Principles of Helicopter Aerodyn

Helicopters are highly capable and useful rotating-wing aircraft that have a variety of civilian and military applications. Their usefulness lies in their unique ability to take off and land vertically, to hover stationary relative to the ground, and to fly forward, backward, or sideways. These unique flying qualities, however, come at a price, including complex aerodynamic problems, significant vibrations, high levels of noise, and relatively large power requirements compared to a fixed-wing aircraft.

This book, written by an internationally recognized teacher and researcher in the field, provides a thorough, modern treatment of the aerodynamic principles of helicopters and other rotating-wing vertical lift aircraft. The first part of the text begins with a unique technical history of helicopter flight, and then covers basic methods of rotor aerodynamic analysis and related issues associated with helicopter performance and aerodynamic design. The second part is devoted to more advanced topics in helicopter aerodynamics, including airfoil flows, unsteady aerodynamics, dynamic stall, and rotor wakes. Every chapter is extensively illustrated, and concludes with a full bibliography and homework problems. The text is complemented by a companion solutions manual to the problems.

Advanced undergraduate and graduate students, as well as practicing engineers and researchers, will welcome this thorough and up-to-date text on rotating-wing aerodynamics.

Dr. J. Gordon Leishman is an Associate Professor of Aerospace Engineering at the University of Maryland, and a former aerodynamicist at Westland Helicopters. He has written extensively on topics in helicopter aerodynamics.

Cambridge Aerospace Series

Editors
Michael J. Rycroft
and
Robert F. Stengel

1. J. M. Rolfe and K. J. Staples (eds.): *Flight Simulation*
2. P. Berlin: *The Geostationary Applications Satellite*
3. M. J. T. Smith: *Aircraft Noise*
4. N. X. Vinh: *Flight Mechanics of High-Performance Aircraft*
5. W. A. Mair and D. L. Birdsall: *Aircraft Performance*
6. M. J. Abzug and E. E. Larrabee: *Airplane Stability and Control*
7. M. J. Sidi: *Spacecraft Dynamics and Control*
8. J. D. Anderson: *A History of Aerodynamics*
9. A. M. Cruise, J. A. Bowles, C. V. Goodall, and T. J. Patrick: *Principles of Space Instrument Design*
10. G. A. Khoury and J. D. Gillett (eds.): *Airship Technology*
11. J. Fielding: *Introduction to Aircraft Design*
12. J. G. Leishman: *Principles of Helicopter Aerodynamics*

Principles of Helicopter Aerodynamics

J. GORDON LEISHMAN
University of Maryland

TO MY STUDENTS

in appreciation of all they have taught me

PUBLISHED BY THE PRESS SYNDICATE OF THE UNIVERSITY OF CAMBRIDGE
The Pitt Building, Trumpington Street, Cambridge, United Kingdom

CAMBRIDGE UNIVERSITY PRESS
The Edinburgh Building, Cambridge CB2 2RU, UK
40 West 20th Street, New York, NY 10011-4211, USA
10 Stamford Road, Oakleigh, VIC 3166, Australia
Ruiz de Alarcón 13, 28014 Madrid, Spain
Dock House, The Waterfront, Cape Town 8001, South Africa

http://www.cambridge.org

First published 2000
Reprinted with corrections 2001
First paperback edition 2002

Printed in the United States of America

Typeface Times Roman 10/12 pt. *System* LaTeX 2_ε [TB]

A catalog record for this book is available from the British Library.

Library of Congress Cataloging in Publication Data

Leishman, J. Gordon.
Principles of helicopter aerodynamics / J. Gordon Leishman.
p. cm.
Includes bibliographical references (p.).
ISBN 0-521-66060-2 (hardcover)
1. Helicopters – Aerodynamics.
TL716.L43 2000
629.133′352 – dc21 99-38291
CIP

ISBN 0 521 66060 2 hardback
ISBN 0 521 52396 6 paperback

Contents

List of Figures

List of Tables

Preface

This book is a college-level analytical and applied level exposition of the aerodynamic principles of helicopters and other rotating-wing vertical lift aircraft. It is written for students who have no background in rotating-wing aerodynamics but have had at least two semesters of basic aerodynamics at the undergraduate level and possibly one course at graduate level. The material covered has grown mainly out of two graduate level courses in "Helicopter Aerodynamics" that I have taught at the University of Maryland since 1988. I have also taught a somewhat more general senior-level undergraduate course in "Helicopter Theory" about every other year, which is centered around the first half of this book. These courses have been offered as part of the formal curriculum in the Center for Rotorcraft Education and Research, which was originally founded in 1982, partly through the efforts of Professor Alfred Gessow. It is now nearly fifty years since Alfred Gessow and Gary Myers's well-known book *The Aerodynamics of the Helicopter* was first published. I am pleased to record in the preface to this book that in his status as Professor Emeritus, Alfred Gessow continues to be active in activities at the University of Maryland and also within government and professional organizations. As a testimony to his lifelong dedication to education and research in helicopter technology, the Rotorcraft Center at the University of Maryland has been recently named in his honor.

In the institutions where formal courses in helicopter technology have been taught, either at the undergraduate or graduate level, my experience is that they have been well received and very popular with the students. What is often most attractive to students is the highly multidisciplinary nature of helicopter engineering problems. Therefore, an introductory course in helicopters provides a good capstone to the aerospace engineering curriculum. Another factor for most students who have taken a helicopter course is the realization that so much more remains to be learned about fundamental aerodynamics, especially as it applies to rotating-wing aircraft. This is reflected in the experience levels in predicting the aerodynamics and overall behavior of helicopters before their first flight, which are less than desirable. Consequently, it is fair to say that the various aerodynamic problems associated with helicopters and other new forms of rotary-wing aircraft probably provide the scientists and engineers of the future with some of the most outstanding research challenges to be found in the field of theoretical and applied aerodynamics.

Serious work on this book project started about three years ago and was motivated primarily by my students. It has grown out of a relatively informal collection of classroom notes and research papers and an overwhelming need to synthesize both older and newer information on the subject of helicopter aerodynamics into one coherent volume. With the ever increasing content of new research material on helicopters and the increasing proportion of recent research material being included in the course, especially in the second semester, the development of a formal textbook was really a logical step. Ever since 1980, when Wayne Johnson's excellent book *Helicopter Theory* was published, progress in understanding helicopter aerodynamics and other related fields has been remarkable. A glance at the content of recent proceedings of the Annual Forum of the American Helicopter Society or the European Rotorcraft Forum shows the scope and depth of new work being conducted today, which continues despite tightly controlled budgets. This has been fueled, in part, by the great advances in computer technology, which have fostered ambitious new analytical and numerical approaches to solving helicopter technology problems. Some of these approaches now come under the banner of computational fluid dynamics (CFD) using numerical solutions to the Euler and Navier–Stokes equations. While these techniques are not yet mature, CFD methods have begun to provide new insight into the complicated aerodynamic problems associated with helicopters that were previously intractable with existing mathematical methods or were limited by available experimental techniques. As these new numerical methods continue to mature and become better validated, the first decade of the twenty-first century will see an increasing use of CFD tools in the design of new and improved helicopters. The past twenty years have also marked a revolution in the experimental studies of helicopter aerodynamics, where advances in flow diagnostic and other instrumentation has allowed measurements on rotors to be made with a fidelity that was considered impossible just a few years ago. The complex nature of the problems found on helicopters means that both experiment and theory must continue to go hand in hand to forge a better understanding of the whole. This will result in the development of new rotating-wing aircraft with better performance, lower vibration, better reliability and maintainability, and lower direct and operational costs. The modern spirit of international cooperation in research and development makes the years ahead in the twenty-first century very exciting.

As I have already mentioned, a significant part of the content of this book has been focused toward students at the graduate level who are learning the principles of helicopter aerodynamics and are exploring the tools available to approach the vigorous, multidisciplinary research and development of the modern helicopter. In planning the content of this book, I have organized the material in two main parts. The first part will be appropriate for a one-semester course in helicopter aerodynamics for senior undergraduate and first-year graduate students. This material essentially represents a thorough introduction to fundamental helicopter problems, basic methods of rotor analysis, issues associated with helicopter performance, and conceptual design issues. I have attempted to follow the spirit of Gessow & Myers's book, where theory is supported throughout by liberal references to experimental observations and measurements. In this regard, rediscovering the less well-known early NACA and RAE technical literature on the subject of helicopter aerodynamics proved to be one of the most satisfying aspects of writing this book. The rapid progress made in understanding the problems of the helicopter during the period between 1930 and 1950, and the ingenuity shown in both the experimental and analytical work, are quite remarkable.

The second part of the book gives a more advanced treatment of more detailed aspects of helicopter aerodynamics, again with emphasis on physical concepts and basic methods of analysis. This part will appeal more to those students who plan to conduct research in helicopter aerodynamics and related fields or who are already practicing engineers in industry and government laboratories. However, I do hope that practicing engineers will relish the opportunity to revisit the first part of the book to review the basics and also to review the inherent assumptions and limitations of the fundamental concepts and methods. These are so often taken for granted, but they form the backbone of many modern forms of helicopter analysis. Because I have always found a need to bring "industrial practice" into the classroom, I have tried to incorporate engineering practice as well as some of my own industrial experience into the second part of the book. For this, I thank my former colleagues at Westland Helicopters for sharing their knowledge with me.

Like most textbooks, the final product has turned out to be not exactly what was originally planned. Along the way, topics have been added, parts of the text rearranged, and some other topics deleted. Also, in light of the reviews of the preliminary manuscript, new figures have been added, many were modified, and others deleted. While over 400 figures were originally prepared, less than 275 finally made it to the finished book. In the interests of space and publication costs, two chapters have been left out completely. These were "Helicopter Interactional Aerodynamics" and "Advanced Computational Aerodynamics of Rotorcraft." Both topics are referred to, albeit briefly, throughout the book, but to include them would have made the final size of the book prohibitively long. There have also been such rapid recent advances in these areas in the past few years that they are best left for a second edition. The list of key references for each chapter is extensive but by no means complete, and the reader is encouraged to follow through with the references contained in each publication, as required. Problems are provided at the end of each chapter and have been drawn on time proven homework and examination questions. Solutions have been provided in a companion instructors' solution manual.

Chapter 1 introduces the helicopter through its technical history. This chapter grew out of my personal research into the Berliner helicopter experiments that were conducted during the early 1920s at the College Park airport, which is close to the University of Maryland. Whereas in the technical development of fixed-wing aircraft it is possible to point to several key historical events, it becomes quickly apparent that things are much less clear in the development of the helicopter. There are already many authoritative publications that detail the historical development of the helicopter, so I have tried to approach the discussion on more of a technical theme and to put the background and difficulties in understanding the aerodynamics of vertical flight into broader perspective. This introduction is followed in Chapter 2 by an analysis of hovering and axial flight using the Rankine–Froude momentum theory. The basic momentum theory concept was extended to forward flight by Glauert and others from the RAE at Farnborough, inspired not by the helicopter, but by the success of Cierva's autogiro. It is shown that many of the important performance and operational characteristics of the helicopter can be deduced from Glauert's extension of the basic momentum theory. The blade element and combined blade-element momentum theory is discussed in Chapter 3. These ideas were developed in the 1940s and provide the foundation for a more modern treatment of the aerodynamics of rotors. On the basis of certain assumptions, important information on blade design, such as optimum or ideal shapes for the blade planform and blade twist, can be deduced from the combined blade element momentum theory. Because helicopter blades have articulation, in that they can flap and lag about hinges located near the root of each blade, it is not possible to understand the behavior of the helicopter solely from an aerodynamics perspective. Therefore, in Chapter 4 a discussion of rigid blade motion leads naturally into an understanding of the issues associated with rotor response to the changing aerodynamic loads and also to rotor control. Also introduced here are the ideas of rotor trim, that is, the pilot's control inputs required to enable equilibrium flight of the helicopter. Chapter 5 gives an introduction to helicopter performance and operational issues such as climbing and descending flight, including the autorotative state and flight near the ground. The first part of the book concludes in Chapter 6, which reviews issues associated with the conceptual aerodynamic design of helicopters, including the main rotor, the fuselage and empennage, and the tail rotor. Although it might have been more appropriate to place this chapter at the end of the book, it provides a good bridge between the fundamentals and results to many problems that are still more of a research nature for which experimental research is incomplete and predictive capabilities are not yet mature.

Chapter 7 starts the second part of the book with an important practical review of basic airfoil aerodynamics, including boundary layer and viscous aerodynamics, and the role of compressibility. This is followed by applications of these concepts to understanding some of the special requirements and characteristics of rotor airfoils. Again, liberal reference to experimental measurements makes the present treatment relatively unique amongst textbooks on helicopters. Chapters 8 and 9 comprise a comprehensive discussion on unsteady aerodynamics, with emphasis on the relevance to helicopter problems. Classical techniques of unsteady aerodynamics, including Theodorsen's and Loewy's theory, the indicial response method, and dynamic inflow, are reviewed in Chapter 8. Extensions of some of these methods to the compressible flow problems found on helicopters are described, with validation with experimental measurements where possible. Indicial methods are treated in some detail because they form the foundation for many modern methods of helicopter analysis and are not covered in any previous helicopter text. Chapter 9 discusses the problem of dynamic stall, which is known to be a barrier to attaining high speed forward flight with a conventional helicopter. Engineering methods of dynamic stall prediction are also reviewed, along with some examples of the general predictive capability to be found with these models. The physical nature of helicopter rotor wakes, both in hovering and forward flight, are discussed in Chapter 10. Nearly all of what we know about rotor wakes comes from empirical observations, and this has led to the development of well validated mathematical models of the rotor wake using vortex techniques. Chapter 10 concludes with a brief discussion on interactional aerodynamics. Although many the problems of rotor aerodynamics can be studied by considering the rotor in isolation to the fuselage, tail rotor, and empennage, aerodynamic interactions between the components lead to many problems that are not yet fully understood.

A word about systems of units is in order. In the preparation of this book, I have found it necessary to use both the British (Imperial) and metric (SI) systems. Any preference for one over the other is done simply for the sake of convenience, and I think there is really little reason to change units for the sake of standardization in the text. As aerospace engineers, and particularly helicopter engineers, we are used to working with both systems and even with mixed units in the same breath, and so most readers will have no problems with this approach. Many recent reports and publications still use the older Imperial system of units, especially those from the United States. For the foreseeable future, students will have to learn to become fully conversant in both systems. Where I have felt it would be helpful, units in both systems are stated. For convenience, a table of conversion factors is also included in the appendix.

Having said all this and made my excuses, I hope this edition of the book will be judged, especially by the practicing engineer, on what it contains and not on what else could potentially have been included. Like everything in aviation, the final product is always a compromise and is always under a continuous state of revision, development, and improvement. As a final comment, it seems appropriate to quote Igor Sikorsky who has said: "At that time [1908] aeronautics was neither an industry nor even a science ... it was an art, I might say, a passion. Indeed at that time it was a miracle." As we stand now at the turn of the new millenium, I'm sure that Igor Sikorsky would have agreed that the past century has indeed seen many miracles, both in aeronautics and in helicopter technology. The new century will almost certainly see more.

Acknowledgments

This book is the product of an opportunity afforded to me by the University of Maryland. While the research and writing of this book has been a unique experience, it has not been easy to devote the best part of 2,500 work hours, mainly during evenings and weekends and spread over three years, that were neccessary for its successful completion. I am forever grateful to my wife, Alice Marie Leishman, for her love and understanding and for providing me continuous support, especially during the last six months of the project, where writing, proofreading, drawing figures, and plotting graphs meant many sixteen-hour days, a good number of sleepless nights, and all too short weekends.

Acknowledgment is due to a great many people, both on and off the University of Maryland campus. I am grateful for the counsel of my colleagues, Professors Alfred Gessow, Inderjit Chopra, Roberto Celi and James Baeder, and Dr. Vengalattore Nagaraj. Professor Gessow read substantial parts of the text and offered useful suggestions for improvement. I also appreciate his letting me adapt some examination questions from his earlier courses. Professors Inderjit Chopra and Roberto Celi provided good suggestions for Chapter 4 and kindly allowed me to use some of their own course material on blade motion and rotor trim. Professor James Baeder read Chapter 8 on unsteady aerodynamics, and I am grateful to him for letting me use his finite-difference results to some of the problems. Dr. Vengalattore Nagaraj read Chapters 5 and 6 and offered good suggestions on helicopter design issues.

I am particularly indebted to my graduate students, both past and current, who have both directly and indirectly contributed to the content of this book. The current members of my research group, namely Mahendra Bhagwat, Preston Martin, Jacob Park, and Greg Pugliese, enthusiastically read (and reread) many parts of the text, found the bulk of the typographical errors in the equations, and offered much useful advice and constructive criticism from a student's perspective. I am grateful to Mahendra Bhagwat for running some of the prescribed and free-wake solutions and for allowing me to adapt his analysis of the tip vortex aging problem in Chapter 10. Preston Martin took several of the rotor wake visualization images for Chapter 10 and did much to help me find references and other material for Chapter 7. Jacob Park ran several of the rotor wake calculations, sifted through lots of experimental results on rotor wakes, and organized the data for a good number of the figures in Chapter 10. Gregory Pugliese read many of the draft chapters several times, made endless trips to the library to hunt down reports, cross-checked several thousand references, and helped me turn more than a few of my many crude sketches into professional looking figures. Other former students who have directly or indirectly contributed material used in this book include Dr. Ashish Bagai, Dr. Nai-pei Bi, Dr. Gilbert Crouse, Mark Daghir, Erwin Moederesheim, Joe Tyler, and Berend van der Wall.

Off the university campus, there are a great many people who provided suggestions or read parts of the text or sent me hard to find reports, computer files of experimental data, figures, photographs, or other information that was used or adapted for the book. In particular, I appreciate the contributions of Dr. Bill Bousman, Dr. Mark Chaffin, Prof. Muguru Chandrasekhara, Dr. Colin Coleman, Ms. Susan Gorton, Dr. Richard Green, Mr. Robert Hansford, Ms. Jennifer Henderson, Mr. Markus Krekel, Dr. Kenneth McAlister, Dr. Khanh

Nguyen, Dr. Reinert Müller, Mr. John Perry, and Dr. James Tangler. A special thanks to Dr. Ashish Bagai of Sikorsky Aircraft, who read the final draft of the whole book and helped me smooth off the remaining rough edges. His generosity in allowing me to adapt some of his graphics for Chapter 10 is also gratefully appreciated. I am particularly indebted to the reviewers of my original proposal who were able to make good suggestions on what should be included and what should be left out. In particular, I appreciate the contructive reviews by Edward Smith of Penn State University and Andrew Lemnios of Rensselaer Polytechnic Institute.

Thanks are due to Madelyn Bush and Jack Satterfield of Boeing Helicopters, Kevin Hale of Bell Helicopter Textron, David Long of Kaman Aircraft Corporation, the Public Affairs Department at GKN Westland, and the National Air and Space Museum for giving permission to publish some of the photographs included in Chapter 1.

Finally, I want to express my sincere gratitude to Florence Padgett and her staff at Cambridge University Press for their help in getting this book project off the ground, as well as for their continous support during the writing and publication process of this book. My thanks also to Andrew Wilson and Michie Shaw of TechBooks for their help during the production process.

List of Main Symbols

Listed below alphabetically are the main symbols used in this book. Note that more than one meaning may be assigned to a symbol. Other symbols are defined internally within each chapter.

a	sonic velocity
A	rotor disk area, πR^2
A	axial (chord) force
A_b	blade area, σA
A_{ov}	overlap area
B	tip loss factor
c	blade chord
C	Theodorsen's function
C'	Lowey's function
C_d	drag coefficient
C_{d_0}	zero-lift drag coefficient
C_{d_f}	fuselage drag coefficient
C_{D_v}	vertical drag coefficient
c_f	local skin friction coefficient, $\tau_w/\frac{1}{2}\rho V_\infty^2$
C_H	H-force coefficient, $H/\rho A(\Omega R)^2$
C_l	section lift coefficient, $L/\frac{1}{2}\rho U^2 c$
$C_{l_{\max}}$	maximum lift coefficient
C_{l_α}	lift-curve slope
$\bar{C}_L$	rotor mean lift coefficient, $6(C_T/\sigma)$
C_m	section moment coefficient, $M/\frac{1}{2}\rho U^2 c^2$
$C_{m_{1/4}}$	section moment coefficient about quarter chord
$C_{m_{0.25}}$	section moment coefficient about quarter chord
$C_{m_{ac}}$	section moment coefficient about aerodynamic center
C_{M_x}	roll moment coefficient, $M_x/\rho A R(\Omega R)^2$
C_{M_y}	pitching moment coefficient, $M_y/\rho A R(\Omega R)^2$
C_n	normal force coefficient, $N/\frac{1}{2}\rho U^2 c$
C_{n_α}	slope of normal force coefficient versus α curve
C_p	pressure coefficient
C_P	rotor power coefficient, $P/\rho A(\Omega R)^3$
C_{P_c}	climb power coefficient
C_{P_i}	induced power coefficient
C_{P_0}	profile power coefficient
C_{P_p}	parasitic power coefficient
C_Q	rotor torque coefficient, $Q/\rho A(\Omega R)^2 R$
C_T	rotor thrust coefficient, $T/\rho A(\Omega R)^2$
C_T/σ	blade loading coefficient

C_W	weight coefficient, $W/\rho A(\Omega R)^2$
C_W	work coefficient or torsional damping factor, $\oint C_m \, d\alpha$
C_Y	Y-force coefficient, $Y/\rho A(\Omega R)^2$
D	drag force
DL	rotor disk loading, T/A
$D.F.$	torsional damping factor, $\oint C_m \, d\alpha$
e	flapping hinge offset
f	equivalent flat plate area of fuselage etc., $D/\frac{1}{2}\rho V_\infty^2$
f	effective separation point
F	Prandtl's tip loss function
FM	figure of merit, $C_T^{3/2}/(\sqrt{2}C_P)$
F_x	aerodynamic force parallel to disk plane
F_z	aerodynamic force normal to disk plane
h	perpendicular distance from vortex to evaluation point
h	altitude (height)
h_p	pressure altitude
h_ρ	density altitude
h	plunge displacement
H	rotor drag force
H	Hankel function
i	blade control point index
$I_{b_{i,j}}$	influence coefficient in lifting-line model
I_b	blade moment of inertia about the flapping hinge
I_ζ	blade moment of inertia about the lagging hinge
$\Im$	imaginary or out-of-phase part
J	Bessel function of the first kind
k	reduced frequency, $\Omega c/2V$
k_g	gust reduced frequency
k_x, k_y	longitudinal and lateral inflow gradients, respectively
l	length of influencing vortex filament
L	section lift force
m	mass per unit length of rotor blade
m	maximum camber
m	tandem rotor overlap fraction
$\dot{m}$	mass flow rate through rotor
M	Mach number or free-stream Mach number, U/a
M^*	critical Mach number
M	mass of helicopter
M	moment
M_∞	free-stream Mach number
M_{dd}	drag divergence Mach number
M_{tip}	rotor hover tip Mach number, $\Omega R/a$
M_x	rotor rolling moment, positive to starboard
M_y	rotor pitching moment, positive nose-up
M_β	aerodynamic flapping moment
N	normal force
N_b	number of rotor blades
p	pressure

p	roll rate about x-axis
p	point of maximum camber
$\bar{p}$	nondimensional roll rate, p/Ω
p	Laplace variable
P	rotor power
PL	rotor power loading, T/P
q	non-dimensional pitch rate, $\dot{\alpha}c/V$
q	pitch rate about y-axis
q_∞	free-stream dynamic pressure, $\frac{1}{2}\rho V_\infty$
$\bar{q}$	nondimensional pitch rate, q/Ω
Q	rotor torque
r	radial distance
r_0	blade root cutout
r_c	vortex core radius
r_v	vortex release point along blade span
$\vec{r}$	position vector of point p in space
$\bar{r}$	nondimensional radial coordinate, r/r_c
R	blade radius
Re	Reynolds number based on chord
Re_x	Reynolds number based on distance x
$\Re$	real part
s	relative distance traveled by airfoil in semichords, $2Vt/c$
S	Sears's function, referenced with respect to airfoil midchord
S	reference or surface area
S_{ref}	reference area
S'	Sears's function, referenced with respect to airfoil leading edge
SFC	specific fuel consumption
t	time
t	maximum airfoil thickness
T	temperature
T	rotor thrust
T/A	rotor disk loading
U	resultant velocity at blade element, $\sqrt{U_P^2 + U_T^2}$
U_P	out-of-plane velocity normal to rotor disk plane
U_R	radial velocity along blade at disk plane
U_T	in-plane velocity parallel to rotor disk plane
v_h	hover induced velocity
v_r, v_θ, v_z	radial, tangential, axial velocity components
V	velocity
$\vec{V}$	velocity vector
V_∞	magnitude of free-stream velocity
$\vec{V}_\infty$	free-stream velocity vector
V_c	climb velocity
V_g	gust convection velocity
V_{mp}	speed to fly for minimum power
V_{mr}	speed to fly for maximum range
w	slipstream velocity
w	velocity normal to chord

w_g	upwash velocity normal to chord
W	gross weight of helicopter
W	work done on fluid
W_F	weight of fuel
x, y, z	Cartesian coordinate system
$\bar{x}$	nondimensional distance along airfoil chord
X_n, Y_n	recurrence functions
Y	rotor side force
Y	Bessel function of the second kind
α	blade section angle of attack
α	rotor disk angle of attack
α_0	zero-lift angle of attack
α_e	effective angle of attack
α_{ind}	induced angle of attack
α_{TPP}	tip-path-plane angle of attack
β	Glauert compressibility factor, $\sqrt{1 - M^2}$
β	blade flapping angle (positive up)
β_0	blade coning angle
β_{1c}	longitudinal flapping angle
β_{1s}	lateral flapping angle
γ	ratio of specific heats
γ	blade Lock number, $\rho C_{l_\alpha} c R^4 / I_b$
γ_b	bound vortex sheet strength
γ_w	vortex sheet strength in wake
Γ_b	bound vortex strength
Γ_v	vortex circulation strength
δ	boundary layer thickness
δ	effective viscosity coefficient multiplier
ζ	blade lag angle
η	efficiency factor
η	flap deflection angle
θ	blade pitch angle
θ_0	blade collective pitch
θ_{1c}	lateral cyclic pitch
θ_{1s}	longitudinal cyclic pitch
θ_{tip}	geometric pitch at blade tip
θ_{tw}	linear blade twist rate
$\theta_{0.75}, \theta_{75}$	pitch angle at 75% radius
κ	induced power factor
κ_{int}	induced power factor from interference
λ	rotor inflow ratio (positive downward through disk)
λ	gust speed ratio, $V/(V + V_g)$
λ_c	climb inflow ratio
λ_g	wavelength of sinusoidal gust
λ_i	rotor induced inflow ratio
Λ	sweep angle
μ	rotor advance ratio, $V_\infty \cos\alpha / \Omega R$
μ	dynamic viscosity

ν	kinematic viscosity coefficient
ν_β	blade rotating flap frequency (nondimensional)
ν_ζ	blade rotating lag frequency (nondimensional)
ρ	density of air
σ	rotor solidity, $N_b c/\pi R$
σ_e	equivalent rotor solidity
τ	time constant
τ_w	wall shear stress
ϕ	inflow angle of attack, $\tan^{-1}(U_P/U_T)$
ϕ	general indicial response function
ϕ	Wagner function
ψ	blade/rotor azimuth angle
ψ	general sharp-edged gust function
ψ	Küssner function
ψ_b	blade azimuth location
ψ_w	wake age
ω	angular velocity
ω_g	gust frequency
Ω	rotational frequency of rotor
χ	wake skew angle

Subscripts and Superscripts

∞	free-stream conditions
0	profile
$1c$	first harmonic cosine component
$1s$	first harmonic sine component
$1/4$	quarter chord
$3/4$	three-quarter chord
ac	aerodynamic center
b	associated with bound vortex
c	climb
c	associated with camber
c	circulatory part
cg	with respect to the center of gravity
CP	control plane
FP	flight path
h	hover
HT	horizontal tail
i	induced
l	lower surface
LE	leading edge
max	maximum
MR	main rotor
n	normal to leading edge
nc	noncirculatory part
NFP	no feathering plane
p	parasitic
p	component resulting from pitch rate

q	roll rate or airfoil pitch rate
qs	quasi-steady part
t	associated with thickness
tip	blade tip component
TPP	tip path plane
TR	tail rotor
u	upper surface
VT	vertical tail
α	component resulting from angle of attack

Other Symbols

∇^2	$= \dfrac{\partial^2}{\partial r^2} + \dfrac{1}{r}\dfrac{\partial}{\partial r} + \dfrac{\partial^2}{\partial z^2}$
$\rightarrow$	vector quantity
[ ]	matrix
{ }	column vector

Abbreviations

AIAA	American Institute of Aeronautics and Astronautics (formally IAS)
ARL	Army Research Laboratories
ARC R & M	Aeronautical Research Committee, Reports & Memoranda
BEMT	blade element momentum theory
BET	blade element theory
BVI	blade vortex interaction
CF	centrifugal force
DERA	Defence Evaluation Research Agency
FAI	Fédération Aeronautique Internationale
GTOW	gross takeoff weight
HWA	hot wire anemometry
HOT	higher order term
IAS	Institute of Aeronautical Sciences (now AIAA)
ICAN	International Commission for Air Navigation
ICAO	International Civil Aviation Organization
LDV	laser Doppler velocimetry
LE	leading edge
NACA	National Advisory Committee for Aeronautics (now NASA)
NASA	National Aeronautics and Space Administration (formally NACA)
NASM	National Air and Space Museum
NTIS	National Technical Information Service
NPL	National Physical Laboratory
ONERA	Office National d'Etude et de Recherches Aérospatiales
RAE	Royal Aircraft Establishment (now DERA)
RAeS	Royal Aeronautical Society
PDE	partial differential equation
rpm	revolutions per minute
SAE	Society of Automotive Engineers
SL	sea level
TE	trailing edge
TPP	tip path plane

TsAGI	Central AeroHydrodynamic Institute
USAAVLABS	U.S. Army Aviation Laboratories
USAAMRDL	U.S. Army Aviation Air Mobility Research & Development Laboratories
USARTL	U.S. Army Research and Technology Laboratories
VTOL	vertical takeoff and landing
2-D	two-dimensional
3-D	three-dimensional

CHAPTER 1

Introduction: A History of Helicopter Flight

The idea of a vehicle that could lift itself vertically from the ground and hover motionless in the air was probably born at the same time that man first dreamed of flying.

Igor Ivanovitch Sikorsky

1.1 Introduction

The science of aerodynamics is the fundament of all flight. It is the role of aerodynamics in the engineering analysis and design of rotating-wing vertical lift aircraft that is the subject of this book. Igor Sikorsky's vision of a rotating-wing aircraft that could safely hover and perform other desirable flight maneuvers under full control of the pilot was only to be achieved some thirty years after fixed-wing aircraft (airplanes) were flying successfully. This rotating-wing aircraft we know today as the helicopter. Although the helicopter is considered by some to be a basic and somewhat cumbersome looking aircraft, the modern helicopter is indeed a machine of considerable engineering sophistication and refinement and plays a unique role in modern aviation provided by no other aircraft.

In the introduction to this book, the technical evolution of the helicopter is traced from a cumbersome, vibrating contraption that could barely lift its own weight into a modern and efficient aircraft that has become an indispensable part of modern life. Compared to fixed-wing flight, the development of which can be clearly traced to Lilienthal, Langley, and the first fully controlled flight of a piloted powered aircraft by the Wright Brothers in 1903, the origins of successful helicopter flight are less clear. Nonetheless, there are many parallels in the development of the helicopter when compared to fixed-wing aircraft. However, the longer and perhaps more tumultuous gestation period of the helicopter is directly attributable to the greater depth of scientific and aeronautical knowledge that was required before all the various technical problems could be understood and overcome. Besides the need to understand the basic aerodynamics of vertical flight and improve upon the aerodynamic efficiency of the helicopter, other technical barriers included the need to develop suitable high power-to-weight engines and high-strength, low-weight materials for the rotor blades, hub, fuselage, and transmission.

A helicopter can be defined as any flying machine using rotating wings (i.e., rotors) to provide lift, propulsion, and control forces that enable the aircraft to hover relative to the ground without forward flight speed to generate these forces. The thrust on the rotor(s) is generated by the aerodynamic lift forces created on the spinning blades. To turn the rotor, power from an engine must be transmitted to the rotor shaft. It is the relatively low amount of power required to lift the machine compared to other vertical take off and landing (VTOL) aircraft that makes the helicopter unique. Efficient hovering flight with low power requirements comes about by accelerating a large mass of air at a relatively low velocity; hence we have the large diameter rotors that are one obvious characteristic of helicopters. In addition, the helicopter must be able to fly forward, climb, cruise at speed, and then descend and come back into a hover for landing. This demanding flight capability comes at a price, including mechanical and aerodynamic complexity and higher power requirements than for a fixed-wing aircraft of the same gross weight. All of these factors influence the design, acquisition, and operational costs of the helicopter.

Besides generating all of the vertical lift, the rotor is also the primary source of control and propulsion for the helicopter, whereas these functions are separated on a fixed-wing aircraft. For forward flight, the rotor disk plane must be tilted so that the rotor thrust vector is inclined forward to provide a propulsive component to overcome rotor and airframe drag. The orientation of the rotor disk to the flow also provides the forces and moments to control the attitude and position of the aircraft. The pilot controls the magnitude and direction of the rotor thrust vector by changing the blade pitch angles (using collective and cyclic pitch inputs), which changes the blade lift and the distribution of thrust over the rotor disk. By incorporating articulation into the rotor design through the use of mechanical flapping and lead/lag hinges that are situated near the root of each blade, the rotor disk can be tilted in any direction in response to these blade pitch inputs. As the helicopter begins to move into forward flight, the blades on the side of the rotor disk that advance into the relative wind will experience a higher dynamic pressure and lift than the blades on the retreating side of the disk, and so asymmetric aerodynamic forces and moments will be produced on the rotor. Articulation helps allow the blades to naturally flap and lag so as to help balance out these asymmetric aerodynamic effects. However, the mechanical complexity of the rotor hub required to allow for articulation and pitch control leads to high design and maintenance costs. With the inherently asymmetric flow environment and the flapping and pitching blades, the aerodynamics of the rotor become relatively complicated and lead to unsteady forces. These forces are transmitted from the rotor to the airframe and can be a source of vibrations, resulting in not only crew and passenger discomfort, but also considerably reduced airframe component lives and higher maintenance costs. However, with a thorough knowledge of the aerodynamics and careful design, all these adverse factors can be minimized or overcome to produce a highly reliable and versatile aircraft.

1.2 Early Attempts at Vertical Flight

There are many authoritative sources that record the development of helicopters and other rotating-wing aircraft such as autogiros. These include Gregory (1944), Lambermont (1958), Gablehouse (1967), Gunston (1983), Apostolo (1984), Boulet (1984), Lopez & Boyne (1984), Taylor (1984), Everett-Heath (1986), Fay (1987) and Spenser (1999), amongst others. Boulet (1984) takes a unique approach in that he gives a first-hand account of the early helicopter developments through interviews with the pioneers, constructors, and pilots of the machines. A remarkably detailed history of early helicopter developments is given by Liberatore (1950, 1988, 1998). For original publications documenting early technical developments of the autogiro and helicopter, see Warner (1920), von Kármán (1921), Balaban (1923), Moreno-Caracciolo (1923), Klemin (1925), Wimperis (1926), and Seiferth (1927).

As described by Liberatore (1998), the early work on the development of the helicopter can be placed into two distinct categories: inventive and scientific. The former is one where intuition is used in lieu of formal technical training, whereas the latter is one where a trained, systematic approach is used. Prior to the nineteenth century there were few scientific investigations of flight or the science of aerodynamics. The inherent mechanical and aerodynamic complexities in building a practical helicopter that had adequate power and control, and did not vibrate itself to pieces, resisted many ambitious efforts. The history of flight documents literally hundreds of failed helicopter projects, which, at most, made only brief uncontrolled hops into the air. Some designs provided a contribution to new knowledge that ultimately led to the successful development of the modern helicopter. Yet, it was not until the more scientific contributions of engineers such as Juan de la Cierva, Heinrich Focke, Raoul Hafner, Igor Sikorsky, Arthur Young, and others did the design of a truly safe and practical helicopter become a reality.

Six fundamental technical problems can be identified that limited early experiments with helicopters. These problems are expounded by Sikorsky (1938, and various editions) in his autobiography. In summary, these problems were:

1. Understanding the aerodynamics of vertical flight. The theoretical power required to produce a fixed amount of lift was an unknown quantity to the earliest experimenters, who were guided more by intuition than by science.[1]
2. The lack of a suitable engine. This was a problem that was not to be overcome until the beginning of the twentieth century, through the development of internal combustion engines.
3. Keeping structural weight and engine weight down so the machine could lift a pilot and a payload. Early power plants were made of cast iron and were heavy.[2]
4. Counteracting rotor torque reaction. A tail rotor was not used on most early designs; these machines were either coaxial or laterally side-by-side rotor configurations. Yet, building and controlling two rotors was even more difficult than for one rotor.
5. Providing stability and properly controlling the machine, including a means of defeating the unequal lift produced on the advancing and retreating blades in forward flight. These were problems that were only to be fully overcome with the use of blade articulation, ideas that were pioneered by Cierva, Breguet, and others, and with the development of blade cyclic pitch control.
6. Conquering the problem of vibrations. This was a source of many mechanical failures of the rotor and airframe, because of an insufficient understanding of the dynamic and aerodynamic behavior of rotating wings.

The relatively high weight of the structure, engine, and transmission was mainly responsible for the painfully slow development of the helicopter until about 1920. However, by then gasoline powered piston engines with higher power-to-weight ratios were more widely available, and the antitorque and control problems of achieving successful vertical flight were at the forefront. This resulted in the development of a vast number of prototype helicopters. Many of the early designs were built in Great Britain, France, Germany, Italy, and the United States, who led the field in several technical areas. However, with all the various incremental improvements that had been made to the basic helicopter concept during the pre–World War II years, it was not until the late inter war period that significant technical advances were made and more practical helicopter designs began to appear. The most important advances of all were in engine technology, both piston and gas turbines, the latter of which revolutionized both fixed-wing and rotating-wing flight.

A time-line documenting the evolution of rotating-wing aircraft through 1950 is shown in Fig. 1.1. The ideas of vertical flight can be traced back to early Chinese tops, a toy first used about 400 BC. Everett-Heath (1986) and Liberatore (1998) give a detailed history of such devices. The earliest versions of the Chinese top consisted of feathers at the end of

[1] The first significant application of aerodynamic theory to helicopter rotors came about in the early 1920s.

[2] Aluminum was not available commercially until about 1890 and was inordinately expensive. It was not used in aeronautical applications until about 1915.

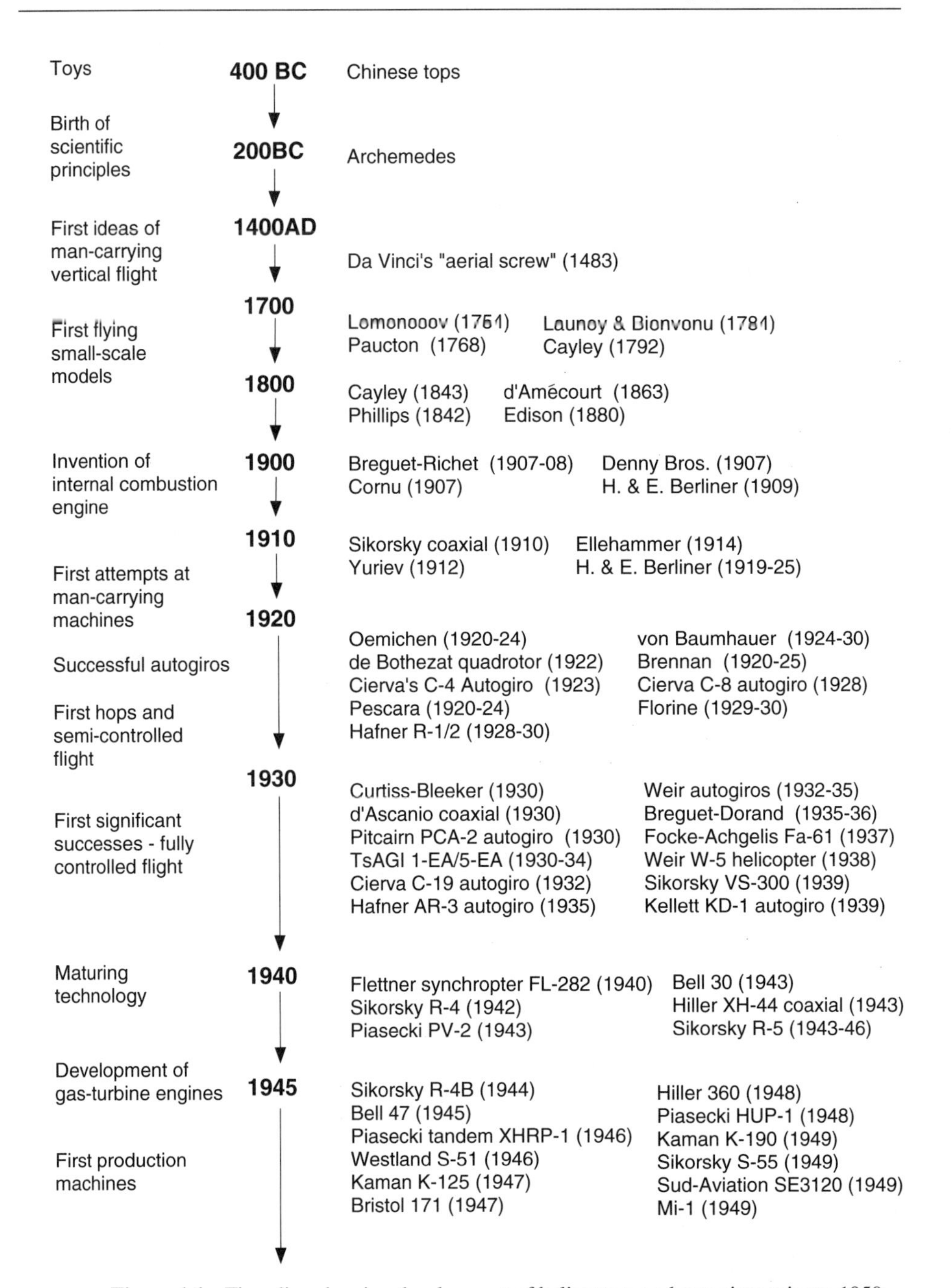

Figure 1.1 Time-line showing development of helicopters and autogiros prior to 1950.

a stick, which was rapidly spun between the hands to generate lift and then released into flight. More than 2,000 years later in 1784, Launoy & Bienvenu used a coaxial version of the Chinese top in a model consisting of a counterrotating set of turkey feathers, powered by a string wound around its shaft and tensioned by a crossbow. It is also recorded that Mikhail Lomonosov of Russia had developed, as early as 1754, a small coaxial rotor modeled after the Chinese top but powered by a wound-up spring device. In 1786, the French mathematician

A. J. P. Paucton published a paper entitled "Théorie de la vis D'Archimèdes," where he proposed a human-carrying flying machine, with one rotor to provide lift and another for propulsion.

Amongst his many intricate drawings, Leonardo da Vinci shows what is a basic human-carrying helicopterlike machine, an obvious elaboration of an Archimedes water-screw. His sketch of the "aerial-screw" device, which is shown in Fig. 1.2, is dated to 1483 but was first published nearly three centuries later. The device comprises a helical surface that da Vinci describes should be "rotated with speed that said screw bores through the air and climbs high." He realized that the density of air is much less than that of water, and so da Vinci describes how the device needed to be relatively large to accomplish this feat (the number "8" in his writing to the left of the sketch indicates that the size of the rotor is 8 *braccia* or arm lengths). He also describes in some detail how the machine should be built using wood, wire, and linen cloth. Although da Vinci worked on various concepts of

Figure 1.2 Leonardo da Vinci's aerial screw machine, dated to 1483. Original drawing is MS 2173 of Manuscript (codex) B, folio 83 verso, in the collection of the Bibliothèque L'Institut de France (Paris).

engines, turbines, and gears, he did not unite the ideas of his aerial-screw machine to an engine nor did he appreciate the problems of torque reaction. See Hart (1961) or Giacomelli (1930) for further details of da Vinci's aeronautical work.

Sir George Cayley is famous for his work on the basic principles of flight, which dates from the 1790s – see Pritchard (1961). By the end of the eighteenth century, Cayley had constructed several successful vertical-flight models based on Chinese tops driven by wound-up clock springs. He designed and constructed a whirling-arm device in 1804, which was probably one of the first scientific attempts to study the aerodynamic forces produced by lifting wings. Cayley (1809–10) published a three-part paper that was to lay down the foundations of aerodynamics – see Anderson (1997). In a later paper, published in 1843, Cayley gives details of a vertical flight aircraft design that he called an "Aerial Carriage," which had two pairs of lateral side-by-side rotors. Also, in the 1840s, another Englishman, Horatio Phillips, constructed a steam-driven vertical flight machine, where steam generated by a miniature boiler was ejected out of the blade tips. Although impractical, Phillips's machine was significant in that it marked the first time that a model helicopter had flown under the power of an engine rather than stored energy devices such as wound-up springs.

In the early 1860s, Ponton d'Amécourt of France flew a number of small helicopter models. He called his machines *hélicoptères*, which is a word derived from the Greek adjective *elikoeioas*, meaning spiral or winding, and the noun *pteron*, meaning feather or wing – see Wolf (1974) and Liberatore (1998). In 1863, d'Amécourt built a steam propelled model helicopter, but it could not generate enough lift to fly. However, the novelist Jules Verne was still impressed with d'Amécourt's attempts, and in 1886 he wrote "The Clipper of the Clouds" where the hero cruised around the skies in a giant helicopterlike machine that was lifted by thirty-seven small coaxial rotors and pulled through the air by two propellers.

Other notable vertical flight models that were constructed at about this time include the coaxial design of Bright in 1861 and the twin-rotor steam-driven model of Dieuaide in 1877. Wilhelm von Achenbach of Germany built a single rotor model in 1874, and he was probably the first to use the idea of a tail rotor to counteract the torque reaction from the main rotor. Later, Achenbach conducted experiments with propellers, the results of which were published by NACA – see Achenbach (1923). About 1869 a Russian helicopter concept was developed by Lodygin, using a rotor for lift and a propeller for propulsion and control. Around 1878, Enrico Forlanini of Italy also built a flying steam-driven helicopter model. This model had dual counterrotating rotors, but like many other model helicopters of the time, it was underpowered and had no means of control.

In the 1880s, Thomas Alva Edison experimented with small helicopter models in the United States. He tested several rotor configurations driven by a guncotton engine, which was an early form of internal combustion engine. Later, Edison used an electric motor for power, and he was one of the first to realize from his experiments the need for a large diameter rotor with low solidity to give good hovering efficiency [Liberatore (1998)]. Unlike other experimenters of the time, Edison's more scientific approach to the problem proved that both high aerodynamic efficiency of the rotor and high power from an engine were required if successful vertical flight was to be achieved. In 1910, Edison patented a rather cumbersome looking full-scale helicopter concept with boxkite-like blades, but there is no record that it was ever constructed.

In 1907, about four years after the Wright brothers' first successful powered flights in fixed-wing airplanes at Kitty Hawk in the United States, Paul Cornu of France constructed a vertical flight machine that carried a human off the ground for the first time. Boulet (1984) gives a good account of the work. The airframe was very simple, with a rotor at each end

Figure 1.3 The Cornu helicopter, circa 1907. (Courtesy NASM, Smithsonian Institution, SI Neg. No. 74-8533.)

(Fig. 1.3). Power was supplied to the rotors by a gasoline motor and belt transmission. Each rotor had two blades, and the rotors rotated in opposite directions to cancel torque reaction. A primitive means of control was achieved by placing a wing in the slipstream below the rotors. The machine was reported to have made several tethered flights of a few seconds at low altitude. Also in France, the Breguet brothers had begun to conduct helicopter experiments about 1907. Their complicated quadrotor "Gyroplane" carried a pilot off the ground, albeit briefly, but like the Cornu machine it was underpowered, and it lacked stability and a proper means of control.

In the early 1900s, Igor Sikorsky and Boris Yur'ev independently began to design and build vertical-lift machines in Czarist Russia. By 1909, Sikorsky had built a nonpiloted coaxial prototype. This machine did not fly because of vibration problems and the lack of a powerful enough engine. Sikorsky (1938) stated that he had to await "better engines, lighter materials, and experienced mechanics." His first design was unable to lift its own weight, and the second, even with a more powerful engine, only made short (nonpiloted) hops. Sikorsky abandoned the helicopter idea and devoted his skills to fixed-wing (conventional airplane) designs at which he was very successful. Although he never gave up his vision of the helicopter, it was not until the 1930s after he had emigrated to the United States that he pursued his ideas again (see Section 1.4). Good accounts of the life and work of Igor Sikorsky are documented by Bartlett (1947), Delear (1969), Sikorsky (1964, 1971), Sikorsky & Andrews (1984), Finne (1987), and Cochrane et al. (1989).

Unbeknown to Sikorsky, Boris Yur'ev had also tried to build a helicopter in Russia around 1912, but with a single rotor and tail rotor configuration. Like Sikorsky's machine, the aircraft lacked a powerful enough engine. Besides being one of the first to use a tail rotor design, Yur'ev was one of the first to propose the concept of cyclic pitch for rotor control. (Another early design was patented by Gaetano Crocco of Italy in 1906). Good accounts of Yur'ev's machine are given by Gablehouse (1967) and Liberatore (1998). There is also evidence of the construction of a primitive coaxial helicopter by Professor Zhukovskii (Joukowski) and his students at Moscow University in 1910 – see Gablehouse (1967). Joukowski is well known for his theoretical contributions to aerodynamics, and he published several papers on the subject of rotating wings and helicopters; see also Margoulis (1922) and Tokaty (1971).

Figure 1.4 Danish aviation pioneer Jens Ellehammer flew a coaxial rotor helicopter design in 1914.

About 1914, the Danish aviation pioneer Jens Ellchammer designed a coaxial rotor helicopter. Boulet (1984) gives a good description of the machine, which is shown in Fig. 1.4. The rotor blades themselves were very short; six of these were attached to each of two large circular aluminum rings. The lower disk was covered with fabric and was intended to serve as a parachute in the event the blades or engine failed. A cyclic pitch mechanism was used to provide control, this being another one of many early applications of the concept. The pilot was supported in a seat that could be moved forward and sideways below the rotor, allowing for additional kinesthetic control. The aircraft made many short hops into the air but never made a properly controlled free flight.

An Austrian, Stephan Petroczy, with the assistance of the well-known aerodynamicist Theodore von Kármán, built and flew a coaxial rotor helicopter during 1917–1920. Interesting design features of this machine included a pilot/observer position above the rotors, inflated bags for landing gear, and a quick-opening parachute. While the machine never really flew freely, it accomplished numerous limited tethered vertical flights. The work is summarized in a report by von Kármán (1921) and published by the NACA. It is significant that von Kármán also gives results of laboratory tests on the "rotors," which were really oversize propellers. With the work of William F. Durand [see Warner (1920) and the analysis by Munk (1923)] these were some of the first attempts to scientifically study rotor performance and the power required for vertical flight.

In the United States, Emile and Henry Berliner (a father and son) were interested in vertical flight aircraft. As early as 1909, they had designed and built a helicopter based on pioneering forward flight experiments with a wheeled test rig. In 1918 the Berliners patented a single-rotor helicopter design, but there is no record that this machine was built. Instead, by about 1919, Henry Berliner had built a counterrotating coaxial rotor machine, which made brief uncontrolled hops into the air. By the early 1920s at the College

Figure 1.5 This Berliner helicopter with side-by-side rotors made short flights at College Park airport in Maryland in 1922. (Courtesy of College Park Airport Museum.)

Park airport, the Berliners were flying an aircraft with side-by-side rotors (Fig. 1.5). The rotors were oversized wooden propellers, but with special airfoil profiles and twist distributions. Differential longitudinal tilt of the rotor shafts provided yaw control. On later variants, lateral control was aided by cascades of wings located in the slipstream of the rotors. All variants used a conventional elevator and rudder assembly at the tail, also with a small vertically thrusting auxiliary rotor on the rear of the fuselage. The Berliner's early flights with the coaxial rotor and side-by-side rotor machines are credited as some of the first rudimentary piloted helicopter developments in the United States. However, because true vertical flight capability with these machines was limited, the Berliners abandoned the pure helicopter in favor of a hybrid machine they called a "helicoplane." This still used the rotors for vertical lift but incorporated a set of triplane wings and a larger oversized rudder. The Berliner's final hybrid machine of 1924 was a biplane wing configuration with side-by-side rotors. See also Berliner (1908, 1915).

In Britain during the 1920s, Louis Brennan worked on a helicopter concept with an unusually large single two-bladed rotor. Fay (1987) gives a good account of Brennan's work. Brennan, who was an inventor of some notoriety, had a different approach to solving the problem of torque reaction by powering the rotor with propellers mounted on the blades (Fig. 1.6). Control was achieved by the use of "ailerons" inboard of the propellers. In 1922, the machine lifted off inside a balloon shed. Further brief low altitude flights outdoors were undertaken through 1925, but the machine crashed, and further work stopped because of increasing interest in the autogiro (see Section 1.3).

During the early 1920s, Raul Pescara, an Argentinean living and working in Spain and France, was building and attempting to fly a coaxial helicopter with biplane-type rotors (Fig. 1.7). As described by Boulet (1984), each rotor had a remarkable five sets of biplane blades that were mounted rigidly to the rotor shaft. Pescara's work focused on the need

Figure 1.6 The Brennan helicopter suspended in the balloon shed at RAE Farnborough, circa 1922.

Figure 1.7 Pescara's helicopter hovering in a hanger about 1923. (Courtesy NASM Smithsonian Institution, SI Neg. No. 83-16343.)

Figure 1.8 Between 1924 and 1930, A. G. von Baumhauer made attempts to fly a single main rotor helicopter with a separately powered tail rotor. (Courtesy of NASM, Smithsonian Institution, Neg. No. 77-721.)

for complete control of the machine, which was achieved through cyclic-pitch changes that could be obtained by warping the blades periodically as they rotated. This was one of the first successful applications of cyclic pitch. Yaw was controlled by differential collective pitch between the two rotors. Early versions of his machine were underpowered, which may not be surprising considering the high drag of the bracing wires of his rotor, and the aircraft did not fly. With a later version of his helicopter using a more powerful engine, some successful flights were accomplished, albeit under limited control. However, most flights resulted in damage or serious crashes followed by long periods of rebuilding. By 1925, Pescara had abandoned his helicopter projects.

Between 1924 and 1930, a Dutchman named A. G. von Baumhauer designed and built one of the first single-rotor helicopters with a tail rotor to counteract torque reaction. Boulet (1984) gives a good description of the machine. Figure 1.8 shows that the fuselage consisted essentially of a tubular truss, with an engine mounted on one end. The other end carried a smaller engine mounted at right angles to the main rotor shaft, which turned a conventional propeller to counter the main rotor torque reaction. The main rotor had two blades, which were restrained by cables so that the blades flapped about a hinge like a seesaw or teeter board. Control was achieved by a swashplate and cyclic-pitch mechanism, which was another very early application of this mechanism. Unfortunately, the main and tail rotors were in no way connected, and this caused considerable difficulties in achieving proper control. Nevertheless, the machine was reported to have made numerous short, semicontrolled flights.

In the late 1920s, the Austrian engineer Raoul Hafner designed and built a single-seat helicopter called the R-2 Revoplane – see Everett-Heath (1986) and Fey (1987). The flights were mostly unsuccessful despite some brief tethered flights of up to a minute. His early machines used a single-rotor configuration with a pair of fixed wings located in the rotor downwash to provide antitorque. For rotor control, Hafner's machine is notable in that it used a swashplate for blade pitch, which was another early application of the mechanism. Hafner later emigrated to England, where he and Juan de la Cierva independently continued work on blade articulation for autogiros, and later, for helicopters.

1.3 The Era of the Autogiro

The Spanish engineer Juan de la Cierva had built and flown another type of rotating-wing aircraft as early as 1923 – see Juan de la Cierva (1926, 1930). This aircraft had a rotor that could freely turn on a vertical shaft but was not powered directly. Instead, the rotor disk was inclined backward at a small angle of attack, and as the machine was pulled forward by a propeller, the rotor was turned by the action of the airflow on the blades. This aerodynamic phenomenon, called autorotation, had been understood by Crocco and Yur'ev before 1910. Juan de la Cierva called his rotating-wing aircraft an "Autogiro." The name Autogiro was coined by de la Cierva as a proprietary name, but when spelled with a small "a" it is used as a generic name for this class of rotorcraft.

De la Cierva's first Autogiro was a coaxial design. The problem of asymmetric lift was well known to de la Cierva, and his idea of using a counter rotating coaxial design was that the lower rotor would counteract the asymmetry of lift produced on the upper rotor, thereby balancing the rolling moment on the aircraft. However, the aerodynamic interference produced between the rotors resulted in different rotor speeds, spoiling the required aerodynamic roll balance. De la Cierva conducted basic wind tunnel experiments on model rotors at Quatro Vientos near Madrid, and he was one of the first to establish a scientific understanding of their aerodynamic behavior. He built two more machines with single rotors before he achieved final success in January 1923 with the C-4. Based on his tests with small models, this machine incorporated blades with mechanical hinges at the root, which de la Cierva used as a means of equalizing the lift on the two sides of the rotor in forward flight – see de la Cierva & Rose (1931). The blades could flap up or down about these hinges, responding to the changing airloads during each blade revolution. Although the principle of flapping blades had actually been suggested by Charles Renard in 1904 and also patented by Louis Breguet in 1908, de la Cierva is credited with the first successful practical application.

In 1925, de la Cierva was invited to Britain by the Weir Company. His C-6 Autogiro was demonstrated at the Royal Aircraft Establishment (RAE), and these flights stimulated early theoretical work on rotating-wing aerodynamics at the RAE, mainly by Glauert and Lock. Early theoretical developments were also conducted by Munk in the United States. De la Cierva himself was to write two books for the fledgling rotorcraft industry, albeit formally unpublished, called *Engineering Theory of the Autogiro*, and *Theory of Stresses in Autogiro Rotor Blades*. In later models of his Autogiro, de la Cierva added a lag hinge to the blades, which alleviated stresses caused by in-plane Coriolis forces, and completed the development of the articulated rotor hub. A control stick was also connected to the rotor hub, which allowed the rotor disk to be tilted for control purposes (orientable direct rotor control). While this allowed the ailerons to be dispensed with, the rudder and elevator on the machine were retained. The Cierva Autogiro Company went on to build many more versions of the Autogiro, including the successful C-19 (Fig. 1.9), which is described by de la Cierva (1935).

Although the autogiro was still not a direct-lift machine and could not hover, it required only minimal forward airspeed to maintain flight. Juan de la Cierva proved that his autogiros were very safe and essentially stall-proof, and because of their low speed handling capability, they could be landed in confined areas. Takeoffs required a short runway, but this was

Figure 1.9 Cierva's C-19 successful two-seater autogiro, circa 1931.

rectified with the advent of the "jump" takeoff technique. To this end, Hafner introduced the "spider" cyclic-pitch control system to autogiros in 1934 – see Fay (1987). This provided a means of increasing collective pitch and also tilting the rotor rotor disk without tilting the rotor shaft with a control stick as in de la Cierva's direct control system. Hafner used this design in his third autogiro, the AR-3, which flew in 1935. In the "jump" takeoff technique, the blades are set to flat pitch and the rotor rpm is increased above the normal flight rpm using the engine. This is followed by the rapid application of collective pitch, while simultaneously declutching the rotor and thereby avoiding any torque reaction on the fuselage. This technique lifts the aircraft rapidly off the ground, powered only by the stored kinetic energy in the rotor system. As forward speed builds, the rotor settles into its normal autorotative state – see also Prewitt (1938). With its jump takeoff capability and low handling speeds, the autogiro was to closely rival the helicopter in performance capability.

Several other British companies including Weir, Avro, de Havilland, and Westland went on to build variants of the de la Cierva autogiro designs. The first Weir designs were developments of de la Cierva's models and used the orientable direct rotor control system. The Weir W-1 through W-4 models were all autogiros and were some of the first machines to use a clutch to help bring up the rotor rpm prior to takeoff. The de Havilland and Westland companies built a few larger prototype autogiros. The Westland C-29 was a five-seat cabin autogiro built in 1934. The aircraft was never flown because it exhibited serious ground resonance problems, and the project was canceled with the untimely death of Juan de la Cierva in 1936. However, de la Cierva's work was carried on by designers from Weir, and another Westland designed autogiro called the CL-20 was flown just before World War II – see Mondey (1982).

The Kellett and Pitcairn companies entered into licensing agreements with de la Cierva, resulting in the first flight of an autogiro in the United States in 1928. Pitcairn went on to design and patent many improvements into the de la Cierva rotor system [see Smith (1985)], but it became clear that it was a true helicopter with power delivered to the rotor shaft that was required. The autogiro was extensively tested in the United States by the NACA. Gustafson (1971) gives an authoritative account of the early NACA technical work on autogiros and

Figure 1.10 The Jerome–de Bothezat helicopter, which made limited controlled flights in 1922. (Courtesy NASM, Smithsonian Institution, SI Neg. No. 87-6022.)

helicopters.[3] In Russia, the TsAGI built autogiros derived from the de la Cierva designs. Kuznetsov and Mil built the 2-EA, which was derived from the Cierva C-19 – see Everett-Heath (1988). Later developments of this design led to the first Russian helicopters built with the assistance of Vittorio Isacco, who had earlier led basic helicopter developments in Italy during the 1920s.

1.4 The First Successes with Helicopters

In 1922, a Russian émigré to the United States by the name of Georges de Bothezat built one of the largest helicopters of the time. De Bothezat had been a student of Professor Joukowski in Russia and had written one of the first technical manuscripts on rotating-wing aerodynamics – see de Bothezat (1919). De Bothezat's machine was a quad-rotor with a rotor located at each end of a truss structure of intersecting beams, placed in the shape of a cross (Fig. 1.10). Ivan Jerome was the codesigner. Each rotor had six wide chord blades. Control of the machine was achieved by collective, differential collective, and cyclic blade pitch variations, and the design likely derived directly from those of Yur'ev. In 1922, the ungainly Jerome–de Bothezat quad-rotor flew successfully many times, albeit at low altitudes and forward speeds. However, because of insufficient performance and the increasing military interest in autogiros at the time, the project was canceled.

In 1920, Étienne Oemichen of France built a quad-rotor machine in a similar style to that of de Bothezat's, but with a number of additional rotors for control and propulsion. His machine typified the cumbrous complexity of the various helicopters of the time. His initial design was underpowered, and it had to have a hydrogen balloon attached to provide additional lift and stability. However, he went on to design a "pure" helicopter that was flown between 1923 and 1924. By 1924, Oemichen's machine proved "perfectly maneuverable and stable," and he was awarded a prize by the FAI for demonstrating the first helicopter to fly a standard closed 1 km circuit. See also NACA (1921) and Oemichen (1923).

[3] In addition, the entire first issue of the *Journal of the American Helicopter Society*, 1 (1), Jan. 1956, was devoted to the early autogiro and helicopter developments in the United States.

Figure 1.11 Corradino d'Ascanio's coaxial machine, circa 1930, which used servo-tabs on the blades to change their lift. (Courtesy John Schneider.)

In 1930, Corradino d'Ascanio of Italy built a highly successful coaxial helicopter. This machine held various speed and altitude records for the time. Figure 1.11 shows that this relatively large machine had two, two-bladed, counterrotating rotors. Following the work of de la Cierva, the blades had hinges that allowed for flapping and a feathering capability to change blade pitch. Control was achieved by auxiliary wings or servo-tabs on the trailing edges of the blades, a concept that was later adopted by others, including Bleeker and Kaman in the United States. D'Ascanio designed these servo-tabs so that they could be deflected periodically by a system of cables and pulleys, thereby cyclically changing the lift on the blade. For vertical flight, the tabs on all the blades moved collectively to increase the rotor thrust. Three small propellers mounted to the airframe were used for additional pitch, roll, and yaw control.

In 1930, Maitland Bleeker of the United States followed Brennan's approach to the torque reaction problem by delivering power to propellers that were mounted on each rotor blade. Power was supplied through a system of chains and gears from an engine mounted at the center of the machine. Like d'Ascanio's machine, Bleeker's helicopter was controlled by auxiliary aerodynamic surfaces he called "stabovators," which were fastened to the trailing edges of each blade. Both collective and cyclic pitch were incorporated into the design. Bleeker's machine accomplished numerous precarious hovers, but vibrations and control problems caused the project to be abandoned in 1933. Liberatore (1998) gives one of the best accounts of the project.

During 1929–1930, the Russian born engineer Nicolas Florine built a tandem rotor helicopter in Belgium. The rotors turned in the same direction but were tilted in opposite directions to cancel torque reaction. Boulet (1984) describes the mechanical aspects of the machine. Florine's first aircraft was destroyed in 1930, but he had a second design flying successfully by 1933, which made a flight of over 9 minutes and exceeded d'Ascanio's record of the time. Yet, Florine's designs suffered many setbacks, and work was discontinued in the mid 1930s.

During the period 1930–1936, the famous French aviation pioneers Louis Breguet and René Dorand made notable advances in helicopter development. Figure 1.12 shows their

Figure 1.12 The successful Breguet-Dorand coaxial helicopter, circa 1936. (Courtesy NASM, Smithsonian Institution, SI Neg. No. A-42078-c.)

machine of 1935, which was relatively large with coaxial rotors. Boulet (1984) and Kretz (1987) give an excellent account of the work. Each rotor had two tapered blades that were mounted with flap and lag hinges and were controlled in cyclic pitch using a swashplate design. Yaw control was achieved by differential torque on one rotor with respect to the other. Horizontal and vertical tails were used for increased stability. For its time, the aircraft had held several records, including a duration of 62 minutes and distance flown of 44 km (27 mi). Autorotations[4] were attempted with the Breguet–Dorand machine, some moderately successful, but the aircraft crashed and further work stopped prior to the outbreak of World War II.

In 1937, Heinrich Focke in Germany, in conjunction with Georg Wulf and later Gerd Achgelis, demonstrated a successful lateral side-by-side, two-rotor machine, called the Fa-61. The details of the machine are described by Focke (1938, 1965) and Boulet (1984).

[4] All aircraft must possess safe flight characteristics after a loss of power. While a fixed-wing aircraft can glide, the helicopter can take advantage of autorotation with the rotor unpowered as a means of maintaining rotor rpm, lift, and control in the event of engine failure. However, to get the rotor to autorotate, the helicopter must descend (at a relatively high rate) so that the relative wind comes upward through the rotor disk. The pilot, in effect, gives up altitude (potential energy) at a controlled rate for kinetic energy to drive the rotor and, with care, can "autorotate" the aircraft safely onto the ground. The ability to autorotate is a distinguishing feature of a successful helicopter.

Figure 1.13 The Focke Fa-61 flew successfully in 1938. (Courtesy NASM, Smithsonian Institution, SI Neg. No. A-2316.)

This machine was constructed from some rotor components provided by the Weir company. Figure 1.13 shows that the rotors were mounted on outriggers and were inclined slightly inward to provide lateral stability. The blades were tapered in planform and were attached to the rotor hub by both flapping and lagging hinges. Longitudinal control was achieved by tilting the rotors fore and aft by means of a swashplate mechanism, while yaw control was gained by tilting the rotors differentially. The rotors had no variable collective pitch, instead using a slow and clumsy system of changing rotor speed to change the rotor thrust. A vertical rudder and horizontal tail provided additional directional stability.

The Fa-61 machine is significant in that it was the first helicopter to demonstrate successful autorotations. Provision was made in the design for a fixed low collective pitch setting to keep the rotor from stalling during the descent. It also set records for duration (80 minutes), climb to altitude (3,427 m, 11,427 feet), forward speed (122 km/hr, 76 miles per hour), and distance (230 km, 143 miles). The machine gained a certain amount of notoriety prior to the outbreak of World War II when the famous test pilot Hanna Reitsch flew it inside a Berlin sports arena. The Fa-61 aircraft was used as a basis to develop the Fa-266 (Fa-233E), which first flew in 1940. This was a fairly large aircraft, with two three-bladed rotors, and could carry up to four crew. It went into limited production during the Second World War. Boulet (1984) gives a good account of the later helicopter work of Focke. After the war, some of the German machines were used as a basis to develop helicopters in Russia, for which a comprehensive account is given by Everett-Heath (1988).

With the assistance of Juan de la Cierva, the Weir Company had formed an aircraft department in Scotland in 1932. The W-5 was their first true helicopter design, since the W-1 through W-4 all had been autogiros. Initially, the W-5 was a coaxial design, but concerns about stability and control led to the redevelopment as a side-by-side configuration, which flew successfully in June 1938. Control was achieved with cyclic pitch but there was no collective pitch; vertical control was obtained by altering the rotor speed, a cumbersome feature used also on the Fa-61. The Weir W-5 (and later the W-6) and the Fa-61 were ahead of Sikorsky's VS-300. The W-6, which first flew in 1939, was a much larger version of the W-5 but still used the lateral side-by-side rotor configuration.

During the period 1938–1943, Antoine Flettner, also of Germany, developed several helicopter designs. Flettner's success came with using a side-by-side intermeshing rotor configuration, which became known as a *synchropter*. This rotor idea was first patented by Bourcart in 1903 and by Mees in 1910. In the synchropter design, the rotor shafts were close together but arranged so that they were at a significant outward angle with the overlapping rotors turning in opposite directions. A gearing system ensured the exact phasing of the rotors. In 1939, Flettner's Fl-265 synchropter was the first helicopter to demonstrate transition into autorotation and then back again into powered flight. Flettner built several other machines, including the Fl-282 Hummingbird.[5] After World War II, in the United States, the Kellett Aircraft Company (which also built autogiros as a licensee to Pitcairn) adopted Flettner's synchropter configuration but used three-bladed instead of two-bladed intermeshing rotors. The aircraft flew very successfully, but it never went into production. The synchropter concept was also adopted by Charles Kaman, whose company Kaman Aircraft Corp. was later to put the type into successful production.

As described earlier, Igor Sikorsky had experimented in Czarist Russia with primitive vertical lift aircraft as early as 1907 – see Sikorsky (1938) and Finne (1987). After Sikorsky had emigrated to the United States, he went on to design and build giant flying boats. In 1935, Sikorsky was issued a patent, which showed a relatively modern looking single rotor/tail rotor helicopter design with flapping hinges and a form of cyclic pitch control. Although Sikorsky encountered many technical challenges, his first helicopter, the VS-300, was flying by May 1940. A good summary of the technical design is given by Sikorsky (1941, 1942, 1943). His first machine had one main rotor and three auxiliary tail rotors, with longitudinal and lateral control being obtained by means of pitch variations on the two vertically thrusting horizontal tail rotors. The main lifting rotor of the VS-300 was used in the later VS-300A, but only the vertical tail rotor was retained out of the original three auxiliary rotors for antitorque and directional (yaw) control purposes. In this configuration, longitudinal and lateral control was achieved by tilting the main rotor by means of cyclic-pitch control, while directional control was achieved by pitch variations of the tail rotor. This configuration was to become the standard for most modern helicopters. By 1941, Sikorsky had started production of the R-4 (Fig. 1.14) and in 1943 Sikorsky developed the R-5, which, although still only a two-seater helicopter, was much larger and had better performance than the R-4.

In 1946 Westland Helicopters in Great Britain obtained a license to build the Sikorsky machines, the first being designated as the WS-51 after the S-51, which was the first commercial helicopter designed by Sikorsky. This period was the start of a long relationship between the two companies which continues today. Westland already had a history as a successful fixed-wing manufacturer but decided to specialize in helicopters in 1946. After significantly reengineering the Sikorsky machine, Westland called the aircraft the Dragonfly. The Widgeon later followed, and this was a very modern looking and powerful version of the Dragonfly with a larger passenger cabin.

During 1942, the Weir Company, prompted by the success of Sikorsky's R-5, proposed a rather large single-rotor machine called the W-9, which was rather unique in its use of jet thrust to counteract rotor torque reaction – see Everett-Heath (1986). However, because the rotor lacked any collective pitch control, rotor thrust was controlled by changing rotor speed as in the pre-war Weir W-5/6 models. The W-9 crashed during a test flight in 1946, and the project was subsequently abandoned. The Weir and Cierva companies later went

[5] The Fl-282 was probably one of the first helicopters in production; however, like the Focke helicopters, numbers were limited because of the Second World War.

Figure 1.14 The famous Sikorsky R-4B, circa 1944.

on to design the W-11 Air-Horse, which was an unorthodox three-rotor helicopter. The final helicopter of the Weir–Cierva line was the W-14 Skeeter, which was a small two-seater training helicopter, but this saw limited production.

It is significant to note that while helicopters were becoming more and more successful, the development of the autogiro continued in Europe and the United States well into the 1950s. Considerable development work was undertaken by The Pitcairn Company [see Pitcairn (1930) and Smith (1985)] and the Kellett Aircraft Companies in the United States. Harold Pitcairn patented over 100 concepts in rotor blade design and rotor control, some of which were later licensed to Sikorsky. The Pitcairn and Kellett autogiros or "Mailwings" flew on part of Eastern Airlines' network delivering mail for the United States Post Office. Even into the 1960s small single- and two-seat autogiro designs were being developed in the United States by Umbaugh and McCulloch for the private market.

In Great Britain during the 1940s and 1950s, the autogiro concept was pursued to some significant end by the Fairey Aviation Company. The Fairey Girodyne compound aircraft used a propeller set on the end of a stub wing to provide both propulsion and antitorque – see Everett-Heath (1986). The Fairey Company went on to develop the Jet Girodyne in which the rotor system was driven by tip jets. This ultimately led to the Rotodyne, which was the world's biggest giroplane with a cabin big enough for forty passengers – see Hislop (1958). The aircraft set a world speed record for a convertiplane in 1959 before the project was canceled. Smaller single- and two-seat autogiros were later built in Britain by the Wallis company.

Besides Igor Sikorsky, there were several other helicopter design pioneers in the United States during the 1940s. These included Arthur Young, Frank Piasecki, Stanley Hiller, and Charles Kaman. In the late 1930s, Arthur Young began a series of experiments with model helicopters that were ultimately to lead to the design of the renowned Bell-47 helicopter. After much research, Young invented the teetering rotor with a stabilizer bar; see Young (1948, 1979). The bar had bob weights attached to each end and was directly linked to the rotor blades through the pitch control linkages. The idea was that if the rotor was disturbed in pitch or roll, the gyroscopic inertia of the bar could be used to introduce cyclic pitch into the main rotor system, increasing the effective damping to disturbances and giving stability to the rotor system – see also Kelly (1954).

Figure 1.15 A version of the Bell-47, which was the world's first commercially certified helicopter. (By permission of Bell Helicopter Textron.)

The prototype Bell-30 was built in 1942 and had a single main teetering rotor with Young's stabilizer bar. The first untethered flights took place in 1943, and the machine was soon flying at speeds in excess of 70 mph. The Bell-30 led to the famous Bell Model 47 (Fig. 1.15), which was the world's first commercially certified helicopter. During its nearly thirty-year manufacturing period over 5,000 were produced in the United States alone, and many were also license built in more than twenty other countries. Tipton (1989), Brown (1995) and Spenser (1999) give good historical overviews of the Bell machines. Schneider (1995) gives a brief biography of Arthur Young and his novel teetering rotor design.

In 1943, Frank Piasecki designed and flew a small helicopter that was called the PV-2. This was the second successful helicopter to fly in the United States after Sikorsky's VS-300. Piasecki immediately turned to larger helicopters, and in 1943 the Piasecki Helicopter Corporation built a tandem rotor helicopter called the PV-3 Dogship. Details are given by the Piasecki Aircraft Corp. (1967) and Spenser (1999). This aircraft was popularly called the "Flying Banana" because of its distinctive curved fuselage shape. Despite its nickname, however, the aircraft was very successful and larger and more powerful versions quickly followed, including the H-16 and H-21 "Workhorse" models of 1952. The Vertol Company incorporated Piasecki's corporation in 1956 and went on to develop the Vertol 107 and two highly successful military tandem rotor models, the CH-46 and CH-47 (Fig. 1.16). The company finally became Boeing Helicopters. An overview of the Boeing–Vertol machines produced up to the mid-1970s is given by Grina (1975). In the late 1980s, the company produced a demonstrator of an advanced technology tandem rotor helicopter called the Model 360. The only other company in the United States to build a tandem helicopter design was Bell, who manufactured the XSL-1 during the 1950s. In 1998, Boeing announced the launch of the CH-47F and the CH-47SD "Super-D" Chinook. See also The Boeing Company (1999).

The British company, Bristol Helicopters, had designed and built a tandem helicopter during the late 1940s under the leadership of the helicopter pioneer Raoul Hafner – see Hobbs (1984) and Everett-Heath (1986). The Bristol Type-173 had a long, slim fuselage

Figure 1.16 The Boeing-Vertol CH-47 – one of the largest and most widely used military transport helicopters. (By permission of the Boeing Company.)

with two three-bladed rotors at each end, similar to the Piasecki machines. The Bristol Type-192 Belvedere was an improved tandem rotor design, which followed in 1958 with more powerful engines. The other Bristol design of note was the single rotor Type 171 Sycamore – see Hafner (1949).

In the United States, Charles Kaman adopted Antoine Flettner's synchropter rotor design. One of Kaman's innovations was the use of torsionally compliant solid spar spruce rotor blades with servo-flaps. The servo-flaps were mounted at the three-quarter rotor radius, some distance behind the elastic axis of the blade. When these flaps were deflected cyclically, the aerodynamic moments caused the blades to twist, changing their angle of attack and thus introducing a cyclic rotor control capability – a system first used by d'Ascanio. The first Kaman helicopter, the K-125A, flew in 1947. An improved version, the K-225, became the first helicopter to fly powered by a gas turbine. A family of larger machines, known as the H-43 Huskie and its derivatives, were produced through 1964. While Kaman reverted to conventional single-rotor helicopter designs in the later 1950s, the servo-flap concept continued to be a trademark of the Kaman helicopters. The H-2 Seasprite first flew in 1959 and has been produced in considerable numbers. Kaman has recently returned to the synchropter concept with the design of the K-Max (Fig. 1.17), which first flew in 1991. See Kaman Aircraft Corp. (1999) for further details of their helicopter line.

Stanley Hiller is another pioneer in the development of the modern helicopter [see Straubel (1964) and Spenser (1992, 1999)]. Hiller built several helicopter prototypes, including the coaxial XH-44 of 1943. Hiller's later machines used a conventional main rotor and tail rotor configuration. His main breakthrough was the "Rotormatic" main rotor design, where the cyclic pitch controls were connected to a set of small auxiliary blades set at ninety degrees to the main rotor blades. These auxiliary blades provided damping in pitch and roll helping to augment the hovering stability of the machine. It is significant to note that both Hiller and Young designed in stability-producing mechanisms for their helicopters, whereas the Sikorsky machines had none and so they had a reputation for being harder to fly. While the Hiller machines are probably less well known than those of Sikorsky

Figure 1.17 A modern synchropter design, the Kaman K-Max. (By permission of Kaman Aircraft Corporation.)

or Bell, the Hiller company went on to build many thousands of helicopters, including the 360 model and later the UH-12A and H-23.

1.5 Maturing Technology

The early 1950s saw helicopters quickly maturing into safe, successful, and viable aircraft that were easier to fly and more comfortable for crew and passengers alike. This era is marked by significant mass production of helicopters by various manufacturers in the United States. The Sikorsky S-55 and S-58 models were probably two of the greatest single advances in helicopter design. These aircraft had a large cabin under the rotor, and to give a wide allowable center of gravity position, the engine was placed in the nose. Westland also maintained their relationship with Sikorsky and built versions called the S-55 Whirlwind and S-58 Wessex. The 1960s saw the development of the Sikorsky S-61 Sea King, the S-64 Sky Crane, and the larger CH-53 models. Later, the S-70 (UH-60) Blackhawk was to become the mainstay of the Sikorsky company, and the machine is expected to remain in production well into the twenty-first century. The civilian S-76 has been successful in its role as an executive transport and air ambulance, amongst other roles. In the 1970s, Sikorsky and Boeing teamed to build the military RAH-66 Comanche (Fig. 1.18), which will be a scout/attack helicopter for the new millennium. The latest Sikorsky machine, the civilian medium lift S-92 Helibus, flew for the first time in 1998. For more information, see Sikorsky Aircraft (1999).

The success with the Model-47 led Bell Helicopter to develop the UH-1 Huey, which was delivered starting in 1959. The Bell 212 was a two-engine development of the UH-1D and proved to be a successful military and civilian machine. The Huey-Cobra (Fig. 1.19) also grew out of the UH-1 series, retaining the same rotor components, but having a more

Figure 1.18 The Sikorsky/Boeing RAH-66 Comanche, which first flew in 1997. (By permission of the Boeing Company.)

streamlined fuselage with the crew seated in tandem. The type is still in production in 1999 as the AH-1W Super-Cobra, which uses an advanced composite four-bladed rotor. The Bell 412 is basically a 212 model, but with a four-bladed composite rotor replacing the two-bladed teetering rotor. Bell also conquered the civilian market with its Model 206 Jet-Ranger and variants, which first flew in 1966 and has become one of the most widely used helicopters. The OH-58 military version was sold in considerable numbers and with sustained improvements over the years, with the OH-58D having an advanced four-bladed

Figure 1.19 The Bell AH-1 Huey-Cobra. (By permission of Bell Helicopter Textron.)

Figure 1.20 The McDonnell–Douglas (now Boeing) AH-64 Apache. (By permission of the Boeing Company.)

rotor with mast-mounted sight. One of most recent civilian models is the Bell 427, which is an eight-place light twin. See also Bell Helicopter Textron (1999).

Hughes built the military TH-55 and later the Hughes-500 series, which has seen extensive civilian use in various models. However, the AH-64 Apache, which was designed in 1976, proved to be the biggest success story for the Hughes company, which became part of McDonnell-Douglas in 1984. The AH-64D Longbow model (Fig. 1.20) is still in production over twenty years later. It is also produced under license in the UK by GKN-Westland. McDonnell-Douglas has also produced a line of light commercial helicopters including the MD 500 and 600 series, and it most recently has marketed the MD-900 Explorer. This aircraft uses a new bearingless rotor design and the 'No Tail Rotor' (NOTAR) circulation control antitorque concept (see Chapter 6).

Although the bulk of helicopters produced are for the military, several manufacturers produce training helicopters or helicopters aimed at the general aviation market, including Robinson, Schweizer, and Enstrom. In the United States, Robinson produces the R-22 two-seat and R-44 four-seat helicopters. Both are powered by piston engines. Schweizer produces an updated version of the two-seat Hughes 300 for the training market, and a larger derivative, designated as the Model-330, has a gas turbine.

European manufacturers such as Aerospatiale, Agusta, MBB, and Westland have produced many successful helicopter designs since the 1960s. Agusta and Westland have also license-produced helicopters designed in the United States, such as those of Sikorsky and Bell. The Aerospatiale (Sud-Aviation) Alouette was one of the most successful European helicopters, and in 1955 it was one of the first machines to be powered by a gas turbine. The

Aerospatiale Super Frelon was a large transport machine, first flown in 1962. In the early 1970's the Aerospatiale/Westland SA330 Puma became Europe's best selling transport helicopter. The Aerospatiale/Westland Gazelle was a successful successor to the Alouette, first flown in 1967, and it introduced the fenestron tail rotor.[6] The Dauphin, first flown in 1972, used an improved fenestron tail rotor and a composite main rotor hub. Messerschmitt-Bölkow-Blohm (MBB) introduced the BO105 in 1967 with a hingeless titanium rotor, with the larger and more capable BK117 machine first flying in 1979. In the 1990s, Aerospatiale and MBB joined resources to form Eurocopter, which produces a large number of civilian and military helicopter models – see Eurocopter (1999).

In 1952, Agusta purchased a license to build the Bell Model-47, and through 1965 it built several variants of the Bell machine to its own specifications. Agusta also began to design its own machines, with the large three-engined A-101 flying in 1964, but it never went into production. The Agusta A-109 was one of the most aerodynamically attractive helicopters. First flown in 1971, this high-speed transport and multirole helicopter has been very successful and is used in both civilian and military roles. The A-129 Mangusta, first flown in 1983, is a military version of the A109 with a different fuselage.

Westland Helicopters (now GKN-Westland) has been a key player in British aviation since the 1930s – see Mondey (1982). The earliest helicopters built by Westland were under license from Sikorsky, but these were significantly modified to meet British airworthiness standards. During 1959–1960, Westland took over the operation of the Bristol, Saunders-Roe, and Fairey companies. Saunders-Roe (SARO) had previously taken over the Cierva Company in 1951. The Westland/SARO/Cierva Skeeter was a small two-seat trainer, which led to the bigger and relatively successful Wasp in 1962. The Westland Wessex was a development of the Sikorsky S-58, which was built in many configurations through 1970. The Sea King and Commando were derived from the S-61, which were steadily improved upon since the first models flew in the late 1960s. The latest versions of the Sea King sold through 1990 have used composite rotor blades and various airframe improvements. Westland designed its own line of helicopters, starting with the military Lynx, which first flew in 1971. The Westland WG-30 was a larger multirole transport version of the Lynx. Although this aircraft saw some civilian use, production was limited. New versions of the Lynx (Super Lynx) are fitted with the Westland/RAE British Experimental Rotor Program (BERP) blade, which has improved airfoil sections and special tip shapes (see Chapter 6). A Lynx with the BERP rotor currently holds the absolute straight line speed record for a single-rotor helicopter at some 250 kts (400 km/h; 287 mi/h). The BERP blade design is also used on the Westland-Agusta built EH Industries EH-101 (Fig. 1.21), which is a medium-lift helicopter that entered production in 1996 in both civilian and military variants. Westland also has a license agreement to build the WS-70 Blackhawk and AH-64 Apache. See also GKN-Westland (1999) for more information on the current lineage.

Significant numbers of helicopters have also been built in the former Soviet Union. In the 1930s, the TsAGI Technical Institute in Moscow built a series of autogiros based on the de la Cierva designs. Everett-Heath (1988) gives a good account of the early work. Later, work with the Focke-Achgelis company of Germany resulted in a number of prototype helicopter designs with a lateral side-by-side rotor configuration. The Mil, Kamov, and Yak companies all went on to build successful helicopter lines. An overview of the early Russian machines is given by Free (1970). Mikhail Mil adopted the single main rotor tail rotor configuration, with the Mi-1 flying in 1950. The Mi-2 was a turbine-powered

[6] The fenestron is a ducted tail rotor design, fully integrated into the fuselage and vertical fin. The name "fenestron" comes from the French for "little window."

Figure 1.21 A civil variant of the EH Industries EH-101. (By permission of GKN-Westland Helicopters.)

version. The more efficient Mi-3 and larger Mi-4 machines quickly followed. The Mi-4 looked very much like the S-55, but it was much bigger and more capable. The Mi-2 was also built in significant numbers in Poland, with the Mi-4 being produced in China. The Mi-6 of 1957 was one of the largest helicopters ever built, with a rotor diameter of 35 m (115 ft) and a gross weight of over 42,500 kg (93,700 lb). This was followed by the smaller Mi-8 (similar to the Mi-4), which went into civilian service. The Mi-10 of 1961 was a flying crane development of the Mi-6, with a tall, wide, quadricycle landing gear. However, the credit for the world's largest and heaviest helicopter goes to the Russian Mil Mi-12. This aircraft had a lateral side-by-side rotor configuration, with the span of the aircraft from rotor tip to rotor tip exceeding that of the wing span of the Boeing 747. Power was provided by four gas turbines, installed as pairs at the end of each wing pylon. The Mi-24 assault/transport helicopter was designed in 1972, and it has been produced in large numbers. The Mi-26 entered service in 1982 and is the largest helicopter currently flying. The Mi-28 is an attack helicopter, similar in configuration to the AH-64 Apache. The latest Mil design, the Mi-38, is planned as a successor to the Mi-8/17 and is similar in size and weight to the Westland/Agusta EH-101.

The Kamov company built a series of very successful light and medium weight coaxial rotor helicopter designs, including the Ka-15 and Ka-18 in 1956 and the Ka-20 in 1961. Kamov was the only company to ever put the coaxial helicopter design into mass production. The Ka-25 and most of the later models were all gas turbine powered. The Ka-27 and the civilian model Ka-32 have been in production since 1972. One of the most recent Kamov designs is the Ka-50 (Fig. 1.22), which is a lightweight attack helicopter of considerable performance. One exception to the Kamov coaxial line was the Ka-22 convertiplane of 1961. Another new design is the Kamov Ka-62, which is a conventional light utility helicopter design incorporating a fenestron that, at the time of writing, was in the prototype stage. See the Kamov Company (1999) for more information. Alexander Yakolev built many successful fixed-wing designs, but with the assistance of Mil designed the large tandem Yak-24 helicopter in the early 1950s. This helicopter was produced from about 1952 to 1959, but it was not very successful. Further information on Russian helicopter developments is given by Anoschenko (1968) and Everett-Heath (1988).

Figure 1.22 The Kamov Ka-50 helicopter, which has a coaxial rotor design. (Photo courtesy of Jari Juvonen.)

1.6 Tilt-Wings and Tilt-Rotors

The conventional helicopter is limited in forward flight performance by the aerodynamic lift and propulsion limitations of the main rotor. These rotor limits arise because of compressibility effects on the advancing blade, as well as stall on the retreating blade. In addition, the relatively high parasitic drag of the rotor hub and other airframe components leads to a relatively poor overall lift-to-drag ratio of the helicopter. This generally limits performance of conventional helicopters to level-flight cruise speeds in the range of 150 kts (278 km/h; 172 mi/h), with dash speeds up to 200 kts (370 km/h; 230 mi/h). Although somewhat higher flight speeds are possible with compound designs, which use auxiliary propulsion devices and wings to offload the rotor (see Section 6.7.1), this is always at the expense of much higher power required and fuel burned than would be necessary with a fixed-wing aircraft of the same gross-weight and cruise speed.

The need for a machine that could combine the benefits of vertical takeoff and landing (VTOL) capability with the high speed cruise of a fixed-wing aircraft has led to the evolution of *tilt-wing* and *tilt-rotor* concepts. A good history of the many VTOL designs, including tilt-wings and tilt-rotors, is given by Hirschberg (1997). However, this potential capability comes at an even greater price than for a conventional helicopter, including increased mechanical complexity, increased weight, and the susceptibility for the rotors and wing to exhibit various aeroelastic problems.

The tilt-wing is basically a convertiplane concept, but it never became a viable rotating-wing concept to replace or surpass the helicopter. The idea is that the wing can be tilted from its normal flying position with the propellers providing forward thrust, to a vertical position with the propellers providing vertical lift. Several companies seriously considered the tilt-wing concept in the 1950s, with Boeing, Hiller, Vought-Hiller-Ryan, and Canadair all producing flying prototype aircraft. The Boeing-Vertol VZ-2 first flew in 1957 and went on to make many successful conversions from hover into forward flight. However, the flow separation produced by the wing as it stalled during the conversion flight regime resulted in some difficult piloting, and these issues were never satisfactorily resolved. The Hiller X-18 was a large tilt-wing aircraft compared to the VZ-2. The aircraft used two large diameter counterrotating propellers (from the earlier Ryan Pogo concept)–see Straubel (1964). The aircraft underwent flight testing in 1960, but the program was canceled in 1961 after the aircraft suffered a loss of control. In the 1980s, the Ishida Co. developed the TW-68 tilt-wing aircraft as a private venture, but the company went into bankruptcy before the aircraft could be completed.

The tilt-rotor aircraft takes off and lands vertically with the rotors pointed vertically upward like a helicopter. For forward flight, the wing-tip-mounted rotors are progressively tilted to convert the aircraft into something that looks like a fixed-wing turboprop airplane. In this mode, the tilt-rotor is able to achieve considerably higher flight speeds (about 300 kts; 555 km/h; 344 mi/h) than would be possible with a helicopter. Therefore, the tilt-rotor combines some attributes of the conventional helicopter with those of a fixed-wing aircraft. Because the rotors of a tilt-rotor cannot be as large as those of a helicopter, the hovering efficiency of the tilt-rotor is not as high as that of a helicopter.[7]

The tilt-rotor concept was first demonstrated in a joint project between the Transcendental Aircraft Corporation and Bell in 1954. The first aircraft, the Model 1-G, had two three-bladed fully articulated rotors. Various technical problems were encountered, especially in the conversion from helicopter mode to fixed-wing flight. Bell later led the development of the XV-3 in 1951, which had two fully articulated three-bladed rotors. The XV-3 was damaged in an accident in 1956 after an aeroelastic problem with the rotor. The second XV-3 used a two-bladed teetering rotor system, and the aircraft was successfully flown in 1958. However, several aeromechanical problems were again encountered, including pylon whirl flutter.

By the late 1960s, Bell had developed the Model 266 tilt-rotor and later the Model 300. Various wind-tunnel tests of scaled models led to an improved understanding of the rotor and wing aeroelastic issues involved with tilt-rotors, especially during the conversion mode, and Bell continued to develop the Model 301. This aircraft later became the XV-15, which fully demonstrated the viability of the tilt-rotor concept; but the aircraft was never designed with production in mind. However, in 1983 the much larger V-22 Osprey tilt-rotor program was begun (see Fig. 1.23). This joint Bell/Boeing project has resulted in several test and preproduction aircraft, and in 1997 the decision was made to put the aircraft into production for the United States Navy and Marines. In 1997, Bell announced the development of the Model 609 civilian tiltrotor, which will be capable of transporting 9 passengers at 275 kts (509 km/h; 315 mi/h) over 750 nm (1,390 km; 860 mi) sectors. See Bell Helicopter Textron (1999) for further information.

1.7 Chapter Review

During the past sixty years since the first successful flights, helicopters have matured from noisy, unstable, rickety vibrating contraptions that could barely lift the pilot and a small payload into sophisticated machines of quite extraordinary capability. Igor Sikorsky's dream has certainly been fulfilled, perhaps in many ways that he could not have imagined. In 1999, there were about 40,000 helicopters flying worldwide. Its roles encompass air ambulance, sea and mountain rescue, police surveillance, corporate services, and oil rig servicing in the civilian world and troop transport and anti-tank gunships in military use. In rescue operations alone, the helicopter has saved the lives of well over a million people.

[7] In the design of the Bell-Boeing V-22 Osprey, the rotor diameter was also limited by the need to operate and hangar the aircraft on board an aircraft carrier.

Figure 1.23 The Bell/Boeing V-22 Osprey tilt rotor. (By permission of the Boeing Company.)

Sustained scientific research and development in many aeronautical disciplines has allowed for dramatic increases in helicopter performance, overall lifting and cruise efficiencies, and mechanical reliability. Continuous aerodynamic improvements to rotor efficiency have allowed the helicopter to lift a payload that is more than its empty weight and to fly in level flight at speeds in excess of 200 kts (370 km/h; 229 mi/h). Since the 1980s, there has been an accelerating scientific effort to understand and overcome some of the most difficult technical problems associated with helicopter flight, particularly in regard to aerodynamic limitations associated with the main rotor. The improved design of the helicopter and the increasing viability of other vertical lift aircraft such as the tilt-rotor continue to advance as a result of the revolution in computer-aided design and manufacturing and the advent of new lightweight composite materials. The helicopter today is a safe, versatile, and reliable aircraft, and it will continue to be an indispensable part of modern life well into the twenty-first century.

1.8 Questions

1.1. Give an overview of the main technological problems that were encountered when trying to design and fly a successful helicopter prior to 1900. In your discussion, consider issues such as overall levels of aeronautical knowledge, stability and control requirements, and the availability of suitable power-plants and construction materials.

1.2. The development of the autogiro formed the basis for the design of the modern helicopter. Yet, the autogiro has been much less of a commercial success than the helicopter. What are the physical design features that distinguish a helicopter from an autogiro? Discuss the relative merits of the helicopter versus the autogiro from the standpoint of maximum speed capability, capital costs, and cruise efficiency.

1.3. Most modern helicopters are of the single main-rotor and tail-rotor (conventional) configuration. For the same overall aircraft gross weight, what might be the relative advantages of a tandem rotor helicopter over a conventional helicopter? Also, compare the potential relative merits of a coaxial rotor design over a tandem design.

1.4. Although tilt-wing and tilt-rotor concepts date back more than forty years, it is only in the past few years that a civilian tilt-rotor aircraft has been proposed. Discuss the technical, economic, and any other reasons for this long gestation period.

1.5. For a truly heavy lift helicopter, say in excess of 100,000 kg (220,000 lb) gross weight, a single-rotor design would be considered impractical. Discuss the reasons why this might be so.

1.6. So-called 'compound' helicopter designs have enjoyed periodic popularity over the last 50 years, but none have entered production. What is a 'compound' helicopter? Briefly, suggest some of the advantages and disadvantages of a compound helicopter design. Explain the difference between 'lift' compounding and 'thrust' compounding.

Bibliography

Achenbach, W. 1923. "Variation in the Number of Revolutions of Air Propellers," NACA Technical Note 131.

Anderson, J. D., Jr. 1997. *A History of Aerodynamics: And Its Impact on Flying Machines*, Cambridge University Press, Cambridge, England.

Anoshchenko, N. D. (ed.) 1968. "History of Aviation and Cosmonautics, Vol. 5," NASA TT F-11851.

Apostolo, G. 1984. *The Illustrated Encyclopedia of Helicopters*, Bonanza, New York.

Balaban, K. 1923. "Evolution of the Helicopter," NACA TN 196.

Bartlett, R. M. 1947. *Sky Pioneer: The Story of Igor I. Sikorsky*, Charles Scribner's & Sons, New York.

Bell Helicopter Textron. 1999. http://www.belltextron.com/

Berliner, E. 1908. "The Berliner Helicopter," *Aeronautics*, Vol. 3.

Berliner, E. 1915. "Elements of a Girocopter," *Aeronautics*, Vol. 10. See also *Scientific American*, 27 May 1921.

[The] Boeing Company. 1999. Helicopter Division. http://www.boeing.com/rotorcraft/

Boulet, J. 1984. *The History of the Helicopter as Told by Its Pioneers 1907–1956*, Editions France-Empire, Paris.

Brown, D. A. 1995. *The Bell Helicopter Textron Story*, Aerofax, Arlington, TX.

Cayley, G. 1809–10. "On Aerial Navigation," *Nicholson's Journal of Natural Philosophy, Chemistry and the Arts*, Nov. 1809, Feb. 1810 & Mar. 1810.

Cochrane, D., Hardesty, V., and Lee, R. 1989. *The Aviation Careers of Igor Sikorsky*, University of Washington Press, Seattle & London.

de Bothezat, G. 1919. "The General Theory of Blade Screws," NACA TR 29.

de la Cierva, J. 1926. "The Development of the Autogiro," *J. Royal Aeronaut. Soc.*, 30 (181), pp. 8–29.

de la Cierva, J. 1930. "The Autogiro," *J. Royal Aeronaut. Soc.*, 39 (239), pp. 902–921.

de la Cierva, J. and Rose, D. 1931. *Wings of Tomorrow: The Story of the Autogiro*, Brewer, Warren & Putnam, New York.

de la Cierva, J. 1935. "New Developments of the Autogiro," *J. Royal Aeronaut. Soc.*, 39, Dec., pp. 1125–1143.

Delear, F. J. 1969. *Igor Sikorsky: His Three Careers in Aviation*, Dodd, Mead & Co., New York.

[The] Eurocopter Company. 1999. http://www.eurocopter.com/

Everett-Heath, J. 1986. *British Military Helicopters*, Arms and Armour Press, London.

Everett-Heath, J. 1988. *Soviet Helicopters – Design, Development and Tactics*, Jane's Information Group, London.

Fay, J. 1987. *The Helicopter: History, Piloting and How It Flies*, Hippocrene, New York.

Finne, K. N. 1987. *Igor Sikorsky: The Russian Years*, edited by C. J. Brobrow and Von Hardesty, Smithsonian Institution Press, Washington DC.

Focke, H. 1938. "The Focke Helicopter," NACA TM 858.

Focke, E. H. H. 1965. "German Thinking on Rotary Wing Development," *J. Royal Aeronaut. Soc.*, 69 (653), pp. 293–305.

Free, F. W. 1970. "Russian Helicopters," *The Aeronaut. J.*, 74 (717), Sept., pp. 767–785.

Gablehouse, C. 1967. *Helicopters and Autogiros: A Chronicle of Rotating Wing Aircraft*, Lippincott, Philadelphia & New York.

Giacomelli, R. 1930. "The Aerodynamics of Leonardo da Vinci," *J. Royal Aeronant. Soc.*, 34 (240), pp. 1016–1038.

GKN-Westland. 1999. http://www.gkn-westland.co.uk/index.html/

Gregory, H. F. 1944. *Anything a Horse Can Do: The Story of the Helicopter*, Introduction by Igor Sikorsky, Reynal & Hitchcock, New York.

Grina, K. I. 1975. "Helicopter Development at Boeing Vertol Company," *The Aeronaut. J.*, 79 (777), pp. 401–416.

Gunston, B. 1983. *Helicopters of the World*, Crescent, New York.

Gustafson, F. B. 1971. "A History of NACA/NASA Rotating-Wing Aircraft Research, 1915–1970," Limited-Edition Reprint (VF-70) from *Vertiflite*, American Helicopter Soc., Alexandria, VA.

Hafner, R. 1949. "The Bristol 171 Helicopter," *J. Royal Aeronaut. Soc.*, 53, April, pp. 324–333.

Hart, I. B. 1961. *The World of Leonardo da Vinci*, MacDonald, London.

Hirschberg, M. J. 1997. "V/STOL: The First Half-Century," *Vertica*, 43 (2), pp. 34–54.

Hislop, G. S. 1958. "The Fairey Rotodyne," Meetings of The Helicopter Assoc. of Great Britain & The Royal Aeronaut. Soc., London, 7th Nov.

Hobbs, J. 1984. *Bristol Helicopters: A Tribute to Raoul Hafner*, Frenchay Publications, Bristol, UK.

[The] Kaman Corporation. 1999. http://www.kaman.com/

[The] Kamov Company. 1999. http://www.kamov.com/. See also http://www.alaincharles.com/kamov.html

Kelly, B. 1954. "Helicopter Stability and Young's Lifting Rotor," *SAE J.*, 53 (12), pp. 685–690.

Klemin, A. 1925. "An Introduction to the Helicopter," NACA TM 340.

Kretz, M. 1987. "René Dorand: The Life of a Pioneer," 13th European Rotorcraft Forum, Arles, France, Sept. 8–11.

Lambermont, P. (with Price, A.) 1958. *Helicopters and Autogiros of the World*, Cassell & Co., London.

Liberatore, E. K. (ed.) 1950. Rotary Wing Handbooks and History, United States Air Force Air Materiel Command. 18 volumes.

Liberatore, E. K. 1988. "An Eclectic Bibliography of Rotorcraft," *Vertiflite*, 34 (1), pp. 50–59.

Liberatore, E. K. 1998. *Helicopters before Helicopters*, Krieger Publishing, Malabar, FL.

Lopez, D. S. and Boyne, W. J. (eds.) 1984. *Vertical Flight: The Age of the Helicopter*, Smithsonian Institution Press, Washington DC.

Margoulis, W. 1922. "Propeller Theory of Professor Joukowski and His Pupils," NACA TM 79.

Mondey, D., 1982. *Planemakers 2: Westland*, Jane's Publishing, London.

Moreno-Caracciolo, M. 1923. "The Autogiro" NACA TM 218.

Munk, M. M. 1923. "Analysis of Dr. Schaffran's Propeller Model Tests," NACA TN 153.

NACA. 1921. "The Oehmichen–Peugeot Helicopter," NACA TM 13.

Oemichen, E. 1923. "My Experiments with Helicopters," NACA TM 199.

Piasecki Aircraft Corp. 1967. *The Piasecki Story of Vertical Lift: Pioneers in Progress for Over Forty Years*, Lakehurst, NJ.

Pitcairn, H. F. 1930. *The Autogiro: Its Characteristics and Accomplishments*, Annual Report of the Board of Regents to the Smithsonian Institution.

Prewitt, R. H. 1938. "Possibilities of the Jump-Off Autogiro," *J. Aeronaut. Sci.*, 6 (1), pp. 10–14.

Pritchard, J. L. 1961. *Sir George Cayley: The Inventor of the Airplane*, Max Parrish & Co., London.

Schneider, J. 1995. "Arthur Young – Inventor, Developer and Metaphysicist, the Developer of the Bell Teetering Rotor," *Vertiflite*, 41 (2), March/April, pp. 36–39.

Seiferth, R. 1927. "Testing a Windmill Airplane (Autogiro)," NACA TM 394.

Sikorsky, I. I. 1938 (Rev. 1958, 1967). *The Story of the Winged-S: An Autobiography*, Dodd, Mead & Co., New York.

Sikorsky, I. I. 1941. "Development of the VS-300 Helicopter," Paper presented at the Rotating Wing Aircraft Meeting of the Inst. of the Aeronaut. Sci., Jan. 29.

Sikorsky, I. I. 1942. "Technical Development of the VS-300 Helicopter During 1941," *J. Aeronaut. Sci.*, 9 (8), pp. 309–311.

Sikorsky, I. I. 1943. "Progress of the Vought–Sikorsky Helicopter Program in 1942," *Aeronaut. Eng. Rev.*, 2 (4), pp. 41–43.

Sikorsky, I. I. 1964. "Recollections and Thoughts of a Pioneer," Lecture Presented to the Wings Club, New York, Nov. 16.

Sikorsky, I. I. 1971. "Sixty Years of Flying," *The Aeronaut. J.*, 75 (731), Nov., pp. 761–768.

Sikorsky, S. and Andrews, A. 1984. *Straight Up*, Sikorsky Aircraft, United Technologies Corp.

Sikorsky Aircraft Corp. 1999. http://www.sikorsky.com/

Smith, F. K. 1985. *Legacy of Wings: The Story of Harold Pitcairn*, Jason Aronson, New York.

Spenser, J. P. 1992. *Vertical Challenge: The Hiller Aircraft Story*, University of Washington Press, Seattle & London.

Spenser, J. P. 1999. *Whirlybirds: A History of the U.S. Helicopter Pioneers*, University of Washington Press, Seattle & London.

Straubel, J. F. 1964. *One Way Up*, Hiller Aircraft Co., Inc., Palo Alto, CA.

Taylor, M. J. 1984. *History of the Helicopter*, Hamlyn Publishing, London.

Tipton, R. S. 1989. *They Filled the Skies*, Bell Helicopter Textron, Fortworth, TX.

Tokaty, G. A. 1971. *A History and Philosophy of Fluid Mechanics*, Foulis & Co., Henley-on-Thames, England.

von Kármán, Th. 1921. "Tests Made With Captive Helicopters." In the report: "Recent European Developments in Helicopters," NACA TN 47.

Warner, E. P. 1920. "The Problem of the Helicopter," NACA TN 4.

Wimperis, H. E. 1926. "The Rotating Wing in Aircraft," ARC R & M 1108.

Wolf, A. L. 1974. "The Vision of D'Amecourt," *Vertiflite*, 20 (5), pp. 2–6.

Young, A. M. 1948. "The Helicopter and Stability," *American Helicopter Magazine*, March, pp. 6–25.

Young, A. M. 1979. *The Bell Notes: A Journey from Metaphysics to Physics*, Delacorte Press/Seymour Lawrence, New York.

CHAPTER 2

Fundamentals of Rotor Aerodynamics

The essential theory of flight can be reduced to a comparatively simple statement, though it becomes a highly complicated affair as it is presented in figures and formulae.

Juan de la Cierva (1931)

2.1 Introduction

The rotor of a helicopter provides three basic functions: 1. the generation of a vertical lifting force (thrust) in opposition to the aircraft weight; 2. the generation of a horizontal propulsive force for forward flight; and 3. a means of generating forces and moments to control the attitude and position of the helicopter. All three of these functions must be under full control of the pilot. Unlike a fixed-wing aircraft where these functions are separated, the helicopter rotor alone must provide all three. To ensure the rotor can meet these demanding roles, the rotor designer requires considerable knowledge of both the aerodynamic environment in which the rotor operates as well as how the aerodynamic loads affect the blade dynamic response and overall rotor behavior.

The lifting capability of any part of a rotating blade is related to its local angle of attack and local dynamic pressure. The blade position can be defined in terms of an azimuth angle, ψ, which is defined as zero when the blade is pointing downstream, as shown in Fig. 2.1. In hovering flight, the velocity variation along the blade is azimuthally axisymmetric and radially linear, with zero flow velocity at the hub and the velocity reaching a maximum, V_{tip}, at the blade tip. The local dynamic pressure at any blade element is proportional to the square of the distance from the rotational axis. Based on elementary considerations, the average rotor thrust will depend on the square of the rotor tip speed, $V_{\text{tip}} = \Omega R$, that is, $T \propto V_{\text{tip}}^2$. Also, the rotor power, P, will depend on the cube of the tip speed, that is, $P \propto V_{\text{tip}}^3$ (see Question 2.1).

In forward flight, a component of the free stream, V_∞, adds to or subtracts from the rotational velocity at each part of the blade; that is, V_{tip} now becomes $\Omega R + V_\infty \sin\psi$. As shown by Fig. 2.1(b), while the distribution of velocity along the blade remains linear, it is no longer axisymmetric and varies in magnitude with respect to blade azimuth angle. However, it will also be evident that forward flight speed, blade pitch angle, and any blade flapping, as well as the distribution of induced inflow through the rotor, will all affect the blade section angle of attack and, therefore, the blade lift distribution, rotor thrust, and rotor power consumption. This is the complication with the helicopter rotor that makes its aerodynamic analysis relatively difficult.

Unlike on a fixed wing, which has a relatively uniform lift loading over its span, the high dynamic pressure found at the tips of a helicopter blade produces a concentration of aerodynamic forces there. As a consequence, strong vortices form and trail from each blade tip. Figure 2.2 shows an example of the physical nature of the vortical wake generated by a helicopter rotor in hovering flight. Here, the blade tip vortices are rendered visible by

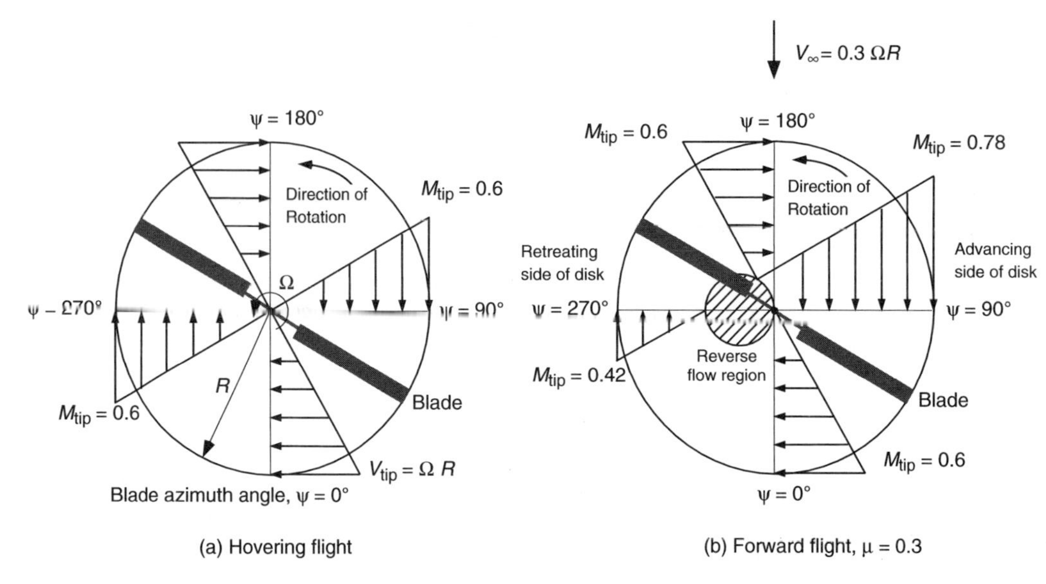

Figure 2.1 Distribution of incident velocity normal to the leading edge of rotor blade. (a) Hovering flight. (b) Forward flight at $\mu = 0.3$.

natural condensation of water vapor in the air.[1] It will be seen that the vortices are convected downward below the rotor and form a series of interlocking, almost helical trajectories. For the most part, the net flow velocity at the plane of the rotor and in the rotor wake itself is comprised of the velocities induced by these tip vortices. For this reason, predicting the strengths and locations of the tip vortices plays an important role in determining rotor performance and in designing the rotor.

All helicopters spend considerable time in hover, which is a flight condition where they are specifically designed to be operationally efficient. In hover, the main purpose of the rotor is to provide a vertical lifting force in opposition to the aircraft weight. However, in forward flight the rotor must also provide a propulsive force to overcome the aircraft drag. This is obtained by tilting the plane of the rotor forward, while increasing the overall rotor thrust so that the vertical component of thrust (lift) remains equal to the aircraft weight. In forward flight the rotor blades encounter an asymmetric velocity field, which is a maximum on the blade that advances into the relative wind and a minimum on the blade that retreats away from the relative wind (see Fig. 2.1(b)). The local dynamic pressure and the blade airloads, therefore, become periodic primarily at the rotational speed of the rotor (i.e., once per revolution or 1/rev). Because of the articulation built into helicopter blade designs, the rotor blades will begin to flap about their hinges causing the rotor disk to tilt. This inherent tendency can be compensated for by the pilot by using cyclic pitch inputs to the blades. This changes the magnitude and phasing of the 1/rev aerodynamic lift forces over the disk, and so it can be used to maintain a desirable orientation of the rotor disk to meet propulsion and control requirements. The inherent coupling among blade pitch inputs, the aerodynamic forces, and the blade flapping response is discussed in detail in Chapter 4.

The overall aerodynamic complexity of the helicopter in forward flight can be appreciated from Fig. 2.3. The flow field in which the rotor operates is considerably more complex than

[1] This effect is only obtained under conditions when the air temperature is close to the dew point; it is produced by the small amount of cooling that takes place inside the low pressure vortex cores.

Figure 2.2 Hovering helicopter showing the vortical rotor wake through natural condensation of water vapor inside the tip vortex cores (Courtesy of the US Navy, Patuxent River NAS).

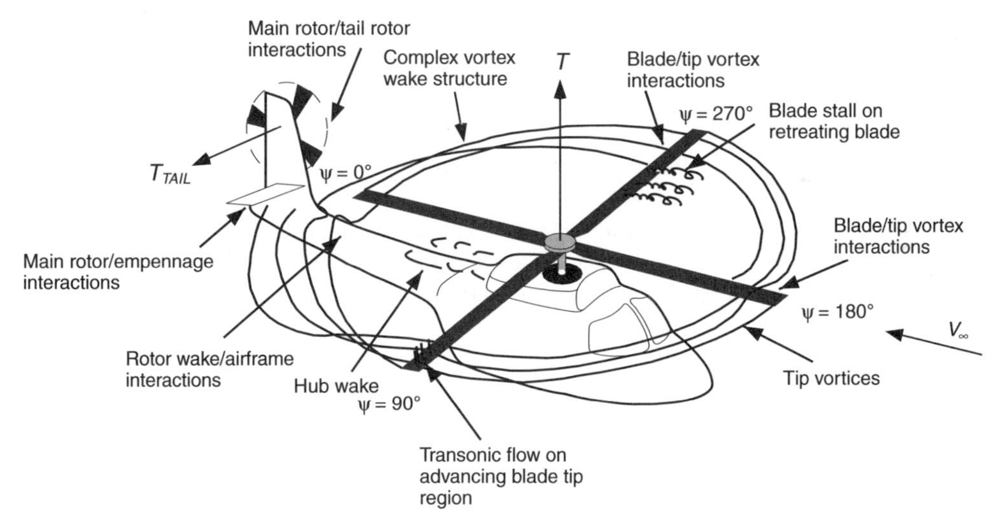

Figure 2.3 Schematic showing the flow structure and some aerodynamic problem areas on a helicopter in forward flight.

that of a fixed-wing aircraft, mainly because of the individual wakes trailed from each blade. For a fixed-wing aircraft, the tip vortices trail downstream of the aircraft. However, for a helicopter in forward flight, the blade tip vortices can remain close to the rotor and to the following blades for several rotor revolutions. As a result of the low disk loading (thrust carried per unit area of the rotor disk) and generally low average flow velocity through the rotor disk, these vortices remain close enough to produce a strongly three-dimensional induced velocity field. As following blades encounter this induced velocity field, fluctuating airloads are produced on the blades.[2] Besides affecting the rotor performance, these time-varying airloads can be a source of high rotor vibrations and strongly focused obtrusive noise.

At higher forward flight speeds, the inherently asymmetric nature of the flow over the rotor disk gives rise to a number of aerodynamic problems that ultimately limit the rotor performance. The most obvious is that the blade tips on the advancing side of the rotor disk can start to penetrate into supercritical and transonic flow regimes, with the associated formation of compressibility zones and, ultimately, strong shock waves. In addition to the occurrence of wave drag and the possibilities of shock-induced flow separation, both phenomena requiring much more power to drive the rotor, the periodic formation of shock waves is another source of obtrusive noise. The increased power demands placed on the rotor system when compressibility effects manifest will eventually limit forward flight speed. Although compressibility effects can be relieved to some extent by the use of swept tip blades, the problems of increased power requirements and noise are only delayed to moderately higher forward flight speeds and are not eliminated.

On the retreating side of the disk, that is, where the blades are retreating away from the relative wind because of the forward flight velocity of the helicopter, the local velocity and dynamic pressure at the blade are relatively low, and the blades are required to operate at higher angles of attack to maintain lift. If these angles of attack become too large, then the retreating blade will stall. This results in a loss of overall lifting and propulsive capability from the rotor and sets an intrinsic barrier to further increases in forward flight speed. It is interesting to note that while stall on a fixed-wing aircraft occurs at low flight speeds, a helicopter encounters the problem of stall at relatively high flight speeds. Because of the inherent time-dependent nature of the flow environment on the rotor blades in forward flight, retreating blade stall is highly unsteady in nature and is referred to as dynamic stall. The unsteady loads produced during dynamic stall are an additional source of vibration on the helicopter, which can significantly limit its forward flight and maneuvering capability. The various aerodynamic interactions that can exist between the rotor wake and the airframe, including the tail rotor and empennage, are also worthy of note (see Fig. 2.3). While fundamentally very complicated, they can lead to various significant aerodynamic interference effects that cannot be ignored in the design of the helicopter.

2.2 Momentum Analysis in Axial Flight

The helicopter, or any other rotating-wing vehicle, must operate in a variety of flight regimes. These include hover, climb, descent, or forward flight. In addition, the aircraft may undergo maneuvers, which may comprise a combination of these basic flight regimes. In hover or axial flight, the flow is axisymmetric, and the flow through the rotor is either upward or downward. This is the easiest flow regime to analyze and, at least in principle, should be the easiest to predict by means of mathematical models. It has been found, however, that even with modern mathematical models of the rotor flow, accurate prediction of hovering performance is by no means straightforward. Although it must be remembered that the

[2] This phenomenon is called blade vortex interaction or BVI – see Chapters 8 and 10.

actual physical flow about the rotor will comprise a complicated vortical wake structure, as previously shown in Fig. 2.2, the basic performance of the rotor can be analyzed by a simpler approach known as *momentum theory*. The momentum theory approach allows one to derive a first-order prediction of the rotor thrust and power, and the principles also form a foundation for more elaborate treatments of the rotor aerodynamics problem.

2.2.1 *Flow near a Hovering Rotor*

Hover is a very unique flight condition. Here, the rotor has zero forward speed and zero vertical speed (no climb or descent). The rotor flow field is, therefore, azimuthally axisymmetric. A set of velocity measurements near a sub-scale hovering rotor and in its wake is shown in Fig. 2.4. Note that the fluid velocity is increased smoothly as it is entrained into and through the rotor disk plane. There is no jump in velocity across the disk, although because a thrust is produced, there must be a jump in pressure over the disk. The existence of a wake boundary or slipstream is apparent, with the flow velocity outside this boundary being relatively quiescent. The blade tip vortices trail behind and below each blade and are convected along this wake boundary. Inside the wake boundary, the flow velocities are substantial and may be distributed nonuniformly across the slipstream. Note also the contraction in the diameter of the wake below the rotor corresponding to an increase in the slipstream velocity.

With the physical picture of the hovering rotor flow now apparent, it is possible to approach a mathematical solution to this problem. Consider the application of the three basic conservation laws (conservation of mass, momentum, and energy) to the rotor and its flow field. The conservation laws will be applied in a quasi-one-dimensional integral formulation to a control volume surrounding the rotor and its wake. This approach permits us to perform a first level analysis of the rotor performance (e.g., its thrust and power), but without actually having to consider the details of the flow environment, that is without having to consider what is happening locally at each blade section.

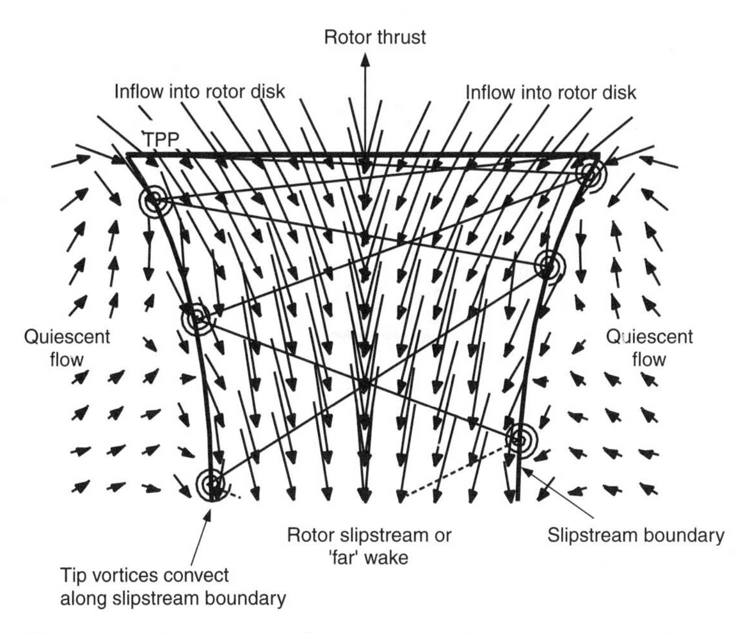

Figure 2.4 Measurements of the velocity field near and below a hovering two-bladed rotor. Data source: Leishman et al. (1995).

This approach, which is called *momentum theory*, was first developed by Rankine (1865) for use in the analysis of marine propellers. The theory was developed further by W. Froude (1878) and R. E. Froude (1889) and Betz (1920a,b; 1922). Momentum theory has also been further generalized by Glauert – see Durand (1935). The main difference between the Froude and Rankine theories is in the treatment of the rotor disk as a series of elementary rings, versus the treatment of the disk as a whole. In either case, one fundamental assumption in the basic momentum theory is that the rotor can be idealized as an infinitestimally thin *actuator disk* over which a pressure difference exists. The concept is equivalent to considering an infinite number of blades of zero thickness. The actuator disk supports the thrust force that is generated by the rotation of the rotor blades about the shaft and their action on the air. Power is required to generate this thrust, which is supplied in the form of a torque to the rotor shaft. Work done on the rotor leads to a gain in kinetic energy of the rotor slipstream, and this is an unavoidable energy loss that is called *induced* power.

2.2.2 *Conservation Laws of Fluid Mechanics*

In the general approach to the problem, it will be assumed that the flow through the rotor is one dimensional, quasi-steady, incompressible, and inviscid. Consider an *ideal fluid*, that is, one that generates no viscous shear between fluid elements. Therefore, induced losses are the sole source of losses in the fluid, with other losses resulting from the action of viscosity being assumed negligible. Furthermore, assume that the flow is quasi-steady, in that the flow properties at a point do not change with time. Finally, assume that the flow is one dimensional, and so the properties across any plane parallel to the rotor plane are constant; that is, the fluid properties change only with axial (vertical) position relative to the rotor.

Let the control volume surrounding the rotor and its wake have surface area S, as shown in Fig. 2.5. This figure represents the helicopter rotor in an axial climb with velocity V_c, for which the hovering condition is obtained in the limit as $V_c \rightarrow 0$. Let $d\vec{S}$ be the unit normal area vector, which by convention always points out of the control volume. A general

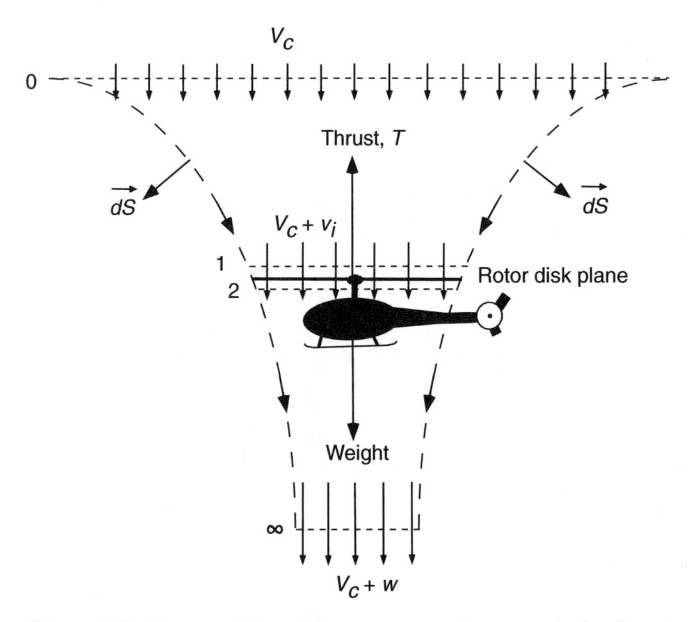

Figure 2.5 Flow model used for momentum theory analysis of a rotor in axial flight.

equation governing the conservation of fluid mass applied to this finite control volume can be written as

$$\oint\!\!\oint_S \rho \vec{V} \cdot d\vec{S} = 0, \tag{2.1}$$

where $\vec{V}$ is the local velocity and ρ is the density of the fluid. This equation states that the mass flow into the control volume must equal the mass flow out of the control volume across surface S. Note that this is a scalar equation. Similarly, an equation governing the conservation of fluid momentum can be written as

$$\vec{F} = \oint\!\!\oint_S p d\vec{S} + \oint\!\!\oint_S (\rho \vec{V} \cdot d\vec{S})\vec{V}. \tag{2.2}$$

For an unconstrained flow, the net pressure force on the fluid inside the control volume is zero. Therefore, the net force on the fluid, $\vec{F}$, is simply equal to the rate of change with time of the fluid momentum across the control surface, S. Although this is a vector equation, it can be simplified considerably by the assumptions of quasi-one-dimensional flow. Because the force on the fluid is supplied by the rotor, by Newton's third law the fluid must exert an equal and opposite force on the rotor. This reaction force is the rotor thrust, T. Thirdly, an equation governing the conservation of energy in the flow can be written as

$$W = \oint\!\!\oint_S \frac{1}{2}(\rho \vec{V} \cdot d\vec{S})|\vec{V}|^2. \tag{2.3}$$

This equation states simply that the work done on the fluid by the rotor manifests as a gain in kinetic energy of the fluid in the rotor slipstream, per unit time. It is also a scalar equation.

2.2.3 *Application to a Hovering Rotor*

These general equations of fluid mass, momentum, and energy conservation may now be applied to the specific problem of a hovering rotor. This corresponds to the condition $V_c = 0$ in Fig. 2.5. Let cross section 0 denote the plane far upstream of the rotor, where in the hovering case the fluid is quiescent (i.e., $V_0 = V_c = 0$). Cross sections 1 and 2 are the planes just above and below the rotor, respectively, and the "far" wake (the slipstream well downstream of the rotor) is denoted by cross section ∞. At the plane of the rotor, assume that the velocity (the induced velocity or velocity imparted to the mass of air contained in the control volume at the rotor disk) is v_i. In the far wake (the slipstream), the velocity will be increased over that at the plane of the rotor, and this velocity is denoted by w. The rotor disk area is denoted by A.

From the assumption that the flow is quasi-steady, and by the principle of conservation of mass, the mass flow rate, $\dot{m}$, must be constant within the boundaries of the rotor wake (control volume). Therefore,

$$\dot{m} = \iint_\infty \rho \vec{V} \cdot d\vec{S} = \iint_2 \rho \vec{V} \cdot d\vec{S}, \tag{2.4}$$

and the one-dimensional incompressible flow assumption reduces this equation to

$$\rho A_\infty w = \rho A_2 v_i = \rho A v_i. \tag{2.5}$$

The principle of conservation of fluid momentum gives the relationship between the rotor thrust, T, and the net time rate of change of momentum out of the control volume (Newton's second law). The rotor thrust is equal and opposite to the force on the fluid, which is given by

$$\vec{F} = T = \iint_{\infty} \rho(\vec{V} \cdot d\vec{S})\vec{V} - \iint_{0} \rho(\vec{V} \cdot d\vec{S})\vec{V}. \tag{2.6}$$

Because in hovering flight the velocity well upstream of the rotor is quiescent, the second term on the right-hand side of the above equation is zero. Therefore, for the hover problem, the rotor thrust can be written as the scalar equation

$$T = \iint_{\infty} \rho(\vec{V} \cdot d\vec{S})\vec{V} = \dot{m}w. \tag{2.7}$$

From the principle of conservation of energy, the work done on the rotor is equal to the gain in energy of the fluid per unit time. The work done per unit time, or the power consumed by the rotor, is Tv_i, and this results in the equation

$$Tv_i = \iint_{\infty} \frac{1}{2}\rho(\vec{V} \cdot d\vec{S})\vec{V}^2 - \iint_{0} \frac{1}{2}\rho(\vec{V} \cdot d\vec{S})\vec{V}^2 \tag{2.8}$$

In hover, the second term on the right-hand side of the above equation is zero so that

$$Tv_i = \iint_{\infty} \frac{1}{2}\rho(\vec{V} \cdot d\vec{S})\vec{V}^2 = \frac{1}{2}\dot{m}w^2. \tag{2.9}$$

From Eqs. 2.7 and 2.9 it is clear that

$$v_i = \frac{1}{2}w \tag{2.10}$$

or that $w = 2v_i$. This, therefore, gives a simple relationship between the induced velocity in the plane of the rotor and the "far" wake or slipstream velocity.

Rotor Slipstream

Because the flow velocity increases in the wake below the rotor, continuity considerations require that the area of the slipstream must decrease. This is apparent from the observations in Figs. 2.2 and 2.4. It follows from the conservation of fluid mass that

$$\rho A v_i = \rho A_{\infty} w = \rho A_{\infty}(2v_i) = 2\rho A_{\infty} v_i, \tag{2.11}$$

so that in hover the ratio of the cross-sectional area of the far wake to the area of the rotor disk is

$$\frac{A_{\infty}}{A} = \frac{1}{2}. \tag{2.12}$$

In other words, based on ideal fluid flow assumptions, the far wake below the rotor contracts to an area that is exactly half of the rotor disk area. Alternatively, by considering the radius of the far rotor wake, r_{∞}, relative to that of the rotor, R, it is easy to show that

$$r_{\infty} = \frac{R}{\sqrt{2}}. \tag{2.13}$$

Therefore, the ratio of the radius of the wake to the radius of the rotor is $1/\sqrt{2} = 0.707$. This is called the *wake contraction ratio*. In practice, it has been found experimentally that the wake contraction ratio is not as much as the theoretical value given by the momentum theory; typically it is only about 0.78 compared to 0.707. This is mainly a consequence of the viscosity of the fluid, the reality that a nonuniform inflow will be produced over the disk, and a small swirl component of velocity in the rotor wake induced by the spinning rotor. These effects serve to reduce the net change of the fluid momentum in the vertical direction, and they will decrease the rotor thrust for a given shaft torque (power supplied) or will increase the rotor power required to produce a given thrust. Behaviors directly attributable to the viscosity of the fluid are termed "nonideal" effects, and these will be considered in detail later.

Rotor Power

It has been shown previously using Eq. 2.7 that momentum theory can be used to relate the rotor thrust to the induced velocity at the rotor disk by using the equation

$$T = \dot{m}w = \dot{m}(2v_i) = 2(\rho A v_i)v_i = 2\rho A v_i^2. \tag{2.14}$$

Rearranging this equation gives the induced velocity at the plane of the rotor disk in hover as

$$v_h \equiv v_i = \sqrt{\frac{T}{2\rho A}} = \sqrt{\left(\frac{T}{A}\right)\frac{1}{2\rho}}. \tag{2.15}$$

The ratio T/A is known as the *disk loading*. Note that $v_h \equiv v_i$ is used to represent the induced velocity in hover. This value will be used later as a reference when the axial climb and descending flight conditions are considered.

The power required to hover (or the time rate-of-work done by the rotor on the fluid per unit thrust) is given by

$$P = T v_i \equiv T v_h = T\sqrt{\frac{T}{2\rho A}} = \frac{T^{3/2}}{\sqrt{2\rho A}}. \tag{2.16}$$

This power, called *ideal power*, is entirely induced in nature because the contribution of viscous effects have not been considered in the present level of analysis. Alternatively, one can write

$$P = T v_i = \left(2\rho A v_i^2\right)v_i = 2\rho A v_i^3. \tag{2.17}$$

From this equation it is noted that the power required to hover will increase with the cube of the induced velocity (or inflow) at the disk. Obviously, to make a rotor hover at a given thrust with minimum induced power, the induced velocity at the disk must be small. Therefore, the mass flow through the disk must be large, and this consequently requires a large rotor disk area. This is a fundamental design feature of all helicopters.

Pressure Variation

The pressure variation through the rotor flow field in the hover state can be found from the application of Bernoulli's equation along a streamline above and below the rotor disk. Remember that there is a pressure jump across the disk as a result of energy addition by the rotor, so that Bernoulli's equation cannot be applied there. Applying Bernoulli's equation up to the disk between stations 0 and 1 produces

$$p_0 = p_\infty = p_1 + \frac{1}{2}\rho v_i^2. \tag{2.18}$$

Below the disk, between stations 2 and ∞, the application of Bernoulli's equation gives

$$p_2 + \frac{1}{2}\rho v_i^2 = p_\infty + \frac{1}{2}\rho w^2. \tag{2.19}$$

Because the jump in pressure Δp is assumed to be uniform across the disk, this pressure jump must be equal to the disk loading, T/A, that is,

$$\Delta p = p_2 - p_1 = \frac{T}{A}. \tag{2.20}$$

Therefore, one can write

$$\frac{T}{A} = p_2 - p_1 = \left(p_\infty + \frac{1}{2}\rho w^2 - \frac{1}{2}\rho v_i^2\right) - \left(p_\infty - \frac{1}{2}\rho v_i^2\right) = \frac{1}{2}\rho w^2, \tag{2.21}$$

from which it is seen that the rotor disk loading is equal to the dynamic pressure in the rotor slipstream. One can also determine the pressure just above the disk and just below the disk in terms of the disk loading. Just above the disk, the use of Bernoulli's equation gives

$$p_1 = p_\infty - \frac{1}{2}\rho v_i^2 = p_\infty - \frac{1}{2}\rho\left(\frac{w}{2}\right)^2 = p_\infty - \frac{1}{4}\left(\frac{T}{A}\right), \tag{2.22}$$

and just below the disk one gets

$$p_2 = p_0 + \frac{1}{2}\rho w^2 - \frac{1}{2}\rho\left(\frac{w}{2}\right)^2 = p_0 + \frac{3}{4}\left(\frac{T}{A}\right). \tag{2.23}$$

Therefore, the static pressure is reduced by $\frac{1}{4}(T/A)$ above the rotor disk and is increased by $\frac{3}{4}(T/A)$ below the disk.

2.2.4 *Disk Loading and Power Loading*

A parameter used frequently in helicopter analysis that appears in the preceding equations is the disk loading, T/A, which is denoted by DL. Because for a single-rotor helicopter in hover, the rotor thrust, T, is equal to the weight of the aircraft, W, the disk loading is sometimes written as W/A or $W/\pi R^2$. Disk loading is measured in pounds per square foot (lb ft^{-2}) in British (Imperial) units or Newtons per square meter (N m^{-2}) in the SI system. One may also use kilograms per square meter (kg m^{-2}) in the SI system. The direct use of the kilogram (kg) as a unit of force is frequently found in engineering practice, particularly in the aerospace field. For the purposes of computing the disk loading for multirotor helicopters such as tandems and coaxials, the convention is to assume that each rotor carries an equal proportion of the aircraft weight. Values of rotor disk loading for a selection of rotating-wing aircraft are given in the appendix.

The power loading is defined at T/P, which is denoted by PL. Power loading is measured in pounds per horsepower (lb hp^{-1}) in Imperial units or Newtons per kilo Watt (N kW^{-1}) or kilograms per kilo Watt (kg kW^{-1}) in the SI system. Remember that the induced (ideal) power required to hover is given by $P = Tv_i$. This means that the *ideal power loading* will be inversely proportional to the induced velocity at the disk. To see this, recall that the inflow velocity at the disk v_i can be written in terms of the disk loading as

$$v_i = \sqrt{\frac{T}{2\rho A}} = \sqrt{\frac{DL}{2\rho}} = \frac{P}{T} = (PL)^{-1}. \tag{2.24}$$

According to Fig. 2.6, the ratio T/P (power loading) decreases quickly with increasing disk loading (note the logarithmic scale on the ordinate). Therefore, rotors that have a low disk loading will require a low power per unit thrust (high ideal power loading) and will

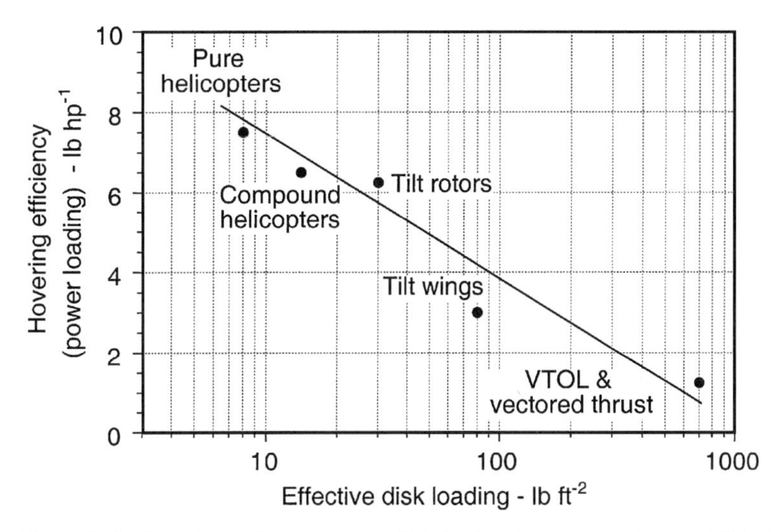

Figure 2.6 Hovering efficiency versus disk loading for a range of vertical lift aircraft.

tend to be more efficient; that is, the rotor will require less power (and consume less fuel) to generate any given amount of thrust. Calculation of the actual power loading and rotor efficiency, however, requires the consideration of viscous losses.

Helicopters operate with low disk loadings in the region of 24–48 kg m^{-2} or 5–10 lb ft^{-2}; thus they can provide a large amount of lift for a relatively low power [more than 5 kg kW^{-1} ($\sim$50 N kW^{-1}) or 10 lb hp^{-1}]. Therefore, the helicopter is a very efficient aircraft in hover compared to other vertical takeoff and landing (VTOL) aircraft. Tilt-rotors can be considered a hybrid helicopter/fixed-wing aircraft and have somewhat higher rotor disk loadings. Therefore, they are somewhat less efficient in hover than a conventional helicopter for the same gross weight but still are much more efficient than other VTOL aircraft without rotating wings.

2.2.5 *Induced Inflow Ratio*

The induced inflow velocity, v_i, at the rotor disk can be written as

$$v_h \equiv v_i = \lambda_h \Omega R, \tag{2.25}$$

where the nondimensional quantity λ_h is called the induced inflow ratio in hover. The angular or rotational speed of the rotor is denoted by Ω and R is the rotor radius; so the product is simply the tip speed, V_{tip}. The inflow ratio is normally the preferable quantity to use when comparing results from different rotors because it is a nondimensional quantity. For rotating-wing aircraft, it is the convention to nondimensionalize all velocities by the blade tip speed in hover (i.e., $V_{\text{tip}} = \Omega R$).

2.2.6 *Thrust and Power Coefficients*

Nondimensional coefficients are normally employed in helicopter rotor analysis. The rotor thrust coefficient is defined on the basis of the rotor disk area, A, and the tip speed, ΩR, by

$$C_T = \frac{T}{\rho A V_{\text{tip}}^2} = \frac{T}{\rho A \Omega^2 R^2}. \tag{2.26}$$

Therefore, the inflow ratio, λ_i, is related to the thrust coefficient in hover by

$$\lambda_h = \lambda_i = \frac{v_i}{\Omega R} = \frac{1}{\Omega R}\sqrt{\frac{T}{2\rho A}} = \sqrt{\frac{T}{2\rho A(\Omega R)^2}} = \sqrt{\frac{C_T}{2}}. \tag{2.27}$$

This is based on the one-dimensional flow assumption made in the preceding analysis, which means that this value of inflow is assumed to be distributed uniformly over the disk. The rotor power coefficient is defined as

$$C_P = \frac{P}{\rho A V_{\text{tip}}^3} = \frac{P}{\rho A \Omega^3 R^3} \tag{2.28}$$

so that based on momentum theory the power coefficient for the hovering rotor is

$$C_P = C_T \lambda_i = \frac{C_T^{3/2}}{\sqrt{2}}. \tag{2.29}$$

Again, this is calculated on the basis of uniform inflow and no viscous losses, and so it is called the *ideal power*. The corresponding rotor shaft torque coefficient is defined as

$$C_Q = \frac{Q}{\rho A V_{\text{tip}}^2 R} = \frac{Q}{\rho A \Omega^2 R^3}. \tag{2.30}$$

Note that because power is related to torque by $P = \Omega Q$, then $C_P = C_Q$.

It is important to note that the US customary definition of the thrust, torque, and power coefficients is different to that used in some parts of the world, where a factor of one half is used in the denominator giving

$$C_T = \frac{T}{\frac{1}{2}\rho A(\Omega R)^2}, \quad C_Q = \frac{Q}{\frac{1}{2}\rho A(\Omega R)^2 R} \quad \text{and} \quad C_P = \frac{P}{\frac{1}{2}\rho A(\Omega R)^3} \tag{2.31}$$

This means that the values of thrust, torque, and power coefficients are all a factor of two greater than the values obtained with the US customary definition. The US definition is used throughout this book.

Figure 2.7 shows a comparison of the simple momentum theory with thrust and power measurements made for a hovering rotor using Eq. 2.29. Note that the momentum theory underpredicts the power required, but the predicted trend that $C_P \propto C_T^{3/2}$ is essentially correct. These differences between the momentum theory and experiment occur because viscous effects have been totally neglected so far.

2.2.7 *Nonideal Effects on Rotor Performance*

In hovering flight the induced power can be approximately described by a simple modification to the momentum result in Eq. 2.29, namely

$$C_{P_i} = \frac{\kappa C_T^{3/2}}{\sqrt{2}}, \tag{2.32}$$

where κ is called an *induced power correction factor* or simply an *induced power factor*. This is an empirical coefficient derived from rotor measurements or flight tests, and it encompasses a number of nonideal, but physical effects, such as nonuniform inflow, tip losses, wake swirl, less than ideal wake contraction, finite number of blades, etc. For preliminary

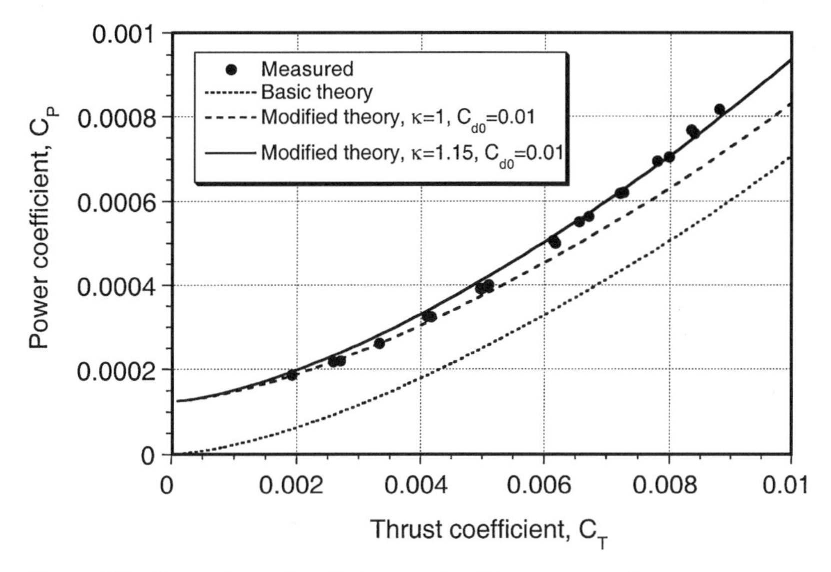

Figure 2.7 Comparison of predictions made with momentum theory to measured power for a hovering rotor. Data source: Bagai & Leishman (1992).

design, most helicopter manufacturers use their own measurements and experience to estimate values of κ, a typical value being about 1.15. Values of κ can also be computed using more advanced blade element methods (see Chapter 3).

Proper estimates for the profile power consumed by the rotor requires a knowledge of the drag coefficients of the airfoils that make up the rotor blades; that is, a strip or blade element analysis is required. The airfoil drag coefficient will be a function of both Reynolds number, Re, and Mach number, M, which obviously vary along the span of the blade. A result for the profile power can be obtained from an element-by-element analysis of sectional drag forces and by radially integrating the sectional drag force along the length of the blade using

$$P_0 = \Omega N_b \int_0^R D y dy, \tag{2.33}$$

where N_b is the number of blades and D is the drag force per unit span at a section a distance y from the rotational axis. The drag force can be expressed conventionally as

$$D = \frac{1}{2}\rho(\Omega y)^2 c C_d, \tag{2.34}$$

where c is the blade chord. If the section profile drag coefficient, C_d, is assumed to be constant ($= C_{d_0}$) and independent of Re and M (which is not an unrealistic assumption), and the blade is not tapered (a rectangular blade), then the profile power will be

$$P_0 = \frac{1}{8}\rho N_b \Omega^3 c C_{d_0} R^4. \tag{2.35}$$

Converting to a power coefficient by dividing through by $\rho A(\Omega R)^3$ gives

$$C_{P_0} = \frac{1}{8}\left(\frac{N_b c R}{A}\right) C_{d_0} = \frac{1}{8}\left(\frac{N_b c R}{\pi R^2}\right) C_{d_0} = \frac{1}{8}\left(\frac{N_b c}{\pi R}\right) C_{d_0} = \frac{1}{8}\sigma C_{d_0}. \tag{2.36}$$

The grouping $N_b c R/A$ (or $N_b c/\pi R$) is known as the *rotor solidity*, which is the ratio of blade area to rotor disk area, and is represented by the symbol σ. Typical values of σ for a helicopter rotor range between 0.07 and 0.12.

Armed with these estimates of the induced and profile power losses, it is possible to recalculate the rotor power requirements. These alternative results are shown in Fig. 2.7, as denoted by the "modified theory," which has been calculated by assuming $C_{d_0} = 0.01$. In the first case, it has been assumed that $\kappa = 1.0$ (ideal induced losses), and in the second case, $\kappa = 1.15$ (nonideal losses). The solidity for this particular rotor is 0.1. Note the need to account for nonideal induced losses to give agreement with the measured data. The overall level of correlation thus obtained gives considerable confidence in the modified momentum theory approach for basic rotor performance studies, at least in hover.

2.2.8 *Figure of Merit*

There is a difficulty in defining an efficiency factor for a helicopter rotor because many parameters are involved. The power loading parameter discussed previously is one measure of rotor efficiency because a helicopter of a given weight should be designed to hover with the minimum power and fuel consumption; that is, the ratio T/P should be made as large as possible. However, the power loading is a dimensional quantity and so a standard nondimensional measure of hovering thrust efficiency called the *figure of merit* has been adopted. This quantity is calculated using the simple momentum theory as a reference and was introduced in the 1940s by Richard H. Prewitt of Kellett Aircraft. The figure of merit is equivalent to a static thrust efficiency and defined as the ratio of the ideal power required to hover to the *actual* power required, that is,

$$FM = \frac{\text{Ideal power required to hover}}{\text{Actual power required to hover}}. \tag{2.37}$$

The *ideal* power is given by the simple momentum result in Eq. 2.29. For the ideal case, the figure of merit must always be unity because the momentum theory assumes no viscous losses; hence the ideal power is entirely induced in origin. In reality, viscous effects manifest as both induced and profile contributions, and these are always present in the power measurements. Therefore, for a real rotor the figure of merit must always be less than one. The figure of merit or *FM* can be used as a gauge as to how efficient a hovering rotor is in terms of generating thrust for a given power. However, it should only be used as a comparative measure between two rotors when the rotors are also compared at the same disk loading. The figure of merit can also be written as

$$FM = \frac{P_{\text{ideal}}}{P_{\text{meas}}} = \frac{C_T^{3/2}/\sqrt{2}}{C_{P_{\text{meas}}}} = \frac{C_T^{3/2}}{\sqrt{2}\, C_{P_{\text{meas}}}}, \tag{2.38}$$

where the measured value of power, $C_{P_{\text{meas}}}$, will include both induced effects and all of the other physical factors that have their origin from viscosity.

A representative plot of measured figure of merit versus rotor thrust is shown in Fig. 2.8. It will be apparent that the *FM* reaches a maximum and then remains constant or drops slightly. This is because of the higher profile drag coefficients ($> C_{d_0}$) obtained at higher rotor thrust and higher blade section angles of attack. For some rotors, especially those with older and less efficient airfoils, the curve can exhibit a peak in *FM*, followed by either a progressive or abrupt decrease thereafter. Therefore, the *FM* behavior in the high thrust range will, to some extent, be a function of airfoil shape and airfoil stall type (i.e., gradual or

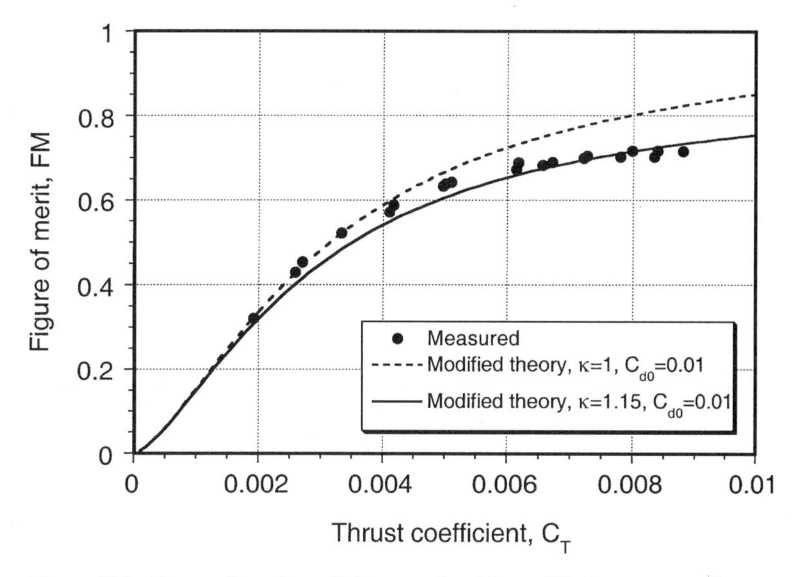

Figure 2.8 Figure of merit predictions made with modified momentum theory compared to measured results for a hovering rotor. Data source: Bagai & Leishman (1992).

abrupt). In practice, FM values between 0.7 and 0.8 represent a good hovering performance for a helicopter rotor.

Using the modified form of the momentum theory with the nonideal approximation for the power, the figure of merit can be written as

$$FM = \frac{\text{Ideal power}}{\text{Induced power} + \text{Profile power}} = \frac{\dfrac{C_T^{3/2}}{\sqrt{2}}}{\dfrac{\kappa C_T^{3/2}}{\sqrt{2}} + \dfrac{\sigma C_{d_0}}{8}}, \tag{2.39}$$

with the results being shown in Fig. 2.8. Note that at low operating thrusts the figure of merit is small. This is because in Eq. 2.39 the profile drag term in the denominator is large compared to the numerator. As the value of C_T increases, however, the relative importance of the profile power term decreases and FM increases. This continues until the induced power dominates the profile term, and the figure of merit approaches a value of $1/\kappa$. In practice, the profile drag contribution decreases this value somewhat. Therefore, the FM curve reaches a firm plateau region, which represents the maximum attainable FM. Note again, that to be meaningful, the figure of merit must only be used as a gauge of rotor efficiency when two or more rotors are compared at the same disk loading. This is because increasing T/A will increase the induced power relative to the profile power, producing a higher figure of merit and a potentially misleading comparison.

2.2.9 *Worked Example*

A tilt-rotor aircraft has a gross weight of 60,500 lb ($\approx$27,500 kg). The rotor diameter is 38 ft (11.58 m). On the basis of the momentum theory, estimate the power required for the aircraft to hover at sea level on a standard day[3]. Assume that the figure of merit of the rotors is 0.75 and transmission losses amount to 5%.

[3] At sea-level on a standard day, the density of air will be 0.00238 slugs ft^{-3} or 1.225 kg m^{-3}.

A tilt-rotor has two rotors, which are each assumed to carry half of the total aircraft weight, that is, $T = 30{,}250$ lb. For each of the rotors, the disk area is $A = \pi(38/2)^2 = 1134.12$ ft^2. The induced velocity in the plane of the rotor is

$$v_i = \sqrt{\frac{T}{2\rho A}} = \sqrt{\frac{30250}{2 \times 0.00238 \times 1134.12}} = 74.86 \text{ ft s}^{-1}.$$

The ideal power per rotor will be $Tv_i = 30{,}250 \times 74.86 = 2{,}264{,}515$ lb ft s^{-1}. This result is converted into horsepower (hp) by dividing by 550 to give 4,117.3 hp per rotor. Remember that the figure of merit accounts for the aerodynamic efficiency of the rotor. Therefore, the actual power required per rotor to overcome induced and profile losses will be $4{,}117.3/0.75 = 5{,}489.7$ hp, followed by multiplying the result by two to account for both rotors, that is, $2 \times 5{,}489.7 = 10{,}979.4$ hp. Transmission losses account for another 5%, so that the total power required to hover is $1.05 \times 10{,}979.4 = 11{,}528.4$ hp.

The problem can also be worked in SI units. In this case, $T = 13{,}750 \times 9.81 = 134{,}887.5$ N. The disk area is $A = \pi(11.58/2)^2 = 105.32$ m^2. The induced velocity in the plane of the rotor is

$$v_i = \sqrt{\frac{T}{2\rho A}} = \sqrt{\frac{134{,}887.5}{2 \times 1.225 \times 105.35}} = 22.9 \text{ ms}^{-1}.$$

The ideal power per rotor will be $Tv_i = 134{,}887.5 \times 22.9 = 3{,}088.9$ kW. The actual power required per rotor to overcome induced and profile losses will be $3{,}088.9/0.75 = 4{,}118.5$ kW followed by multiplying the result by two to account for both rotors, that is, 8,237 kW. Transmission losses mean that the total power required to hover will be 8,648.9 kW.

2.2.10 *Induced Tip Loss*

The formation of a trailed vortex at the tip of each blade produces a high local inflow over the tip region and effectively reduces the lifting capability there. This is referred to as a *tip loss*, in that it represents a loss relative to the finite value of lift that would otherwise be produced without the influence of any tip vortices in the flow. In performance or preliminary design work, a simple tip loss factor B can be used to account for this physical effect such that the product BR corresponds to an effective blade radius, $R_e < R$. A tip loss essentially corresponds to a reduction in the rotor disk area by a factor B^2, that is,

$$A_e = \pi R_e^2 = \pi(BR)^2 = B^2(\pi R^2) = B^2 A. \tag{2.40}$$

This "tip-loss" effect will manifest as a higher effective disk loading (i.e., T/A_e) and an increase in the average induced velocity by a factor B^{-1} for a given thrust, with a corresponding increase in induced power. This is the essence of Prandtl's approach to modeling tip-loss effects, where an analogy was made between the helical vortex wake below the rotor and a system of parallel vortex sheets – see Durand (1935). Prandtl showed that when accounting for the tip loss, the effective blade radius is given by

$$\frac{R_e}{R} \approx 1 - \left(\frac{1.386}{N_b}\right)\frac{\lambda_i}{\sqrt{1+\lambda_i^2}}, \tag{2.41}$$

where N_b is the number of blades. For helicopter rotors λ_i is typically less than 0.07; thus λ_i^2 will be small and the preceding equation can be simplified to

$$\frac{R_e}{R} \approx 1 - \left(\frac{1.386}{N_b}\right)\lambda_i. \tag{2.42}$$

A more general tip-loss equation is sometimes used where

$$B = 1 - \frac{1.386\lambda}{N_b} \tag{2.43}$$

and where the inflow ratio is $\lambda = (V_c + v_i)/\Omega R$, where V_c is the climb velocity. For hovering flight with the assumption of uniform inflow it has already been shown that

$$\lambda_i = \sqrt{\frac{C_T}{2}}, \tag{2.44}$$

so that to a good approximation the tip-loss factor is

$$\frac{R_e}{R} = B = 1 - \left(\frac{1.386}{\sqrt{2}}\right)\frac{\sqrt{C_T}}{N_b} \approx 1 - \frac{\sqrt{C_T}}{N_b}. \tag{2.45}$$

A graph of this result is shown in Fig. 2.9, where the factor B is shown to decrease with decreasing number of blades and also with increasing rotor thrust. The former effect results from blade-to-blade interference, whereas the latter effect has its origin in the spacing of the vortex sheets below the rotor (helical pitch of the wake). In practice, values of B for helicopter rotors are found to range from about 0.95 to 0.98, depending on the number of blades.

Gessow & Myers (1952) suggest an empirical tip-loss factor based on blade geometry alone where

$$B = 1 - \frac{c}{2R} \tag{2.46}$$

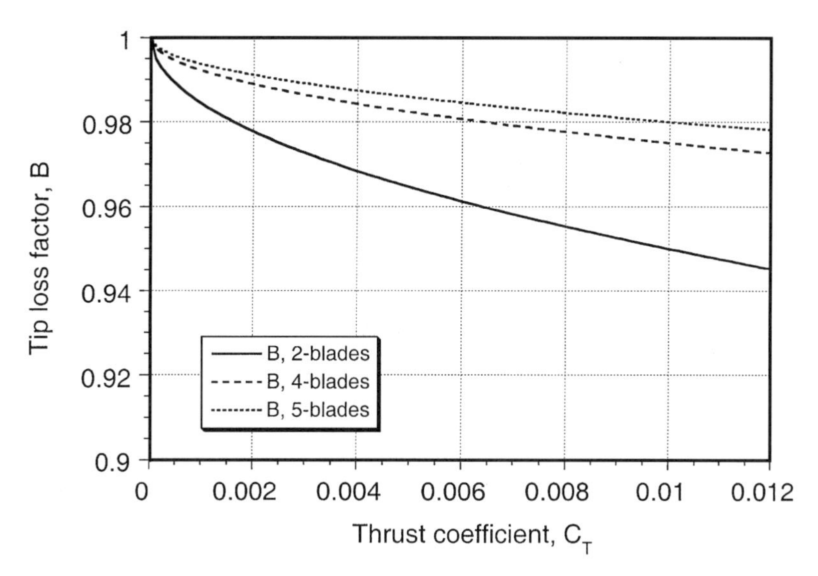

Figure 2.9 The effect of thrust and number of blades on Prandtl's tip loss factor.

and where c is the tip chord, although it would seem that this result is not general enough to deal with other than rectangular blade tips. Sissingh (1939) has proposed the alternative geometric expression

$$B = 1 - \frac{c_0(1 + 0.7\tau_r)}{1.5R}, \tag{2.47}$$

where c_0 is the root chord of the main blade and τ_r is the blade taper ratio (i.e., the ratio of the tip chord to the root chord). The need to determine B by one of these equations can be avoided if a more general numerical approach to solving for tip-loss effects is used (see Chapter 3).

2.2.11 *Rotor Solidity and Blade Loading Coefficient*

It will be seen from Eq. 2.39 that the solidity, σ, appears in the expression for figure of merit, FM. The solidity represents the ratio of the lifting area of the blades to the area of the rotor. If FM is plotted for rotors with different values of σ, the behavior is typified by Fig. 2.10. The data have been taken from the classic experiments of Knight & Hefner (1937), for which the thrust and power were measured for rotors with different numbers of blades. Results predicted by means of the modified momentum theory are also shown. From the measurements at zero thrust it was deduced that $C_{d_0} = 0.011$, and κ was estimated to be 1.25 and independent of the number of blades. While this value of κ is perhaps on the high side for an actual helicopter rotor, it must be recognized that these measurements were made with untwisted blades and so, as will be shown in Chapter 3, this will produce higher induced flow near the tip and, therefore, the rotor is less efficient with a higher value of κ.

From the results shown in Fig. 2.10, it will be noted that higher values of FM are obtained with the *lowest* possible solidity at the same design C_T (same aircraft gross weight or disk loading). This is hardly an unexpected result from Eq. 2.39 (all other terms such as κ being assumed constant) and simply means that the viscous drag on the rotor is being minimized by reducing the net blade area. However, the minimization of rotor solidity, σ,

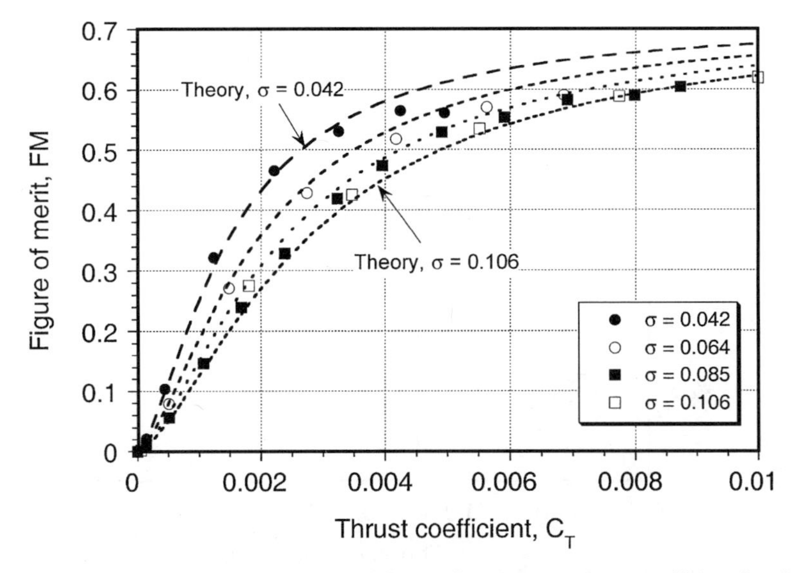

Figure 2.10 Measured and predicted figure of merit versus thrust coefficient for a hovering rotor with different values of solidity. Data source: Knight & Hefner (1937).

must be done with extreme caution. This is because reducing blade area must always result in higher blade section angles of attack (and higher lift coefficients) to obtain the same values of C_T and disk loading. Therefore, the lowest allowable value of σ must ultimately be limited by the onset of blade stall. This effect is well shown by the results in Fig. 2.10 for the lowest solidity of 0.042, where a progressive departure occurs from the theoretical predictions for $C_T > 0.003$.

An alternative format that emphasizes the local lift loading on the blades is to plot the figure of merit versus blade loading coefficient, C_T/σ. Note that the blade loading coefficient can be written as

$$\frac{C_T}{\sigma} = \frac{T}{\rho A(\Omega R^2)}\left(\frac{A}{A_b}\right) = \frac{T}{\rho A_b(\Omega R)^2}, \tag{2.48}$$

where A_b is the area of the blades. These results are shown in Fig. 2.11. Note that reducing the value of σ results in higher values of C_T/σ for the same operational value of C_T. Although Fig. 2.11 shows the rotor operates at higher values of FM with increased blade loading coefficient, the maximum value is clearly limited by the occurrence of blade stall. Typically, for a helicopter rotor, the maximum realizable value of blade loading coefficient is about 0.12, but the influence of Reynolds number on blade stall must also be considered. The maximum attainable value of C_T/σ will also depend on the distribution of local lift coefficients along the blade, which in turn depends on both the blade twist and the planform. As will be shown in Chapter 3, the local lift coefficients can be related to the blade loading by means of the blade element theory, and so the blade twist and blade planform can be designed to delay the effects of stall to higher values of C_T/σ. Nevertheless, it is clear that to maximize the figure of merit, the blade sections must always operate close to their maximum lift coefficients (i.e., at the highest possible blade loading without the occurrence of blade stall). It can also be concluded that a rotor that uses airfoils with higher values of maximum lift coefficient can be designed to have lower solidity. This has the benefits of lower blade and hub weight. Consequently, there is considerable research emphasis

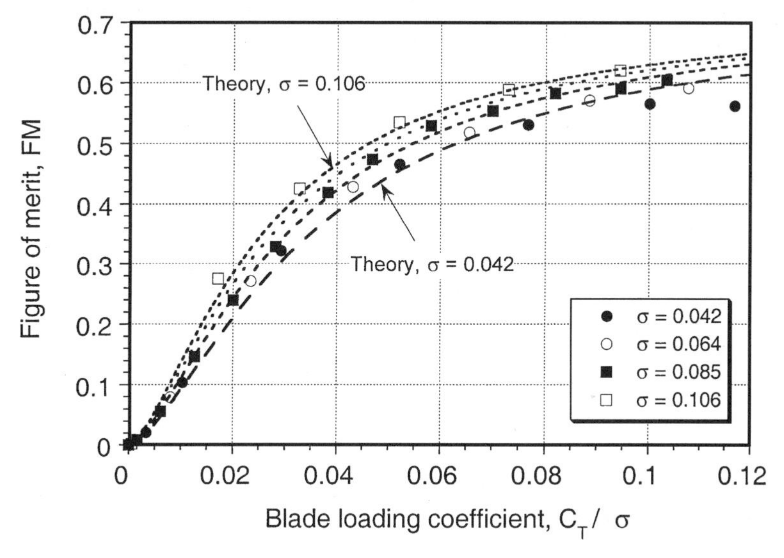

Figure 2.11 Measured and predicted figure of merit versus blade loading coefficient for a hovering rotor with different solidities. Data source: Knight & Hefner (1937).

in the helicopter industry to design airfoils with high values of maximum lift coefficient (see Chapter 7).

2.2.12 *Power Loading*

The ratio T/P or the power loading has been previously defined as a rotor efficiency parameter. Helicopters and other rotorcraft are generally designed to hover with the lowest possible power required (and hence lowest fuel burn) for a given gross weight, that is, at a high power loading. Helicopters spend a good proportion of their flight time in hover or low speed forward flight, and the use of the hover condition as an initial design point is clear. However, as will be shown, optimizing the rotor for maximum hovering efficiency can also have some trade-offs in terms of efficient high speed forward flight performance.

Power loading is the ratio of the thrust produced to the power required to hover, that is,

$$PL = \frac{T}{P} = \frac{W}{P} = \frac{C_T}{(\Omega R) C_P}. \tag{2.49}$$

This quantity should be as close as possible to the ideal value for best hovering efficiency. Because $T \propto (\Omega R)^2$ but $P \propto (\Omega R)^3$, maximizing the power loading requires a low tip speed (ΩR). On the basis of simple momentum theory considerations the ratio P/T is given by

$$\frac{P}{T} = \frac{T^{3/2}}{T\sqrt{2\rho A}} = \sqrt{\frac{T}{2\rho A}} = \sqrt{\frac{DL}{2\rho}} = v_i = (PL)^{-1} \tag{2.50}$$

which, as shown previously in Eq. 2.24, is related to the disk loading. Therefore, to maximize the power loading, that is, to minimize the ratio P/T, the disk loading should be low (i.e., the disk area should be large for a given gross weight to give a low induced velocity and the tip speed should be low). Generally, the tip speed is set on the basis of various performance requirements for a given rotor size. As will be shown in Chapter 6, this may include autorotative requirements and rotor noise constraints.

When using the modified momentum theory, the ratio P/T is given by

$$\frac{P}{T} = \kappa \sqrt{\frac{T}{2\rho A}} + \frac{P_0}{T}. \tag{2.51}$$

Alternatively, we can write

$$\begin{aligned} \frac{P}{T} &= \Omega R \frac{C_P}{C_T} = \Omega R \left(\kappa \sqrt{\frac{C_T}{2}} + \frac{C_{P_0}}{C_T} \right) \\ &= \frac{\Omega R}{C_T} \left(\kappa \frac{C_T^{3/2}}{\sqrt{2}} + \frac{\sigma C_{d_0}}{8} \right), \end{aligned} \tag{2.52}$$

which is a result that depends on the rotor solidity, σ. Equation 2.52 can also be written in terms of the figure of merit, that is,

$$\frac{P}{T} = \frac{\sqrt{DL}}{\sqrt{2\rho}FM} \propto \frac{\sqrt{DL}}{FM} \quad \text{or} \quad PL(\text{actual}) \propto \frac{FM}{\sqrt{DL}} \tag{2.53}$$

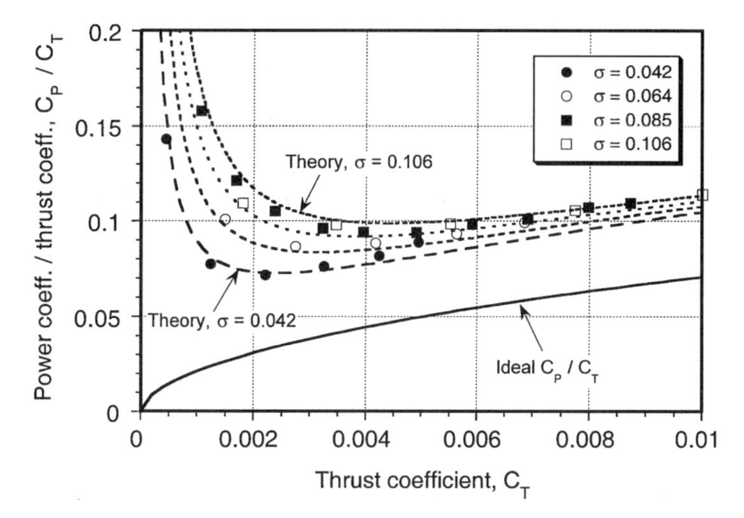

Figure 2.12 Measured variation of C_P/C_T versus thrust coefficient for a hovering rotor with different solidities. Data source: Knight & Hefner (1937).

This means that the best efficiency (maximum power loading) of the rotor is obtained when the disk loading is a minimum and the figure of merit is a maximum.

A representative example of the measured C_P/C_T for a rotor (which is inversely proportional to power loading at a fixed tip speed) versus thrust is shown in Fig. 2.12 for several values of rotor solidity. By differentiating Eq. 2.52 with respect to C_T, it is easy to show that the operating C_T to give the lowest ratio of P/T is

$$C_T \text{ (for maximum } PL) = \frac{1}{2}\left(\frac{\sigma C_{d_0}}{\kappa}\right)^{2/3}, \tag{2.54}$$

which depends on both the profile and induced effects. This is equivalent to a figure of merit of $2/3\kappa$. Using this result, the disk loading for maximum power loading will be

$$\frac{T}{A} = \frac{1}{2}\rho(\Omega R)^2\left(\frac{\sigma C_{d_0}}{\kappa}\right)^{2/3}, \tag{2.55}$$

and for design purposes this would determine the optimum radius of the rotor for a given gross weight. However, it is apparent that the conditions for the most efficient operation of the rotor are relatively insensitive to the operating state in that the C_P/C_T curve is fairly flat over the normal range of thrust coefficients. Therefore, there is some latitude in selecting the rotor radius, which, as mentioned previously, may be constrained because of other (nonaerodynamic) factors.

2.3 Axial Climb and Descent

2.3.1 *Axial Climb*

Adequate climbing flight performance is an important operational consideration, and sufficient power reserves must be designed into the helicopter to ensure this performance is maintained over a wide range of gross weights and operational density altitudes. Again, we can apply the three conservation laws to a control volume surrounding the climbing rotor and its flow field, as shown previously in Fig. 2.5, but this time $V_c > 0$. As before, consider the problem to be quasi-one-dimensional in that the flow properties will be assumed to vary only in the vertical direction through the disk and at each cross section the flow properties are distributed uniformly. In contrast to the hover case where the climb velocity is zero, the relative velocity far upstream relative to the rotor will now be V_c. At the plane of the rotor, the velocity will be $V_c + v_i$, and the slipstream velocity is $V_c + w$. By the conservation of mass, the mass flow rate is constant within the boundaries of the wake and so

$$\dot{m} = \iint_{\infty} \rho \vec{V} \cdot d\vec{S} = \iint_{2} \rho \vec{V} \cdot d\vec{S}. \tag{2.56}$$

Therefore, substituting the values for this problem results in

$$\rho A_{\infty}(V_c + w) = \rho A(V_c + v_i). \tag{2.57}$$

The application of the momentum equation gives

$$T = \iint_{\infty} \rho(\vec{V} \cdot d\vec{S})\vec{V} - \iint_{0} \rho(\vec{V} \cdot d\vec{S})\vec{V}. \tag{2.58}$$

Now, in a steady climb the velocity far upstream of the rotor is finite, so that both terms on the right-hand side of the above equation are nonzero. Therefore, in this case

$$T = \dot{m}(V_c + w) - \dot{m} V_c = \dot{m} w. \tag{2.59}$$

Note that this is the same equation obtained for the rotor thrust in the hover case (Eq. 2.7). Because the work done by the climbing rotor is now $T(V_c + v_i)$, then

$$\begin{aligned} T(V_c + v_i) &= \iint_{\infty} \frac{1}{2}\rho(\vec{V} \cdot d\vec{S})\vec{V}^2 - \iint_{0} \frac{1}{2}\rho(\vec{V} \cdot d\vec{S})\vec{V}^2 \\ &= \frac{1}{2}\dot{m}(V_c + w)^2 - \frac{1}{2}\dot{m} V_c^2 = \frac{1}{2}\dot{m} w(2V_c + w). \end{aligned} \tag{2.60}$$

From these latter two equations, it is readily apparent that $w = 2v_i$, which is, again, the same result as for the hover case.

The relationship between the rotor thrust and the induced velocity at the rotor disk in hover is

$$v_h \equiv v_i = \sqrt{\frac{T}{2\rho A}} \tag{2.61}$$

and for the climbing rotor it has been shown from Eq. 2.59 that

$$T = \dot{m} w = \rho A(V_c + v_i) w = 2\rho A(V_c + v_i) v_i, \tag{2.62}$$

so that

$$\frac{T}{2\rho A} = v_h^2 = (V_c + v_i) v_i = V_c v_i + v_i^2. \tag{2.63}$$

Dividing through by v_h^2 results in

$$\left(\frac{v_i}{v_h}\right)^2 + \frac{V_c}{v_h}\left(\frac{v_i}{v_h}\right) - 1 = 0, \tag{2.64}$$

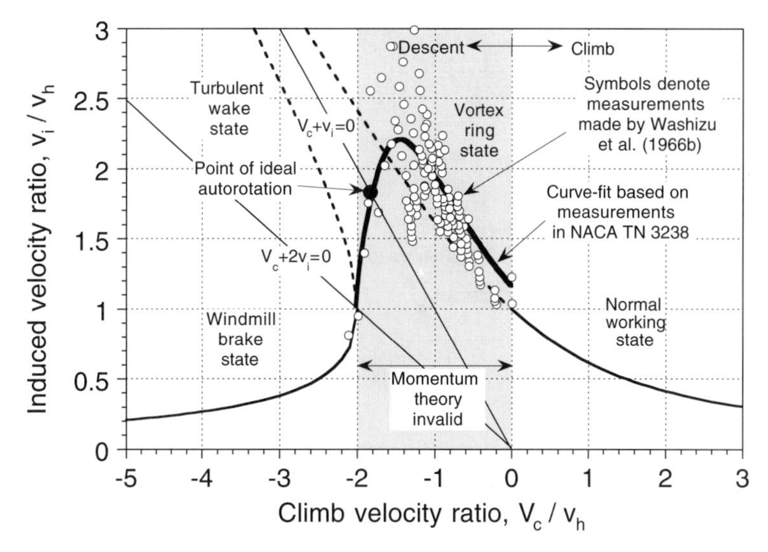

Figure 2.13 Induced velocity variation as a function of climb and descent velocity based on momentum theory (complete induced velocity curve).

which is a quadratic equation in v_i/v_h. This has the solution

$$\frac{v_i}{v_h} = -\left(\frac{V_c}{2v_h}\right) \pm \sqrt{\left(\frac{V_c}{2v_h}\right)^2 + 1}. \tag{2.65}$$

Although there are two possible solutions, v_i/v_h must always be positive in the climb so as not to violate the assumed flow model. Therefore, the only valid solution becomes

$$\frac{v_i}{v_h} = -\left(\frac{V_c}{2v_h}\right) + \sqrt{\left(\frac{V_c}{2v_h}\right)^2 + 1}. \tag{2.66}$$

The results from this analysis are shown in Fig. 2.13, which is presented in a form first suggested by Hafner (1947). The other root of the quadratic equation lies below the V_c/v_h axis and is physically invalid. It will be apparent that as the climb velocity increases the induced velocity at the rotor decreases. This is called the *normal working state* of the rotor. The branch of the induced velocity curve denoted by the broken line gives a solution to Eq. 2.66 for negative values of V_c (i.e., a descent). However, just as the rotor begins to descend there can be two possible flow directions; this violates the assumed flow model and so this solution is also physically invalid.

2.3.2 *Axial Descent*

The climb model cannot be used in a descent (where $V_c < 0$) because now V_c is directed upward, and so the slipstream will be above the rotor. This will be the case whenever $|V_c|$ is more than twice the average induced velocity at the disk. For cases where the descent velocity is in the range $-2v_i < V_c < 0$, the velocity at any plane through the rotor slipstream can be either upward or downward. Under these circumstances, a more complicated recirculating (and usually more turbulent) flow pattern may exist at the rotor, and momentum theory cannot be used because no definitive control volume can be established. The operating state where $-2v_i < V_c < 0$ will be discussed in detail later in Section 2.3.3.

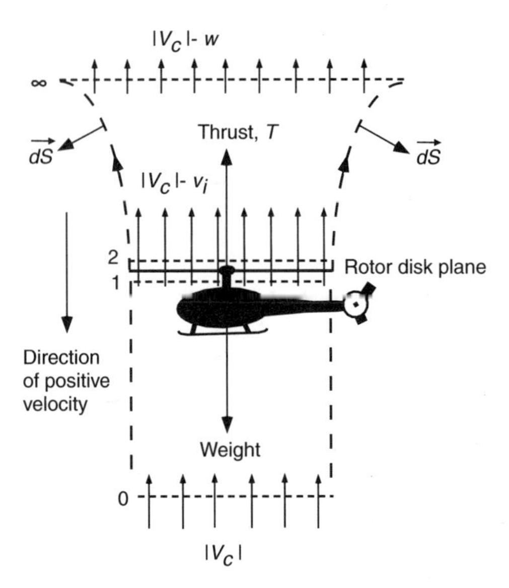

Figure 2.14 Flow model used for momentum theory analysis of a rotor in a vertical descent.

The assumed flow model and control volume surrounding the descending rotor is shown in Fig. 2.14. To proceed, the assumption must be made that $|V_c| > 2v_h$ so that a well-defined slipstream will always exist above the rotor. Far upstream (well below) the rotor, the magnitude of the velocity is the descent velocity, which is equal to $|V_c|$. Note that to avoid any ambiguity, it will be assumed that the velocity is measured as positive when directed in a downward direction. At the plane of the rotor, the velocity is $|V_c| - v_i$. In the far wake (above the rotor), the velocity is $|V_c| - w$. By the conservation of mass, the fluid mass flow rate, $\dot{m}$, through the rotor disk is

$$\dot{m} = \iint_{\infty} \rho \vec{V} \cdot d\vec{S} = \iint_{2} \rho \vec{V} \cdot d\vec{S}. \tag{2.67}$$

Therefore,

$$\dot{m} = \rho A_{\infty}(V_c + w) = \rho A(V_c + v_i). \tag{2.68}$$

Conservation of fluid momentum gives in this case

$$T = -\left[\iint_{\infty} \rho(\vec{V} \cdot d\vec{S})\vec{V} - \iint_{0} \rho(\vec{V} \cdot d\vec{S})\vec{V}\right], \tag{2.69}$$

with the negative sign arising because the flow direction is now reversed compared to the climb case. In a steady descent, the velocity far upstream of (below) the rotor must be finite so that both terms on the right-hand-side of the above equation are nonzero. Therefore, in this case

$$T = (-\dot{m})(V_c + w) - (-\dot{m})V_c = -\dot{m}w. \tag{2.70}$$

Note that T is not negative because $\dot{m}$ is negative by convention. Because the work done by the rotor in descent is $T(v_i + V_c)$, this gives

$$T(v_i + V_c) = \iint_0 \frac{1}{2}\rho(\vec{V} \cdot d\vec{S})\vec{V}^2 - \iint_\infty \frac{1}{2}\rho(\vec{V} \cdot d\vec{S})\vec{V}^2$$
$$= \frac{1}{2}\dot{m}V_c^2 - \frac{1}{2}\dot{m}(V_c + v_i)^2 = -\frac{1}{2}\dot{m}w(2V_c + m), \quad (2.71)$$

which is a negative quantity. Therefore, the rotor is now extracting power from the airstream, and this operating condition is known as the *windmill brake state*. Using Eqs. 2.70 and 2.71 it is seen, again, that $w = 2v_i$. Note, however, that the net velocity in the slipstream is *less* than V_c and so from continuity considerations the wake boundary expands above the descending rotor disk. For the descending rotor

$$T = -\dot{m}w = -\rho A(V_c + v_i)w = -2\rho A(V_c + v_i)v_i. \quad (2.72)$$

Therefore, we can write

$$\frac{T}{2\rho A} = v_h^2 = -(V_c + v_i)v_i = -V_c v_i - v_i^2. \quad (2.73)$$

Dividing through by v_h^2 gives

$$\left(\frac{v_i}{v_h}\right)^2 + \frac{V_c}{v_h}\left(\frac{v_i}{v_h}\right) + 1 = 0, \quad (2.74)$$

which is a quadratic equation in v_i/v_h. This has the solution

$$\frac{v_i}{v_h} = -\left(\frac{V_c}{2v_h}\right) \pm \sqrt{\left(\frac{V_c}{2v_h}\right)^2 - 1}. \quad (2.75)$$

Again, like the climb case, there are two possible solutions for v_i/v_h. However, because $|V_c| > 2v_h$ the only valid solution is

$$\frac{v_i}{v_h} = -\left(\frac{V_c}{2v_h}\right) - \sqrt{\left(\frac{V_c}{2v_h}\right)^2 - 1}, \quad (2.76)$$

which is valid for $V_c/v_h \leq -2$. Therefore, it is noted from Fig. 2.13 that as the descent velocity increases the induced velocity also decreases and asymptotes smoothly to zero at high descent rates. The other solution to the quadratic, which is denoted by the broken line in Fig. 2.13, violates the assumed flow model and so is a nonphysical solution.

2.3.3 *The Region* $-2 \leq V_c/v_h \leq 0$

In the region $-2 \leq V_c/v_h \leq 0$, momentum theory is invalid because the flow can take on two possible directions and a well-defined slipstream ceases to exist. However, the velocity curve can still be defined approximately on the basis of flight tests or other experiments. Unfortunately descending flight accentuates interactions of the tip vortices with other blades, and so the flow becomes rather unsteady and turbulent, and experimental measurements of rotor thrust and power are difficult to make. Also, the average induced velocity cannot be measured directly. Instead, it is obtained indirectly from the measured rotor power and thrust – see Gessow (1948, 1954) and Brotherhood (1949). The measured rotor power can be written in the assumed form as

$$P_{\text{meas}} = T(V_c + v_i) + P_0, \tag{2.77}$$

where P_0 is the profile power, and where v_i is recognized as an averaged induced velocity through the disk. Using the result that $P_h = T v_h$, we get

$$\begin{aligned} \frac{V_c + v_i}{v_h} &= \frac{P_{\text{meas}} - P_0}{P_h} = \frac{P_{\text{meas}} - P_0}{T\sqrt{T/2\rho A}} \\ &= \frac{C_{P\,\text{meas}} - C_{P_0}}{C_T^{3/2}/\sqrt{2}} = \frac{\sqrt{2}C_{P_i}}{C_T^{3/2}}. \end{aligned} \tag{2.78}$$

Therefore, in addition to the measured rotor power, to obtain an estimate for the averaged induced velocity ratio it is necessary to know the rotor profile power. As shown previously in Section 2.2.7, one simple estimate for the profile power of a rectangular rotor is $C_{P_0} = \sigma C_{d_0}/8$, where C_{d_0} is the mean (average) drag coefficient of the airfoil sections comprising the rotor, and σ is the rotor solidity.

Because of the high levels of turbulence near the rotor in this operating state, the derived measurements of the average induced velocity contain a relatively large amount of scatter. The curve shown in Fig. 2.13 is taken from Gessow (1954) and is a composite of flight and wind tunnel measurements made by Lock et al. (1926), Brotherhood (1949), and Drees & Hendal (1951). Note, however, that the curve follows the other branch of the induced velocity curve derived on the basis of momentum theory up to about $V_c/v_h \approx -1.5$, after which it drops off precipitously and joins the branch of the curve defined on the basis of the momentum result in a descent. The higher values of measured power in hover and at low rates of descent are simply a result of higher induced power losses, which, as explained previously, are not predicted directly by the simple momentum theory.

Because the nature of the induced velocity curve is not analytically predictable in the range $-2 \leq V_c/v_h \leq 0$, the experimental estimates can be used to find a "best-fit" approximation for v_i at any rate of descent. Various authors, including Young (1978) and Johnson (1980), suggest a linear approximation to the measured curve. Following Young (1978), then one approximation is

$$\frac{v_i}{v_h} = \kappa - \frac{V_c}{v_h} \quad \text{for} \quad 0 \leq \frac{V_c}{v_h} \leq -1.5 \tag{2.79}$$

and

$$\frac{v_i}{v_h} = \kappa\left[7 + 3\frac{V_c}{v_h}\right] \quad \text{for} \quad -1.5 \leq \frac{V_c}{v_h} \leq -2, \tag{2.80}$$

where κ is the measured induced power factor in hover. A slighly better approximation to the measured curve is the quartic

$$\frac{v_i}{v_h} = \kappa + k_1\left(\frac{V_c}{v_h}\right) + k_2\left(\frac{V_c}{v_h}\right)^2 + k_3\left(\frac{V_c}{v_h}\right)^3 + k_4\left(\frac{V_c}{v_h}\right)^4 \tag{2.81}$$

with $k_1 = -1.125$, $k_2 = -1.372$, $k_3 = -1.718$, and $k_4 = -0.655$, which is valid for the full range $0 \leq V_c/v_h \leq -2$.

2.3.4 *Power Required*

Note that because the climb and descent velocity changes the induced velocity at the rotor, the induced power will also be affected. In a climb or descent the power ratio is

$$\frac{P}{P_h} = \frac{V_c + v_i}{v_h} = \frac{V_c}{v_h} + \frac{v_i}{v_h}. \tag{2.82}$$

The two terms on the right-hand side of the above equation are the work done to change the potential energy of the rotor and the work done on the air, respectively. The latter induced loss appears as an increase in dynamic pressure and gain in kinetic energy of the slipstream. Using Eq. 2.66 and substituting and rearranging gives the power ratio for a *climb* as

$$\frac{P}{P_h} = \frac{V_c}{2v_h} + \sqrt{\left(\frac{V_c}{2v_h}\right)^2 + 1}. \tag{2.83}$$

In a *descent*, Eq. 2.76 is applicable, and substituting in Eq. 2.82 and rearranging gives the power ratio as

$$\frac{P}{P_h} = \frac{V_c}{2v_h} - \sqrt{\left(\frac{V_c}{2v_h}\right)^2 - 1}. \tag{2.84}$$

Both results are shown in Fig. 2.15, which is usually called the *universal power curve* – a form first suggested by Lock (1947). This graph shows the total rotor power ratio, P/P_h, plotted versus the climb ratio, V_c/v_h. Note that the power required to climb is always greater than the power required to hover. However, as the climb velocity increases the induced power becomes a progressively smaller percentage of the total power required to climb. It is also significant to note that in a descent, at least above a certain rate, the rotor extracts power from the air and uses less power than required to hover (i.e., the rotor operates like a windmill).

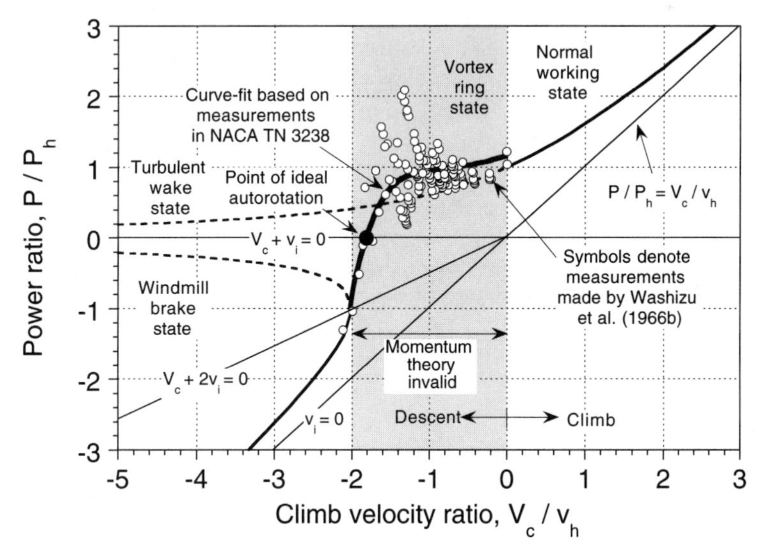

Figure 2.15 Total power required as a function of climb and descent velocity (universal power curve).

2.3.5 *Working States of the Rotor in Axial Flight*

Figures 2.13 and 2.15 show the lines $V_c + v_i = 0$ and $V_c + 2v_i = 0$, and these lines are used to demarcate four axial operating states of the rotor. For points above the line $V_c + v_i = 0$, the rotor is absorbing power from the rotor shaft. Below this line, the rotor is extracting power from the relative airstream. To understand the complicated physical nature of the rotor wake under these conditions, Fig. 2.16 shows flow visualization images of the wake at various descent velocities. Unlike some of the earlier work by Lock (1928) where smoke was used to visualize the gross flow structure, the individual blade tip vortices were rendered visible here by means of shadowgraphy, which is a density gradient method (see Chapter 10). Because the flow is nominally axisymmetric, only one side of the rotor is shown for clarity.

Figure 2.16(a) shows an image of the flow in the *normal working state*. Here, the tip vortices follow smooth helicoidal-like trajectories. The flow is highly periodic and free of any significant disturbances. A schematic of the mean flow is also shown. For low rates of descent, the tip vortex filaments are convected closer to the plane of the rotor than for the hover case, but they also move radially outward away from the rotor. At slightly higher descent rates, the tip vortices become very close to the rotor plane, and considerable unsteadiness (aperiodicity) becomes apparent. This can be seen in Fig. 2.16(b) by the contortions in the tip vortex trajectories and the lack of any distinct slipstream boundary. This is close to the flow condition known as the *vortex ring state*, where the accumulation of tip vortices in the rotor plane begins to resemble a concentric set of vortex rings (see adjacent schematic). In the vortex ring state, the rotor can experience highly unsteady flow with regions of concurrent upward and downward velocities, and the flow can periodically break away from the rotor disk. This flow state appears to have been first recognized by de Bothezat (1919), and such flows have been visualized and measured in rotor experiments by Lock et al. (1926), Castles & Gray (1951), Drees & Hendal (1951), Washizu et al. (1966a,b), and Azuma & Obata (1968). Some interesting in-flight flow visualization on a helicopter during vertical descent were performed by Brotherhood (1949). Glauert (1926b) and Nikolsky & Seckel (1949a,b) were some of the first to conduct a mathematical analysis of the problem. From a piloting perspective, entry into autorotative flight requires transition through the vortex ring state. The vortex ring state is known to exhibit considerable unsteadiness at the rotor, and this can lead to significant blade flapping and a loss of rotor control. If the vortex ring state occurs on the tail rotor, such as during sideways flight or hovering in a crosswind, then directional (yaw) control may be seriously impaired.

As the descent velocity increases further, the wake above the rotor becomes more turbulent and aperiodic and is representative of the flow conditions known as the *turbulent wake state*. This shown by Fig. 2.16(c). At even higher descent velocities, the wake is again observed to develop a more definite slipstream boundary that expands downstream (above) of the rotor. When in this state, the wake structure is found to return to a more regular helical structure, as shown by Fig. 2.16(d), and as previously described is known as the *windmill brake state*.

2.3.6 *Autorotation*

Note that from the universal power curve in Fig. 2.15 there is a value of V_c/v_h for which zero net power is required for the rotor (i.e., $P = T(V_c + v_i) = 0$ or $P/P_h = 0$). This condition is called *ideal autorotation*, and for a given thrust it is a self-sustained operating state where the energy to drive the rotor comes from the descent velocity (which is upward

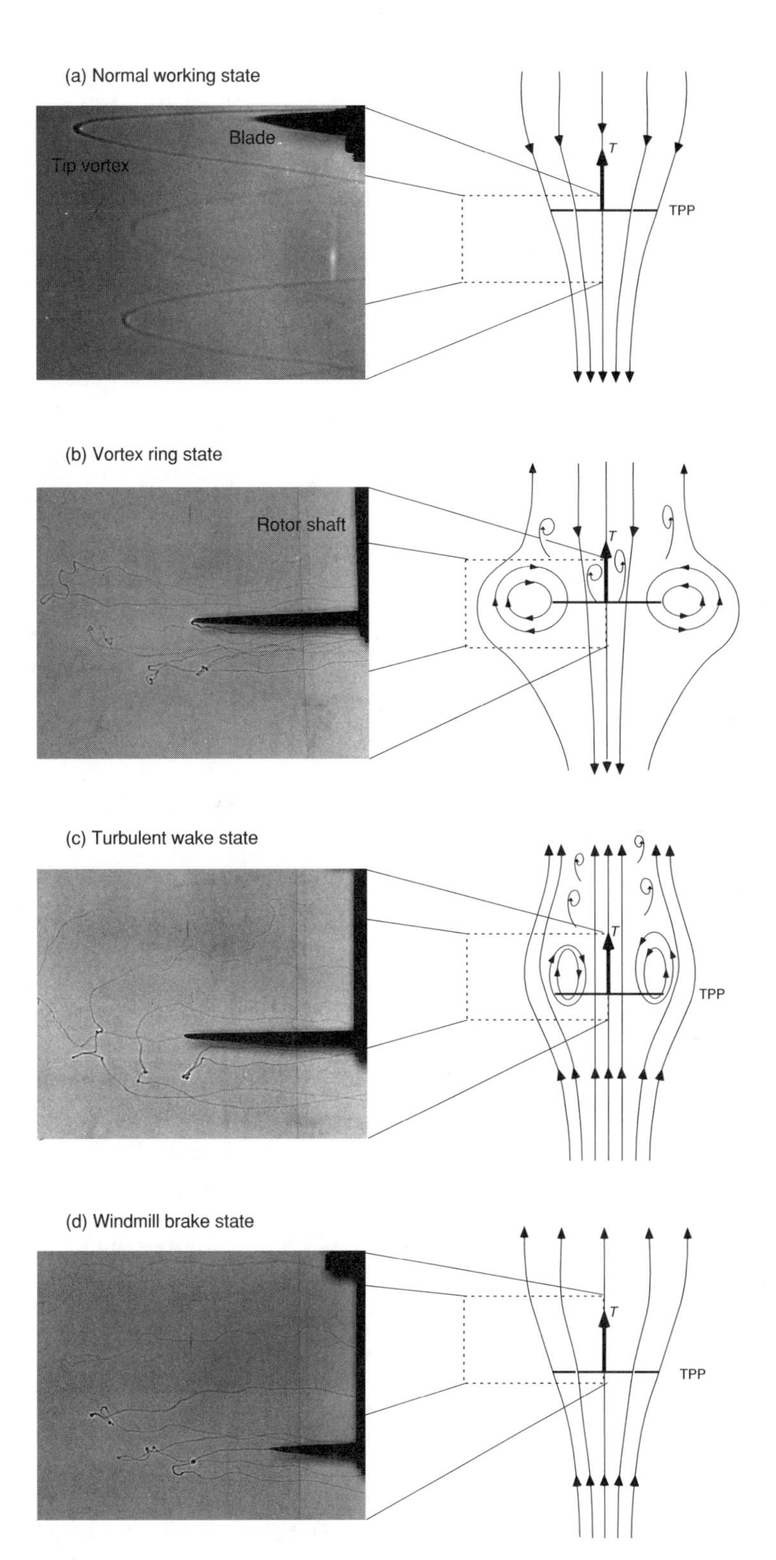

Figure 2.16 The physical nature of the rotor wake in simulated axial descent. (a) Normal working state. (b) Approaching the vortex ring state. (c) Turbulent wake state. (d) Windmill brake state. Source: University of Maryland.

relative to the rotor). An autorotation is a maneuver that can be used to recover the helicopter in the event of an engine failure, transmission problems, or a loss of the tail rotor. It requires that the pilot let the helicopter descend at a sufficiently high rate where the energy to drive the rotor can be obtained by giving up potential energy (altitude) at a controlled rate, thereby averting a ballistic fall.

On the basis of Eq. 2.80, the power curve crosses the ideal autorotation line $V_c + v_i = 0$ at

$$\frac{V_c}{v_h} = -\left(\frac{7\kappa}{1+3\kappa}\right), \tag{2.85}$$

which gives $V_c/v_h = -1.75$ for an ideal rotor ($\kappa = 1$). In practice, a real autorotation in axial flight occurs at a slightly higher rate than this, because in addition to induced losses at the rotor, there are also profile losses to overcome. In a real autorotation one can write

$$P = T\,(V_c + v_i) + P_0 = 0. \tag{2.86}$$

Therefore, in a stable autorotation an energy balance must exist where the decrease in potential energy of the rotor TV_c just balances the sum of the induced (Tv_i) *and* profile (P_0) losses of the rotor. Using Eq. 2.86, we see that this condition is achieved when

$$\frac{V_c + v_i}{v_h} = -\frac{P_0}{Tv_h} = -\frac{P_0\sqrt{2\rho A}}{T^{3/2}}. \tag{2.87}$$

Also, using the definition of figure of merit (and assuming the induced and profile losses do not vary substantially from the hover values), we find that

$$\frac{P_0\sqrt{2\rho A}}{T^{3/2}} = \left(\frac{1}{FM} - \kappa\right). \tag{2.88}$$

Using Eq. 2.88 and 2.80 we obtain the real (actual) autorotation condition:

$$\frac{V_c}{v_h} = -\frac{FM^{-1} - \kappa}{1+3\kappa} - \frac{7\kappa}{1+3\kappa}. \tag{2.89}$$

The first term on the right-hand side of the latter equation will vary in magnitude from between -0.04 to -0.09, depending on the rotor efficiency. Compared to the second term, the extra rate of descent required to overcome profile losses is relatively small. Therefore, on the basis of the foregoing, it is apparent that a real autorotation will occur for values of V_c/v_h between -1.85 and -1.9, that is, with the rotor operating in the turbulent wake state. Under these autorotative conditions, the flow above the rotor is known from experimental tests to be turbulent and resemble that from behind a bluff body (see Question 2.15).

It is found that with a helicopter, autorotations must be performed at relatively high rates of descent. Using the result that $v_h \approx 14.49\sqrt{T/A}$, where T/A is in lb/ft^2, gives $V_c \approx 26.81\sqrt{T/A}$ ft s^{-1}, which for a representative disk loading of 5 lb ft^{-2} leads to a rate of descent of about 3,600 ft min^{-1}. However, it will be shown in Section 2.4 that with some additional forward speed, the power required at the rotor is considerably lower than in the hover case, so that the autorotative rate of descent can be reduced by about half, although this is still a relatively high rate. Consequently, in an autorotation the proper recovery of the aircraft throughout the maneuver requires a high level of skill from the pilot. As the aircraft approaches the ground, the rotational kinetic energy stored in the rotor can be used to arrest the rate of descent. To do this, the pilot will progressively increase the collective pitch and flare the helicopter using cyclic pitch, so that the aircraft will settle onto the ground with minimum vertical and forward velocity. Clearly, the autorotative characteristics of the helicopter are an important design issue and will be addressed again in more detail in Chapters 3, 5, and 6.

2.4 Momentum Analysis in Forward Flight

Under forward flight conditions the rotor moves through the air with an edgewise component of velocity that is parallel to the plane of the rotor disk. Because helicopter rotors are required to produce both a lifting force (to overcome the aircraft weight) and a propulsive force (to propel the aircraft forward), the rotor disk must be tilted forward at an angle of attack relative to the oncoming flow. Under these conditions the axisymmetry of the flow through the rotor is lost. Despite the potentially more complicated nature of the flow in forward flight, the simple momentum theory can be extended to encompass these conditions on the basis of certain assumptions.

The following treatment of rotor performance in forward flight was first derived by Glauert (1926a, 1928). An adaptation of Glauert's flow model is shown in Fig. 2.17, where the analysis is performed with respect to an axis aligned with the rotor disk. The mass flow rate, $\dot{m}$, through the actuator disk is now

$$\dot{m} = \rho A U, \tag{2.90}$$

where U is the resultant velocity at the disk and is given by

$$U = \sqrt{(V_\infty \cos\alpha)^2 + (V_\infty \sin\alpha + v_i)^2} = \sqrt{V_\infty^2 + 2V_\infty v_i \sin\alpha + v_i^2}. \tag{2.91}$$

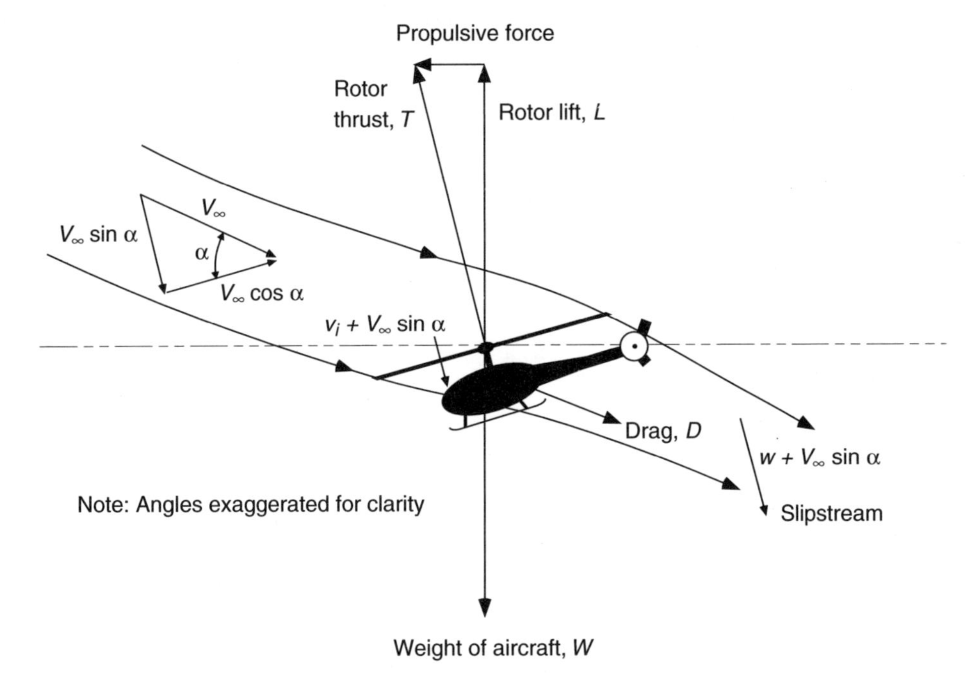

Figure 2.17 Glauert flow model for momentum analysis of a rotor in forward flight.

The application of the conservation of momentum in a direction normal to the disk gives

$$T = \dot{m}(w + V_\infty \sin\alpha) - \dot{m} V_\infty \sin\alpha = \dot{m} w, \tag{2.92}$$

and by the application of conservation of energy, we obtain

$$\begin{aligned} P = T(v_i + V_\infty \sin\alpha) &= \frac{1}{2}\dot{m}(V_\infty \sin\alpha + w)^2 - \frac{1}{2}\dot{m} V_\infty^2 \sin^2\alpha \\ &= \frac{1}{2}\dot{m}(2V_\infty w \sin\alpha + w^2). \end{aligned} \tag{2.93}$$

Using Eqs. 2.92 and 2.93 we get

$$2wv_i + 2V_\infty w \sin\alpha = 2V_\infty w \sin\alpha + w^2 \tag{2.94}$$

or simply $w = 2v_i$, which is the same result shown previously for the axial flight cases. Therefore,

$$T = 2\dot{m} v_i = 2\rho A v_i \sqrt{V_\infty^2 + 2V_\infty v_i \sin\alpha + v_i^2}. \tag{2.95}$$

Note that for hovering flight, $V_\infty = 0$, so that Eq. 2.95 reduces to

$$T = 2\rho A v_i^2, \tag{2.96}$$

which confirms that the forward flight result above reduces to the hover result (Eq. 2.14), as required. In high speed forward flight $V_\infty \gg v_i$, and Eq. 2.95 reduces to

$$T = 2\rho A v_i V_\infty. \tag{2.97}$$

This result is exactly the lift on an elliptically loaded fixed-wing with a circular planform of radius R and can be proved with the aid of classical fixed-wing lifting-line theory (see Question 2.16).

2.4.1 *Induced Velocity in Forward Flight*

In forward flight, Eq. 2.95 shows that the rotor thrust is given by

$$T = 2\dot{m} v_i = 2(\rho A U) v_i \tag{2.98}$$

or

$$T = 2\rho A v_i \sqrt{(V_\infty \cos\alpha)^2 + (V_\infty \sin\alpha + v_i)^2}. \tag{2.99}$$

Recall from Eq. 2.96 that for hovering flight

$$v_h^2 = \frac{T}{2\rho A}. \tag{2.100}$$

Then the induced velocity in forward flight can be written as

$$v_i = \frac{v_h^2}{\sqrt{(V_\infty \cos\alpha)^2 + (V_\infty \sin\alpha + v_i)^2}}. \tag{2.101}$$

The idea of a *tip speed ratio* or *advance ratio*, μ, is now used. By using the velocity parallel to the plane of the rotor, then we define $\mu = V_\infty \cos\alpha / \Omega R$. The inflow ratio is

$\lambda = (V_\infty \sin\alpha + v_i)/\Omega R$. This leads to the expression

$$\lambda = \frac{V_\infty \sin\alpha}{\Omega R} + \frac{v_i}{\Omega R} = \mu \tan\alpha + \lambda_i. \tag{2.102}$$

Also, Eq. 2.101 becomes

$$\lambda_i = \frac{\lambda_h^2}{\sqrt{\mu^2 + \lambda^2}}. \tag{2.103}$$

But, it is also known from the hover case that

$$\lambda_h = \sqrt{\frac{C_T}{2}}. \tag{2.104}$$

Therefore,

$$\lambda_i = \frac{C_T}{2\sqrt{\mu^2 + \lambda^2}}. \tag{2.105}$$

Finally, the solution for the inflow ratio, λ, is

$$\lambda = \mu \tan\alpha + \frac{C_T}{2\sqrt{\mu^2 + \lambda^2}}. \tag{2.106}$$

Note that λ appears on both sides of the above equation, and so for the general case a numerical procedure is normally used to solve for λ.

2.4.2 *Special Case, $\alpha = 0$*

If the disk angle of attack is zero ($\alpha = 0$), an exact analytical solution for λ can be determined. This is a physically unrealistic situation because the rotor must always be tilted slightly forward to produce a propulsive force. However, this special solution serves to illustrate the basic form of the induced part of the inflow through the rotor disk in forward flight and also provides a check for the numerical solution (considered next). With $\alpha = 0$, the induced velocity ratio in forward flight is

$$\lambda_i = \frac{C_T}{2\sqrt{\mu^2 + \lambda_i^2}} = \frac{\lambda_h^2}{\sqrt{\mu^2 + \lambda_i^2}} \tag{2.107}$$

Squaring both sides of the above equation and rearranging gives

$$\lambda_i^4 + \mu^2\lambda_i^2 - \lambda_h^4 = 0 \tag{2.108}$$

Dividing by λ_h^4 gives

$$\left(\frac{\lambda_i}{\lambda_h}\right)^4 + \left(\frac{\mu}{\lambda_h}\right)^2 \left(\frac{\lambda_i}{\lambda_h}\right)^2 - 1 = 0, \tag{2.109}$$

which is a quadratic in $(\lambda_i/\lambda_h)^2$. This quadratic has the solution

$$\frac{\lambda_i}{\lambda_h} = \left[\sqrt{\frac{1}{4}\left(\frac{\mu}{\lambda_h}\right)^4 + 1} - \frac{1}{2}\left(\frac{\mu}{\lambda_h}\right)^2\right]^{1/2}. \tag{2.110}$$

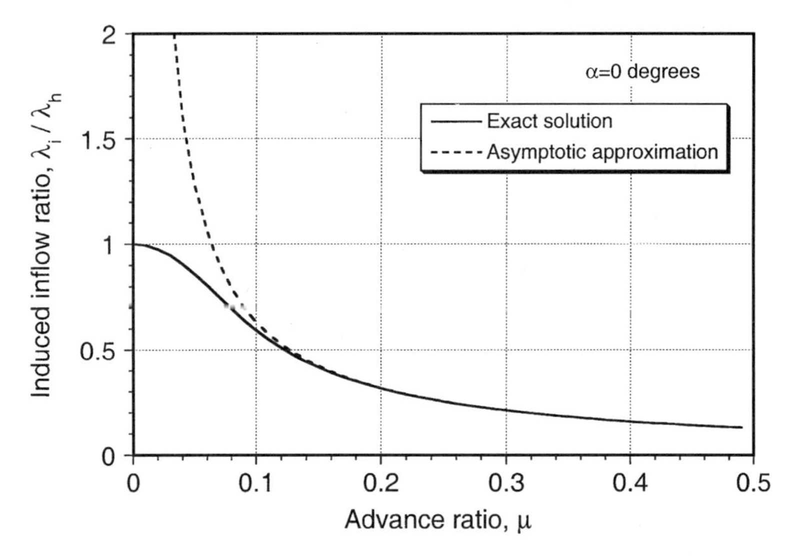

Figure 2.18 Induced inflow ratio at the rotor disk λ_i/λ_h as a function of forward speed μ for $\alpha = 0$, $\lambda_h = 0.05$.

This result is shown in nondimensional form in Fig. 2.18. Note that the induced velocity decreases quite quickly with increasing forward flight velocity. The asymptotic approximation is obtained by letting $\mu \gg \lambda$, so that in high speed forward flight Eq. 2.107 gives

$$\frac{\lambda_i}{\lambda_h} \rightarrow \frac{\lambda_h}{\mu} \quad \text{or that} \quad \lambda_i = \frac{C_T}{2\mu}, \tag{2.111}$$

which is denoted by the broken line in Fig. 2.18. For most practical purposes, this approximation is satisfactory for $\mu/\lambda_h > 2$, which is normally the case for $\mu > 0.10$.

2.4.3 *Numerical Solution to Inflow Equation*

Because α can never be zero in any practical case, Eq. 2.106 must be solved numerically. There are two common numerical approaches: 1. a simple fixed-point iteration and 2. a Newton–Raphson iteration. The algorithm for the fixed-point iteration is extremely simple. It consists of a loop to iteratively compute new estimates of λ until a termination criterion has been met. Equation 2.106 can be written as the iteration equation

$$\lambda_{n+1} = \mu \tan\alpha + \frac{C_T}{2\sqrt{\mu^2 + \lambda_n^2}}, \tag{2.112}$$

where n is the iteration number. The starting value for λ_0 is usually the hover value (i.e., $\lambda_0 = \lambda_h = \sqrt{C_T/2}$). The error estimator is

$$\epsilon = \left\| \frac{\lambda_{n+1} - \lambda_n}{\lambda_{n+1}} \right\|. \tag{2.113}$$

Normally, convergence is said to occur if $\epsilon < 0.0005$ or 0.05%. One normally finds that between 10 and 15 iterations are required with the fixed-point iteration approach. However, under some conditions, especially at lower advance ratios, a larger number of iterations may be necessary.

One can also use a Newton–Raphson procedure to solve for λ. The advantage here is that for the price of computing a simple first derivative, the convergence is much more rapid. In this case, the iteration scheme is

$$\lambda_{n+1} = \lambda_n - \left[\frac{f(\lambda)}{f'(\lambda)}\right]_n, \tag{2.114}$$

where n is the iteration number. Equation 2.106 may be rearranged in the form $f(\lambda) = 0$ giving

$$f(\lambda) = \lambda - \mu \tan\alpha - \frac{C_T}{2\sqrt{\mu^2 + \lambda^2}} = 0. \tag{2.115}$$

Differentiating this expression to find $f'(\lambda)$ gives

$$f'(\lambda) = 1 + \frac{C_T}{2}(\mu^2 + \lambda^2)^{-3/2}\lambda. \tag{2.116}$$

While the Newton–Raphson method can be sensitive to the starting value (initial conditions), in most cases the hover value $\lambda_h = \lambda_0$ works well, with only 3–4 iterations being required compared to up to 15 iterations using fixed-point iteration (see Questions 2.17 and 2.18). However, the computing costs with either method are essentially trivial.

Results for the inflow ratio λ/λ_h as computed using the iterative scheme are plotted in Fig. 2.19 for several different values of α and over a range of values of μ/λ_h typical of a helicopter. Note that the induced part of the total inflow decreases with increasing advance ratio and the total inflow becomes dominated by the $\mu \tan\alpha$ term at higher advance ratios.

2.4.4 *Validity of the Inflow Equation*

The inflow equation as given by Eq. 2.106 is widely employed for practical calculations involving rotors in climbs and descents in both axial flight and forward flight. However, a nonphysical solution will always be obtained if there is a descent (upward) component of

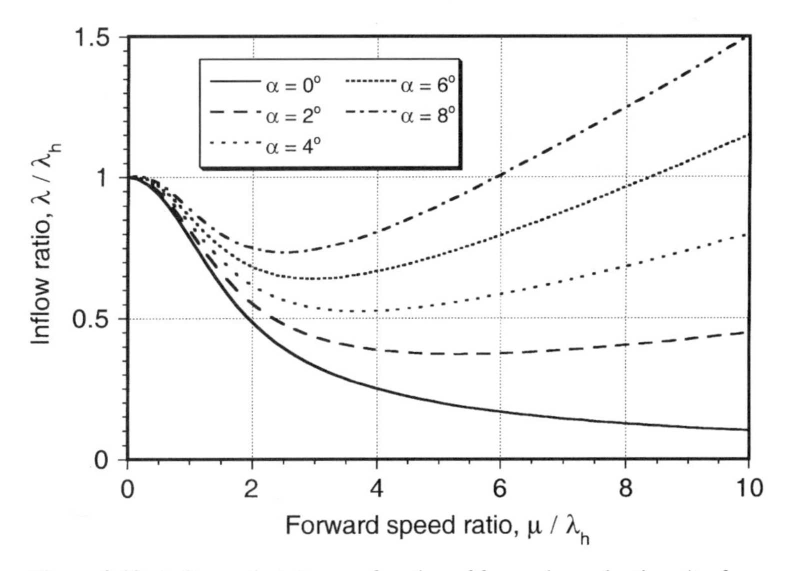

Figure 2.19 Inflow ratio λ/λ_h as a function of forward speed ratio μ/λ_h for several rotor angles of attack.

velocity normal to the rotor disk that is between 0 and $2v_i$ (i.e., if $-2v_i \leq V_\infty \sin\alpha \leq 0$). Under these conditions there can always be two possible directions for the flow and there can be no well-defined slipstream boundary as was assumed in the physical model. Therefore, the momentum theory cannot be applied under these conditions.

With the numerical solution several things may occur if these restrictions are ignored. First, the simple fixed-point iteration method may not converge. This will always be the case for axial flight ($\mu = 0$) when $-2 \leq V_c/v_i \leq 0$. Second, the fixed-point iteration method may converge, but to a nonphysical solution. In such cases, the results should not be used. The Newton–Raphson method will generally converge under all conditions but again the solutions will be nonphysical in the range $-2v_i \leq V_\infty \sin\alpha \leq 0$. The Newton–Raphson method is also sensitive to the initial conditions and may converge to different nonphysical solutions if different initial conditions are used. Generally it is assumed that $\lambda_0 = \sqrt{C_T/2}$, and this will work well when the rotor is in the normal working state. However, this initial condition will cause the method to fail for descents (i.e., $V_c < 0$). Here, convergence of the Newton–Raphson method to the proper physical result can generally be ensured if $\lambda_0 = |\lambda_c|$.

2.4.5 *Rotor Power in Forward Flight*

The rotor power in forward flight is given by the equation

$$P = T(V_\infty \sin\alpha + v_i) = TV_\infty \sin\alpha + Tv_i \tag{2.117}$$

assuming no viscous losses at this stage. The first term on the right-hand side of the above equation is the power required to propel the rotor forward and also to climb. The second term is the induced power. As for the axial flight case, one may reference the rotor power in forward flight to the hover result and so

$$\begin{aligned} \frac{P}{P_h} &= \frac{P}{Tv_h} = \frac{T(V_\infty \sin\alpha + v_i)}{Tv_h} \\ &= \frac{V_\infty \sin\alpha + v_i}{v_h} = \frac{\lambda}{\lambda_h}. \end{aligned} \tag{2.118}$$

Therefore, the form of the power curve simply mimics the inflow curves shown previously in Fig. 2.19. The results depend on the disk angle of attack, which must be tilted forward slightly for propulsion (see Fig. 2.17). Recall that

$$\lambda = \mu \tan\alpha + \frac{C_T}{2\sqrt{\mu^2 + \lambda^2}} = \mu \tan\alpha + \frac{\lambda_h^2}{\sqrt{\mu^2 + \lambda^2}}. \tag{2.119}$$

Therefore,

$$\frac{\lambda}{\lambda_h} = \frac{P}{P_h} = \frac{\mu}{\lambda_h}\tan\alpha + \frac{\lambda_h}{\sqrt{\mu^2 + \lambda^2}}. \tag{2.120}$$

The first term on the right-hand side of the above equation is the extra power to meet propulsion and climb requirements, while the second term is the induced power. As mentioned previously, the evaluation of the propulsive power requires a knowledge of the rotor disk angle of attack, which in turn requires a knowledge of the aircraft drag, D. Assuming straight-and-level flight, the disk angle of attack, α, can be calculated from a simple force equilibrium (see Fig. 2.17). For vertical equilibrium $T\cos\alpha = W$ and for horizontal equilibrium $T \sin\alpha = D\cos\alpha \approx D$. Therefore,

$$\tan\alpha = \frac{D}{W} = \frac{D}{L} \approx \frac{D}{T}, \tag{2.121}$$

which is expressed in terms of the helicopter lift-to-drag ratio. Therefore, the power equation in straight-and-level flight can be written as

$$\frac{P}{P_h} = \frac{\mu}{\lambda_h}\left(\frac{D}{T}\right) + \frac{\lambda_h}{\sqrt{\mu^2 + \lambda^2}}. \tag{2.122}$$

Since D will be proportional to the square of the forward flight velocity, the first term of the above equation is proportional to the cube of the flight speed. This component of the power required is called *parasite power*. The angle of attack of the disk relative to the oncoming flow will change in a climb or descent, thereby altering the power required. In this case, the power can be written as

$$\frac{P}{P_h} = \lambda_c \cos\alpha + \frac{\mu}{\lambda_h}\tan\alpha + \frac{\lambda_h}{\sqrt{\mu^2 + \lambda^2}}, \tag{2.123}$$

where λ_c is climb velocity ratio. In each case, the disk angle of attack can be solved for on the basis of a free-flight force equilibrium.

2.4.6 *Other Applications of the Momentum Theory*

The momentum theory analysis has found use in the analysis of other helicopter rotor designs, including coaxials, tandems, and ducted fans. The former are now discussed, with the ducted fan problem being discussed in relation to a fan-in-fin rotor or fenestron in Chapter 6 (Section 6.6.7).

The Coaxial Rotor

One advantage of the coaxial rotor design is that the net size of the rotor(s) is reduced (for a given aircraft gross weight) because each rotor provides vertical thrust. In addition, no tail rotor is required for antitorque purposes, so that all power can be devoted to useful vertical lift. However, the two rotors and their wakes interact with one another, producing a somewhat more complicated flow field than is found for a single rotor. Coleman (1993) gives a good summary of coaxial helicopter rotors and a comprehensive list of relevant citations on performance, wake characteristics, and methods of analysis.

Following Payne (1959), consider a simple momentum analysis of the hovering coaxial rotor problem. Assume that the rotor planes are sufficiently close together and that each rotor provides an equal fraction of the total system thrust, $2T$. The effective induced velocity of the rotor system will be

$$(v_i)_e = \sqrt{\frac{2T}{2\rho A}}. \tag{2.124}$$

Therefore, the induced power is

$$(P_i)_{tot} = 2T(v_i)_e = \frac{(2T)^{3/2}}{\sqrt{2\rho A}}. \tag{2.125}$$

However, if one treats each rotor separately then the induced power for either rotor will be $T v_i$ and for the two separate rotors

$$P_i = \frac{2T^{3/2}}{\sqrt{2\rho A}}. \tag{2.126}$$

If the interference-induced power factor κ_{int} is considered to be the ratio of Eqs. 2.125 and 2.126 then

$$\kappa_{\text{int}} = \frac{(P_i)_{\text{tot}}}{P_i} = \left(\frac{(2T)^{3/2}}{\sqrt{2\rho A}}\right)\left(\frac{2T^{3/2}}{\sqrt{2\rho A}}\right)^{-1} = \sqrt{2} \tag{2.127}$$

which is a 41% increase in induced power relative to the power required to operate the two rotors in complete isolation.

This simple momentum analysis of the problem has been shown to be overly pessimistic when compared with experimental measurements for closely spaced coaxial rotors – see Harrington (1951) and the review by Coleman (1993). The main reason for the overprediction of induced power is related to the actual (finite) spacing between the two rotors. Generally, on coaxial designs the rotors are spaced sufficiently far apart that the lower rotor operates in the fully developed slipstream of the upper rotor. This is justified from the flow visualization results of Taylor (1950), for example. Based on ideal flow considerations, this means that half of the area of the lower rotor operates in an effective climb velocity induced by the upper rotor. In this case, the net effect found from momentum theory is that $\kappa_{\text{int}} = 1.28$ compared to 1.41 when the rotors have no vertical separation (see Questions 2.19 and 2.20). This is closer to the values deduced from experiments for which $\kappa_{\text{int}} \approx 1.16$ [see, for example, Dingeldein (1954)] but still overpredicts the value. Figure 2.20 shows the power versus thrust relationship of single and coaxial rotors operating in hover; the measurements are taken from Dingeldein (1954). The power for the single two-bladed rotor was calculated

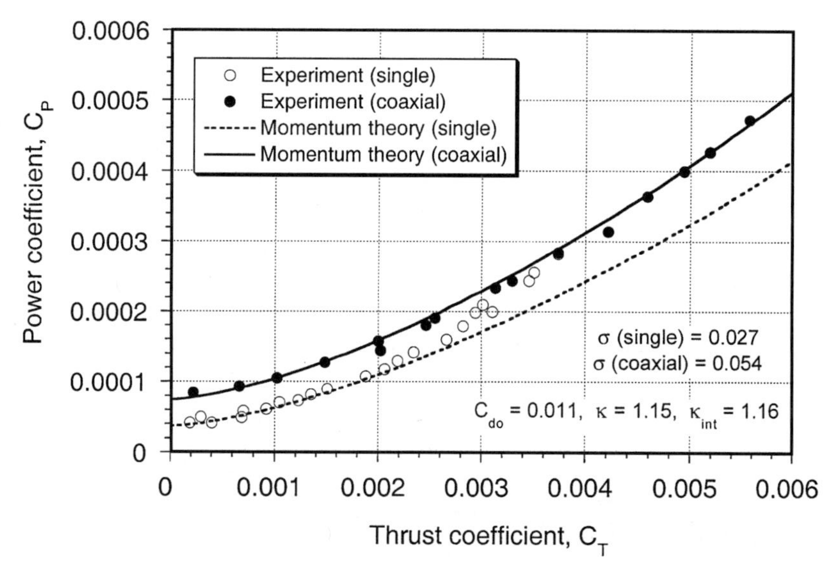

Figure 2.20 Hover performance predictions for a coaxial rotor compared to measurements. Data source: Dingeldein (1954).

from momentum theory using

$$P = \frac{\kappa T^{3/2}}{\sqrt{2\rho A}} + \rho A(\Omega R)^3 \left(\frac{\sigma C_{d_0}}{8} \right) \tag{2.128}$$

and for the coaxial using

$$P = \frac{\kappa_{\text{int}}\, \kappa (2T)^{3/2}}{\sqrt{2\rho A}} + \rho A(\Omega R)^3 \left(\frac{2\sigma C_{d_0}}{8} \right) \tag{2.129}$$

with $\kappa = 1.15$ and $\kappa_{\text{int}} = 1.16$. The agreement between momentum theory and the measurements is good and confirms that the coaxial rotor operates basically as two isolated rotors but with a mutually induced interference effect. See also Andrew (1981), Saito & Azuma (1981), and Zimmer (1985) for more details on coaxial rotor performance.

Tandem Rotors

The basic momentum analysis can also be extended to overlapping tandem rotors. Tandem rotor designs are sometimes used for heavy-lift helicopters because, like the coaxial design, all of the rotor power can be used to provide useful lift. However, like a coaxial design, the induced power of partly overlapping tandem rotors is found to be higher than that of the two isolated rotors. This is because one of the rotors must operate in the slipstream of the other rotor, resulting in a higher induced power for the same thrust. The tandem rotor problem is discussed extensively by Stepniewski (1955) and Stepniewski & Keys (1984). Other results for twin rotor performance are given by Sweet (1960) and Fail & Squire (1947).

The analysis of overlapping rotors from the perspective of momentum theory is normally based on the ideas of overlapping areas – see Payne (1959). Let $A_{\text{ov}} = mA$ be the overlap area according to the inset shown in Fig. 2.21. The rotors are assumed to have no vertical

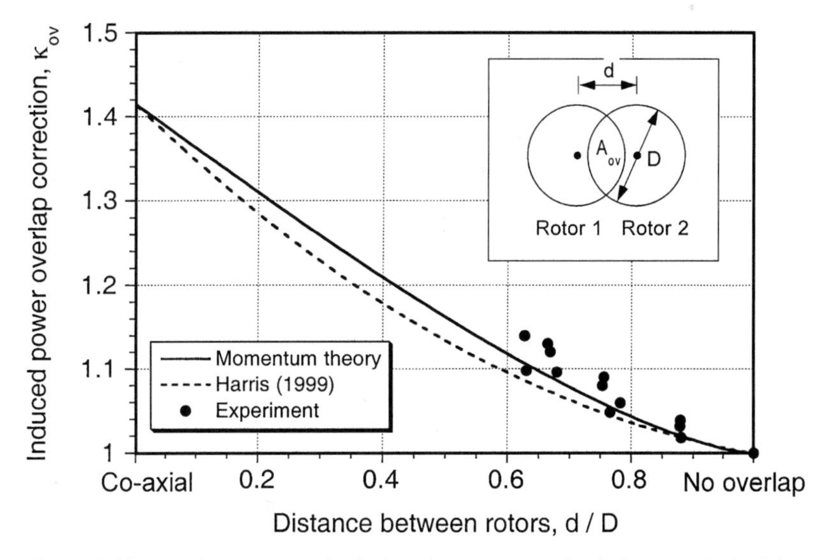

Figure 2.21 Tandem rotor overlap induced power correction in hover as derived from momentum theory and compared to measurements. Data source: Stepniewski & Keys (1984).

spacing. By means of the geometry of the problem it can be shown that

$$m = \frac{2}{\pi}\left[\theta - \frac{d}{D}\sin\theta\right], \quad \text{where } \theta = \cos^{-1}\left(\frac{d}{D}\right). \tag{2.130}$$

Let T_1 and T_2 be the thrusts on the two rotors, which may be unequal. Therefore, $m(T_1+T_2)$ is the thrust on the overlapped region. Based on uniform inflow assumptions then the induced power of the rotors consumed by each of the areas is

$$P_1 = \frac{(1-m)T_1^{3/2}}{\sqrt{2\rho A}}, \quad P_2 = \frac{(1-m)T_2^{3/2}}{\sqrt{2\rho A}}, \quad \text{and} \quad P_{ov} = \frac{m(T_1+T_2)^{3/2}}{\sqrt{2\rho A}},$$

where the total power is $(P_i)_{tot} = P_1 + P_2 + P_{ov}$. If the rotors are isolated, then $m = 0$ and the total system power will be

$$P_i = P_1 + P_2 = \frac{\left(T_1^{3/2} + T_2^{3/2}\right)}{\sqrt{2\rho A}}. \tag{2.131}$$

Proceeding as for the coaxial case, the induced power factor can be considered as the ratio

$$\frac{(P_i)_{tot}}{P_i} = \kappa_{ov} = \frac{(1-m)T_1^{3/2} + (1-m)T_2^{3/2} + m(T_1+T_2)^{3/2}}{T_1^{3/2} + T_2^{3/2}}. \tag{2.132}$$

Therefore, the total induced power required to hover for a tandem overlapping design can be expressed as

$$P_i = \kappa_{ov}\,\kappa T\sqrt{\frac{T}{4\rho A}}, \tag{2.133}$$

where A is the disk area of any one of the two rotors and T is the total system thrust as generated by both rotors. If it is assumed that each rotor carries an equal fraction of the total thrust ($T_1 = T_2$) then

$$\frac{(P_i)_{tot}}{P_i} = \kappa_{ov} = 1 + (\sqrt{2}-1)m = 1 + 0.4142\,m. \tag{2.134}$$

where m is given by Eq. 2.130. Harris (1999) suggests an approximation to κ_{ov} where

$$\kappa_{ov} \approx \left[\sqrt{2} - \frac{\sqrt{2}}{2}\left(\frac{d}{D}\right) + \left(1 - \frac{\sqrt{2}}{2}\right)\left(\frac{d}{D}\right)^2\right]. \tag{2.135}$$

The assumptions made in the derivation of the previous equations are not rigorous, as it can be shown that if each rotor carries a different fraction of the total thrust then κ_{ov} is only slightly higher than if an equal fraction is assumed (see Question 2.21). Note that in Eq. 2.134 as $m \rightarrow 1$, which is a coaxial, then $\kappa_{ov} \rightarrow \sqrt{2}$, as before. When the rotors are completely separated with no overlap ($m = 0$), then $\kappa_{ov} \rightarrow 1$ as required for isolated performance. The result of Eq. 2.134 is plotted in Fig. 2.21 in terms of the spacing ratio d/D, where d is the spacing between the two rotor axes and D is the rotor diameter. As d becomes larger and the rotors are more separated, the induced power overlap correction factor approaches unity.

The dependency of κ_{ov} on rotor overlap has been measured experimentally using small-scale rotor models. The amount of these data is relatively scarce compared to single-rotor data, but the available results have been collated by Stepniewski & Keys (1984) from several sources including Sweet (1960) and Boeing-Vertol experiments. Their results showing the relationship between κ_{ov} and the overlap d/D are reproduced in Fig. 2.21. The momentum theory result gives a good agreement with the measurements, although Harris's approximate result underpredicts κ_{ov}. For tandem rotor designs, such as the CH-46 and CH-47 models, d/D is approximately 0.65, giving κ_{ov} of about 1.13. Unfortunately, because the interference effects are related to the vertical spacing between the rotors as well as the degree of overlap, the results shown in Fig. 2.21 indicate some variance. Also, Dingeldein (1954) shows results that suggest κ_{ov} to be less than unity when the rotors are just separated such that $d/D > 1$; that is, a favorable interference effect exists. Apart from this anomaly, the correlation of Eq. 2.134 with the measured data is sufficiently good to enable the momentum theory to be used for at least preliminary design of tandem rotor performance.

In forward flight, both coaxials and tandem rotors systems appear to behave very much like two single rotors but with one of the rotors operating in the fully developed downwash of the other rotor – see Dingeldein (1954). Stepniewski & Keys (1984) discuss induced power interference effects and tandem rotor performance in forward flight. See also Chapter 5 for forward flight performance predictions of coaxial and tandem rotors.

2.5 Chapter Review

With certain assumptions and approximations, the application of the conservation laws of fluid mechanics in integral form has permitted an understanding of the factors that influence the basic performance of the helicopter rotor. The momentum theory has allowed a quantification of the thrust and power of a lifting rotor, and it has been shown how these quantities are related to the downwash (inflow) velocity through the rotor. The momentum method has permitted a preliminary evaluation of rotor performance in hover, climb, and descent. It has been shown that the disk loading is a key parameter governing rotor performance, and the need for a low disk loading is essential to give a helicopter good hovering efficiency. The ideas of power loading and figure of merit have been introduced as quantities that can help in designing the rotor as well as to compare the relative performance of two different rotor designs. To account for various nonideal effects that have their origin in the viscosity of the fluid, it has been shown how the basic momentum theory can be modified empirically to give a methodology that is in substantially better agreement with experimental measurements of rotor performance, while still retaining the simplicity of the overall method. The basic momentum theory has also been extended to forward flight, for which numerical solutions are generally required to solve for the inflow. These numerical techniques have been examined, along with the identification of limitations in their use. Finally, the ideas embodied in momentum theory have been extended to coaxial and partly overlapping rotors.

Despite the advantages of momentum theory in providing clarity of insight into the basic aspects of the rotor problem, it has many limitations. For example, it provides no information about the distribution of loads over the blade or as to how the rotor blades should be designed (planform, twist, thickness, airfoils, etc.) to produce a given performance. However, this is a limitation that can be overcome by more advanced methods based on a blade element analysis of the rotor blades, which is considered in the next chapter.

2.6 Questions

2.1. Show that for a hovering rotor with blades that operate at constant lift and drag coefficients, the thrust on the rotor is proportional to the square of the tip speed, and the power is proportional to the cube of tip speed.

2.2. The induced velocity, v_i, at the rotor plane in hover is known to be a function of the rotor tip speed, V, the rotor thrust, T, the rotor disk area, A, and air density, ρ. By means of dimensional analysis show that the functional form for v_i can be written as $\lambda_i = f(C_T)$.

2.3. Using Bernoulli's equation instead of the general energy equation, show that the induced velocity in the fully contracted wake of a rotor climbing with a vertical velocity is twice the induced velocity in the rotor plane.

2.4. The simple momentum theory assumes that the jump in pressure across the disk of a hovering rotor is the same everywhere. By considering an elemental annulus of the rotor disk, prove that this is consistent with a distribution of lift across the rotor disk that varies linearly from zero at the rotational axis of the rotor.

2.5. One difference between an actuator disk and a real rotor is that the aerodynamically active parts of real rotor blades do not extend all the way to the rotational axis – the so-called root cut-out. This is because of the need to attach the blades to the hub and also provide flap and lead/lag hinges, as well as a feathering (pitch) bearing. For a root cut-out of 25% of radius and a tip loss factor $B = 0.97$, estimate the additional effect of the root cut-out on the induced power factor of a hovering rotor.

2.6. For a hovering rotor, momentum theory gives a result for the ideal contraction ratio of the wake. In practice, experimental observations show that the wake contraction is not as much as the ideal value. Why? Show by means of simple momentum theory that a wake contraction that is less than the ideal will result in a higher induced power for the same total thrust. Find an expression for the induced power ratio (actual power relative to ideal at a constant thrust) in terms of a measured wake radial contraction ratio of 0.78.

2.7. Calculate and plot the values of the slipstream velocities in the fully contacted wake of a hovering rotor at sea-level as a function of disk loading, T/A. Assuming a figure of merit of 0.75, compute and plot the power loading versus disk loading.

2.8. A helicopter with a gross weight of 3,000 lb (1,363.6 kg), a main rotor radius of 13.2 ft (4.0 m), and a rotor tip speed of 680 ft/s (207.3 m/s) has 275 hp (205 kW) delivered to the main rotor shaft. For hovering conditions at sea-level, compute: (a) the rotor disk loading, (b) the ideal power loading, (c) the thrust and torque coefficients, (d) the figure of merit and actual power loading.

2.9. For the helicopter in the previous question, the tail rotor radius is 2.3 ft (0.701 m) and the tail rotor is located 15.3 ft (4.66 m) from the main rotor shaft. Calculate the thrust and power required by the tail rotor for hovering conditions at sea-level. Assume that the *FM* of the tail rotor is 0.70.

2.10. A helicopter is hovering in a steady cross wind at a gross weight of 3,000 lb (1,360.8 kg). This helicopter has 275 hp (205 kW) delivered to the main rotor shaft. The tail rotor radius is 2.3 ft (0.701 m) and has an induced power factor of 1.15. The tail rotor is located 15.3 ft (4.66 m) from the main rotor shaft. Determine the crosswind conditions (velocity and direction) in which the tail rotor effectiveness may be reduced or lost. If the center of gravity is assumed to lie on the rotor shaft axis, determine the feasible yawing angular velocity that the pilot can demand that may also result in a loss of tail rotor effectiveness.

2.11. An inventor claims to have built a "flying car" that can hover, where the lifting force is provided by two ducted fans. The car weighs 2,200 lb (1,000 kg) and has a 200 hp (149.14 kW) engine. The ducted fans are 7 ft (2.13 m) in diameter. Is hovering flight possible? [Hint: A ducted fan can be considered to have an effective disk area that is twice that of an unducted rotor.]

2.12. A preliminary design of a tandem rotor helicopter with a gross weight of 19,500 lb (8,845.2 kg) suggests a rotor diameter of 45 ft (13.72 m), a blade chord of 20 inches (0.508 m), three blades, and a rotor tip speed of 700 ft/s (213.43 m/s). Estimate the total shaft power required to hover if the induced power factor for the front rotor is 1.20 and that for the rear rotor is 1.15. The rotor airfoil to be used has a zero lift drag coefficient of 0.01. Estimate the installed power if transmission losses amount to 5% and the aircraft must demonstrate a vertical rate of climb of 1,000 ft/min (304.9 m/min) at sea-level.

2.13. A tilt-rotor has a gross weight (mass) of 25,500 kg. The rotor diameter is 12 m. On the basis of the simple momentum theory, estimate the power required for the aircraft to hover at sea-level on a standard day. Assume that the figure of merit of the rotors is 0.80 and transmission losses amount to 5%. If each of the two turbo-shaft engines delivers 4,500 kW, estimate the maximum vertical rate of climb at sea-level.

2.14. Given a helicopter of weight, $W = 6{,}000$ lb (2,727.3 kg), calculate the power required in hover and up to 600 ft/min axial rate-of-climb. The radius of the main rotor is 20 ft and the rotor has a figure of merit of 0.75. Assume sea-level conditions. Plot your results in the form of power required versus climb velocity. Discuss the factors that will determine the maximum vertical climb rate of a helicopter.

2.15. Find the effective drag coefficient, C_D, acting on a rotor that is in vertical autorotation at $V_c/v_h \approx -1.85$. Compare your result to the published drag coefficients of standard bluff-body shapes, and comment.

2.16. Using classical lifting-line theory, show that the lift on an elliptically loaded fixed-wing with a circular planform of radius R in a free-stream of velocity V_∞ is given by $L = 2\rho A v_i V_\infty$ where A is the wing area and v_i is the induced downwash over the wing.

2.17. Compare the results from the iterative solution of the general inflow equation to the exact analytical inflow equation for the case of axial flight (i.e., $\mu = 0$). Use both the fixed-point iterative method and the Newton–Raphson method. Determine the conditions (if any) under which these iterative methods fail.

2.18. Starting from the momentum theory result for the induced flow at the rotor disk in forward flight, program numerical solutions to this equation using both the fixed-point iterative method and the Newton–Raphson method. Plot some example graphs of inflow ratio versus advance ratio for a series of rotor angles of attack. Explore the effect of the initial guess for λ on the solution.

2.19. In a coaxial rotor design, the rotors are spaced sufficiently far apart such that the lower rotor operates in the fully developed slipstream of the upper rotor. Show by means of the momentum theory that the induced power factor resulting from interference is 1.28 compared to 1.41 when the rotors have no vertical separation. Assume that the thrusts of both rotors are equal.

2.20. In a coaxial rotor design, the rotors are spaced sufficiently far apart such that the lower rotor operates in the fully developed slipstream of the upper rotor. Show that by means of the momentum theory and on the basis of equal torque (power) that the induced power factor resulting from interference is 1.22 compared to 1.41 when the rotors have no vertical separation.

2.21. Using the momentum analysis for an overlapping rotor configuration with unequal rotor thrusts, find an expression for the overlap induced power factor κ_{ov} in terms of the overlap area $A_{ov} = mA$ where

$$m = \frac{2}{\pi}\left[\cos^{-1}\left(\frac{d}{D}\right) - \left(\frac{d}{D}\right)\sqrt{1-\left(\frac{d}{D}\right)^2}\right],$$

and where the rotors are assumed to have no vertical spacing. Assume that the total system thrust is expressed as $2T = \tau T_1 + (2-\tau)T_2$. Plot the results as a function of horizontal rotor separation distance d/D for several values of τ, and comment on the results.

Bibliography

Andrew, M. J. 1981. "Co-Axial Rotor Aerodynamics in Hover," *Vertica*, 5, pp. 163–172.

Azuma, A. and Obata, A. 1968. "Induced Flow Variation of the Helicopter Rotor Operating in the Vortex Ring State," *J. of Aircraft*, 5 (4), pp. 381–386.

Bagai, A. and Leishman, J. G. 1992. "A Study of Rotor Wake Developments and Wake/Body Interactions in Hover Using Wide-Field Shadowgraphy," *J. American Helicopter Soc.*, 37 (4), pp. 48–57.

Betz, A. 1920a. *Z. Flugtech. Motorluftschiffahrt*, 11, pp. 105.

Betz, A. 1920b. "Development of the Inflow Theory of the Propeller," NACA TN 24.

Betz, A. 1922. "The Theory of the Screw Propeller," NACA TN 83.

Brotherhood, P. 1949. "Flow Through a Helicopter Rotor in Vertical Descent," ARC R & M 2735.

Castles, W. and Gray, R. B. 1951. "Empirical Relation between Induced Velocity, Thrust and Rate of Descent of a Helicopter Rotor as Determined by Wind Tunnel Tests on Four Model Rotors, NACA TN 2474.

Coleman, C. P. 1993. "A Survey of Theoretical and Experimental Coaxial Rotor Aerodynamic Research," 19th European Rotorcraft Forum, Como, Italy.

de Bothezat, G. 1919. "The General Theory of Blade Screws," NACA Report 29.

de la Cierva, J. and Rose, D. 1931. *Wings to Tomorrow: The Story of the Autogiro*, Brewer, Warren & Putnam, New York.

Dingeldein, R. C. 1954. "Wind Tunnel Studies of the Performance of a Multirotor Configurations," NACA Technical Note 3236.

Drees, J. M. and Hendal, W. P. 1951. "Airflow Patterns in the Neighborhood of Helicopter Rotors," *Aircraft Engineering*, 23 (266), pp. 107–111.

Durand, W. F. (ed.) 1935. *Aerodynamic Theory*, Vol. IV, Divisions J-M, Springer-Verlag, Berlin.

Fail, R. and Squire, H. B. 1947. "24-ft Wind Tunnel Tests of Model Multi-Rotor Helicopters," RAE Report No. Aeronaut. 2207.

Froude, R. E. 1889. "On the Part Played in Propulsion by Differences of Fluid Pressure," *Trans. Inst. Naval Architects*, 30, p. 390.

Froude, W. 1878. "On the Elementary Relation Between Pitch, Slip and Propulsive Efficiency," *Trans. Inst. Naval Architects*, 19, pp. 47–57.

Gessow, A. 1948. "Flight Investigation of Effects of Rotor Blade Twist on Helicopter Performance in the High Speed and Vertical Autorotative Descent Conditions," NACA TN 1666.

Gessow, A. and Meyers, G. C. 1952. *Aerodynamics of the Helicopter*, Macmillan Co. (republished by Frederick Ungar Publishing, NY, 1967), pp. 73–75.

Gessow, A. 1954. "Review of Information on Induced Flow of a Lifting Rotor," NACA TN 3238.

Glauert, H. 1926a. "The Analysis of Experimental Results in the Windmill Brake and Vortex Ring States of an Airscrew," ARC R & M 1026.

Glauert, H. 1926b. "A General Theory of the Autogiro," ARC R & M 1111.

Glauert, H. 1928. "On the Horizontal Flight of a Helicopter," ARC R & M 1157.

Hafner, R. 1947. "Rotor Systems and Control Problems of the Helicopter," Anglo-American Aeronaut. Conf., Sept.

Harrington, R. D. 1951. "Full-Scale Tunnel Investigation of the Static Thrust Performance of a Coaxial Helicopter Rotor," NACA Technical Note 2318.

Harris, F. D. 1999. "Twin Rotor Hover Performance," *J. American Helicopter Soc.*, 44 (1), pp. 34–37.

Johnson, W. 1980. *Helicopter Theory*, Princeton University Press, Princeton, NJ, pp. 105–106.

Knight, M. and Hefner, R. A. 1937. "Static Thrust of the Lifting Airscrew," NACA TN 626.

Leishman, J. G., Baker, A., and Coyne, A. 1995. "Measurements of Rotor Tip Vortices Using Three-Component Laser Doppler Velocimetry," American Helicopter Soc. Aeromechanics Specialists Meeting, Fairfield, CT, Oct. 11–13.

Lock, C. N. H., Bateman, H., and Townend, H. C. H. 1926. "An Extension of the Vortex Theory of Airscrews with Applications to Airscrews of Small Pitch, Including Experimental Results," ARC R & M 1014.

Lock, C. N. H. 1928. "Photographs of Streamers Illustrating the Flow Around an Airscrew in the Vortex Ring State," ARC R & M 1167.

Lock, C. N. H. 1947. "Note on the Characteristic Curve for an Airscrew or Helicopter," ARC R & M 2673.

Nikolsky, A. A. and Seckel, E. 1949a. "An Analytical Study of the Steady Vertical Descent in Autorotation of Single-Rotor Helicopters," NACA TN 1906.

Nikolsky, A. A. and Seckel, E. 1949b. "An Analysis of the Transition of a Helicopter from Hovering to Steady Autorotative Vertical Descent," NACA TN 1907.

Payne, P. R. 1959. *Helicopter Dynamics and Aerodynamics*, The MacMillan Company, New York. pp. 90–98.

Rankine, W. J. M. 1865. "On the Mechanical Principles of the Action of Propellers," *Trans. Inst. Naval Architects*, 6, pp. 13–39.

Saito, S. and Azuma, S. 1981. "A Numerical Approach to Co-Axial Rotor Aerodynamics," 7th European Rotorcraft Forum, Garmisch-Partenkirchen, Germany, Sept. 8–11.

Sissingh, G. 1939. "Contribution to the Aerodynamics of Rotating-Wing Aircraft," NACA TM 921.

Stepniewski, W. Z. 1955. "A Simplified Approach to the Aerodynamic Rotor Interference of Tandem Helicopter," American Helicopter Soc. West Coast Region Meeting, Sept. 21–22.

Stepniewski, W. Z. and Keys, C. N. 1984. *Rotary-Wing Aerodynamics*, Dover, New York. Vol. II, Chapter 5.

Sweet, G. E. 1960. "Hovering Measurements for Twin-Rotor Configurations with and without Overlap," NASA Technical Note D-534.

Taylor, M. K. 1950. "A Balsa-Dust Technique for Air-Flow Visualization and Its Application to Flow Through Model Helicopter Rotors in Static Thrust," NACA Technical Note 2220.

Washizu, K., Azuma, A., Koo, J. and Oka, T. 1966a. "Experimental Study on the Unsteady Aerodynamics of a Tandem Rotor Operating in the Vortex Ring State," 22nd Annual Forum of the American Helicopter Soc., Washington, DC

Washizu, K. Azuma, A., Koo, J., and Oka, T. 1966b. "Experimental on a Model Helicopter Rotor Operating in the Vortex Ring State," *J. of Aircraft*, 3 (3), pp. 225–230.

Young, C. 1978. "A Note of the Velocity Induced by a Helicopter Rotor in the Vortex Ring State," RAE Technical Report 78125.

Zimmer, H. 1985. "The Aerodynamic Calculation of Counter Rotating Coaxial Rotors," 11th European Rotorcraft Forum, London, Sept. 10–13.

CHAPTER 3

Blade Element Analysis

Whatever progress the airplane might make, the helicopter will come to be taken up by advanced students of aeronautics.

Thomas Edison

3.1 Introduction

The blade element theory (BET) forms the basis of most modern analyses of rotor aerodynamics because it provides estimates of the radial and azimuthal distributions of blade aerodynamic loading. The BET theory assumes that each blade section acts as a quasi-two-dimensional airfoil to produce aerodynamic forces (and moments). Rotor performance can be obtained by integrating the sectional airloads at each blade element over the length of the blade and averaging the result over a rotor revolution. Therefore, unlike the simple momentum theory, the BET can be used as a basis to help design the rotor blades in terms of blade twist, planform distribution, and airfoil shape, to provide a given overall rotor performance.

The idea of the BET was first suggested by Drzwiecki (1892, 1909) for the analysis of airplane propellers.[1] In the elementary blade element theory, there is no mutual influence of adjacent blade elements sections; these sections are idealized as two-dimensional (2-D) airfoils. However, the effects of a nonuniform "induced inflow" across the blade (its source from the rotor wake) is accounted for through a modification to the angle of attack at each blade element. Unless we make some simple analytic assumption for the distribution of induced velocity over the disk, such as a uniform or linear distribution, the blade element calculation is quite a formidable undertaking because it must more precisely represent the highly nonuniform velocity field induced by the vortical wake trailed from each blade, as well as account for the influence of all the blades and possibly airframe components. However, if the induced velocity can be calculated, or even approximated, then the net thrust and power and other forces and moments acting on the rotor can be readily obtained.

In an extension to the basic approach, the BET and momentum theories were linked together by Reissner (1910, 1937, 1940), de Bothezat (1919), and Glauert (see Durand, 1935) to define the induced velocity or induced angle of attack distribution. A similar approach for hovering helicopter rotors was developed by Gustafson & Gessow (1946), Gessow (1948) and Gessow & Myers (1952). The BET was extended to explicitly include the influence of the vortical wake (and the rest of the rotor) through an induced angle of attack component as calculated by means of the Biot–Savart law. These basic ideas seem to have been first pursued by Joukowski – see Tokaty (1971), Glauert (1922), Biennen & von Kármán (1924), and Lock et al. (1925). Betz (1919, and appendix therein by Prandtl) and

[1] At the beginning of the twentieth century, there was considerable scientific debate between Stefan Drzwiecki and Louis Bréguet about the proper theoretical aerodynamic analysis of propellers and helicopter rotors – see Liberatore (1998).

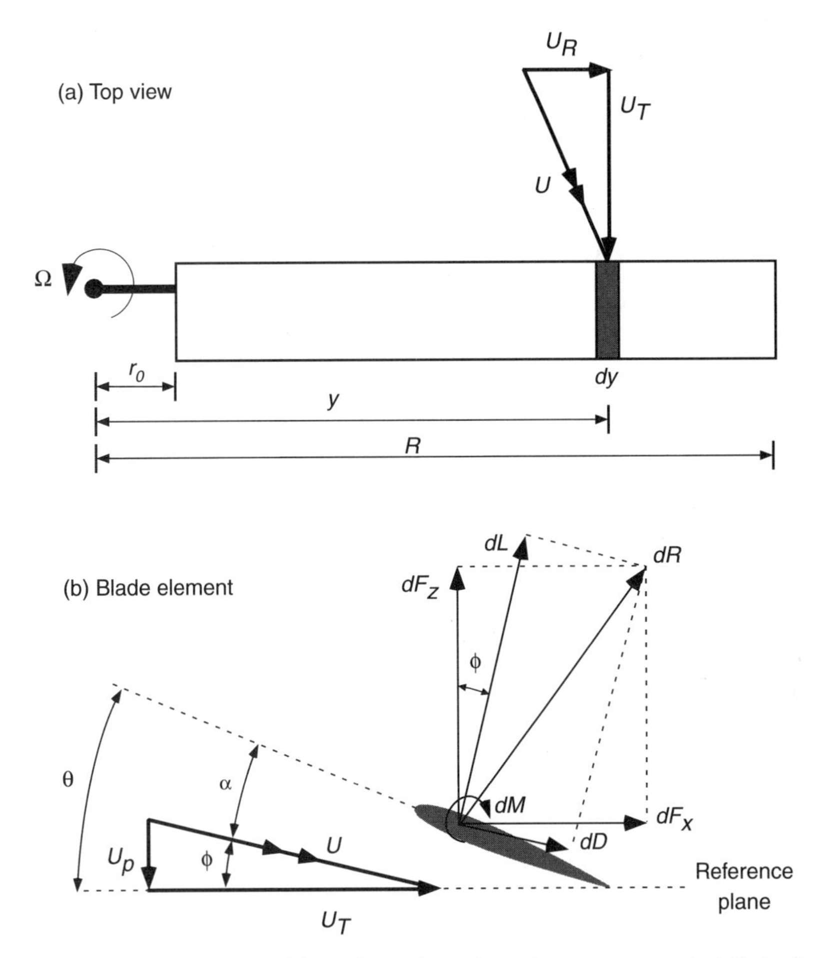

Figure 3.1 Incident velocities and aerodynamic environment at a typical blade element.

Goldstein (1929) developed a prescribed vortex wake theory for lifting propellers. This work was later extended by Theodorsen (1948). The early work with the technique as it applies to propellers is reviewed by Glauert – see Durand (1935). Knight & Hefner (1937) were one of the first to apply blade element and prescribed vortex wake principles to the calculation of the helicopter rotor problem. Coleman et al. (1945), Castles & De Leeuw (1954), and Castles & Durham (1965) later extended this work to helicopters operating in forward flight.

Figure 3.1 shows a sketch of the flow environment and aerodynamic forces at representative blade element on the rotor. The aerodynamic forces are assumed to arise solely from the velocity and angle of attack normal to the leading edge of the blade section. The effect of the radial component of velocity, U_R, on the lift is usually ignored in accordance with the independence principle; see Jones & Cohen (1957). However, the U_R component will affect the drag on the blade in forward flight and should be included in this case. The measured 2-D aerodynamic characteristics of the airfoil as a function of angle of attack can be assumed for the purposes of calculating the resultant lift and pitching moment on each blade element. Such results are available in the published literature for a large number of airfoils and over a wide range of operating conditions. The induced angle of attack, ϕ, arises because of the velocity induced by the rotor and its wake. Therefore, the induced velocity serves to modify the direction of the relative flow velocity vector and, therefore, alters the angle of attack at each blade element from its 2-D value. This induced velocity also inclines the local lift vectors, which by definition act perpendicular to the resultant velocity vector at the blade element and, therefore, provide a source of induced drag (drag due to lift) and the source of induced power required at the rotor shaft.

3.2 Blade Element Analysis in Hover and Axial Flight

The resultant local flow velocity at any blade element at a radial distance y from the rotational axis has an out-of-plane component $U_P = V_c + v_i$ normal to the rotor as a result of climb and induced inflow and an in-plane component $U_T = \Omega y$ parallel to the rotor because of blade rotation, relative to the disk plane. The resultant velocity at the blade element is, therefore,

$$U = \sqrt{U_T^2 + U_P^2}. \tag{3.1}$$

The relative inflow angle (or induced angle of attack) at the blade element will be

$$\phi = \tan^{-1}\left(\frac{U_P}{U_T}\right) \approx \frac{U_P}{U_T} \text{ for small angles.} \tag{3.2}$$

Thus, if the pitch angle at the blade element is θ, then the aerodynamic or effective angle of attack is

$$\alpha = \theta - \phi = \theta - \frac{U_P}{U_T}. \tag{3.3}$$

The resultant incremental lift dL and drag dD per unit span on this blade element are

$$dL = \frac{1}{2}\rho U^2 c C_l \, dy \quad \text{and} \quad dD = \frac{1}{2}\rho U^2 c C_d \, dy, \tag{3.4}$$

where C_l and C_d are the lift and drag coefficients, respectively. The lift dL and drag dD act perpendicular and parallel to the resultant flow velocity, respectively. Note that the quantity c is the *local* blade chord. Using Fig. 3.1 these forces can be resolved perpendicular and parallel to the rotor disk plane giving

$$dF_z = dL \cos\phi - dD \sin\phi \quad \text{and} \quad dF_x = dL \sin\phi + dD \cos\phi. \tag{3.5}$$

Therefore, the contributions to the thrust, torque, and power of the rotor are

$$dT = N_b dF_z, \quad dQ = N_b dF_x y, \quad \text{and} \quad dP = N_b dF_x \Omega y, \tag{3.6}$$

where N_b is the number of blades comprising the rotor. Note that in the hover or axial flight condition, the aerodynamic environment is (ideally) axisymmetric, and the airloads are independent of the blade azimuth angle. Substituting the results for dF_x and dF_z from Eq. 3.5 gives

$$dT = N_b(dL \cos\phi - dD \sin\phi), \tag{3.7}$$

$$dQ = N_b(dL \sin\phi + dD \cos\phi)y, \tag{3.8}$$

$$dP = N_b(dL \sin\phi + dD \cos\phi)\Omega y. \tag{3.9}$$

For helicopter rotors, the following simplifying assumptions can be made:

1. The out-of-plane velocity U_P is much smaller than the in-plane velocity U_T, so that $U = \sqrt{U_T^2 + U_P^2} \approx U_T$. This is a valid approximation except near the blade root, but the aerodynamic forces are small here anyway.
2. The induced angle ϕ is small, so that $\phi = U_P/U_T$. Also, $\sin\phi = \phi$ and $\cos\phi = 1$.
3. The drag is at least one order of magnitude less than the lift, so that the contribution $dD \sin\phi$ (or $dD\phi$) is negligible.

Applying these simplifications to the preceding equations results in

$$dT = N_b dL, \tag{3.10}$$

$$dQ = N_b(\phi dL + dD)y, \tag{3.11}$$

$$dP = N_b\Omega(\phi dL + dD)y. \tag{3.12}$$

Proceeding further, it is convenient to introduce nondimensional quantities by dividing lengths by R and velocities by ΩR. Hence, $r = y/R$, and $U/\Omega R = \Omega y/\Omega R = y/R = r$. Also, $dC_T = dT/\rho A(\Omega R)^2$, $dC_Q = dQ/\rho A(\Omega R)^2 R$, and $dC_P = dP/\rho A(\Omega R)^3$. The inflow ratio can be written as

$$\lambda = \frac{V_c + v_i}{\Omega R} = \frac{V_c + v_i}{\Omega y}\left(\frac{\Omega y}{\Omega R}\right) = \frac{U_P}{U_T}\left(\frac{y}{R}\right) = \phi r. \tag{3.13}$$

Therefore,

$$\begin{aligned} dC_T &= \frac{N_b dL}{\rho A(\Omega R)^2} = \frac{N_b\left(\frac{1}{2}\rho U_T^2 c C_l\, dy\right)}{\rho(\pi R^2)(\Omega R)^2} \\ &= \frac{1}{2}\left(\frac{N_b c}{\pi R}\right) C_l \left(\frac{y}{R}\right)^2 d\left(\frac{y}{R}\right) \\ &= \frac{1}{2}\left(\frac{N_b c}{\pi R}\right) C_l r^2\, dr. \end{aligned} \tag{3.14}$$

As described in Chapter 2, the grouping $N_b c/\pi R$ is known as the rotor solidity, σ. For a rectangular blade (c = constant), this is the ratio of the rotor blade area to the rotor disk area:

$$\sigma = \frac{\text{Blade area}}{\text{Disk area}} = \frac{A_b}{A} = \frac{N_b c R}{\pi R^2} = \frac{N_b c}{\pi R}.$$

Therefore, the rotor thrust increment is

$$dC_T = \frac{1}{2}\sigma C_l r^2\, dr. \tag{3.15}$$

This is one of the most fundamental equations for rotating-wing analysis by means of the BET. By a similar approach, it can be shown that the rotor power increment is

$$\begin{aligned} dC_P \equiv dC_Q &= \frac{dQ}{\rho A(\Omega R)^2 R} = \frac{N_b(\phi dL + dD)y}{\rho(\pi R^2)(\Omega R)^2 R} \\ &= \frac{1}{2}\left(\frac{N_b c}{\pi R}\right)(\phi C_l + C_d) r^3\, dr \\ &= \frac{1}{2}\sigma\, (\phi C_l + C_d) r^3\, dr. \end{aligned} \tag{3.16}$$

3.2.1 Integrated Rotor Thrust and Power

To find the total C_T and C_Q, the incremental thrust and power quantities derived above must be integrated along the blade from the root to the tip. For a rectangular blade, the thrust coefficient is

$$C_T = \frac{1}{2}\sigma \int_0^1 C_l r^2 \, dr, \tag{3.17}$$

where the limits of integration are $r = 0$ at the root to $r = 1$ at the tip. For the corresponding torque or power coefficient

$$C_Q \equiv C_P = \frac{1}{2}\sigma \int_0^1 (\phi C_l + C_d) r^3 \, dr = \frac{1}{2}\sigma \int_0^1 (\lambda C_l r^2 + C_d r^3) \, dr \tag{3.18}$$

using the general result that $\phi = \lambda/r$ from Eq. 3.13.

To evaluate C_T and C_Q it is necessary to predict the spanwise variation in the inflow, λ, as well as the sectional aerodynamic forces, C_l and C_d. If two-dimensional aerodynamics are assumed, then $C_l = C_l(\alpha, Re, M)$ and $C_d = C_d(\alpha, Re, M)$, where Re and M are the local Reynolds number and Mach number, respectively (see Chapter 7). Also, $\alpha = \alpha(V_c, \theta, v_i)$, and $v_i = v_i(r)$. Because these effects cannot, in general, be expressed as simple analytic results, it is necessary to numerically solve the integrals for C_T and C_Q. However, with certain assumptions and approximations, it is possible to find closed-form analytical solutions. These solutions are very useful because they serve to illustrate the fundamental form of the results in terms of the operational and geometric parameters of the rotor.

3.2.2 Thrust Approximations

Based on steady linearized aerodynamics, the local blade lift coefficient can be written as

$$C_l = C_{l_\alpha}(\alpha - \alpha_0) = C_{l_\alpha}(\theta - \alpha_0 - \phi) \tag{3.19}$$

where C_{l_α} is the two-dimensional lift-curve-slope of the airfoil section(s) comprising the rotor, and α_0 is the corresponding zero-lift angle. For an incompressible flow, C_{l_α} would have a value close to the thin-airfoil result of 2π per radian. Although C_{l_α} will take a different value at each blade station because it is a function of local incident Mach number and Reynolds number, an average value for the rotor can be assumed without serious loss of accuracy (i.e., C_{l_α} = constant). Also, unless otherwise stated, it will be assumed that symmetric airfoils are used throughout so that $\alpha_0 = 0$. Therefore, C_{l_α} can be taken outside of the integral sign giving

$$C_T = \frac{1}{2}\sigma \int_0^1 C_l r^2 \, dr = \frac{1}{2}\sigma C_{l_\alpha} \int_0^1 (\theta - \phi) r^2 \, dr, \tag{3.20}$$

and with $\phi = \lambda/r$, the thrust coefficient can be written as

$$C_T = \frac{1}{2}\sigma C_{l_\alpha} \int_0^1 (\theta r^2 - \lambda r) \, dr. \tag{3.21}$$

3.2.3 *Untwisted Blades, Uniform Inflow*

For a blade with zero twist, θ = constant = θ_0. Also, for uniform inflow velocity (as assumed in simple momentum theory), λ = constant. Therefore, in this case the thrust coefficient is given by

$$C_T = \frac{1}{2}\sigma C_{l_\alpha} \int_0^1 (\theta_0 r^2 - \lambda r)\, dr = \frac{1}{2}\sigma C_{l_\alpha} \left[\frac{\theta_0 r^3}{3} - \frac{\lambda r^2}{2} \right]_0^1$$

or

$$C_T = \frac{1}{2}\sigma C_{l_\alpha} \left[\frac{\theta_0}{3} - \frac{\lambda}{2} \right]. \tag{3.22}$$

To find the direct relationship between C_T and the blade pitch, one can use the relationship between C_T and λ as given by the simple momentum theory in Chapter 2 $\left(\text{i.e., } \lambda_i \equiv \lambda_h = \sqrt{C_T/2}\right)$. Therefore,

$$C_T = \frac{1}{2}\sigma C_{l_\alpha} \left[\frac{\theta_0}{3} - \frac{1}{2}\sqrt{\frac{C_T}{2}} \right]. \tag{3.23}$$

This equation must be solved iteratively. Alternatively, solving for the pitch angle, θ_0, in terms of thrust gives

$$\theta_0 = \frac{6C_T}{\sigma C_{l_\alpha}} + \frac{3}{2}\sqrt{\frac{C_T}{2}}. \tag{3.24}$$

The first term is the blade pitch required to produce thrust, and the second term is the additional pitch required to compensate for the inflow resulting from that thrust.

Figure 3.2 shows results for the thrust versus collective pitch for four rotors of different solidity. The experimental data are taken from the experiments of Knight & Hefner (1937), who tested subscale rotors with no blade twist. For the purposes of these calculations, C_{l_α} was

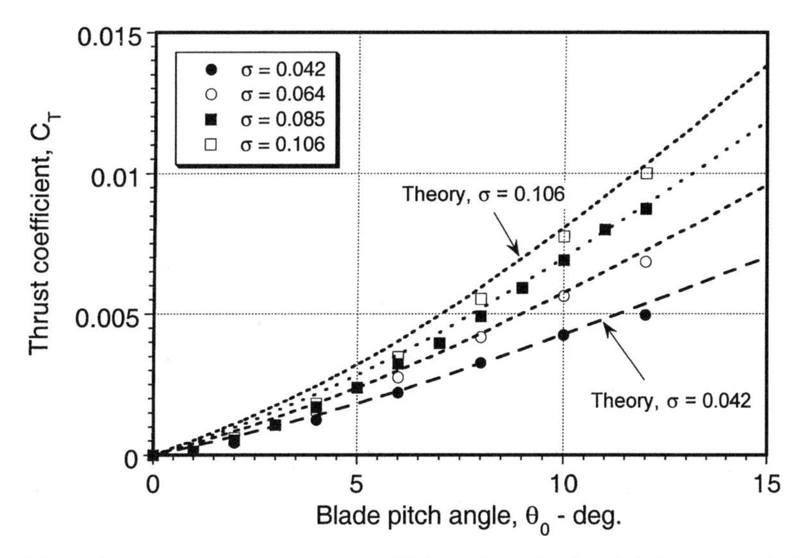

Figure 3.2 Variation in rotor thrust coefficient with collective pitch for rotors with different solidities. Data source: Knight & Hefner (1937).

set to a value of 5.73/radian, which represents a small reduction of the two-dimensional thin-airfoil result of 2π/radian because of finite airfoil thickness and Reynolds number effects. The agreement between Eq. 3.23 and the measurements is found to be good, although there is a slight overprediction of the thrust because the nonuniformity of the inflow and nonideal effects such as tip loss have not been accounted for.

3.2.4 *Linearly Twisted Blades, Uniform Inflow*

All helicopter rotor blades use some amount of twist, although in different forms and with different amounts. As will be shown, the use of blade twist provides the rotor with several important performance advantages. Many helicopter rotor blades are designed with a linear twist, so that θ takes the form $\theta(r) = \theta_0 + r\theta_{\text{tw}}$, where θ_{tw} is the blade twist rate in the appropriate angular units. Using this variation in $\theta(r)$, we get

$$C_T = \frac{1}{2}\sigma C_{l_\alpha} \int_0^1 [(\theta_0 + r\theta_{\text{tw}})r^2 - \lambda r]\, dr$$

$$= \frac{1}{2}\sigma C_{l_\alpha} \left[\frac{\theta_0 r^3}{3} + \frac{\theta_{\text{tw}} r^4}{4} - \frac{\lambda r^2}{2}\right]_0^1,$$

$$C_T = \frac{1}{2}\sigma C_{l_\alpha} \left[\frac{\theta_0}{3} + \frac{\theta_{\text{tw}}}{4} - \frac{\lambda}{2}\right]. \tag{3.25}$$

If the reference blade pitch angle (or collective pitch) is taken at 3/4-radius (known as $\theta_{0.75}$, or sometimes just as θ_{75}), then $\theta(r) = \theta_{75} + (r - 0.75)\theta_{\text{tw}}$ and

$$C_T = \frac{1}{2}\sigma C_{l_\alpha} \int_0^1 \{[(\theta_{75} + (r - 0.75)\theta_{\text{tw}}]r^2 - \lambda r\}\, dr$$

$$= \frac{1}{2}\sigma C_{l_\alpha} \int_0^1 (\theta_{75} r^2 + \theta_{\text{tw}} r^3 - 0.75\theta_{\text{tw}} r^2 - \lambda r)\, dr$$

$$= \frac{1}{2}\sigma C_{l_\alpha} \left[\frac{\theta_{75}}{3} + \frac{\theta_{\text{tw}}}{4} - \frac{\theta_{\text{tw}}}{4} - \frac{\lambda}{2}\right]$$

$$= \frac{1}{2}\sigma C_{l_\alpha} \left[\frac{\theta_{75}}{3} - \frac{\lambda}{2}\right]. \tag{3.26}$$

Comparing this latter equation with Eq. 3.22 shows an interesting result, namely that a blade with linear twist has the same thrust coefficient as one of constant pitch when θ is set to the pitch of the twisted blade defined at 3/4-radius [see also Gessow & Myers (1952)].

3.2.5 *Torque/Power Approximations*

According to the BET, the rotor incremental power coefficient can be written as

$$dC_P = dC_Q = \frac{\sigma}{2}(\phi C_l + C_d) r^3\, dr. \tag{3.27}$$

Using the result that $\lambda = \phi r$ and expanding gives

$$dC_P = \frac{\sigma}{2}\phi C_l r^3\, dr + \frac{\sigma}{2} C_d r^3\, dr$$

$$= \frac{\sigma}{2} C_l \lambda r^2\, dr + \frac{\sigma}{2} C_d r^3\, dr$$

$$= dC_{P_i} + dC_{P_0}, \tag{3.28}$$

where dC_{P_i} is the induced power and dC_{P_0} is the profile power. Recall that the incremental thrust coefficient can be written as $dC_T = \frac{1}{2}\sigma C_l r^2\,dr$ so that $dC_{P_i} = \lambda dC_T$. Therefore,

$$dC_P = \lambda dC_T + dC_{P_0} \tag{3.29}$$

and the total power coefficient is

$$C_P = \int_{r=0}^{r=1} \lambda dC_T + \int_0^1 \frac{1}{2}\sigma C_d r^3\,dr. \tag{3.30}$$

By assuming uniform inflow and $C_d = C_{d_0}$ = constant, then after integration we obtain

$$C_P = \lambda C_T + \frac{1}{8}\sigma C_{d_0}. \tag{3.31}$$

But in hover $\lambda = \sqrt{C_T/2}$, and so

$$C_P = \frac{C_T^{3/2}}{\sqrt{2}} + \frac{1}{8}\sigma C_{d_0}. \tag{3.32}$$

The first term in this equation will be recognized as the simple momentum theory result given previously in Chapter 2. The second term is the extra power predicted by the BET that is required to overcome profile drag of the rotor blades.

3.2.6 *Tip-Loss Factor*

The ideas of a so-called tip-loss factor have been introduced in Chapter 2 and can be used to account for the effects on the rotor thrust and induced power because of the locally high induced velocities produced at the blade tips by the trailed tip vortices. The basic BET permits a finite lift to be produced at the blade tip, which, of course, is physically unrealistic. The Prandtl tip-loss factor, B, is used to represent this loss of blade lift and can be considered as an "effective" blade radius, $R_e = BR$. Usually the value of B is found to take a value between 0.95 and 0.98 for most helicopter rotors.

When the tip loss is included in the calculation of rotor thrust using the BET, one approach is to consider the outer portion of the blade, $R - R_e$, to be incapable of carrying lift. This is the approach suggested by Gessow (1948) and Gessow & Myers (1952) and followed by Payne (1959) and Johnson (1980) and others. In this case the result for the lift is given by integrating the segment lift over the effective blade span as

$$C_T = \int_0^B \frac{\sigma}{2} C_l r^2\,dr = \int_0^B \frac{\sigma C_{l_\alpha}}{2}(\theta r^2 - \lambda r)\,dr. \tag{3.33}$$

For untwisted blades ($\theta = \theta_0$) and uniform inflow, this becomes

$$C_T = \frac{1}{2}\sigma C_{l_\alpha} B^2 \left[\frac{\theta_0 B}{3} - \frac{\lambda}{2}\right], \tag{3.34}$$

which can be compared to the result given previously in Eq. 3.22 with no tip losses (or Eq. 3.34 with $B = 1$). For a rotor with a twist distribution of the form $\theta = \theta_{\text{tip}}/r$ (as will be shown, this is known as ideal twist), then

$$\begin{aligned} C_T &= \frac{\sigma C_{l_\alpha}}{2} \int_0^B (\theta_{\text{tip}} - \lambda) r\,dr = \frac{\sigma C_{l_\alpha}}{2}\left[(\theta_{\text{tip}} - \lambda)\frac{r^2}{2}\right]_0^B \\ &= \frac{\sigma C_{l_\alpha}}{4} B^2 [\theta_{\text{tip}} - \lambda]. \end{aligned} \tag{3.35}$$

In either case, because B is between 0.95 and 0.98, one finds a 6 to 10% reduction in rotor thrust because of tip-loss effects for a given blade pitch setting.

Strictly speaking, the tip-loss equation deduced from the Prandtl theory as first discussed in Chapter 2 should be applied to the calculation of an increased inflow (for the same total thrust). Therefore, to assume that the outboard part of the blade, $R - R_e$, is ineffective in carrying lift is not the correct interpretation of Prandtl's theory. This fact has also been pointed out by Bramwell (1976). The correct interpretation is to consider that for the same thrust the induced inflow will be increased to a value

$$v_h = \sqrt{\frac{T}{2\rho A_e}} = \sqrt{\frac{T}{2\rho(AB^2)}} = \frac{1}{B}\sqrt{\frac{T}{2\rho A}}; \tag{3.36}$$

that is, v_h (or λ_h) is increased by a factor B^{-1} compared to the case with no assumed tip losses. For untwisted blades and uniform inflow with tip losses alone, the thrust becomes

$$C_T = \frac{1}{2}\sigma C_{l_\alpha}\left[\frac{\theta_0}{3} - \frac{\lambda}{2B}\right], \tag{3.37}$$

which can be compared to Eq. 3.34 with the alternative interpretation of tip loss. For ideal twist and uniform inflow, the thrust now becomes

$$C_T = \frac{\sigma C_{l_\alpha}}{4}\left[\theta_{\text{tip}} - \frac{\lambda}{B}\right] \tag{3.38}$$

compared to the result in Eq. 3.35, which will overpredict the effects of tip loss.

Because of tip-loss effects, a real rotor will always have a higher overall average induced velocity compared to that given by momentum theory and so the induced power will also be increased relative to the simple momentum result. Using the BET, the induced power can be written as

$$C_{P_i} = \int_{r=0}^{r=1} \lambda dC_T = \int_0^1 \frac{1}{2}\sigma\lambda C_l r^2\, dr, \tag{3.39}$$

where λ must be calculated for each element. For untwisted rectangular blades and uniform inflow then

$$C_P = \frac{1}{2}\sigma C_{l_\alpha}\frac{\lambda}{B}\left[\frac{\theta_0}{3} - \frac{\lambda}{2B}\right] + \frac{1}{8}\sigma C_{d_0}. \tag{3.40}$$

Assuming ideal twist and uniform inflow with tip losses only, one can integrate Eq. 3.39 to get

$$C_P = \frac{\sigma C_{l_\alpha}}{4}\left[\frac{\lambda}{B}\left(\theta_{\text{tip}} - \frac{\lambda}{B}\right)\right] + \frac{1}{8}\sigma C_{d_0}. \tag{3.41}$$

Figure 3.3 shows the variation in power required versus collective pitch for four rotors with increasing solidity. Again, the data are taken from Knight & Hefner (1937). The calculations have used a tip-loss factor, B, of 0.97 and $C_{d_0} = 0.011$. The result in Eq. 3.40 with the assumption of uniform inflow correlates well with the measured data but shows a slight underprediction at the higher collectives. For the most part, this is because of the higher drag coefficients associated with boundary layer thickening and the onset of blade stall, the effects of which tend to increase the profile power above the value assumed with a constant profile drag coefficient.

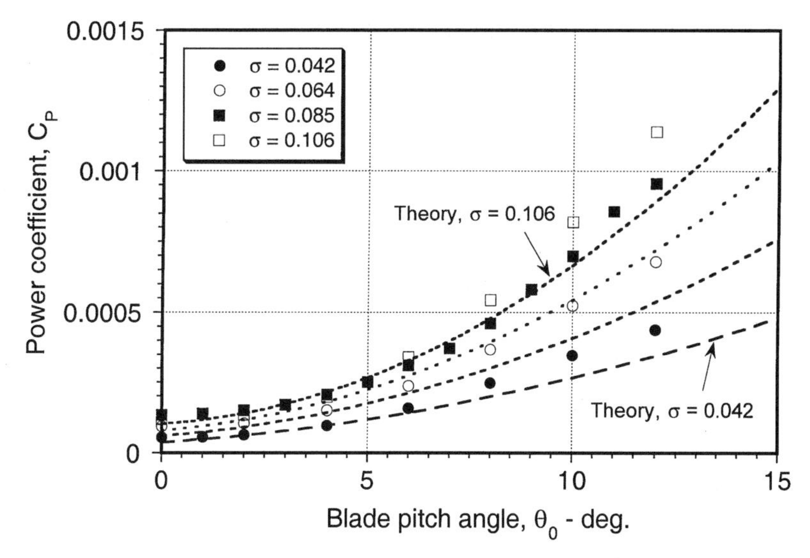

Figure 3.3 Variation in rotor power coefficient with collective pitch for rotors with different solidities. Data source: Knight & Hefner (1937).

3.3 Blade Element Momentum Theory (BEMT)

The blade element momentum theory (BEMT) for hovering rotors is a hybrid method that was first proposed for helicopter use by Gustafson & Gessow (1946) and Gessow (1948) and combines the basic principles from both the blade element and momentum approaches. With certain assumptions, the BEMT allows the inflow distribution along the blade to be estimated.

Consider first the application of the conservation laws to an annulus of the rotor disk, as shown in Fig. 3.4. This annulus is at a distance y from the rotational axis and has a width dy. The area of this annulus is, therefore, $dA = 2\pi y\ dy$. The incremental thrust,

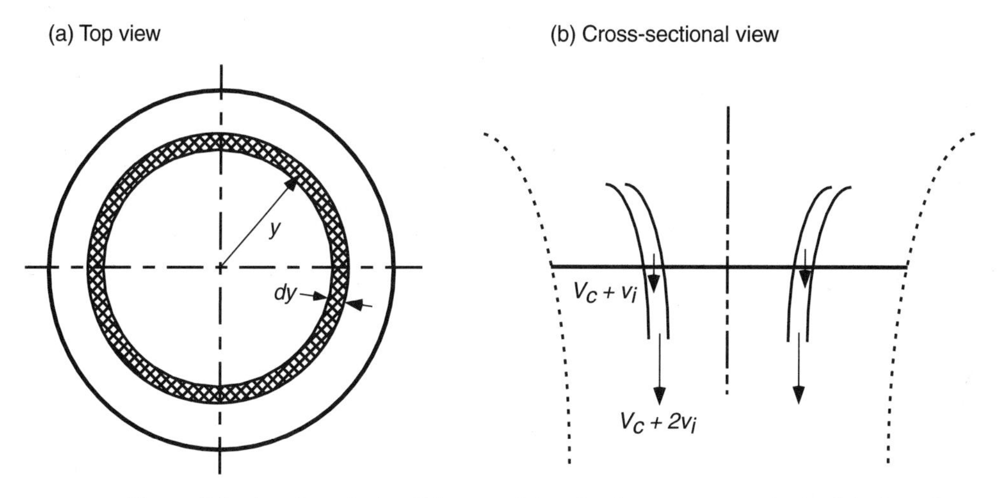

Figure 3.4 Annulus of rotor disk as used for a local momentum analysis of the hovering rotor. (a) Plan view. (b) Side view.

dT, on this annulus may be calculated on the basis of simple momentum theory and with the assumption that successive rotor annuli have no mutual effects on each other. As might be expected, this approach has good validity except near the blade tips. The removal of this two-dimensional restriction requires a considerably more advanced treatment of the problem using vortex wake theory. However, a good approximation to the tip-loss effect on the inflow distribution can be made using Prandtl's "circulation-loss" function, which will be discussed in Section 3.3.7.

On the basis of simple one-dimensional momentum theory developed in Chapter 2, one may compute the incremental thrust on the rotor annulus as the product of the mass flow rate through the annulus and twice the induced velocity at that section. In this case the mass flow rate over the annulus of the disk is

$$d\dot{m} = \rho dA(V_c + v_i) = 2\pi\rho(V_c + v_i)y\,dy, \tag{3.42}$$

so that the incremental thrust on the annulus is

$$dT = 2\rho\,(V_c + v_i)\,v_i dA = 4\pi\rho\,(V_c + v_i)\,v_i y\,dy. \tag{3.43}$$

This is also known as the Froude–Finsterwalder equation.[2] It is more convenient to work in nondimensional quantities so that

$$dC_T = \frac{dT}{\rho(\pi R^2)(\Omega R)^2} = \frac{2\rho(V_c + v_i)v_i dA}{\rho\pi R^2(\Omega R)^2}$$

$$= \frac{2\rho(V_c + v_i)v_i(2\pi y\,dy)}{\rho\pi R^2(\Omega R)^2} = 4\frac{(V_c + v_i)}{\Omega R}\left(\frac{v_i}{\Omega R}\right)\left(\frac{y}{R}\right)d\left(\frac{y}{R}\right)$$

or simply

$$dC_T = 4\lambda\lambda_i r\,dr. \tag{3.44}$$

Therefore, the incremental thrust coefficient on the annulus can be written as

$$dC_T = 4\lambda\lambda_i r\,dr = 4\lambda\,(\lambda - \lambda_c)r\,dr \tag{3.45}$$

because $\lambda_i = \lambda - \lambda_c$. The induced power consumed by the annulus is

$$dC_{P_i} = \lambda dC_T = 4\lambda^2\lambda_i r\,dr = 4\lambda^2\,(\lambda - \lambda_c)r\,dr. \tag{3.46}$$

Consider first the hovering state where $\lambda_c = 0$. The incremental thrust and power of the annulus are given by

$$dC_T = 4\lambda^2 r\,dr \quad \text{and} \quad dC_{P_i} = 4\lambda^3 r\,dr.$$

The total thrust coefficient of the rotor is

$$C_T = \int_{r=0}^{r=1} dC_T = 4\int_0^1 \lambda^2 r\,dr \tag{3.47}$$

and the corresponding induced power coefficient is

$$C_{P_i} = \int_{r=0}^{r=1} \lambda dC_T = 4\int_0^1 \lambda^3 r\,dr. \tag{3.48}$$

[2] Tokaty (1971) gives a good historical overview of the origins of the blade element and blade element momentum theories.

These results are valid for any radial form of induced velocity distribution across the disk. However, assume for illustrative purposes that the inflow can be expressed in the simple form

$$\lambda(r) = \lambda_{\text{tip}} r^n \quad \text{for } n \geq 0. \tag{3.49}$$

Then using Eq. 3.47 leads to

$$C_T = 4\int_0^1 \lambda^2 r\, dr = 4\lambda_{\text{tip}}^2 \int_0^1 r^{2n+1}\, dr = \frac{4\lambda_{\text{tip}}^2}{2n+2}. \tag{3.50}$$

Solving for λ_{tip} gives

$$\lambda_{\text{tip}} = \sqrt{n+1}\sqrt{\frac{C_T}{2}}. \tag{3.51}$$

The induced power coefficient is

$$C_{P_i} = 4\int_0^1 \lambda^3 r\, dr = 4\lambda_{\text{tip}}^3 \int_0^1 r^{3n+1}\, dr = \frac{4\lambda_{\text{tip}}^3}{3n+2}, \tag{3.52}$$

and substituting Eq. 3.51 gives

$$C_{P_i} = \frac{2(n+1)^{3/2} C_T^{3/2}}{(3n+2)\sqrt{2}}. \tag{3.53}$$

It has been shown previously that the induced power can be written as

$$C_{P_i} = \frac{\kappa C_T^{3/2}}{\sqrt{2}}, \tag{3.54}$$

and so in this case the induced power factor, κ, is

$$\kappa = \frac{2(n+1)^{3/2}}{(3n+2)}. \tag{3.55}$$

Note that for $n = 0$ (uniform inflow) then $\kappa = 1$ (ideal case). For $n > 0$, then $\kappa > 1$, and as the value of n increases κ increases as the inflow becomes more heavily biased toward the tips.

Although it is clear from the foregoing that uniform inflow will give the lowest induced power, the question is now how the blade properties (pitch angle and chord) should be adjusted to achieve this objective. One solution can be obtained using a hybrid blade element momentum approach to solve directly for the inflow distribution. From the BET it has been shown previously that the incremental thrust produced on an annulus of the disk is

$$dC_T = \frac{1}{2}\sigma C_l r^2\, dr = \frac{\sigma C_{l_\alpha}}{2}(\theta r^2 - \lambda r)\, dr. \tag{3.56}$$

Equating the incremental thrust coefficients from the momentum and blade element theories (Eqs. 3.45 and 3.56) one finds that

$$\frac{\sigma C_{l_\alpha}}{2}(\theta r^2 - \lambda r) = 4\lambda(\lambda - \lambda_c) r, \tag{3.57}$$

which gives

$$\frac{\sigma C_{l_\alpha}}{8}\theta r - \frac{\sigma C_{l_\alpha}}{8}\lambda = \lambda^2 - \lambda_c \lambda \tag{3.58}$$

or

$$\lambda^2 + \left(\frac{\sigma C_{l_\alpha}}{8} - \lambda_c \right) \lambda - \frac{\sigma C_{l_\alpha}}{8} \theta r = 0. \tag{3.59}$$

This quadratic equation in λ has the solution

$$\lambda(r, \lambda_c) = \sqrt{\left(\frac{\sigma C_{l_\alpha}}{16} - \frac{\lambda_c}{2} \right)^2 + \frac{\sigma C_{l_\alpha}}{8} \theta r} - \left(\frac{\sigma C_{l_\alpha}}{16} - \frac{\lambda_c}{2} \right). \tag{3.60}$$

Consider hovering conditions where $\lambda_c = 0$, then Eq. 3.60 simplifies to

$$\lambda(r) = \frac{\sigma C_{l_\alpha}}{16} \left[\sqrt{1 + \frac{32}{\sigma C_{l_\alpha}} \theta r} - 1 \right]. \tag{3.61}$$

Equation 3.61 allows us to solve for the inflow as a function of radius for any given blade pitch, blade twist distribution, planform (chord distribution), and airfoil section (through the effect of lift-curve-slope and zero-lift angle). When the inflow is obtained, the rotor thrust and induced power may then be found by integration across the rotor disk using Eqs. 3.47 and 3.48.

For an untwisted blade of constant chord and uniform airfoil section, the distribution of inflow as predicted by Eq. 3.61 is shown in Fig. 3.5 and is compared with the uniform inflow at the same thrust coefficient. It is clear that for the untwisted blade the distribution of inflow is concentrated toward its tips. Clearly, the combination of blade geometric parameters for the results shown in Fig. 3.5 is nonideal because as shown by Eq. 3.55 the induced power will be higher than the minimum possible with uniform inflow. The question is now how the blade geometric properties can be adjusted to give a more uniform inflow and, therefore, to minimize the induced power.

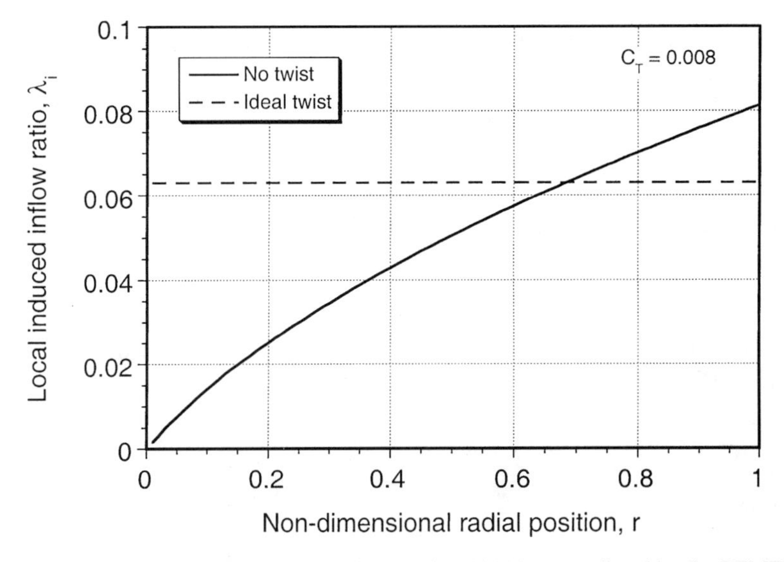

Figure 3.5 Distribution of inflow for untwisted blade as predicted by the BEMT.

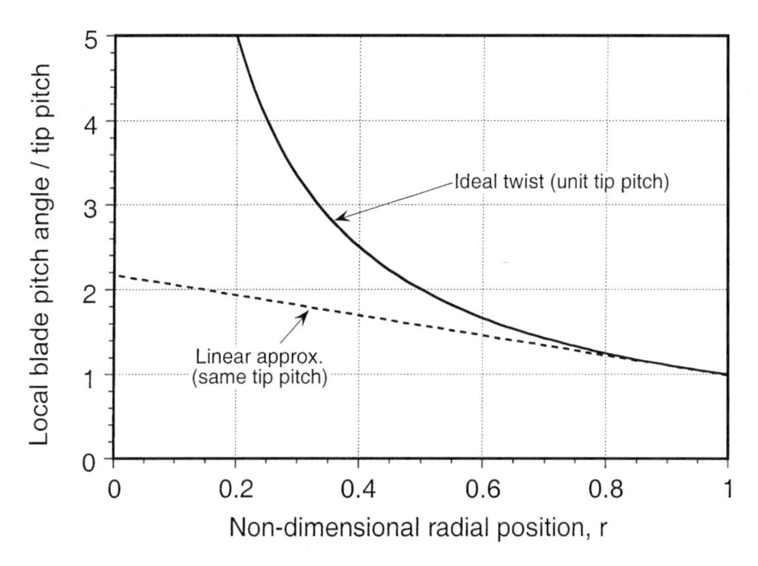

Figure 3.6 Radial distribution of blade twist in ideal case.

3.3.1 *Ideal Twist*

Gessow (1948) showed that if $\theta r = \text{constant} = \theta_{tip}$ there is a special solution to Eq. 3.61 that gives uniform inflow, that is,

$$\theta(r) = \frac{\theta_{tip}}{r}. \tag{3.62}$$

This twist distribution is called *ideal twist* and is shown in Fig. 3.6. This result is of considerable interest for blade design purposes because the uniform inflow case must always correspond to the minimum induced power for the rotor when operating in hover or in axial climb. Unfortunately, the hyperbolic form of pitch angle or twist distribution given by Eq. 3.62 is physically unrealizable as $r \rightarrow 0$. However, because of the root cutout, the blade pitch variation here does not matter anyway. A linear twist distribution is reasonably close to the ideal case over the outer part of the blade.

A careful examination of airplane propellers will show that they are highly twisted and, in fact, closely conform to the ideal twist given by Eq. 3.62. For helicopter rotors, such high amounts of twist over the blade span is nonoptimum for the entire flight envelope. This is because the rotor must also operate in forward flight, and high values of blade twist can lead to less efficient lift and propulsion generation from the advancing side of the disk at high advance ratios. Nevertheless, because helicopters spend a substantial part of their operating time in hover and low speed forward flight, there are considerable performance benefits to be realized by incorporating some blade twist.

With ideal twist the performance of the rotor can now be recalculated. If $\theta r = \text{constant} = \theta_{tip}$, then

$$C_T = \frac{\sigma C_{l_\alpha}}{2} \int_0^1 (\theta_{tip} - \lambda) r \, dr = \frac{\sigma C_{l_\alpha}}{2} \left(\frac{\theta_{tip}}{2} - \frac{\lambda}{2} \right). \tag{3.63}$$

Using Eq. 3.13 gives $\lambda = r\phi = \phi_{tip} = \text{constant}$, and so the preceding equation can also be written as

$$C_T = \frac{\sigma C_{l_\alpha}}{4} (\theta_{tip} - \phi_{tip}) = \frac{\sigma C_{l_\alpha}}{4} \alpha_{tip}. \tag{3.64}$$

Using Eqs. 3.61 and 3.62 gives

$$\lambda(r) = \frac{\sigma C_{l_\alpha}}{16}\left[\sqrt{1 + \frac{32\theta_{\text{tip}}}{\sigma C_{l_\alpha}}} - 1\right] = \text{constant} = \sqrt{\frac{C_T}{2}}. \tag{3.65}$$

Then a solution for the blade pitch angle can be found from

$$\theta_{\text{tip}} = \frac{4C_T}{\sigma C_{l_\alpha}} + \sqrt{\frac{C_T}{2}} \tag{3.66}$$

or

$$\theta_{\text{tip}} = \frac{4C_T}{\sigma C_{l_\alpha}} + \lambda. \tag{3.67}$$

The first term in the previous equation can be thought of as a mean blade pitch term, and the second term is an additional blade pitch angle required because of the induced inflow produced by the generation of thrust.

One can also prove that the ideal twist distribution gives a special result, namely a linear (triangular) lift distribution over the blade, which is consistent with uniform disk loading. Recall that the incremental thrust distribution is given by

$$dC_T = \frac{\sigma C_{l_\alpha}}{2}(\theta_{\text{tip}} - \lambda) r \, dr. \tag{3.68}$$

Because λ is a constant, the thrust distribution on the blade varies in proportion to r (i.e., it is a linear distribution). Another way of writing the preceding equation is just

$$dC_T = \frac{\sigma C_{l_\alpha}}{2}\alpha_{\text{tip}} r \, dr = \frac{\sigma}{2} C_{l_{\text{tip}}} r \, dr. \tag{3.69}$$

Therefore, the total thrust on the rotor with ideal blade twist is

$$C_T = \frac{\sigma}{4} C_{l_{\text{tip}}} = \frac{\sigma}{4} C_{l_\alpha} \alpha_{\text{tip}}. \tag{3.70}$$

3.3.2 *BEMT – A Numerical Approach*

In most applications, the equations of the BEMT are solved numerically. In this approach, the blade must be discretized into a series of small elements of span Δr. The inflow is now obtained using the discretized equation

$$\lambda(r_n) = \frac{\sigma C_{l_\alpha}}{16}\left[\sqrt{1 + \frac{32}{\sigma C_{l_\alpha}}\theta(r_n) r_n} - 1\right], \tag{3.71}$$

where $n = 1, N$ is the element location and r_n and $\theta(r_n)$ are the radius and pitch angle at the midspan of each of the N elements, respectively.[3] When the inflow is determined using Eq. 3.71, the incremental thrust at each segment can be found using

$$\Delta C_{T_n} = \frac{\sigma C_{l_\alpha}}{2}\left(\theta(r_n)\, r_n^2 - \lambda(r_n)\, r_n\right) \Delta r. \tag{3.72}$$

[3] Typically, a minimum of 20 elements must be used to ensure an adequate resolution of the inflow and spanwise loading.

The total thrust is then obtained by numerically integrating over the blade. The simplest approach is to use the rectangle rule, which is equivalent to considering the inflow and thrust to be constant over each segment. In this case

$$C_T = \sum_{n=1}^{N} \Delta C_{T_n}, \tag{3.73}$$

which will be adequate for most purposes. The torque (power) can be obtained by a similar process (see Questions).

It is normally desirable to compare results for different rotors at the same value of C_T. Therefore, for rotors with different twist or planforms, the blade (collective) pitch, θ_0, required to obtain the required value of thrust coefficient, say $C_{T_{\text{req}}}$, must be obtained interatively starting from an assumed value for θ_0. For this, one can use as a basis the simple relationships found previously between the blade pitch and C_T. For example, for a blade with linear twist and with the assumption of uniform inflow, it was shown that

$$C_T = \frac{1}{2}\sigma C_{l_\alpha}\left[\frac{\theta_0}{3} + \frac{\theta_{\text{tw}}}{4} - \frac{1}{2}\sqrt{\frac{C_T}{2}}\right], \tag{3.74}$$

or rearranging and solving for θ_0 gives

$$\theta_0 = \frac{6C_T}{\sigma C_{l_\alpha}} - \frac{3}{4}\theta_{\text{tw}} + \frac{3}{2}\sqrt{\frac{C_T}{2}}. \tag{3.75}$$

Therefore, one simple iterative scheme of correcting the collective pitch from iteration j to $j+1$ is

$$\theta_0^{(j+1)} = \theta_0^{(j)} + \left[\frac{6\left(C_{T_{\text{req}}} - C_T^{(j)}\right)}{\sigma C_{l_\alpha}} + \frac{3\sqrt{2}}{4}\left(\sqrt{C_{T_{\text{req}}}} - \sqrt{C_T^{(j)}}\right)\right] \tag{3.76}$$

starting from the initial value

$$\theta_0^{(0)} = \frac{6C_{T_{\text{req}}}}{\sigma C_{l_\alpha}} - \frac{3}{4}\theta_{\text{tw}} + \frac{3\sqrt{2}}{4}\sqrt{C_{T_{\text{req}}}}. \tag{3.77}$$

Using this approach, the collective pitch is found to converge rapidly to provide the required value of C_T for any arbitrary rotor geometry, typically within two to four iterations.

3.3.3 *Distributions of Inflow and Airloads*

Using the numerical implementation of the BEMT, the distribution of inflow and lift on a blade with any distribution of twist (and planform) can be readily determined. Representative results for λ and dC_T/dr are shown in Figs. 3.7(a) and (b), respectively, for a rectangular blade with different linear twist rates and at a constant value of thrust. For reference purposes, the distribution corresponding to ideal twist is also shown, which gives a uniform inflow and a linear variation of lift from the root to the tip. Notice that the lift on the untwisted blade is parabolic. With moderate values of twist, the lift distribution becomes more linear, with an off-loading at the tip and a greater loading being produced inboard, and more closely resembles that obtained in the ideal twist case. For very high values of blade twist, the tip is off-loaded to the point that the inflow is more nonuniform inflow again, and the rotor thereby becomes less aerodynamically efficient.

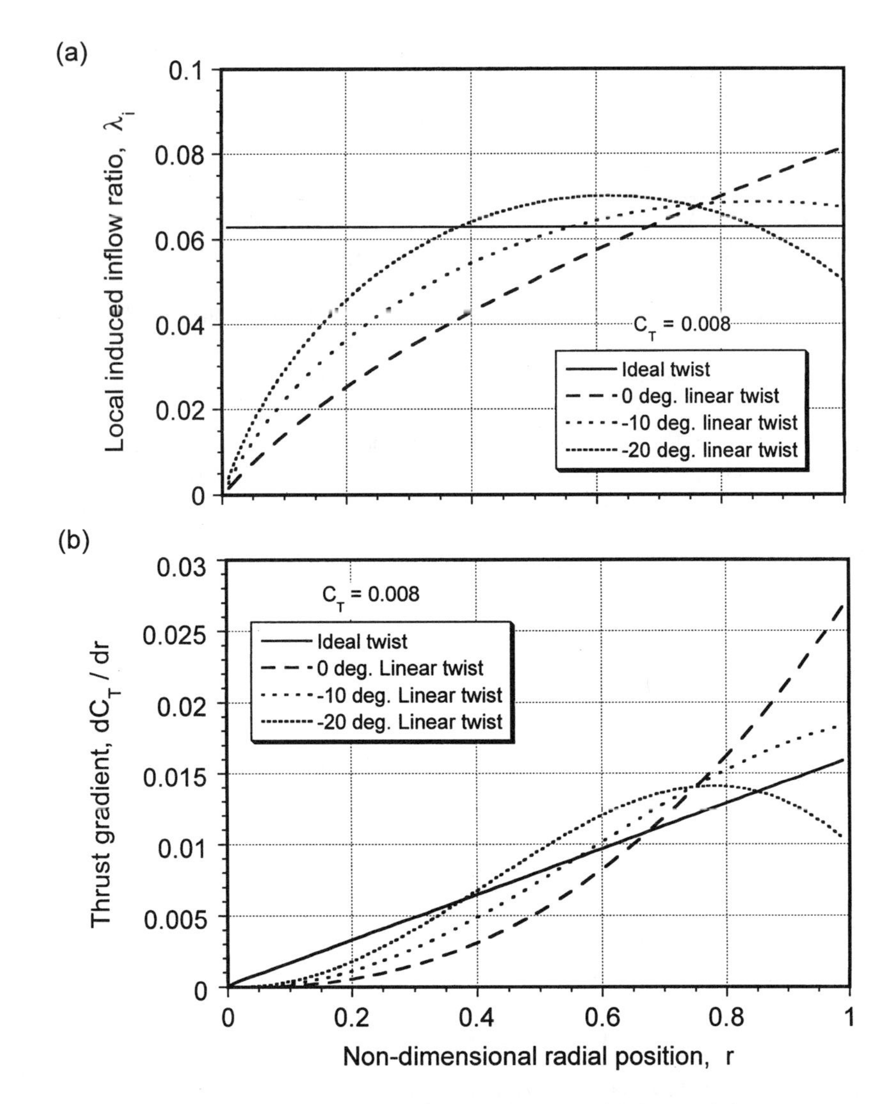

Figure 3.7 BEMT predictions of spanwise distributions of inflow and thrust on a rotor blade for different linear twist rates. Results are compared at a constant total thrust. (a) Inflow ratio. (b) Thrust per unit span.

The corresponding local lift coefficient distribution over the blade is found using

$$C_l(r_n) = C_{l_\alpha}\left(\theta(r_n) - \frac{\lambda(r_n)}{r_n}\right), \tag{3.78}$$

with results being shown in Fig. 3.8 for different amounts of linear twist. For a rotor with constant blade chord and the ideal twist distribution $\theta = \theta_{\text{tip}}/r$, the local lift coefficient can be determined analytically. The incremental thrust coefficient is given by the standard BET equation

$$dC_T = \frac{\sigma C_{l_\alpha}}{2}\left(\theta - \frac{\lambda}{r}\right) r^2\, dr = \frac{\sigma C_{l_\alpha}}{2}\left(\frac{\theta_{\text{tip}}}{r} - \frac{\lambda}{r}\right) r^2\, dr. \tag{3.79}$$

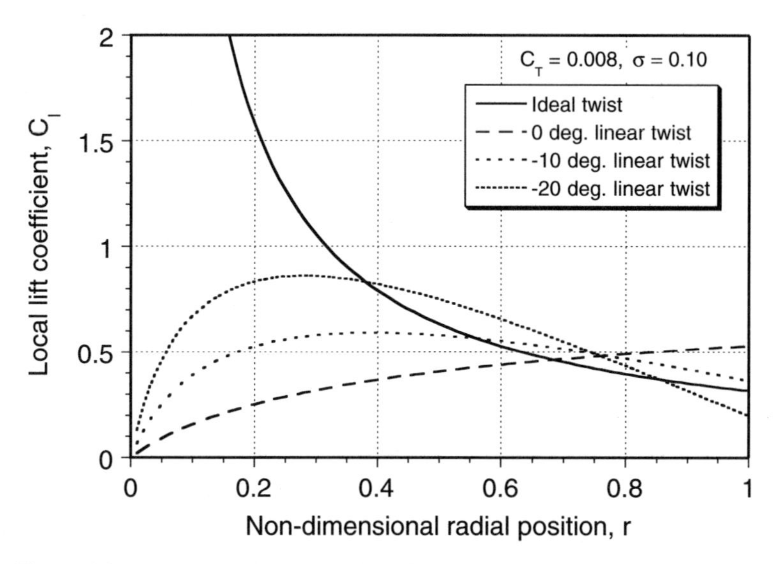

Figure 3.8 BEMT predictions of lift coefficient distribution on a rotor blade for different linear twist rates. Results are compared at a constant total rotor thrust.

In the ideal case λ is a constant, so that the angle of attack along the blade must be

$$\alpha(r) = \frac{\theta_{\text{tip}} - \lambda}{r} = \frac{\alpha_{\text{tip}}}{r}. \tag{3.80}$$

This confirms that the blade loading distribution is triangular, that is,

$$dC_T = \frac{\sigma C_{l_\alpha}}{2} \alpha_{\text{tip}} r \, dr. \tag{3.81}$$

As previously shown, besides uniform inflow the ideal twist distribution also gives uniform disk loading. For the ideal rotor (with constant chord, ideal twist, uniform inflow) integration along the blade leads to

$$C_T = \frac{\sigma C_{l_\alpha}}{4} (\theta_{\text{tip}} - \lambda) = \frac{\sigma C_{l_\alpha}}{4} \alpha_{\text{tip}}. \tag{3.82}$$

Remember that simple momentum theory gives a value of $\lambda = \sqrt{C_T/2}$ in hover. Therefore, the blade pitch angle is $\theta_{\text{tip}} = \alpha_{\text{tip}} + \lambda$ so that

$$\theta_{\text{tip}} = \frac{4C_T}{\sigma C_{l_\alpha}} + \sqrt{\frac{C_T}{2}}. \tag{3.83}$$

The local blade angle of attack can now be written as

$$\alpha(r) = \frac{\alpha_{\text{tip}}}{r} = \frac{4C_T}{\sigma C_{l_\alpha}} \left(\frac{1}{r} \right). \tag{3.84}$$

The corresponding local blade lift coefficient is

$$C_l(r) = C_{l_\alpha} \alpha = \frac{4C_T}{\sigma} \left(\frac{1}{r} \right), \tag{3.85}$$

which is plotted in Fig. 3.8. This result shows that for a rectangular blade with ideal twist (minimum possible induced power), the lift coefficients become very large at the root end of the blade; thus drag coefficients will increase and rotor performance will ultimately be

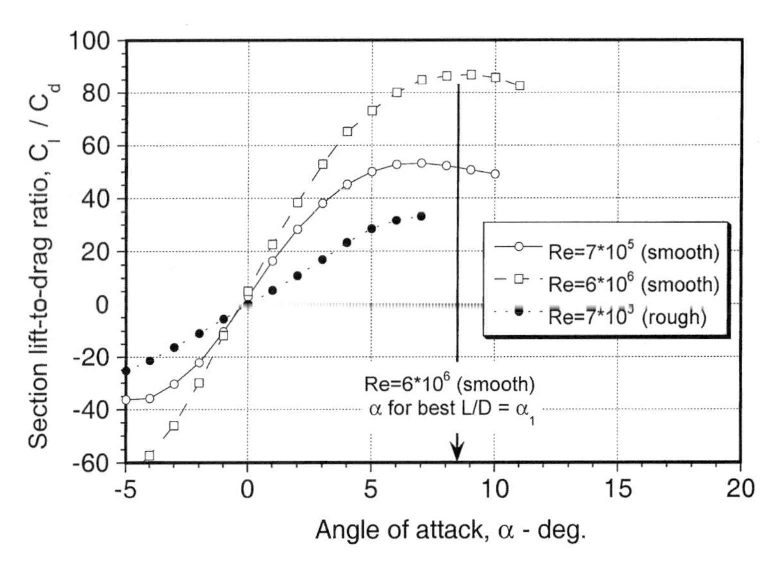

Figure 3.9 Representative lift-to-drag ratio of a 2-D airfoil at low angles of attack and low Mach numbers. Data source: Loftin & Smith (1949).

limited by the onset of stall. Besides the fact that C_l (and C_d) increases rapidly when moving inboard from the tip, there can only be one blade station on the blade operating at its best lift-to-drag ratio. Therefore, anything less than the best C_l/C_d must always result in a higher profile power consumption from the rotor than would otherwise be possible.

This effect can be seen more clearly if a representative lift-to-drag ratio of a two-dimensional airfoil is examined, as shown in Fig. 3.9. In this case, the best C_l/C_d of the airfoil section occurs at an angle of attack between 5 and 10 degrees (denoted as α_1), and this operating condition should be the design goal at all blade stations to achieve minimum profile power.[4] A hovering rotor that is designed to minimize both the induced power and the profile power is called an *optimum hovering rotor.*

3.3.4 *The Optimum Hovering Rotor*

Minimum induced power requires uniform inflow over the disk. The corresponding condition for minimum profile power requires that each blade station operate at the angle of attack for maximum C_l/C_d (i.e., at $\alpha = \alpha_1$). For minimum induced power then $\theta = \theta_{\text{tip}}/r$, and if each element of the blade is to operate at α_1 then

$$dC_T = \frac{\sigma C_{l_\alpha}}{2}\left(\frac{\theta_{\text{tip}}}{r} - \frac{\lambda}{r}\right) r^2\, dr = \frac{\sigma C_{l_\alpha}}{2}\alpha_1 r^2\, dr. \tag{3.86}$$

Also, it has been shown previously from the BEMT that for an annulus of the disk

$$dC_T = 4\lambda^2 r\, dr. \tag{3.87}$$

Equating Eqs. 3.86 and 3.87 and solving for λ gives

$$\lambda = \sqrt{\frac{\sigma r C_{l_\alpha}\alpha_1}{8}}, \tag{3.88}$$

[4] Note that the angle of attack for the best C_l/C_d is a function of Reynolds number and surface finish (roughness). It will also be a function of Mach number.

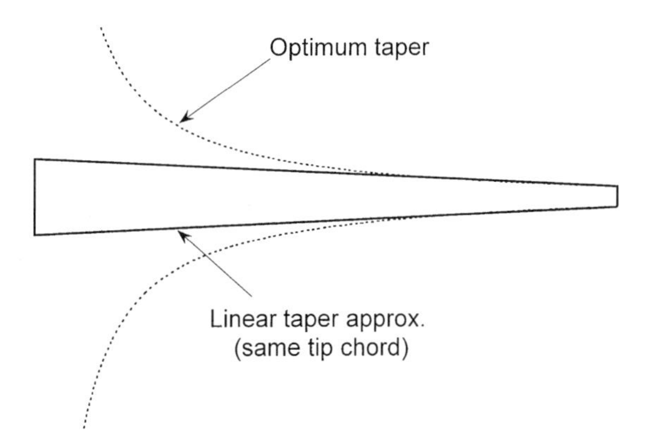

Figure 3.10 Radial distribution of blade chord for optimum rotor and linear approximation.

which is constant over the disk, as required. If it is assumed that α_1 is the same for all the airfoils along the blade span and independent of Reynolds number and Mach number, then for uniform inflow the product σr in Eq. 3.88 must be a constant, that is,

$$\left(\frac{N_b}{\pi R}\right) cr = \text{constant}, \tag{3.89}$$

which requires a local chord distribution over the blade to be given by

$$c(r) = \frac{c_{\text{tip}}}{r} \quad \text{or} \quad \sigma(r) = \frac{\sigma_{\text{tip}}}{r}. \tag{3.90}$$

Therefore, for each section of the blade to operate at the optimum lift-to-drag ratio, the local blade solidity, $\sigma(r)$, or blade chord, $c(r)$, must vary hyperbolically with span, as shown by Fig. 3.10. Clearly this distribution is physically unrealizable but can be adequately approximated by a linear taper over the outer part of the blade. Because of the root cutout, the chord variation as $r \to 0$ does not matter anyway. Furthermore, as seen from Fig. 3.9, operating the airfoils at angles of attack that are few degrees less or greater than α_1 does not result in a serious degradation of C_l/C_d. Therefore, the use of some planform taper, while not exactly hyperbolic or applied over the whole blade, will generally always have a beneficial effect on hovering rotor performance.

The blade twist required for uniform inflow *and* for a constant angle of attack along the blade can be found from

$$\alpha = \left(\theta - \frac{\lambda}{r}\right) = \alpha_1 = \text{constant} \tag{3.91}$$

or

$$\theta(r) = \alpha_1 + \frac{\lambda}{r} = \alpha_1 + \sqrt{\frac{\sigma_{\text{tip}} C_{l_\alpha} \alpha_1}{8}} \left(\frac{1}{r}\right). \tag{3.92}$$

In this case, the rotor thrust distribution is still linear across the disk from the root to the tip (corresponding to the uniform pressure jump), with the total thrust being given by

$$C_T = \frac{1}{2} C_{l_\alpha} \int_0^1 \sigma \alpha_1 r^2 \, dr = \left(\frac{\sigma_{\text{tip}} C_{l_\alpha}}{4}\right) \alpha_1. \tag{3.93}$$

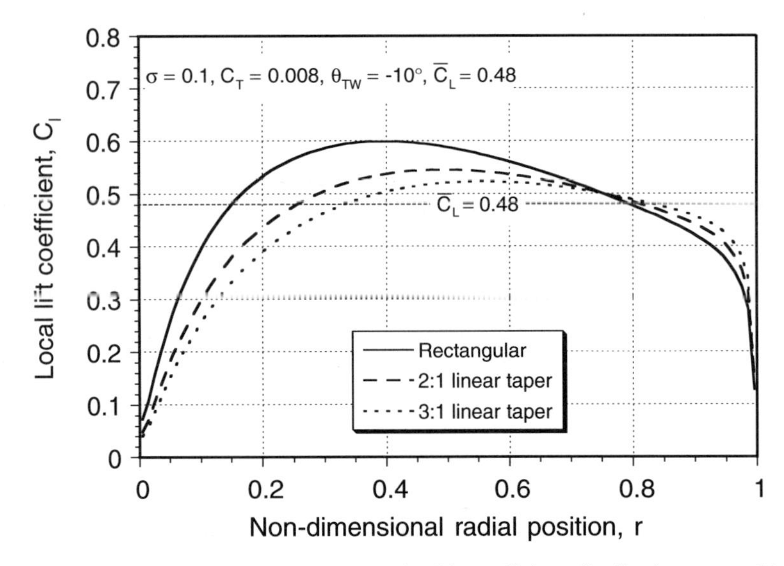

Figure 3.11 Effect of blade taper on the lift coefficient distribution over a blade with constant twist.

The local solidity (chord) of the optimum rotor can be written as

$$\sigma(r) = \left(\frac{4C_T}{C_{l_\alpha}\alpha_1}\right)\frac{1}{r} \tag{3.94}$$

and the blade twist as

$$\theta(r) = \alpha_1 + \frac{\lambda}{r} = \alpha_1 + \sqrt{\frac{C_T}{2}}\left(\frac{1}{r}\right), \tag{3.95}$$

which is hyperbolic. This is the "optimum" hovering rotor referred to by Gessow (1948). Note that the final optimum blade design depends on the rotor thrust, C_T.

It is common for linear blade twist and taper planform variations to be employed on helicopter rotor blades, and this is found sufficiently close to the optimum values defined on the basis of the BEMT. For example, Fig. 3.11 shows the effects on the spanwise distribution of lift coefficient for different values of linear taper. The results are compared at the same thrust-weighted solidity.[5] Without taper the lift coefficients at the root of the blade are relatively high and cannot operate at the best C_l/C_d ratio for the airfoil section. As C_T is increased, the rotor performance will be limited by the onset of stall at the blade root. From Fig. 3.11 we see that with the introduction of a moderate amount of taper the values of C_l become much more uniform, with the values of C_l being significantly reduced at the blade root along with a mild increase in C_l at the tip. This reduces the profile power, and so the rotor can be operated at the same thrust but with an improved figure of merit.

The point is expounded further by the results shown in Fig. 3.12, where the figure of merit is plotted versus the twist rate for a rectangular and for a linearly tapered blade. A mild amount of taper clearly gives a notable increase in *FM*. In either case, the rotor ultimately becomes limited by the onset of stall, but the use of a tapered blade clearly allows the rotor to operate with a higher stall margin. This also allows a higher possible collective pitch, a higher attainable rotor thrust, and better overall hovering efficiency.

[5] For a linearly tapered blade, this means that the solidity at 75% blade radius is the same in all cases – see Section 3.3.16.

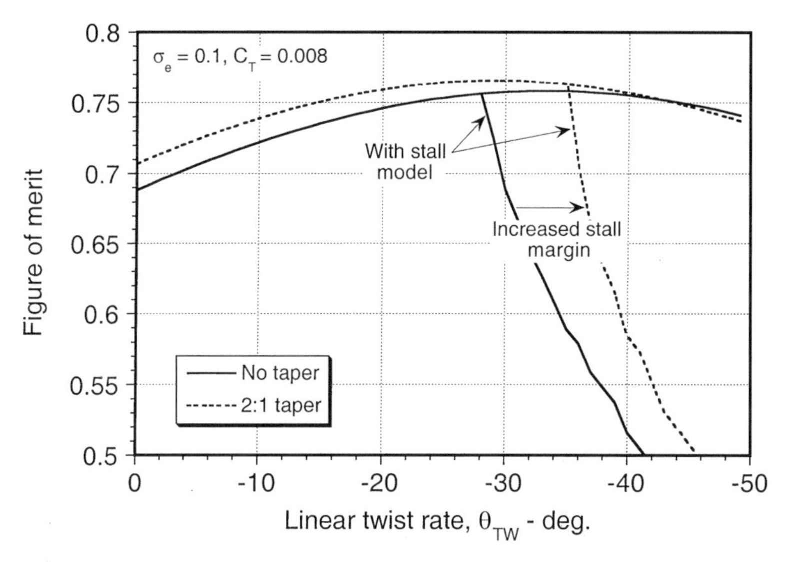

Figure 3.12 Effect of blade twist and taper on rotor figure of merit.

3.3.5 *Circulation Theory of Lift*

Another interesting result of the foregoing is that with a linear lift loading distribution, the local blade circulation is constant. The link between lift per unit length of span and the local circulation, $\Gamma(y)$, is given by the Kutta–Joukowski theorem, so that

$$dL = \rho(\Omega y)\Gamma\, dy = \frac{1}{2}\rho(\Omega y)^2 c C_l\, dy \tag{3.96}$$

or

$$C_l = \frac{2\Gamma}{\Omega y c}. \tag{3.97}$$

Therefore, using Eq. 3.69

$$dC_T = \frac{N_b \rho(\Omega y)\Gamma\, dy}{\rho A(\Omega R)^2} = \frac{N_b \Gamma r\, dr}{\Omega \pi R^2} = \frac{\sigma C_{l_{\text{tip}}}}{2} r\, dr, \tag{3.98}$$

so that solving for the circulation gives

$$\Gamma(r) = \frac{\sigma C_{l_{\text{tip}}} \Omega \pi R^2}{2N_b} = \frac{c R C_{l_{\text{tip}}} \Omega}{2} = \text{constant}, \tag{3.99}$$

confirming that ideal twist gives uniform inflow, a linear distribution of lift, *and* a constant circulation over the blade.

A link can also be made between the momentum theory of lift and the vortex theory (considered in Chapter 10). With a uniform circulation along the blade span, Helmholtz's theorem requires a single vortex of the same strength to trail from the blade tips. This vortex strength can be related to the blade loading as follows. The lift per unit span along the blade is

$$dL = \rho(\Omega y)\Gamma\, dy. \tag{3.100}$$

Because Γ is constant, then the lift on one blade is

$$L = \rho\Omega\Gamma \int_0^R y\, dy = \rho\Omega\Gamma R^2/2. \tag{3.101}$$

The total rotor thrust is $T = N_b L$, and in coefficient form we have

$$C_T = \frac{N_b \Gamma}{2(\pi R^2)\Omega}. \tag{3.102}$$

Using the result for the solidity that $\sigma = N_b c/\pi R$, we get

$$\frac{C_T}{\sigma} = \frac{\Gamma}{2\Omega c R} \tag{3.103}$$

or

$$\Gamma = 2\Omega c R \left(\frac{C_T}{\sigma}\right). \tag{3.104}$$

Remember that this result is for an "ideal" rotor operating in hover. However, the result provides a simple connection between the rotor operating state and the strength (circulation) of the tip vortex filaments trailed into the rotor wake.

3.3.6 *Power Estimates*

For a real rotor, the nonuniformity of λ over the disk means that the induced power must be calculated by numerically integrating the equation

$$C_{P_i} = \int_{r=0}^{r=1} \lambda\, dC_T \tag{3.105}$$

using the actual induced velocity distribution computed using the BEMT. This equation also allows the calculation of the induced power factor, κ, for different twist distributions, that is,

$$\kappa = \frac{C_{P_i}}{C_T^{3/2}/\sqrt{2}}. \tag{3.106}$$

For a more accurate calculation of the rotor profile power, one must consider the variation in sectional drag coefficient with blade section angle of attack – see Bailey & Gustafson (1944). For most airfoils the sectional drag coefficient below stall[6] can be approximated as

$$C_d = C_{d_0} + d_1\alpha + d_2\alpha^2, \tag{3.107}$$

as shown in Fig. 3.13. Clearly, the coefficients of this expression are also a function of airfoil section, Mach number, Reynolds number, and surface finish, and the behavior cannot easily be generalized. In practice, it is found sufficiently accurate to use 2-D airfoil measurements for the Reynolds and Mach numbers corresponding to those at 75% radius on the rotor. Particular caution should be exercised when analyzing subscale rotors. If measurements for the airfoil in question are not available, then catalogs of low Reynolds number airfoil data such as those compiled by Althaus (1972) and Miley (1982) are often useful in estimating the anticipated effects.

[6] Measurements of 2-D section drag are often made by measuring the velocity in the wake of the airfoil, followed by a momentum balance using the conservation laws in integral form. This, however, is a technique valid only below stall where there are no viscous losses from rotation of the fluid (see Question 7.9).

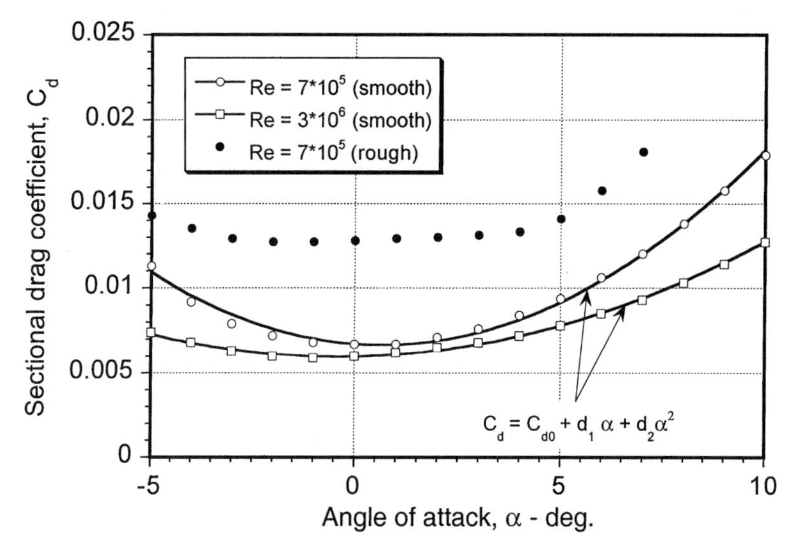

Figure 3.13 Representative 2-D drag coefficient variation for an airfoil as a function of angle of attack. Data source: Loftin & Smith (1949).

The profile part of the rotor power is given by

$$C_{P_0} = \frac{\sigma}{2} \int_0^1 C_d r^3 \, dr. \tag{3.108}$$

Proceeding on the basis that Eq. 3.107 is valid, then substituting gives

$$\begin{aligned} C_{P_0} &= \frac{\sigma}{2} \int_0^1 [C_{d_0} + d_1(\theta - \phi) + d_2(\theta - \phi)^2] r^3 \, dr \\ &= \frac{\sigma}{2} \int_0^1 \left[C_{d_0} + d_1 \left(\theta - \frac{\lambda}{r} \right) + d_2 \left(\theta - \frac{\lambda}{r} \right)^2 \right] r^3 \, dr. \end{aligned} \tag{3.109}$$

Now, for ideally twisted blades, $\lambda =$ constant and $\theta r = \theta_{\text{tip}}$ so that

$$\begin{aligned} C_{P_0} &= \frac{\sigma C_{d_0}}{8} + \frac{\sigma d_1}{6}(\theta_{\text{tip}} - \lambda) + \frac{\sigma d_2}{4}(\theta_{\text{tip}} - \lambda)^2 \\ &= \frac{\sigma C_{d_0}}{8} + \left(\frac{2 d_1}{3 C_{l_\alpha}} \right) C_T + \left(\frac{4 d_2}{\sigma C_{l_\alpha}^2} \right) C_T^2, \end{aligned} \tag{3.110}$$

with the latter equation being obtained using the result that

$$\theta_{\text{tip}} - \lambda = \frac{4 C_T}{\sigma C_{l_\alpha}}. \tag{3.111}$$

By including the higher-order approximation for the airfoil drag, improvements in power prediction can be obtained. Results from Knight & Hefner's experiments are shown again in Fig. 3.14, but with the modified drag expression. Note the substantially better agreement than that shown previously in Fig. 3.3 when a constant drag coefficient was assumed.

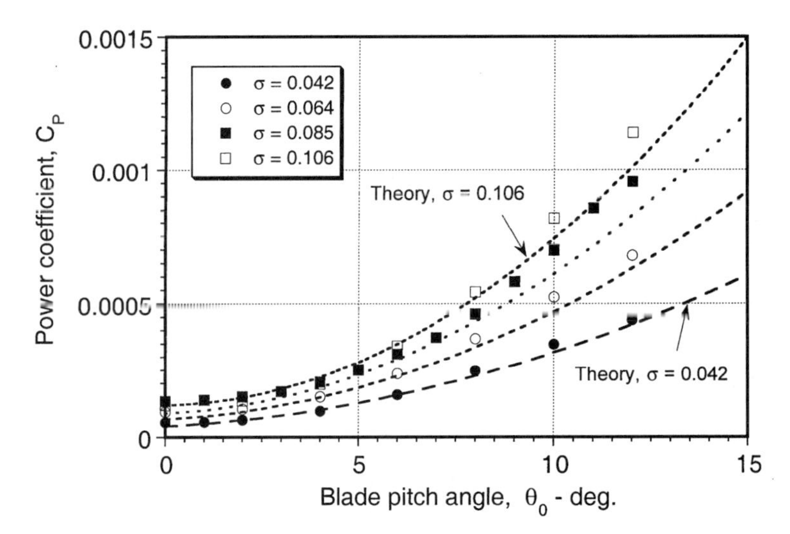

Figure 3.14 Predicted power coefficient collective pitch using higher-order profile drag variation. Data source: Knight & Hefner (1937).

3.3.7 *Prandtl's Tip-Loss Function*

The idea of a tip-loss factor, B, has already been introduced. Instead of assuming a value for B it is possible to compute tip losses on the basis of a method devised by Prandtl – see Betz (1919). Prandtl provided a solution to the problem of the loss of lift near the tips resulting from the induced effects associated with a finite number of blades. This theory contains all the elements of a model that was developed later by Goldstein (1929) and Lock (1930) but has several simplifications that make it attractive for helicopter rotor analysis. Prandtl replaced the curved helical vortex sheets of the rotor wake by a series of two-dimensional sheets, the assumption here being that the radius of curvature at the blade tips is large. This is a satisfactory assumption for helicopter rotors but is less so for propellers. Prandtl's final result can be expressed in terms of an induced velocity correction factor, F, where

$$F = \left(\frac{2}{\pi}\right)\cos^{-1} e^{-f}, \tag{3.112}$$

where the exponent f is given in terms of the number of blades and the radial position of the blade element, r, by

$$f = \frac{N_b}{2}\left(\frac{1-r}{r\phi}\right) \tag{3.113}$$

and ϕ is the induced inflow angle $[= \lambda(r)/r]$.

Prandtl's F function is plotted in Fig. 3.15 versus the local inflow angle, ϕ, for a two-bladed rotor. The basic effect of the F function is to increase the induced velocity over the tip region and reduce the lift generated there. The Prandtl function is, therefore, sometimes referred to as a circulation loss function. Prandtl's F function is plotted versus the radial position of the blade element in Fig. 3.16 for two- and four-bladed rotors, and for different values of the inflow angle, ϕ. The gradual reduction in the function as the tip is approached represents the basic induced effects of the vortex wake. Note that F is a function of the number of blades and always has larger values for rotors with fewer blades. This is because the circulation is distributed over fewer blades and the induced effects in the wake are

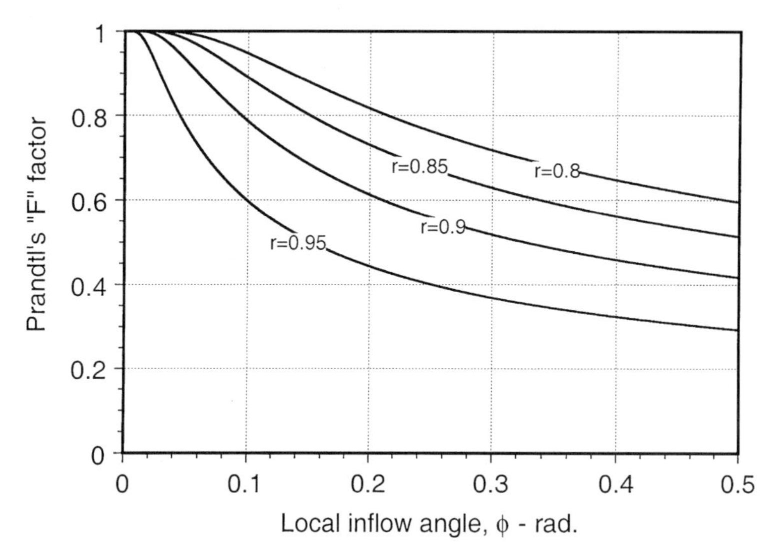

Figure 3.15 Variation of Prandtl tip loss function versus local flow angle for a two-bladed rotor.

higher. For the limiting case where $N_b \to \infty$, which approximates an actuator disk, the circulation is distributed uniformly over the disk and $F \to 1$.

In the application of the Prandtl tip-loss method, the function F can also be interpreted as a reduction factor applied to the change in fluid velocity as it passes through the control volume. The tip-loss effect can be incorporated into the blade element momentum theory as follows. For hovering flight Eq. 3.44 is now modified by the use of the Prandtl factor, F, to give

$$dC_T = 4F\lambda^2 r\, dr. \tag{3.114}$$

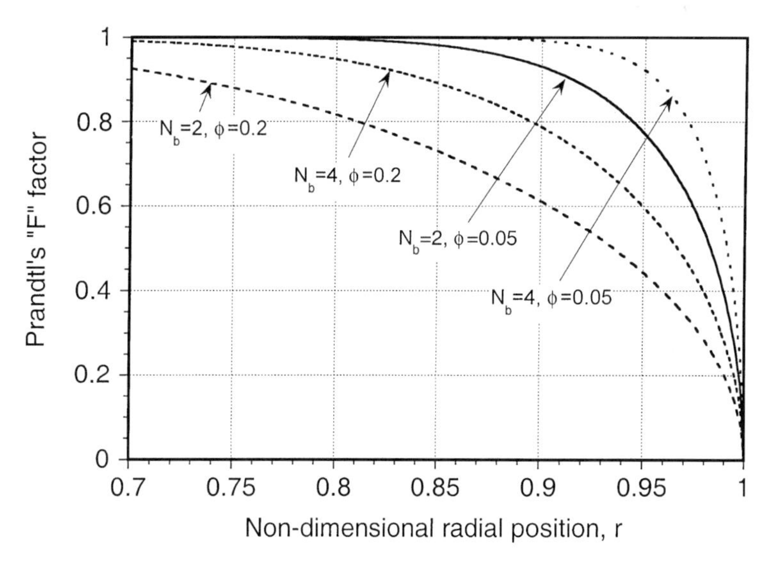

Figure 3.16 Variation of Prandtl tip loss function versus radial position for two- and four-bladed rotors.

Also, from the BET it has been shown previously that

$$dC_T = \frac{\sigma C_{l_\alpha}}{2}(\theta r^2 - \lambda r)\,dr. \tag{3.115}$$

Equating the incremental thrust coefficients from the momentum and blade element theories gives

$$\frac{\sigma C_{l_\alpha}}{2}(\theta r^2 - \lambda r) = 4F\lambda^2 r \tag{3.116}$$

or

$$\lambda^2 + \left(\frac{\sigma C_{l_\alpha}}{8F}\right)\lambda - \frac{\sigma C_{l_\alpha}}{8F}\theta r = 0. \tag{3.117}$$

This quadratic has the solution

$$\lambda(r) = \frac{\sigma C_{l_\alpha}}{16F}\left[\sqrt{1 + \frac{32F}{\sigma C_{l_\alpha}}\theta r} - 1\right]. \tag{3.118}$$

Because F is a function of λ, this equation cannot be solved immediately because λ is initially unknown. Therefore, it is solved iteratively by first calculating λ using $F = 1$ (corresponding to an infinite number of blades) and then finding F from Eq. 3.112 and re-calculating λ from the numerical solution to Eq. 3.118. Convergence is rapid and is obtained in three or four iterations.

The resulting effect of the application of Prandtl's F function on the blade thrust distribution is shown in Fig. 3.17 for an untwisted blade. These comparisons have been performed at the same value of thrust obtained without using the tip-loss effect. Note that the primary effect of the tip loss is to reduce the thrust production over the immediate tip region. This loss has to be compensated by a slightly greater blade pitch, which increases the angles of attack further inboard to produce the same total rotor thrust. Generally it is found that the use of a larger number of blades tends to improve the efficiency of the rotor (increases

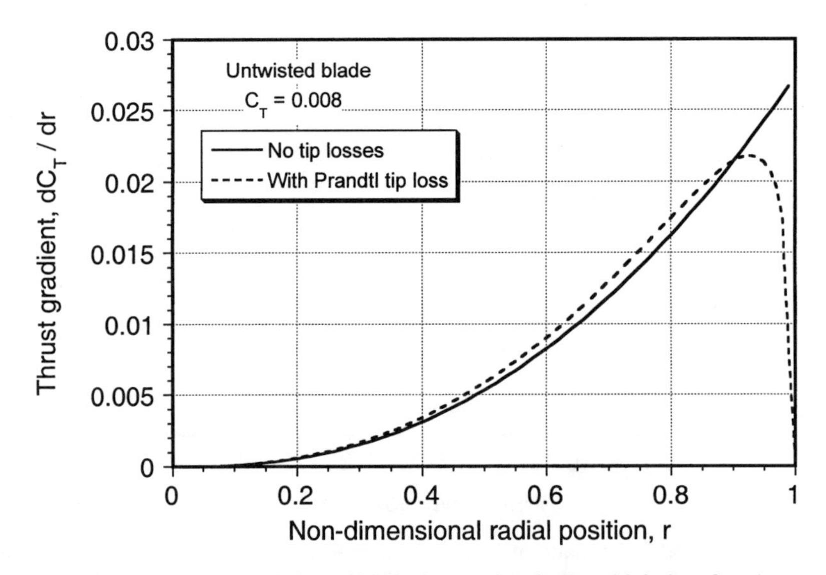

Figure 3.17 Spanwise variation of blade thrust using the Prandtl tip loss function.

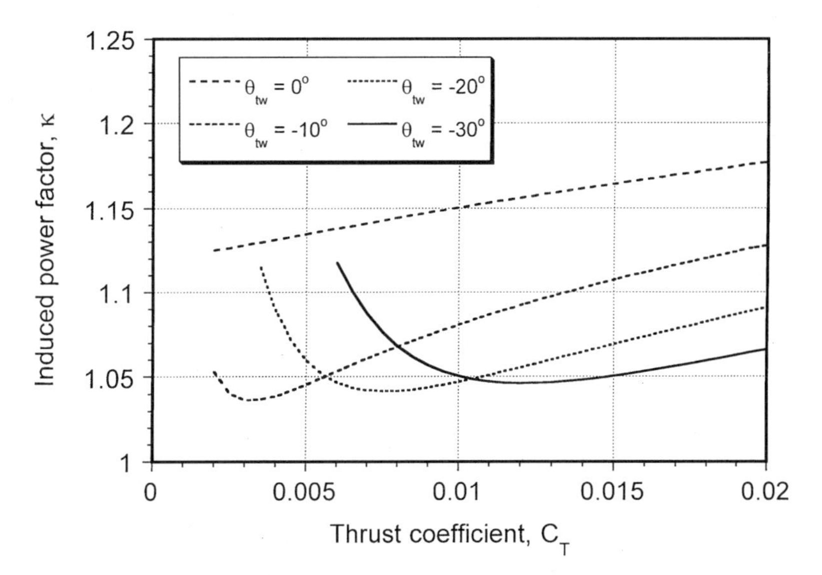

Figure 3.18 BEMT predictions of induced power factor for a four-bladed rotor with linear twist using Prandtl's tip loss function.

the figure of merit) for a given solidity and disk loading because the induced tip-loss effects tend to decrease with increasing number of blades (see Questions 3.2 & 3.3).

3.3.8 *Figure of Merit*

On the basis of the predictions of profile power from the blade element analysis, the figure of merit can be written as

$$FM = \frac{C_{P_{\text{ideal}}}}{C_{P_i} + C_{P_0}}, \tag{3.119}$$

where $C_{P_{\text{ideal}}} = C_T^{3/2}/\sqrt{2}$ and C_{P_i} can be written in terms of the induced power factor, κ, as $C_{P_i} = \kappa C_T^{3/2}/\sqrt{2}$. From the numerical approach to the BEM theory, and incorporating Prandtl's tip loss function, we can readily obtain the results for C_{P_i}, C_{P_0}, κ, and FM as a function of collective pitch or C_T and the results compared for different blade twist rates and/or planforms. It has been shown previously that blade twist primarily affects the induced power. Results showing the effects of blade twist on the induced power factor, κ, versus rotor thrust coefficient are given in Fig. 3.18. The values of κ have been computed for a four-bladed rotor. Note that there are significant reductions in induced power (and gains in FM) to be made by the addition of nose-down blade twist. Up to -20 degrees of twist is considered optimal, with further increases in nose-down twist rate giving only minor returns within the operational values of C_T for a helicopter. (The use of high blade twist rates is nonoptimal for forward flight because such blades tend to generate lower lift on the advancing blade compared to a design with less twist.)

The profile power also affects the FM. Using the higher-order drag variation, we have

$$FM = \frac{\dfrac{C_T^{3/2}}{\sqrt{2}}}{\dfrac{\kappa C_T^{3/2}}{\sqrt{2}} + \left[\dfrac{\sigma C_{d_0}}{8} + \left(\dfrac{2 d_1}{3 C_{l_\alpha}}\right) C_T + \left(\dfrac{4 d_2}{\sigma C_{l_\alpha}^2}\right) C_T^2\right]}. \tag{3.120}$$

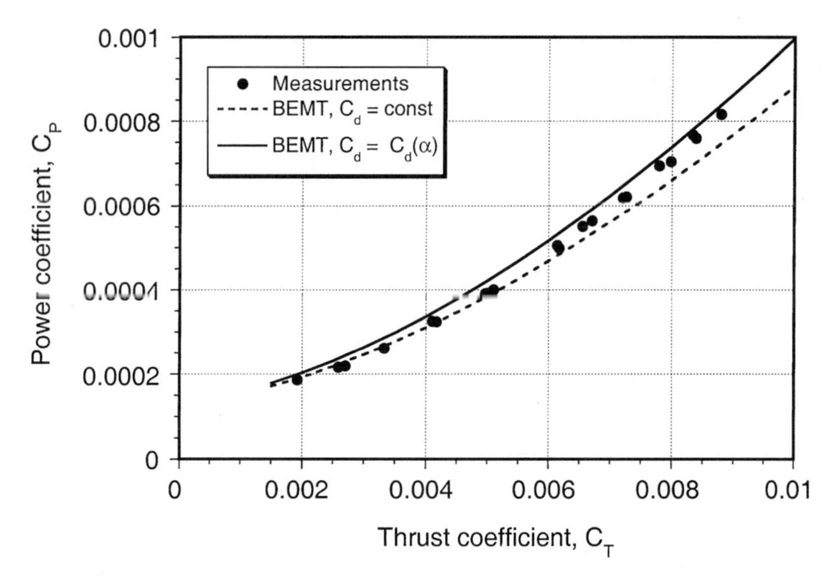

Figure 3.19 Comparison of the BEMT with and without higher-order profile drag terms versus measured thrust and power data for a four-bladed hovering rotor. Data source: Bagai & Leishman (1992).

Note that the leading term of the profile power is affected by the rotor solidity, σ. Therefore, it is apparent that to maximize the figure of merit the profile part of the power must be kept as low as possible and also the onset of stall must be delayed to high angles of attack.

3.3.9 *Further Comparisons of BEMT with Experiment*

Results showing predictions made by the complete BEMT with measured thrust and power data for a hovering rotor are given in Fig. 3.19 for a four-bladed rotor of solidity 0.1 and with -13 degrees of linear twist. Compared to the modified momentum theory now only a model for the profile drag losses must be assumed; all the induced losses resulting from nonuniform inflow and Prandtl tip losses are now calculated directly from the more complete BEMT. Figure 3.19 shows that, with the assumption that $C_d =$ constant, the complete BEMT underpredicts the power, and consequently it will tend to overpredict the figure of merit. The addition of the higher-order drag terms using Eq. 3.107 rectifies this, and the agreement with the measured data is much better.

Figure 3.20 shows a comparison of the complete BEMT with the measurements of thrust and power for a hovering rotor made by Knight & Hefner (1937). Again, the higher-order drag curve using Eq. 3.107 gives excellent agreement with the measurements. The only discrepancy is at higher values of C_T where the rotor begins to stall and Eq. 3.107 is insufficient to model the drag behavior.

Measurements of thrust and power in the axial climb condition are relatively rare. Flight test results have been given by Gustafson & Gessow (1945) and others, although it is difficult to estimate fuselage vertical drag and also to separate wake distortion effects resulting from the fuselage from the measured rotor thrust and shaft power in flight tests. These factors make comparisons of flight test results with the BEMT relatively difficult. Such issues are reviewed by Harris (1987), but as shown by Prouty (1986) and others, reasonable climb performance estimation can be expected from the momentum theory if appropriate empirical corrections are taken into account.

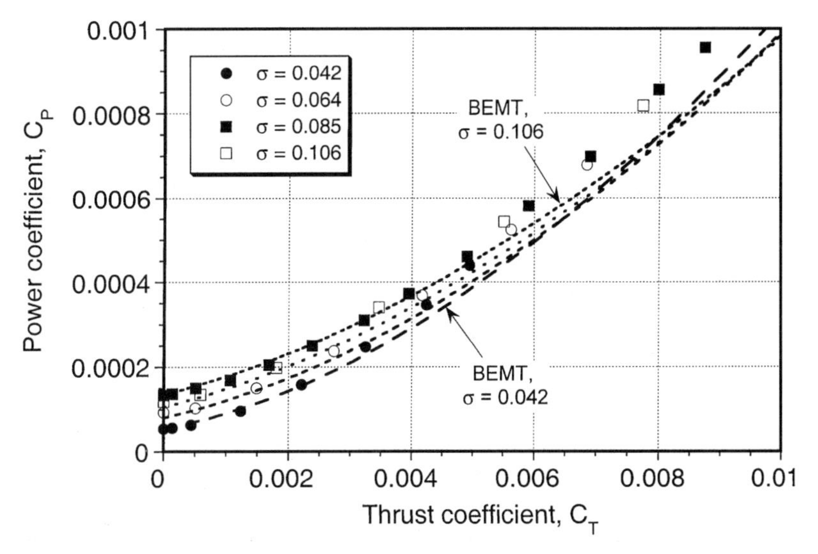

Figure 3.20 Comparison of the BEMT with measured thrust and power data for hovering rotors of different solidity. Data source: Knight & Hefner (1937).

Felker & McKillip (1994) have measured isolated rotor performance on model rotors tested on a track facility. The thrusting rotor was mounted horizontally on a carriage and moved at constant velocity in still air, thereby simulating a steady climb. The thrust and power were measured with a balance system for constant collective pitch at several axial climb velocities. The measured thrust and power coefficient as a function of climb velocity, λ_c/λ_h or V_c/v_h, are shown in Fig. 3.21 and are compared with results obtained using the BEMT. Because data were acquired at a constant collective pitch, increasing climb velocity reduces the blade element angles of attack resulting in a decrease in rotor thrust. Note that although the BEMT predicts the thrust behavior well, it tends to underpredict the thrust somewhat at the higher climb velocities. For the lower collective pitch, the maximum underprediction is about 7%; and at the higher collective pitch it is about 4%. Since estimated uncertainties in the measurements are about 5%, the agreement of the BEMT with the measurements is good. The corresponding predictions of power are shown in Fig. 3.21(b), where the work done by the rotor, $C_T\lambda_c$, has been added to the measurements. Again the agreement between the BEMT and experiment is good, except at the higher climb velocities where there is a maximum underprediction of no more than 10%.

3.3.10 *Compressibility Corrections*

A questionable assumption made up until now is that the lift-curve-slope of the blade airfoil sections and their drag coefficients are independent of Mach number. To examine the effects of compressibility on C_T, consider a correction to the lift-curve-slope of each blade element according to Glauert's rule in which

$$C_{l_\alpha}(M) = \frac{C_{l_\alpha}|_{M=0.1}}{\sqrt{1-M^2}}, \tag{3.121}$$

where $C_{l_\alpha}|_{M=0.1}$ is the measured 2-D lift-curve-slope of the airfoil at $M = 0.1$. The local

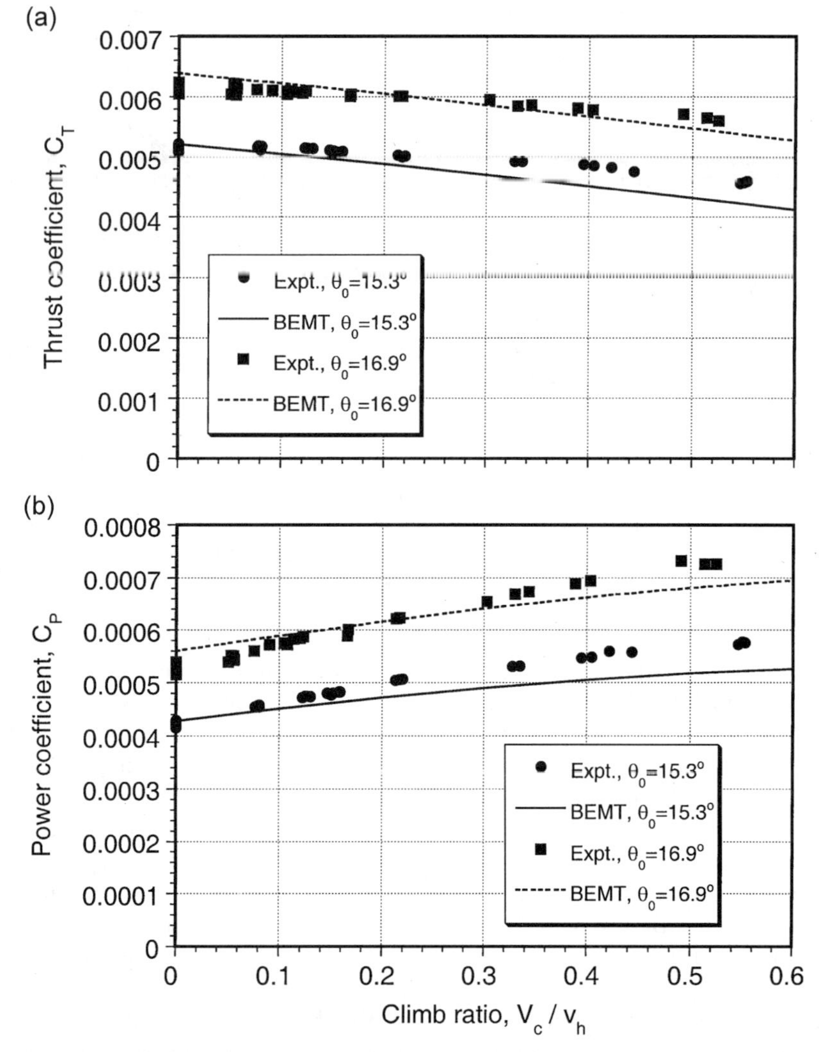

Figure 3.21 Comparison of the BEMT with measured thrust and power data for a climbing rotor. (a) Thrust versus climb ratio. (b) Power versus climb ratio. Data source: Felker & McKillip (1994).

blade Mach number will be

$$M(y) = \frac{U_T}{a} = \frac{\Omega y}{a}. \tag{3.122}$$

Therefore, the lift-curve-slope correction in terms of the tip Mach number will be

$$\frac{1}{\sqrt{1-M^2}} = \frac{1}{\sqrt{1-\left(\frac{\Omega}{a}\right)^2 y^2}} = \frac{1}{\sqrt{1-M_{\text{tip}}^2 r^2}}. \tag{3.123}$$

where M_{tip} is the tip Mach number. Remembering that

$$dC_T = \frac{1}{2}\sigma C_l r^2 \, dr \tag{3.124}$$

then we may write

$$dC_T = \frac{1}{2}\sigma \frac{C_{l_\alpha}|_{M=0.1}}{\sqrt{1 - M_{\text{tip}}^2 r^2}}(\theta r^2 - \lambda r)\, dr. \tag{3.125}$$

If we still assume ideal twist, and that the inflow remains uniformly distributed, then

$$dC_T = \frac{1}{2}\sigma C_{l_\alpha}|_{M=0.1}(\theta_{\text{tip}} - \lambda)\frac{r}{\sqrt{1 - M_{\text{tip}}^2 r^2}}\, dr. \tag{3.126}$$

The total blade thrust is then

$$\begin{aligned} C_T &= \frac{1}{2}\sigma C_{l_\alpha}|_{M=0.1}(\theta_{\text{tip}} - \lambda)\int_0^1 \frac{r}{\sqrt{1 - M_{\text{tip}}^2 r^2}}\, dr \\ &= \frac{1}{2}\sigma C_{l_\alpha}|_{M=0.1}(\theta_{\text{tip}} - \lambda)\left[\frac{1}{1 + \sqrt{1 - M_{\text{tip}}^2}}\right] \\ &= \frac{1}{4}\sigma K C_{l_\alpha}|_{M=0.1}(\theta_{\text{tip}} - \lambda), \end{aligned} \tag{3.127}$$

where the factor K is given by

$$K = \frac{2}{1 + \sqrt{1 - M_{\text{tip}}^2}}. \tag{3.128}$$

We see that as $M_{\text{tip}} \to 0$ then $K \to 1$ and the incompressible result given previously in Eq. 3.63 is obtained. If it assumed that an averaged compressible lift-curve-slope can be used for the entire rotor then

$$C_{l_\alpha} = \frac{C_{l_\alpha}|_{M=0.1}}{\sqrt{1 - (r_e M_{\text{tip}})^2}}, \tag{3.129}$$

where r_e is an effective radius. It can be shown that $r_e = 1/\sqrt{2} = 0.707$ when $M_{\text{tip}} \to 0$, increasing to a value of $r_e = 0.75$ for $M_{\text{tip}} = 0.8$. This is in close agreement with Payne (1959), who suggests using an effective lift-curve-slope at 70% radius, which is probably accurate enough for $M_{\text{tip}} < 0.6$. Peters & Ormiston (1975) suggest using the value of the lift-curve-slope at 75% radius, which is realistic for the normal operational tip Mach numbers of most helicopters. Generally, the effects of compressibility will increase the rotor thrust coefficient by about 10% for a given collective pitch setting. However, tip relief effects tend to reduce the effect somewhat (see Section 5.3.2).

Compressibility effects become more important for helicopter rotors when they are operated in forward flight, and especially where the advancing blade tip Mach numbers approach transonic conditions. In forward flight conditions, the idea of a mean lift-curve-slope corrected for compressibility becomes less applicable, and compressibility corrections must be included inside the integral sign in the thrust integral and for other integrated quantities – that is, compressibility modeling must be considered separately at each blade element and the net effect obtained by numerical integration.

3.3.11 *Equivalent Chords and Weighted Solidity*

For blades that are nonrectangular in planform the local solidity $\sigma(r)$ varies along the blade span. In this case, to find the thrust coefficient one must use

$$C_T = \frac{1}{2}\int_0^1 \sigma(r) r^2 C_l \, dr, \tag{3.130}$$

where the local solidity of the blade appears *inside* the integral sign. Because the chord varies along the blade span, the rotor solidity must be written as

$$\sigma_{\text{rotor}} = \frac{\text{Blade area}}{\text{Rotor area}} = \int_0^1 \sigma(r)\, dr. \tag{3.131}$$

For a rectangular blade, the rotor solidity and the local solidity are, of course, identical.

As discussed by Gessow & Myers (1952), the purpose of weighted solidities is to help compare the performance of several rotors that may have different blade planforms, for example, rotors with different amounts of taper. The main idea is to generate an equivalent rectangular rotor blade that takes into account the fundamental aerodynamic effects of varying blade chord. The concept, however, is not applicable strictly when applied to other than simple planform shapes. The weighted solidity concept is similar to the mean geometric and aerodynamic chords used in fixed-wing analysis, where an equivalent rectangular wing is derived. In helicopter analyses, two forms of weighted solidities may be used, namely, thrust weighted solidity and power (or torque) weighted solidity. Before defining these quantities, it is instructive to briefly recall the concepts of mean chords used in fixed-wing studies.

3.3.12 *Mean Wing Chords*

For a fixed (nonrotating) wing, the lift L on the wing of area S and total lift coefficient C_L is given by

$$L = \frac{1}{2}\rho V_\infty^2 S C_L = \frac{1}{2}\rho V_\infty^2 \int_{-s}^{s} c C_l \, dy, \tag{3.132}$$

where C_l is the local section lift coefficient, c is the local chord, and s is the semi-span – see, for example, Houghton & Carpenter (1993). We note that the wing area can be written as

$$S = \int_{-s}^{s} c \, dy = 2\int_0^s c \, dy = 2\bar{c}\int_0^s dy, \tag{3.133}$$

where $\bar{c}$ is known as the standard mean chord or the geometric mean chord. Using this definition then

$$C_L \bar{c}\int_0^s dy = \int_0^s c C_l \, dy. \tag{3.134}$$

Now, if an elliptically loaded elliptical chord wing is assumed, then C_l is constant along the wing, and so $C_l = C_L$. This gives

$$\bar{c} = \frac{\int_0^s c \, dy}{\int_0^s dy} = \frac{1}{s}\int_0^s c \, dy = \int_0^1 c d\left(\frac{y}{s}\right), \tag{3.135}$$

which is the definition of mean chord used for fixed wings.

3.3.13 *Thrust Weighted Solidity*

Now consider the rotor case. The rotor thrust coefficient can be written as

$$C_T = \frac{1}{2}\int_0^1 \sigma r^2 C_l \, dr = \frac{1}{2}\sigma_e \int_0^1 r^2 C_l \, dr. \tag{3.136}$$

Assuming constant C_l, as in the case of the fixed wing, gives

$$\int_0^1 \sigma r^2 \, dr = \sigma_e \int_0^1 r^2 \, dr = \frac{\sigma_e}{3}. \tag{3.137}$$

Therefore, based on this assumption the equivalent thrust weighted solidity is

$$\sigma_e = 3\int_0^1 \sigma(r) r^2 \, dr \tag{3.138}$$

or the equivalent chord is

$$c_e = \frac{3\pi R}{N_b}\int_0^1 \sigma(r) r^2 \, dr. \tag{3.139}$$

This parameter takes into account the primary aerodynamic effect of varying planform, weighting the effects at the tip more heavily than stations further inboard. McVeigh & McHugh (1982) suggest a modification to the weighted solidity definition to take account of tip sweep. In this case, Eq. 3.138 is modified to read

$$\sigma_e = 3\int_0^1 \sigma(r) r^2 \cos^2 \Lambda(r) \, dr, \tag{3.140}$$

where σ is now measured perpendicular to the local 1/4-chord line and Λ is the local sweep angle of the 1/4-chord from the blade reference axis. The proper validity of this latter expression, however, has not been confirmed.

3.3.14 *Power/Torque Weighted Solidity*

The rotor power or torque coefficient can be written as

$$\begin{aligned} C_P = C_Q &= \int_{r=0}^{r=1} \lambda dC_T + \frac{1}{2}\int_0^1 \sigma(r) r^3 C_d \, dr \\ &= \int_{r=0}^{r=1} \lambda dC_T + \frac{1}{2}\sigma_e \int_0^1 r^3 C_d \, dr. \end{aligned} \tag{3.141}$$

Assuming constant C_d and uniform inflow then the equivalent power/torque weighted solidity is

$$\sigma_e = 4\int_0^1 \sigma(r) r^3 \, dr. \tag{3.142}$$

Torque weighted solidity is analogous to the activity factor used in propeller design. It is also used in wind turbine design but rarely in helicopter design. Again, McVeigh & McHugh (1982) suggest a modification to the weighted solidity if swept tip blades are used so that

$$\sigma_e = 4\int_0^1 \sigma(r) r^3 \cos^2 \Lambda(r) \, dr. \tag{3.143}$$

3.3.15 *Weighted Solidity of the Optimum Rotor*

With the optimum taper distribution, the profile power coefficient can be established. However, to do this properly one must compare the results at the same *weighted solidity*. In terms of thrust weighted solidity, σ_e, the optimum planform leads to

$$\sigma_e = 3\int_0^1 \sigma r^2\, dr = 3\int_0^1 \left(\frac{\sigma_{\text{tip}}}{r}\right) r^2\, dr = \frac{3}{2}\sigma_{\text{tip}}. \tag{3.144}$$

The profile power coefficient at the same weighted solidity is now

$$C_{P_0} = \int_0^1 \frac{\sigma}{2} r^3 C_{d_0}\, dr = \frac{(2/3)\sigma_e C_{d_0}}{6} = \frac{\sigma_e C_{d_0}}{9}, \tag{3.145}$$

which is about 11% lower than for a rotor with rectangular blades. The figure of merit for the optimum rotor becomes

$$FM = \frac{\dfrac{C_T^{3/2}}{\sqrt{2}}}{\dfrac{\kappa C_T^{3/2}}{\sqrt{2}} + \dfrac{\sigma C_{d_0}}{9}}, \tag{3.146}$$

which is a good 2 to 5% higher than for a rectangular rotor with ideal twist, and this difference can translate into substantial payload gains (see Question 6.8).

3.3.16 *Weighted Solidities of Tapered Blades*

The equivalent solidities for rotors with taper or other planform variation can now be computed. One must be cautious, however, in that the rotor planforms must not be too radically different. This is because the definition of weighted solidity, although based on the blade planform, has a hidden assumption in that the C_l distribution is assumed nominally constant. If this is not the case, then the definition breaks down. Consider a linearly tapered blade that can be described by $\sigma(r) = \sigma_0 + \sigma_1 r$, where σ_0 and σ_1 are constants. The thrust weighted solidity is given by

$$\begin{aligned}
\sigma_e &= 3\int_0^1 \sigma r^2\, dr = 3\int_0^1 (\sigma_0 + \sigma_1 r) r^2\, dr \\
&= 3\left[\frac{\sigma_0 r^3}{3}\right]_0^1 + 3\left[\frac{\sigma_1 r^4}{4}\right]_0^1 \\
&= \sigma_0 + \frac{3}{4}\sigma_1 = \text{ Solidity at 3/4 (75\%) blade radius.}
\end{aligned}$$

The corresponding power/torque weighted solidity is given by

$$\begin{aligned}
\sigma_e &= 4\int_0^1 \sigma r^3\, dr = 4\int_0^1 (\sigma_0 + \sigma_1 r) r^3\, dr \\
&= 4\left[\frac{\sigma_0 r^4}{4}\right]_0^1 + 4\left[\frac{\sigma_1 r^5}{5}\right]_0^1 \\
&= \sigma_0 + \frac{4}{5}\sigma_1 = \text{ Solidity at 4/5 (80\%) blade radius.}
\end{aligned}$$

Similar results can be obtained for blades with taper extending only over part of the span (see Question 3.12). In this case, the integral for the equivalent weighted solidity must be split into two parts, where the inboard portion of the blade has a constant chord c_r from the root to a point $r = r_1$ (i.e., $\sigma = \sigma_r$) and then a linear taper from r_1 to the tip at $r = 1$ where the chord is c_t (i.e., $\sigma = (\frac{\sigma_t - \sigma_r}{1 - r_1})(r - 1) + \sigma_t$). Then for this blade the equivalent thrust weighted solidity is given by

$$\sigma_e = (\sigma_r - \sigma_t)r_1^3 + \left(\frac{\sigma_t - \sigma_r}{1 - r_1}\right)\left[-\frac{1}{4} + r_1^3 - \frac{3}{4}r_1^4\right] + \sigma_t. \tag{3.147}$$

3.3.17 *Mean Lift Coefficient*

Another parameter that is useful in rotor analyses is the mean lift coefficient, $\bar{C}_L$. The mean lift coefficient is defined to give the same thrust coefficient as

$$C_T = \frac{1}{2}\int_0^1 \sigma r^2 C_l \, dr \tag{3.148}$$

when the entire blade is assumed to be operating at the same local lift coefficient as $\bar{C}_L$ (i.e., with an optimum rotor). Thus,

$$C_T = \frac{1}{2}\int_0^1 \sigma r^2 \bar{C}_L \, dr = \frac{1}{2}\bar{C}_L \int_0^1 \sigma r^2 \, dr = \frac{1}{6}\sigma \bar{C}_L \tag{3.149}$$

or

$$\bar{C}_L = 6\left(\frac{C_T}{\sigma}\right). \tag{3.150}$$

Therefore, the quantity $\bar{C}_L / C_{l_\alpha}$ can be viewed as a mean angle of attack of the blades. Typically, $\bar{C}_L$ is found to be in the range 0.5 to 0.8 for helicopter rotors, and so the mean angles of attack for hovering flight vary from about 5 to 8 degrees. Remember that, like the weighted solidity factors, the mean lift coefficient is derived on the basis of an optimum rotor (i.e., uniform inflow and uniform C_l). In practice, however, for any rotor the mean lift coefficient remains a good overall indicator of the average working state of the blades. Note that the figure of merit can be written in terms of $\bar{C}_L$ as

$$FM = \frac{\lambda_h C_T}{\kappa \lambda_h C_T + \sigma C_{d_0}/8} = \frac{1}{\kappa + \frac{3}{4}[(C_{d_0}/\bar{C}_L)/\lambda_h]}, \tag{3.151}$$

confirming that the use of airfoils with a high average lift-to-drag ratio is required for good hover performance (see Question 3.7).

3.4 Blade Element Analysis in Forward Flight

For a rotating-wing aircraft the rotor is unique because, unlike a propeller, it must provide both a lifting force (in opposition to the aircraft weight) and a propulsive force (to overcome the rotor and airframe drag) in forward flight. In forward flight, the rotor moves almost edgewise through the air, and the blade sections must encounter a periodic variation in local velocity. This gives rise to a number of complications in the aerodynamics of the rotor including significant compressibility effects, unsteady effects, nonlinear aerodynamics, and the possibility of stall, reverse flow, and the complex induced velocity from the rotor wake. All these effects are difficult to model. However, using the BET with certain simplifying assumptions, the leading terms of the rotor aerodynamic forces can be obtained – see Section 4.12. These solutions are very instructive and also provide closed-form expressions that can be used as checks when analyzing the results from more comprehensive mathematical models of the rotor aerodynamics.

3.4.1 *Blade Forces*

The same blade element assumptions and approximations previously used for the axisymmetric flight case can also be considered as valid in forward flight. As before, the velocity at the blade element with a pitch angle θ is decomposed into an out-of-plane (perpendicular) velocity component, U_P, and a tangential (in-plane) component, U_T, perpendicular to the leading edge of the blade, both relative to the rotor disk plane, as shown previously in Fig. 3.1. The resultant velocity is

$$U = \sqrt{U_T^2 + U_P^2} \approx U_T \tag{3.152}$$

because U_P is generally small relative to U_T over most of the blade. The relative inflow angle (or induced angle of attack) at the blade section is

$$\phi = \tan^{-1}\left(\frac{U_P}{U_T}\right) \approx \frac{U_P}{U_T} \quad \text{for small angles.} \tag{3.153}$$

Therefore, the aerodynamic angle of attack of the blade element is given by

$$\alpha = \theta - \phi. \tag{3.154}$$

The resultant incremental lift dL per unit span on the blade element is given by

$$\begin{aligned} dL &= \frac{1}{2}\rho U^2 c C_l \, dy = \frac{1}{2}\rho U_T^2 c C_{l_\alpha}(\theta - \phi)\, dy \\ &= \frac{1}{2}\rho U_T^2 c C_{l_\alpha}\left(\theta - \frac{U_P}{U_T}\right) dy \\ &= \frac{1}{2}\rho c C_{l_\alpha}\left(\theta U_T^2 - U_P U_T\right) dy \end{aligned} \tag{3.155}$$

and the incremental drag is

$$dD = \frac{1}{2}\rho U^2 c C_d \, dy, \tag{3.156}$$

which act perpendicular and parallel to the resultant flow velocity at the section, respectively. Resolving these forces perpendicular and parallel to the rotor disk gives

$$\begin{aligned} dF_z &= dL\cos\phi - dD\sin\phi \approx dL \\ &= \frac{1}{2}\rho c C_{l_\alpha}\left(\theta U_T^2 - U_P U_T\right) dy \end{aligned} \tag{3.157}$$

and

$$\begin{aligned} dF_x &= dL\sin\phi + dD\cos\phi \approx \phi dL + dD \\ &= \frac{1}{2}\rho c C_{l_\alpha}\left(\theta U_P U_T - U_P^2 + \frac{C_d}{C_{l_\alpha}} U_T^2\right) dy. \end{aligned} \tag{3.158}$$

In forward flight the blade element velocity components are periodic at the rotor rotational frequency. As for the hover case, there is an in-plane velocity component because of blade

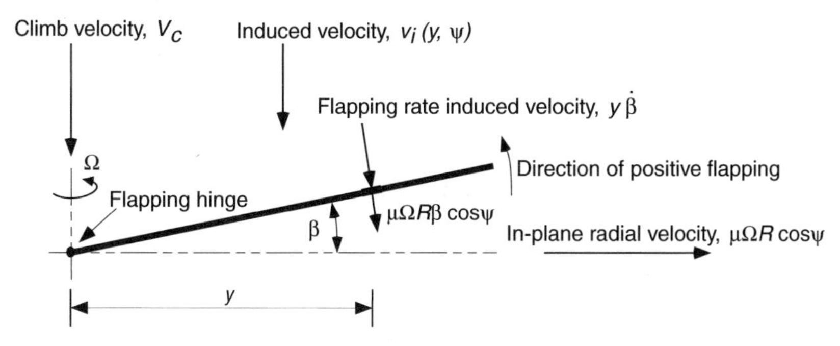

Figure 3.22 Perturbation velocities on the blade resulting from blade flapping velocity and rotor coning.

rotation about the rotor shaft, but now there is a further free-stream (translational) part such that

$$U_T(y, \psi) = \Omega y + V_\infty \sin\psi = \Omega y + \mu\Omega R \sin\psi. \tag{3.159}$$

The out-of-plane component consists of three parts. The first is a component comprising the inflow velocity, as in the hover case. The other two components result from perturbations in velocity at the blade element that are produced by blade motion (i.e., flapping). With reference to Fig. 3.22, a perturbation in velocity $y\dot{\beta}$ is produced as a result of the blade flapping velocity about a hinge, with another perturbation $\mu\Omega R\beta\cos\psi$ produced because of blade flapping displacements (coning). The origin of blade flapping is considered in detail in Chapter 4. Therefore, the velocity perpendicular to the disk can be written as

$$U_P(y, \psi) = (\lambda_c + \lambda_i)\Omega R + y\dot{\beta}(\psi) + \mu\Omega R\beta(\psi)\cos\psi. \tag{3.160}$$

Also, there is a radial velocity component parallel to the axis of the blade and this is given by

$$U_R(\psi) = \mu\Omega R\cos\psi. \tag{3.161}$$

In the BET the aerodynamic effects resulting from the radial velocity are neglected. This is in accordance with the independence principle of sweep, which states that the aerodynamics result only because of the velocity components and angle of attack perpendicular to the leading edge of the blade (see Section 9.7). However, the effects of the radial velocity component along the blade may need to be considered when estimating the rotor drag.

In nondimensional form, the three preceding equations can be written as

$$\frac{U_T}{\Omega R} = \left(\frac{y}{R} + \mu\sin\psi\right) = (r + \mu\sin\psi), \tag{3.162}$$

$$\frac{U_P}{\Omega R} = \left(\lambda + \frac{y\dot{\beta}}{\Omega R} + \mu\beta\cos\psi\right) = \left(\lambda + \frac{r\dot{\beta}}{\Omega} + \mu\beta\cos\psi\right), \tag{3.163}$$

$$\frac{U_R}{\Omega R} = \mu\cos\psi. \tag{3.164}$$

3.4.2 *Induced Velocity Field*

Besides accounting for the effects of blade pitch and flapping motion, the blade element method in forward flight requires an estimate of the induced velocity field. This is not known a priori because it is based on a knowledge of the rotor wake, which in turn depends on the rotor thrust, the blade flapping and overall trim state (i.e., blade pitch angles), and the distribution of airloads over the blades. The effects of the individual tip vortices tend to produce a highly nonuniform inflow over the rotor disk, and the calculation of these effects is a formidable undertaking (see Chapter 10). Nevertheless, the performance of the rotor can be analyzed with the aid of simpler models that represent the basic effects on the inflow resulting from the rotor wake. These models are called "inflow" models and can be formulated on the basis of experimental results or more advanced vortex theories. Because of their simplicity, inflow models have found great utility in many problems in rotor aerodynamics, aeroelasticity, and flight dynamics.

Linear Inflow Models

A remarkable in-flight experiment to measure the time-averaged induced velocity over the rotor disk in forward flight was made by Brotherhood & Stewart (1949). Based on measurements of the angular displacements of smoke 'streamers' introduced upstream of the rotor, the longitudinal inflow variation was determined to be approximately linear. Some interesting aspects of the vortical wake can also be seen in the experiments. Similar results documenting the approximately linear longitudinal variation in the inflow were deduced by Heyson & Katsoff (1957). Since then, many experiments have confirmed the complicated nature of the inflow. During the transition from hover into forward flight, that is, within the range $0.0 \leq \mu \leq 0.1$, the induced velocity in the plane of the rotor is the most nonuniform, it being strongly affected by the presence of discrete tip vortices that sweep downstream near the rotor plane. In higher speed forward flight ($\mu > 0.15$), the time-averaged longitudinal inflow becomes more linear and can be approximately represented by the variation

$$\lambda_i = \lambda_0 \left(1 + k_x \frac{x}{R} \right) = \lambda_0 \left(1 + k_x r \cos \psi \right), \tag{3.165}$$

which is a form first suggested by Glauert (1926). The coefficient λ_0 is the mean (average) induced velocity at the center of the rotor as given by the standard (uniform) momentum theory where

$$\lambda_i = \lambda_0 = \frac{C_T}{2\sqrt{\mu^2 + \lambda_i^2}}. \tag{3.166}$$

Glauert suggested that $k_x = 1.2$, so that there is a small upwash at the leading edge of the rotor and an increase in downwash relative to the average value all along the trailing edge. A variation of Glauert's result is to consider both a longitudinal and lateral variation in the inflow. In this case

$$\begin{aligned} \lambda_i &= \lambda_0 \left(1 + k_x \frac{x}{R} + k_y \frac{y}{R} \right) \\ &= \lambda_0 \left(1 + k_x r \cos \psi + k_y r \sin \psi \right). \end{aligned} \tag{3.167}$$

Here k_x and k_y can be viewed as weighting factors and represent the deviation of the inflow from the uniform value predicted by the simple momentum theory. From experiments with trimmed rotors, it is generally found that the inflow is heavily biased toward the rear of the disk and weakly biased toward the retreating side.

Various attempts have been made to directly calculate k_x (and k_y). One estimate can be found using an adaptation of linear momentum theory, as described by Payne (1959) and Johnson (1980) (see also Question 3.14). Weighting factors can also be estimated using

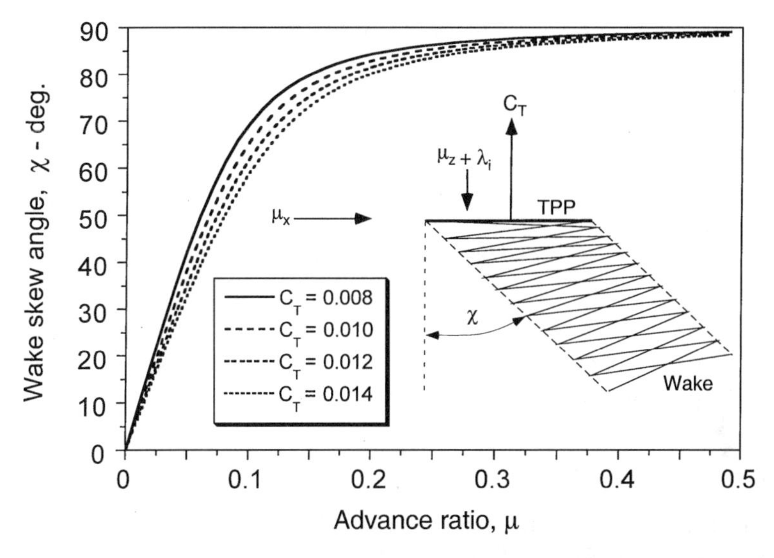

Figure 3.23 Typical variation in rotor wake skew angle with thrust and advance ratio.

rigid cylindrical vortex wake theories – see Coleman et al. (1945) and Johnson (1980) for a summary. One approximation for k_x is

$$k_x = \tan\left(\frac{\chi}{2}\right), \tag{3.168}$$

where χ is the *wake skew angle* and is given by

$$\chi = \tan^{-1}\left(\frac{\mu_x}{\mu_z + \lambda_i}\right), \tag{3.169}$$

and where μ_x and μ_z are advance ratios defined parallel and perpendicular to the rotor disk (see Fig. 3.23). It is apparent that the skew angle increases rapidly with advance ratio, and for $\mu > 0.2$ the wake is relatively flat. Note that for high speed forward flight, k_x approaches unity according to the above expression and does not represent the small region of upwash that is usually measured at the leading edge of the disk. Another simple linear inflow model frequently employed in basic rotor analyses is attributed to Drees (1949). In this model the coefficients of the linear part of the inflow are obtained from another variation of vortex theory – see Johnson (1980) for a good summary. The inflow coefficients are given by

$$k_x = \frac{4}{3}\left(\frac{1 - \cos\chi - 1.8\mu^2}{\sin\chi}\right), \tag{3.170}$$

$$k_y = -2\mu. \tag{3.171}$$

Drees's model gives $k_x = 0$ for $\mu = 0$, a maximum value of 1.11 at $\mu \approx 0.2$, and k_x slowly decreasing thereafter. Like the other linear inflow models, Drees's model is easy to implement in rotor analyses and gives a reasonably good description of the rotor inflow. Various other authors have suggested values for k_x, which are summarized in Table 3.1. Payne (1959) suggests a value for k_x based on the numerical results of Castles & De Leeuw (1954), which approaches 4/3 at high advance ratios, and in light of the experimental evidence appears to be one of the better representations of the longitudinal inflow distribution. Overall, the Drees (1949), Payne (1959), and Pitt & Peters (1981) models are found to give

Table 3.1. *Various Estimated Values of First Harmonic Inflow*

Author(s)	k_x	k_y
Coleman et al. (1945)	$\tan(\chi/2)$	0
Drees (1949)	$(4/3)(1 - \cos\chi - 1.8\mu^2)/\sin\chi$	-2μ
Payne (1959)	$(4/3)(\mu/\lambda/(1.2 + \mu/\lambda))$	0
White & Blake (1979)	$\sqrt{2}\sin\chi$	0
Pitt & Peters (1981)	$(15\pi/23)\tan(\chi/2)$	0
Howlett (1981)	$\sin^2\chi$	0

the best representation of the inflow gradient as functions of wake skew angle and advance ratio when compared to experimental evidence.

Inflow Model of Mangler & Squire

Another inflow model that has found some use in rotor analyses is that developed by Mangler (1948) and Mangler & Squire (1950). The method uses the incompressible, linearized, Euler equations to relate the pressure field across the disk to an inflow. This theory is partially summarized by Bramwell (1976) and Stepniewski & Keys (1984). Mangler & Squire assume that the loading on the rotor disk can be expressed as a linear combination of two fundamental forms: Type I, which is an elliptical loading, and Type III, which is a loading that vanishes at the edges and center of the disk. The pressure loading can be written as

$$\Delta p_m \propto r^{m-1}\sqrt{1-r^2}, \qquad m = 1, 3, \tag{3.172}$$

with r being the distance from the rotational axis. These two pressure distributions correspond to extreme forms of the disk loading, which in the real case will typically comprise a combination of these two loadings, depending on the blade twist and flight conditions.[7] The resulting inflow can be described by the Fourier series

$$\lambda_i = \left(\frac{2C_T}{\mu}\right)\left[\frac{c_0}{2} + \sum_{n=1}^{\infty}(-1)^n c_n(r,\alpha)\cos n\psi\right], \tag{3.173}$$

where α = the disk angle of attack. The coefficients in this equation depend on the assumed form of the disk loading. For Type I loading then

$$c_0 = \frac{3}{4}\nu \quad \text{and} \quad c_1 = -\frac{3\pi}{16}\sqrt{1-\nu^2}\left(\frac{1-\sin\alpha}{1+\sin\alpha}\right)^{1/2}, \tag{3.174}$$

where $\nu^2 = 1 - r^2$. For even values of $n \geq 2$,

$$c_n = (-1)^{\frac{n-2}{2}}\left(\frac{3}{4}\right)\left(\frac{\nu+n}{n^2-1}\right)\left(\frac{1-\nu}{1+\nu}\right)^{\frac{n}{2}}\left(\frac{1-\sin\alpha}{1+\sin\alpha}\right)^{\frac{n}{2}}, \tag{3.175}$$

and for odd values of $n \geq 3$, $c_n = 0$. The inflow obtained from this loading is approximately linearly distributed longitudinally across the disk. For Type III loading, the coefficients

[7] Type I ($m = 1$) elliptic loading is consistent with Glauert's high speed approximation to the rotor problem.

are

$$c_0 = \frac{15}{8}\nu(1-\nu^2), \tag{3.176}$$

$$c_1 = -\frac{15\pi}{256}(5-9\nu^2)\sqrt{1-\nu^2}\left(\frac{1-\sin\alpha}{1+\sin\alpha}\right)^{1/2}, \tag{3.177}$$

$$c_3 = \frac{45\pi}{256}(1-\nu^2)^{3/2}\left(\frac{1-\sin\alpha}{1+\sin\alpha}\right)^{3/2}. \tag{3.178}$$

For even values of $n \geq 2$,

$$c_n = (-1)^{\frac{n-2}{2}}\left(\frac{15}{8}\right)\left[\left(\frac{\nu+n}{n^2-1}\right)\frac{9\nu^2+n^2-6}{n^2-9} + \frac{3\nu}{n^2-9}\right] \times \left(\frac{1-\nu}{1+\nu}\right)^{\frac{n}{2}}\left(\frac{1-\sin\alpha}{1+\sin\alpha}\right)^{\frac{n}{2}}, \tag{3.179}$$

and for odd values of $n \geq 5$, $c_n = 0$. Typical results from the Mangler & Squire theory are shown in Fig. 3.24, where the longitudinal and lateral inflow distributions across the disk are plotted for $\alpha = 0$ and -15 degrees. For Type I disk loading, the inflow is exactly linear for $\alpha = 0$ and is only a weak function of α. Type III disk loading gives a more nonuniform distribution of inflow, with zero at the center of the disk. Both forms of the assumed disk loading give a lateral distribution of inflow that is symmetric with respect to the longitudinal axis of the rotor. Bramwell (1976) uses the high speed approximation to the inflow and replaces the leading $2C_T/\mu$ term in Eq. 3.173 by $4\lambda_0$, where λ_0 is the mean inflow from momentum theory. This implies that the Mangler & Squire theory is valid through hover, but clearly it is not because it violates the high speed assumptions made in the original work that $v_i \ll V_\infty$. Therefore, the theory should be used only for advance ratios greater than 0.10. In applying the Mangler & Squire theory, it can usually be assumed that the loading on the rotor is described by a linear combination of Type I and Type III loadings, that is,

$$\Delta p = w_1 \Delta p_1 + w_3 \Delta p_3, \qquad w_1 + w_2 = 1. \tag{3.180}$$

This is the main disadvantage of theory, which requires the aerodynamic loading on the rotor to be known or assumed a priori. Based on downwash measurements behind a rotor, Fail & Eyre (1954) show that the downwash behind the advancing side of the rotor corresponds to Type I loading, whereas from behind the retreating blade the downwash corresponds closely to Type III loading.

The results from the various inflow models described above are compared with inflow measurements over the rotor disk in Fig. 3.25. These data are taken from Elliott et al. (1988a,b,c), which were measured one chord above the rotor plane using a laser Doppler velocimetry system. Because the experiment used both a rotor and a fuselage, the measured results cannot be considered entirely representative of an isolated rotor. The rotor was trimmed such that the TPP was perpendicular to the rotor shaft, with a forward shaft tilt of -3 degrees. Figure 3.25 shows that the inflow along the longitudinal axis of the rotor is reasonably well described by linear inflow models such as that of Drees, or with the Type I loading in the Mangler & Squire model. The main discrepancies are at the leading and trailing edges of the disk, and also near the rotor hub.

Using a simple average of Type I and Type III loadings ($w_1 = w_3 = 0.5$), Mangler & Squire's theory gives a somewhat better description of the longitudinal inflow, especially over the back of the rotor. Strictly speaking the coefficients of the Mangler & Squire theory

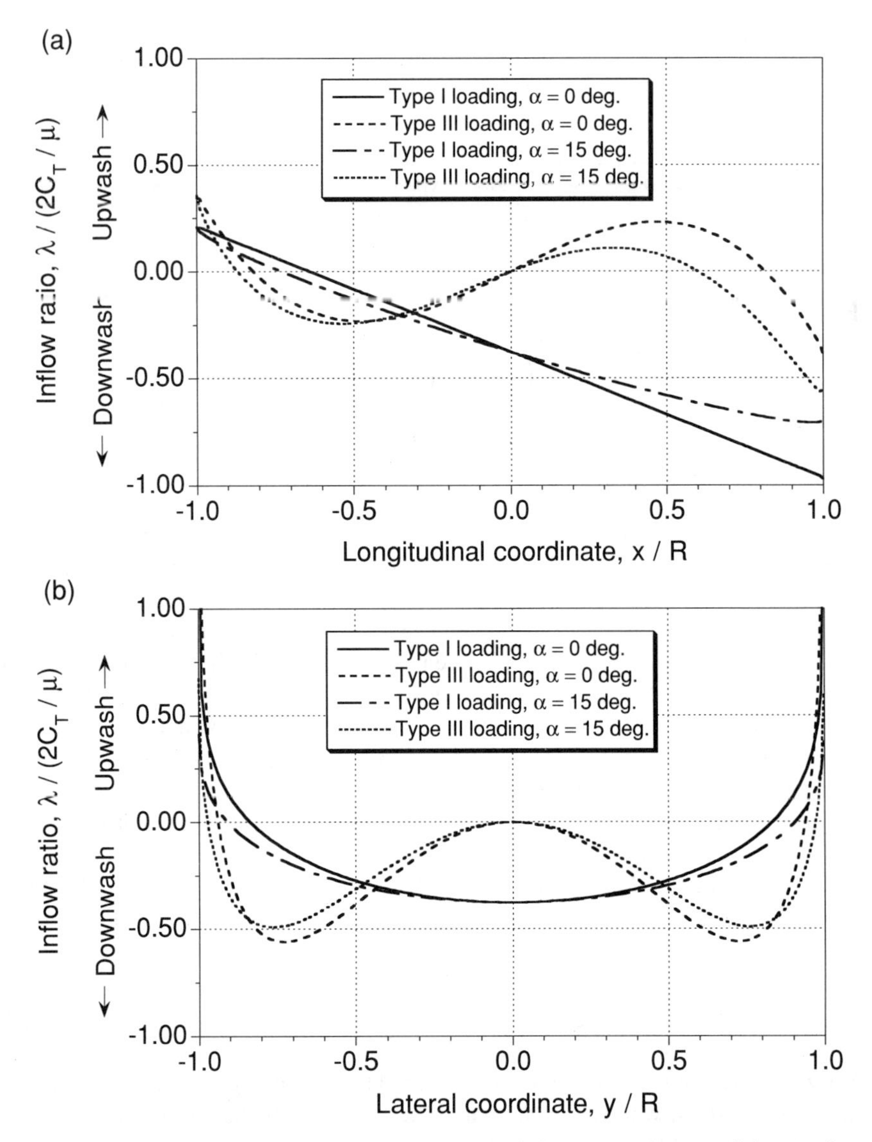

Figure 3.24 Variation in the longitudinal and lateral inflow across the rotor disk according to the theory of Mangler & Squire. (a) Longitudinal inflow for Types I and III loading. (b) Lateral inflow for Types I and III loading.

are obtained by solving for the rotor loading (Δp) at the given trim state by means of the BET. Vortex theory (which will be described in Chapter 10) does much better at the leading and trailing edges of the disk and agrees with the linear inflow model over the remainder of the disk. The somewhat larger discrepancies found between all of the theories and the measurements with increasing advance ratio are almost certainly a result of perturbations to the velocity field caused by the rotor hub and the fuselage below the rotor. Because the mean rotor inflow decreases slightly with increasing advance ratio and the fuselage perturbations increase, the net effect is that a larger fraction of the total inflow is affected by the presence of the fuselage. Such effects are not easy to calculate but at least must be carefully recognized as a contributing factor to most rotor inflow measurements. The

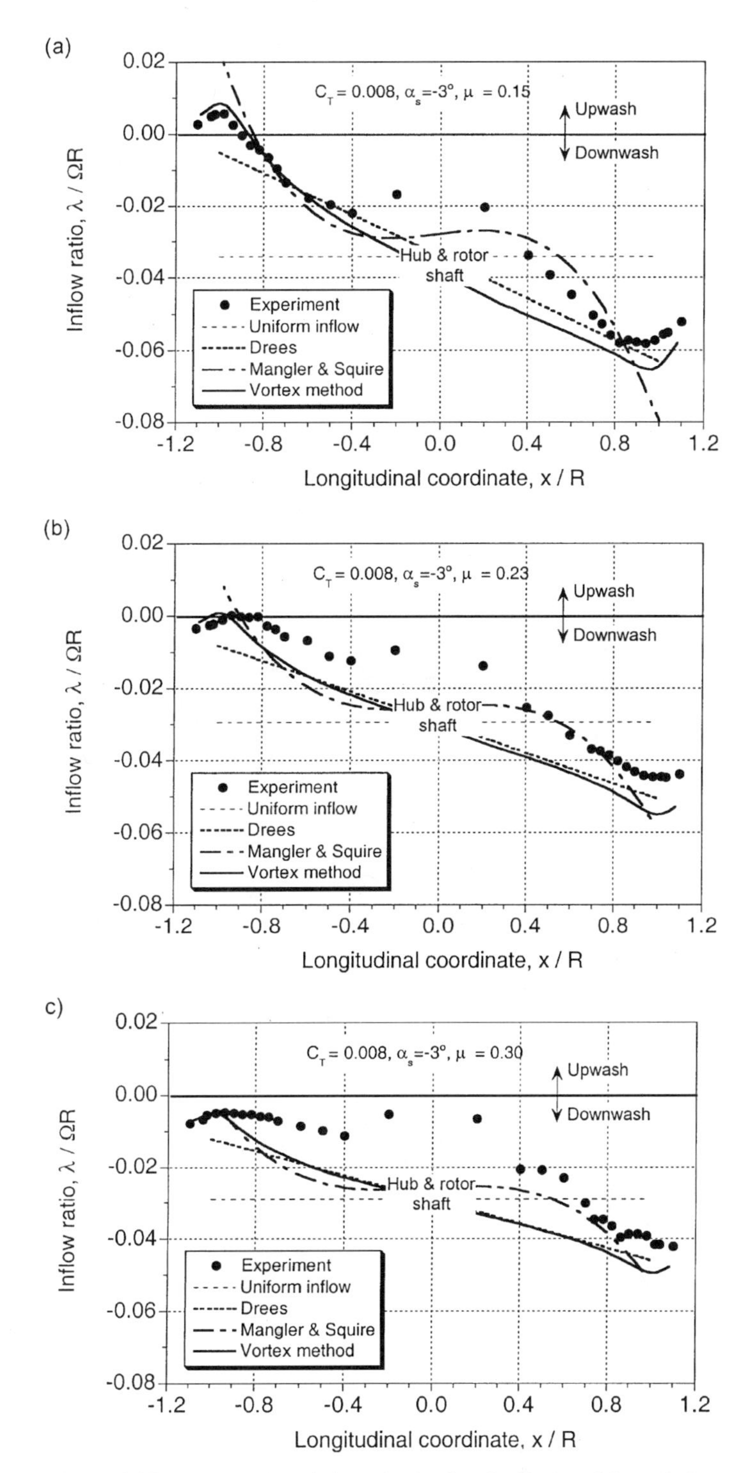

Figure 3.25 Measured variations in the longitudinal and lateral time-averaged inflow across the disk in forward flight compared to inflow models. Measurements made one chord above TPP. (a) Longitudinal inflow at $\mu = 0.15$. (b) Longitudinal inflow at $\mu = 0.23$. (c) Longitudinal inflow at $\mu = 0.30$. (d) Lateral inflow at $\mu = 0.15$. (e) Lateral inflow at $\mu = 0.23$. (f) Lateral inflow at $\mu = 0.30$.

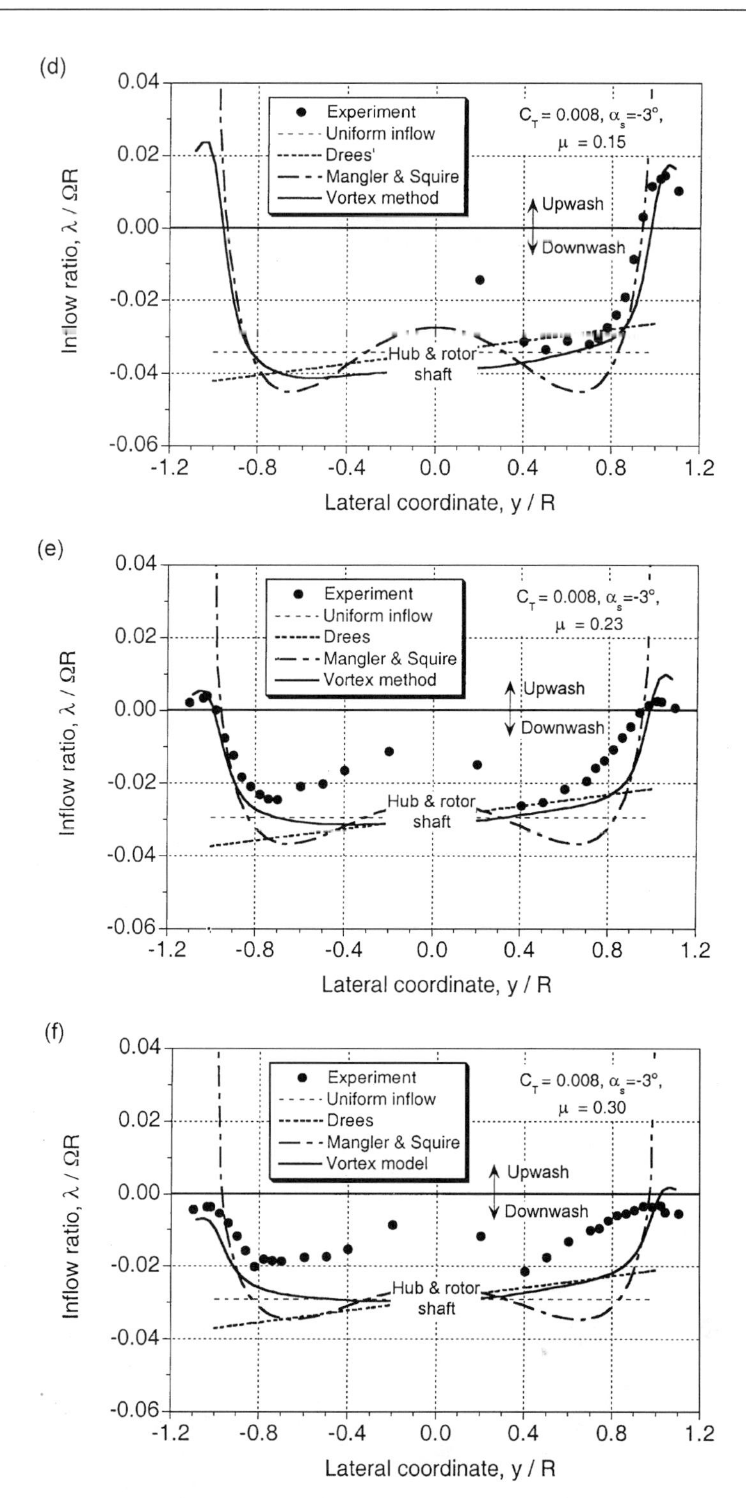

Figure 3.25 *(Continued)*

lateral distribution of inflow across the disk is found to be relatively uniform compared to the longitudinal variation but decreases rapidly near the edges of the disk. Outside the edges of the disk, there is an upwash velocity. Drees's model is certainly not unrealistic, but the Mangler & Squire theory gives a much better description of the inflow near the edges of the disk and agrees closely with the results from vortex theory. Again, the influence of the fuselage and rotor hub are responsible for some of the discrepancies seen between theory and experiment.

Ormiston (1972) describes a more general inflow formulation in the spirit of Mangler & Squire's approach, where the rotor loading and inflow are solved more consistently. For a simple uniform bound circulation distribution on the blades, the inflow results appear substantially similar to those obtained with Mangler & Squire's theory. Other more recent developments of the actuator disk inflow model are described by Peters et al. (1987) and Peters & He (1989, 1995). In this theory, the acceleration potential and inflow distribution are expressed in the form of an infinite series of shape functions. When these are substituted into the incompressible, linearized Euler equations, a set of first-order ordinary differential equations are obtained. An aerodynamic loading model is required to solve the equations, which is based on a standard blade element approach. These methods are still in a state of ongoing development but have been shown to give improvements in predictive capability for many rotor problems, especially those involving rotor aeroelasticity, and they avoid the need to explicitly model the complicated nature of the true vortex wake.

3.5 Chapter Review

The blade element approach is a powerful tool for the aerodynamic analysis of helicopter rotors. It forms the basis for nearly all modern computational methods used for performance, airloads, and aeroelastic analyses. The basic ideas consist of representing the airloads on two-dimensional sections of the blades and integrating their effect to find the performance of the rotor as a whole. This allows tremendous flexibility in the rotor analysis and also allows the effects of airfoil shape to be examined, along with the effects of Reynolds number and Mach number, and even some elementary effects associated with nonlinear aerodynamics and stall.

On the basis of certain assumptions, it has been shown how the combined blade element momentum theory (BEMT) can provide analytic results about how to design the rotor in terms of optimum blade planform and blade twist distribution to enable maximum hovering efficiency. Whereas the simple momentum theory shows that the rotor should be designed for low disk loading, the blade element approach allows the trade-offs associated with the interrelated effects of disk loading, blade tip speed, blade loading, blade twist, and blade planform to be examined. While the BEMT is by no means complete, it paves the way for initial studies in rotor design to meet specified requirements.

In forward flight, the blade element theory (BET) allows for the calculation of the nonaxisymmetric airloads over the rotor disk. Besides the effects of blade motion (which are considered next in Chapter 4), this chapter has also introduced the ideas that more accurately representing the effects of the rotor wake is one key to the successful use of the BET. The use of other than uniform inflow models provides a better representation of the effects of the rotor wake and a better overall physical picture of the rotor aerodynamics problem. While inflow models are not completely rigorous, especially at low forward flight speeds, they give reasonable descriptions of the induced velocity field over the rotor disk and may be well suited for the analysis of many rotor problems. Their computational simplicity allows for easy integration into blade element based rotor models and may allow for the inclusion of unsteady aerodynamic effects as well (see Chapter 8). However, it must be remembered that in many situations where the individual tip vortices come close to the disk (especially in low speed forward flight or in descents) the induced velocity distribution is considerably more complicated than can be expressed by these simple inflow models. The use of prescribed- or free-vortex methods provides the fidelity necessary under these conditions, albeit at much greater computational expense (see Chapter 10).

However, before the rotor problem in forward flight can be fully enunciated, it is necessary to consider the effects of the rotor blade motion on the aerodynamic forces. Because the aerodynamics and blade motion are intrinsically coupled, they must be solved as a system. This then allows the calculation of rotor trim, that is, the control inputs required to orient the rotor in a direction to meet vehicle lift, propulsion, and control requirements. These issues are considered in the next chapter.

3.6 Questions

3.1. A rotor has a rectangular blade planform, a solidity of 0.1, ideal blade twist, and operates in hover at a thrust coefficient of 0.008. By numerical means, using the combined blade element momentum theory (BEMT), compare the predicted inflow ratio across the span of the blade with the exact (analytic) result. Show also the radial distributions of thrust, induced torque, and lift coefficient, and compare with the solutions obtained using the exact theory. Finally, compare the numerical calculations of the integrated thrust and induced power with the exact results for this rotor over the range $0.0 \leq C_T \leq 0.010$.

3.2. Using the BEMT, implement Prandtl's tip-loss function. For a rotor of solidity 0.1 and with no blade twist, show the effects of the tip-loss effect on the induced inflow, thrust distribution, lift coefficient distribution, and torque distribution across the blade span. Neglect profile drag. Compare the results for rotors with two and four blades at a thrust coefficient of 0.008. Also, calculate the induced power factor for the two rotors, with and without the tip-loss effect, and comment on your results.

3.3. Using the BEMT, show the effect of increasing linear twist on the variations in inflow, thrust, induced power, profile power, and lift coefficient across the span of a rotor with four blades of rectangular planform and solidity 0.1, and operating at a thrust coefficient of 0.008. Assume $C_{d_0} = 0.01$. Include Prandtl tip-loss effects. All of the results should be compared at a constant thrust (disk loading).

3.4. For the same rotor parameters as used in the previous question, compute the power required to hover versus thrust for different values of linear blade twist. Also compute the figure of merit and induced power factor versus thrust coefficient. Assume that the profile drag is given by $C_d = 0.01 + 0.025\alpha + 0.65\alpha^2$. Show your results graphically. Include the Prandtl tip-loss function. Comment on your results.

3.5. Tapering the blade planform can have a powerful impact on the profile power of the rotor. Using the BEMT, show the effect of increasing taper on the variations in inflow, thrust, power, and lift coefficient across the span of a rotor with two blades of linearly tapered planform and with a thrust weighted solidity of 0.1. The blades have 10 degrees of linear nose-down twist. Consider blade taper ratios of 1:1, 2:1 and 3:1. Include Prandtl tip-loss effects. Remember that all of the results should be compared at a constant thrust (disk loading) and thrust weighted solidity. Determine an optimum combination of twist and taper that will give a maximum figure of merit.

3.6. Explain and discuss the potential benefits to be gained by using blade twist, planform taper, low solidity, large radius, and low rotational speed for the main rotor of a heavy lift helicopter that is designed to operate primarily in hover. Discuss any disadvantages associated with these design factors.

3.7. Assuming an optimum hovering rotor, rewrite an expression for the figure of merit in terms of rotor tip speed, disk loading, and airfoil lift-to-drag ratio. Comment on your result.

3.8. A manufacturer has developed a growth version of an existing helicopter that has a higher gross weight. This new aircraft has a main rotor that has the same rotor disk area, but with a higher solidity. Explain qualitatively, quantitatively, or both, what benefits and/or trade-offs may be expected with reference to the hover, forward flight, and maneuver capabilities of the new helicopter.

3.9. Given that the inflow distribution over a rotor with rectangular untwisted blades is approximately triangular, and assuming no tip-loss effects: (a) Compute the variation with radius of inflow angle and section angle of attack for such a rotor. (b) Derive the relation between the blade element lift coefficients and C_T/σ. (c) Derive an expression for the hovering power of such a rotor in terms of C_T/σ and C_{d_0}.

3.10. The inflow distribution over a rotor operating in hover with no tip losses is assumed to be linear; that is, $\lambda = \lambda_{\text{tip}} r$, where λ_{tip} is the value of the inflow at the tip. By considering a momentum balance on successive annuli of the rotor disk, show that the induced power factor for such a rotor is $4\sqrt{2}/5$.

3.11. What is the purpose of using "equivalent" solidities in helicopter rotor performance studies? Explain the potential pitfalls in using such "equivalent" factors with rotors.

3.12. A blade has an inboard portion of constant chord c_r from the root to a point $r = r_1$, and then a linear taper from r_1 to the tip at $r = 1$ where the chord c_t. Show that the equivalent thrust weighted solidity for this blade is given by

$$\sigma_e = (\sigma_r - \sigma_t) r_1^3 + \left(\frac{\sigma_t - \sigma_r}{1 - r_1} \right) \left[-\frac{1}{4} + r_1^3 - \frac{3}{4} r_1^4 \right] + \sigma_t .$$

3.13. For helicopter operations at high speed, it is possible that a "reverse flow" region can exist at the root end of the blade on the retreating side of the rotor disk. Show that the reverse flow region on the rotor disk is a circle of diameter μ with a center located at $(r, \psi) = (\mu/2, 3\pi/2)$.

3.14. By considering a local momentum analysis of a uniformly loaded rotor disk in high-speed forward flight, show that the induced inflow can be written in the form

$$v_i = v_0 \left[1 + K r \cos\psi \right],$$

where the inflow gradient $K = \left[\sqrt{1 - (y/R)^2} \right]^{-1}$ and v_0 is the velocity at the center of the disk.

Bibliography

Althaus, D. 1972. "Stuttgarter Profilkatalog I," Institut für Aerodynamik und Gasdynamik der Universität Sutttgart.

Bagai, A. and Leishman, J. G. 1992. "A Study of Rotor Wake Developments and Wake/Body Interactions in Hover Using Wide-Field Shadowgraphy," *J. American Helicopter Soc.*, 37 (4), Oct., pp. 48–57.

Bailey, F. J. and Gustafson, F. B. 1944. "Charts for Estimation of the Characteristics of a Helicopter Rotor in Forward Flight," NACA ACR L4H07.

Betz, A. 1919. *Schraubenpropeller mit geringstem Energieverlust*, Göttinger Nachrichten, p. 193.

Bienen, T. and von Karman, T. 1924. *Z. Ver. Deutsch. Ing.*, 68, p. 1237.

Bramwell, A. R. S. 1976. *Helicopter Dynamics*, Edward Arnold, London.

Brotherhood, P. and Stewart, W. 1949. "An Experimental Investigation of the Flow through a Helicopter Rotor in Forward Flight," ARC R & M 2734.

Castles, W., Jr. and De Leeuw, J. H. 1954. "The Normal Component of the Induced Velocity in the Vicinity of a Lifting Rotor and Some Examples of Its Application," NACA Report 1184.

Castles, W., Jr. and Durham, H. L., Jr. 1956. "Distribution of Normal Component of Induced Velocity in Lateral Plane of a Lifting Rotor," NACA TN 3841.

Coleman, R. P., Feingold, A. M., and Stempin, C. W. 1945. "Evaluation of the Induced velocity Fields of an Idealized Helicopter Rotor," NACA ARR L5E10.

de Bothezat, G. 1919. "The General Theory of Blade Screws," NACA TR 29.

Drzwiecki, S. 1892. Bulletin de l'Association Technique Maritime.

Drzwiecki, S. 1909. *Théorie Générale de l'Hélice*, Paris.

Drees, J. M. 1949. "A Theory of Airflow Through Rotors and Its Application to Some Helicopter Problems," *J. Helicopter Assoc. Great Britain*, 3 (2), July–Sept.

Durand, W. F. (ed.) 1935. *Aerodynamic Theory*, Vol. IV, Divisions J-M, Julius Springer, Berlin.

Elliott, J. W., Althoff, S. L., and Sailey, R. H. 1988a. "Inflow Measurements Made with a Laser Velocimeter on a Helicopter Model in Forward Flight, Volume I: Rectangular Planform at an Advance Ratio of 0.15," NASA TM 100545.

Elliott, J. W., Althoff, S. L., and Sailey, R. H. 1988b. "Inflow Measurements Made with a Laser Velocimeter on a Helicopter Model in Forward Flight, Volume II: Rectangular Planform at an Advance Ratio of 0.23," NASA TM 100546.

Elliott, J. W., Althoff, S. L., and Sailey, R. H. 1988c. "Inflow Measurements Made with a Laser Velocimeter on a Helicopter Model in Forward Flight, Volume III: Rectangular Planform at an Advance Ratio of 0.30," NASA TM 100547.

Fail, R. A. and Eyre, R. C. W. 1954. "Downwash Measurements behind a 12-ft Diameter Helicopter Rotor in the 24-ft Wind Tunnel," ARC R & M 2810.

Felker, F. F. and McKillip, R. M. 1994. "Comparisons of Predicted and Measured Rotor Performance in Vertical Climb and Descent," 50th Annual Forum of the American Helicopter Soc., Washington DC, May 11–13.

Gessow, A. 1948. "Effect of Rotor-Blade Twist and Plan-Form Taper on Helicopter Hovering Performance," NACA Technical Note 1542.

Gessow, A. and Meyers, G. C. 1952. *Aerodynamics of the Helicopter*, Macmillan Co. (republished by Frederick Ungar Publishing, New York, 1967), Chapter 4.

Glauert, H. 1922. "An Aerodynamic Theory of the Airscrew," ARC R & M 786.

Glauert, H. 1926. "A General Theory of the Autogiro," ARC R & M 1111.

Goldstein, L. 1929. "On the Vortex Theory of Screw Propellers," *Proc. of the Royal Soc.*, Series A 123, p. 440.

Gustafson, F. B. and Gessow, A. 1945. "Flight Tests on the Sikorsky HNS-1 (Army YR-4B) Helicopter," NACA MR L5D09a.

Gustafson, F. B. and Gessow, A. 1946. "Effect of Rotor Tip Speed on Helicopter Rotor Performance and Maximum Forward Speed," NACA ARR No. L6A16.

Harris, F. D. 1987. "Rotary Wing Aerodynamics: Historical Perspectives and Important Issues," American Helicopter Soc. Specialists' Meeting on Aerodynamics and Aeroacoustics, Fort Worth, TX.

Heyson, H. H. and Katsoff, S. 1957. "Induced Velocities Near a Lifting Rotor with Nonuniform Disk Loading," NACA Technical Report 1319.

Houghton, E. L. and Carpenter P. W. 1993. *Aerodynamics for Engineering Students*, John Wiley & Sons, New York.

Howlett, J. J. 1981. "UH-60A Blackhawk Engineering Simulation Program: Vol. 1 – Mathematical Model," NASA CR-66309.

Johnson, W. 1980. *Helicopter Theory*, Princeton University Press, Princeton, NJ, Chapter 4.

Jones, R. T. and Cohen, D. 1957. "Aerodynamics of Wings at High Speeds," Vol. VII of *High Speed Aerodynamics and Jet Propulsion*, Section A, Chapter 1, pp. 36–48, Aerodynamic Components of Aircraft at High Speeds, A. F. Donovan & H. R. Lawrence (eds.), Princeton University Press, Princeton, NJ.

Knight, M. and Hefner, R. A. 1937. "Static Thrust of the Lifting Airscrew," NACA TN 626.

Liberatore, E. K. 1998. *Helicopters before Helicopters*, Krieger Publishing, Malabar, FL, pp. 60–61, 249.

Lock, C. N. H., Bateman, H., and Townend, H. 1925. "An Extension of the Vortex Theory of Airscrews with Applications to Airscrews of Small Pitch and Including Experimental Results," ARC R & M 1014.

Lock, C. N. H. 1930. "The Application of Goldstein's Theory to the Practical Design of Airscrews," ARC R & M 1377.

Loftin, L. K. and Smith, H. A. 1949. "Aerodynamic Characteristics of 15 NACA Airfoil Sections at Seven Reynolds Numbers from 0.7×10^6 to 9×10^6," NACA Technical Note 1945.

Mangler, K. W. 1948. "Calculation of the Induced Velocity Field of a Rotor," Royal Aircraft Establishment, Report Aero 2247.

Mangler, K. W., and Squire, H. B. 1950. "The Induced Velocity Field of a Rotor," ARC R & M 2642.

McVeigh, M. A. and McHugh, F. J. 1982. "Recent Advances in Rotor Technology at Boeing Vertol," 38th Annual Forum of the American Helicopter Soc., Anaheim, CA, May 4–7.

Miley, S. J. 1982. "A Catalog of Low Reynolds Number Airfoil Data for Wind Turbine Applications," US Department of Energy, Wind Energy Division, RFP-338, UC-60.

Ormiston, R. A. 1972. "An Actuator Disk Theory for Rotor Wake Induced Velocities," AGARD-CP-111, Aerodynamics of Rotary Wings.

Payne, P. R. 1959. *Helicopter Dynamics and Aerodynamics*, Pitman & Sons, London.

Peters, D. A. and Ormiston, R. A. 1975. "Flapping Response Characteristics of Hingeless Rotor Blades by a Generalized Harmonic Balance Method," NASA TN D-7856.

Peters, D. A., Boyd, D. D., and He, C. J. 1987. "Finite-State Induced-Flow Model for Rotors in Hover and Forward Flight," 43rd Annual Forum of the American Helicopter Soc., St. Louis, MO, May 18–20.

Peters, D. A. and He, C. J. 1989. "Correlation of Measured Induced Velocities with a Finite-State Wake Model," 45th Annual Forum of the American Helicopter Soc., Boston, MA, May 22–24.

Peters, D. A. and He, C.-J. 1995. "Finite-State Induced Flow Models Part II: Three-Dimensional Rotor Disk," *J. of Aircraft*, 32 (2), March–April, pp. 323–333.

Pitt, D. M. and Peters, D. A. 1981. "Theoretical Prediction of Dynamic Inflow Deriviatives," *Vertica*, 5, pp. 21–34.

Prouty, R. W. 1986. *Helicopter Performance, Stability, and Control*, Prindle, Weber & Schmitdt, PWS Engineering, Boston. pp. 97–101.

Reissner, H. 1910. *Z. Flugtech. Motorluftschiffahrt*, 1, pp. 257–309.

Reissner, H. 1937. "On the Vortex Theory of the Screw Propeller," *J. Aeronaut. Sci.*, 5 (1), pp. 1–6.

Reissner, H. 1940. "A Generalized Vortex Theory of the Screw Propeller and Its Application," NACA TN 750.

Stepniewski, W. Z. and Keys, C. N. 1984. *Rotary-Wing Aerodynamics*, Dover, New York, Part 1, Chapter 5.

Theodorsen, T. 1948. *Theory of Propellers*, McGraw-Hill, New York.

Tokaty, G. A. 1971. *A History and Philosophy of Fluid Mechanics*, Foulis & Co., Henley-on-Thames, England.

White, F. and Blake, B. B. 1979. "Improved Method of Predicting Helicopter Control Response and Gust Sensitivity," 35th Annual Forum of the American Helicopter Soc., Washington DC, May 21–23.

CHAPTER 4

Rotating Blade Motion

It is sufficient to say that it [the dissymmetry of lift on the rotor] is completely eliminated by the articulation of the rotor blades, which no longer set up a powerful inherent force which must be overcome in order to control the craft.

Juan de la Cierva (1931)

4.1 Introduction

The interdependent coupling of the aerodynamic forces produced on the blades and the rotor blade motion is the key to understanding the behavior and control of the rotor system. A distinctive feature of helicopter rotors is that articulation in the form of flapping and lead/lag hinges is incorporated at the root of each blade. These may be mechanical hinges, or modern rotor hub designs may use hingeless flexures that allow motion about a "virtual" hinge location. In either case, the hinges allow each blade to independently flap and lead/lag with respect to the hub plane under the action of varying aerodynamic lift and drag loads (see Fig. 4.1). The idea of a flapping hinge for a rotor was first patented by Bartha & Madzer in 1912. However, the flapping hinge was first successfully used by Juan de la Cierva on his autogiros. The addition of a lead/lag hinge allows in-plane motion of the blade in response to the Coriolis forces that are produced when the radius of gyration of the blade changes by virtue of flapping. A pitch bearing is also incorporated into the blade design to allow the blades to *feather*, providing an ability to change their pitch. This can be done collectively (together), thereby changing the magnitude of the rotor thrust, or cyclically, that is, with respect to blade azimuth, thereby changing the phasing of the aerodynamic loads over the disk. The latter allows the rotor disk to tilt so as to reorient the rotor thrust vector and provide the pilot with pitch and roll control.

In hovering flight, the flow field is azimuthally axisymmetric and so each blade encounters the same aerodynamic environment. The blades flap up and lag back with respect to the hub and reach a steady equilibrium position under the action of these steady (nonazimuthally varying) aerodynamic and centrifugal forces. The blades will cone up to form a static balance between the blade aerodynamic forces and the centrifugal forces, and the rotor disk plane takes on some orientation in space. The centrifugal forces are dominant, and so the coning angles on helicopter rotors always remain relatively small (a few degrees). In addition, the aerodynamic drag forces on the blades cause them to lag back. Because the drag forces are only a fraction of the lift forces and are overpowered by the centrifugal forces, the lag angle displacements are even smaller than the coning angles.

In forward flight, the asymmetry of the onset flow and dynamic pressure over the disk produces aerodynamic forces that are functions of blade azimuth position (i.e., cyclically varying airloads are now produced). The flapping hinge allows each blade to freely flap up and down in a periodic manner with respect to azimuth angle under the action of these varying aerodynamic loads. The blades reach an equilibrium condition again when the local changes in angle of attack and the aerodynamic loads produced as a result of blade flapping are sufficient to compensate for the local changes in the airloads resulting from variations

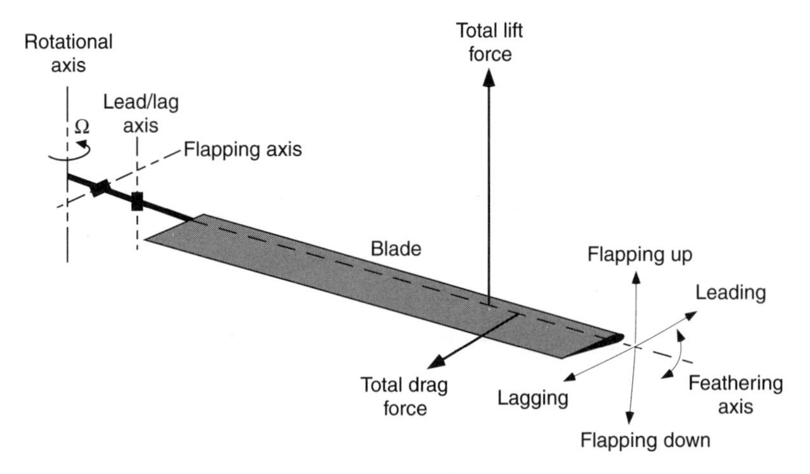

Figure 4.1 Schematic showing flapping, lead/lag, and feathering motion of a rotor blade.

in dynamic pressure. The rotor disk, therefore, naturally takes up a new orientation in inertial space.

Some of the first theoretical studies of blade flapping and rotor response were undertaken in Great Britain by Lock (1927), Glauert (1928), and Squire (1936), motivated initially by the success of the autogiro. In the United Sates, Wheatly (1934), Sissingh (1939, 1941), Bailey (1941), and Wald (1943) were responsible for most of the initial work in understanding rotor blade flapping behavior in forward flight. Nikolsky (1944, 1951) gave one of the first formal treatments of the various problems in rotor dynamics and the effect on rotor performance. Bramwell (1976) and Johnson (1980) provide the most thorough textbook expositions of rotating blade dynamics and rotor response. Loewy (1969), Reichert (1973), Friedmann (1977), and Chopra (1990) give good reviews of the complicated coupled dynamics and various aeroelastic problems associated with helicopter rotors.

4.2 Types of Rotors

There are basically four types of helicopter rotor hubs in use. These are the teetering design, the articulated design, the hingeless design, and the bearingless design. A teetering rotor has two blades that are hinged at the rotational axis (i.e., on the shaft) and uses no independent flap or lead/lag hinges. The well-known Bell (Young) design shown in Fig. 4.2(a) is an example of a simple teetering rotor design. The blades are connected together, so that as one blade flaps up the other flaps down like a seesaw or teeter board. A separate pitch or feathering bearing on each blade allows for cyclic and collective pitch capability. The teetering design has the advantage of being mechanically simple with a low parts count, and it is easy to maintain. One disadvantage of the design is that it has a relatively high parasitic drag in forward flight.

A variation of the teetering design is the underslung teetering hub, an example being used on the Bell-Huey series of helicopters. Here, the blades are given a pre-cone angle so that a downward component of the centrifugal forces produced by blade rotation eliminates the upward bending moment at the hub resulting from the aerodynamic loads. However, any pre-cone displacement will move the center of gravity of the rotor system above the axis of the teetering hinge, and this can introduce an undesirable coupling into the rotor system through Coriolis effects. To prevent this, the rotor is underslung below the teetering hinge

(a) Teetering hub design (Bell UH-1)

(b) Articulated hub (Sikorsky S-55)

(c) Articulated hub (Boeing CH-46)

(d) Bearingless hub (Aerospatiale Dauphin)

Figure 4.2 Various types of rotor hubs. (a) Bell teetering hub design. Note the bob weights attached to the stabilizer bar. (b) An articulated hub, in this case with coincident flapping and lag hinges. (c) An articulated hub, with the lag hinge outboard of both the flap and feathering bearing. (d) A bearingless hub, where the mechanical hinges and bearings are replaced by flexures.

to minimize the center of gravity movement relative to the shaft. A problem, however, that has arisen with underslung teetering rotor designs is mast bumping, which occurs when the rotor is lightly loaded such as in a push-over maneuver and excessive blade flapping causes the blade hub to contact the rotor shaft. While snubbers are used to help prevent this, repetition of the problem can result in damage to the rotor shaft. Another variation of the teetering design, called the gimballed hub, is used on tilt-rotor aircraft such as the V-22, where the three blades and the hub are attached to the rotor shaft by means of a gimbal or universal joint.

A large number of helicopters use conventional or fully articulated rotor hubs. Here, mechanical flap and lead/lag hinges are provided on each blade along with a feathering bearing. Different helicopters use various sequences of hinges and bearings, and this affects the dynamics of the rotor system. Many Sikorsky helicopters use coincident flap and lead/lag hinges, with the feathering bearing located further outboard (see Fig. 4.2(b)). The Boeing CH-46 and CH-47 machines use a lead/lag hinge outboard of the feathering bearing (see Fig. 4.2(c)). In both cases, because of the relatively low drag and aerodynamic damping in the lead/lag plane, mechanical dampers are fitted at the lag hinges. Needless to say, the articulated rotor design is mechanically complicated with a high parts count and is expensive to maintain. It is also heavy and produces relatively high drag in forward flight. Nevertheless, the fully articulated design is the classic approach to providing blade articulation, and has proven extremely robust and reliable.

A hingeless rotor design eliminates the flap and lead/lag hinges by using a flexure to accommodate blade motion. A feathering bearing is still used to allow for pitch changes on each blade. The advantage of a hingeless design is that it is mechanically simple. However, because blade articulation is achieved by the elastic flexing of a structural beam, the design of such rotors is rather complicated. In addition, because it is not possible to completely isolate the flapping motion from the lead/lag motion, there is usually significant flap/lag coupling to contend with. Advantages of the hingeless hub design include mechanical simplicity with a low parts count and low aerodynamic drag. In addition, the relatively stiff hub design gives the helicopter outstanding maneuvering capability in response to control inputs.

Bearingless hub designs are a relatively new innovation. In addition to eliminating the mechanical flap and lead/lag hinges, the bearingless hub also eliminates the feathering bearing. All three degrees of motion are obtained by bending, flexing, and twisting of the hub structure. Needless to say, obtaining the required stiffnesses is only possible by means of new high strength composite materials such as glass, carbon, and kevlar, which can be arranged so that load paths, stiffnesses, and couplings can all be controlled. Designing a bearingless hub is difficult and requires a finite-element based structural dynamic analysis. While retaining all the advantages of a hingeless design, bearingless hubs can be particularly susceptible to aeromechanical instabilities as a result of low in-plane damping. Examples of bearingless helicopter hub designs are found on the Aerospatiale Dauphin (see Fig. 4.2(d)), the Eurocopter BK-117, and the McDonnell-Douglas MD-640 Explorer.

4.3 Equilibrium about the Flapping Hinge

Consider now the mathematical analysis of a rotating flapping blade. Figure 4.3 shows a rigid rotor blade flapping about a hinge located at a distance eR from the rotational axis. The equilibrium position of the blade is determined by the balance of aerodynamic and centrifugal forces (CF). Because the centrifugal forces are much larger than the aerodynamic forces, the flapping angle β is usually quite small (between 3–6 degrees is typical for a helicopter rotor).

Assume first for simplicity that the flapping hinge is at the rotational axis, that is, $e = 0$. For other than teetering rotor designs, the flapping hinge is offset from the rotational axis, but this does not alter the basic physics of the problem. The rotational speed about the axis is Ω radians/s and constant. Assume a uniform mass per unit length of the blade, m. Consider a small element of the blade of length dy. The mass of this element is $m dy$. The contribution of this small element to the centrifugal force acting in a direction parallel to the plane of rotation is

$$d(F_{CF}) = (m dy) y \Omega^2 = m\Omega^2 y \, dy. \tag{4.1}$$

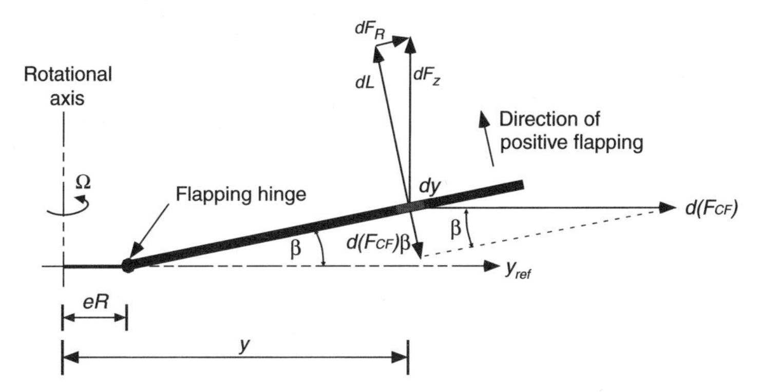

Figure 4.3 Equilibrium of blade aerodynamic and centrifugal forces about the flapping hinge.

Therefore, the total centrifugal force acting on the blade is

$$F_{CF} = \int_0^R m\Omega^2 y\, dy = \frac{m\Omega^2 R^2}{2} = \frac{M\Omega^2 R}{2}, \tag{4.2}$$

where the mass of the blade is $M(= mR)$. Note that the centrifugal forces increase linearly in proportion to blade mass and length and are also proportional to Ω^2. If the blade is coned up at some angle β, then the contribution of this small element to the CF acting in a direction perpendicular to the blade is

$$d(F_{CF}) \sin\beta = (m\,dy) y\Omega^2 \sin\beta \approx my\Omega^2\beta\, dy. \tag{4.3}$$

The moment about the flapping hinge (at the rotational axis) as a result of the centrifugal forces produced by all the elements is, therefore,

$$\begin{aligned} M_{CF} &= \int_0^R m\Omega^2 y^2 \beta\, dy = m\Omega^2\beta \int_0^R y^2\, dy = \frac{m\Omega^2 \beta R^3}{3} \\ &= \frac{M\Omega^2\beta R^2}{3} = \frac{2}{3} F_{CF} R\beta. \end{aligned} \tag{4.4}$$

The aerodynamic moment about the flap hinge, M_β, depends on the distribution of lift across the blade, that is,

$$M_\beta = -\int_0^R Ly\, dy. \tag{4.5}$$

Note that the negative sign indicates that the aerodynamic moment is in the opposite direction to the centrifugal moment. The rotating blade will reach an equilibrium position where the CF moment about the hinge is equal and opposite to the aerodynamic moment about the hinge. Therefore, the equilibrium equation can be written as

$$M_\beta + M_{CF} = 0 \tag{4.6}$$

and so the equilibrium or coning angle β_0 will be given by

$$\beta_0 = \frac{\displaystyle\int_0^R Ly\, dy}{\left(\dfrac{M\Omega^2 R^2}{3}\right)}. \tag{4.7}$$

This result is valid for any form of the aerodynamic loading over the blade. Also, the flapping hinge has been assumed to lie at the rotational axis of the rotor. The important point to remember is that the coning angle increases in proportion to rotor thrust and decreases inversely with centrifugal forces. In other words, increasing either blade mass or rotor speed will decrease the coning angle.

In practice the flapping hinge is not normally at the rotational axis (except for teetering rotors) but is offset by a small distance, eR. Typically, $e < 0.15$. In this case, the aerodynamic moment about the flap hinge, M_β, is

$$M_\beta = -\int_{eR}^R Ly\, dy. \tag{4.8}$$

Also, the moment about the hinge is

$$M_{CF} = \int_{eR}^R m\Omega^2 y^2 \beta\, dy = \frac{m\Omega^2\beta R^3(1-e^3)}{3} = \frac{M\Omega^2\beta R^2(1+e)}{3} + O(e^2). \tag{4.9}$$

where $M = mR(1-e)$. Therefore, in this case the coning angle will be given by

$$\beta_0 = \frac{\displaystyle\int_{eR}^{R} Ly\,dy}{\left(\dfrac{M\Omega^2 R^2(1+e)}{3}\right)}. \tag{4.10}$$

4.4 Equilibrium about the Lead/Lag Hinge

The equilibrium of the blade about the lead/lag hinge is also determined by a balance of centrifugal and aerodynamic moments. In this case, the aerodynamic moments are generated by the aerodynamic drag of the blade as it rotates. The lag angle, ζ, is defined as positive in the lagging direction (see Fig. 4.4). The centrifugal force on the blade element is

$$d(F_{CF}) = m\Omega^2 y\,dy \tag{4.11}$$

and the component of this force perpendicular to the axis of the blade, which tends to rotate the blade toward zero lag angle, is

$$d(F_{CF}) = m\Omega^2 y\,dy\zeta. \tag{4.12}$$

Therefore, with a zero hinge offset the lag moment will be

$$M_\zeta = \int_0^R m\Omega^2 \zeta y^2\,dy. \tag{4.13}$$

The aerodynamic forces acting on the blade in the plane of rotation include both induced drag and profile drag components. For the time being, to keep the analysis as simple as possible, it will be assumed that the resultant of all these drag forces is denoted by F_D and acts at a distance y_D from the lead/lag hinge. The equation of equilibrium is therefore

$$F_D y_D = M_\zeta = \int_0^R m\Omega^2 \zeta y^2\,dy = \frac{m\Omega^2 R^3 \zeta}{3} = \frac{M\Omega^2 R^2 \zeta}{3} \tag{4.14}$$

or

$$\zeta = \frac{3F_D y_D}{M\Omega^2 R^2}. \tag{4.15}$$

Like the coning angle, this equation indicates that, for a constant aerodynamic loading, the mean lag angle is inversely proportional to blade mass and Ω^2.

Again, in practice, like the flapping hinge, the lead/lag hinge is also offset from the rotational axis. Note that the lag hinge offset may be different from the flap hinge offset. In

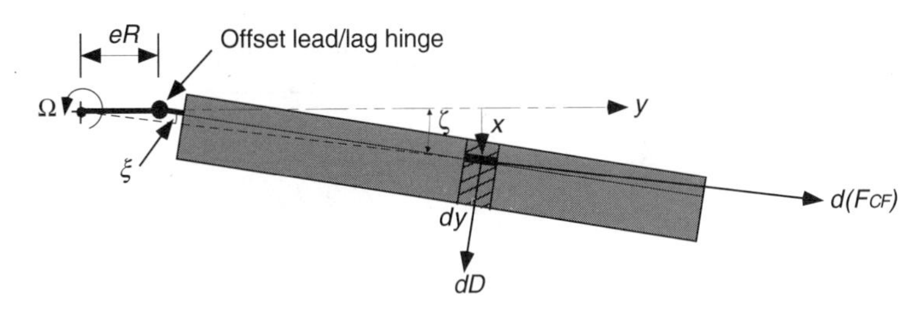

Figure 4.4 Blade with an offset lead/lag hinge.

this case, the centrifugal force as a result of lag about the offset hinge is

$$F_{CF} = \int_{eR}^{R} m\Omega^2(\zeta - \xi) y \, dy. \tag{4.16}$$

From the geometry of the problem shown in Fig. 4.4, it is noted that

$$\xi = \zeta \left(1 - \frac{eR}{y}\right). \tag{4.17}$$

Therefore, the centrifugal moment about the hinge is

$$M_\zeta = \int_{eR}^{R} m\Omega^2(\zeta - \xi) y^2 \, dy = \int_{eR}^{R} m\Omega^2 \zeta e R y \, dy$$
$$= \frac{M\Omega^2 \zeta e R(1+e)}{2} + O(e^2), \tag{4.18}$$

which shows that the centrifugal force acts at a distance $R(1+e)/2$ from the rotational axis (i.e., at the center of gravity of the blade, y_{cg}).

The resultant of the aerodynamic forces because of blade drag is denoted by F_D and acts at a distance y_D from the hinge axis. The shear force on the lead/lag hinge because of the aerodynamic and centrifugal forces must be equal to the shaft torque, Q, divided by the hinge offset, eR. The equation of force equilibrium is therefore

$$F_D \cos\zeta - \int_{eR}^{R} m\Omega^2 y \sin\xi \, dy = \frac{Q}{eR}. \tag{4.19}$$

This equation indicates that, for a given rotor, the mean drag angle of the blades is essentially a function of Q/Ω^2.

4.5 Equation of Motion for Flapping Blade

In hovering flight, the solution for β is a constant ($= \beta_0$) independent of ψ, and this angle is called the *coning angle*. Under forward flight conditions as a result of the cyclically (azimuthally) varying airloads, the blade flaps up and down in a periodic manner with respect to azimuth. Consider Fig. 4.5, which shows the line of action of the aerodynamic forces, the centrifugal forces, and the inertial forces acting on a small element of the blade span. Define the moment to be positive in a direction such as to reduce β. The centrifugal moment about the hinge is

$$d(M_{CF}) = (m dy) y^2 \Omega^2 \beta = m y^2 \Omega^2 \beta \, dy. \tag{4.20}$$

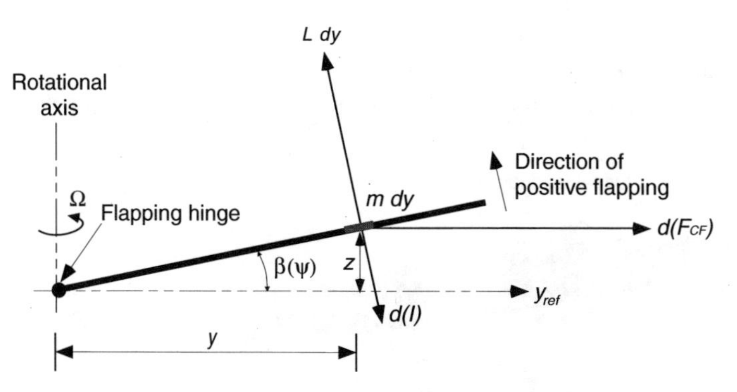

Figure 4.5 Forces acting on an element of a flapping blade.

The inertial moment about the hinge is

$$d(I) = (m\,dy)y^2\ddot{\beta} = my^2\ddot{\beta}\;dy \tag{4.21}$$

and the aerodynamic moment about the hinge is

$$d(M_\beta) = -Ly\;dy, \tag{4.22}$$

where the sign of the latter term is noted. It will be assumed that there is no hinge offset, i.e., $e = 0$. Therefore, the equation of motion can be derived by summing the moments about the flap hinge giving

$$\int_0^R d(M_{CF}) + \int_0^R d(I) + \int_0^R d(M_\beta) = 0. \tag{4.23}$$

Introducing the relevant expressions for $d(M_{CF})$, $d(I)$, and $d(M_\beta)$ gives

$$\int_0^R m\Omega^2\beta y^2\;dy + \int_0^R m\ddot{\beta}y^2\;dy - \int_0^R Ly\;dy = 0. \tag{4.24}$$

Collecting terms results in

$$\left(\int_0^R my^2\;dy\right)(\ddot{\beta} + \Omega^2\beta) = \int_0^R Ly\;dy. \tag{4.25}$$

The mass moment of inertia of the blade about the flap hinge is

$$I_b = \int_0^R my^2\;dy. \tag{4.26}$$

Thus the equation of flapping motion in Eq. 4.25 can be written as

$$I_b\ddot{\beta} + I_b\Omega^2\beta = \int_0^R Ly\;dy. \tag{4.27}$$

Noting that $\psi = \Omega t$ results in the following transformations:

$$\dot{\beta} = \frac{\partial\beta}{\partial t} = \Omega\frac{\partial\beta}{\partial\psi} = \Omega\overset{*}{\beta} \quad\text{and}\quad \ddot{\beta} = \frac{\partial^2\beta}{\partial t^2} = \Omega^2\frac{\partial^2\beta}{\partial\psi^2} = \Omega^2\overset{**}{\beta}, \tag{4.28}$$

we can write the flapping equation as

$$I_b\Omega^2\frac{\partial^2\beta}{\partial\psi^2} + I_b\Omega^2\beta = \int_0^R Ly\;dy \tag{4.29}$$

or in short-hand notation

$$\overset{**}{\beta} + \beta = \frac{1}{I_b\Omega^2}\int_0^R Ly\;dy. \tag{4.30}$$

Now, consider the aerodynamic forces. At any blade element of chord c at a radial distance y from the rotational axis it can be easily shown from Section 3.4.1 that with uniform inflow the aerodynamic force per unit length is

$$L = \frac{1}{2}\rho U_T^2 cC_{l_\alpha}\left(\theta - \frac{\dot{\beta}y}{U_T} - \frac{v_i}{U_T}\right) \tag{4.31}$$

so the aerodynamic moment about the hinge will be

$$\int_0^R Ly\,dy = \frac{1}{2}\rho\Omega^2 cC_{l_\alpha}\int_0^R y^3\left(\theta - \frac{\dot{\beta}}{\Omega} - \frac{v_i}{\Omega y}\right)dy$$
$$= \frac{1}{8}\rho\Omega^2 cC_{l_\alpha}R^4\left[\theta - \frac{\dot{\beta}}{\Omega} - \frac{4\lambda_i}{3}\right]. \tag{4.32}$$

The Lock number is defined as

$$\gamma = \frac{\rho C_{l_\alpha} cR^4}{I_b}, \tag{4.33}$$

which can be viewed as a measure of the ratio of inertial forces to aerodynamic forces. For a helicopter rotor, the value is approximately 8. Therefore, the equation governing the behavior of the flapping blade becomes

$$\overset{**}{\beta} + \beta = \frac{\rho C_{l_\alpha} cR^4}{I_b}\left(\frac{1}{8}\right)\left[\theta - \overset{*}{\beta} - \frac{4\lambda_i}{3}\right] \tag{4.34}$$

or

$$\overset{**}{\beta} + \left(\frac{\gamma}{8}\right)\overset{*}{\beta} + \beta = \frac{\gamma}{8}\left[\theta - \frac{4\lambda_i}{3}\right]. \tag{4.35}$$

This is the flapping equation for a *centrally hinged* blade, that is, one that is hinged at the rotational axis. A more general form is to leave the aerodynamic force (moment) on the right-hand side unintegrated, in which case

$$\overset{**}{\beta} + \beta = \gamma\bar{M}_\beta, \tag{4.36}$$

where

$$\bar{M}_\beta = \frac{1}{\rho C_{l_\alpha} cR^4\Omega^2}\int_0^R Ly\,dy. \tag{4.37}$$

Note the similarity of Eq. 4.36 with the equation of motion of a single degree-of-freedom spring–mass–damper system, that is, $m\ddot{x} + c\dot{x} + kx = F$, where m is the mass, c is the damping, and k is the spring stiffness. This system has an undamped natural frequency of $\omega_n = \sqrt{k/m}$. Therefore, the undamped natural frequency of the flapping blade about a hinge located at the rotational axis is $\omega_n = \Omega$ rad/s or once per revolution (1/rev).

Consider first the case where the rotor operates in a vacuum, in which case there are no aerodynamic forces present. The flapping equation reduces to

$$\overset{**}{\beta} + \beta = 0. \tag{4.38}$$

This equation has the general solution

$$\beta = \beta_{1c}\cos\psi + \beta_{1s}\sin\psi, \tag{4.39}$$

where β_{1c} and β_{1s} are arbitrary coefficients. Thus, in the absence of aerodynamic forces, the rotor takes up an arbitrary orientation in inertial space. In effect, the rotor acts like a gyroscope. The introduction of aerodynamic forces produces an aerodynamic flapping moment about the hinge, which causes the rotor to *precess* to a new orientation until the aerodynamic damping causes equilibrium to be obtained once again.

In forward flight, the aerodynamic forces provide the forcing to the flapping blade at multiples of the rotor frequency. This aerodynamic excitation is primarily at 1/rev. The blade flapping motion with respect to the rotor hub can be represented as an infinite Fourier series of the form

$$\begin{aligned}\beta(\psi) &= \beta_0 + \beta_{1c}\cos\psi + \beta_{1s}\sin\psi + \beta_{2c}\cos 2\psi + \beta_{2s}\sin 2\psi + \cdots \\ &= \beta_0 + \sum_{n=1}^{\infty}\left(\beta_{nc}\cos n\psi + \beta_{ns}\sin n\psi\right),\end{aligned} \tag{4.40}$$

where the time scale has been nondimensionalized with respect to rotor time such that the dimensionless period is 2π radians. The Fourier coefficients may be evaluated from the flapping displacements using

$$\beta_0 = \frac{1}{2\pi}\int_0^{2\pi} \beta \, d\psi,$$

$$\beta_{nc} = \frac{1}{\pi}\int_0^{2\pi} \beta\cos n\psi \, d\psi,$$

$$\beta_{ns} = \frac{1}{\pi}\int_0^{2\pi} \beta\sin n\psi \, d\psi.$$

Assuming uniform inflow and linearly twisted blades, we can evaluate the aerodynamic flapping moment about the hinge analytically using the blade element theory where

$$\bar{M}_\beta = \frac{1}{\rho c_{l_\alpha} c R^4 \Omega^2}\int_0^R y\,dF_z \tag{4.41}$$

$$= \frac{1}{2}\int_0^1 r\left[\left(\frac{U_T}{\Omega R}\right)^2\theta - \left(\frac{U_P}{\Omega R}\right)\left(\frac{U_T}{\Omega R}\right)\right]dr. \tag{4.42}$$

After substituting the results for U_T and U_P obtained in Section 3.4.1, and also assuming uniform inflow and linearly twisted blades, we get

$$\begin{aligned}\bar{M}_\beta &= \theta\left(\frac{1}{8} + \frac{\mu}{3}\sin\psi + \frac{\mu^2}{4}\sin^2\psi\right) \\ &\quad + \theta_{\text{tw}}\left(\frac{1}{10} + \frac{\mu}{4}\sin\psi + \frac{\mu^2}{6}\sin^2\psi\right) - \lambda\left(\frac{1}{6} + \frac{\mu}{4}\sin\psi\right) \\ &\quad - \overset{*}{\beta}\left(\frac{1}{8} + \frac{\mu}{6}\sin\psi\right) - \beta\mu\cos\psi\left(\frac{1}{6} + \frac{\mu}{4}\sin\psi\right).\end{aligned} \tag{4.43}$$

And this result can then be substituted into Eq. 4.36 to compute β.

Some interesting characteristics of the resulting flapping equation are:

1. In forward flight, that is, when $\mu \neq 0$, the equation has periodic coefficients. This does not allow an analytical closed-form solution of the flapping equation.
2. The flap damping term, which is the coefficient associated with the $\overset{*}{\beta}$ term, is

$$\frac{\gamma}{8}\left(1 + \frac{4}{3}\mu\sin\psi\right),$$

which is of aerodynamic origin (all the terms multiplied by the Lock number γ come from the aerodynamics) and is usually very high. For the case of hover, and for the natural frequency of 1/rev, the corresponding damping ratio is $\gamma/16$. For a Lock number $\gamma = 8$ this means that the damping is 50% of the critical value; thus the blade flapping motion is stable and well damped.

3. Finally, note that the equation has been derived with respect to the plane defined by the hub of the rotor, and so both the flap angle β and the pitch control angle θ will generally be functions of the azimuth angle ψ.

The general flapping equation of motion cannot be solved analytically in closed form for the general case of $\mu \neq 0$. Therefore, two options present themselves:

1. *Solve the equation numerically*. The equation can be integrated for given values of collective pitch θ_0, lateral cyclic θ_{1c}, longitudinal cyclic θ_{1s}, and inflow λ_i. The selection of the initial conditions for the flapping is not very important because even extreme initial conditions will simply cause a numerical transient that will disappear after a few iterations because of the high damping. The main problem with the numerical solution is that it does not provide any insight into how the blade flapping response is affected by the various parameters.
2. *Find a periodic solution*. In this case the problem is to find a steady-state, periodic solution, in the form of a Fourier series. Obviously this solution is not adequate for transient situations such as during a maneuver, but it allows the identification of the relationships between the flapping response and the various problem parameters.

Assuming the solution for the blade flapping motion to be given by the first harmonics only, that is,

$$\beta(\psi) = \beta_0 + \beta_{1c}\cos\psi + \beta_{1s}\sin\psi, \tag{4.44}$$

and harmonically matching constant and periodic (sine and cosine) terms on both sides of the derived flapping equation gives

$$\beta_0 = \gamma\left[\frac{\theta_0}{8}(1+\mu^2) + \frac{\theta_{\text{tw}}}{10}\left(1+\frac{5}{6}\mu^2\right) + \frac{\mu}{6}\theta_{1s} - \frac{\lambda}{6}\right], \tag{4.45}$$

$$\beta_{1s} - \theta_{1c} = \frac{\left(-\frac{4}{3}\mu\beta_0\right)}{\left(1+\frac{1}{2}\mu^2\right)}, \tag{4.46}$$

$$\beta_{1c} + \theta_{1s} = \frac{-\frac{8}{3}\mu\left[\theta_0 - \frac{3}{4}\lambda + \frac{3}{4}\mu\theta_{1s} + \frac{3}{4}\theta_{\text{tw}}\right]}{\left(1-\frac{1}{2}\mu^2\right)}. \tag{4.47}$$

Note that by setting $\mu = 0$ (hovering flight) in the above equations, the following results are obtained:

$$\beta_{1s} - \theta_{1c} = 0 \quad \text{or} \quad \beta_{1s} = \theta_{1c},$$
$$\beta_{1c} + \theta_{1s} = 0 \quad \text{or} \quad \beta_{1c} = -\theta_{1s}.$$

This shows that there is an equivalence between pitching motion and flapping motion. If the cyclic pitch motion is assumed to be

$$\theta = \theta_0 + \theta_{1c}\cos\psi + \theta_{1s}\sin\psi, \tag{4.48}$$

then the flapping response will be

$$\beta = \beta_0 + \theta_{1c}\cos\left(\psi - \frac{\pi}{2}\right) + \theta_{1s}\sin\left(\psi - \frac{\pi}{2}\right). \tag{4.49}$$

Therefore, because of the dynamic behavior of the blade, the flapping response lags the blade pitch (aerodynamic) inputs by 90°, which is the resonant condition.[1] Strictly speaking this is for a rotor with a flapping hinge at the rotational axis. The effect of a hinge offset on the flapping problem will be considered later.

4.6 Physical Description of Blade Flapping

4.6.1 *Coning Angle*

The coefficient β_0 is the average or mean part of the flapping motion that is independent of time or blade azimuth, ψ. In hovering flight, $\beta(\psi) = \beta_0$, which is called the coning angle. The presence of a coning angle has been pointed out previously to be the angle that results from the moment balance about the flapping hinge as a result of the centrifugal and aerodynamic forces. Because the centrifugal loads remain constant for a given rotor speed, the coning angle varies with both the magnitude and distribution of lift across the blade. For example, a higher gross weight requires a higher blade lift to hover, which tends to increase the aerodynamic moment about the hinge resulting in a higher coning angle. Also, it is already known from previous discussion in Chapter 3 that the inflow velocity has an effect on the blade spanwise loading. As the magnitude of the inflow increases, for a given overall rotor thrust the blade must become more highly lift loaded toward the tips. This produces a higher aerodynamic moment about the flap hinge and, therefore, a higher coning angle. Because the time-averaged inflow (induced velocity) changes with forward speed, the rotor coning angle will mimic the variation in mean inflow through the disk with forward speed, as discussed in Chapter 2.

4.6.2 *Longitudinal Flapping*

The coefficient β_{1c} represents the amplitude of the pure cosine flapping motion (see Fig. 4.6). This represents a longitudinal or fore–aft tilt of the rotor tip path plane. In forward flight, the rotor disk has a natural tendency to tilt back (longitudinally) because of the dissymmetry in lift produced between the advancing and retreating sides of the disk. As a result of the higher dynamic pressure on the advancing side of the disk, the blade lift is increased over that obtained at $\psi = 0°$ and $\psi = 180°$. Therefore, as the blade rotates into the advancing side of the disk, the excess lift causes the blade to flap upward. Over the front of the disk the dynamic pressure reduces progressively, and the blade reaches a maximum *displacement* with $\dot{\beta} = 0$ at $\psi = 180°$. As the blade rotates into the retreating side of the rotor disk, the deficiency in dynamic pressure causes the blade to flap downward. This downward flapping motion increases the angle of attack at the blade element, which tends to increase blade lift over the lift that would have been obtained without flapping motion. This upward and downward flapping of the blade tends to reduce and increase the angle of attack at the blade elements by an amount

$$\Delta\alpha(y, \mu) = -\tan^{-1}\left(\frac{y\dot{\beta}}{\Omega y + \mu\Omega R}\right) = -\tan^{-1}\left(\frac{r\overset{*}{\beta}}{r + \mu}\right). \tag{4.50}$$

For example, as a result of the flapping upward ($\dot{\beta} > 0$), the blade lift tends to decrease relative to the lift that would have been produced if there was no flapping hinge.

[1] For a single degree-of-freedom system excited at its natural frequency, there will always be a phase lag of 90° between the input and the output.

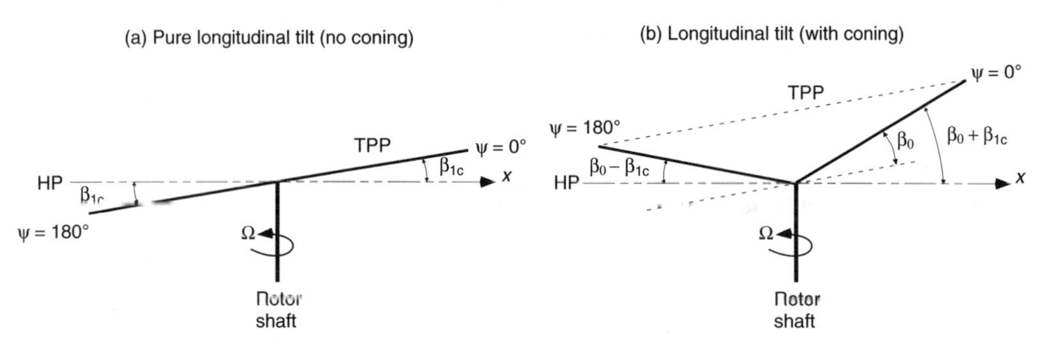

Figure 4.6 Pure longitudinal flapping of the rotor.

The upshot of all this flapping motion is that the rotor blades reach an equilibrium condition again when the local changes in angle of attack and aerodynamic loads as a result of blade flapping become sufficient to compensate for local changes in the airloads resulting from variations in dynamic pressure between the advancing and retreating sides of the disk. For the situation described previously, in the final equilibrium condition the disk will be tilted back longitudinally with respect to the hub (i.e., a $-\beta_{1c}$ flapping motion). Remember that the forcing function in this case is phased such that the maximum aerodynamic force occurs at $\psi = 90^\circ$, but because of the dynamic behavior of the rotor blade, the maximum flapping displacement occurs 90° later at $\psi = 180^\circ$. This is of fundamental importance because it allows an understanding of what happens to the rotor flapping under the action of aerodynamic loads that are phased differently with respect to the rotor azimuth.

4.6.3 *Lateral Flapping*

The coefficient β_{1s} represents the amplitude of the pure sine motion (see Fig. 4.7). This represents the lateral or left–right tilt of the tip path plane. In addition to the natural tendency for the disk to tilt back with a change in forward flight speed, the disk also has a tendency to tilt laterally to the right. This effect arises because of blade flapping displacement (coning). For the coned rotor, the blade angle of attack is decreased when the blade is at $\psi = 0^\circ$ and increased when $\psi = 180^\circ$. Again, another source of periodic force is produced, but now this is phased 90° out of phase compared to the effect discussed previously. Because of the 90° force/displacement lag of blade, this results in a lateral tilt of the rotor disk to the right when viewed from behind (i.e., a $-\beta_{1s}$ motion). Note that in the hypothetical case with no coning, the blades see the same increase in angle of attack at $\psi = 0^\circ$ and 180° and there will be no lateral tilt.

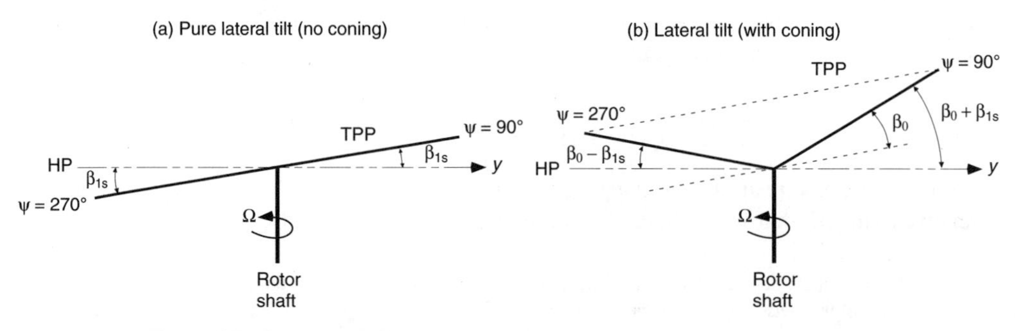

Figure 4.7 Pure lateral flapping of the rotor.

The coefficients β_{2c}, β_{2s}, etc. represent the amplitudes of the higher harmonics of the blade motion. In practice, these are found to be very small and of no substantial significance but appear as a slight warping of the rotor tip path plane. For rotor trim and performance evaluation it is considered acceptable engineering practice to neglect all harmonics above the first. However, the effects of higher harmonics of flapping on the vibration and aeroelastic stability characteristics of the rotor are important.

4.7 Dynamics of Blade Flapping with a Hinge Offset

The analysis of a blade with a hinge offset is similar to the foregoing, but there are some important quantifiable differences. The blade is assumed to be rigid and hinged at a distance eR from the rotational axis. The forces acting on an element of the blade are:

1. Inertia force $m(y-eR)\ddot{\beta}dy$ acting at a distance $(y-eR)$ from the hinge.
2. Centrifugal force $my\Omega^2\, dy$ acting at a distance $(y-eR)\beta$ from the hinge.
3. Aerodynamic lift forces $L\, dy$ acting at a distance $(y-eR)$ from the hinge.

Taking moments about the flap hinge gives the equation of motion

$$\int_{eR}^{R} m(y-eR)^2\ddot{\beta}\, dy + \int_{eR}^{R} m\Omega^2 y(y-eR)\beta\, dy - \int_{eR}^{R} L(y-eR)\, dy = 0. \quad (4.51)$$

In this case, the mass moment of inertia about the flap hinge is

$$I_b = \int_{eR}^{R} m(y-eR)^2\, dy, \quad (4.52)$$

so that the flapping equation becomes

$$I_b\left\{\ddot{\beta} + \Omega^2\left(1 + \frac{eR\int_{eR}^{R} m(y-eR)\,dy}{I_b}\right)\beta\right\} = \int_{eR}^{R} L(y-eR)\,dy. \quad (4.53)$$

Dividing through by Ω^2 gives

$$I_b(\overset{**}{\beta} + \nu_\beta^2\beta) = \frac{1}{\Omega^2}\int_{eR}^{R} L(y-eR)\,dy, \quad (4.54)$$

where ν_β is the nondimensional flap frequency in terms of the rotational speed, that is,

$$\nu_\beta^2 = 1 + \frac{eR\displaystyle\int_{eR}^{R} m(y-eR)\,dy}{I_b}. \quad (4.55)$$

Again, the physical analogy with a spring–mass–damper system can be drawn. Evaluation of Eq. 4.55 shows that the undamped natural frequency of the system is now

$$\nu_\beta = \omega_n = \sqrt{1 + \frac{3e}{2(1-e)}} \approx \sqrt{1 + \frac{3}{2}e}, \quad (4.56)$$

see also Question 4.1. Typically, the value of e varies from 4 to 6% for an articulated blade, so that the natural frequency of the rotor is only slightly greater than Ω or 1/rev. This also means that the phase lag between the forcing and the rotor flapping response must be less than 90°. In this case, the flapping equation is

$$\overset{**}{\beta} + \nu_\beta^2\beta = \gamma\bar{M}_\beta, \quad (4.57)$$

where ν_β is the rotating flap frequency in terms of rotational speed. (Remember that $\nu_\beta = 1$ for a hinge at the rotational axis.) Therefore, in hover the flapping response to cyclic pitch inputs is given by

$$\beta_{1c}\left(\nu_\beta^2 - 1\right) + \beta_{1s}\frac{\gamma}{8} = \frac{\gamma}{8}\theta_{1c}, \tag{4.58}$$

$$\beta_{1s}\left(\nu_\beta^2 - 1\right) - \beta_{1c}\frac{\gamma}{8} = \frac{\gamma}{8}\theta_{1s}. \tag{4.59}$$

This gives for the longitudinal flapping angle

$$\beta_{1c} = \frac{-\theta_{1s} + \left(\nu_\beta^2 - 1\right)\dfrac{8}{\gamma}\theta_{1c}}{1 + \left[\left(\nu_\beta^2 - 1\right)\dfrac{8}{\gamma}\right]^2} \tag{4.60}$$

and for the lateral flapping angle

$$\beta_{1s} = \frac{\theta_{1c} + \left(\nu_\beta^2 - 1\right)\dfrac{8}{\gamma}\theta_{1s}}{1 + \left[\left(\nu_\beta^2 - 1\right)\dfrac{8}{\gamma}\right]^2}. \tag{4.61}$$

In this case, the forcing frequency (1/rev) is less than the natural flapping frequency (off resonance condition) and it can be shown that the phase lag, ϕ, will be less than 90° as given by

$$\phi = \tan^{-1}\left(\frac{\gamma\left(1 - \dfrac{8e}{3}\right)}{8\left(\nu_\beta^2 - 1\right)}\right) \approx \tan^{-1}\left(\frac{\gamma/8}{\nu_\beta^2 - 1}\right), \tag{4.62}$$

see also Question 4.7. For hingeless rotors, which have a relatively high effective hinge offset, the phase lag is about 75–80°.

4.8 Blade Feathering and the Swashplate

The blade pitch (or feathering) motion can be described as the Fourier series

$$\theta(r,\psi) = \theta_{tw}r + \theta_0 + \theta_{1c}\cos\psi + \theta_{1s}\sin\psi + \cdots + \theta_{nc}\cos n\psi$$
$$+ \theta_{ns}\sin n\psi + \cdots, \tag{4.63}$$

where r is the nondimensional radial position on the blade. Blade pitch motion comes from two sources, namely

1. Commanded input from the helicopter control system. This is done by means of a swashplate, the orientation of which is controlled by the pilot. The control inputs produced by the swashplate consist of the collective pitch θ_0 and the first harmonics of the Fourier series: the lateral cyclic θ_{1c} and the longitudinal cyclic θ_{1s}. The collective pitch controls the average blade pitch angle and, therefore, the blade lift and average rotor thrust. The cyclic pitch controls the orientation or tilt of the rotor disk and so the direction of the rotor thrust vector.
2. Elastic deformations (twist) of the blade and control system. The elastic (torsional) deformations of the blade, although small, are significant. However, these can be neglected at the present level of analysis.

The swashplate is the key to effecting pitch control to the rotor blades. The swashplate has rotating and nonrotating disks concentric with the rotor shaft. A set of bearings between the two disks allows the upper disk to rotate while the lower is non-rotating. Both, disks can be slid up and down the shaft in response to collective inputs and the swashplate can also be tilted to an arbitrary orientation in response to cyclic inputs from the pilots' controls. The blade pitch motion itself is induced about a pitch bearing. A pitch horn is attached to the blade outboard of the pitch bearing. A pitch link is attached to the pitch horn and the upper (rotating) part of the swashplate in such a way that as the upper plate rotates the vertical displacement of the pitch link produces blade pitch motion. The novel part of the helicopter swashplate is the ability to tilt the plate to an arbitrary orientation, which requires a gimbal or spherical bearing between the swashplate and the rotor shaft. This allows a first harmonic blade pitch input with any phase angle. The earliest swashplate mechanisms are documented in patents by Crocco of Italy in 1906 and Yurev of Russia in 1910. In 1929, Raoul Hafner was one of the first to use the swashplate in the form known today.

A photograph of the rotor hub, pitch links, and swashplate mechanism of the AH-64 helicopter is shown in Fig. 4.8. The individual parts can be seen by cross-referencing with a simple schematic of the swashplate shown in Fig. 4.9. The vertical motion of the swashplate results in a vertical motion of the pitch link and a collective pitch change to the blades. This either increases or decreases the rotor thrust. The fore and aft tilt of the swashplate translates into a once-per-revolution cyclic pitch on each rotor blade. This produces a once-per-revolution aerodynamic forcing, which causes the rotor blades to flap and the rotor to precess to a new orientation in space, thereby tilting the thrust vector. The system is fundamentally simple; an aerodynamic forcing is applied at or close to the natural frequency of the flapping blade and the blades respond so that a unit of cyclic pitch input results in (almost) a unit of flapping response.

Figure 4.8 Photograph of the hub, pitch links, and swashplate mechanism on an AH-64.

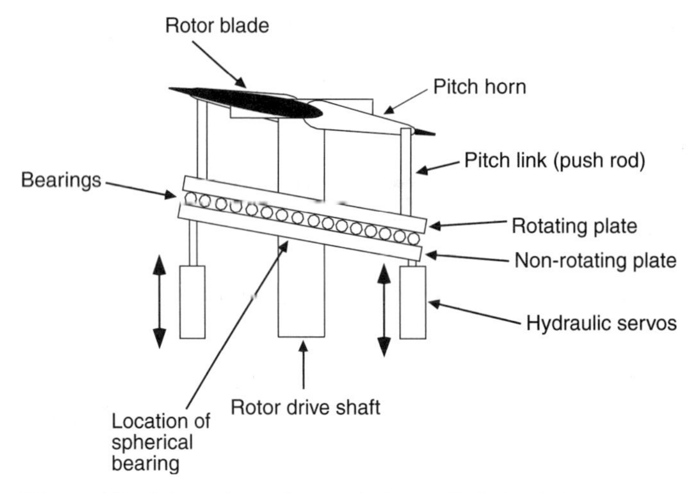

Figure 4.9 Schematic of the swashplate and blade pitch attachments for a two-bladed rotor.

Note that for a centrally hinged rotor system the term θ_{1c} controls the lateral orientation of the rotor disk and θ_{1s} controls the longitudinal orientation. Remember that for a rotor with a centrally located flap hinge there is an exact 90° force/displacement phase lag. In the case of pure θ_{1c} (cosine) cyclic motion, the maximum applied aerodynamic force occurs at $\psi = 0°$ and so the maximum flapping displacement occurs 90° later at $\psi = 90°$. Therefore, the application of a θ_{1c} pitch displacement causes the rotor to tilt laterally to the left, which is equivalent to a β_{1s} flapping motion, and therefore this is called *lateral cyclic*. By a similar argument, the application of a θ_{1s} pitch displacement causes the disk to tilt back longitudinally, which is equivalent to a $-\beta_{1c}$ flapping displacement, and this is called *longitudinal cyclic*.

For a coaxial rotor system, both rotors must be tilted. Therefore, the pitch angles of both sets of rotors must be connected together so that they tilt in unison, see Fig. 4.10. Nevertheless, the different aerodynamic environments found on both rotors result in considerable differential flapping. This requires significant spacing between the rotors to ensure that the blades do not collide. The much larger rotor mast and exposed rotor flight control linkages, therefore, result in a high parasitic drag in forward flight. This is a major disadvantage to the coaxial rotor configuration.

4.9 Review of Rotor Reference Axes

There are several physical planes that can be used to describe the equations of motion of the rotor blades. These reference planes or axes systems have evolved as a matter of convenience, and each has advantages over others for certain types of analysis. It is always possible to transform an analysis from one reference axes to another. Figure 4.11 schematically shows the various reference planes and axes that are often used in helicopter analyses. These planes are:

1. Hub Plane (HP): The rotor hub plane is perpendicular to the rotor shaft. In the HP, an observer would see both blade flapping and feathering (pitch changes) during forward flight. While this plane is the most complicated for analysis of the rotor, it has the advantage of being linked to a physical part of the aircraft. The HP is often used for blade dynamic analyses.

Figure 4.10 Hub of a coaxial rotor system (Kamov Ka-26).

2. No Feathering Plane (NFP): The NFP is a plane where an observer sees no variation in cyclic pitch, that is, both θ_{1c} and θ_{1s} are zero. However, the observer will still see a cyclic variation in blade flap angle. Normally, this plane is used for performance analyses.
3. Tip Path Plane (TPP): This is the plane whose boundary is described by the blade tips. Therefore, an observer will see no variation in flapping, that is, both β_{1c} and

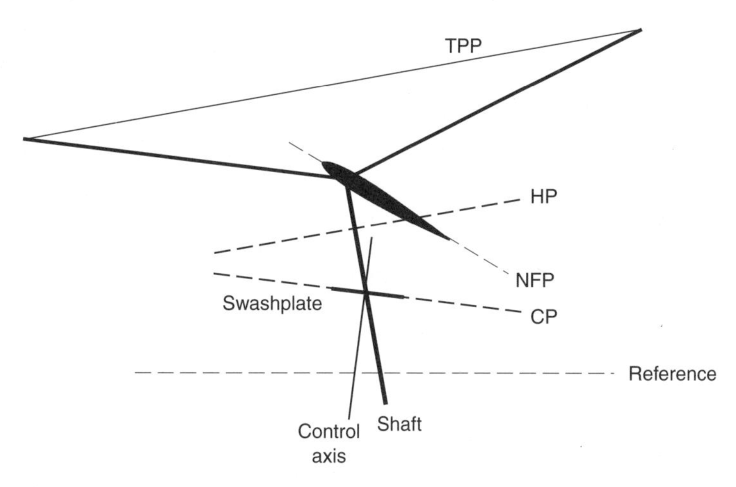

Figure 4.11 Schematic of rotor reference axes and planes.

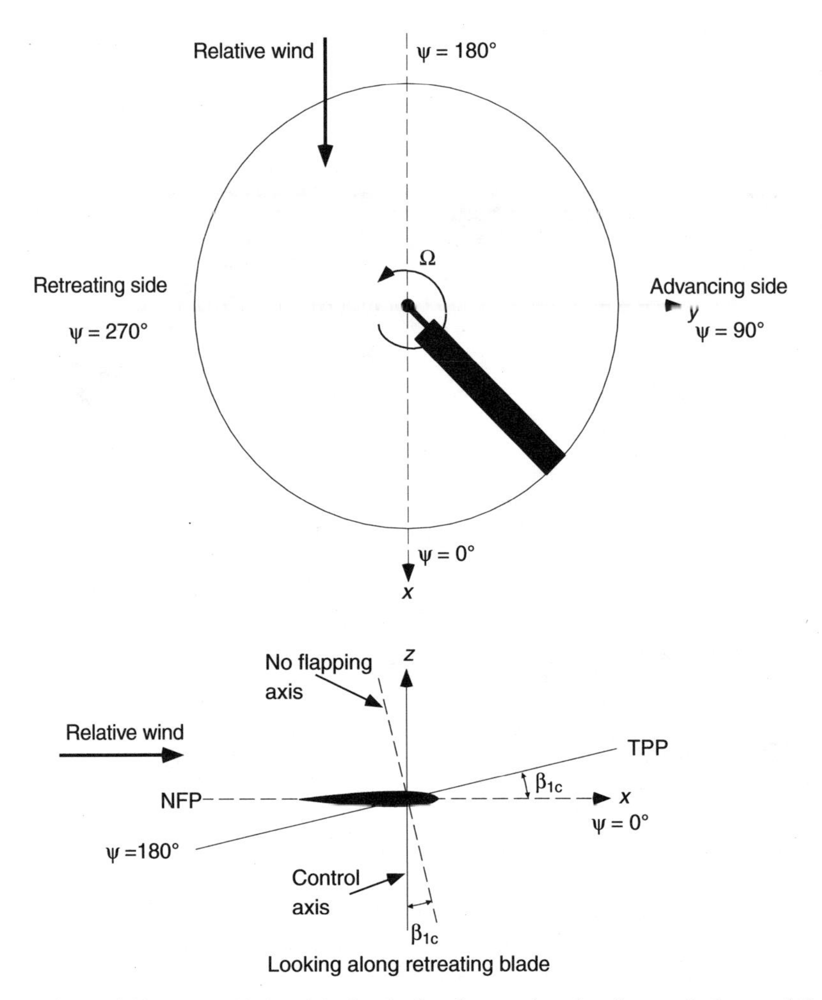

Figure 4.12 Rotor blade with simple flapping motion showing equivalence of flapping and feathering.

β_{1s} will be zero. This plane is commonly used for aerodynamic analyses, such as rotor wake models.

4. Control Plane (CP): This plane represents the commanded cyclic pitch plane and is sometimes known as the swashplate plane.

Consider now a rotor in forward flight, and assume that the rotor has a simple flapping motion with only β_{1c} – that is, the rotor tilts forward by the amount β_{1c} as shown in Fig. 4.12. Note that the relative amount of blade flapping versus feathering depends on the axis system to which the blade motion is being referenced. At $\psi = 0°$ we see that with respect to the NFP, the blade flapping is full up, but the blade pitch is neutral (zero in this case). At $\psi = 90°$, with respect to the NFP the blade flapping is now zero, again with the pitch being neutral. However, at this azimuth the blade pitch is a maximum with respect to the TPP.

Now consider a more general case, where the disk is tilted both forward and to the left. Again, as shown by Fig. 4.13, we see that the amount of blade feathering versus flapping depends on the axis system used to view the blade motion. Therefore, in light of this and

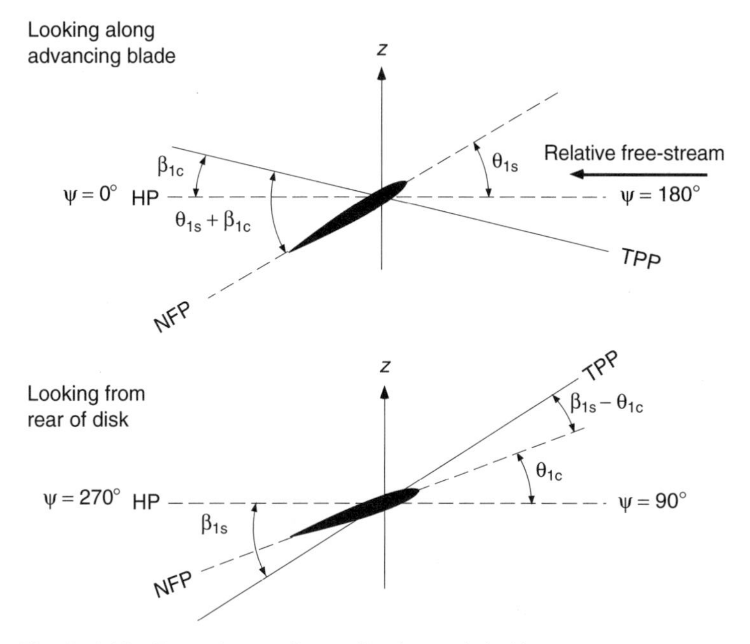

Figure 4.13 General case of rotor flapping and pitching.

the previous example, it is apparent that the amount of blade pitch (feathering) with respect to the TPP is equal to the amount of blade flapping with respect to the NFP. Fore and aft (longitudinal) flapping or β_{1c} with respect to the NFP is equivalent to lateral feathering (pitch) with respect to the TPP. Therefore, there is an equivalence of flapping and feathering motion. Strictly speaking, this equivalence is only true for a teetering type of rotor (where the flapping hinge is exactly at the rotational axis), but the equivalence is strong for all types of helicopter rotors.

In the transformation from one plane to another, it will be apparent that

$$\beta_{1c} + \theta_{1s} = \text{constant} = (\beta_{1c})_{\text{NFP}} = (\theta_{1s})_{\text{TPP}} . \tag{4.64}$$

Also,

$$\beta_{1s} - \theta_{1c} = \text{constant} = (\beta_{1s})_{\text{NFP}} = -(\theta_{1c})_{\text{TPP}} . \tag{4.65}$$

Using the above relationships, it is possible to transfer any analysis from one reference system to another, that is, from the TPP to the NFP or vice versa. For example, the angles of attack in the TPP are related to those in the NFP by

$$(\alpha)_{\text{TPP}} = (\alpha)_{\text{NFP}} - (\beta_{1c} + \theta_{1s}) . \tag{4.66}$$

4.10 Dynamics of a Lagging Blade with Hinge Offset

The blade is assumed to be rigid and undergoes a simple lagging in the plane of rotation about a hinge located at a distance eR from the rotational axis. The various forces acting on the blade are now:

1. The inertia force $m\ddot{x}\, dy = m(y - eR)\ddot{\zeta}\, dy$ acting at a distance $(y - eR)$ about the lag hinge.

2. The centrifugal force $m\Omega^2 y\, dy$ acting at a distance eRx/y from the rotational axis.
3. The aerodynamic drag force D acting at a distance $y - eR$ from the hinge axis.

Taking moments about the lag hinge gives

$$\int_{eR}^{R} m(y-eR)^2 \ddot{\zeta}\, dy - \int_{eR}^{R} m\Omega^2 y(y-eR)\frac{eR}{y}\zeta\, dy - \int_{eR}^{R} D(y-eR)\, dy = 0. \tag{4.67}$$

The mass moment of inertia I_ζ about the lag hinge is

$$\int_{eR}^{R} m(y-eR)^2\, dy = I_\zeta. \tag{4.68}$$

Therefore the equation of motion for lagging about the lead/lag hinge can be written as

$$I_\zeta\left(\overset{*}{\zeta} + \nu_\zeta^2 \zeta\right) = \frac{1}{\Omega^2}\int_{eR}^{R} D(y-eR)\, dy, \tag{4.69}$$

where ν_ζ is the nondimensional lag frequency in terms of the rotational speed, determined with

$$\nu_\zeta^2 = \frac{eR \displaystyle\int_{eR}^{R} m(y-eR)\, dy}{I_\zeta}. \tag{4.70}$$

Note that $\nu_\zeta = 0$ is zero if there is no hinge offset. Evaluation of Eq. 4.70 gives the lag frequency with a hinge offset as

$$\nu_\zeta = \sqrt{\frac{3}{2}\left(\frac{eR}{R-eR}\right)} \approx \sqrt{\frac{3}{2}e} \tag{4.71}$$

Because the centrifugal restoring moment about the lag hinge is much smaller than in flapping, the corresponding uncoupled natural frequency of the lag motion is much smaller. Typically, for articulated rotors, the uncoupled rotating lag frequency varies from about 0.2Ω to 0.3Ω. (See also Question 4.9.)

Note that the lag motion of the blade is lightly damped. Besides the fact that the lead/lag displacements about the hinge are small, they produce aerodynamic forces through a change in velocity normal to the leading edge of the blade. This has a much smaller effect than aerodynamic forces produced by changes in angle of attack induced by flapping. Also, the drag forces acting on the blades are almost two orders of magnitude less than the lift forces. Therefore, the blade lag motion is very susceptible to various types of aeroelastic and aeromechanical instabilities. One important example of this is *ground resonance*, where the blade lag motion and lateral motion of the fuselage become coupled to produce a catastrophic aeromechanical instability. This is the reason why most rotors have mechanical lag dampers, which provide artificial damping to suppress the occurrence of such so-called aeromechanical phenomena.

4.11 Coupled Flap–Lag Motion

It will be apparent from the foregoing that, in practice, the blade flap and lead/lag motion are coupled. Now consider the analysis of the problem where the blade simultaneously undergoes two types of motion with both flap and lead/lag about their respective hinges. For simplicity, it will be assumed that the flap and lead/lag hinges are coincident. This type of model is a good representation of a blade with a high torsional stiffness. In this case, note that the flap and lead/lag motions are coupled as a result of Coriolis and aerodynamic forces. Coriolis effects, although subtle, introduce an important coupling between blade flapping or out-of-plane motion and lead/lag or in-plane motion. On a rotor, Coriolis forces will appear whenever there is a radial lengthening or shortening of the blade about the rotational axis, which will be a result of blade flapping or bending. Coriolis effects produce forces in the plane of rotation of the rotor. A more complete description of the various forces acting on the rotor can now be obtained.

For the flap motion, the forces acting on the blade element are:

1. Inertia force $m(y - eR)\ddot{\beta}\, dy$ acting at a distance $(y - eR)$ about the hinge.
2. Centrifugal force $m\Omega^2 y\, dy$ acting at a distance $(y - eR)\beta$ from the hinge.
3. Coriolis force $2m(y - eR)\Omega\dot{\zeta}\, dy$ acting at a distance $(y - eR)\beta$ from the hinge.
4. Aerodynamic lift forces $L\, dy$ acting at a distance $(y - eR)$ from the hinge.

For the lag motion, the various forces acting on the blade element are:

1. Inertia force $m(y - eR)\ddot{\zeta}\, dy$ acting at a distance $(y - eR)$ about the hinge.
2. Centrifugal force $m\Omega^2 y\, dy$ acting at a distance $(y - eR)eR/y\zeta$ from the hinge.
3. Coriolis force $2m(y - eR)\beta\Omega\dot{\beta}\, dy$ acting at a distance $(y - eR)\beta$ from the hinge.
4. Aerodynamic drag forces $D\, dy$ acting at a distance $(y - eR)$ from the hinge.

Taking the moment of the forces about the flap hinge gives

$$\int_{eR}^{R} m(y - eR)^2 \ddot{\beta}\, dy + \int_{eR}^{R} m\Omega^2 y(y - eR)\beta\, dy - \int_{eR}^{R} 2m\Omega(y - eR)^2 \beta\dot{\zeta}\, dy - \int_{eR}^{R} L(y - eR)\, dy = 0, \tag{4.72}$$

so that the coupled equation of motion becomes

$$I_b \left(\ddot{\beta} + \nu_\beta^2 \Omega^2 \beta - 2\Omega\dot{\zeta}\beta \right) = \int_{eR}^{R} L(y - eR)\, dy, \tag{4.73}$$

where

$$\nu_\beta^2 = 1 + \frac{eR \displaystyle\int_{eR}^{R} m(y - eR)\, dy}{I_b}. \tag{4.74}$$

This reduces to the correct result that $\nu_\beta \rightarrow 1$ as $e \rightarrow 0$. Taking the moment of the forces about the lead/lag hinge gives

$$\int_{eR}^{R} m(y - eR)^2 \ddot{\zeta}\, dy + \int_{eR}^{R} m\Omega^2 eR(y - eR)\zeta\, dy + \int_{eR}^{R} 2m\Omega(y - eR)^2 \beta\dot{\beta}\, dy - \int_{eR}^{R} D(y - eR)\, dy = 0, \tag{4.75}$$

so that the coupled equation of motion for the blade becomes

$$I_\zeta \left(\ddot{\beta} + \nu_\zeta^2 \Omega^2 \zeta + 2\Omega \dot{\beta} \beta\right) = \int_{eR}^{R} D(y - eR)\, dy, \tag{4.76}$$

where

$$\nu_\zeta^2 = \frac{eR \displaystyle\int_{eR}^{R} m(y - eR)\, dy}{I_\zeta}, \tag{4.77}$$

which, again, reduces to the correct result that $\nu_\zeta \rightarrow 0$ as $e \rightarrow 0$. It is of particular significance to note how the flap–lag equations are coupled through the Coriolis acceleration terms.

4.12 Introduction to Rotor Trim

The trim solution involves the calculation of the rotor control settings, rotor disk orientation (pitch angles and blade flapping), and overall helicopter orientation for the prescribed flight conditions. To control the position of the helicopter in free-flight requires the adjustment of the forces and moments about all three axes. For a conventional helicopter there are three independent controls used for this purpose:

1. *Collective pitch:* This input (θ_0) increases the main rotor blade pitch angles by the same amount and, therefore, changes the magnitude of the rotor thrust. The collective is changed by a lever that is held in the pilot's left hand, with an upward pulling motion required for an increase in thrust.
2. *Lateral and longitudinal cyclic pitch:* These inputs impart a once-per-revolution cyclic pitch change to the blades. Lateral cyclic (θ_{1c}) is applied such that the rotor disk can be tilted left and right. This changes the orientation of the rotor thrust vector, producing both a side force and a rolling moment. Longitudinal cyclic (θ_{1s}) imparts a once-per-revolution cyclic pitch change to the blades such that the rotor disk can be tilted fore and aft. Like the lateral cyclic, this changes the orientation of the rotor thrust vector, in this case producing both a longitudinal force and pitching moment. Both lateral and longitudinal cyclic are controlled by the pilot using a cyclic stick (similar to the conventional stick on a fixed-wing aircraft), which is held in the pilot's right hand.
3. *Yaw*: This is controlled by using the tail rotor thrust. The pilot has a set of pedals, which are operated by the pilot's feet just like a rudder on a fixed-wing aircraft. By pushing the pedals in the required direction, the collective pitch on the tail rotor is changed, producing a change in tail rotor thrust and, therefore, causing the nose to yaw right or left. On a tandem rotor system, yaw is controlled by differential lateral cyclic on the two rotors.

As will be appreciated, there is a considerable amount of cross-coupling of the forces and moments on the helicopter when the pilot applies the controls. The relevant equations describing the behavior of the helicopter are complicated, and there is no exact mathematical solution to this problem. For example, a change in rotor thrust produced by pulling up on the collective will require a higher power and will create a larger torque reaction on the fuselage. This, in turn, will require a yaw correction to be made to keep the helicopter tracking in a straight line. These and other so-called cross-coupling effects are unavoidable, but much can be done to minimize their effects and reduce the workload for the pilot by appropriate mixing of the control inputs. On early helicopters this was done mechanically, but now these effects can be handled electronically by an on-board flight control system.

There are two types of trim solution that are of interest to helicopter engineers: "propulsive," or "free-flight," trim and "wind-tunnel" trim. Wind-tunnel trim is used when testing model rotors in the wind tunnel and is somewhat different from free-flight trim because only force equations are used. For propulsive trim, the solution simulates the free-flight conditions of the helicopter. For a specified helicopter gross weight, center of gravity, and forward speed, the trim solution must numerically evaluate the rotor controls, namely the collective pitch angle θ_0; the cyclic pitch angles θ_{1c} and θ_{1s}; the rotor disk orientation – which is described by β_0, β_{1c}, and β_{1s}; and the vehicle orientation – which is described by the inertial angles (pitch angle, θ, roll angle, ϕ) and the aerodynamic angles (angles of attack, α, and sideslip, β, and the tail rotor collective pitch).

There are many forms of trim solutions possible, and several levels of approximations and assumptions may be used. In all cases, the free-flight trim solution is obtained from a set of vehicle equilibrium equations. These are usually simplified by using small-angle assumptions. Some trim solutions neglect the lateral equilibrium equation, and this is justified because any lateral tilt of the rotor disk does not substantially change the rotor or fuselage aerodynamics. The trim solution can be approached from two perspectives. One approach is to use the blade element theory with certain assumptions for the wake inflow to calculate analytic results for the blade flapping and control angles. This gives a good first estimate of the rotor trim state. Another approach is to use a less restrictive solution for the blade aerodynamics (which may include nonuniform inflow and nonlinear aerodynamics) and to calculate the rotor trim state numerically. In either case, an iterative approach is required.

Figure 4.14 shows the forces and moments acting on the helicopter in free-flight. In the treatment of the trim problem, it is useful to decompose each of the components according to its origin. For example, the moment can be written in terms of the contributions from the main rotor (MR), fuselage (F), horizontal tail (HT), vertical tail (VT), tail rotor (TR), and other sources (O) as

$$M = M_{MR} + M_F + M_{HT} + M_{VT} + M_{TR} + M_O \tag{4.78}$$

and similarly for the other force and moment components. The hub plane (HP) is used as the reference plane. The flight path (FP) angle is θ_{FP}. It will be assumed that there is no side-slip angle, and so the fuselage side force (Y_F) can be assumed negligible. Also, it will be assumed for simplicity that there are no contributions from the horizontal and vertical tails. For vertical force equilibrium

$$W - T_{MR}\cos\alpha_F\cos\phi_F + D\sin\theta_{FP} - H_{MR}\sin\alpha_F + Y_{MR}\sin\phi_F + Y_{TR}\sin\phi_F = 0. \tag{4.79}$$

For longitudinal force equilibrium a balance of forces results in

$$D\cos\theta_{FP} + H_{MR}\cos\alpha_F - T_{MR}\sin\alpha_F\cos\phi_F = 0. \tag{4.80}$$

For lateral force equilibrium the tail rotor thrust, T_{TR}, must be included to give

$$Y_{MR}\cos\phi_F + T_{TR}\cos\phi_F + T_{MR}\cos\alpha_F\sin\phi_F = 0. \tag{4.81}$$

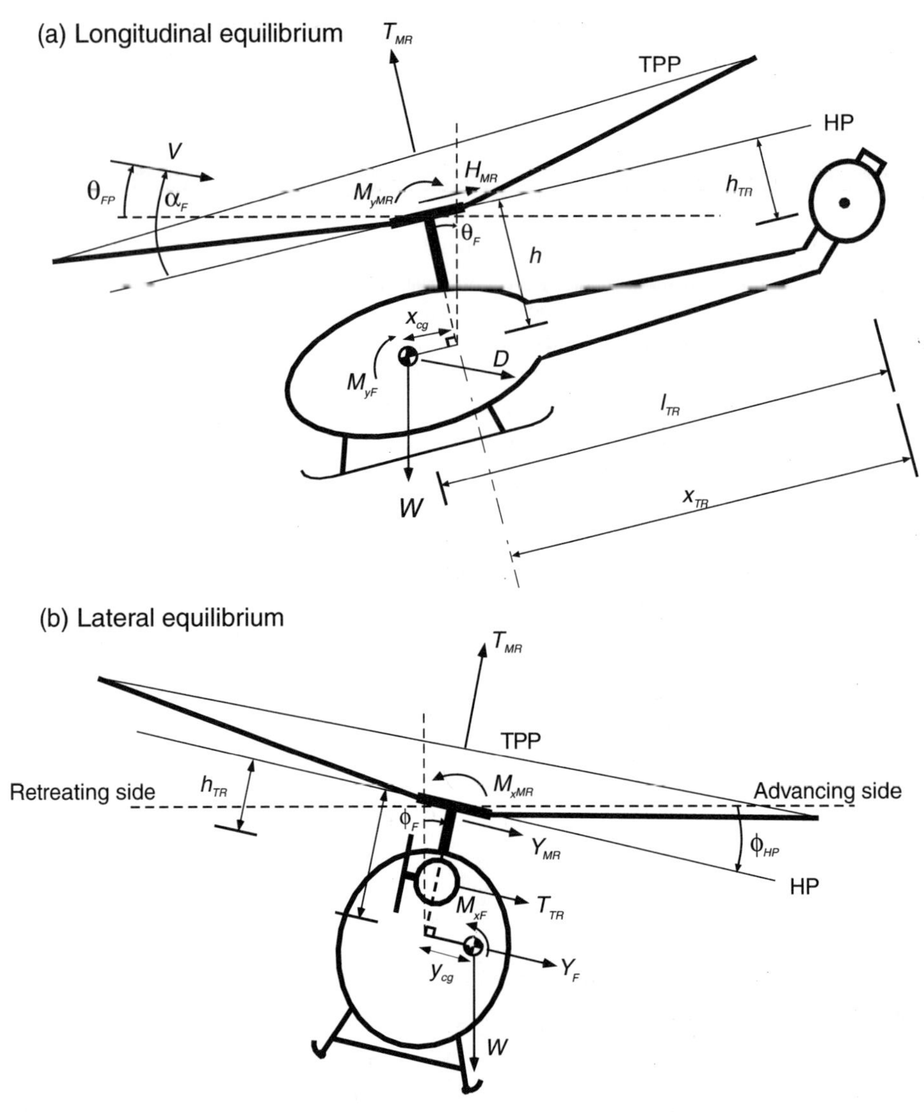

Figure 4.14 Forces and moments acting on a helicopter in free-flight. (a) Longitudinal forces and moments. (b) Lateral forces and moments.

For pitching moment equilibrium about the hub

$$M_{yMR} + M_{y_F} - W(x_{cg}\cos\alpha_F - h\sin\alpha_F) - D(h\cos\alpha_F + x_{cg}\sin\alpha_F) = 0. \quad (4.82)$$

For rolling moment equilibrium about the hub

$$M_{xMR} + M_{x_F} + T_{TR}h_{TR} + W(h\sin\phi_F - y_{cg}\cos\phi_F) = 0. \quad (4.83)$$

Finally, torque equilibrium about the shaft gives

$$Q_{MR} - Y_{TR}l_{TR} = 0. \quad (4.84)$$

Using small-angle assumptions, the equilibrium equations can be reduced to the set

$$\begin{aligned} W - T_{MR} &= 0, \\ D + H_{MR} - T_{MR}\alpha_s &= 0, \\ Y + T_{TR} + T_{MR}\phi_F &= 0, \\ M_{y_{MR}} + M_{y_F} + W(h\alpha_F - x_{cg}) - hD &= 0, \\ M_{x_{MR}} + M_{x_F} + W(h\phi_F - y_{cg}) + T_{TR}h_{TR} &= 0, \\ Q_{MR} - T_{TR}l_{TR} &= 0. \end{aligned}$$

The rotor thrust is simply the average of the blade lift during one revolution multiplied by the number of blades. Mathematically, this is stated as

$$T_{MR} = \frac{N_b}{2\pi}\int_0^{2\pi}\int_0^R dF_z d\psi. \tag{4.85}$$

The thrust coefficient can therefore be written as

$$C_{T_{MR}} = \frac{\sigma C_{l_\alpha}}{2}\frac{1}{2\pi}\int_0^{2\pi}\int_0^1 \left[\left(\frac{U_T}{\Omega R}\right)^2\theta - \left(\frac{U_P}{\Omega R}\right)\left(\frac{U_T}{\Omega R}\right)\right] dr d\psi. \tag{4.86}$$

Because of the complexity of the expressions for U_P, U_T, and θ, this equation must usually be solved numerically. However, by assuming uniform inflow, that is, $\lambda(r,\psi) = \lambda =$ constant, $C_d = C_{d_0}$, $c =$ constant, and linear twist, we can obtain the result for the rotor thrust coefficient analytically. The result can be shown to be

$$C_{T_{MR}} = \frac{\sigma C_{l_\alpha}}{2}\left[\frac{\theta_0}{3}\left(1 + \frac{3}{2}\mu^2\right) + \frac{\theta_{tw}}{4}(1+\mu^2) + \frac{\mu}{2}\theta_{1s} - \frac{\lambda}{2}\right]. \tag{4.87}$$

In addition, the rotor torque, side force, drag force, and moments about the respective axes can be computed by a similar process. The rotor drag force (also known as the H-force) is given by

$$H_{MR} = \frac{N_b}{2\pi}\int_0^{2\pi}\int_0^R [dF_x \sin\psi + dF_R\cos\psi]\, d\psi \tag{4.88}$$

or, in coefficient form,

$$\begin{aligned} C_{H_{MR}} = \frac{\sigma C_{l_\alpha}}{2}\frac{1}{2\pi}\int_0^{2\pi}\int_0^1 \Bigg\{ &\sin\psi\left[\left(\frac{U_P}{\Omega R}\right)\left(\frac{U_T}{\Omega R}\right)\theta - \left(\frac{U_P}{\Omega R}\right)^2\right] \\ &- \beta\cos\psi\left[\left(\frac{U_T}{\Omega R}\right)^2\theta - \left(\frac{U_P}{\Omega R}\right)\left(\frac{U_T}{\Omega R}\right)\right] + \frac{C_{d_0}}{C_{l_\alpha}}\left(\frac{U_T}{\Omega R}\right)^2\sin\psi\Bigg\} dr d\psi. \end{aligned} \tag{4.89}$$

The rotor side force (also known as the Y-force) is given by

$$Y_{MR} = \frac{N_b}{2\pi}\int_o^{2\pi}\int_0^R [-dF_x\cos\psi + dF_R\sin\psi]\, d\psi. \tag{4.90}$$

The rotor torque is given by

$$Q_{MR} = \frac{N_b}{2\pi}\int_0^{2\pi}\int_0^R y dF_x d\psi. \tag{4.91}$$

The rotor rolling moment is given by

$$M_{x_{MR}} = \frac{N_b}{2\pi} \int_0^{2\pi} \int_0^R y \, dF_z \sin\psi \, d\psi. \tag{4.92}$$

Finally, the rotor pitching moment is given by

$$M_{y_{MR}} = -\frac{N_b}{2\pi} \int_0^{2\pi} \int_0^R y \, dF_z \cos\psi \, d\psi. \tag{4.93}$$

Closed-form expressions for the latter quantities can also be derived.

Additional equations may be necessary. For example, two more equations should be added to determine the trim value of main rotor inflow λ_{MR} and tail rotor inflow λ_{TR}. These inflow equations should be solved together with all the other trim equations. For this purpose it is convenient to rewrite them so that all the terms are on one side of the equal sign. Using simple momentum theory for the main rotor (MR) gives

$$\lambda_{MR} - \mu_{MR} \tan\alpha_{MR} - \frac{C_{T_{MR}}}{2\sqrt{\mu_{MR}^2 + \lambda_{MR}^2}} = 0, \tag{4.94}$$

where $\mu_{MR} \cos\alpha_{MR} = V/\Omega_{MR} R_{MR}$ and α_{MR} is the disk angle of attack, for which setting $\alpha_{MR} = \alpha$ is usually reasonable. For the tail rotor (TR),

$$\lambda_{TR} - \mu_{TR} \tan\alpha_{TR} - \frac{C_{T_{TR}}}{2\sqrt{\mu_{TR}^2 + \lambda_{TR}^2}} = 0, \tag{4.95}$$

where $\mu_{TR} = V/\Omega_{TR} R_{TR}$ and α_{TR} is the disk angle of attack of the tail rotor (this will be zero if there is no side-slip angle).

The vehicle equilibrium equations, along with the inflow equations, can then written in the form $\mathbf{F}(\mathbf{X}) = 0$, where $\mathbf{X}$ is a vector of rotor trim unknowns defined as

$$\mathbf{X} = [\theta_0 \; \theta_{1c} \; \theta_{1s} \; \lambda_{MR} \; \lambda_{TR} \; \theta_F \; \phi_F]^T, \tag{4.96}$$

which can be solved numerically using a nonlinear algebraic equation solver. In the trim process, the rotor collective and cyclic pitch controls are being adjusted to control the orientation of the rotor to provide trim, control, and propulsion requirements. For wind-tunnel trim, the rotor cyclic controls are calculated from a prescribed shaft angle α_F. In a wind-tunnel situation, the model rotor is controlled by a conventional swashplate, and the operator remotely "flies" the model rotor very much in the way of an actual helicopter. However, because the rotor is rigidly attached to a support system, the operator must ensure that large amounts of cyclic flapping do not occur as the wind speed is increased. This is done by controlling the orientation of the swashplate to eliminate cyclic flapping with respect to the rotor shaft.

Representative variations in the control input angles, θ_0, θ_{1c}, and θ_{1s}, versus forward speed, as computed from a free-flight trim solution, are shown in Fig. 4.15. The results must be considered representative only but are typical of the inputs that would be required on any single rotor helicopter. An example helicopter, which resembles the UH-60, is used for these calculations. Experimental measurements of control angles are also shown, which are taken from Ballin (1987). Remember that as the rotor transitions from hover to forward flight, a primary effect is the dissymmetry in lift between the advancing and retreating blades. The excess of lift on the advancing blade causes the blade to flap upward, reaching a maximum displacement about 90° later at the front of the disk. Conversely, the reduced

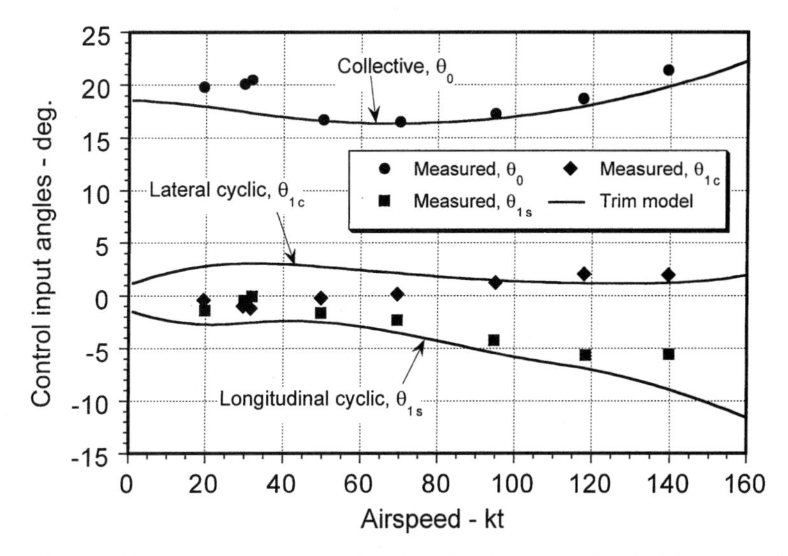

Figure 4.15 Representative variations in collective and cyclic pitch inputs to trim a rotor in forward flight. Propulsive trim calculation. Data source: Ballin (1987).

dynamic pressure on the retreating blade causes the blade to flap downward over the rear of the disk. The net effect is that the rotor disk naturally wants to tilt back (i.e., a negative longitudinal flapping or $-\beta_{1c}$). Therefore, to prevent this longitudinal flapping, the cyclic pitch controls must be adjusted to bring the rotor disk back to an orientation that will meet propulsion (to overcome fuselage drag and other airframe aerodynamic forces and moments) and control requirements. Note that the θ_{1s} component of cyclic pitch (longitudinal cyclic) controls the longitudinal flapping; thus an increasingly negative value of θ_{1s} (of magnitude approximately equal to $-\beta_{1c}$) must be imposed by using forward stick.

The effects of blade coning cause a lateral flapping response. When a blade passes over the front of the disk, the effects of the forward flight velocity cause an increase in angle of attack because it acts as an upwash. At the rear of the disk, the effects of the free stream cause a decrease in angle of attack. Therefore, there is a once-per-revolution (1/rev) forcing function generated as a result of blade coning, and the response to this forcing causes a negative lateral disk tilt (to starboard). The amount of lateral tilt is proportional to the rotor coning angle, which in turn depends on inflow for a given rotor thrust. To counteract this effect, a positive value of lateral cyclic (θ_{1c}) must be applied using left stick. A small additional amount of lateral cyclic must also be applied to counter the thrust and moments produced by the tail rotor, which will vary with the antitorque requirements. Also, some lateral cyclic is required to compensate for the nonuniformity of the longitudinal inflow over the rotor disk. Because of these coupled effects, the amount of lateral cyclic required for trim does not show a strong trend in one direction unlike the longitudinal cyclic. See Harris (1972) for a systematic study of inflow effects on rotor flapping response.

4.13 Chapter Review

A basic understanding of the dynamic behavior of the rotor blades in response to the changing aerodynamic loads has been the objective of this chapter. The blades have two primary degrees of freedom: flapping and lagging, which take place about either mechanical or virtual hinges near the blade root. A third degree of freedom allows cyclic pitch or feathering of the blade. Despite the fact that helicopter blades are relatively flexible, the basic physics of the blade dynamics can be explained by assuming them as rigid. In hovering flight the airloads do not vary with azimuth, and so the blades flap up and lag back with respect to the hub and reach a steady equilibrium position under a simple balance of aerodynamic and centrifugal forces. However, in forward flight the fluctuating airloads cause continuous flapping motion and give rise to aerodynamic, inertial, and Coriolis forces on the blades that result in a dynamic response. The flapping hinge allows the effects of the cyclically varying airloads to reach an equilibrium with airloads produced by the blade flapping motion. It has also been shown that the flapping motion is highly damped by the aerodynamic forces.

A rotor blade has a flapping natural frequency that is equal to its rotational frequency (or nearly so). Because the rotor is excited by the aerodynamic loads primarily at 1/rev, this means that there is a 90° phase lag between the aerodynamic forcing and the blade flapping response. It has been shown how collective and cyclic pitch can be used to change the magnitude and phasing of the aerodynamic loads over the disk. This is the key to changing the flapping of the rotor and the orientation of the rotor plane, thereby effecting a means of control for the helicopter. The ideas of controlling the orientation of the rotor are formally embodied in a procedure known as "trim." While there are many forms of solution to the trim problem, the basic procedure is the same: to adjust the control inputs to give resultant forces and moments on the helicopter that will satisfy a specified equilibrium and propulsive force condition. The evaluation of the various forces and moments acting on the helicopter is by no means trivial. However, under certain assumptions, and with sufficient labor, many of the results can be obtained analytically in closed form. This allows the identification of the relationships between the solution and various problem parameters and will be useful for many problems related to the preliminary design of the helicopter.

4.14 Questions

4.1. A rigid rotor blade with a uniform mass distribution has its flapping hinge located at a distance e from the rotational (shaft) axis. If the shaft is rotating at an angular velocity Ω, calculate the natural frequency of the blade about the flapping hinge. Show all the steps in your derivation. Assume $e/R \ll 1$.

4.2. Derive expressions for the mean coning angle, β_0, of a hovering rotor blade that has a uniform mass per unit length m and a concentrated mass M at its tip. Zero hinge offset can be assumed. Assume that the thrust on the blade varies linearly with radius.

4.3. (a) Name and explain the functions of the three principal degrees-of-freedom of a rotor blade. (b) Describe the differences between (i) an articulated rotor system, (ii) a hingeless rotor system, and (iii) a bearingless rotor system. What are the principal advantages of one versus the other?

4.4. A rotor in a given flight condition has the following flapping motion with respect to the control axis (control plane): $\beta(\psi) = 6^\circ - 4^\circ \cos\psi - 4^\circ \sin\psi$. (a) Sketch a side view and rear view of the rotor. (b) How much is the TPP inclined in the fore and aft direction? Forward or backward? (c) How much is the TPP inclined laterally? Is the advancing or retreating blade high? (d) What angle does the blade make with the control plane at $\psi = 0, 90, 180, 270$ degrees. (e) At what azimuth angle is the flapping angle greatest? What is the flapping angle at this point?

4.5. If the flapping motion of a centrally hinged blade with uniform mass distribution in forward flight is represented as $\beta = \beta_0 + \beta_{1c} \cos\psi + \beta_{1s} \sin\psi$, prove that the thrust moment (that is the flapping moment created by the blade thrust) about the flapping hinge is constant around the disk (i.e., independent of ψ).

4.6. By considering the blade thrust, blade weight, centrifugal force, and blade inertia of a centrally flapping untwisted untapered blade, find an expression for the coning angle of the blade in forward flight. Assume uniform inflow.

4.7. Starting from the equation of motion of a flapping blade with a hinge offset operating in hover, show that the flapping response lags the aerodynamic forcing by an angle

$$\phi = \tan^{-1}\left[\frac{\gamma}{8}\left(\frac{1}{\nu_\beta^2 - 1}\right)\right],$$

where all the symbols have their usual meanings.

4.8. A tail rotor uses pitch-flap coupling via a δ_3 hinge to reduce cyclic flapping. This is achieved by placing the pitch link/pitch horn connection to lie off the flap hinge axis. Find a relationship between the coning angle and the collective pitch, and the effect on the flapping frequency. Assume zero flapping hinge offset and uniform inflow.

4.9. Calculate and plot as a function of hinge offset the natural frequency about the lag hinge of a blade that is twice as heavy per unit length at the root compared to the tip, where e is the lag hinge offset.

4.10. Explain why, in reality, the variation of inflow through a helicopter rotor in forward flight is highly nonuniform. If the time-averaged inflow is assumed linearly distributed longitudinally and laterally over the disk, qualitatively describe the effects of this nonuniform inflow on the rotor flapping response, compared to the uniform inflow case with the same mean value of inflow.

Bibliography

Bailey, F. J. 1941. "A Simplified Theoretical Method of Determining the Characteristics of a Lifting Rotor in Forward Flight," NACA Report 716.

Ballin, M. G. 1987. "Validation of a Real-Time Engineering Simulation of the UH-60 Helicopter," NASA TM-88360.

Bramwell, A. R. S. 1976. *Helicopter Dynamics*, Edward Arnold, London.

Chopra, I. 1990. "Perspectives in Aeromechanical Stability of Helicopter Rotors," *Vertica*, 14 (4), pp. 457–508.

de la Cierva, J. and Rose, D. 1931. *Wings of Tomorrow: The Story of the Autogiro*, Brewer, Warren & Putnam, NY.

Friedmann, P. P. 1977. "Recent Developments in Rotary-Wing Aeroelasticity," *J. of Aircraft*, 14 (11), pp. 1027–1041.

Glauert, H. 1928. "On the Horizontal Flight of a Helicopter," ARC R & M 1157.

Harris, F. D. 1972. "Articulated Rotor Blade Flapping Motion at Low Advance Ratios, "*J. American Helicopter Society* 17 (1), pp. 41–48.

Johnson, W. 1980. *Helicopter Theory*, Princeton University Press, Princeton, NJ, Chapter 12.

Lock, C. N. H. 1927. "Further Development of Autogyro Theory," ARC R & M 1127.

Loewy, R. G. 1969. "Review of Rotary Wing V/STOL Dynamics and Aeroelastic Problems," *J. American Helicopter Soc.*, 14 (3), pp. 3–24.

Nikolsky, A. 1944. *Notes on Helicopter Design Theory*, Princeton University Press, Princeton, NJ.

Nikolsky, A. 1951. *Helicopter Analysis*, John Wiley & Sons, Inc., New York.

Reichert, G. 1973. "Basic Dynamics of Rotors," AGARD-LS-63.

Sissingh, G. 1939. "Contributions to the Aerodynamics of Rotating Wing Aircraft," NACA TM 921.

Sissingh, G. 1941. "Contributions to the Aerodynamics of Rotating Wing Aircraft, Part 2." NACA TM 990.

Squire, H. B. 1936. "The Flight of a Helicopter," ARC R & M 1730.

Wald, Q. 1943. "A Method for Rapid Estimation of Helicopter Performance," *J. Aeronaut. Sci.*, 10 (4), April pp. 131–135.

Wheatly, J. B. 1934. "An Aerodynamic Analysis of the Autogyro Rotor and Comparison between Theory and Experiments," NACA TR 487.

CHAPTER 5

Basic Helicopter Performance

The future of the helicopter . . . therefore lies not in competition with the airplane, but in its ability to perform certain functions which the airplane cannot undertake.

Dr. Alexander Klemin (1925)

5.1 Introduction

The aerodynamic tools described in the previous chapters can now be used to analyze the basic performance of the helicopter. By the term helicopter *performance* we mean the estimation of the installed engine power required for a given flight condition, determination of maximum level flight speed, or the estimation of the endurance or range of the helicopter. Both the momentum and blade element methods will allow for good estimates of total rotor thrust, power, and figure of merit in hover and can be used with confidence to predict overall rotor performance. In addition, the performance of the helicopter in autorotational flight and when operating near the ground will be considered in this chapter.

5.2 Hovering and Axial Climb Performance

It has already been shown in Chapter 3 that the rotor power in hover can be estimated using the modified momentum theory to get

$$P = P_0 + P_i = \rho A(\Omega R)^3 \left(\frac{\sigma C_{d_0}}{8} \right) + \frac{\kappa W^{3/2}}{\sqrt{2\rho A}}, \tag{5.1}$$

where $W(\approx T)$ is the weight, and the induced power factor, κ, is estimated on the basis of the best available method. While the above equation assumes a rectangular blade design, the profile power part can easily be adjusted to take into account blade taper. Note that the hover power required is a function of aircraft gross weight. Another important point to note is that the rotor power required is a function of the air density. This makes it difficult to compare aircraft performance unless the data are corrected to standard conditions. For decreasing density (increasing altitude and/or temperature), a greater power is required for the same aircraft gross takeoff weight (GTOW). To remedy this, the International Standard Atmosphere (ISA) has been established, which gives the height in a standard atmosphere corresponding to the density of the air in which the aircraft is operating. The standards[1] are defined in Minzner et al. (1959), ICAO (1964) and NASA (1966), where a "standard day" at sea level is an air temperature of 15°C (59°F) with a barometric pressure of 760 mm (29.92 inches) of mercury or 1013.2 millibars. Up to 11 km altitude, the pressure, p, and the temperature, T, are related by

$$\frac{dp}{p} = \frac{dT}{T}\frac{1}{R}\left(\frac{dh}{dT}\right), \tag{5.2}$$

[1] The ICAN standard may be used in many European countries.

where the standard lapse rate dT/dh is 6.51°K (or 6.51°C) per km of altitude or 3.57°R (or 3.57°F) per 1,000 ft of altitude. In practical terms, the lapse rate in the standard atmosphere can be expressed by $T = 15 - 0.001981h$, where the altitude h is expressed in feet and the temperature T in °C. Integrating Eq. 5.2 from sea level using the standard lapse rate gives a relationship between temperature and pressure, and using the equation of state, the corresponding density of the air can be determined. The *density altitude*, h_ρ, (in feet) is defined using the ambient density in the relation

$$h_\rho = \frac{288.16}{0.001981}\left[1 - \frac{\rho}{\rho_0}\right]^{0.235}, \tag{5.3}$$

where the subscript 0 refers to sea-level conditions. (The result can be converted from feet to meters by multiplying by 0.3048.) *Pressure altitude*, h_p, is the height at which a given pressure is found in the standard atmosphere. The following expression is used:

$$h_p = \frac{288.16}{0.001981}\left[1 - \left(\frac{p}{p_0}\right)^{0.235}\right]. \tag{5.4}$$

The aircraft altimeter is calibrated according to Eq. 5.4. The advantage of using pressure altitude is that it is a function of the ambient pressure alone, and the value of h_p can be read directly off the altimeter in the aircraft. This is done by setting the value in the altimeter Kohlsman (reference) window to the standard sea-level pressure of 29.92 inches or 1013.2 millibars. Density altitude must be obtained from measurements of pressure altitude corrected for nonstandard ambient temperature. The density ratio can be obtained from

$$\frac{\rho}{\rho_0} = \frac{288.16}{(T + 273.16)}\left(1 - \frac{0.001981\, h_p}{288.16}\right)^{5.256}, \tag{5.5}$$

where h_p is in feet and T is in °C. Pressure altitude and density altitude are identical if the temperature conforms to standard conditions. For example, on a standard day the density ratio will be 1.0 at sea level and 0.864 at 4,000 ft. On a very hot day with a pressure (indicated) altitude of 4,000 ft but with an outside air temperature of 30°C, the density ratio will be 0.82. Performance charts in a helicopter flight manual may be expressed in terms of either pressure altitude or density altitude, although pressure altitude is the more common form. Because the aircraft altimeter will read pressure altitude, a simple altitude conversion chart is provided to the pilot to allow rapid conversions to density altitude for a given ambient temperature, if required.

The basic effect of density altitude on hovering performance is illustrated in Fig. 5.1. The results are for a representative helicopter closely resembling the UH-60 at a GTOW of 16,000 lb (≈ 7,250 kg). The basic parameters of this aircraft are given in Table 5.1. The empty weight of the aircraft is 11,000 lb (≈ 5,000 kg). Notice from Fig. 5.1 that as much as 20% greater power will be required to hover at a density altitude of 9,000 ft versus that required at sea-level conditions. The rotor efficiency or figure of merit is not substantially affected by density altitude. While the power required to hover increases with increasing density altitude, the engine power available also decreases. A good approximation to the power available at any altitude can be found by multiplying the power available at sea level by the ratio of the density of that altitude to the density at sea level.

Representative results for the example helicopter that show the decrease in excess power with aircraft gross weight for various density altitudes are given in Fig. 5.2. At a given altitude the decrease in excess power is almost proportional to the extra aircraft weight. The

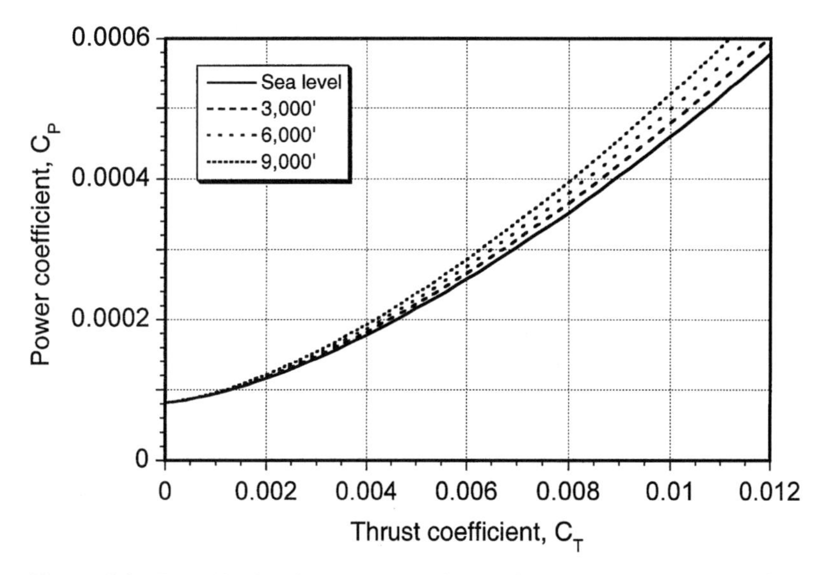

Figure 5.1 Example showing power required to hover versus rotor thrust for various density altitudes.

excess power available ultimately becomes zero, and this defines the hover ceiling. For a maximum GTOW of 20,000 lb ($\approx$ 9,100 kg) the hover ceiling for the example helicopter is about 7,000 ft ($\approx$ 2,100 m), which means that the aircraft cannot hover above this altitude at this weight. (See also Question 5.1.)

The power required for any vertical rate of climb, V_c, can be estimated by solving for the induced velocity using the simple momentum result given previously in Chapter 2, namely

$$\frac{v_i}{v_h} = -\frac{V_c}{2v_h} + \sqrt{\left(\frac{V_c}{2v_h}\right)^2 + 1} \approx -\frac{V_c}{2v_h} + 1 \quad \text{for low rates of climb,} \tag{5.6}$$

where v_h is the hover induced velocity. The maximum rate of climb is then obtained by solving for V_c using

$$\frac{P_h + \Delta P}{P_h} = \frac{V_c}{2v_h} + \sqrt{\left(\frac{V_c}{2v_h}\right)^2 + 1} \approx \frac{V_c}{2v_h} + 1 \quad \text{for low rates of climb,} \tag{5.7}$$

where ΔP is the excess power available over and above that required for hover. Note that

Table 5.1. *Parameters for Example Helicopter*

Parameter	Symbol	Main Rotor	Tail Rotor
Radius (ft)	R	27.0	5.5
Chord (ft)	c	1.7	0.80
Solidity	σ	0.082	0.19
Number of blades	N_b	4	4
Tip speed (ft s^{-1})	ΩR	725	685
Induced power factor	κ	1.15	1.15
Profile drag coefficient	C_{d_0}	0.008	0.008

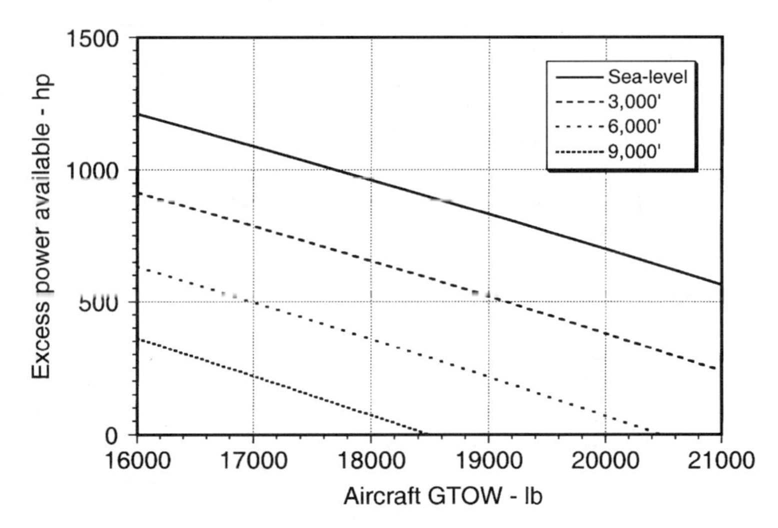

Figure 5.2 Excess hover power available as a function of GTOW and density altitude.

the climb velocity does not depend on excess rotor thrust but on an excess of power. It will be apparent for low to moderate rates of climb that $\Delta P \approx T V_c/2$, or that V_c is given approximately by $2\Delta P/W$ in axial flight. Normally it is convention in performance work for the rate of climb to be expressed in terms of feet/min, mainly because the rate of climb indicator in the aircraft will be calibrated this way.

Representative results for the maximum possible climb velocity for the example helicopter at a GTOW = 16,000 lb (≈7,256 kg) are shown in Fig. 5.3. Note the significant decrease in the maximum possible climb velocity with increasing density altitude. High values of density altitude can be encountered during "hot and high" operations, and it is important for the pilot to recognize that a potentially serious performance degradation may occur under these conditions.

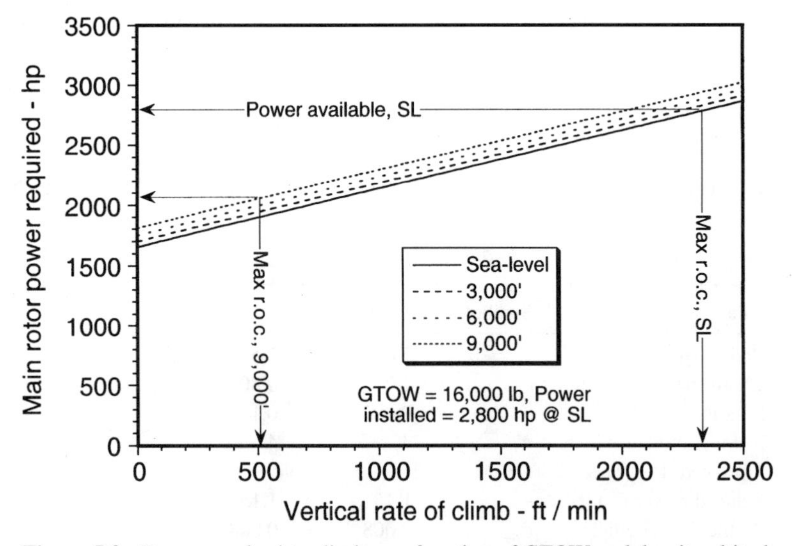

Figure 5.3 Power required to climb as a function of GTOW and density altitude.

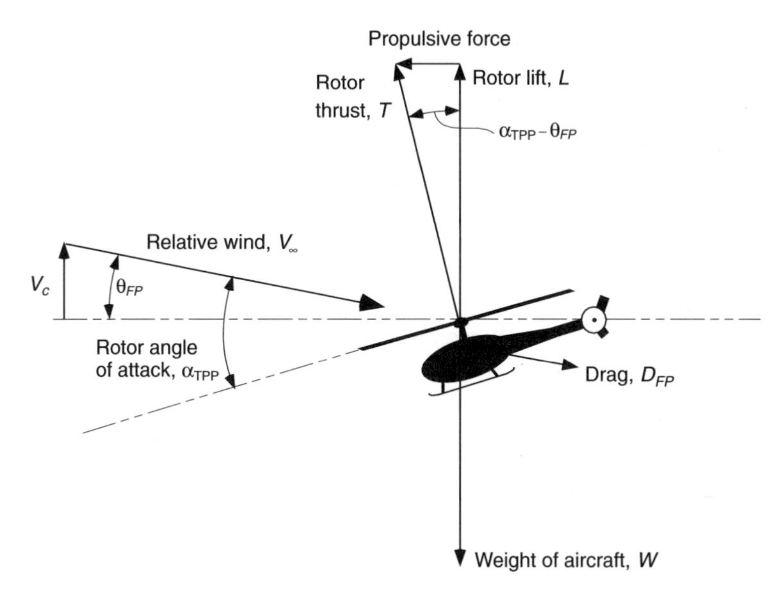

Figure 5.4 Equilibrium of forces on a helicopter in forward flight.

5.3 Forward Flight Performance

For a helicopter in forward flight, the total power required, P, can be expressed by the equation

$$P = P_i + P_0 + P_p + P_c, \tag{5.8}$$

where P_i is the induced power, P_0 is the profile power required to overcome viscous losses, P_p is the parasite power required to overcome the drag of the aircraft, and P_c is the climb power required to increase the gravitational potential of the aircraft. Consider the equilibrium of forces on a single rotor helicopter in a climbing forward flight situation, as shown in Fig. 5.4. In the figure θ_{FP} is the flight path angle, so that for small angles the climb velocity, $V_c = V_\infty \theta_{FP}$. For small angles, satisfying vertical equilibrium gives the equation

$$T \cos(\alpha_{\text{TPP}} - \theta_{FP}) = W \approx T. \tag{5.9}$$

Satisfying horizontal equilibrium leads to

$$T \sin(\alpha_{\text{TPP}} - \theta_{FP}) = D_{FP} \cos \theta_{FP}. \tag{5.10}$$

Assuming D_{FP} is independent of the angle of climb, then this latter equation simplifies to

$$T(\alpha_{\text{TPP}} - \theta_{FP}) = D. \tag{5.11}$$

Rearranging and solving for the disk angle of attack, α_{TPP}, gives

$$\alpha_{\text{TPP}} = \theta_{FP} + \frac{D}{W}. \tag{5.12}$$

Now, consider the power to undertake a climb (and also to propel the rotor forward). This part of the power is

$$T V_\infty \sin \alpha_{\text{TPP}} \approx T V_\infty \alpha_{\text{TPP}} = W V_\infty \left(\theta_{FP} + \frac{D}{W} \right) \tag{5.13}$$

$$= W V_\infty \theta_{FP} + D V_\infty = W V_c + D V_\infty. \tag{5.14}$$

The term WV_c is known as the climb power, P_c. The term DV_∞ is known as the *parasite power*, P_p, because this is energy lost to viscous effects.

5.3.1 *Induced Power*

It is already known from the simple one-dimensional momentum theory described in Chapter 2 that the induced power of the rotor, P_i, can be approximated as

$$P_i = \kappa T v_i. \tag{5.15}$$

If the forward velocity is sufficiently high, say $\mu > 0.1$, then the induced velocity can be approximated by the asymptotic result given previously (Glauert's formula) so that

$$v_i = \frac{T}{2\rho A V_\infty} = \frac{W}{2\rho A V_\infty}. \tag{5.16}$$

Therefore, the power equation can be written more simply as

$$P = P_0 + WV_c + P_p + \frac{\kappa W^2}{2\rho A V_\infty}, \tag{5.17}$$

where κ is the now familiar empirical correction to account for a multitude of aerodynamic phenomena, mainly those resulting from tip losses and nonuniform inflow. The value of κ cannot necessarily be assumed independent of advance ratio, but the use of a mean value is usually sufficiently accurate for preliminary power predictions. The previous equation shows the origin of the terms that comprise the basic power requirements of the helicopter in forward flight. Note that in coefficient form the induced power can be written as

$$C_{P_i} = \frac{\kappa C_W^2}{2\sqrt{\lambda^2 + \mu^2}} \approx \frac{\kappa C_W^2}{2\mu} \text{ for larger } \mu. \tag{5.18}$$

5.3.2 *Blade Profile Power*

Glauert (1926) and Bennett (1940) were among the first to establish estimates of profile power using the blade element theory. The profile power coefficient with a uniform chord is

$$C_{P_0} = C_{Q_0} = \frac{\sigma C_{d_0}}{4\pi} \int_0^{2\pi} \int_0^1 \left(\frac{U}{\Omega R} \right)^3 dr\, d\psi, \tag{5.19}$$

where U is the resultant velocity at the element, and where C_{d_0} is the profile (viscous) drag coefficient of the airfoils that make up the rotor blades. Neglecting radial flow component U_R such that $U = U_T = \Omega R(r + \mu \sin\psi)$ gives

$$C_{P_0} = \frac{\sigma C_{d_0}}{4\pi} \int_0^{2\pi} \int_0^1 (r + \mu \sin\psi)^3\, dr\, d\psi. \tag{5.20}$$

Expanding and integrating gives

$$\begin{aligned} C_{P_0} &= \frac{\sigma C_{d_0}}{4\pi} \int_0^{2\pi} \int_0^1 (r^3 + 3r^2\mu \sin\psi + 3r\mu^2 \sin^2\psi + \mu^3 \sin^3\psi)\, dr d\psi \\ &= \frac{\sigma C_{d_0}}{8}(1 + 3\mu^2). \end{aligned} \tag{5.21}$$

The inclusion of the radial component of the velocity at the blade element means $U^2 = U_T^2 + U_R^2$, where $U_R = \Omega R \mu \cos \psi$. The problem of calculating C_{P_0} can now only be solved numerically. The results from the analysis of Glauert (1926) and Bennett (1940) show that the profile drag can be approximated as

$$C_{P_0} = \frac{\sigma C_{d_0}}{8}(1 + K\mu^2), \tag{5.22}$$

where the numerical value of K varies from 4 in hover to 5 at $\mu = 0.5$, depending on the assumptions and/or approximations that are made. In practice, usually average values of K are used, independent of advance ratio. Bennett (1940) used an average value of $K = 4.6$, while Stepniewski (1973) suggests $K = 4.7$. Either value will be acceptable for basic performance studies.

Johnson (1980) summarizes how various other assumptions affect the profile power, including the effects of reverse flow and modified drag coefficients in yawed and reverse flow (also, see Question 5.4). The validity of some of these assumptions, however, is questionable and they do not necessarily lead to more accurate models of the profile power. For example, because the blade clearly stalls in the reverse flow region, the assumption that $C_d = C_{d_0} =$ constant is clearly invalid there. Generally, however, the effects of the various other assumptions are small, except for radial flow, which must be included to give an accurate calculation of power. In this case numerical evaluation of the integral is required, but the advantage of this numerical approach is that more realistic models of the airfoil drag as a function of angle of attack (including stall) can be included.

Gessow & Crim (1956) have estimated the additional effects of compressibility on the overall rotor profile power requirements. They suggest a power increment of the form

$$\frac{\Delta C_{P_0}}{\sigma} = 0.007(\Delta M_{dd}) + 0.052(\Delta M_{dd})^2 \tag{5.23}$$

be applied when the tip of the advancing blade exceeds the drag divergence Mach number M_{dd} by the amount ΔM_{dd}. Such information must be drawn from tests on 2-D airfoils (see Chapter 7). Generally, it is found that the drag on an airfoil remains nominally constant and independent of Mach number until M_{dd} is reached. Although M_{dd} is also a function of angle of attack, the relationship between M_{dd} and the rotor mean lift coefficient can be approximated empirically – see Stepniewski & Keys (1984) and Prouty (1986). Similar techniques can be developed to account for the effects of stall. See also Gustafson & Myers (1946), Gustafson & Gessow (1947), and Amer (1955). While these techniques are by no means exact, they allow one to make a relatively simple estimate of compressibility and stall effects on rotor performance predictions.

A more detailed analysis of compressibility effects would represent the actual nonlinear airfoil characteristics as functions of Mach number through stall at each blade element, followed by integrating numerically to find the effects on rotor thrust and power. Some allowance for so-called tip-relief effects would normally be included in any power calculation. Tip relief accounts for the relaxation of compressibility effects at the edge of a finite wing, and simple approximations for the effect can be developed based on transonic similarity rules. The effect was first noted in experiments on propellers, which showed that losses in propulsion efficiency did not occur until the tip Mach numbers well exceeded the drag divergence Mach number of the blade tip sections estimated from two-dimensional considerations. One practical analysis of tip-relief effects is given by Prouty (1971, 1986). See also LeNard (1972) and LeNard & Boehler (1973).

5.3.3 *Parasitic Power*

The parasitic power, P_p, is a power loss as a result of viscous shear effects and flow separation (pressure drag) on the airframe, rotor hub, etc. Because helicopter airframes are much less aerodynamic than their fixed-wing counterparts (for the same gross weights), this source of drag can be very significant. We can write the parasitic power as

$$P_p = \left(\frac{1}{2}\rho V_\infty^2 S_{\text{ref}} C_{D_f}\right) V_\infty, \tag{5.24}$$

where S_{ref} is some reference area and C_{D_f} is the drag coefficient of the fuselage based on this reference area. In nondimensional form, this becomes

$$C_{P_p} = \frac{1}{2}\left(\frac{S_{\text{ref}}}{A}\right)\mu^3 C_{D_f} = \frac{1}{2}\left(\frac{f}{A}\right)\mu^3, \tag{5.25}$$

where A is the rotor disk area and f is known as the *equivalent wetted area* or *equivalent flat-plate area*. This parameter accounts for the drag of the hub, fuselage, landing gear, etc., in aggregate. The concept of equivalent wetted area comes from noting that while the fuselage drag coefficient can be written in the conventional way as

$$C_{D_f} = \frac{D_f}{\frac{1}{2}\rho V_\infty^2 S_{\text{ref}}}, \tag{5.26}$$

where S_{ref} is a reference area, the definition of S_{ref} may not be unique. Thus an equivalent wetted area is used, which is defined as

$$f = \frac{D_f}{\frac{1}{2}\rho V_\infty^2}. \tag{5.27}$$

This avoids any confusion that may arise through the definition of S_{ref}. It is found that values of f range from about 10 ft^2 (0.93 m^2) on smaller helicopters to as much as 50 ft^2 (4.65 m^2) on large utility helicopter designs. The concept of equivalent flat-plate area is discussed further in Chapter 6.

5.3.4 *Climb Power*

The climb power is equal to the time rate of increase of potential energy. If the potential energy is denoted as E, then $E = Wh$. The rate of increase of potential energy is $W\dot{h} = TV_c = WV_c$, where W is the aircraft weight and V_c is the climb velocity. The climb power coefficient can be written as $C_{P_c} = \lambda_c C_W$. The effect of the fuselage vertical drag is normally taken into account when estimating the climb power, and this is discussed further in Chapter 6.

5.3.5 *Tail Rotor Power*

The power required from the tail rotor typically varies between 5 and 10% of the main rotor power. It is calculated in a similar way to the main rotor power, with the thrust required being set equal to the value necessary to balance the main rotor torque reaction on the fuselage. The use of vertical tail surfaces to produce a side force in forward flight can help to reduce the power fraction required for the tail rotor, albeit at the expense of some increase in parasitic and induced drag. If the distance from the main rotor shaft to the tail rotor shaft is x_{TR}, the tail rotor thrust required will be

$$T_{TR} = \frac{(P_i + P_0 + P_p)}{\Omega\, x_{TR}}, \tag{5.28}$$

where Ω is the angular velocity of the main rotor. This assumes that there is no off-loading of the tail rotor by the fin. The interference between the main rotor and the tail rotor, and between the tail rotor and the vertical fin, is usually neglected in preliminary analysis. However, the effects of the main rotor wake may be accounted for by an increase in the induced power factor, κ, to take into account the generally higher nonuniform inflow at the tail rotor location. The loss of tail rotor efficiency because of the vertical fin can be approximately accounted for by results discussed in Chapter 6. Although the tail rotor power consumption is relatively low, interference effects may increase the power required by up to 20%, depending on the tail rotor and fin configuration. The tail rotor power required is initially high in hover, but it quickly decreases as airspeed builds up and the main rotor torque requirements decrease. In high speed forward flight, tail rotor power required increases again as the main rotor torque increases to overcome parasitic drag. However, this is usually offset to some extent by the aerodynamic side force produced on a vertical fin. Because of the relatively low amount of power consumed by the tail rotor, for first estimates of performance the power required can be expressed as a fraction of the total main rotor power.

5.3.6 *Total Power*

In light of the forgoing, the total main rotor power coefficient in forward flight can be written in the form

$$C_P = C_Q = \frac{\kappa C_W^2}{2\sqrt{\lambda^2 + \mu^2}} + \frac{\sigma C_{d_0}}{8}(1 + K\mu^2) + \frac{1}{2}\left(\frac{f}{A}\right)\mu^3 + \lambda_c C_W. \tag{5.29}$$

For larger values of μ, $\lambda < \mu$, so that Glauert's formula allows this equation to be simplified to

$$C_P = C_Q = \frac{\kappa C_W^2}{2\mu} + \frac{\sigma C_{d_0}}{8}(1 + K\mu^2) + \frac{1}{2}\left(\frac{f}{A}\right)\mu^3 + \lambda_c C_W. \tag{5.30}$$

Representative results of net power required for the example helicopter in straight-and-level flight is shown in Fig. 5.5. A gross takeoff weight of 16,000 lb (7,256 kg) and an operating altitude of 5,200 ft (1,585 m) were assumed. The rotor disk angle of attack was calculated at each airspeed to satisfy the horizontal force equilibrium equation, which, although not a complete trim calculation, provides reasonably acceptable results. An analysis of the predicted components of the total rotor power are also shown, including that of the tail rotor. The equivalent flat-plate area, f, of the helicopter is 23.0 ft (2.137 m^2). For both the main and tail rotors, it is assumed that $\kappa = 1.15$ and $C_{d_0} = 0.008$. The distance between main and tail rotor shafts, x_{TR}, is 32.5 ft (9.9 m).

Note from Fig. 5.5 that the induced part of the power initially decreases with increasing airspeed but increases again as the disk is progressively tilted forward to meet propulsion requirements. It can be seen that the power required for high speed forward flight increases dramatically at higher airspeeds because it is proportional to μ^3. The airframe drag makes the major contribution to the total power required in high speed flight, and much can be done to expand the flight envelope by designing a more streamlined airframe. Unfortunately,

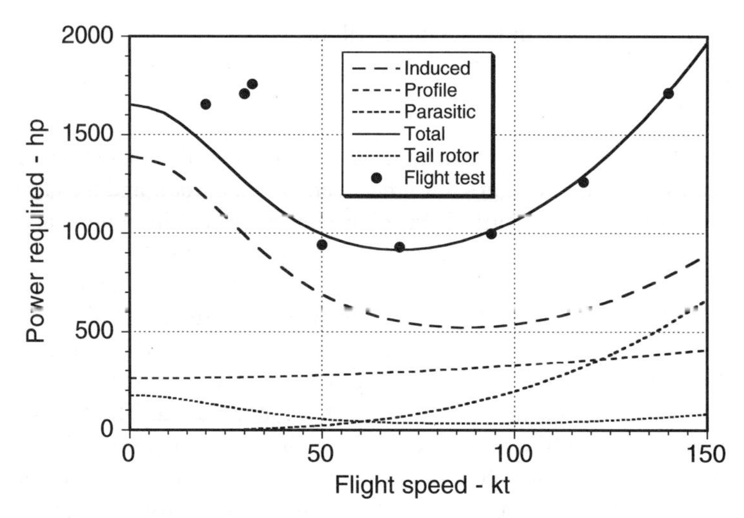

Figure 5.5 Predictions of main rotor power in forward flight. Data source: Ballin (1987).

because of various design constraints, this is not always an easy process. However, noticeable improvements can be made by fairing in the hub to the fuselage.

5.3.7 *Effect of Gross Weight*

Clearly the power required in forward flight will also be a function of GTOW. Representative results showing the effect of GTOW on the rotor power required are given in Fig. 5.6 for the example helicopter at sea-level (SL) conditions. Note that with increasing GTOW, the excess power available becomes progressively less, but it is particularly affected at lower airspeed where the induced power requirement constitutes a greater fraction of the total power. In this case, the power available at SL is 2,800 hp and for a gas turbine this stays relatively constant with airspeed. The airspeed obtained at the intersection of

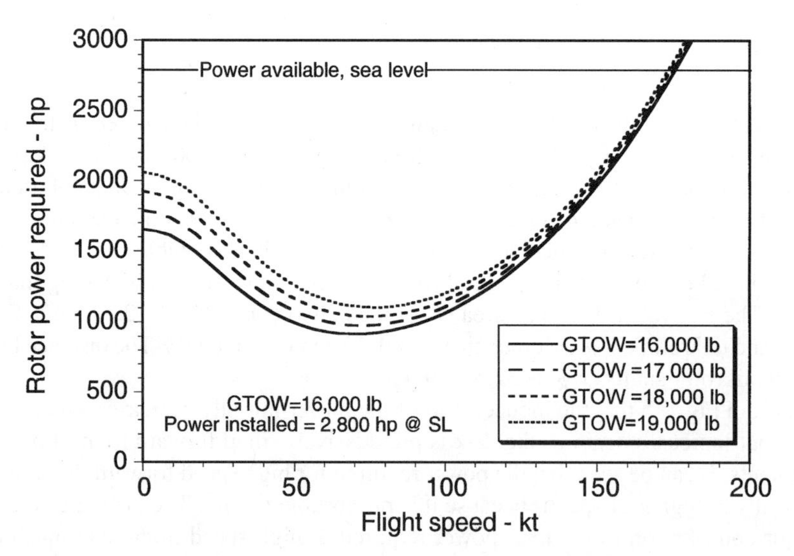

Figure 5.6 Predictions of main rotor power in forward flight at different gross take off weights.

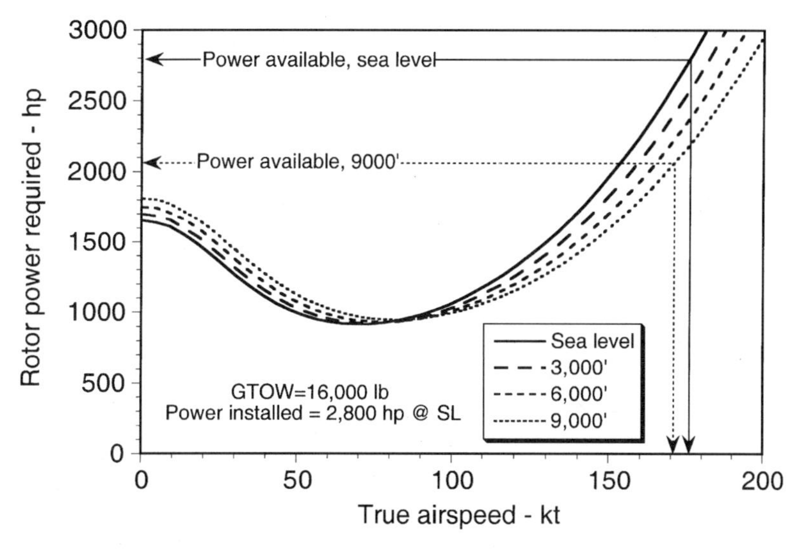

Figure 5.7 Predictions of main rotor power in forward flight at different density altitudes.

the power required curve with the power available curve gives the maximum level flight speed; however, the maximum speed is likely to be limited by the onset of rotor stall and compressibility effects before this point is reached.

5.3.8 *Effect of Density Altitude*

As discussed at the beginning of this chapter, an important operational consideration is the effect of altitude on overall helicopter performance. As shown in Fig. 5.7, increasing density altitude increases the power required in hover and lower airspeeds. At higher airspeeds, the result of lower air density results in a lower power requirement because of the reduction of parasitic drag. However, a higher density altitude will also affect the engine power available. As shown in Fig. 5.7, at 9,000 ft the power available is about 25% less than that available at sea-level conditions, resulting in a large decrease in the excess power available at any airspeed relative to that at sea-level conditions.

5.3.9 *Lift-to-Drag Ratios*

The lift-to-drag ratio (L/D) of the aircraft or just the rotor alone can also be calculated from the power required curves. This is useful for comparison of the forward flight efficiency with another rotor, helicopter, or a fixed-wing aircraft. Because the rotor provides both propulsive and lifting forces, the lift force is $T\cos\alpha_{\text{TPP}}$. The effective drag can be calculated from the power expended (i.e., $D = P/V_\infty$). For the rotor alone the power is $P = P_i + P_0$, and for the complete helicopter $P = P_i + P_0 + P_P + P_{TR}$. Therefore, for the rotor alone the lift-to-drag ratio is

$$\frac{L}{D} = \frac{T\cos\alpha_{\text{TPP}}}{(P_i + P_0)/V_\infty} \approx \frac{W V_\infty}{P_i + P_0}. \tag{5.31}$$

For the complete helicopter the lift-to-drag ratio is

$$\frac{L}{D} = \frac{T\cos\alpha_{\text{TPP}}}{(P_i + P_0 + P_p + P_{TR})/V_\infty} \approx \frac{W V_\infty}{P_i + P_0 + P_p}. \tag{5.32}$$

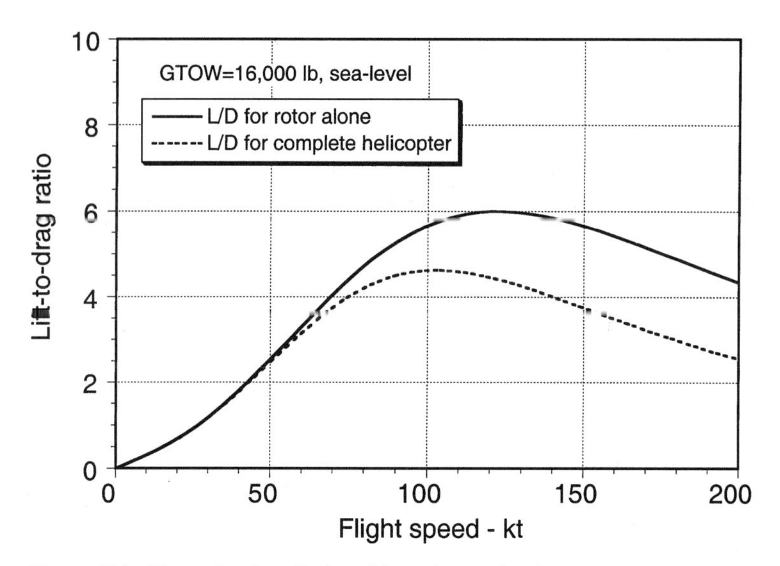

Figure 5.8 Example of equivalent lift-to-drag ratios for rotor and complete helicopter.

Representative results for the lift-to-drag ratio for the example helicopter in forward flight are shown in Fig. 5.8. It is apparent that the L/D increases rapidly as induced power requirements decrease, reaches a maximum, then drops off sharply as the parasitic power requirements rapidly increase. The maximum lift-to-drag ratio of the rotor is about 6, which is typical of most modern helicopters and is also typical of a low aspect ratio fixed wing. The maximum lift-to-drag ratio of the complete helicopter is about 4.5, which gives some idea as to the considerable effect of the airframe drag in the overall cruise efficiency of the aircraft. A typical L/D ratio for a fixed-wing aircraft of the same gross weight will be about 3 to 4 times this value. Note that the maximum lift-to-drag ratio for this example occurs at about 100 kts, which will be the airspeed to fly for maximum range. Also, note that the lift-to-drag ratio of the rotor and helicopter will also be a function of density altitude.

5.3.10 *Climb Performance*

The general power equation can be used to estimate the climb velocity, V_c, that is possible at any given airspeed. Rearranging in terms of V_c leads to

$$V_c = \frac{P - \left(P_0 + P_p + \dfrac{\kappa T^2}{2\rho A V_\infty} \right)}{T}. \tag{5.33}$$

It is realistic to assume that for low rates of climb or descent the rotor induced power, P_i, the profile power, P_0, and the airframe drag, D, remain nominally constant. In this case we can easily solve for the climb velocity to get

$$V_c = \frac{P - P_{\text{level}}}{T} = \frac{\Delta P}{T}. \tag{5.34}$$

Note that P_{level} is simply the net power required to maintain level flight conditions at the same forward speed. If the installed power available is P_a, then it will be seen that the power

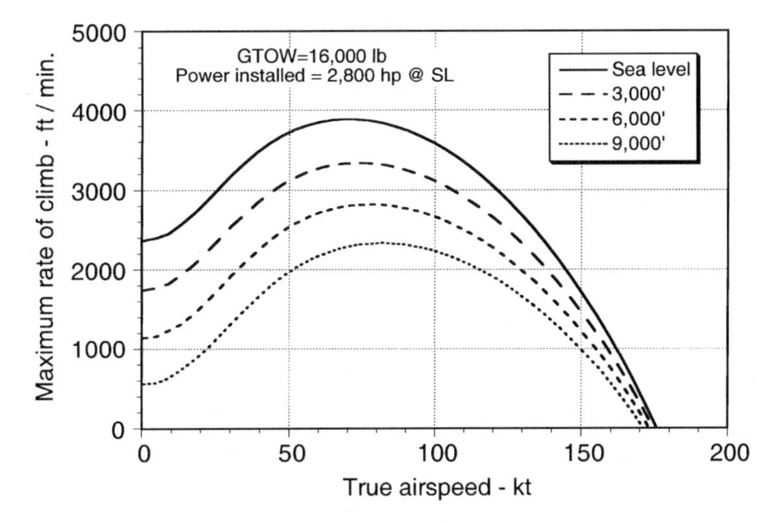

Figure 5.9 Maximum possible rate of climb as a function of airspeed for different density altitudes.

available to climb varies with forward flight speed. The climb velocity can then be obtained from

$$V_c = \frac{P_a - (P_i + P_0 + P_p)}{T}, \tag{5.35}$$

or more simply,

$$V_c = \frac{\Delta P}{W}, \tag{5.36}$$

where ΔP is the excess power available at that combination of airspeed and altitude.

Calculations of the maximum rate of climb as a function of flight speed and density altitude is shown in Fig. 5.9 for the example helicopter. These curves mimic the excess power available curves because the climb (or descent) velocity is determined simply by the excess (or decrease) in power required, ΔP, relative to steady level flight conditions. As shown by Fig. 5.9, the climb performance is substantially affected by density altitude.

5.3.11 *Speed for Minimum Power*

On the basis of the preceding analysis, it is clear that many of the climb and autorotative characteristics of the helicopter in forward flight may be estimated from curves of excess power available relative to that required for straight-and-level flight at the same airspeed. The maximum possible rate of climb is obtained at the speed for minimum power in level flight (this speed is often denoted by V_{mp}). It can be seen from Fig. 5.6 or Fig. 5.7 that this situation occurs at a fairly low airspeed, usually in the range 60 to 80 kts.

The speed V_{mp} will also be the optimum speed to fly for minimum autorotative rate of descent. At this airspeed the power required by the rotor is a minimum, and for an autorotation at this airspeed the pilot needs to give up the least amount of potential energy (altitude) per unit time. Thus, on the basis of the power required curve it can be deduced that in the event of a mechanical failure in the hover condition, it is beneficial for the pilot to translate some of the stored potential energy into translational (forward flight) kinetic energy because the autorotative rate of descent under these conditions will be lower. However, if a mechanical failure occurs close to the ground, this transition may not be possible. For this reason, the normal operational envelope of a helicopter is restricted to an acceptable combination of altitude and airspeed that always allows safe autorotational landings to be performed (see Section 5.4.2).

In addition to being the speed to fly for maximum rate of climb and minimum autorotative rate of descent, V_{mp} also determines the speed to give the best endurance, that is, to obtain maximum time on station for surveillance, search, etc. To obtain maximum endurance the fuel burn per unit time must be a minimum. Typically fuel burn curves versus V are shallow enough that this airspeed can also be estimated where the power required is a minimum.

To estimate the value of V_{mp}, the approximation previously derived in Eq. 5.29 for the helicopter power can be used, namely

$$C_P = C_Q = \frac{\kappa C_W^2}{2\mu} + \frac{\sigma C_{d_0}}{8}(1 + K\mu^2) + \frac{1}{2}\left(\frac{f}{A}\right)\mu^3 + \lambda_c C_W. \tag{5.37}$$

At lower airspeeds the rotor profile power is small and builds slowly with the square of airspeed. Therefore, the minimum power is essentially determined by the variation in induced power and the parasitic power at these low air speeds; that is, it can be assumed that

$$C_P \approx \frac{\kappa C_W^2}{2\mu} + \frac{1}{2}\left(\frac{f}{A}\right)\mu^3. \tag{5.38}$$

Differentiating this expression with respect to μ gives

$$\frac{dC_P}{d\mu} = -\frac{\kappa C_W^2}{2\mu^2} + \frac{3}{2}\left(\frac{f}{A}\right)\mu^2, \tag{5.39}$$

which equals zero for a minimum. Therefore,

$$\frac{\kappa C_W^2}{2\mu^2} = \frac{3}{2}\left(\frac{f}{A}\right)\mu^2 \tag{5.40}$$

or

$$\mu = \left(\frac{\kappa C_W^2}{3f/A}\right)^{1/4} \tag{5.41}$$

for maximum endurance. Also, recall that $\lambda_h = \sqrt{C_W/2}$ so that

$$\mu = \lambda_h \left(\frac{4\kappa}{3f/A}\right)^{1/4} = \sqrt{\frac{C_W}{2}}\left(\frac{4\kappa}{3f/A}\right)^{1/4} \tag{5.42}$$

or

$$V_{mp} = V_h \left(\frac{4\kappa}{3f/A}\right)^{1/4} = \sqrt{\frac{W}{2\rho A}}\left(\frac{4\kappa}{3f/A}\right)^{1/4}. \tag{5.43}$$

Note that, in general, this speed will increase with increasing density altitude, that is, with both increases in altitude and temperature. In addition, it will be apparent that this airspeed is a function of aircraft weight.

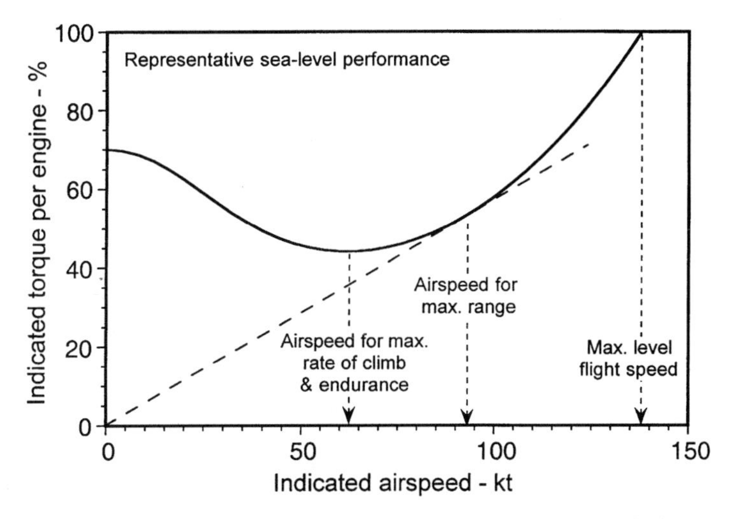

Figure 5.10 Determination of speed to fly for maximum rate of climb, maximum range, and maximum level flight speed from representative torque (power) curve.

5.3.12 *Speed for Maximum Range*

The range of the aircraft is the distance it can fly for a given takeoff weight and for a given amount of fuel. In all cases, contingency fuel allowances must be taken into account, depending on the mission profile. The speed to fly to give the best range is obtained when the ratio P/V is a minimum or the ratio V/P is a maximum, that is, when the aircraft is operated at the best lift-to-drag ratio. For no-wind conditions this speed is graphically obtained from a line drawn through the origin and tangent to the P/V curve, as shown in Fig. 5.10. As can be seen, this airspeed is usually at a somewhat higher airspeed than that required for maximum endurance.

The speed for maximum range, V_{mr}, is also determined essentially by the variation in induced power and the parasitic power. Therefore, the ratio of P/V can be approximated by the expression

$$\frac{C_P}{\mu} \approx \frac{\kappa C_W^2}{2\mu^2} + \frac{1}{2}\left(\frac{f}{A}\right)\mu^2. \tag{5.44}$$

Differentiating this latter expression with respect to μ gives

$$\frac{d(C_P/\mu)}{d\mu} = -\frac{\kappa C_W^2}{\mu^3} + \left(\frac{f}{A}\right)\mu, \tag{5.45}$$

which equals zero for a minimum. Therefore,

$$\mu = \left(\frac{\kappa C_W^2}{f/A}\right)^{1/4} \tag{5.46}$$

for maximum range, or

$$V_{\text{mr}} = V_h\left(\frac{4\kappa}{f/A}\right)^{1/4} = \sqrt{\frac{W}{2\rho A}}\left(\frac{4\kappa}{f/A}\right)^{1/4}. \tag{5.47}$$

Note that, like the speed for maximum range, this speed will increase with increasing density altitude and aircraft weight.

5.3.13 *Range–Payload and Endurance–Payload*

Range–payload and endurance–payload curves provide information of the effects of aircraft range and endurance when trading off payload for fuel. The engine characteristics must be taken into account to determine both the maximum endurance and maximum range. Fuel flow curves must be derived versus indicated airspeed and gross weight. Generally, fuel flow curves versus airspeed (at a given density altitude) are fairly flat and this alternative form of the performance curves closely follows the power curves. McCormick (1995) lays down the basic range analysis for an aircraft, which can be adapted for the helicopter. The fuel flow rate, W_F, with respect to distance, R, will be

$$\frac{dW_F}{dR} = \frac{P \times (SFC)}{V}, \tag{5.48}$$

where *SFC* is the specific fuel consumption of the engine(s). The power required varies with gross weight and density altitude, and the *SFC* depends on power and density altitude. Because the weight decreases as fuel is burned, Eq. 5.48 must be integrated numerically to find the range. Fuel burned during takeoff, climb, and descent is factored into the calculation, along with a mandated fuel reserve. However, because the fuel weight is normally a small fraction of the total gross weight and the specific range data are fairly linear with respect to weight, Eq. 5.48 can be realistically evaluated at the point in the cruise where the aircraft weight is equal to the initial gross weight (gross takeoff weight, W_{GTOW}) less half the initial fuel weight, that is, at the point where $W = W_{GTOW} - W_F/2$, where W_F is the initial fuel weight. In this case the range, R, of the helicopter is given by

$$R = W_F \left[\frac{V}{P \times (SFC)} \right]_{W_{GTOW} - W_F/2} \tag{5.49}$$

less an allowance for the other contingency factors described previously. By a similar process, the estimated endurance, E, will be given by

$$E = W_F \left[\frac{1}{P \times (SFC)} \right]_{W_{GTOW} - W_F/2} \tag{5.50}$$

Generally, it is sufficiently accurate to estimate endurance by dividing the useable fuel on board by the average fuel flow rate.

An example of a payload–range curve is shown in Fig. 5.11. The detailed nature of the range and endurance curves are obtained by numerical calculation and then confirmed by the manufacturer by flight testing. Some helicopters may be able to be fitted with long-range fuel tanks, for which ferry operations over considerable distances are possible. All of this payload–range information is presented to the pilot in the form of performance charts, and these are included in the aircraft flight manual. The pilot can use these charts for flight planning at the combination of aircraft weight and atmospheric conditions. Stepniewski & Keys (1984) give good examples of payload–range and payload–endurance curves and provide step-by-step procedures for determining curves for specific mission profiles.

5.3.14 *Factors Affecting Maximum Attainable Forward Speed*

Conventional helicopters are relatively low speed machines compared to their fixed-wing counterparts. The maximum flight speed will be determined by a combination of one or more of the following: 1. installed engine power, 2. airframe parasitic drag,

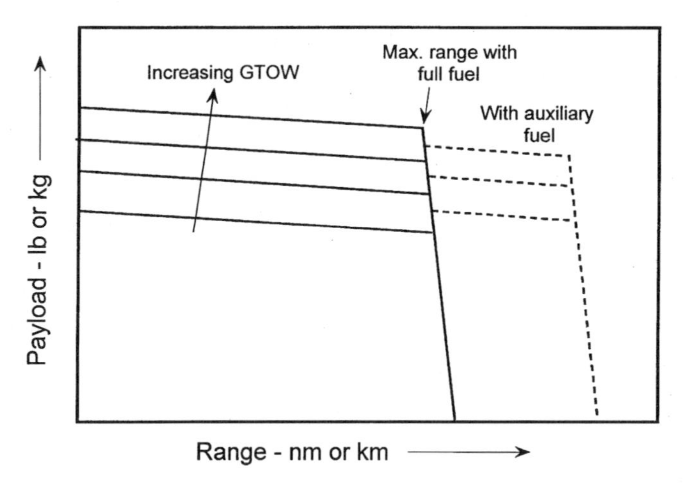

Figure 5.11 A representative payload–range curve for a helicopter.

3. gearbox (transmission) torque limits, and 4. rotor stall and/or compressibility effects. Early helicopters were powered by piston engines and were mostly limited in performance because of the lack of installed power. Piston engines have relatively poor power-to-weight ratios and become extremely heavy when large amounts of power are required. A piston engine, on average, weighs about 2 lb/hp (1.22 kg/kW), whereas a modern gas turbine weighs about half as much. Above a certain aircraft gross weight, it is inefficient to use piston engines on helicopters. As a result, on modern helicopters turbine engines are almost universally used because of their superior power-to-weight ratios. However, turbines have a high acquisition cost and are not usually found on small training helicopters. On turbine powered helicopters, performance limits are dictated by allowable transmission torque. When it is necessary to transmit large amounts of torque to the rotor shaft, the transmission system becomes relatively heavy, and so there is usually a torque limit imposed to minimize overall transmission weight. In this case helicopter performance charts are usually presented in terms of indicated engine torque versus indicated airspeed (e.g., Fig. 5.10).

The minimization of airframe drag is a major issue in the design of a modern helicopter. Over the past twenty years or so there have been progressive improvements in reducing airframe drag and improving forward flight speeds and reducing fuel burn. A major drag producer at high forward speed is the rotor hub, especially because the blade hinges and controls are exposed to the airstream. Careful contouring of the fuselage in this region can significantly help reduce hub drag and control the extent and intensity of the separated wake behind the hub. More recently, there has been a shift to the use of hingeless or bearingless rotor hubs. Besides being mechanically simpler than conventional articulated rotor hubs, these types of hub designs are also aerodynamically cleaner and have a much lower equivalent flat-plate area.

On many helicopters, the maximum forward flight performance is limited by the aerodynamics of the rotor itself. This is because of the occurrence of one of two possible factors. First, high power (or torque) is required to overcome compressibility effects generated on the advancing side of the rotor disk. Second, retreating blade stall can produce sufficiently high blade loads and vibration levels to limit the flight speed. Compressibility effects manifest as wave drag as a result of the onset of supercritical (transonic) flow and the generation of shock waves. The intensity of the supercritical flow may also progress to a point where the shock waves are sufficiently strong to promote rapid thickening of the local boundary layer, and it may even produce shock induced separation and stall. The approach of the rotor into these conditions is usually accompanied by a relatively gradual increase in power required with mild increases in vibration. However, the occurrence of retreating blade stall is often quite sudden in its occurrence and is accompanied by high vibration levels.

An expansion of the flight boundary of helicopters to high flight speeds is limited by not only aerodynamic constraints, but by aeroelastic and structural constraints as well. Usually, high stresses or intolerable fatigue loadings of the various structural components are limiting factors, particularly on the hub and pitch links. These vibratory stresses result from the generation of unsteady aerodynamic loads on the rotor system. The complex nature of these loads reflects the need to understand the highly unsteady aerodynamic flow field produced within the rotor disk, which is discussed in detail in Chapter 8.

5.3.15 *Performance of Coaxials and Tandems*

The hovering performance of coaxial and tandem rotors has been previously discussed in Chapter 2. By accounting for the induced interference effects between the rotors, it has been shown that the relationship between power and thrust can be adequately estimated using the momentum theory. The power required for a coaxial rotor system operating in forward flight at a constant thrust coefficient and over a range of advance ratios is shown in Fig. 5.12. The experimental data are taken from Dingeldein (1954). The predictions were made using the extension of simple momentum theory to forward flight, with the effects of profile losses accounted for through the blade element theory, in the same manner used previously for the single rotor. The single rotor had a solidity of 0.027 and the coaxial had a solidity of 0.054. An equivalent flat-plate parasitic drag area of 10 ft^2 was used to define the propulsive force component, along with $C_{d_0} = 0.01$, $\kappa = 1.15$, and $\kappa_{\text{int}} = 1.16$, the latter being derived from the hover case and assumed to be valid also for forward flight.

For both the single and coaxial rotors, the predictions compare favorably with the measurements, although there is a slight overprediction of power at the highest advance ratios. It is clear that for the coaxial configuration, there is a higher overall power requirement

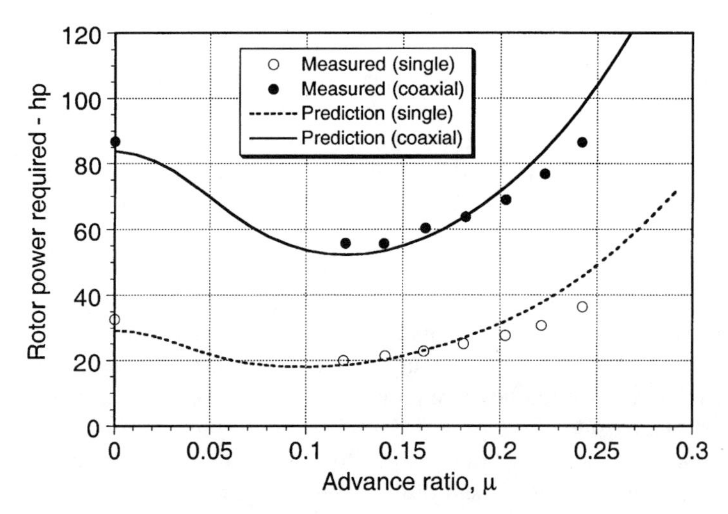

Figure 5.12 Predictions of power in forward flight for single and coaxial rotor systems compared to measurements. Data source: Dingeldein (1954).

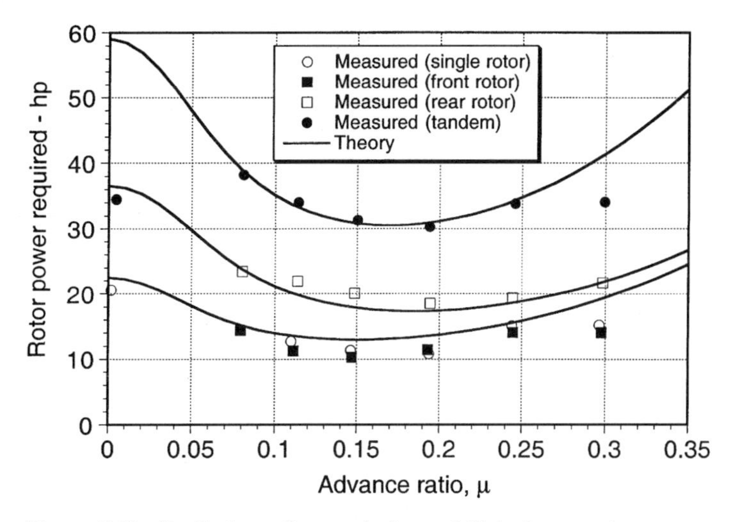

Figure 5.13 Predictions of power in forward flight for a tandem rotor system compared to measurements. Data source: Dingeldein (1954).

than for an equivalent single rotor. This is because of the interference effects between the two rotors. Also, the higher overall parasitic drag of the two hubs and dual control linkages (see Fig. 4.10) make a coaxial rotor less efficient than a single rotor. However, this negative aspect can be outweighed by the overall compactness of the coaxial rotor design.

The forward flight performance obtained with a tandem rotor configuration is shown in Fig. 5.13, along with a breakdown of the power required separately for the front and rear rotors. There is no rotor overlap for this particular tandem configuration, with the rotor shafts being separated 103% of the rotor diameter. Each rotor had a solidity of 0.054. The equivalent flat-plate area was 2 ft^2. Note that the performance of the front rotor was almost identical to that of the single rotor, suggesting that in this case there is little or no interference produced on the forward rotor by the rear rotor during forward flight. This, however, may not be a general result independent of rotor spacing or relative difference in rotor heights. Prediction of performance by means of the blade element/momentum theory is, therefore, of the quality expected based on previous studies with single rotors. The power required for the rear rotor is considerably higher because it operates in the downwash generated by the front rotor[2] – see Heyson (1954). By computing the downwash from the front rotor, this can be used to redefine the flow environment seen by the rear rotor. The induced power for the combination becomes

$$P_i = T_F v_{i_F} + \kappa_{ov} T_R v_{i_R}. \tag{5.51}$$

An induced power interference factor, κ_{ov}, equal to 1.14 was assumed for the rear rotor, for which the power required can then be estimated. For advance ratios of 0.1 and above, there is a good agreement between the predictions and the measurements. Note that because of the effects of the forward rotor, the minimum power required for the rear rotor is attained at a much higher advance ratio. Combining the results for the two rotors gives the power required for the tandem configuration. Agreement between prediction and experiment is generally good, except for low advance ratios approaching hover where the experimental results show

[2] On tandems such as the CH-46 or CH-47 the rear rotor is placed substantially higher than the front rotor to minimize these interference effects.

a favorable interference effect between the two rotors. Based on the results shown previously in Chapter 2, this favorable effect seems unique to this tandem configuration and would not be expected for overlapping tandem rotors.

5.4 Autorotation Revisited

The autorotation maneuver has been discussed in Chapter 2 and is defined as a self-sustained rotation of the rotor without the application of any shaft torque from the engine (i.e., $C_Q \approx 0$). Under these conditions, the power to drive the rotor comes from the relative airstream through the rotor as the helicopter descends through the air. Autorotation is used as a means of recovering safe flight of the helicopter in the event of a catastrophic mechanical failure, such as engine, transmission, or tail rotor failure. Under established autorotative conditions, there is an energy balance where the decrease in aircraft potential energy per unit time is equal to the power required to sustain the rotor speed. In other words, the pilot gives up altitude at a controlled rate in return for energy to turn the rotor to keep it producing thrust. Recall from Chapter 2 that an autorotation takes place in the turbulent wake state.

Consider now the flow environment encountered at a blade element on the rotor during a vertical descent, as shown in Fig. 5.14. Although most autorotations are conducted with some forward speed, for simplicity first consider a vertical autorotative descent with no forward speed. In an autorotation, the inflow angle ϕ must be such that there is no net in-plane force and, therefore, no contribution to rotor torque; that is, for this station on the blade,

$$dQ = (D - \phi L)y\,dy = 0. \tag{5.52}$$

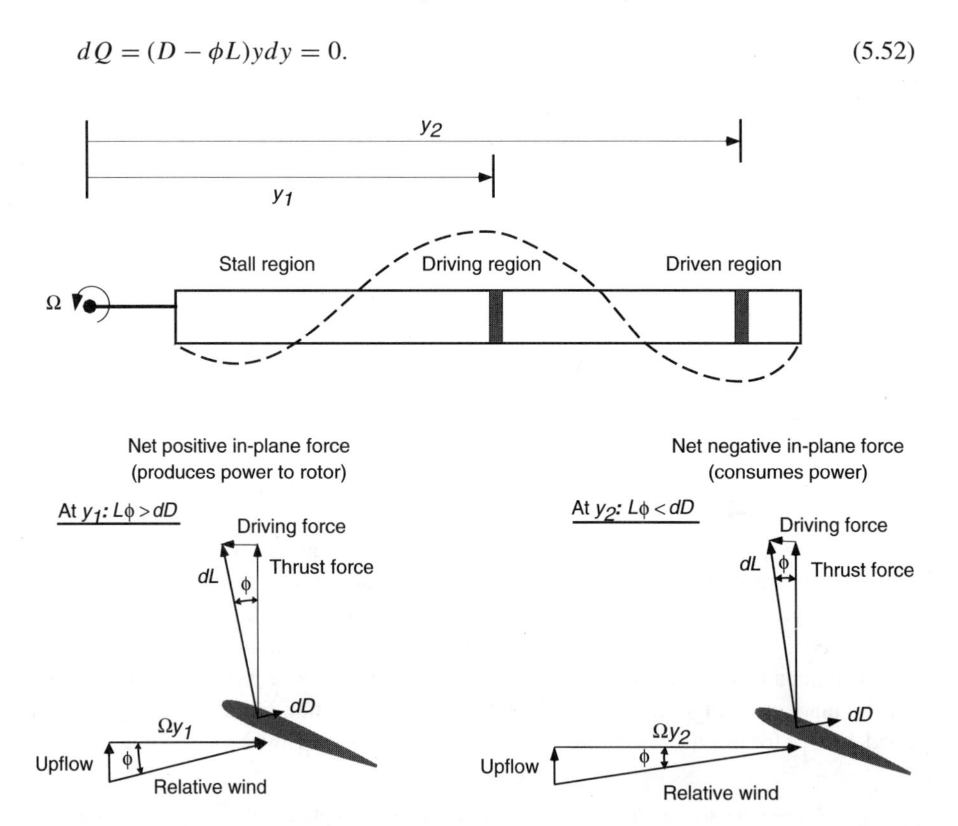

Figure 5.14 In autorotation, different parts of the rotor consume power or produce power.

However, this is a condition that can only exist at two radial stations on the blade. In general, some stations on the rotor will absorb power from the relative airstream and some will consume power such that the net power at the rotor shaft is approximately zero. With the assumption of uniform inflow over the disk, the induced angle of attack is given by

$$\phi = \tan^{-1}\left(\frac{|V_c + v_i|}{\Omega y}\right). \tag{5.53}$$

It follows that the induced angles of attack over the inboard stations of the blade are high, and near the tip ϕ is low. Therefore, one finds that at the inboard part of the blade the net angle of attack results in a forward inclination of the sectional lift vector and is such that in this region there is now a negative induced drag component that is greater than the profile drag. Therefore, the blade element absorbs power from the airstream. At the tip of the blade where ϕ is low, these sections consume power because, as a result of the forward inclination of the lift vector, the propulsive component is insufficient to overcome the profile drag. The effect is summarized in Fig. 5.14.

The consequence of this behavior is that the rotor rpm (Ω) will adjust itself until equilibrium is obtained. This is a stable equilibrium point because if Ω increases, ϕ will decrease and the region of accelerating torque will decrease inboard, and this tends to decrease rotor rpm. Conversely, if the rotor rpm decreases then ϕ will increase and the region of accelerating torque will grow outward. Therefore, when fully established in the autorotative state the rotor speed is quite stable.

Consider the autorotation diagram shown in Fig. 5.15, where the blade section C_d/C_l is plotted versus angle of attack at the blade section. This is a form originally used by Wimperis

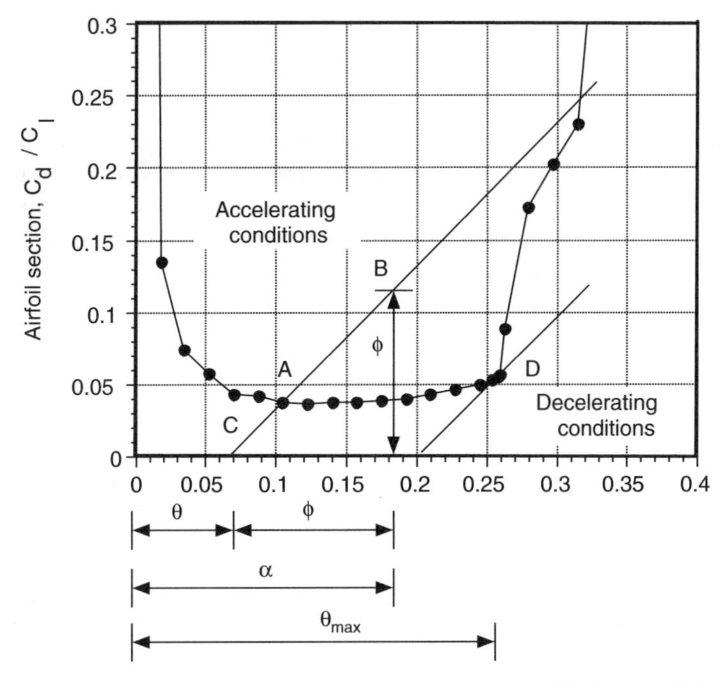

Figure 5.15 Autorotative diagram used to describe equilibrium conditions at the blade element.

(1928). Both Nikolsky (1944) and Gessow & Myers (1952) describe rotor equilibrium at the blade element in terms of this interpretation, which is relatively useful for further understanding the phenomenon. For a single section in equilibrium recall that

$$C_d - \phi C_l = 0 \quad \text{or} \quad \frac{C_d}{C_l} = \phi = \alpha - \theta, \tag{5.54}$$

where θ is the blade pitch angle. For a given value of blade pitch angle, θ, and inflow angle, ϕ, the previous equation represents a series of points that form a straight line, which is plotted on Fig. 5.15. The intersection of this line with the measured C_d/C_l data at point A corresponds to the equilibrium condition where $\phi = C_d/C_l$. Above this point, say at point B, $\phi > C_d/C_l$; so this represents an accelerating torque condition. Point C is where $\phi < C_d/C_l$; this represents a decelerating torque condition. Note that above a certain pitch angle, say θ_{max}, equilibrium is not possible and so for operation at point D stall will occur causing the rotor rpm to decay.

Establishing stable autorotation requires a certain level of skill from the pilot. The rotor rpm and the rate of descent can be controlled by the pilot by means of judicious adjustment of the collective pitch setting. The cyclic pitch is used to control the airspeed. The collective controls the mean blade pitch (and hence the mean aerodynamic angles of attack at the blade sections) and, therefore, the blade mean lift and drag and rotor rpm. In an autorotation the collective pitch is always held at a low value to ensure that the blade sections never reach high enough angles of attack to stall. However, the inboard parts of the rotor blades always operate at high angles of attack during the autorotative descent. Therefore, the pilot must ensure that the collective pitch angle is kept low enough to prevent stall propagating out from the blade root region, which will tend to quickly decrease rotor rpm because of the high profile drag associated with stall.

The initial stages of the autorotative maneuver are the most critical. During these initial stages, the pilot must sharply lower the collective pitch setting from the normal flight value in order to prevent blade stall and a rapid decay in rotor rpm. Both military and civil certification requirements impose a finite time delay (usually a few seconds) to account for normal pilot reaction time before the collective pitch is lowered; thus there is always some safety margin imposed in all helicopter designs. However, in most cases the pilot must still react sufficiently quickly to ensure that the rotor rpm does not decay below acceptable margins. How quickly the rotor rpm decays is a function of the power required at the time of failure and the rotational inertia of the rotor. Newman (1994) gives a good analysis of the effect of rotor inertia on the autorotation problem.

5.4.1 *Autorotation in Forward Flight*

The autorotative energy balance in forward flight is basically the same as for vertical flight. However, because of the forward flight velocity there is a loss of axial symmetry in the induced velocity and angles of attack over the rotor disk. This tends to move the distribution of parts of the rotor disk that consume power and absorb power, as shown in Fig. 5.16. However, the basic physics of the autorotational problem remains unchanged. Estimates of the autorotative rate of descent in forward flight can be made using the power equation given in terms of the momentum and blade element theories. While an autorotation takes place in the turbulent wake state, which is not strictly amenable to analysis by momentum theory, the power equation gives results that are sufficiently accurate for engineering estimates of the rate of descent as long as averaged flow properties are considered. In an autorotation

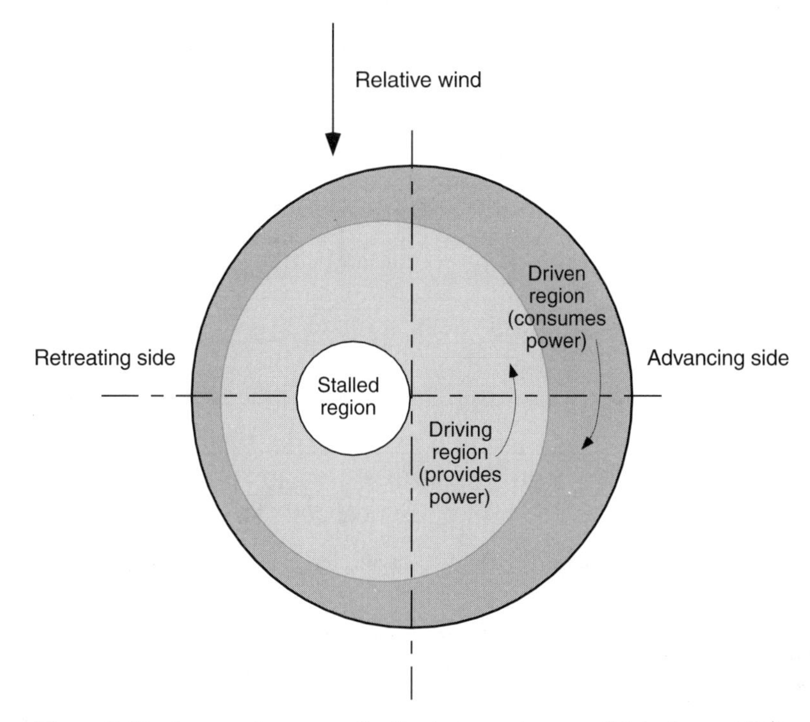

Figure 5.16 Autorotative power distribution over the rotor disk in forward flight.

$C_Q = 0$ so that to a first approximation to the rate of descent we can use

$$C_Q = 0 = \frac{\kappa C_W^2}{2\mu} + \frac{\sigma C_{d_0}}{8}(1 + K\mu^2) + \frac{1}{2}\left(\frac{f}{A}\right)\mu^3 + \lambda_c C_W. \tag{5.55}$$

Rearranging and solving for λ_c gives

$$\lambda_c \equiv \lambda_d = -\frac{\kappa C_W}{2\mu} - \frac{\sigma C_{d_0}}{8C_W}(1 + K\mu^2) - \frac{1}{2C_W}\left(\frac{f}{A}\right)\mu^3. \tag{5.56}$$

The result is normally expressed in ft/min so that $V_d = 60\lambda_d \Omega R$.

Representative results for the autorotative rate of descent in forward flight based on the use of the power curve are shown in Fig. 5.17. The shape of the curve mimics the power (or torque) required curve for steady level flight. [See also Gessow & Myers (1947) and Wheatly (1932).] Note the extremely high rates required at very low airspeeds, but at or near V_{mp}, the rate of descent is reduced to about half the value required in axial flight. Therefore, should a problem arise that requires the pilot to put the aircraft into an autorotation, the airspeed should be immediately increased or decreased to V_{mp} to enable the lowest possible rate of descent. For example, if there is a mechanical problem in high speed forward flight, the pilot would immediately pull back on the stick (application of longitudinal cyclic), to gain some additional altitude (zoom-climb) and lose airspeed, before quickly lowering the collective and entering the autorotation before the rotor rpm begins to decay.

Autorotation performed at or near V_{mp} (airspeed for minimum power under normal flight conditions) will give the pilot more time to try and correct the problem. When established at the minimum rate of descent, the pilot will have the maximum possible time to select a suitable landing site. It is likely that an actual autorotation will have to be performed away from the vicinity of an airport, and maximizing the time to complete the descent is essential.

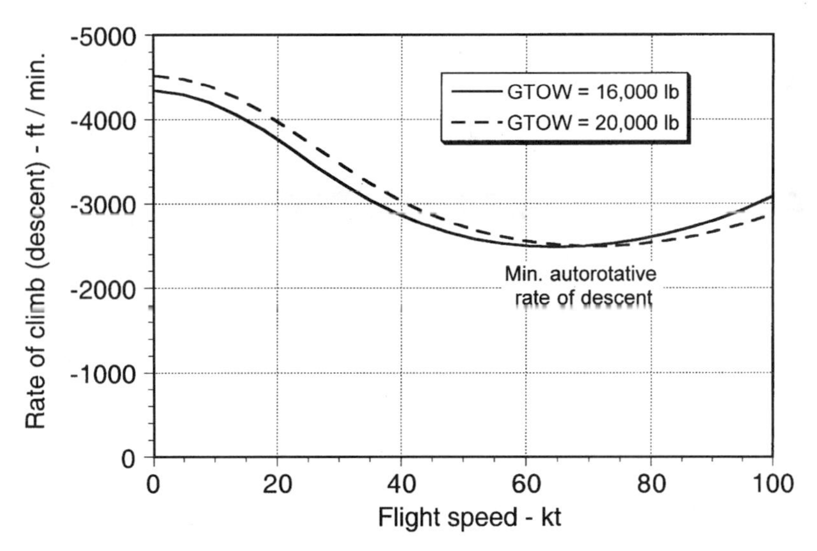

Figure 5.17 Estimates of rate of descent in autorotation.

Although the autorotative rate of descent is high, the actual maneuver is relatively safe from the pilot's perspective when sufficient altitude is available.

The final stages of the autorotation happen more quickly and require a high level of skill from the pilot. At about 50 ft ($\approx$17 m) from the ground, the pilot must begin to decelerate the helicopter. This is done by slowly pulling up on the collective pitch (increasing blade pitch angles and so increasing rotor thrust), while simultaneously pulling back slightly on the cyclic to reduce forward speed. As the collective pitch is increased, the rotor rpm will reduce quickly and so the pilot must ensure that the collective is brought in progressively and at a rate that still allows the rotor rpm to stay at acceptable values to produce thrust. The overall objective is to cushion the rate of descent such that the helicopter touches down with a rate of descent less than about 10 ft/s ($\approx$3 m/s). The pilot must also ensure that the rpm does not decay to the point that excessive blade flapping occurs. There have been several serious accidents over the years during the autorotative maneuver as a result of excessive blade flapping, which can cause the blades to strike the airframe. Performed skillfully, however, the autorotation is a safe and relatively benign maneuver.

5.4.2 *Height–Velocity (HV) Curve*

The flight conditions that will allow safe entry to an autorotation and recovery of the helicopter are summarized in the form of height–velocity or HV curves. Figure 5.18 shows representative examples of the HV curves for single-engine and multiengine helicopters, which are typical of the information included in the aircraft flight manual. The curves that define the "avoid" regions are established through test flights prior to certification of the helicopter. Anywhere outside the avoid region, the pilot should be able to safely recover the helicopter through an autorotative maneuver in the event of a catastrophic mechanical failure.

The actual size and shape of the HV curve depends on many factors, including the characteristics of the machine, its gross weight, and operational density altitude [Pegg (1968)]. As shown in Chapter 2, the disk loading T/A is the primary factor influencing the autorotative rate of descent. The number of engines installed will also affect the shape of curves, and multiple curves may be defined for single- and multiengine operations. Note that

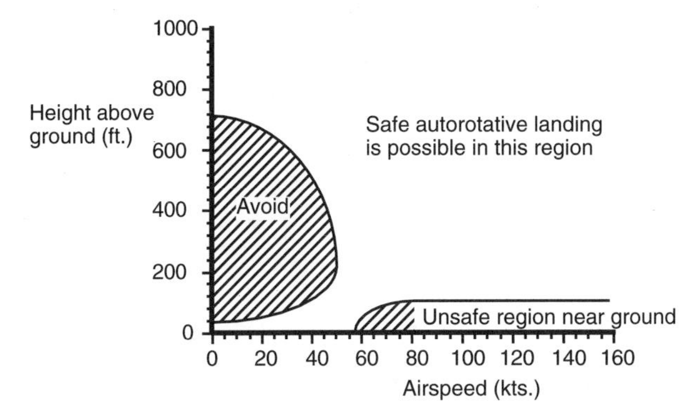

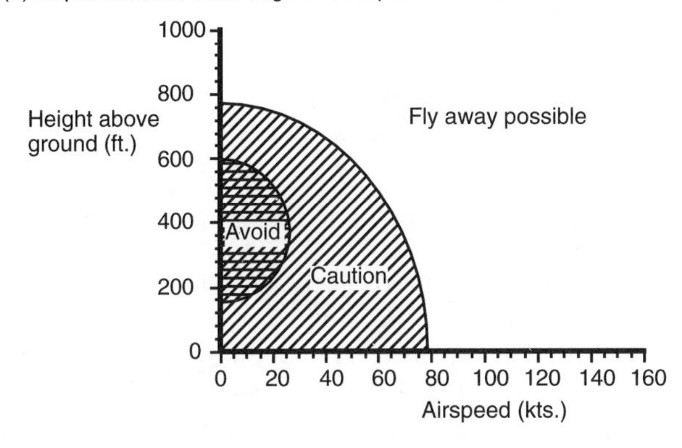

Figure 5.18 Representative height–velocity curves for single-engine and multiengine helicopters. (a) Single-engine. (b) Multi-engine.

there are two unsafe regions defined by the HV curve. The avoid region at low altitude and high airspeeds determines the minimum altitude below which translational kinetic energy cannot be converted into potential energy by means of a zoom-climb prior to entering the autorotation. This boundary is also marked to prevent unsafe operations close to the ground. The most important avoid region is at low airspeeds. The bottom of the HV curve is established in a full-power climb, with some prescribed allowance for pilot reaction time, that is, the time between the power failure and the reduction in collective pitch. Military and civil requirements may differ slightly, so that the limits of the curve will also vary. The top of the curve is established for level flight power conditions, again by including some prescribed pilot reaction time.

Reducing the size of the avoid region is desirable but generally difficult. Helicopters with low disk loading will tend to have much smaller avoid regions; hence the autorotative characteristics of the helicopter usually enter into the basic sizing of the rotor. Increasing the stored rotational kinetic energy by adding blade mass is one possibility, but this will be at the expense of a lower payload. For a multiengine helicopter the unsafe or avoid region shrinks considerably, as shown in Fig. 5.18(b). For twin-engine helicopters, the avoid region diminishes to the point where there is only a slight chance that a fly-away or safe autorotation could not be performed. For three-engine helicopters, the avoid region essentially disappears in the event of a single engine failure. However, there will always be some avoid regions marked on the HV diagram in the event of a tail rotor failure.

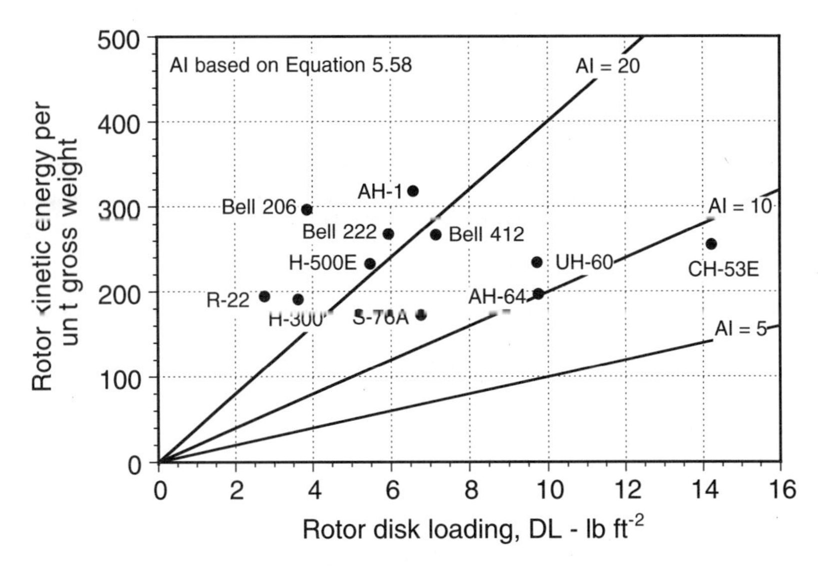

Figure 5.19 Autorotative indices derived for several helicopters. Data source: Various published helicopter specifications.

5.4.3 *Autorotation Index*

It is clear that autorotative performance of a helicopter depends on several factors. These include the rotor disk loading (which affects the descent rate), the stored kinetic energy in the rotor system (which influences the entry and completion of the autorotation), as well as subjective assessments by pilots. To help select the rotor diameter during predesign studies, an "Autorotative Index" is often used. Although various types of indices have been used [see White et al. (1982) for a summary] the autorotation index is basically an energy factor. One form of the index can be defined in terms of ratio of the kinetic energy of the main rotor to the gross weight of the aircraft, that is,

$$AI = \frac{I_R \Omega^2}{2W}, \tag{5.57}$$

which is used by Bell – see Wood (1976). An alternative autorotative index used by Sikorsky [see Fradenburgh (1984)] is

$$AI = \frac{I_R \Omega^2}{2W\,DL}. \tag{5.58}$$

Figure 5.19 shows the autorotative index for several helicopters, which have been calculated using Eq. 5.58 based on published information for each helicopter. These indices are of great use in the sizing of the main rotor or in examining the effects in autorotative characteristics with increasing gross weight or density altitude. Note that the absolute values of the index are of no significance, but the relative values provide a means of comparing the autorotative performance of a new helicopter design against another helicopter with already acceptable autorotative characteristics. An index of about 20 would normally be considered acceptable for single-engine helicopters, whereas a multiengine helicopter can have an index as low as 10 and still have safe autorotative characteristics.

The autorotative characteristics of the helicopter may also be expressed in terms of equivalent hover time – see Wood (1976). This is the time that the stored kinetic energy in the rotor system can supply sufficient power to hover before the rotor rpm decays to the point that stall occurs. This "equivalent time" parameter seems to correlate well with pilot opinion of autorotative characteristics. Based on the results of Wood (1976), a design goal for a new helicopter will be to give an equivalent hover time of at least 1.5 seconds.

5.5 Ground Effect

Rotor performance is affected by the presence of the ground or any other boundary that may alter or constrain the flow into the rotor or constrain the development of the rotor wake. "Ground effect" is of concern both in actual flight operations as well as in the wind-tunnel or hover tower testing of rotors. Consider a rotor hovering in close proximity to the ground, as shown in Fig. 5.20. Because the ground must be a streamline to the flow, the rotor

(a) Rotor hovering out-of-ground effect (OGE)

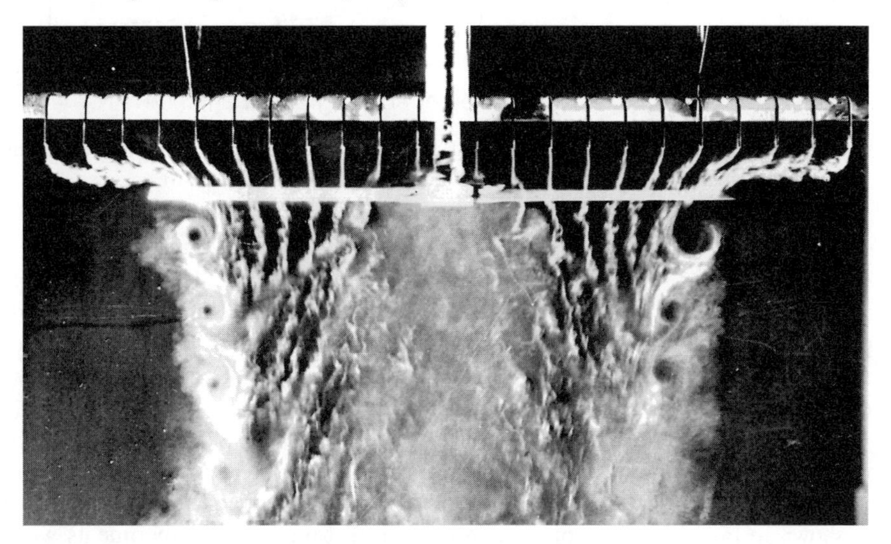

(b) Rotor hovering in-ground effect (IGE)

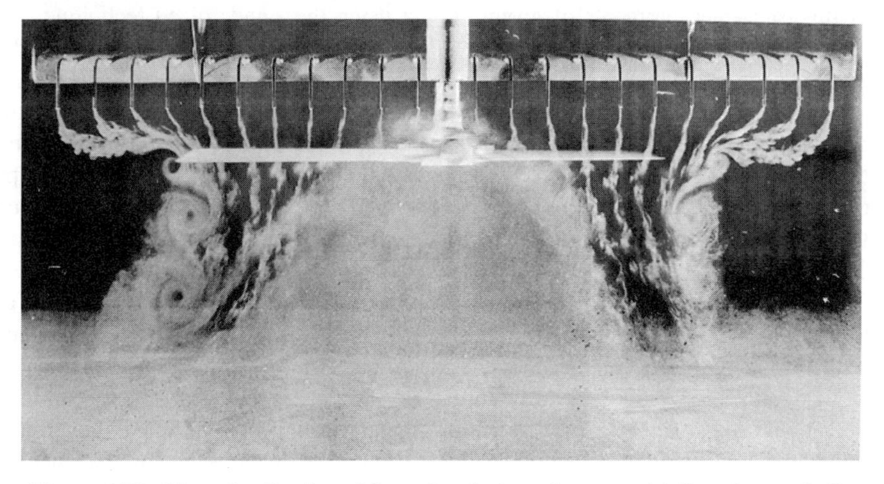

Figure 5.20 Flow visualization of the wake of a hovering rotor. (a) Out-of-ground effect (OGE). (b) In-ground effect (IGE). Reproduced from Fradenburgh (1972).

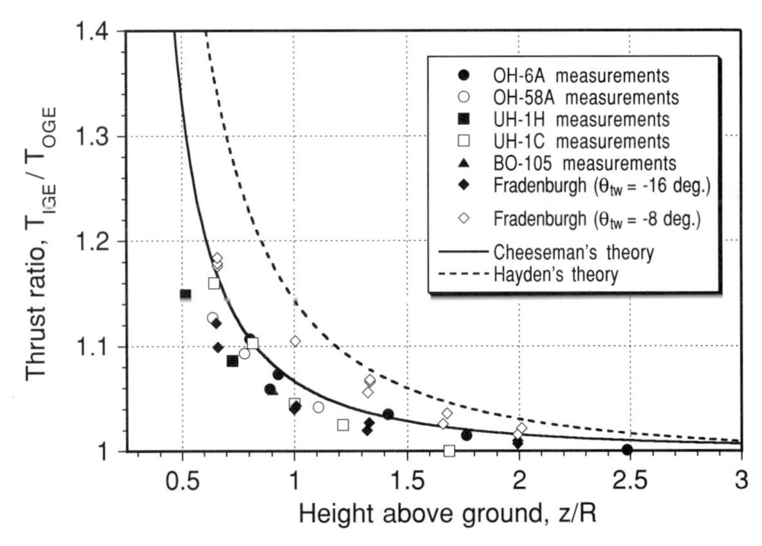

Figure 5.21 Increase in rotor thrust versus distance from the ground for a variety of helicopters. Data sources: Fradenburgh (1972) and Hayden (1976).

slipstream tends to rapidly expand as it approaches the surface. This alters the slipstream velocity, the induced velocity in the plane of the rotor, and, therefore, the rotor thrust and power. Similar effects are obtained both in hover and forward flight, but the effects are strongest in the hover state. Other visualizations of the flow of rotors operating in-ground effect are shown by Taylor (1950) and Light & Norman (1989).

When the rotor is operating in-ground effect, the rotor thrust is found to be increased for a given power. The effect has long been recognized but the aerodynamics are still not fully understood. A representative plot of the thrust ratio in hover versus height from the ground is shown in Fig. 5.21. This plot has been derived from several experiments with rotors operating at different blade loadings, as discussed by Zbrozek (1947) and others, including Betz (1937), Knight & Hefner (1941), Cheeseman & Bennett (1955), Fradenburgh (1960, 1972), and Stepniewski & Keys (1984). Hayden (1976) gives a comprehensive summary of flight test measurements of ground effect using the standardized technique of Lewis (1972). The results suggest significant effects on hovering performance for heights less than one rotor diameter. The results are also dependent on blade loading (or mean lift coefficient), blade aspect ratio, and blade twist. Yet, within the bounds dictated by most helicopters a universal behavior seems a good approximation for engineering estimates of the phenomenon. Besides the effects on actual aircraft performance, these results provide useful guidelines for laboratory testing of rotors in hover, where a minimum distance from the ground is required to ensure performance measurements that are free of interference effects that have their source from the ground.

The problem of ground effect can also be viewed as a reduction in power for a given thrust. Most of the power reduction is induced in nature, but there is also some small reduction in profile power because the blade angles are operating at a somewhat lower angle of attack to produce the same thrust. Because of ground effect there is an important operational advantage to be gained, namely that the aircraft will be able to hover in ground effect (IGE) at a higher gross weight or density altitude than would be possible out of ground effect (OGE). The extra thrust or reduction in power that is felt near the ground will also "cushion" the descent of the helicopter when landing. Systematic studies of such ground effects were conducted by Knight & Hefner (1941).

Ground effect in hovering flight has been examined analytically, albeit approximately, by means of the method of images. Cheeseman & Bennett (1955) replaced the rotor by a simple source with an image source to simulate ground effect and obtained some analytic relationships for the effects of the ground. Knight & Hefner (1941) and Rossow (1985) have used a vortex cylinder model. Based on Cheeseman & Bennett's analysis, ground effect on the rotor thrust can be expressed by the equation

$$\left[\frac{T}{T_\infty}\right]_{P=\text{const}} = \frac{1}{1-(R/4z)^2}. \tag{5.59}$$

Figure 5.21 shows that this equation gives good agreement with the experimental measurements. Because ground effect can be expressed in terms of the increase in thrust at a constant power, then $\lambda_{IGE}\, C_{T_{IGE}} = \lambda_{OGE}\, C_{T_{OGE}}$ or

$$\frac{\lambda_{IGE}}{\lambda_{OGE}} = \frac{T_{OGE}}{T_{IGE}} = k_G. \tag{5.60}$$

Alternatively, the influence of ground effect in hover can be viewed as a reduction in the rotor induced velocity (at a constant thrust and height above the ground) by a factor k_G such that

$$\left[\frac{\text{Power required IGE}}{\text{Power required OGE}}\right]_{T=\text{const}} = \left[\frac{P_{IGE}}{P_{OGE}}\right]_{T=\text{const}} = \left[\frac{C_{P\ IGE}}{C_{P\ OGE}}\right]_{T=\text{const}} = k_G. \tag{5.61}$$

Using a relatively simple model, Betz (1937) has suggested the effect on the rotor power at a constant thrust to be modeled by the equation

$$\left[\frac{P_{IGE}}{P_{OGE}}\right]_{T=\text{const}} = k_G = \frac{2z}{R}. \tag{5.62}$$

Hayden (1976) has used flight test measurements to find the influence of the ground in hover. The profile part of the total power was assumed to be isolated from the induced effect such that only the induced effect is influenced by the ground, that is,

$$P = P_0 + k_G (P_i)_\infty, \tag{5.63}$$

where k_G is derived from a curve fit to the experimental data using

$$k_G = \frac{1}{A + B(2R/z)^2} \tag{5.64}$$

with $A = 0.9926$ and $B = 0.0379$. As shown by Fig. 5.21, when viewed in terms of an increase in thrust for a given power Hayden's result is found to slightly overpredict the rotor thrust. In all cases it is apparent that the effects of the ground on the rotor power become negligible for rotors hovering greater than three rotor radii above the ground.

The effects of the ground on forward flight performance are also significant, but the flow state near the rotor tends to be even more complicated. The typical behavior is shown in Fig. 5.22, which is adapted from the work of Curtiss et al. (1984, 1987). At low forward speeds, a small region of flow recirculation is formed upstream of the rotor near the ground. This phenomenon has negligible effect on performance but may throw loose surface material up into the air that may be reingested by the rotor. As forward speed increases, this

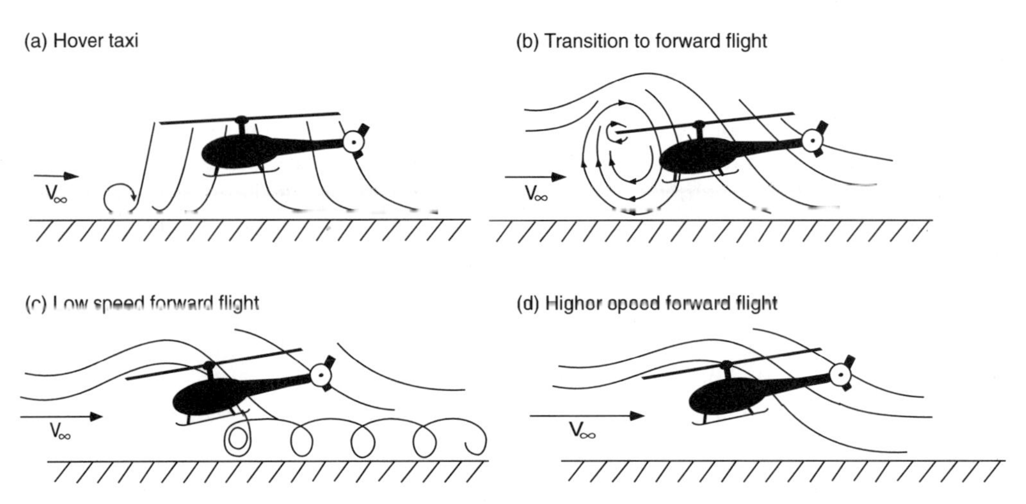

Figure 5.22 Flow characteristics for a rotor in forward flight near the ground. (a) Hover taxi. (b) Transition to forward flight. (c) Low speed forward flight, (d) Higher speed forward flight. Adapted from Curtiss et al. (1984, 1987).

recirculation develops into a small vortical flow region between the ground and the leading edge of the rotor. This vortex has been documented both in wind-tunnel experiments and helicopter operations in the field. For helicopters operating in a dusty or snowy environment the flow recirculation upstream of the rotor becomes particularly evident. This phenomenon increases the inflow through the forward part of the rotor disk, and, as a consequence, the induced power requirements will increase above that required for hover IGE. This will require the pilot to increase the collective pitch to maintain altitude as the aircraft transitions into forward flight. Above a critical advance ratio, which depends on aircraft weight (rotor thrust) and proximity to the ground, a well-defined horseshoe vortex (ground vortex) is formed under the leading edge of the rotor near the ground. With further increases in airspeed, this phenomenon disappears as the rotor wake is skewed back by the flow. Ground effect is usually considered negligible for $V_\infty > 2v_h$ or for advance ratios greater than 0.10.

A representative set of results of rotor power in forward flight for operations in both IGE and OGE is shown in Fig. 5.23, which is reproduced from the wind-tunnel measurements of Sheridan & Weisner (1977). The advantages of ground effect are apparent for hover and very low airspeeds, where the effects indicate a considerable power reduction compared to flight OGE. Note that for operation IGE the power increases rapidly as the helicopter transitions from the hover state. This is because of the formation and influence of flow recirculation at the leading edge of the rotor disk, which causes the rotor to experience a higher induced inflow than hover in OGE, and so power requirements will increase slightly. Similar results showing the same effects have been obtained by Cheeseman & Bennett (1955) on the basis of flight tests with the S-51 and similar helicopters. The problem of ground interference in forward flight has also been studied theoretically by Cheeseman & Bennett (1955) and Heyson (1960) using the method of images. Curtiss and colleagues (1984, 1987) have also conducted experimental measurements and flow visualization of the ground effect phenomenon using sub-scale rotors.

Similar interference effects are noted in wind-tunnel tests on rotors, where the effects of the tunnel floor (as well as the ceiling and side walls, if present) can alter the induced flow through the rotor. Ganzer & Rae (1960) and Lehman & Besold (1971) have studied the effects experimentally. If the objective is to simulate free-air conditions, then the presence

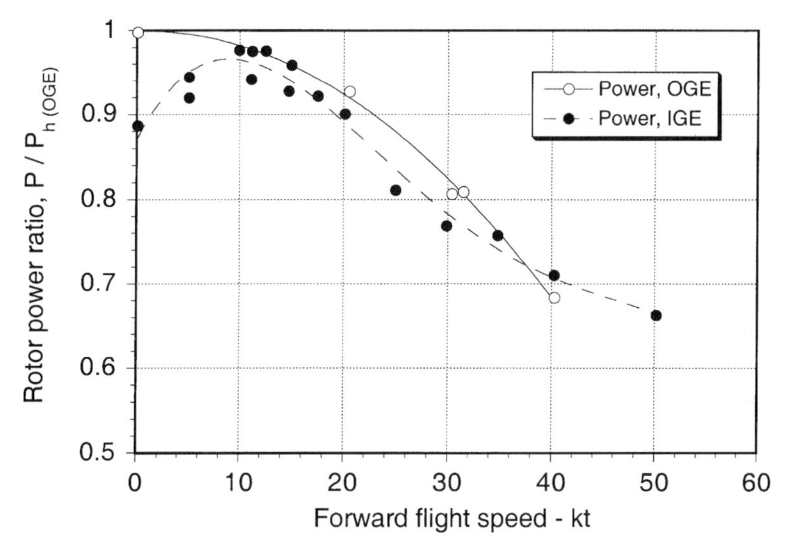

Figure 5.23 Measurements of rotor power versus forward speed when operating near the ground. Data source: Sheridan & Weisner (1977).

of the walls cannot be easily discounted, especially at low airspeeds. For tunnel dimensions that are at least twice the diameter of the rotor, the effects of the tunnel walls are small for advance ratios greater than 0.1. Generally, it must always be assumed that the effects of the tunnel walls will lead to some flow recirculation at lower advance ratios (say, $\mu < 0.075$) making reliable free-air measurements of rotor performance difficult or impossible, even if suitable corrections could be derived – see also Philippe (1990). Measurement of the wall pressure signatures is one of the better ways to allow the investigator to monitor interference effects and help define the lowest allowable wind speed or advance ratio that can be tested without the results being contaminated by wall interference effects.

5.6 Chapter Review

This chapter has addressed some elementary analysis and predicted results that define the overall performance characteristic of the helicopter in hover, climb, and forward flight. It has been shown that these characteristics can be derived, in part, by using relatively simple models that have their origin in the momentum and blade element theories. Estimates of rotor profile power can be made on the basis of blade element theory, perhaps allowing for yawed and reversed flow effects and also compressibility losses at high speeds. Airframe drag has been discussed, and the modeling of these effects has been introduced through the ideas of an equivalent flat-plate parasitic area. The resulting models give good approximations to the rotor power required over the substantial part of the operational flight envelope. The results can be used to estimate performance as functions of aircraft weight and operational factors such as density altitude. Performance issues such as the speed to fly for maximum range or endurance have been discussed, and it has been shown how these results follow directly from a knowledge of the power curves.

However, the performance of modern helicopters is limited by other aerodynamic factors such as retreating blade stall and compressibility effects. These are difficult to model without resorting to more complete types of analyses that model the aerodynamics at a more fundamental level. These phenomena and methods of approximation will be considered in the following chapters. Finally, helicopter performance when operating near the ground or in a wind tunnel has been discussed. The complexity of the recirculating flow is such that this particular aerodynamic problem is not easily amenable to solution, even by using vortex methods, and the problem must be modeled semiempirically.

5.7 Questions

5.1. For the helicopter specified by the parameters listed in the table below, estimate the power required to hover for several values of the density altitude at a gross weight of 16,000 lb. The distance between main and tail rotor shafts, x_{tr}, is 32.5 ft. Assume that transmission power losses amount to 10%. If the power available (installed power) is 3,000 hp at sea level on a standard day, estimate the hover ceiling (as a density altitude) at this gross weight.

Parameter	Symbol	Main Rotor	Tail Rotor
Rotor radius (ft)	R	27.0	5.5
Blade chord (ft)	c	1.7	0.80
Solidity	σ	0.082	0.19
Number of blades	N_b	4	4
Tip speed (ft s^{-1})	$V_{\text{tip}}(=\Omega R)$	725	685
Induced power factor	κ	1.15	1.15
Profile drag coefficient	C_{d_0}	0.008	0.008

5.2. Sketch a representative total power curve for a helicopter in forward flight at sea level. Draw in and label the breakdown of the constituent parts comprising this power curve, and explain the source of each part. Show also how you can use this type of power required curve (at a given altitude and gross weight) to determine: the vertical rate of climb, the speed for maximum endurance, the speed for maximum range, the maximum forward speed, and the maximum rate of climb in forward flight.

5.3. A helicopter is operating in level forward flight at 210 ft/s under the following conditions: shaft power supplied = 655 hp, $W = 6{,}000$ lb, $\rho = 0.00200$ slugs/ft^3. The rotor parameters are $R = 19$ ft, $\sigma = 0.08$, $\Omega R = 700$ ft/s, $k = 1.15$, $C_{d_0} = 0.01$. (i) How much power is required to overcome induced losses? (ii) How much power is required to overcome profile losses? (iii) What is the equivalent flat plate area, f? (iv) If the installed power is 800 hp, estimate the maximum rate of climb possible at this airspeed.

5.4. By means of the blade element theory in forward flight show that by including the effects of reverse flow on the drag, the profile power coefficient can be written as

$$C_{P_0} = \frac{\sigma C_{d_0}}{8}\left(1 + 3\mu^2 + \frac{3}{8}\mu^4\right).$$

Neglect the radial component of velocity. Compare this result to the predicted power obtained without reverse flow.

5.5. An understanding of the vortex ring state is necessary to explain certain aspects of helicopter performance. Describe, with the aid of a diagram(s), what is the mechanism behind the vortex ring state. Under what flight conditions is the vortex ring state important to a pilot and why?

5.6. What is meant by autorotation? Explain the circumstances when an autorotation maneuver might be necessary with a helicopter. What characteristics of the helicopter affect the autorotative performance? By means of a blade element diagram, carefully show and explain why: (a) The mean flow velocity must be vertically upwards through the rotor for autorotation to occur. (b) The blade pitch angles must be low in an autorotation compared to hover or climb. Show where and explain why in an autorotation the rotor blades will absorb power from the airstream at some blade locations and consume power at other blade stations.

5.7. Estimate the autorotative rate of descent in forward flight at sea-level conditions for a small light-weight helicopter with the following characteristics: Weight = 1370 lb, rotor radius = 12.6 ft, rotor solidity = 0.03, tip speed = 700 ft/s.

5.8. Describe the procedure a pilot would follow if the engine of a single-engine helicopter failed in hover. How might the procedure differ if: (a) two engines were installed and one engine failed, (b) the tail rotor failed. Estimate the autorotative rates of descent of three substantially different (in terms of gross weight) single-rotor helicopters of your choosing. Comment on your results.

5.9. Draw and explain the main features of the height–velocity diagram for a single-engine/single-rotor helicopter. Explain if and/or how these curve(s) will change for: (a) a single-rotor/twin-engine helicopter, (b) a tandem-rotor helicopter, (c) a single-rotor system with a high overall rotational inertia, (d) a higher density altitude.

5.10. An understanding of "ground effect" is necessary to explain certain trends in helicopter behavior. Describe the mechanism of ground effect in hover. How does ground effect influence the performance of the helicopter during the transition from hover to forward flight?

Bibliography

Amer, K. B. 1955. "Effect of Blade Stalling and Drag Divergence on Power Required by a Helicopter Rotor at High Forward Speed." 11th Annual National Forum of the American Helicopter Soc., Washington DC.

Bennett, J. A. J. 1940. "Rotary Wing Aircraft," *Aircraft Engineering*, 12 (Nos. 131–138), Jan.–Aug. pp. 7–9, 40–42, 65–67, 109–112, 139–141, 174–176, 208–209, 237–241, 246.

Betz, A. 1937. "The Ground Effect on Lifting Propellers," NACA TM 836.

Cheeseman, I. C. and Bennett, W. E. 1955. "The Effect of the Ground on a Helicopter Rotor in Forward Flight," ARC R & M 3021.

Curtis, H. C., Sun, M., Putman, W. F., and Hanker, E. J. 1984. "Rotor Aerodynamics in Ground Effect at Low Advance Ratios," *J. American Helicopter Soc.*, 29 (1), pp. 48–55.

Curtiss, H. C., Erdman, W., and Sun, M. 1987. "Ground Effect Aerodynamics," *Vertica*, 11 (1/2), pp. 29–42.

Dingeldein, R. C. 1954. "Wind Tunnel Studies of the Performance of Multirotor Configurations," NACA Technical Note 3236.

Fradenburgh, E. A. 1960. "The Helicopter and the Ground Effect Machine," *J. American Helicopter Soc.*, 5 (4), pp. 26–28.

Fradenburgh, E. A. 1972. "Aerodynamic Factors Influencing Overall Hover Performance," AGARD-CP-111.

Fradenburgh, E. A. 1984. "A Simple Autorotational Flare Index," *J. American Helicopter Soc.*, 29 (3), pp. 73–74.

Ganzer, V. M. and Rae, W. H. 1960. "An Experimental Investigation of the Effect of Wind Tunnel Walls on the Aerodynamic Performance of a Helicopter Rotor," NASA TN D-415.

Gessow, A. and Myers, G. C., Jr. 1947. "Flight Tests of a Helicopter in Autorotation, Including a Comparison with Theory," NACA TN 1267.

Gessow, A. and Myers, G. C. 1952. *Aerodynamics of the Helicopter*, Macmillan Co. (republished by Frederick Ungar Publishing, New York, 1967), p. 121.

Gessow, A. and Crim, A. D. 1956. "A Theoretical Estimate of the Effects of Compressibility on the Performance of a Helicopter Rotor in Various Flight Conditions," NACA TN 3798.

Glauert, H. 1926. "On the Horizontal Flight of a Helicopter," ARC R & M 1730.
Gustafson, F. B. and Myers, G. C., Jr. 1946. "Stalling of Helicopter Blades," NACA Rep. No. 840.
Gustafson, F. B. and Gessow, A. 1947. "Effect of Blade Stalling on the Efficiency of a Helicopter Rotor as Measured in Flight," NACA TN 1250.
Hayden, J. S. 1976. "The Effect of the Ground on Helicopter Hovering Power Required," 32th Annual National V/STOL Forum of the American Helicopter Soc., Washington DC, May 10–12.
Heyson, H. H. 1954. "Preliminary Results from Flow Field Measurements around Single and Tandem Rotors in the Langley Full-Scale Tunnel," NACA TN 3242.
Heyson, H. H. 1960. "Ground Effect for Lifting Rotors in Forward Flight," NASA TN D-234.
International Civil Aviation Organization (ICAO). 1964. *Manual of the ICAO Standard Atmosphere*.
Johnson, W. 1980. *Helicopter Theory*, Princeton University Press, Princeton, NJ, pp. 216–221.
Klemin, A. 1925. "An Introduction to the Helicopter," NACA TM 340.
Knight, M. and Hefner, R. A. 1941. "Analysis of Ground Effect on the Lifting Airscrew," NACA TN 835.
Lehman, A. F. and Besold, J. A. 1971. "Test Section Size Influence on Model Helicopter Rotor Performance," USAAVLABS TR 71–6.
LeNard, J. M. 1972. "A Theoretical Analysis of the Tip Relief Effect on Helicopter Rotor Performance," USAAMRDL Technical Report 72–7.
LeNard, J. M. and Boehler, G. D. 1973. "Inclusion of Tip Relief in the Prediction of Compressibility Effects on Helicopter Rotor Performance," USAAMRDL Technical Report 73–71.
Lewis, R. B. 1972. "Army Helicopter Performance Trends," *J. American Helicopter Soc.*, 17 (2), pp. 15–23.
Light, J. S. and Norman, T. 1989. "Tip Vortex Geometry of a Hovering Helicopter Rotor in Ground Effect," 45th Annual Forum of the American Helicopter Soc., Boston, MA, May 22–24.
McCormick, B. W. 1995. *Aerodynamics, Aeronautics, and Flight Mechanics*, John Wiley & Sons, Inc., New York, Chapter 7.
Minzner, R. A., Champion, K. S. W., and Pond, H. L. 1959. "The ARDC Model Atmosphere," AF CRC-TR-59-267.
NASA. 1966. "U.S. Standard Atmosphere Supplements," NASA-CR-88870.
Newman, S. 1994. *The Foundations of Helicopter Flight*, Edward Arnold, London, pp. 122–127.
Nikolsky, A. 1944. *Notes on Helicopter Design Theory*, Princeton University Press, Princeton, NJ, p. 55.
Pegg, R. J. 1968. "An Investigation of the Helicopter Height–Velocity Diagram Showing Effects of Density Altitude and Gross Weight," NASA TN D-4536.
Philippe, J. J. 1990. "Considerations on Wind-Tunnel Testing Techniques for Rotorcraft," AGARD-R-781.
Prouty, R. W. 1971. "Tip Relief for Drag Divergence," *J. American Helicopter Soc.*, 16 (4), pp. 61–62.
Prouty, R. W. 1986. *Helicopter Performance, Stability, and Control*, PWS Engineering Publishing, Boston, MA. pp. 177–187.
Rossow, V. J. 1985. "Effect of Ground and/or Ceiling Planes on Thrust of Rotors in Hover," NASA Technical Memorandum 86754.
Sheridan, P. F. and Wiesner, W. 1977. "Aerodynamics of Helicopter Flight Near the Ground," 33rd Annual Forum of the American Helicopter Soc., Washington DC, May 9–11.
Stepniewski, W. Z. 1973. "Basic Aerodynamics and Performance of the Helicopter," AGARD Lecture Series LS-63.
Stepniewski, W. Z. and Keys, C. N. 1984. *Rotary-Wing Aerodynamics*, Dover Publications, New York. Part II, Chapters 2 and 3.
Taylor, M. K. 1950. "A Balsa-Dust Technique for Air-Flow Visualization and Its Application to Flow through Model Helicopter Rotors in Static Thrust," NACA Technical Note 2220.
Wheatly, J. B. 1932. "Lift and Drag Characteristics and Gliding Performance of an Autogiro as Determined in Flight," NACA Report No. 434.
White, G. T., Logan, A. H., and Graves, J. D. 1982. "An Evaluation of Helicopter Autorotation Assist Concepts," 38th Annual Forum of the American Helicopter Soc., Anaheim, CA, May 4–7.
Wimperis, H. E. 1928. "The Rotating Wing in Aircraft," ARC R & M 1108.
Wood, T. L. 1976. "High Energy Rotor System," 32nd Annual V/STOL Forum of the American Helicopter Soc., Washington DC, May 10–12.
Zbrozek, J. 1947. "Ground Effect on the Lifting Rotor," ARC R & M 2347.

CHAPTER 6

Conceptual Design of Helicopters

We built the first helicopter by what we hoped was intelligent guess. It was time of crystal ball.

Igor Alexis Sikorsky (former Chief Aerodynamicist at Sikorsky Aircraft and cousin to Igor Sikorsky) (1957)

6.1 Introduction

There are many fundamental issues, some of them conflicting, in the aerodynamic design of the modern helicopter. Helicopter designers are concerned with performance, loads, vibration levels, external and internal noise, stability and control, and handling qualities. Specific emphasis in this chapter is placed on the general elements and features contributing to the aerodynamic design of helicopter main rotors and related components. However, many of the issues and arguments are applicable to tilt-rotors and tail rotors as well. The first part of this chapter outlines the basic sizing and overall optimization methodology of helicopter main rotors. This ultimately leads to the consideration of rotor airfoil sections and the need for good airfoil designs, the trade-offs in blade planform design, and the role of tip shape. While helicopter design is often synonymous with rotor aerodynamics, the aerodynamics of the fuselage has received considerable attention in recent years and now constitutes an important part of the overall design process. Finally, the chapter concludes with a discussion of the empennage and tail rotor design and the various aerodynamic interactions between these components and the main rotor.

6.2 Design Requirements

Both the preliminary and detailed design processes comprise a highly interactive effort among aerodynamicists, structural dynamicists, aeroelasticians, material specialists, weight engineers, flight dynamicists, and other specialists. The helicopter design starts with a set of specifications, which are defined based on the needs of a potential customer, or more so in the case of military machines, a specific mission requirement. Design technology for the civilian market is driven mostly by customers who emphasize reduced acquisition and operating costs, increased safety, reduced cabin noise and increased passenger comfort, and better overall mechanical reliability and maintainability. Because many of the helicopters in civilian use will operate from heliports and in populated areas, there is also an increasing emphasis on design for reduced external noise. The military have somewhat different requirements. Military planners constantly emphasize the need for operational flexibility and adaptability and the need for long operational life with components that can be continuously upgraded. Vulnerability and survivability of the craft in a combat situation are also issues important to the military. Today, increasing emphasis is being placed on the dual use of military and civilian technology, which is simply the efficient integration of these traditionally separate design technologies. This has benefits for both the customer and the manufacturer.

Prouty (1998) gives a good overview of the basic helicopter design process, particularly for military machines. The general design requirements for a new helicopter will include (not necessarily in order of priority): 1. hover capability, 2. maximum payload, 3. range and/or endurance, 4. cruise or maximum level flight speed, 5. climb performance, 6. "hot and high" performance, and 7. maneuverability and agility. The general objective for the manufacturer is to design the smallest and lightest helicopter to minimum cost. A challenge in minimizing costs is to lower the design cycle time, and this is where the role of analysis and mathematical models becomes useful. However, the design must proceed on the basis of many constraints, which may limit the number of design choices. These constraints may include (again, not necessarily in any order of priority): 1. maximum main rotor disk loading, 2. maximum physical size of the aircraft, 3. one-engine inoperative performance, 4. autorotative capability, 5. noise issues, 6. radar cross section and detectability, 7. civil certification or military acceptance requirements, 8. vulnerability and survivability requirements, and 9. maintenance issues.

The various requirements for a new helicopter design will be initially specified by the customer. These are then negotiated with the manufacturer and written into a sales contract. Often the "customer" will be the military forces, which will invite various competing manufacturers to respond to a "Request for Proposal" or RFP. Less often, the manufacturer will risk its own resources to develop a new design in anticipation of a production contract. In the design of the new helicopter, performance guarantees will be made to the customer based on metrics such as hover capability, payload, range and endurance, and cruise speed. In addition, one-engine inoperative performance may be included in the guaranteed performance. Methods for determining compliance with the specified performance by means of flight testing will be detailed in the contract. Because failure of the manufacturer to achieve the negotiated performance may result in substantial cost penalties, the manufacturer needs to have high levels of confidence that the performance guarantees can indeed be met.

The basic procedure that a manufacturer will follow in establishing a performance guarantee is based on statistical confidence levels of results obtained from both mathematical models and flight tests. Confidence values can be assigned to predictions based on the established accuracy of a given mathematical model. Typically, most models will have been in use by the manufacturer for some time, and based on correlation with experiments and/or flight tests, they will allow the manufacturer to establish good statistical bounds on the confidence level. For example, methodologies validated with reference to a prototype or a similar aircraft will allow a high confidence level to be accessed. In contrast, a completely new aircraft with an advanced rotor design or blade tip shape may have more uncertainties in the design, and confidence levels in any predictive methodology will be lower. This, however, is where the benefits of fundamental research and development become useful in the design process.

6.3 Design of the Main Rotor

The main rotor is the most important component of the helicopter. Proper design of the rotor is critical to meeting the performance specifications for the helicopter as a whole. The design of the tail rotor is similar to that of the main rotor, but since it has a different set of constraints it will be discussed separately. There remains a great deal of activity in developing an improved understanding of rotor aerodynamics and in developing new and improved mathematical models that will more faithfully predict the flow physics and help design more aerodynamically efficient rotors of lighter weight. One should always bear in mind that small improvements in rotor efficiencies can potentially result in significant increases in aircraft payload capability, maneuver margins, or forward flight speed.

The preliminary design of the main rotor must encompass the following key aerodynamic considerations:

1. General sizing: This will include a determination of rotor diameter, disk loading, and rotor tip speed. There are several important trade-offs in performance and other characteristics with variations in all of these parameters.
2. Blade planform: This will include chord, solidity, number of blades, and blade twist. The optimal blade planform and twist distribution for hover may not be optimum for high speed forward flight. The consideration of other than a rectangular tip shape may be part of the preliminary design.
3. Airfoil section(s): These play an important role in meeting overall performance requirements. On most modern rotors, the use of different airfoils at various stations along the blade will be a likely design choice.

6.3.1 *Rotor Diameter*

There are several conflicting factors that must be examined when determining the main rotor diameter. As shown previously in Chapters 2 and 5, both autorotational capabilities and hover performance call for a large rotor diameter. The advantages of a larger rotor diameter are lower disk loadings, lower average induced velocities, and lower induced power requirements. It was shown in Section 2.2.12 that based on the modified momentum theory the operating thrust coefficient, C_T, to give the best power loading was

$$C_T \text{ for best } PL = \frac{1}{2}\left(\frac{\sigma C_{d_0}}{\kappa}\right)^{2/3}, \tag{6.1}$$

which depends on airfoil section, rotor solidity, and induced power factor. Using this result, the disk loading for minimum power loading will be

$$DL = \frac{T}{A} = \frac{1}{2}\rho(\Omega R)^2\left(\frac{\sigma C_{d_0}}{\kappa}\right)^{2/3}. \tag{6.2}$$

This equation determines the optimum radius of the rotor to maximize power loading at a given gross weight. It has also been shown in Section 2.2.12 that

$$PL = \frac{T}{P} \propto \frac{FM}{DL}, \tag{6.3}$$

which means for a given disk loading the rotor should also be operated at the highest possible figure of merit. However, as shown in Fig. 2.12, the most efficient power loading (compared to the ideal value) is relatively insensitive to the operating state, in that the power loading curve is fairly flat over the normal range of operational thrust coefficients. Therefore, there is some latitude in selecting rotor radius, which may be constrained because of factors other than pure aerodynamic considerations.

For example, if the helicopter is required to operate off loose terrain such as gravel or sand, the design may call for the disk loading to be limited so that the downwash velocities remain low enough not to stir up any loose surface material. This usually means the use of a relatively high rotor diameter. A large diameter also means a larger inertia and stored rotational kinetic energy, which is essential for good autorotational characteristics. Initial sizing studies of the main rotor must always consider autorotational capability, which will require some minimum rotor inertia to meet the normal military acceptance or civilian certification requirements. Often the initial rotor design is guided by previous rotor designs that are known to meet the autorotative specifications necessary for certification. An autorotative index of the form that was discussed in Section 5.4.3 may be used as a means of quantifying the potential characteristics.

However, the overall helicopter size, weight, cost, gearbox torque limitations, speed, maneuverability requirements, and storage or transportation requirements for the helicopter usually require a much smaller rotor diameter for a given blade area and tip speed. A smaller rotor will have a smaller and lighter hub, and a lower overall parasitic drag, and hence will be more efficient for cruising flight. Smaller rotors also permit a more compact aircraft design, which is useful for several operational reasons. In addition, a smaller rotor diameter minimizes the static droop of the blades. The static droop can increase quickly for larger rotor diameters and may cause problems when starting and stopping the rotor, especially in gusty wind conditions, where "blade sailing" may cause the rotor blades to flex and impact the airframe. Unless a very stiff blade can be manufactured (which is unlikely without incurring a weight penalty), the blade radius is usually kept less than about 40 ft ($\approx$12 m).

Figure 6.1 gives a summary of main rotor radius versus gross aircraft weight for a selection of helicopters (see appendix). Plotting the data on a logarithmic scale accentuates the strong correlation between the quantities. This plot shows that the helicopter weight grows much faster than the rotor size. The well-known "square–cube" scaling law gives a simple explanation for this behavior. For geometrically similar aircraft, the surface area of the aircraft increases with the square of a characteristic length; so the weight of the aircraft should increase with the cube power of a characteristic length, that is, $W \propto l^3$ or $l \propto W^{1/3}$. Therefore, for a helicopter, $R \propto W^{1/3}$, which is the trend shown in Fig. 6.1.

The corresponding trend of rotor disk loading versus gross weight is shown in Fig. 6.2. Remember that disk loading must be kept as low as possible to maximize hovering performance, although with the larger and heavier helicopters this comes at some price. Because of the various other constraints posed in the design, such as higher rotor weight, greater torque requirements, etc., a large rotor diameter is not always practical. Generally, the manufacturer will try to find the smallest rotor diameter that will meet all of the specifications

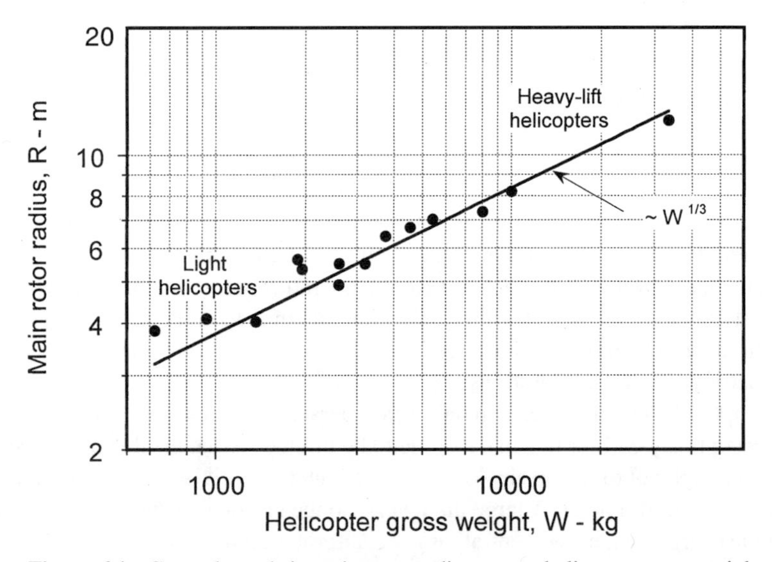

Figure 6.1 General trends in main rotor radius versus helicopter gross weight.

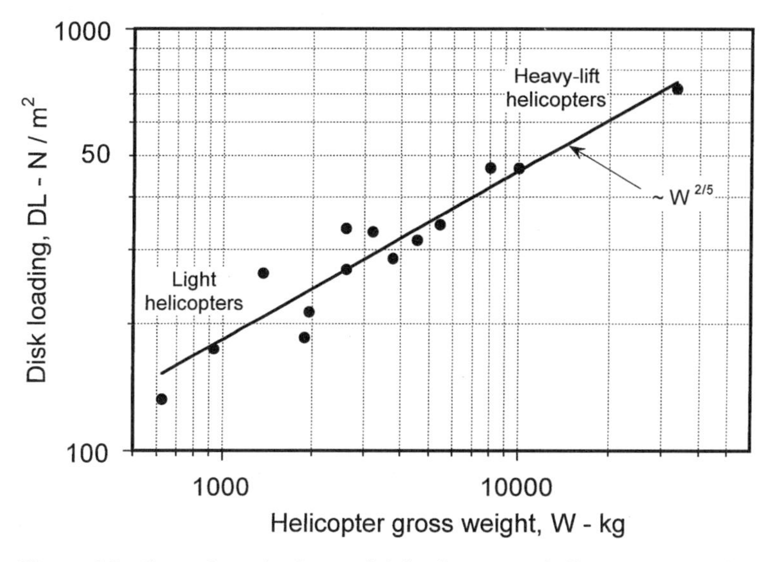

Figure 6.2 General trends of rotor disk loading versus helicopter gross weight.

laid down for that aircraft. The "square–cube" law would suggest that $T/A \propto W^{1/3}$, but the results shown in Fig. 6.2 indicate a slightly more rapid increase such that $T/A \propto W^{2/5}$. Therefore, the corresponding power loading will be proportional to $W^{-1/5}$.

6.3.2 *Tip Speed*

A high rotor tip speed helps to maintain the velocities and decrease the angles of attack on the retreating blade, thereby delaying the onset of blade stall for a given blade area and advance ratio. A high tip speed also gives the rotor a high level of stored rotational kinetic energy for a given radius and reduces design weight. Because $P = \Omega Q$, a high tip speed reduces the rotor torque required for a given power and allows a lighter gearbox and transmission design. However, there are two important factors that work against the use of a high tip speed: compressibility effects and noise. Compressibility effects manifest as increased rotor power requirements. If the drag divergence Mach number of the tip sections is exceeded, the sectional drag (and thus rotor power) increases dramatically. Therefore, reducing the tip speed permits a higher flight speed to be achieved before compressibility effects become important. The use of a thinner airfoil and/or a swept tip shape can help to delay the onset of compressibility effects to a higher advance ratio. Rotor noise also increases rapidly with increasing tip Mach number. At low tip speeds, the noise resulting from the steady and harmonic loading on the blades is dominant. At higher tip speeds, the noise caused by blade thickness effects becomes an important contributor to the overall sound pressure, and the noise signature can become increasingly obtrusive.

Figure 6.3 gives an idea of the range of acceptable tip speeds for conventional helicopters. The challenge is to extend both the stall and compressibility limits to a higher forward flight speed. Yet, this is not easily achieved. Typical design constraints on performance and autorotational characteristics severely limit the range of acceptable rotor tip speeds. However, the development of modern high-lift airfoil sections for rotors has permitted a reduction in rotor tip speed and, therefore, rotor noise levels, without compromising other aspects of the rotor performance. The use of thinner and/or "supercritical-like" airfoils

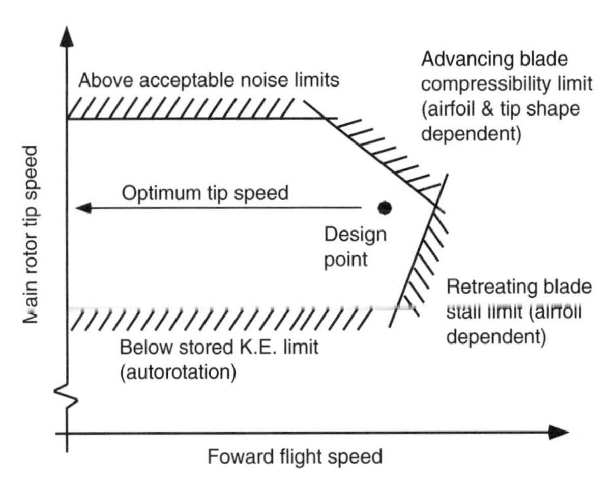

Figure 6.3 Aerodynamic, noise, and autorotative constraints imposed on the selection of rotor tip speed.

and swept tips can also help alleviate compressibility effects. For many rotor designs, the autorotative kinetic energy limit may be an issue. Although this can be improved by adding rotor mass, the corresponding decrease in payload and the increase in blade and hub loads as a result of the higher centrifugal forces are usually unacceptable. In light of the foregoing, these issues tend to set a lower main rotor tip speed limit of about 680 ft/s (207 m/s) for a conventional helicopter.

Systematic experimental measurements of rotor performance (thrust, power, figure of merit) at different operational tip speeds (or tip Mach numbers) and solidities are scarce. Figure 6.4 shows the effects of rotor tip Mach number on the figure of merit as a function of rotor solidity. While there is clearly scatter in the data, the results confirm that operation

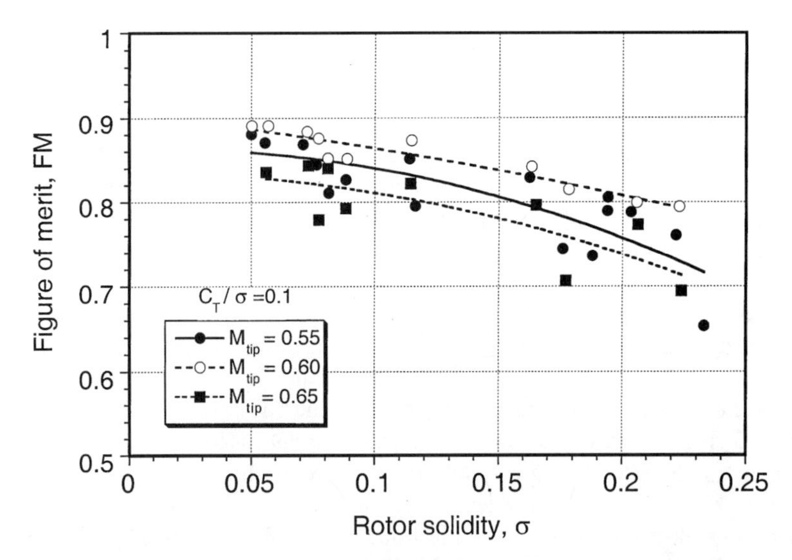

Figure 6.4 Measured figure of merit variation versus rotor solidity for different main rotor tip Mach numbers. Data source: Department of the Army, Engineering Design Handbook (1974).

at lower tip Mach numbers is desirable to maximize hovering performance. At higher tip Mach numbers, performance degrades because of the increasing compressibility losses. However, it must be recalled that operation at low tip speeds will compromise forward flight performance because of the need for the retreating blade to operate at higher angles of attack and closer to the stall, all other factors being equal. In addition, the need to allow sufficient stall margin in the rotor design, especially for helicopters designed for high maneuverability and agility, requires a higher main rotor tip speed.

6.3.3 *Rotor Solidity*

Rotor solidity, σ, has been defined previously and is the ratio of total blade area to the disk area (i.e., $\sigma = N_b c/\pi R$). Values of σ for contemporary helicopters vary from about 0.08 to 0.12. It has been shown in Section 3.3.17 that the mean rotor lift coefficient is given in terms of the blade loading coefficient, C_T/σ, by $\bar{C}_L = 6(C_T/\sigma)$. Typical values of $\bar{C}_L$ for helicopters range from about 0.4 to 0.7.

Selecting the solidity for the rotor design requires a careful consideration of blade stall limits. Certification or acceptance requirements dictate the load factors and bank angles that must be demonstrated without the rotor stalling. Rotors that are designed for high speeds and/or high maneuverability requirements require a higher solidity for a given diameter and tip speed. Obviously, a rotor that uses airfoil sections with high maximum lift coefficients can be designed to have a lower solidity, all other factors being equal. Alternatively, the use of high-lift airfoil sections permits a lower tip speed for the same solidity. Because the rotor noise is considerably reduced when the rotor is operated at lower tip speeds, the development of high-lift airfoils that operate efficiently over the diverse range of conditions found within the rotor environment is an important design goal.

Figure 6.5 shows experimental results for the variation in figure of merit versus rotor solidity. These results simply reaffirm that one way to minimize profile power is to keep the rotor solidity as low as possible. However, this approach must be done with caution because

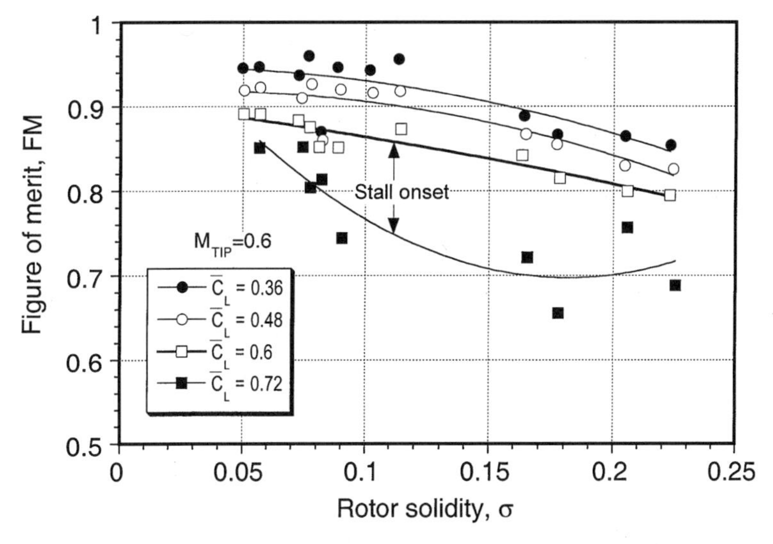

Figure 6.5 Measured figure of merit variation at different mean lift coefficients versus rotor solidity for a constant tip Mach number of 0.6. Data source: Department of the Army, Engineering Design Handbook (1974).

decreasing the solidity reduces the blade lifting area, increases the blade loading coefficients, and elevates the local and mean blade lift coefficients for a given rotor thrust. In other words, decreasing the rotor solidity decreases the *stall margin*, that is, the margin between the normal operational lift coefficients and the maximum lift coefficient of the airfoil sections. For example, consider rotor operation at a nominal mean lift coefficient, $\bar{C}_L$, of 0.6, as shown in Fig. 6.5. Then for a given value of solidity, operation of the rotor at a higher $\bar{C}_L$ results in a substantial degradation in rotor performance. This degradation is alleviated by increasing the solidity, which helps reduce the local lift coefficients and improve the lift-to-drag ratios of the airfoil sections. Because the onset of stall sets the performance limits of a rotor, it is also very important to provide a sufficient stall margin in the rotor design to allow for normal maneuvers and gusts typical of turbulent air. The specified margin, which will vary for different helicopter designs, generally tends to set the lowest allowable solidity to which the rotor can be designed. For example, a highly maneuverable combat helicopter will always require a larger stall margin than a civilian transport machine.

The onset of retreating blade stall also limits rotor performance. In forward flight, the rotor must provide both lifting and propulsive forces, the latter of which depends largely on the parasitic drag of the airframe. The stall inception boundary in forward flight can be observed from strain-gage measurements of blade pitch link and/or blade torsion loads. When stall occurs on the rotor, the aerodynamic pitching moments increase dramatically. Representative measurements of the fluctuating component of the blade root torsion loads are shown in Fig. 6.6 as a function of C_T/σ at two advance ratios of 0.2 and 0.5. At low advance ratios the blade root torsion loads are relatively low, but as C_T/σ reaches about 0.11, there is a more rapid increase as the rotor shows some evidence of stall. As the advance ratio increases, stall inception occurs at progressively lower values of C_T/σ. At $\mu = 0.5$, the stall onset is more sudden and shows large increases in the fluctuating loads because of the onset of dynamic stall on the retreating blade. These fluctuating loads will manifest as high structural stress and increases in vibration levels. Stall will also cause longitudinal flapping, tilting the disk back and setting a natural barrier to further increases in forward speed. Control

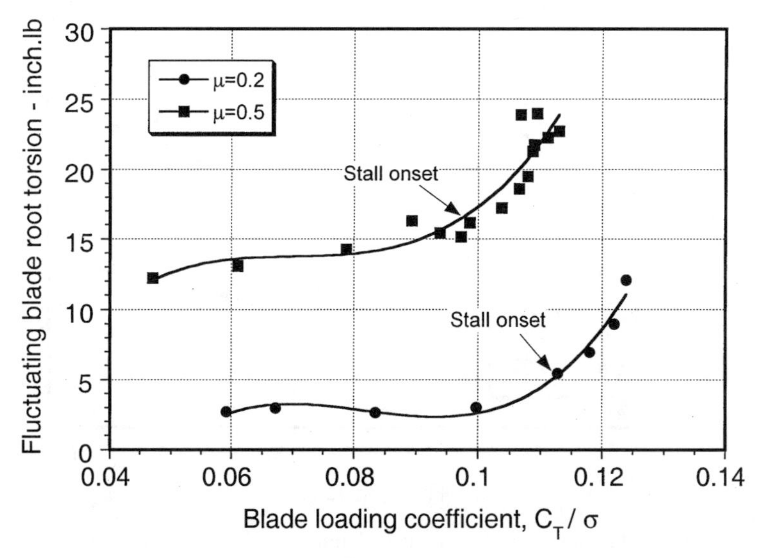

Figure 6.6 Retreating blade stall inception in forward flight deduced from blade root torsion loads. Data source: McHugh (1978).

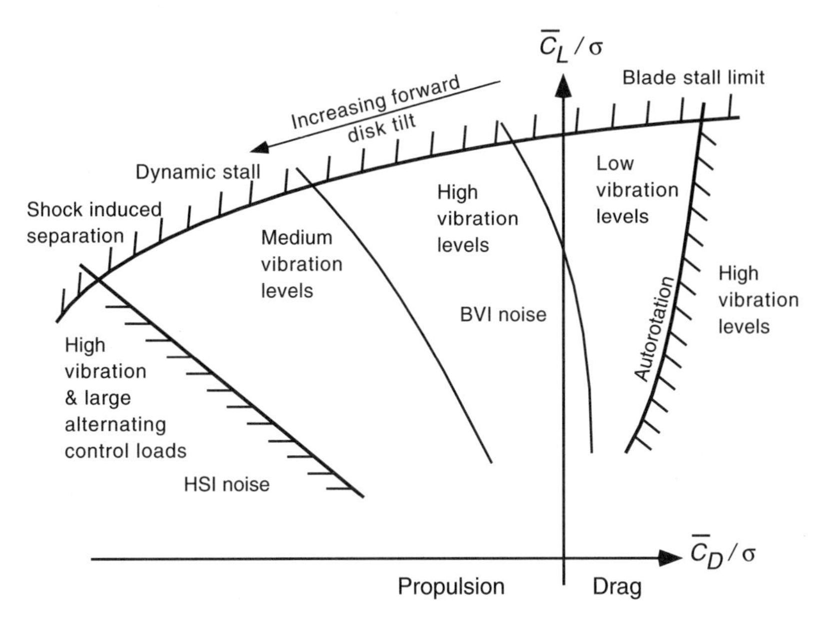

Figure 6.7 Propulsive limits for a conventional helicopter. Adapted from Vuillet (1990).

forces and the higher vibration levels will serve as a warning to the pilot that the limits of the normal operational flight envelope are being exceeded.

As already mentioned, propulsive force considerations also influence the selection of rotor solidity. In forward flight the function of the rotor is to provide a lift force to overcome airframe weight as well as to provide a propulsive force. The overall envelope is summarized by Fig. 6.7, which is a representation of blade loading (or mean lift coefficient) versus propulsive force. The disk must be progressively tilted forward to provide this forward thrust (propulsion) component, and so stall onset will be obtained at progressively lower values of mean lift coefficient. Propulsive requirements can be reduced by minimizing the parasitic drag of the airframe or by thrust and/or lift compounding. Compound helicopter designs are discussed briefly at the end of this chapter. While offering some advantages over a conventional helicopter, they also suffer from many disadvantages as well.

6.3.4 *Number of Blades*

The selection of the number of blades for a rotor (for a given blade area or solidity) is usually based on dynamic rather than aerodynamic criteria. Generally, the larger number of blades the lower the rotor vibration levels. Fewer blades, however, will usually reduce blade and hub weight, minimize hub drag, and may give better reliability and maintainability. Light weight helicopters usually have two blades, whereas heavy helicopters generally have four, five, or even seven blades. As shown in Section 3.3.7, induced tip loss effects are reduced by increasing the number of blades, but the effects on induced power are relatively small for the high aspect ratio blades typical of helicopter rotors.

Knight & Hefner (1937) and Landgrebe (1972) have conducted systematic tests of blade number and solidity on rotor performance. Knight & Hefner tested subscale rotors with 2, 3, 4, and 5 blades with no twist and at fairly low tip speeds. Landgrebe tested subscale rotors with 2, 4, 6, and 8 blades, linear twist, and at full-scale tip Mach numbers. An analysis of these data suggest that hover performance is primarily affected by rotor solidity; the number of blades are secondary. Following Knight & Hefner (1937), the primary rotor behavior can be analyzed by the use of the blade element theory. From Section 3.2.1, the rotor thrust is

$$C_T = \frac{1}{2}\sigma \int_0^1 C_l r^2\, dr = \frac{1}{2}\sigma C_{l_\alpha} \int_0^1 (\theta - \phi) r^2\, dr. \tag{6.4}$$

Also, $\phi(r) = \lambda/r$, and for a linearly twisted blade $\theta(r) = \theta_{75} + \theta_{\text{tw}}(r - 0.75)$. Therefore

$$\begin{aligned} C_T &= \frac{1}{2}\sigma C_{l_\alpha} \int_0^1 [(\theta_{75} + (r - 0.75)\theta_{\text{tw}})r^2 - \lambda r]dr \\ &= \frac{1}{2}\sigma C_{l_\alpha} \left[\frac{\theta_{75}}{3} - \frac{\lambda}{2B}\right]. \end{aligned} \tag{6.5}$$

When including tip losses the corresponding power can be written as

$$\begin{aligned} C_P &= C_{P_i} + C_{P_0} \\ &= \lambda C_T + C_{P_0} \\ &= \frac{\lambda}{2}\sigma C_{l_\alpha} \left[\frac{\theta_{75}}{3} - \frac{\lambda}{2B}\right] + C_{P_0}. \end{aligned} \tag{6.6}$$

Assuming uniform inflow such that $\lambda = \kappa\sqrt{C_T/2}$, and dividing through by σ^3 gives

$$\frac{C_P}{\sigma^3} - \frac{C_{P_0}}{\sigma^3} = \frac{C_{l_\alpha}\kappa}{2\sqrt{2}}\sqrt{\frac{C_T}{\sigma^2}}\left[\frac{\theta_{75}}{3\sigma} - \frac{\kappa}{2\sqrt{2}B}\sqrt{\frac{C_T}{\sigma^2}}\right]. \tag{6.7}$$

Therefore, if the results from rotor experiments are plotted in the form of $(C_P - C_{P_0})/\sigma^3$ versus C_T/σ^2 then this should remove the primary effect of solidity from the measurements. As shown by Fig. 6.8, the results from both the Knight & Hefner and Landgrebe experiments show a good correlation with the result given by Eq. 6.7, confirming that the effect of blade number on rotor performance is indeed secondary. The modified results are based on the

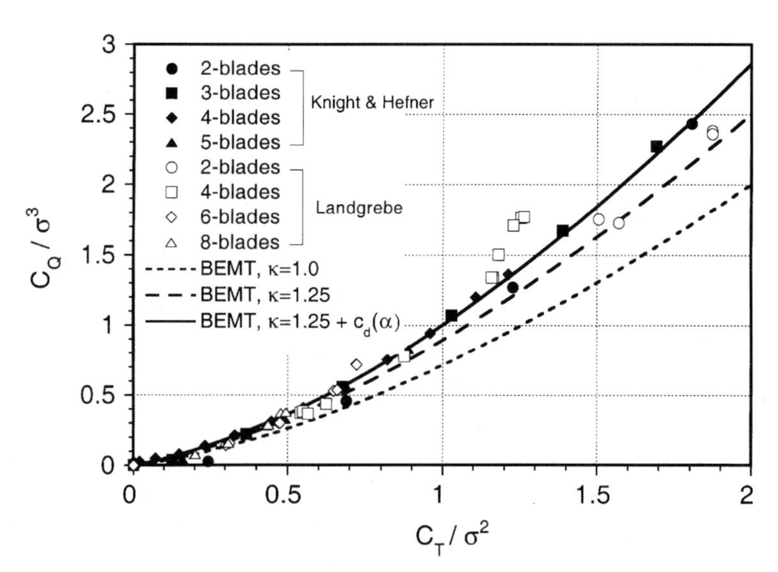

Figure 6.8 Correlation curve of C_Q/σ^3 versus C_T/σ^2 showing negligible effect of blade number on hovering rotor performance.

use of non-uniform inflow and a Prandtl tip-loss factor, and a higher order approximation for the profile power, both effects having been discussed previously in Chapter 3. Note that the latter gives a small but significant improvement in the correlation at high values of thrust where nonlinear drag effects begin to become important.

In subscale rotor experiments the effects of blade number can be masked by other effects, including those associated with the effects of Reynolds number and Mach number. The effects of Reynolds number can be particularly significant on small-scale rotors with small chords and tip Reynolds numbers that are often below 10^6. Increasing the number of blades whilst maintaining the solidity will decrease the chord and decrease the Reynolds numbers. Based on two-dimensional arguments, this will tend to result in slightly higher values of the profile drag coefficient and, therefore, will give a small but finite reduction in the figure of merit.

A larger number of blades also leads to weaker tip vortices (for the same overall rotor thrust), thus potentially reducing the intensity of any airloads produced by blade–vortex interactions (all other factors being assumed equal). However, with more blades, the number of potential blade–vortex interactions over the disk will be increased (see Chapter 10). This can affect both the frequency and directivity of propagated noise, and the problem becomes too complicated to make generalizations. Furthermore, from the perspective of noise generation, a higher number of blades results in a smaller chord and lower "thickness" noise associated with each blade. However, even though the strengths of the tip vortices may be reduced by using a larger number of blades (at a constant thrust), the blades are closer azimuthally. Because the tip vortices are convected vertically only relatively slowly in hover and remain almost in the plane of the rotor, the following blades may more closely interact with the vortices from previous blades. Such a phenomenon has been encountered on hovering rotors with five or seven blades and can significantly affect predictions of induced power – see Clark & Leiper (1970).

6.3.5 *Blade Twist*

Based on the combined blade element momentum theory (BEMT) discussed in Chapter 3, it has been shown that negative (nose down) twist can redistribute the lift over the blade and help reduce the induced power. Therefore, proper use of blade twist can significantly improve the figure of merit in hover. In Chapter 3, the effect of twist on the induced power factor has been shown based on analytical results. Paul & Zincone (1977) show the effects of blade twist on maximum figure of merit deduced from full-scale rotor tests; the results are reproduced in Fig. 6.9. In general, the results confirm the theory in that the figure of merit of highly twisted rotors is nearly always higher than that for moderately twisted rotors when operating at the same disk loading.

In forward flight, rotors with a high nose-down blade twist (say, greater than 15 degrees) may suffer some performance loss. This is because of the reduced angles of attack on the tip of the advancing blade, resulting in a loss of rotor thrust and propulsive force. The problem has been explored by Keys et al. (1987). Results documenting the problem are shown in Fig. 6.10 in terms of power required (relative to hover power) versus advance ratio. Although not severe, some degradation in high speed cruise performance is noted with the higher blade twist. Blades with very large twist rates, while offering performance benefits in hover, may suffer reduced or even negative lift production on the advancing blade tip and are obviously to be avoided. A survey of existing rotor designs shows that most helicopter blades incorporate a negative linear twist between 8 and 15 degrees, with only a few exceptions. This twist range is the best compromise between maximizing the figure of merit of the rotor in hover, whilst simultaneously maintaining good forward flight performance. In the quest

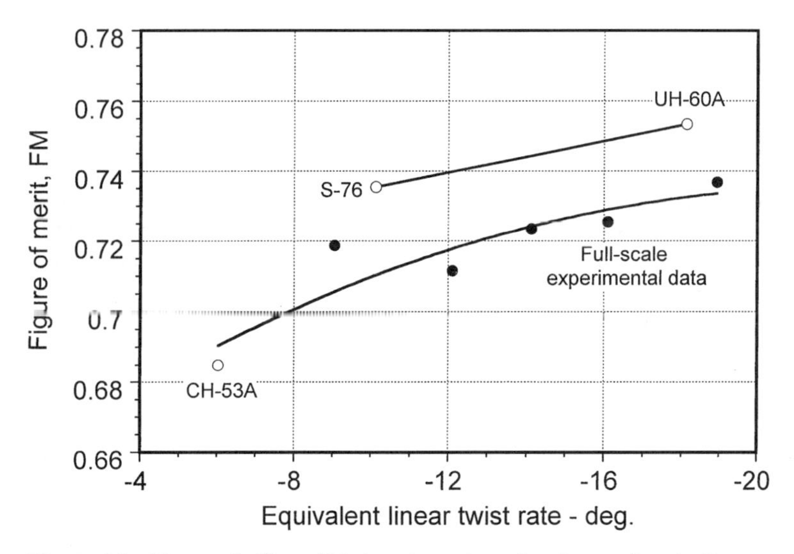

Figure 6.9 Measured effect of blade twist on hovering figure of merit. Data source: Paul & Zincone (1977).

for high speed forward flight capability some manufacturers have used a nonlinear twist or double linear twist, where the effective twist rate is reduced or reversed near the blade tip. This helps give better forward flight performance while retaining most of the hover performance and thus offers another good compromise for the rotor design.

6.3.6 *Blade Planform and Tip Shape*

It is known from the analysis in Chapter 3 that blade planform can also have an important effect on the blade lift distribution and, therefore, on the rotor performance. Usually, small amounts of taper over the blade tip region can help to significantly improve the

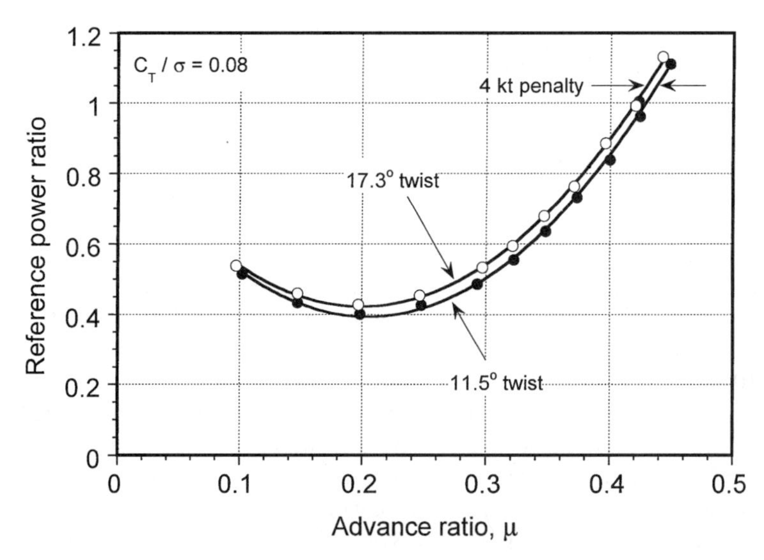

Figure 6.10 Loss of performance with a highly twisted rotor in forward flight. Data source: Keys et al. (1987).

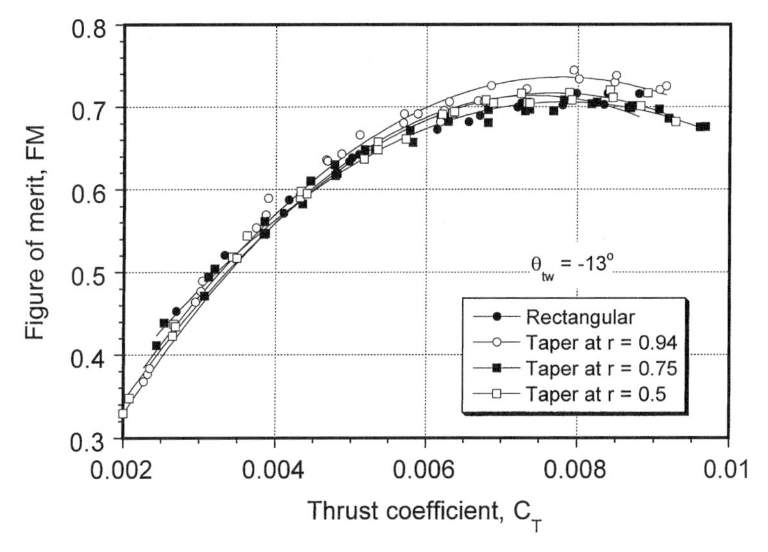

Figure 6.11 Measured effect of blade taper on the figure of merit of a hovering rotor. Data source: Althoff & Noonan (1990).

figure of merit (FM) in hover. This is confirmed in Fig. 6.11, which is taken from the subscale rotor experiments of Althoff & Noonan (1990). The benefits, however, seem to be lost for larger amounts of taper, most likely because of the higher profile drag coefficients associated with operation at small tip chord Reynolds numbers. Nevertheless, the experimental results confirm the previous observations from the BEMT that a maximum figure of merit for the rotor can, in part, be obtained by using both a combination of blade twist and taper. The type and amount of taper and twist, however, needs to be factored into the overall rotor design and operational specifications.

The tips of the blades play a very important role in the aerodynamic performance of the rotor. The blade tips encounter the highest dynamic pressure and highest Mach numbers, and strong trailed tip vortices are produced there. A poorly designed blade tip can have serious implications on the rotor performance. Figure 6.12 shows some blade tip designs that have been used or proposed for helicopter rotors. There are several common designs,

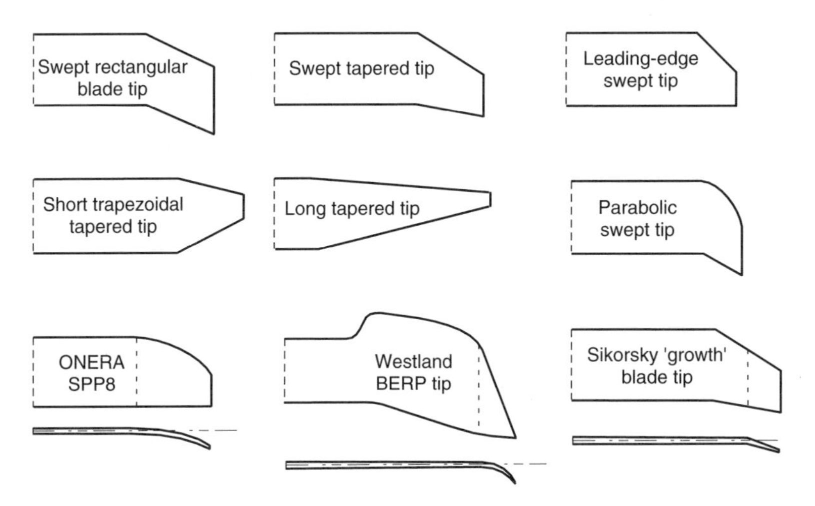

Figure 6.12 Some advanced main rotor blade tip designs. Note: Not to scale.

comprising those with taper, those with sweep, and those with a combination of sweep and taper. Some blade tips may also use an anhedral, which, as shown by Balch (1984) and others, can improve the figure of merit of the rotor. A good analysis of the effects of an anhedral is given by Desopper et al. (1986, 1988), Vuillet et al. (1989), and Tung & Lee (1994). An example of a state-of-the art tip shape is that on the the advanced growth blade designed by Sikorsky for the S-92 Helibus and later models of the UH-60 Blackhawk. This unique tip incorporates all three geometric parameters, sweep, taper, and anhedral. Other special tip designs such as that used on the Westland/RAE BERP blade incorporate somewhat more radical variations in sweep, planform, and anhedral.

Sweeping the leading edge of the blade reduces the Mach number normal to the leading edge of the blade, so allowing the rotor to attain a higher advance ratio before compressibility effects manifest as an increase in power required. The use of sweep also affects tip vortex formation, its location after it has been trailed from the blade, and the overall vortex structure – see, for example, the results of Rorke et al. (1972), Balcerak & Felker (1973), Spivey (1968), Spivey & Morehouse (1970), Carlin & Farrance (1990), Sigl & Smith (1990), and Smith & Sigl (1995). However, the problem of rotor tip vortex formation and the effects of tip shape on the vortex characteristics such as velocity profile and diffusive characteristics is, however, still the subject of ongoing research; systematic studies of the tip vortex characteristics generated by rotors with different tip shapes have not yet been accomplished. It is likely, however, that this is an area where research may lead to blades optimized for lower induced drag.

Blade tips with a constant sweep angle and those with a progressively varying sweep angle have been used. Many modern helicopters (for example, the UH-60 Blackhawk and the AH-64 Apache) use some simple constant sweepback on the blade tip. The amount of sweep is usually kept low enough ($<20^\circ$) so that there are no inertial couplings introduced into the blade dynamics by an aft center of gravity or by aerodynamic couplings caused by a more rearward center of pressure. A blade tip with a progressive sweep angle can be designed on the basis of a simple two-dimensional analysis. One criteria is to choose a sweep angle that is just sufficient to maintain a constant incident Mach number normal to the leading edge. The velocity normal to the leading edge, U_n, is given by

$$U_n = \Omega R (r + \mu \sin \psi) \cos \Lambda, \tag{6.8}$$

where Λ is the local sweep angle of the blade 1/4-chord axis (see inset in Fig. 6.13). The incident Mach number, M_n, is then

$$M_n = \frac{\Omega R}{a} (r + \mu \sin \psi) \cos \Lambda = M_{\text{tip}} (r + \mu \sin \psi) \cos \Lambda, \tag{6.9}$$

where a is the sonic velocity and M_{tip} is the hover tip Mach number. Consider the design point as the advancing blade, that is, where $\psi = 90^\circ$, and ignore any unsteady effects associated with the problem. If the local Mach number, M_n, is to be maintained below the drag divergence Mach number, M_{dd}, then the sweep angle required to do this is given by

$$\Lambda = \cos^{-1}\left(\frac{M_{dd}}{M_{\text{tip}}(r + \mu)}\right). \tag{6.10}$$

For example, assume a typical hover tip Mach number of 0.64, and an airfoil with $M_{dd} = 0.82$. Then the sweep angle required to ensure that M_n at each section remains just below M_{dd} is

$$\Lambda = \cos^{-1}\left(\frac{1.281}{r + \mu}\right). \tag{6.11}$$

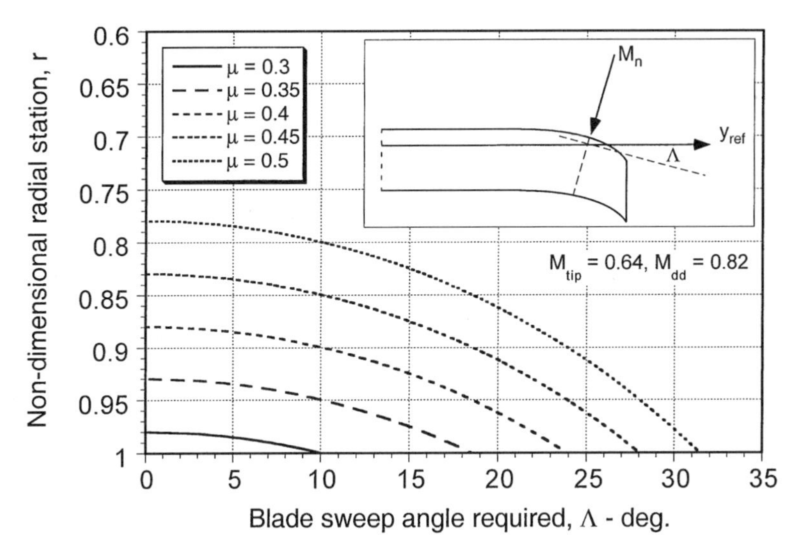

Figure 6.13 Example of sweep angle required to maintain the incident Mach number normal to the blade equal to a constant for a given advance ratio.

The results are plotted in Fig. 6.13, where it will be seen that the 1/4-chord sweep angle is a smooth, almost parabolic, curve. With increasing advance ratio the amount of sweep required increases substantially and the sweep initiation point moves inboard on the blade. (See also Questions 6.9 and 6.10.).

McVeigh & McHugh (1982) have conducted experiments with subscale rotors to study the effects of tip shape on overall rotor performance and cruise lift-to-drag ratio. It was shown that the combined use of improved airfoil sections and tapered tip shapes can help minimize profile power and significantly improve overall rotor cruise efficiency. The effects of tip shapes were examined for four rotors having rectangular, swept, swept-tapered, and tapered tips. All rotors were tested at the same lift, propulsive force, and trim state, which provides a datum for performance comparisons. Figure 6.14 reproduces a sumary of the results from McVeigh & McHugh (1982). The tapered tip is found to give about 10% higher equivalent L/D compared to the rectangular blade, but interestingly enough the rectangular blade gives a better maximum cruise L/D than either of the swept or swept tapered blades. This is because both the advancing and retreating blade characteristics of the tip shape are important, and an integrated performance metric such as L/D does not allow one to distinguish separately between these characteristics. While sweepback alone clearly has the advantage of delaying the onset of compressibility effects to higher advance ratios, the sweep may also promote early flow separation on the retreating blade at lower angles of attack. Therefore, there can also be a performance penalty associated with a swept tip.

Yeager et al. (1987) and Singleton et al. (1990) have examined the performance of a tapered blade versus a blade with a simple swept tip and have found improvements in performance, with lower power requirements up to advance ratios of 0.4. Noonan et al. (1992) have examined the effects of taper alone on forward flight performance. Generally, small amounts of taper are desirable and help to improve rotor performance, at least up to advance ratios of 0.3. Larger amounts of taper yield much smaller improvements. Yeager et al. (1997) have tested a BERP-like (see Section 6.3.8) subscale rotor under similar controlled conditions and have compared the results to a rectangular blade with the same

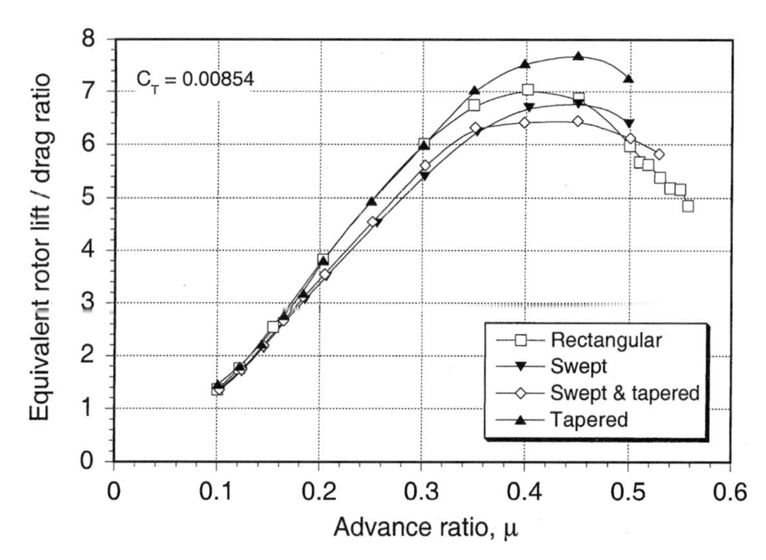

Figure 6.14 Equivalent lift-to-drag ratios for a rotor with different tip shapes. Data source: McVeigh & McHugh (1982).

airfoil section. These and the other results confirm that a sophisticated swept tip design is by no means a panacea; each tip design must be assessed on the performance gains to be expected over the full operational envelope as well as within the overall requirements and constraints imposed by the rotor design process.

6.3.7 *Airfoil Sections*

The choice of airfoil sections for helicopter rotors requires special consideration because significant improvements in rotor performance can be realized with the optimal selection of airfoil shapes. This is of such fundamental importance to helicopter design that Chapter 7 is devoted to a basic understanding of the subject. During each rotor revolution, the airfoil sections on helicopter rotors encounter a wide diversity of operating conditions. For example, for a representative blade station (say at 75% blade radius) the aerodynamic environment in which this section operates in hover and forward flight are different. On the advancing blade in forward flight, the airfoil sections may penetrate into the transonic flow regime. This causes wave drag, and if the shock strengths become sufficiently severe, shock induced flow separation and stall may be produced. On the retreating blade, the tip sections operate at subsonic speeds, but at high angles of attack close to stall. Thus, the blade sections must be thin enough to maximize the drag divergence Mach number, while simultaneously they must have some minimum thickness and incorporate some camber to give a relatively high $C_{l_{\max}}$ but still maintain low pitching moments. No single airfoil profile will meet all the requirements, and the airfoil(s) must be designed to reach a compromise between the requirements of all flight conditions. In general, the goal is to balance the advancing blade requirements (high drag divergence Mach number) with those of the retreating blade (high maximum lift at low Mach numbers), while maintaining a good overall lift-to-drag ratio.

One way to help minimize profile power and maximize figure of merit in hover is to consider the use of a low drag airfoil. The airfoil drag consists of two parts, namely pressure drag and viscous drag or shear stress drag. The pressure drag can be minimized by using relatively thin airfoil sections, although the use of thin airfoils has some disadvantages as well, mainly because of the possibilities of low maximum lift coefficients. The viscous drag component, represented by the coefficient C_{d_0}, can be minimized by carefully controlling the airfoil pressure distribution and maximizing the chordwise extent of laminar flow, at least for a range of low to moderate lift coefficients. Unfortunately, there are other aspects of airfoil performance, such as the need for low pitching moments and high maximum lift, that usually preclude the use of special low drag or laminar flow airfoil sections on rotor blades. Also, environmental factors tend to produce insect accretion and blade erosion at the blade leading edge, which can cause premature boundary layer transition and reduce the run of laminar flow in any case. In some circumstances surface roughness can adversely alter other aspects of the airfoil characteristics.

A powerful parameter affecting profile power is the airfoil thickness. Using two-dimensional airfoil measurements given by Abbott & von Doenhoff (1949), the zero-lift sectional drag coefficient for the NACA symmetric series can be approximated by the equation

$$C_{d_0} \approx 0.007 + 0.025\left(\frac{t}{c}\right), \tag{6.12}$$

where t/c is the thickness-to-chord ratio. The result is valid in the range $0.06 \leq t/c \leq 0.24$. The effects of Mach number compound the behavior of the drag, but at moderate angles of attack below the drag divergence Mach number the effects of compressibility are small and are more sensitive to Reynolds number. Assume, for example, that a blade tapers in thickness from an airfoil with a 12% thickness-to-chord ratio at the root to an 8% ratio at the tip. Therefore, using Eq. 6.13 the drag coefficient can be written as

$$C_{d_0}(r) = 0.007 + 0.025(0.12 - 0.04r) = 0.01 - 0.001r. \tag{6.13}$$

The profile power coefficient can now be estimated using the blade element model where

$$C_{P_0} = \frac{1}{2}\sigma \int_0^1 C_{d_0} r^3 dr = \frac{1}{2}\sigma \int_0^1 (0.01 - 0.001r) r^3 dr. \tag{6.14}$$

Evaluation of this expression gives a value of $\frac{1}{8}\sigma(0.0092)$ compared to the value $\frac{1}{8}\sigma(0.01)$ without the use of thickness variations (i.e., an 8% reduction in profile power). Typically, this would translate into an increase in figure of merit of between 1 and 2%. For a given rotor power or shaft torque, this would offer a 0.5 to 1.5% increase in overall vertical lifting capability (see also Question 6.8). Measurements by McVeigh & McHugh (1982) show a 5% gain in rotor figure of merit and a 25% gain in equivalent lift-to-drag ratio in forward flight through the use of advanced airfoil sections designed for lower profile drag and higher drag divergence Mach numbers.

6.3.8 *The BERP Rotor*

The British Experimental Rotor Program (BERP) blade has been alluded to earlier and is worthy of special consideration. The BERP rotor was designed specifically to meet the conflicting aerodynamic requirements of the advancing and retreating blade conditions, either of which can limit the performance of the rotor in high speed forward flight. The technical details of the BERP research program are described by Perry (1987), Perry et al. (1998), and Wilby (1998). This research paid off in 1986 when a Westland Lynx attained the world absolute speed record for a conventional helicopter, which remains in place at the time of writing.

Figure 6.15 Photograph of the BERP blade as used on the Westland Lynx. (Courtesy of GKN-Westland.)

As shown in Fig. 6.15, the BERP rotor blade is distinctive because of its unique tip shape. However, the aerodynamic improvements shown with the BERP rotor are the result of several innovations in both airfoil design and tip shape design. The BERP blade uses a number of high performance airfoils based on the RAE family (see Fig. 6.16). The main lifting airfoil is the RAE 9645, which is located on the blade from 65% to 85% radius. This airfoil has a maximum lift coefficient of about 1.55, which is high compared to even second-generation helicopter airfoil sections – see Dadone (1976) and Wilby (1998). However, this amount of lift is obtained at the expense of higher pitching moments. These higher pitching moments are offset by the RAE 9648 airfoil, which is reflexed and located inboard of 65% radius where the demands of high maximum lift are not so important. The tip region uses the RAE 9634 airfoil, which has a relatively low thickness-to-chord ratio to give a high drag divergence Mach number. The RAE 9634 airfoil is also cambered in such a way as to give a weak shock wave and low pitching moments.

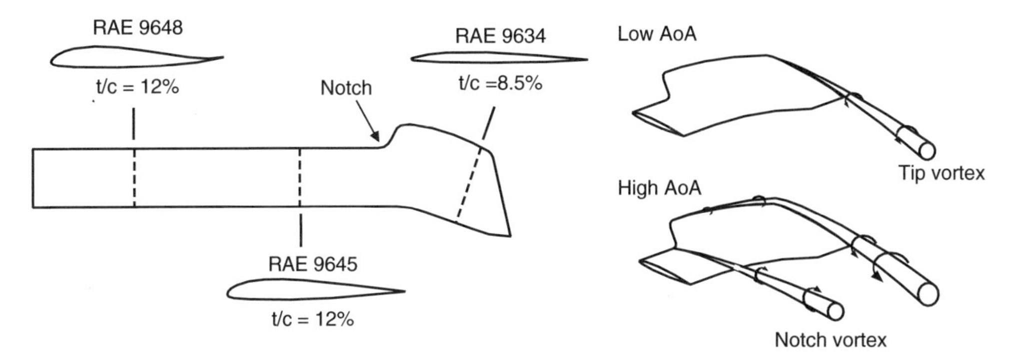

Figure 6.16 Distribution of airfoil sections on the BERP blade and high angle of attack performance. Adapted from Perry (1987).

One of the most recognizable features of the BERP blade is the use of high sweepback over the tip region, which as mentioned previously, is an effective means of reducing compressibility effects and delaying their effects on the rotor to a higher advance ratio. On the BERP blade a progressively increasing sweepback angle is used over the outboard 16% of blade radius. Because the Mach number varies linearly along the blade the amount of sweep necessary can be minimized by keeping the Mach number normal to the leading edge approximately constant, as discussed earlier. What is different for the BERP blade is that the area distribution in the tip region is configured to ensure that the mean center of pressure is located close to the elastic axis of the blade. This is done primarily by offsetting the location of the local 1/4-chord axis forward starting at 86% radius. This offset also produces a discontinuity in the leading edge of the blade, which is referred to as a notch.

It must be recognized that while a swept tip geometry will generally always reduce profile power and delay the onset of increased power requirements to higher advance ratios, it will not necessarily improve the performance of the blade at high angles of attack on the retreating side of the disk. In fact, results have been previously shown in Fig. 6.14 that a swept tip blade may have inferior forward flight characteristics compared to a standard (unswept) blade tip. The BERP blade is specifically designed to perform as a swept tip at high Mach numbers and low angles of attack, but it is also designed to operate at very high angles of attack without stalling. This latter attribute is obtained, in part, through the generation of stable vortex flows that enhance lift and delay the onset of gross flow separation over the tip region to extremely high angles of attack (see Fig. 6.16). This mechanism is promoted by giving the airfoils in this region a small leading-edge nose radius. As the angle of attack of the blade is increased, this vortex begins to develop from a point further forward along the leading edge, following the planform geometry into the more moderately swept region. At a sufficiently high angle of attack, the vortex will initiate close to the forwardmost part of the leading edge near the notch region. Experimental evidence by Duque & Brocklehurst (1990) has shown that a second notch vortex is also formed, which is trailed streamwise across the blade. This vortex acts like an aerodynamic fence and retards the flow separation region from encroaching into the tip region. Numerical calculations by Duque (1992) have shown that in high speed flight this notch also helps to further reduce the strength of shock waves on the tip, acting as a form of "tip relief." Further increases in angle of attack make little change to the flow structure until a very high angle of attack is reached (in the vicinity of 22 degrees) when the flow will finally break down and separate. For a conventional tip planform, a flow breakdown would be expected to occur at about 12 degrees local angle of attack.

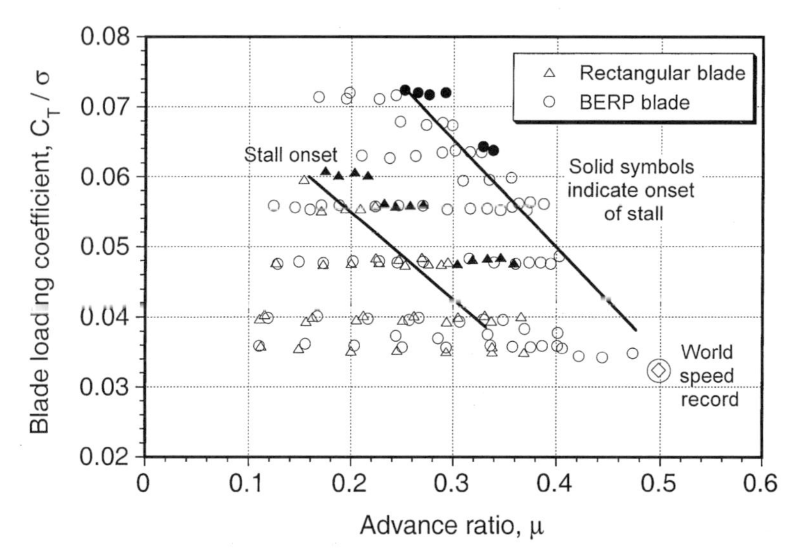

Figure 6.17 Flight envelope of the Lynx aircraft with standard rectangular blades and BERP blades. Data source: Perry et al. (1998).

In summary, the design of the BERP rotor attempts to reduce compressibility effects on the advancing blade while simultaneously delaying the onset of retreating blade stall. This is confirmed by the results shown in Fig. 6.17, which is based on actual flight tests using the Lynx helicopter – see Perry et al. (1998). The results are shown in terms of blade loading coefficient versus advance ratio. Clearly the BERP rotor demonstrates a significant increase in the operational flight envelope. Further results are shown by Perry et al. (1998), who compare the BERP rotor with other rotor blades in terms of hovering performance and the maximum thrust capability in forward flight. Despite some controversy about comparing rotor performance in forward flight on the basis of weighted solidities [see Amer (1989) and Perry (1989)] the performance gains of the BERP rotor are convincing and clearly demonstrate the benefits of a careful synthesis of both improved airfoil design and blade planform shape in the design of better helicopter rotors suitable for high speed forward flight.

6.4 Fuselage Design

The fuselage is the largest airframe component and its aerodynamic characteristics have a significant impact on the performance of the helicopter as a whole. To be efficient, the helicopter fuselage must be fully aerodynamically integrated with the main rotor, tail rotor, and empennage. In isolation, these components exhibit fairly well understood flow phenomena and aerodynamic characteristics. However, when integrated together, they may behave differently and the aerodynamic interactions produced between them may cause an unfavorable behavior that may subtract from the overall performance and handling qualities of the helicopter. For example, the fuselage is immersed in the wake from the main rotor, and so the main rotor downwash velocities will influence the aerodynamic characteristics of the fuselage. The presence of the fuselage also distorts the inflow through the rotor disk and affects the rotor performance. These interactional aerodynamic effects are summarized by Fig. 6.18. Because the helicopter as an entire system must function effectively and predictably throughout the flight envelope, including hover, climb and descent, sideward

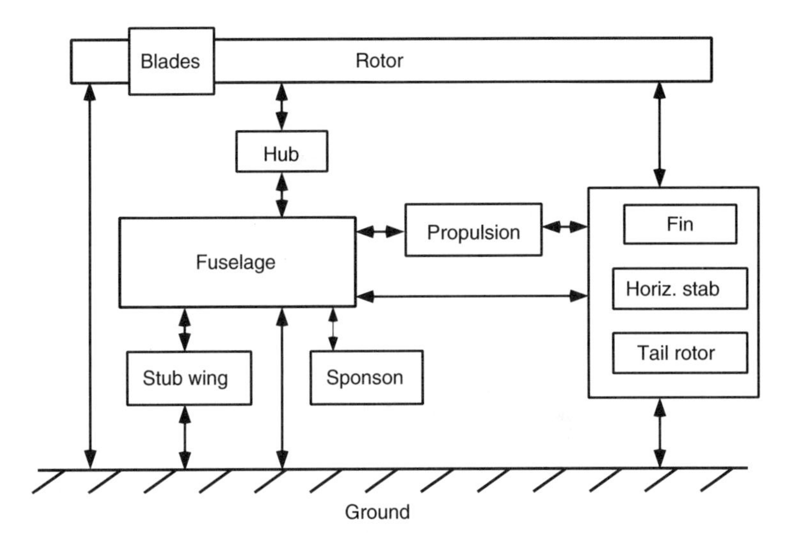

Figure 6.18 Schematic showing that aerodynamic interactions can exist between rotor and airframe components. Adapted from Sheridan & Smith (1979).

flight, and high speed cruise, an understanding of any interactional aerodynamic effects is essential to the design process.

6.4.1 *Fuselage Drag*

The parasitic drag of the fuselage affects cruise speed and fuel consumption. Williams & Montana (1975) and Keys & Wiesner (1975) have emphasized the need for low fuselage drag in the design of helicopters for improved performance. The drag of a helicopter fuselage may be up to one order of magnitude higher than that of a fixed-wing aircraft of the same gross weight. One reason is because of the rotor shaft, hub, and blade attachments, which may account for 30% or more of the total fuselage drag – see Sheehy & Clark (1975) and Sheehy (1977). Another major contributor is the fuselage after-body or bluff-body drag, which may account for 20% of the total fuselage drag. Drag is also caused by flow separation in the region where the main fuselage tapers to the tail boom. Sedden (1982) has shown that large drag penalties result when using fuselage shapes with large rear fuselage upsweep angles, because these can promote flow separation and the formation of two strong trailing vortices. See also Epstein et al. (1994) for details of this phenomenon. There are, however, many constraints that can limit the design of the fuselage shape and consequently the lowest possible drag of the fuselage. Much depends on the operational needs for the machine. For example, as shown in Fig. 6.19, an executive transport helicopter can be designed to have a more streamlined fuselage compared to a utility aircraft, which may need a rear access door. In either case, it is found that the fuselage shape should be more circular than square to keep drag as low as possible. Using a fairing over the top of the fuselage can help reduce drag from the main rotor shaft and control linkages. See also Montana (1975, 1976). Replacing the skids with a retractable undercarriage can reduce drag considerably. However, a fixed wheeled undercarriage will have higher drag than skids.

In view of the complicated interacting viscous dominated flows that can exist over helicopter fuselages, predictive capabilities for pressure and skin friction drag are not yet mature. Current capabilities for design are based on synthesis of component drag using

(a) Unstreamlined fuselage design with large rear upsweep angle

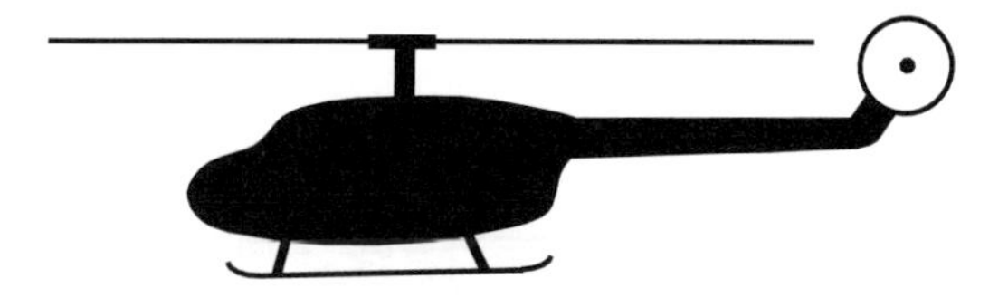

(b) Streamlined fuselage design with shallow rear upsweep angle

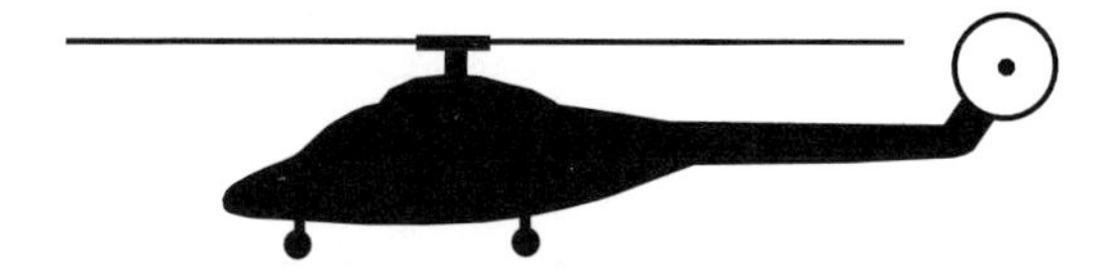

Figure 6.19 Reductions in fuselage parasitic drag are obtained through streamlining. (a) Unstreamlined fuselage. (b) Streamlined fuselage.

experimental data or by using a combination of experimental data and potential flow theory. Classical *panel methods* have found considerable use in routine helicopter fuselage design. These methods are based on the assumption of small disturbance potential flow, with the basic approach being described by Hess & Smith (1962), Rubbert & Saaris (1969), and Hess (1990). A modern treatment of the various types of panel methods is given by Morino (1985) and Katz & Plotkin (1991). Large computer codes that use panel methods are commercially available and are in widespread use in the helicopter industry. Sophisticated computational fluid dynamics methods such as Direct Navier–Stokes (DNS) and Reynolds-Averaged Navier–Stokes (RANS) are still in the development stage, and even if enormous computer memory and storage requirements can be overcome, they are relatively far from being practical for use in routine helicopter fuselage design studies. Another problem with DNS methods is the efficient generation of grids, and especially the proper coupling of structured and unstructured grids. However, rapid progress in grid generation techniques and solution algorithms is being made; see for example Berry et al. (1994), Chaffin & Berry (1994), Duque & Dimanlig (1994), and Duque (1994).

Because of their widespread use in helicopter fuselage design, the theory of panel methods will be briefly reviewed. See also Maskew (1982) for a good summary of the general techniques. In the panel method approach, the fuselage surface is modeled by N small, flat, quadrilateral panels onto which singularities in the form of sources and sinks and/or doublets are placed (see Fig. 6.20). Assume for illustration that only sources are used. If the source strength on the ith "sending" panel is σ_i, then the velocity induced by this panel at the jth control point on the "receiving" panel can be written as $A_{ij}\sigma_i$. A is called an influence coefficient, which depends on the geometry of the "sending" panel, the distance of the "receiving" control point from the panel, and the relative orientation of both panels. For the case where $j = i$, the influence coefficient is exactly $1/2$. The velocity normal to the surface at the jth control point induced by all of the N sources will be

$$v_j = \sum_{i=1}^{N} A_{ij}\sigma_i \quad \text{for } i \neq j. \tag{6.15}$$

The boundary condition of flow tangency is imposed at the centroids of each panel, along with a boundary condition describing the flow away from the fuselage surface. Initial

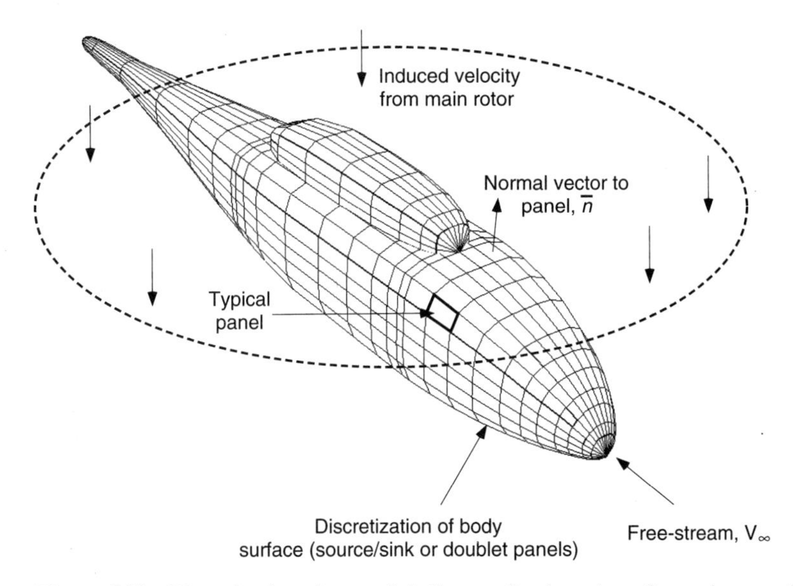

Figure 6.20 Discretization of a generic helicopter fuselage shape for use in a panel method. (Grid courtesy of Mark Chaffin.)

calculations may assume that the rotor is not present in the problem, and the boundary condition is the free-stream flow at infinity. In this case, the component of the free-stream velocity, $\vec{V}_\infty$, normal to the jth panel will be $\vec{V}_\infty \cdot \vec{n}_j$, where $\vec{n}$ is the unit normal vector. Therefore, at the jth panel

$$\sum_{i=1, i\neq j}^{N} A_{ij}\sigma_i + \frac{\sigma_i}{2} + \vec{V}_\infty \cdot \vec{n}_j = 0. \tag{6.16}$$

One equation such as this applies at each of the N panels. This leads to a set of influence coefficients in the form of a set of linear simultaneous equations that can be used to solve for the singularity strengths (σ) using standard numerical methods. The governing equations can be written in matrix form as

$$[A_{ij}]\{\sigma_i\} = -\{\vec{V}_\infty \cdot n_j\}, \tag{6.17}$$

noting that the self-induced influence coefficient, $A_{ii} = 1/2$, runs along the diagonal of the **A** matrix. Once the σ_is are obtained, the velocity components tangential to the panels can be obtained. Normally thousands of panels are required to adequately resolve a helicopter fuselage shape, especially in regions of high curvature such as near the hub. Because the numerical cost of these panel methods is of order N^3, their use is by no means inexpensive, even on a modern computer.

Although panel methods are very flexible and can aid considerably in the design process, they are unable to directly model viscous effects and separated flows typical of helicopter fuselage shapes, and especially hub designs. However, empirical and semi-empirical corrections for the viscous effects can be incorporated into the basic method. For example, a three-dimensional boundary-layer method can be coupled by means of displacement corrections to the fuselage shape, or by means of a transpiration velocity – see Lindhout et al. (1981) and de Bruin (1987). The advantage of the latter approach is that only the right-hand side of the equations must be changed, thereby avoiding the expensive recalculation of the influence coefficients. The separated wake can be modeled as a vortical shear layer, the shape of which is either prescribed or made to be force free – see Polz (1982), Lê et al. (1987), and Katz & Plotkin (1991). A prerequisite, however, is that the separation line be calculated or deduced from experiment. Overall, panel methods and semi-empirical modifications/corrections thereof have been shown to give good predictions of the surface pressure on isolated helicopter fuselages – see Polz & Quentin (1981). Ahmed et al. (1988) and Ahmed (1990) give a good overview of the various predictive models in current use and their general capabilities in predicting pressure distributions and drag on helicopter fuselages. Chaffin & Berry (1994) show that panel methods give good agreement with experimental measurements of fuselage surface pressure, and the predictions are as good as Navier–Stokes approaches except in regions with large-scale flow separation. See also Narramore & Brand (1992).

To supplement numerical predictions of fuselage aerodynamics, semi-empirical drag prediction methods are in widespread use in the helicopter industry. Based on component testing in the wind tunnel, and with some additional engineering judgment, these approaches can give very reliable estimates of fuselage drag. An estimate of the fuselage parasitic equivalent wetted or flat-plate area, f, (see Section 5.3.3) can be determined from a knowledge of the drag coefficients of the various components that make up the aircraft using an equation of the form

$$f = \sum_{n} C_{D_n} S_n, \tag{6.18}$$

where S_n is the area on which the definition of C_D is based. This may be either the wetted area or the projected frontal area of the component. Initial drag estimates will assume no mutual interference effects from individual components, but despite this the approach is found to give a reasonable initial estimation for the fuselage drag. More refined values of f are obtained by component testing of isolated fuselage and rotor/fuselage models in the wind tunnel – see, for example, Bosco (1972), Wilson (1984), Philippe et al. (1985), and Wilson (1990). Component interference effects can then be incorporated into the drag estimate. Further refinements need actual flight tests because wind-tunnel models cannot simulate accurately the effects of full-scale Reynolds numbers or drag producing details such as antennas, leakage through doors, and other gaps. In this case, estimates of drag are derived indirectly from rotor power measurements. The whole process, although somewhat empirical, gives quite reliable estimates of the parasitic drag.

An analysis of the individual items contributing to the parasitic drag of a typical helicopter is given in Table 6.1. These results are not for any one helicopter design and must be considered only as representative. It will be apparent that a major source of the overall drag is the rotor hub and blade attachments, which may account for up to 30–50% of the total parasitic drag on fully articulated blade designs. Sheehy & Clark (1975), Sheehy (1977), Sedden (1979), and Prouty (1986) give good summaries of various sources of published rotor hub drag data. Considerable turbulence may also be produced behind the hub, which can have an influence on the magnitude and frequency of the airloads produced on the tailboom and empennage [see Berry (1997) for a discussion of hub turbulence measurements]. Streamlining the fuselage on the top of the airframe near the hub can help reduce the drag of the rotor shaft, the exposed controls, and the hub. However, it is only with the use of modern hingeless or bearingless rotors that the hub drag can be significantly reduced, with values of f/A for a bearingless rotor being about half those of an articulated design. Cler (1989) describes a fully faired hub design, although this is impractical for many helicopters.

Table 6.1. *Typical Breakdown of Parasitic Drag Components on a Helicopter*

Component	f/A	% of Total
Fuselage	0.00210	30
Nacelles	0.00042	6
Rotor hub & shaft	0.00245	35
Tail rotor hub	0.00028	4
Main landing gear	0.00042	6
Tail landing gear	0.00028	4
Horizontal tail	0.00007	1
Vertical tail	0.00007	1
Rotor/fuselage interference	0.00047	7
Exhaust system	0.00021	3
Miscellaneous	0.00021	3
Total	0.007	100

The equivalent flat-plate areas of helicopter fuselages range from about 10 ft^2 (0.93 m^2) on smaller helicopters to as much as 50 ft^2 (4.65 m^2) on large utility helicopter designs, as shown in Fig. 6.21. The data have been obtained from a variety of sources, including Rosenstein & Stanzione (1981). The results show that values of f/A typically fall between 0.004 for clean helicopter designs and up to 0.025 for first-generation or heavy lift transport helicopters. Both drag curves are approximately proportional to (weight)$^{1/2}$, which follows the "square–cube" law mentioned previously. Obviously, the addition of external hardware or stores on military aircraft tends to dramatically increase the equivalent wetted area of the helicopter, thereby significantly increasing power required and reducing forward flight performance. Conversely, a general drag cleanup of the airframe can result in very substantial increases in climb and forward flight speed capability, as well as in reduced fuel burn and improved overall operational economics.

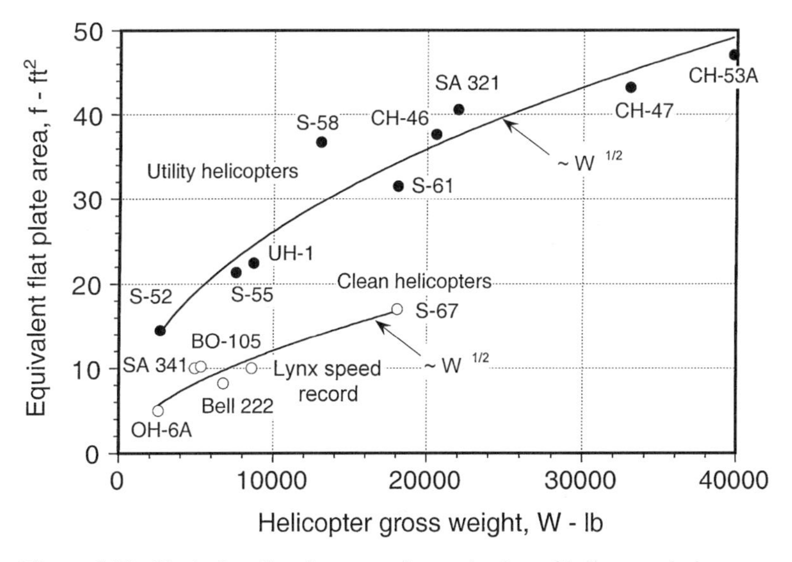

Figure 6.21 Equivalent flat plate areas for a selection of helicopter designs.

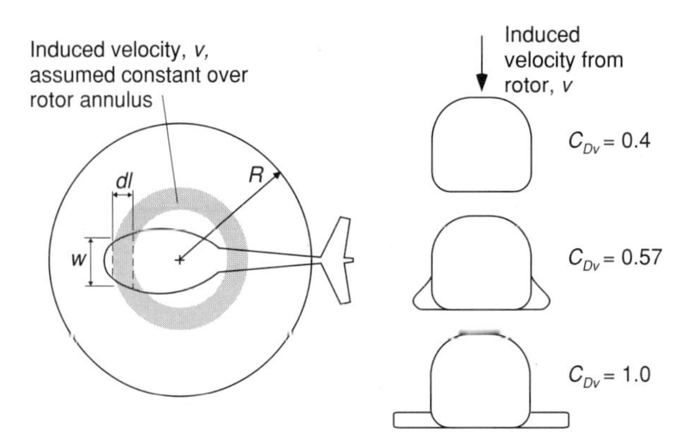

Figure 6.22 Strip analysis of the fuselage for the estimation of vertical drag.

6.4.2 *Vertical Drag or Download*

It is normally assumed that the total thrust, T, required by the main rotor is equal to the aircraft weight, W. However, there is usually an extra increment in power required because of the download or vertical drag, D_v, on the helicopter fuselage that results from the rotor slipstream velocity (i.e., now the rotor thrust will be $T = W + D_v$). See Wilson (1975) for a good summary of the basic problem. Typically, the vertical download on the fuselage can be up to 5% of the aircraft weight, but it can be higher for some rotorcraft designs such as compounds or tilt-rotors that have large wings situated in the rotor downwash field. Because of the bluff-body nature of the flow about the fuselage, estimates of the vertical drag can only be reliably obtained from wind-tunnel testing of isolated fuselage models. Precautions should be taken to properly represent the nonuniformity of the rotor downwash as well as Reynolds number effects on the fuselage; the latter can be very difficult to simulate. Alternatively, the vertical drag on the fuselage is calculated by estimating the drag coefficient of individual two-dimensional fuselage cross sections (i.e., a strip approach), as shown schematically in Fig. 6.22, which is the approach suggested by Stepniewski & Keys (1984). The vertical drag of the entire fuselage can then be obtained by successively testing two-dimensional cross sections of the fuselage to find the pressure distributions and finding the net drag on the fuselage by summation. For example, the incremental vertical drag on any one segment of the fuselage of length dl will be

$$dD_v = \frac{1}{2}\rho v^2 C_{D_v} w \, dl, \tag{6.19}$$

where v is downwash velocity at the element and C_{D_v} is the drag coefficient of that element. The net vertical drag will then be given by

$$D_v = \sum dD_v, \tag{6.20}$$

where it is recognized that C_{D_v} and v will vary between cross sections. Generally, it is found that helicopter fuselage cross-sectional drag coefficients average out at about 0.5, although the addition of sponsons or stub-wings can increase the drag coefficients to over unity. Wilson & Kelly (1983, 1986) give results for a variety of fuselage cross-sectional shapes, which have been estimated from wind-tunnel tests. Drag coefficients for various other bluff-body shapes resembling fuselage cross sections can be found in Delany & Sorensen (1953) and Hoerner (1965).

Note that the strip method of fuselage drag estimation also requires an estimate for the induced velocity below the rotor. A good first estimate can be made by means of the BEMT discussed previously in Chapter 3. However, because the fuselage also influences the rotor in a reciprocal way, accurate estimates of the interference effects on the induced velocity are difficult to obtain. The effect of the fuselage is also known to affect the rotor performance, increasing the thrust as the fuselage is brought closer to the rotor; see for example Sheridan (1978), Sheridan & Smith (1979), and Fradenburgh (1972). This phenomenon, referred to as *thrust recovery*, partly offsets the download obtained on the fuselage. Because of these effects, and the fact that three-dimensional effects have been also neglected, the strip method described above tends to be rather crude, at least in theory. Yet, in practice the method has been shown to give fairly reliable estimates of the extra rotor thrust required to overcome fuselage drag in a vertical climb.

When determining in-ground-effect (IGE) hover capability of the helicopter, the vertical fuselage drag must be corrected to account for a favorable effect that occurs in this regime. The lower drag results from the reduction in pressure drag on the lower surface of the fuselage because of the higher total pressure inside the rotor wake when operating near the ground. Measured results documenting this phenomeon are given by Stepniewski & Keys (1984). The decrease in effective fuselage vertical drag is found to be significant for hovering heights of less than one rotor diameter. Further results documenting this effect are given by Fradenburgh (1972).

6.4.3 *Fuselage Side-Force*

During hovering flight, the tail rotor provides the antitorque thrust to maintain yaw equilibrium. However, in sideward flight or when hovering over a fixed point above the ground in a cross-wind, an aerodynamic side-force will also be produced on the fuselage. For most helicopters this side-force is small enough to have a minimal impact on the handling qualities and operational envelope. For a rotor turning in the conventional direction (counterclockwise when viewed from above), the aerodynamic side-force on the fuselage and tail boom will be opposite to the direction of the tail rotor thrust when the helicopter is in starboard sideward flight or has a starboard crosswind component. On some helicopters this effect can affect directional (yaw) control and in some cases may even limit the operational flight envelope. In particular, this can be an issue on military helicopters, which must often demonstrate high speed sideward flight capability, or for naval helicopters in particular, which must often operate in confined locations in gusty crosswind conditions.

The aerodynamic side-force on the fuselage is found to be accentuated by certain tailboom shapes, which can produce a sizable circulatory lift force when the sideward velocity is combined with the main rotor downwash. To counter this undesirable effect, Brocklehurst (1985) and Wilson et al. (1988) have suggested the use of a strake that runs longitudinally along the one side of the tail boom. This strake forces the flow to separate, spoiling the side-force on the tailboom, thereby restoring or improving the normal operational flight envelope – see Kelly et al. (1993). This design solution, however, will only be required on tailboom cross sections that are relatively deep, typically with height to width ratios greater than about 2.

6.5 Empennage Design

The empennage on a helicopter consists of the vertical and horizontal stabilizer and related fuselage structure. A stabilizer is simply a fixed surface that produces an aerodynamic force. The primary purpose of a stabilizer is to enhance stability about a particular axis, although there are secondary aerodynamic characteristics of stabilizers that are important design considerations for helicopters.

6.5.1 *Horizontal Stabilizer*

The primary purpose of a horizontal stabilizer is to give the helicopter stability in pitch. While a stabilizer is not absolutely necessary, it does negate the inherently negative (adverse) stability derivatives of the fuselage and rotor combination and gives the helicopter better handling qualities. The fuselage itself has a powerful negative stability because of the typically large surface area forward of the center of gravity. The selection of the size and position of the horizontal stabilizer on the tail has proven to be one of the most difficult challenges facing helicopter designers. Because the wake position relative to the empennage is affected by forward flight speed as well as climb or descent velocity, the aerodynamic angles of attack found at the empennage are very nonuniform and their magnitude can be sensitive to the flight conditions. Combined also with the higher total pressure obtained inside the main rotor wake boundary, this can result in substantial changes in the forces on the stabilizer and, therefore, the pitching moments acting on the helicopter. If these changes occur suddenly or unpredictably, undesirable handling qualities can result. Resolving this problem can involve substantial costs because of empennage redesign and associated flight testing time. Good case histories of the stabilizer design are available in the literature [e.g., for the AH-64 Apache as documented by Prouty & Amer (1982) and Prouty (1983)]. Other documented examples are for the AS-360 discussed by Roesch & Vuillet (1981), and for the EH-101 as discussed by Main & Mussi (1990).

Early investigations into the problem of rotor wake/lifting surface interactions include the work of Wheatly (1935), Makofski & Menkick (1956), McKee & Naeseth (1956), and Lynn (1966). Sheridan & Smith (1979) examined specific issues associated with wake induced empennage airloads in their seminal work on rotor/airframe interactions. Leishman & Bi (1994), Foley et al. (1995), and Moedersheim & Leishman (1996) have studied the more detailed aspects of the interactions between a subscale rotor and horizontal lifting surfaces. Torok & Ream (1993) obtained flight test data documenting the interactions found on the RAH-66 T-tail configuration. Frederickson & Lamb (1993) conducted a wind-tunnel test of the RAH-66 Comanche. In all tests, the unsteady airloads and vibration levels can be correlated with the relative position of the main rotor wake boundary.

There are basically three types of horizontal stabilizer design: a forward mounted stabilizer, an aft mounted low stabilizer, and a T-tail design. A forward fixed stabilizer avoids the sudden changes in download caused by wake impingement because it remains inside the wake boundary from hover up until a fairly high forward flight speed. Many Bell helicopter designs use this forward stabilizer position. However, because of the shorter moment arm, the aerodynamic surface must be larger (and heavier) compared to a stabilizer mounted further aft. The stabilizer, however, may have a capability of being used for trim augmentation in that the pitch angle may be linked into the longitudinal cyclic. The download on forward mounted horizontal stabilizers in hovering flight usually represents a significant performance penalty.

There are structural incentives to use a low mounted aft stabilizer, with all the loads being carried directly into the tail boom. However, this stabilizer design tends to produce trim problems associated with transition from low speed flight into hover, where the main rotor wake may suddenly move forward over the empennage location and so produce a nose-down pitching moment on the aircraft. This is shown schematically in Fig. 6.23. Also, the

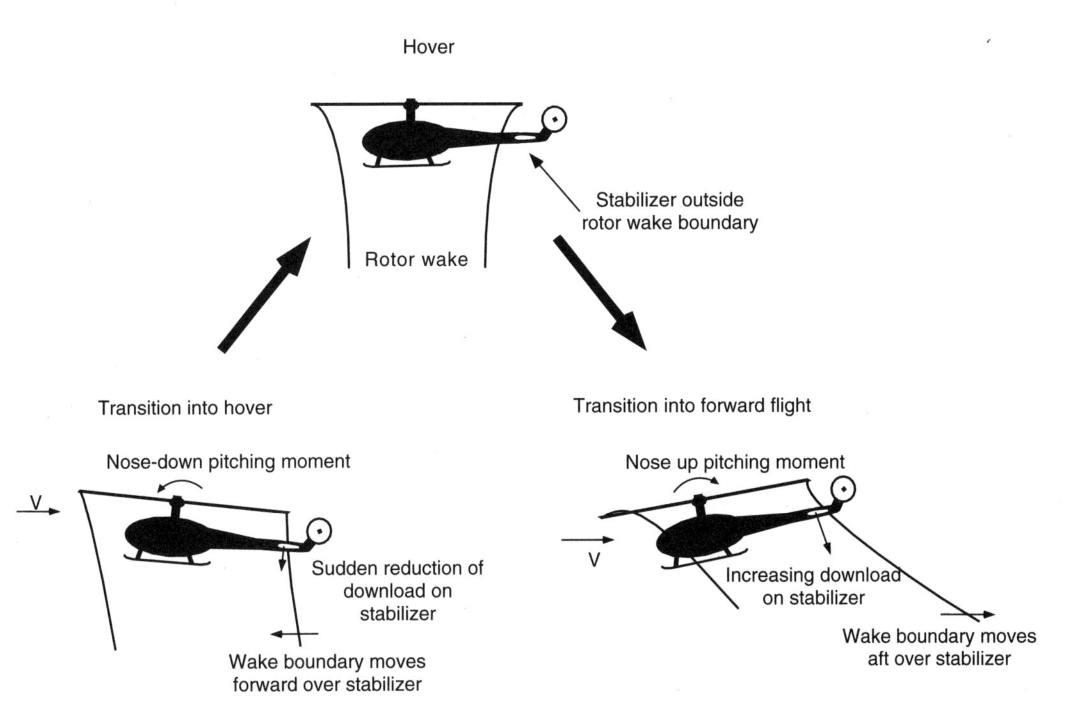

Figure 6.23 Schematic showing interactions between the main rotor wake and the horizontal tail during transition from hover into forward flight.

unsteady separated flow from the upper fuselage and rotor hub tends to reduce the efficiency of this type of stabilizer design so that the lifting area needs to be greater than for one that could be located away from the wake. On military helicopters, ground clearance issues may be important and can preclude this particular design choice. Nevertheless, as evidenced by the large number of helicopters with this low mounted horizontal stabilizer configuration, it is a popular compromise in terms of size, weight, and improved longitudinal stability for helicopters with relatively low disk loading.

In the T-tail design, the horizontal stabilizer is mounted at the top of the vertical fin. This moves the stabilizer away from the rotor wake for most flight conditions, and so it can be smaller in area to give the same overall stability. However, the design is structurally inefficient because of the higher overall weight of the vertical fin required to carry the stabilizer loads, and also because of various low frequency structural vibration modes that can be excited by the main and/or tail rotors. On a low or forward set stabilizer, it may be necessary to have different settings for the port and starboard sides to account for the gradients in downwash in the rotor wake. With a T-tail design, there are often twisting moments that may limit the maximum area of the stabilizer. Often a compromise is drawn by using a stabilizer mounted to only one side of the fin, such as on many of the Sikorsky machines. See Prouty & Amer (1982), Prouty (1983), Hansen (1988), and Main & Mussi (1990) for further information on stabilizer design.

A stabilator is a stabilizer that has a variable incidence capability. It can help to solve the low-speed handling problems mentioned previously that are associated with a fixed stabilizer while also being able to provide desirable modifications to the flying qualities throughout the flight envelope. The stabilator incidence is set automatically based on airspeed and other measurements, although manual override gives the pilot control of the stabilator below certain airspeeds. A stabilator is mechanically complicated and is a structurally inefficient design choice. However, sometimes it can be the best choice to meet the demanding flight envelope of military helicopters. A stabilator design is used on both the AH-64 and UH-60 helicopters.

6.5.2 *Vertical Stabilizer*

The primary purpose of a vertical stabilizer or fin is to provide stability in yaw. While the tail rotor itself provides considerable yaw stability, the vertical stabilizer may also be required to provide sufficient aerodynamic side-force to offset tail rotor thrust in forward flight and to provide sufficient antitorque to allow continued flight in the event of the loss of the tail rotor – see Horst & Reschak (1975). This side-force can be provided by using an airfoil section with a relatively large amount of camber. With sufficient forward speed and some side-slip angle, the side-force can be great enough to allow continued flight without the tail rotor. Alieviating the tail rotor thrust in high speed flight is usually desirable to minimize tail rotor flapping and cyclic loads and to maximize fatigue life.

The vertical stabilizer also forms a structural mount for the tail rotor. Because the flows will strongly interact, the tail rotor can be considered to be an integral part of the fin and empennage assembly. The size of the vertical stabilizer directly and adversely affects tail rotor performance. A smaller stabilizer will reduce the adverse effects on tail rotor efficiency, but this must be balanced against the effects on yaw stability and other design requirements. Because of special complications associated with the design of the tail rotor itself, the aerodynamics of the tail rotors will be discussed separately.

6.5.3 *Modeling*

Computer models have not yet evolved to the point that the empennage design can be conducted without extensive wind-tunnel and actual flight testing. The capabilities of these methods are beyond the state of the art because they are required to model the interactions with the main rotor wake. The problems may involve tip vortex/surface collisions, the cutting of vortex filaments, as well as the formation of three-dimensional unsteady separated flows. Bramwell (1966) used potential flow methods with simple airframe configurations and demonstrated the significance of unsteady aerodynamic effects for the rotor/surface problem. Gangwani (1982) has used a prescribed wake model coupled with a doublet-lattice model of a fixed wing to predict the unsteady bending moments measured on a helicopter tail. Reasonable correlation was obtained with flight test data. Mello & Rand (1991) confirmed that the predicted unsteady loads on the empennage were sensitive to the rotor wake position, confirming a result of wind-tunnel and flight test experience. Curtiss & Quackenbush (1989) considered the effects of the rotor wake induced velocity field at the empennage location on helicopter stability derivatives. The limited correlations obtained with flight test data reiterated the complexity of the main rotor wake/empennage interaction problem. Weinstock (1991) has developed a simple model of the rotor wake/empennage aerodynamic interaction problem for use in flight simulation modeling.

6.6 Design of Tail Rotors

The vast majority of helicopters in production are of the single main rotor with tail rotor configuration. The primary purpose of the tail rotor is twofold. First, the tail rotor provides an antitorque force to counter the torque reaction of the main rotor on the fuselage.

Second, the tail rotor gives yaw stability and provides the pilot with directional control about the yaw axis. The aerodynamics of the tail rotor provide the helicopter with a significant *weathercock stability*. For example, if the aircraft is yawed nose-left, then the tail rotor will experience an effective climb. If the collective pitch is held constant, then this will result in a decrease of thrust (a result of the higher inflow) and a restoring moment about the yawing axis. Similarly, if the helicopter yaws nose-right, the tail rotor experiences an effective descent, with an increase in thrust, and again, a restoring moment is produced. This weathercock stability is a useful characteristic, but it can also make helicopters less maneuverable.

The tail rotor has to operate in a relatively complex aerodynamic environment and must produce thrust with the relative flow coming from essentially any direction. For example, the tail rotor must operate properly in side winds and during yaw maneuvers. In a yawing maneuver, the tail rotor operates either in an effective climb mode or in a descent, depending on the yaw direction. The yawing direction that produces an effective descending condition is the most critical because it is possible for the tail rotor to enter the vortex ring state. This can result in a loss of tail rotor authority, and perhaps even a loss of control under the wrong combination of conditions. These effects are carefully examined during certification of the helicopter to ensure that there is a minimal chance that the machine will inadvertently exhibit undesirable flight characteristics.

As described previously, the tail rotor is also mounted in proximity to a vertical fin or other empennage assembly, and the aerodynamic interactions will affect tail rotor operation. In addition, the operation of the tail rotor will be affected by turbulent separated flow generated by the main rotor hub and fuselage wakes and the energetic main rotor wake itself. This adverse environment means that the aerodynamic design requirements for the tail rotor are different in some respects from those of the main rotor. For these reasons, it is known to be difficult to design a tail rotor that will meet all the aerodynamic, control, stability, weight, and structural requirements. See Lynn (1970), Cook (1978), Byham (1990), and Newman (1994) for a detailed overview of tail rotor design issues.

6.6.1 *Physical Size*

Consider the physical characteristics of typical tail rotors on production helicopters, as listed in the appendix. A comparison of the size of the tail rotor to the main rotor shows that it is roughly one sixth the diameter of the main rotor. Because the ratio decreases with gross weight, the size of the main rotor grows more rapidly than the size of the tail rotor with increasing gross weight, as shown in Fig. 6.24. Note that for machines with fenestrons, the size of the fenestron can be much smaller than that of a conventional tail rotor. Again, the trends (for conventional tail rotors) can be explained with the aid of the "square–cube" law. It has been shown previously that the power (or torque) required is $\propto W^{3/2}$. Also, because the size of the aircraft (and length of the tail moment arm) grows with $W^{1/3}$, the antitorque force requirement should be $\propto W^{7/6}$. Because tail rotor disk loading remains relatively independent of aircraft gross weight (for good efficiency), the size of the tail rotor should be $\propto W^{7/12}$. This trend is in general agreement with the data shown in Fig. 6.24. Note also that the tail rotor tip speed is approximately the same as the main rotor tip speed. This means that the rotational speeds of tail rotors are roughly six times the main rotor frequency, and this has particularly important consequences on rotor noise levels. While the noise energy produced by the tail rotor is only a fraction of that produced by the main rotor, the higher frequencies generated by the tail rotor can be more discernible to the human ear.

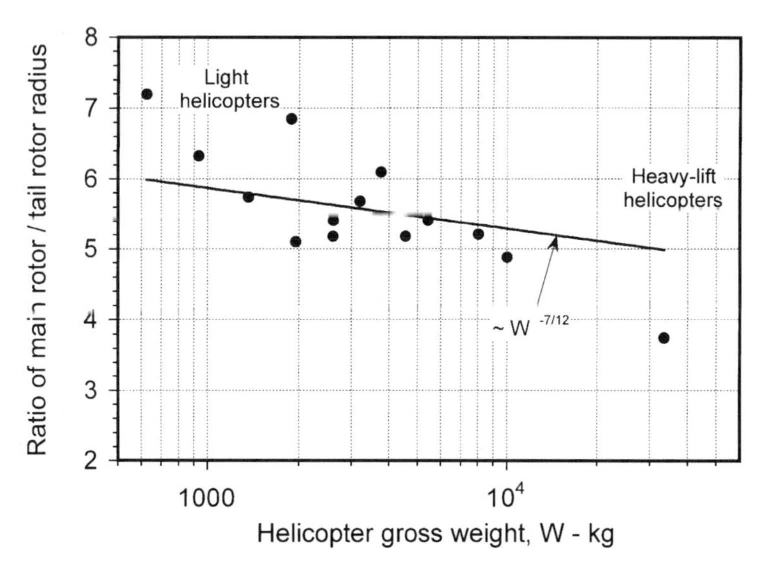

Figure 6.24 General trends of tail rotor size to main rotor size versus helicopter gross weight.

6.6.2 *Thrust Requirements*

The primary purpose of the tail rotor is to provide a sideward force in a direction and of sufficient magnitude to counter the main rotor torque reaction. The tail rotor also provides yaw control. Roughly, the tail rotor consumes up to about 10% of the total aircraft power. This is power that is completely lost, because unless the tail rotor is canted, as on the UH-60 Blackhawk, it provides no useful lifting force. The purpose of the canted tail design is to widen the allowable center of gravity of the aircraft. This, however, introduces an adverse coupling between yaw and pitch, but this effect can be minimized by a flight control system (see also Question 6.13).

The direction of the antitorque force depends on the direction of rotation of the main rotor. For a rotor turning in the conventional direction (counterclockwise direction when viewed from above), the tail rotor thrust is to the right (starboard). The magnitude of this thrust as well as the power consumption depends on the location of the tail rotor from the center of gravity (i.e., the moment arm l_{TR}). The main rotor torque reaction effect, Q_R, is canceled when the tail rotor moment is equal to the yaw reaction torque, that is, $Q_R + I_{zz}\ddot{\Psi} = T_{TR} l_{TR}$, where $\ddot{\Psi}$ is the yaw acceleration and I_{zz} is the mass moment of inertia about the yaw axis.

The tail rotor thrust is controlled by the pilot's feet by pushing on a set of floor mounted pedals. For example, for a rotor turning in the conventional direction, pushing on the left pedal increases tail rotor thrust (positive to starboard) and the helicopter will yaw nose left. The tail rotor must also provide the specified yaw acceleration in the maximum specified crosswind conditions, taking into consideration possible losses in efficiency because of aerodynamic interference effects between the tail rotor and the vertical fin. Keep in mind that when the main rotor thrust or power is increased, for example to climb, the reaction torque, Q_R, on the fuselage is increased. This means that the tail rotor thrust must also increase to balance this torque reaction. Therefore, when the pilot increases the collective to climb, he or she must also apply foot pressure to the appropriate pedal to keep the nose pointed straight in the direction of flight.

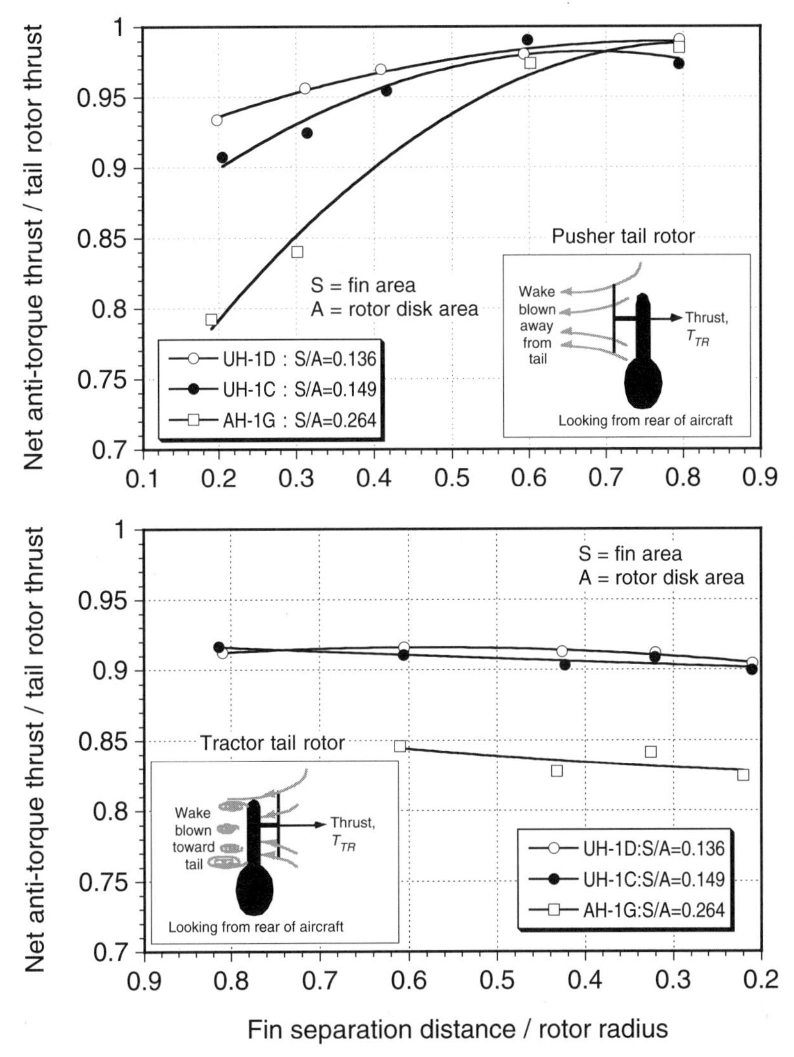

Figure 6.25 Effects of rotor/fin separation distance on net antitorque producing side force. Data source: Lynn (1970).

6.6.3 *Pushers Versus Tractors*

Tail rotors may be either of the pusher or tractor variety and will be located either on the left- or right-hand side of a vertical fin. Both designs suffer from interference effects between the rotor and the fin, these effects being a function of the tail rotor size or disk area, A, fin area, S, and spacing of the rotor plane from the fin. The effects are summarized by Lynn (1970), and representative measurements are reproduced in Fig. 6.25 in terms of net antitorque producing side-force versus fin separation distance. The results are shown for various ratios of fin area to tail rotor disk area, S/A. In the pusher style the wake of the tail rotor is blown away from the vertical tail. This means that the vertical fin distorts the inflow into the tail rotor, the consequences of which are a nonuniform inflow and higher induced power requirement. It can also be a source of $2N_b$ vibratory airloads. In contrast, the tractor design has the vertical fin located inside the high energy region from the tail rotor wake. While this "blockage" or "ground effect" tends to increase the tail rotor thrust, there is also a significant force applied to the vertical tail that is in the opposite direction to the antitorque thrust requirement. It is found, however, that the net effect is a decrease in thrust compared to what would be obtained if the rotor was operating in isolation. In both cases, the interference effects become greater for larger fins and/or smaller rotors. A majority of modern helicopters use a pusher tail rotor design because it has been found through experimentation that this configuration tends to have a higher overall efficiency.

6.6.4 *Design Requirements*

As for the main rotor, the power required to drive the tail rotor depends on the disk loading. Whereas a larger diameter may be preferable for low induced power requirements, this is outweighed by several factors. First, a larger diameter usually means a heavier design and this is undesirable because the adverse effects on the aircraft center of gravity location. Second, to meet certification requirements it is usually desirable that the tail rotor disk loading be high enough so that sideward flight of at least 35 kts is possible without the tail rotor entering into the vortex ring state. Both these constraints dictate the use of a relatively small tail rotor.

Tail rotors typically have two or four blades, with no particular aerodynamic advantage of one number over the other. Only collective pitch is required because there is no need to control the orientation of the tail rotor disk plane. Tail rotor blades usually have some built-in twist to help minimize induced power requirements. However, the amount of twist is small to avoid losses in efficiency and the possibilities of stall when the tail rotor is operating in an effective descent, such as when hovering in crosswinds. Although some designs may use cambered airfoil sections, the tail rotors found on many helicopters use symmetric airfoils because of their reasonable overall performance and low pitching moments. While the higher maximum lift obtained with cambered airfoils can help reduce rotor solidity and thereby minimize tail rotor size and weight, this can be outweighed by their larger pitching moments (which lead to higher control forces) and poorer performance when operating at negative angles of attack. Generally, tail rotors are designed to operate at tip speeds that are comparable to those of the main rotor. Lower tip speeds are desirable to minimize noise. However, for a given thrust, tail rotors operating at lower tip speeds require higher solidity to prevent stall. A lower tip speed also increases the torque requirement. Both of these factors will increase the weight of the drive system.

A secondary effect of the antitorque side-force is the tendency for the helicopter to drift sideways. This is corrected by the main rotor, which is tilted slightly to the left (by means of cyclic pitch inputs) so that a component of the main rotor thrust produces an equal and opposite side-force. This is the reason why, on some occasions, it will be noted that a hovering helicopter will tend to hover with one wheel (or skid) lower than the other. On larger helicopters, the main rotor shaft is physically tilted slightly (as part of the design, thereby introducing a pretilt) so that the pilot does not require as much cyclic pitch input to counter the tail rotor side-force. The tail rotor thrust and the component of the main rotor side-force act together, producing a couple, and thereby, a rolling moment. To reduce this moment, the tail rotor is located vertically up on the tail structure so that the line of action of the thrust vector is close to the center of gravity of the aircraft.

6.6.5 *Aerodynamic Interactions*

The aerodynamic interactions between the main rotor and the tail rotor are discussed by Wiesner & Kohler (1973, 1974), Wilson (1990), Ellin (1993), and Tarttelin &

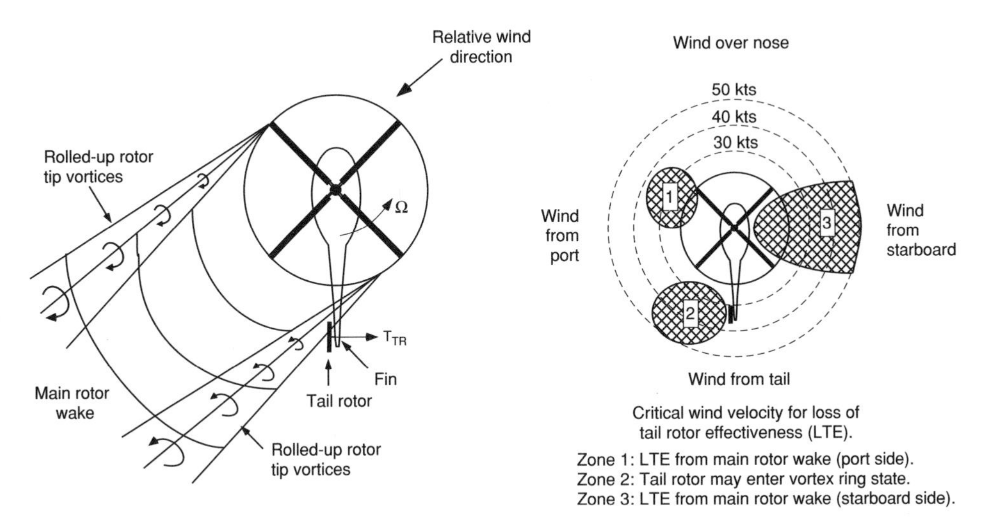

Figure 6.26 The impingement of the main rotor wake on the tail rotor can cause loss of tail rotor effectiveness. Adapted from Vuillet (1990) and other sources.

Martyn (1994). In forward flight there is an asymmetry in dynamic pressure produced laterally across the tail rotor, just as with the main rotor. The main rotor wake is known to roll-up at its edges into a pair of "super"-vortices or vortex bundles, similar to those generated at the tips of a fixed wing. This produces a downwash over the tail rotor with a significant velocity gradient in the longitudinal direction. Finally, there is an increased total pressure in the main rotor wake, which encroaches onto the tail rotor as forward flight speed builds.

Aerodynamic interactions between the main rotor and tail rotor in low speed flight generally prove to be the most severe and the hardest to predict. Critical conditions include right and left crosswinds, where the super-vortices trailed from the main rotor disk can impinge on the tail rotor. An example is shown in Fig. 6.26. Under these conditions the downwash and change in dynamic pressure across the tail rotor disk as produced by the main rotor wake can degrade tail rotor performance. Because of these effects, one normally finds that tail rotors turn in a direction where the top of the tail rotor is retreating aft because this is found to balance the aerodynamic forces and maintain good tail rotor authority over a wider range of flight conditions. Other critical flight conditions include effective rearward flight near the ground or sideward flight to port. In the former condition, the ground vortex (described in Section 5.5) can result in loss of tail rotor effectiveness. In sideward flight (or when hovering in a crosswind), the tail rotor operates in an effective descent and will eventually encounter the vortex ring state where a loss of antitorque and yaw control effectiveness will occur.

6.6.6 *Typical Tail Rotor Designs*

Typical modern tail rotor assemblies are shown in Fig. 6.27. Some tail rotors may be very simple two-bladed teetering assemblies, while others may be relatively sophisticated bearingless or hingeless designs. Common among all tail rotors is the lack of any cyclic pitch; only collective pitch is used because control of the tail rotor disk orientation is not required. Nevertheless, the tail rotor must be provided with flapping so that the blades may be allowed to respond to the changing aerodynamic environment. Lead/lag hinges are not used to save weight and reduce mechanical complexity. Instead, a large amount of δ_3 or pitch-flap

(a)

(b)

Figure 6.27 Representative tail rotor assemblies. (a) UH-60 bearingless tail rotor. (b) AH-64 twin teetering tail rotor.

coupling is built into the tail rotor design. This provides a means of allowing the blades to pitch cyclically in such a way as to minimize blade flapping produced by the changing aerodynamic loads (see also Question 4.8). In addition, one will often see a set of preponderance weights that will be attached to the tail rotor pitch horns. These are a set of weights that lie out of the plane of rotation and rely on centrifugal forces to help keep the control forces to manageable limits. Preponderance weights are usually found on tail rotors that use highly cambered airfoil sections so as to help minimize moments transferred to the control system.

Figure 6.28 The fan-in-fin or fenestron tail rotor, as used on the SA 365 Dauphin.

6.6.7 *Other Antitorque Devices*

Besides the conventional tail rotor, other types of antitorque devices are used on modern helicopters. These are the fenestron (or fan-in-fin or fantail) and the NOTAR concepts.

Fan-in-Fin

Shrouded or ducted fan antitorque designs, which are known as "fenestrons," "fan-in-fin" or "fantail" designs, have been frequently considered over conventional tail rotors, especially for lighter helicopters. A photograph of a typical fan-in-fin design is shown in Fig. 6.28. Details of the design of such antitorque devices are given by Mouille (1970, 1979), Mouille & Dámbra (1986), and Vuilet & Morelli (1986). Fan-in-fin designs typically are found to have lower power requirements than an open tail rotor to produce the same amount of thrust. Alternatively, this means the fan-in-fin design can give the same antitorque and yaw authority with a smaller and lighter design compared to a conventional tail rotor. See also Davidson et al. (1972).

The momentum theory discussed in Chapter 2 can be extended to the analysis of a fan-in-fin design, with the flow model being shown in Fig. 6.29. Far upstream of the fan, the velocity can be assumed to be zero. At the plane of the fan, the induced velocity is v_i. By the principle of conservation of mass, the mass flow rate, $\dot{m}$, is constant through the system so that

$$\dot{m} = \rho A v_i = \rho(a_d A) w, \tag{6.21}$$

where, by virtue of the duct design the area of the flow at the outlet is $a_d A$ with the velocity at the duct outlet being w. This gives the relationship that $w = v_i/a_d$. By means

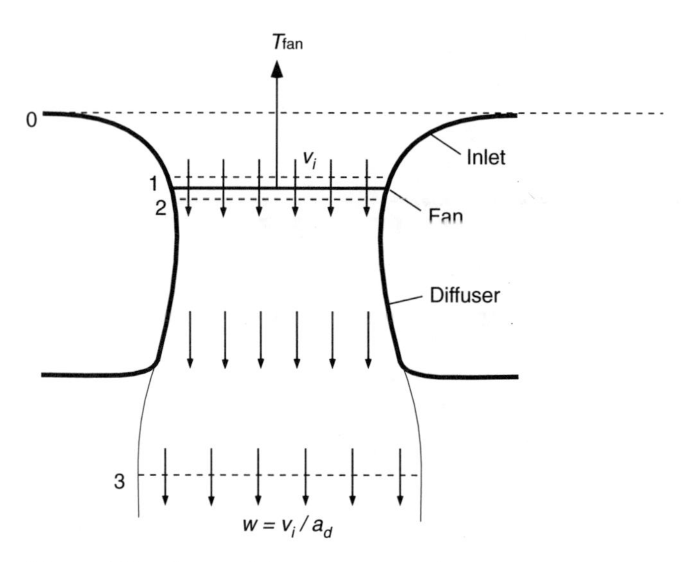

Figure 6.29 Flow model assumed for fan-in-fin analysis using momentum theory.

of conservation of momentum the thrust on the duct and fan is

$$T = T_{\text{duct}} + T_{\text{fan}} = \dot{m}w = (\rho A v_i)w = \frac{\rho A v_i^2}{a_d} \tag{6.22}$$

or

$$v_i = \sqrt{\frac{a_d T}{\rho A}} = \sqrt{\frac{T}{2\rho A_{\text{eff}}}} \tag{6.23}$$

where $A_{\text{eff}} = 2a_d A$. Applying the Bernoulli equation between stations 0 and 1 gives

$$p_0 = p_1 + \frac{1}{2}\rho v_i^2 \tag{6.24}$$

and between stations 2 and 3 gives

$$p_2 + \frac{1}{2}\rho v_i^2 = p_0 + \frac{1}{2}\rho w^2. \tag{6.25}$$

Using Eqs. 6.24 and 6.25 gives for thrust on the fan:

$$T_{\text{fan}} = (p_2 - p_1)A = \frac{1}{2}\rho w^2 A. \tag{6.26}$$

Using Eqs. 6.22 and 6.26 gives

$$\frac{T_{\text{fan}}}{T} = \frac{\frac{1}{2}\rho A w^2}{\rho A v_i w} = \frac{w}{2v_i} = \frac{1}{2a_d}. \tag{6.27}$$

Using Eqs. 6.23 and 6.27, we obtain the induced power consumed by the fan:

$$(P_i)_{\text{fan}} = T_{\text{fan}} v_i = \left(\frac{T}{2a_d}\right)\sqrt{\frac{a_d T}{\rho A}} = \frac{T^{3/2}}{\sqrt{4a_d \rho A}} \tag{6.28}$$

or

$$\frac{(P_i)_{\text{fan}}}{(P_i)_{TR}} = \frac{1}{\sqrt{2a_d}}, \tag{6.29}$$

where $(P_i)_{TR}$ refers to a conventional (unducted) tail rotor. The latter equation gives an interesting result, in that it shows that if the shape of the duct is controlled so that the wake does not contract as much as would occur naturally with a conventional tail rotor (for which in the ideal case, $a_d = 0.5$, $A_{\text{eff}} = A$), less power will be required to produce a given total thrust. In the case where $a_d = 1$ (the assumption of no wake contraction), $A_{\text{eff}} = 2A$, a ducted fan of the same area will consume $1/\sqrt{2}$ of the power of a conventional tail rotor (i.e., 30% less power for the same net thrust). Alternatively, a ducted fan of half the disk area of a conventional tail rotor (i.e., a diameter reduced by a factor $1/\sqrt{2}$) will produce the same thrust and consume the same power. Generally, however, because the length of the duct must be relatively short to minimize structural weight and drag penalties in forward flight, the potential gains in efficiency are not as large as suggested by the above analysis. The lower tip loss effects associated with a fan-in-fin design give a further reduction in induced power requirements and the design offers some overall advantages compared to a conventional tail rotor.

In forward flight the fan-in-fin design is shielded from the external flow and main rotor wake and consequently its performance is usually more predictable. The aerodynamics of "sense of rotation" and the interference effects of the fin assembly, which are important for conventional tail rotors, seem to be less important for the fan-in-fin design. However, the possibility of flow separation at the inlet lip of the shroud must be kept in mind, and usually the lip is carefully contoured to avoid such effects. From a safety perspective, the shrouded nature of the fan-in-fin assembly reduces the possibilities of tail rotor strikes during low altitude operations and also the risk of injury to personnel on the ground. The larger number of blades on a fan-in-fin design increases the frequency of the rotor noise, and this can appear in the helicopter noise spectrum over a range of frequencies to which the human ear is more sensitive. However, at greater distances these higher frequency sounds are more readily absorbed in the atmosphere.

NOTAR Design

The NO TAil Rotor or NOTAR concept uses a different approach to antitorque and yaw control. The design is discussed by Logan (1978) and reviewed by Winn & Logan (1990). Here, antitorque capability comes from a *circulation control concept*, which results in a distributed side-force along the entire tail boom assembly. As shown in Fig. 6.30, a jet (or jets) of air from a pressurized tail boom is blown tangential to the surface out of narrow slots that run lengthwise on one side of the tail boom. In combination with the downwash

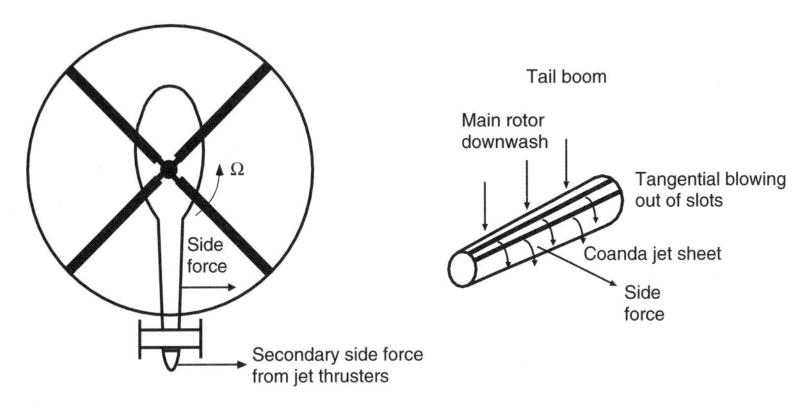

Figure 6.30 The NOTAR antitorque system.

velocity produced by the main rotor, these jets cause the flow to remain attached to the tail boom surface by means of the *Coanda effect*. Downstream of the slots, a powerful suction pressure is produced on one side of the tail boom and a side (antitorque) force results. The magnitude of this side-force depends on the jet velocity out of the slot. This is controlled by adjusting the pressure inside the tail boom by means of a centrifugal compressor. A small auxiliary nozzle at the end of the tail boom provides a pressurized jet to improve yaw control rates and overcome the inherent lag in the circulation control system. The nozzle in the "jet thruster" is rotated by the conventional action of the pilot's foot pedals. In forward flight, the circulation control becomes less effective as the main rotor downwash moves further along the tailboom. Fixed aerodynamic stabilizers and the "jet thruster" combine to produce the necessary antitorque and yaw control. The NOTAR concept has proved attractive to operators because of its low noise, safety for ground personnel, and freedom from blade strikes when operating in confined locations. Also, for a military helicopter this system is attractive because of the absence of a vulnerable tail rotor assembly, and it has a high level of redundancy in the event of any tail boom damage.

6.7 High Speed Rotorcraft

Vertical lift aircraft that use compounding or have tilt-rotor or tilt-wing capability are referred to as high speed rotorcraft. Not only are these craft capable of vertical takeoff and landing, but they can attain much higher speeds than conventional helicopters.

6.7.1 *Compound Helicopters*

A compound helicopter involves a lifting wing in addition to the main rotor (lift compounding) or the addition of a separate source of thrust to the main rotor (thrust compounding); see, for example, the review of Lynn & Drees (1976). The idea is to enhance the basic performance metrics of the helicopter, such as lift-to-drag ratio, propulsive efficiency, and maneuverability. The general benefit is an expansion of the flight envelope compared to a conventional helicopter (see Fig. 6.31). While there are no compound helicopter designs in current production (although many prototypes have been built), except for some Russian designs that may have an element of lift compounding, many helicopter manufacturers have investigated auxiliary propulsion or lifting wings, to off-load the rotor and expand the forward flight envelope. However, there are major drawbacks in this compound design in terms of increased empty weight and loss of payload capability, download penalties in hover, and reduced vertical rate of climb.

One of the first experimental compound helicopter designs was the McDonnell XV-1. This was a pressure jet driven rotor, with a wing and a pusher propeller. After a vertical takeoff, the power was shifted from supplying the tip jets to driving the propeller, and the rotor continued to turn in autorotation. In 1954, the aircraft was flown at speeds approaching 200 mph. The Sikorsky NH-3A was based on the S-61 and used a wing mounted with two turbojets for auxiliary propulsion. It achieved speeds of up to 230 kts – see Fradenburgh & Chuga (1968). The Bell UH-1 compound also had a wing and two turbojets and reached a speed of 275 kts in level flight. The Fairey company built a number of tip jet driven compound machines, with the last and largest being the Rotodyne. This 33,000 lb (15,000 kg) aircraft was powered by two turboprops, which also supplied compressed air for the tip jets. In 1959, the machine set a world speed record in its class of 192 mph. Another prototype compound was the Lockheed AH-56 Cheyenne, which used a wing and a pusher propeller mounted to the tail. Boeing-Vertol flew the tandem Model 347 with relatively large wings,

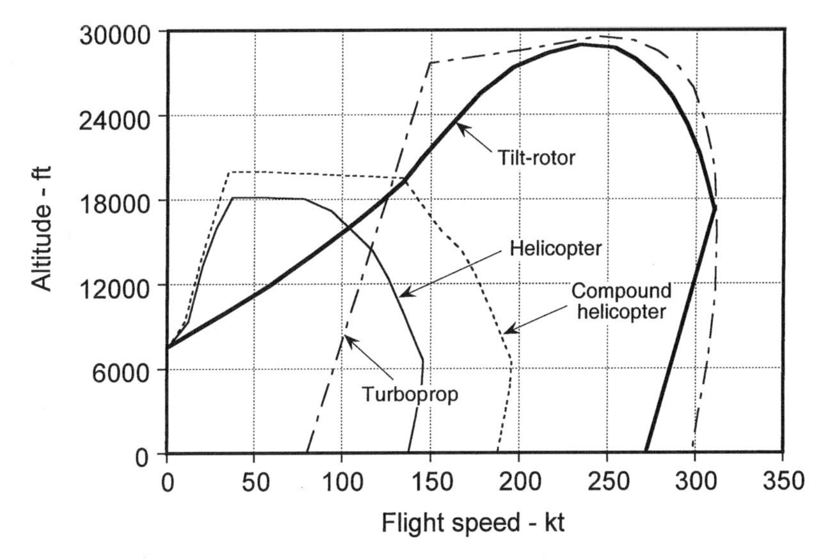

Figure 6.31 Representative flight envelopes of a conventional helicopter versus a compound design, a fixed-wing turboprop, and a tilt-rotor.

the technical development of which is discussed in detail by Stepniewski & Keys (1984). The ideas of compounding have received renewed attention by some manufacturers, with some revived ideas being discussed by Humpherson (1998). It remains to be seen, however, if the compound helicopter design will reemerge as a viable design concept for the twenty-first century.

6.7.2 *Tilt-Rotors*

Based on Bell's experience with the XV-3 (see Chapter 1), the Bell XV-15 tilt-rotor was designed and first flown in 1977. It has proven a reliable research aircraft, from which much knowledge has been gained. The much larger V-22 tilt-rotor flew in 1989, and after a ten-year development phase is now in limited production. In 1998, Bell (and Boeing) announced the Model 609 tilt-rotor for the civilian market.

The aerodynamic design of the tilt-rotor poses many additional challenges over that of the conventional helicopter. The smaller rotor diameter means that in hover the disk loadings of a tilt-rotor are much higher than those of a helicopter of the same gross weight. The efficiency of the rotor is also lowered because of the high effective blade twist, a feature required to ensure good propulsive efficiency in airplane mode. In addition, when in hover the wings of the tilt-rotor operate in the downwash from the rotors, which produces a large download on the aircraft. The fixed wings also influence rotor performance, and the combined effect degrades the hovering efficiency compared to a helicopter of the same gross weight – see McVeigh et al. (1991). Various techniques are used to minimize this effect, including 90-degree deflections of the wing flaps – see Stepniewski & Keys (1984) and Wood & Peryea (1991). In fixed-wing mode, when the rotors act as propellers, they are rather inefficient because the disk loading is too low and profile losses are higher than would be achieved with a conventional propeller. Rosenstein (1986) and McVeigh et al. (1997) give good overviews of the numerous aerodynamic design trade-offs for tilt-rotors.

The tilt-rotor has two sets of controls, one set for helicopter mode and another for fixed-wing mode. In helicopter mode, the V-22's rotors are controlled with full cyclic and collective pitch controls. Yaw is controlled by differential cyclic, just like a tandem rotor machine such as the CH-47. During transition from helicopter mode to fixed-wing mode, the aerodynamic environment encountered by the rotors becomes extremely complex and high loads can be produced. The relationship between the airspeed and the rotor tilt angle must conform to a narrow envelope for safe operation and is controlled with the aid of the fight control system. In airplane mode the rotors act as conventional propellers, with all of the vertical lift force being produced by the wing. In this mode, control is achieved with the use of conventional aerodynamic surfaces, the flaperons, elevator, and rudder. Despite the complexity and high cost of the V-22, it plays a unique role in vertical flight aviation that a conventional helicopter cannot.

6.7.3 *Other High Speed Rotorcraft*

Most of the major US helicopter manufacturers have undertaken studies of *high speed rotorcraft concepts*. Similar studies have been conducted by some European manufacturers. The general requirements call for a rotorcraft capable of efficient hover capability that can cruise at over 400 kts, which is about twice as fast as is attainable with a conventional helicopter. Despite much research, other than tilt-rotors, there is no high speed rotorcraft concept currently flying. Some of the high-speed rotorcraft designs considered have included the following:

1. *The ABC concept*: In the 1960s, Sikorsky Aircraft announced the S-69 research aircraft using the Advancing Blade Concept (ABC). This rigid coaxial rotor system was designed to alleviate the problem of retreating blade stall by allowing a more symmetric distribution of lateral airloads over the rotor. Chaney (1969) gives a good overview of the aerodynamic issues. During flight testing, the machine reached a maximum speed of 263 kts (487 kph). For the time, it was the only rotorcraft to have approached such speeds without lift compounding.
2. *The X-wing concept:* This design employs a unique rotor system that uses circulation control. The blades consist of thick elliptical airfoils, and air is blown at high velocity out of the leading or trailing edges to create lift by means of the Coanda effect – see Cheeseman (1968) and Williams (1976). A complicated pneumatic blowing system was required for cyclic blowing in forward flight. As airspeed builds, the rotor system is slowed down and stopped, forming an X-wing configuration. The full-scale rotor was tested in a wind tunnel in the early 1980s, but it was never flown because of various technical problems and escalating costs.
3. *The variable diameter rotor concept:* This is another Sikorsky concept, which looks similar to a conventional helicopter but the blades telescope in length during flight to increase or decrease the disk area. The design is discussed by Fradenburgh et al. (1973) and Fradenburgh (1992). A system of clutches inside the rotor hub controls the extension and retraction process. This technique allows for low disk loadings and maximum efficiency in hover, while the smaller diameter rotor allows for a higher cruise efficiency. A similar design has been considered for a tilt-rotor, where the rotor diameter is reduced for airplane mode – see Studebaker & Matuska (1993). As of 1999, the system was still under development but full-scale rotor and flight tests were planned.
4. *Rotor/wing concept:* This is a high solidity stoppable rotor system with three blades. The rotor is slowed down and stopped in forward flight, with two blades pointing upstream forming forward swept wing, while the third blade points down-

stream and lies over the fuselage. The concept is similar to the ill-fated X-wing circulation-controlled rotor of the 1980s. See Rutherford et al. (1993) for a history of such stopped rotor designs.

5. *Folding, stowed, or trailed rotor systems:* Here the rotor is stopped, the blades folded back into the fuselage, and the lift is transferred to fixed wings. Various aeroelastic effects are possible during the stopping and stowing process [see Deckert & McCloud (1968)] and these are difficult to contend with. The high capital and maintenance cost of the system, as well as the high weight of the folding and blade storage system, means that this type of machine is not a particularly attractive design option.
6. *Shrouded rotor designs:* The idea here is to use a relatively large ducted fan, integral to a wing, to provide the vertical lift. As the aircraft gains speed, power is transferred from the fan to pure propulsion, shutters cover the fan, and the wing provides all the lift. The concept, however, is not new, having being used on the Ryan XV-5 in the 1960s. This aircraft demonstrated successful transitions from hover to forward flight and back, flying at speeds of up to 330 kts (577 kph).
7. *"Smart" rotor systems:* A "smart" structure is one that involves distributed actuators and sensors, along with a computer to analyze responses and apply displacements or strains to change the characteristics of the structure in an adaptive and desirable way. Smart structures make it possible to alter the properties of rotors in a beneficial way so as to reduce vibrations, improve performance, and enhance other factors. Such alterations would include the use of smart actively controlled aerodynamic surfaces on the rotor, such as with trailing-edge flaps. The field is reviewed by Chopra (1997). While the technology is not yet mature, various concepts are developing quickly, and it is likely that a full-scale helicopter rotor incorporating one or more new smart technologies will fly early in the twenty-first century.

6.8 Chapter Review

This chapter has reviewed many of the aerodynamic issues important in the design of the helicopter. It has been noted that there are several trade-offs in the basic sizing and overall optimization methodology of helicopter main rotors. The final design is always a compromise to meet the needs of a particular set of customer or mission requirements. The use of improved airfoil sections and advanced tip shapes generally helps to improve overall rotor performance, allowing higher figures of merit and better cruise efficiency. The computational tools for rotor design are now at a high level of maturity, although significant empiricism must generally still be relied upon.

The aerodynamics of the helicopter fuselage and empennage are complicated, mainly because of the extensive regions of separated flow that can exist, coupled with the aerodynamic interactions that exist between the main rotor and the fuselage. Predictions of these effects are still beyond the state of the art but can be reliably estimated through component testing of the fuselage and rotor in the wind tunnel. Besides the main rotor and the airframe, the design of the empennage and tail rotor are key elements in the successful design of the helicopter. Because of the various aerodynamic interactions and the trade-offs in stability, the sizing and positioning of the horizontal stabilizer on the tail has proven to be one of the most difficult challenges facing helicopter designers. The tail rotor operates in a complicated flow environment, with its operation being affected strongly by the main rotor wake. Many other factors need to be considered to ensure that the tail rotor operates effectively as an antitorque and directional control device over the full operational flight envelope of the helicopter. Other antitorque devices such as the fenestron and NOTAR have proved viable alternatives to the conventional tail rotor.

Finally, some concepts for compounds and "high speed" helicopters have been reviewed. While many ideas have been put forth over the past fifty years, there are no high speed rotorcraft other than tilt-rotors currently flying. The aerodynamic and aeroelastic problems of high speed helicopters have proved difficult to solve cost effectively. However, with the advent of new technologies such as smart structures to help control aerodynamic forces and vibration levels on the rotor, it is likely that a further expansion in the operational flight envelope of conventional helicopters will occur.

6.9 Questions

6.1. Starting from an expression for the figure of merit for an optimum hovering rotor, show that the figure of merit is a maximum when the blade sections are operating at the highest values of $C_l^{3/2}/C_d$.

6.2. Based on the modified momentum theory, show that the operating thrust coefficient, C_T, to give the lowest power per unit thrust in hover (the best power loading) is

$$C_T \text{ for best } PL = \frac{1}{2}\left(\frac{\sigma C_{d_0}}{\kappa}\right)^{2/3}.$$

Also show that this condition corresponds to a figure of merit of $2/3\kappa$.

6.3. Discuss in detail the reasons why the modern helicopter is still essentially a low speed aircraft. Discuss how a main rotor system might be designed so that a helicopter can achieve higher overall forward flight speeds and a possible expansion of the maneuvering envelope. Identify any potential trade-offs with any design option.

6.4. The design of a highly maneuverable helicopter requires some special considerations that include aerodynamic issues. Discuss the various interdisciplinary design issues that you feel are important.

6.5. The Cierva company designed a heavy-lift helicopter with three main rotors, with a layout such that the rotors were arranged at the corners of an equilateral triangle (two rotors at the back in a lateral arrangement, and one rotor at the front), with all three rotors turning in the same direction. How is torque reaction compensated for, and how would the aircraft be controlled?

6.6. When designing a hovering rotor for maximum aerodynamic efficiency with the disk loading fixed, should a designer choose a high or low value of tip speed? To further increase the rotor figure of merit, should the designer choose a high or low solidity, and why? If the rotor was to be powered by ramjets located at the blade tips, so that it had to operate at a high tip speed, should the designer choose a high or low solidity, and why?

6.7. You are asked to design an optimum rotor for a large helicopter whose primary mission is to lift logs and heavy construction equipment in the mountains. If the gross weight is given as 100,000 lb, rotor diameter is 120 ft, average operating altitude is 15,000 ft, and tip speed is 650 ft/s, calculate the tip chord (assuming seven blades), equivalent rotor solidity, and mean rotor lift coefficient for the helicopter. How much shaft power would be required to hover out of ground effect at the design condition? Choose and obtain data on an appropriate airfoil.

6.8. In a hypothetical helicopter rotor design, the use of blade taper has been shown to increase the figure of merit of the main rotor by 1%. Estimate the percentage increase in vertical lifting and payload capability of the helicopter with all other factors being assumed constant.

6.9. In the design of a swept tip rotor blade, it is desired to introduce only as much sweep angle as necessary to keep the Mach number incident to the leading-edge of the blade equal to a constant, say M_c, over the tip region. Because of unsteady flow effects, the critical azimuth angle for the design is taken to be at $\psi_b = 100°$. Find an expression for the sweep angle as a function of r, the hover tip Mach number M_{tip}, the design mach number M_c, and the advance ratio, μ. If $M_{\text{tip}} = 0.65$ and $M_c = 0.8$, plot the sweep angle required as a function of radial blade station for a series of advance ratios.

6.10. In the design of a swept tip rotor blade, it is desired to introduce only as much sweepback as necessary to keep the Mach number incident to the leading-edge of the blade equal to a constant, say M_c, over the tip region. In addition, the location of the 1/4-chord axis of the blade is to be moved forward by a value $eR = 0.02$ at the point of sweep initiation so as to minimize the torsional moments on the blade that would be associated with the sweepback. Based on 2-dimensional strip considerations, calculate and show the resulting blade geometry for a design advance ratio of 0.5. Assume that the critical azimuth angle for the design is taken to be at $\psi_b = 90°$, and that $M_{\text{tip}} = 0.65$ and $M_c = 0.8$.

6.11. Discuss the relative merits of the compound helicopter design. In your discussion, consider both propulsive and lift compounding as both separate and combined means of compounding.

6.12. It is required to improve the tail rotor authority of an existing helicopter design. The design of the rotor blades is to be changed, but without modifying the blade pitch attachments, hub, pitch links, etc. Define the requirements, outline any conflicting issues, and suggest possible means of tackling this problem.

6.13. A canted tail rotor design has been used on some helicopters. Discuss the relative advantages of such a design.

Bibliography

Abbott, I. H. and von Doenhoff, A. E. 1949. *Theory of Wing Sections, Including a Summary of Airfoil Data*, McGraw-Hill Book Co., Inc., New York. Also Dover Publications, New York, 1959.

Ahmed, S. R., Amtsberg, J., de Bruin, A. C., Cler, A., Falempin, G., Le, T. H., Polz, G., and Wilson, F. T. 1988. "Comparison with Experiment of Various Computational Methods of Airflow on Three Helicopter Fuselage," 14th European Rotorcraft Forum, Milan, Italy, Sept. 20–23.

Ahmed, S. R. 1990. "Fuselage Aerodynamic Design Issues and Rotor/Fuselage Interactional Aerodynamics – Pt. 2: Theoretical Methods," AGARD-R-781.

Althoff, S. L. and Noonan, K. W. 1990. "Effect of Blade Planform Variation on a Small-Scale Hovering Rotor," NASA TM 4146, AVSCOM TM 89-B-009.

Amer, K. B. 1989. "High Speed Rotor Aerodynamics," *J. American Helicopter Soc.*, 34 (1), pp. 63–66.

Balcerak, J. C. and Felker, R. F. 1973. "Effect of Sweep Angle on the Pressure Distributions and Effectiveness of the Ogee Tip in Diffusing a Line Vortex," NASA CR 132355.

Balch, D. A. 1984. "Impact of Main Rotor Tip Geometry on Main Rotor/Tail Rotor Interactions in Hover," 40th Annual Forum of the American Helicopter Soc., Arlington, VA, May 16–18.

Berry, J. D., Chaffin, M. S., and Duque, E. 1994. "Helicopter Fuselage Aerodynamic Predictions: Navier–Stokes and Panel Method Solutions and Comparison with Experiment," American Helicopter Soc. Aeromechanics Specialists Conf., San Francisco, CA, Jan. 19–21.

Berry, J. D. 1997. "Unsteady Velocity Measurement Taken behind a Model Helicopter Rotor Hub in Foward Flight," NASA TM-4738.

Bosco, A. 1972. "Aerodynamics of Helicopter Components Other Than Rotors," AGARD-CP-111.

Bramwell, A. R. S. 1966. "A Theory of the Aerodynamic Interference between a Helicopter Rotor Blade and a Fuselage and Wing in Hovering and Forward Flight," *J. of Sound & Vib.*, 3 (3), pp. 355–383.

Brocklehurst, A. 1985. "A Significant Improvement to the Low Speed Control of the Sea King Helicopter Using a Tail Boom Strake," 11th European Rotorcraft Forum, London, Sept. 10–13.

Byham, G. M. 1990. "An Overview of Conventional Tail Rotors," RAeS Conf. on Helicopter Yaw Control Concepts, London, February–March.

Carlin, G. and Farrance, K. 1990. "Results of Wake Surveys on Advanced Rotor Planforms," 46th Annual Forum of the American Helicopter Soc., Washington DC, May 21–23.

Chaffin, M. S. and Berry, J. D. 1994. "Navier–Stokes and Potential Theory Solutions for a Helicopter Fuselage and Comparison with Experiment," NASA Technical Memorandum 4566, ATCOM Technical Report 94-A-013.

Chaney, M. C. 1969. "The ABC Helicopter," *J. American Helicopter Soc.*, 14 (4), pp. 10–19.

Cheeseman, I. C. 1968. "Circulation Control and It's Application to Stopped Rotor Aircraft," *Aeronaut. J.*, 72, pp. 635–646.

Chopra, I. 1997. "Status of Application of Smart Structures Technology to Rotorcraft Systems," RAeS Proc. of Innovation in Rotorcraft Technology, June 24–25.

Clark, D. R. and Leiper, A. C. 1970. "The Free Wake Analysis, A Method for the Prediction of Helicopter Rotor Hovering Performance," *J. American Helicopter Soc.*, 15 (1), pp. 3–12.

Cler, A. 1989. "High Speed Dauphin Fuselage Aerodynamics," 15th European Rotorcraft Forum, Amsterdam, The Netherlands, Sept. 12–15.

Cook, C. V. 1978. "Tail Rotor Design and Performance," *Vertica*, 2, pp. 163–181.

Curtiss, H. C. and Quackenbush, T. R. 1989. "The Influence of the Rotor Wake on Rotorcraft Stability and Control," 15th European Rotorcraft Forum, Amsterdam, The Netherlands, Sept. 12–15.

Dadone, L. 1976. "US Army Helicopter Design Datcom, Volume 1, Airfoils," US Army Air Mobility Research and Development Laboratory.

Davidson, J. K., Havey, C. T., and Sherrieb, H. E. 1972. "Fan-in-Fin Antitorque Concept Study," USAAMRDL TR 72-44, July 1972.

de Bruin, A. C. 1987. "Three-Dimensional Boundary Layer Calculation Results on the GARTEUR Helicopter Fuselage with Shallow Ramp ($\alpha = 0°$ and $\alpha = -5°$)," NLR TR 87099L.

Deckert. W. H. and McCloud, J. L. 1968. "Considerations of the Stopped Rotor V/STOL Concept," *J. American Helicopter Soc.*, 13 (1), pp. 27–43.

Delany, N. K. and Sorensen, N. E. 1953. "Low Speed Drag of Cylinders of Various Shapes," NACA TN-3038.

Department of the Army. 1974. "Engineering Design Handbook – Helicopter, Part 1 – Preliminary Design," AMCP 706-201, US Army Materiel Command.

Desopper, A., Lafon, P., Philippe, J. J., and Ceroni, P. 1986. "Ten Years of Rotor Flow Studies at ONERA: State of the Art and Future Studies," 42nd Annual Forum of the American Helicopter Soc., Washington DC, June 2–4.

Desopper, A., Lafon, P., Philippe, J. J., and Prieur, J. 1988. "Effect of an Anhedral Sweptback Tip on the Performance of a Helicopter Rotor," 44th Annual Forum of the American Helicopter Soc., Washington DC, June 16–18. Also, 13th European Rotorcraft Forum, Arles, France, Sept. 8–11.

Duque, E. P. N. and Brocklehurst, A. 1990. "Experimental and Numerical Study of the British Experimental Rotor Program Blade," 8th AIAA Applied Aerodynamics Conf., Aug.

Duque, E. P. N. 1992. "Numerical Analysis of the British Experimental Rotor Program Blade," *J. American Helicopter Soc.*, 37 (1), pp. 46–54.

Duque, E. P. N. 1994. "A Structured/Unstructured Embedded Grid Solver for Helicopter Rotor Flows," 50th Annual Forum of the American Helicopter Soc., Washington, DC, May 11–13.

Duque, E. P. N. and Dimanlig, A. C. B. 1994. "Navier–Stokes Simulation of the RAH-66 (Comanche) Helicopter," American Helicopter Soc. Aeromechanics Specialists Conf., San Francisco, CA, Jan. 19–21.

Ellin, A. D. S. 1993. "An In-Flight Investigation of Westland Lynx AH Mk-5 Main Rotor/Tail Rotor Interactions," 19th European Rotorcraft Forum, Como, Italy, Sept. 14–16.

Epstein, R. J., Carbonaro, M. C., and Caudron, F. 1994. "Experimental Investigation of the Flowfield about an Upswept Afterbody," *J. of Aircraft*, 31 (6), pp. 1281–1290.

Fradenburgh, E. A. and Chuga, G. M. 1968. "Flight Program on the NH-3A Research Helicopter," *J. American Helicopter Soc.*, 13 (1), pp. 44–62.

Fradenburgh, E. A. 1972. "Aerodynamic Factors Influencing Overall Hover Performance," AGARD-CP-111.

Fradenburgh, E. A., Murrill, R. J., and Kiely, E. F. 1973. "Dynamic Model Wind Tunnel Tests of a Variable-Diameter, Telescoping-Blade Rotor System (TRAC Rotor)," USAAMRDL TR 73-32.

Fradenbergh, E. A. 1992. "Wind-Tunnel Tests of a Variable Diameter Rotor," 48th Annual Forum of the American Helicopter Soc., Washington DC, June 3–5.

Frederickson, K. C. and Lamb, J. R. 1993. "Experimental Investigation of Main Rotor Wake Induced Vibratory Empennage Airloads for the RAH-66 Comanche Helicopter," 49th Annual Forum of the American Helicopter Soc., St. Louis, MO, May 19–21.

Foley, S. M., Funk, R. B., Fawcett P. A., and Komerath, N. M. 1995. "Rotor-Wake-Induced Flow Separation on a Lifting Surface," *J. American Helicopter Soc.*, 40 (2), pp. 24–27.

Gangwani, S. T. 1982. "A Doublet-Lattice Method for the Determination of Rotor Induced Empennage Vibration Airloads – Analysis, Description, and Program Documentation," NASA CR-165893.

Hansen, K. 1988. "Handling Qualities, Design and Development of the CH-53E, UH-60A and S-76 Helicopters," Proc. of the RAeS Conf. on Handling Qualities and Control, Nov. 15–17, London.

Hess, J. L. and Smith, A. M. O. 1962. "Calculation of Non-Lifting Potential Flow About Arbitrary Three-Dimensional Bodies, " Douglas Aircraft Report ES40622.

Hess, J. L. 1990. "Panel Methods in Computational Fluid Dynamics," *Annual Review of Fluid Mechanics*, 22, pp. 255–274.

Hoerner, S. F. 1965. *Fluid Dynamic Drag*, Published by the Author, Hoerner Fluid Dynamics, Vancouver, WA.

Horst, T. J. and Reschak, R. J. 1975. "Designing to Survive Tail Rotor Loss," 31st Annual National Forum of the American Helicopter Soc., Washington DC, May 13–15.

Humpherson, D. V. 1998. "Compound Interest – A Dividend for the Future, Revisited," 45th Annual Forum of the American Helicopter Soc., Washington DC, May 20–22.

Katz, J. and Plotkin, A. 1991. *Low Speed Aerodynamics – From Wing Theory to Panel Methods*, McGraw-Hill, New York.

Kelley, H. L., Crowell, C. A., Yenni, K. R., and Lance, M. B. 1993. "Flight Investigation of the Effect of Tail Boom Strakes on Helicopter Directional Control," NASA TP-3278, ATCOM TR-93-A-003.

Keys, C. and Wiesner, R. 1975. "Guidelines for Reducing Helicopter Parasitic Drag," in *Rotorcraft Parasitic Drag*, 31st Annual National Forum of the American Helicopter Soc., Washington DC, May 13–15.

Keys, C., Tarzanin, F., and McHugh, F. 1987. "Effect of Twist on Helicopter Performance and Vibratory Loads," 13th European Rotorcraft Forum, Arles, France, Sept. 8–11.

Knight, M. and Hefner, R. A. 1937. "Static Thrust of the Lifting Airscrew," NACA TN 626.

Landgrebe, A. J. 1972. "Wake Geometry of a Hovering Helicopter Rotor and Its Influence on Rotor Performance," *J. American Helicopter Soc.*, 17 (4), pp. 3–15.

Lê, T. H., Ryan, J., and Falempin, G. 1987. "Wake Modeling for Helicopter Fuselage," 13th European Rotorcraft Forum, Arles, France, Sept. 8–11.

Leishman, J. G. and Bi, Nai-pei 1994. "Experimental Investigation of Rotor/Lifting Surface Interactions," *J. of Aircraft*, 31 (4), pp. 846–854.

Lindhout, J. P. F., Moek, G., de Boer, E., and van der Berg, B. 1981. "A Method for the Calculation of 3D Boundary layers on Practical Wing Configurations," *J. of Fluids Eng.*, 103 (1), pp. 104–111.

Logan, A. H. 1978. "Evaluation of a Circulation Control Tail Boom for Yaw Control," USARTL TR 78-10.

Lynn, R. R. 1966. "Wing-Rotor Interactions," *J. of Aircraft*, 3 (4), pp. 326–332.

Lynn, R. R. and Drees, J. M. 1967. "Promise of Compounding," *J. American Helicopter Soc.*, 12 (1), pp. 1–20.

Lynn, R. R. 1970. "Tail Rotor Design," *J. American Helicopter Soc.*, 15 (4), pp. 2–30.

Main, B. J. and Mussi, F. 1990. "EH-101: Development Status Report," 16th European Rotorcraft Forum, Glasgow, Scotland, Sept. 18–20.

Maskew, B. 1982. "Prediction of Subsonic Aerodynamic Characteristics: A Case for Low-Order Panel Methods," *J. of Aircraft*, 19 (2), pp. 157–163.

Makofski, R. A. and Menkick, G. F. 1956. "Investigation of Vertical Drag – Periodic Loads Acting on Flat Panels in a Rotor Slipstream," NACA TN 3900.

McHugh, F. J. 1978. "What Are the Lift and Propulsive Limits at High Speed of the Conventional Rotor?" 34th Forum of the American Helicopter Soc., Washington DC, May.

McKee, J. W. and Naeseth, R. L. 1958. "Experimental Investigation of the Drag on Flat Plates and Cylinders in the Slipstream of a Hovering Rotor," NACA TN 4239.

McVeigh, M. A. and McHugh, F. J. 1982. "Recent Advances in Rotor Technology at Boeing Vertol," 38th Annual Forum of the American Helicopter Soc., Anaheim, CA, May 4–7.

McVeigh, M. A., Grauer, W. K., and Paisley, D. J. 1991. "Rotor–Airframe Interactions on Tilt-Rotor Aircraft," *J. American Helicopter Soc.*, 35 (3), pp. 43–51.

McVeigh, M. A., Liu, J., and O'Toole, S. J. 1997. "V-22 Osprey Aerodynamic Development – A Progress Review," *The Aeronaut. J.*, 101 (1006), pp. 231–244.

Mello, O. A. F. and Rand, O. 1991. "Unsteady Frequency Domain Analysis of Helicopter Non-Rotating Lifting Surfaces," *J. American Helicopter Soc.*, 36 (2), pp. 70–81.

Moedersheim, E. and Leishman, J. G. 1996. "Investigation of Aerodynamic Interactions between a Rotor and a T-Tail Empennage," *J. American Helicopter Soc.*, 43 (1), pp. 37–46.

Montana, P. S. 1975. "Experimental Investigation of Three Rotor Hub Fairing Shapes," David W. Taylor Naval Ship Research and Development Center, Report ASED 333.

Montana, P. S. 1976. "Experimental Evaluation of Analytically Shaped Helicopter Rotor Hub-Pylon Configurations," David W. Taylor Naval Ship Research and Development Center, Report ASED 355

Morino L. (ed.) 1985. *Computational Methods in Potential Aerodynamics*, Springer-Verlag, Berlin.

Mouille, R. 1970. "The Fenestron Shrouded Tail of the SA.341 Gazelle," *J. American Helicopter Soc.*, 15 (4), pp. 31–37.

Mouille, R. 1979. "Ten Years of Aerospatiale Experience with the Fenestron and Conventional Tail Rotor," 35th Annual Forum of the American Helicopter Soc., Washington DC, May 21–23.

Mouille, R. and Dámbra, F. 1986. "The Fenestron, a Shrouded Tail Rotor Concept for Helicopters," 42nd Annual Forum of the American Helicopter Soc., Washington DC, May 21–23.

Narramore, J. C. and Brand, A. G. 1992. "Navier–Stokes Correlations to Fuselage Wind Tunnel Test Data," 48th Annual Forum of the American Helicopter Soc., June 3–5.

Newman, S. 1994. *The Foundations of Helicopter Flight*, Edward Arnold, London, Chapter 11.

Noonan, K. R., Althoff, S. L., Samak, D. K., and Green, M. D. 1992. "Effect of Blade Planform Variation on the Forward Flight Performance of Small-Scale Rotors," NASA Technical Memorandum 4345.

Paul, W. F. and Zincone, R. 1977. "Advanced Technology Applied to the UH-60A and S-76 Helicopters," 3rd European Rotorcraft and Powered Lift Forum, Germany, Sept.

Perry, F. J. 1987. "The Aerodynamics of the World Speed Record," 43rd Annual Forum of the American Helicopter Soc., St. Louis, MO, May 18–20.

Perry, F. J. 1989. "The Contribution of Planform Area to the Performance of the BERP Rotor," *J. American Helicopter Soc.*, 34 (1), pp. 64–66.

Perry, F. J., Wilby, P. G., and Jones, A. F. 1998. "The BERP Rotor – How Does It Work, and What Has It Been Doing Lately?," *Vertifite*, 44 (2), pp. 44–48.

Philippe, J. J., Roesch, P, Dequin, A. M., and Cler, A. 1985. "A Survey of Recent Developments in Helicopter Aerodynamics," AGARD LS-139.

Polz, G. and Quentin, J. 1981. "Separated Flow around Helicopter Bodies," 7th European Rotorcraft Forum, Garmisch-Partenkirchen, Sept. 22–25.

Polz, G. 1982. "The Calculation of Separated Flow at Helicopter Bodies," NASA TM-76715.

Prouty, R. W. and Amer, K. B. 1982. "The YAH-64 Empennage and Tail Rotor – A Technical History," 38th Annual Forum of the American Helicopter Soc., Anaheim, CA, May 4–7.

Prouty, R. W. 1983. "Development of the Empennage Configuration of YAH-64 Advanced Attack Helicopters," USAAVRADCOM-TR-82-D-22.

Prouty, R. W. 1986. *Helicopter Performance, Stability, and Control*, PWS Engineering Publishing, Boston, MA. Chapter 10.

Prouty, R. W. 1998. *Military Helicopter Design Technology*, 2nd ed., Krieger Publishing Co., Malabar, FL.

Roesch, P. and Vuillet, A. 1981. "New Designs for Improved Aerodynamic Stability on Recent Aerospatiale Helicopters," 37th Annual Forum of the American Helicopter Soc., New Orleans, LA, May 17–20.

Rorke, J. B., Moffitt, R. C., and Ward, J. F. 1972. "Wind Tunnel Simulation of Full Scale Vortices," 28th Annual National Forum of the American Helicopter Soc., Washington DC, May.

Rosenstein, H. and Stanzione, K. 1981. "Computer-Aided Helicopter Design," 37th Annual Forum of the American Helicopter Soc., Washington DC, May 17–20.

Rosenstein, H. 1986. "Aerodynamic Development of the V-22 Tilt-Rotor," 12th European Rotorcraft Forum. Garmisch-Partenkirchen, Germany, Sept. 22–25.

Rubbert, P. E. and Saaris, G. R. 1969. "Three-Dimensional Potential Flow Method Predicts V/STOL Aerodynamics," *SAE J.*, 77 (9), Sept., pp. 44–52.

Rutherford, J. W., Bass, S. M., and Larsen, S. D. 1993. "Canard Rotor/Wing: A Revolutionary High-Speed Rotorcraft Concept," Paper 93-1175, AIAA/AHS/ASEE Aerospace Design Conf., Irvine, CA, Feb. 16–19.

Sedden, J. 1979. "An Analysis of Helicopter Rotorhead Drag Based on New Experiments," 5th European Rotorcraft and Powered Lift Forum, Amsterdam, The Netherlands, Sept. 4–7.

Sedden, J. 1982. "Aerodynamics of the Rear Fuselage Upsweep," 8th European Rotorcraft Forum, Aix-en-Province, France, Aug. 31–Sept. 3.

Sheehy, T. W. and Clark, D. R. 1975. "A General Review of Helicopter Rotor Hub Drag Data," in *Rotorcraft Parasitic Drag*, 31st Annual Forum of the American Helicopter Soc., Washington DC, May 13–15.

Sheehy, T. W. 1977. "A General Review of Helicopter Rotor Hub Drag Data," *J. American Helicopter Soc.*, 22 (2), pp. 2–10.

Sheridan, P. F. 1978. "Interactional Aerodynamics of the Single Rotor Helicopter Configuration. Final Report. USARTL-TR-78-23A, U.S. Army.

Sheridan, P. F. and Smith, R. P. 1979. "Interactional Aerodynamics: A New Challenge of Helicopter Technology," 35th Annual National Forum of the American Helicopter Soc., May 21–23.

Sigl, D. G. and Smith, D. E. 1990. "Flow Over and Behind Various Rotor Blade Tip Shapes: Wind Tunnel Test Data Report," NASA CR 182000.

Singleton, J. D., Yeager, W. T., and Wilbur, M. L. 1990. "Performance Data from a Wind-Tunnel Test of Two Main-Rotor Blade Designs for a Utility-Class Helicopter," NASA Technical Memorandum 4183.

Smith, D. G. and Sigl, D. 1995. "Helicopter Tip Shapes for Reduced Blade Vortex Interaction," AIAA paper 95-0192, 33rd Aerospace Sci. Meeting and Exhibit, Reno, NV, Jan. 9–12.

Spivey, R. F. 1968. "Blade Tip Aerodynamics – Profile and Planform Effects," 24th Annual National Forum of the American Helicopter Soc., Washington DC.

Spivey, W. A. and Morehouse, C. G. 1970. "New Insights into the Design of Swept-Tip Rotor Blades." 26th Annual National Forum of the American Helicopter Soc., Washington DC, June 16–18.

Stepniewski, W. Z. and Keys, C. N. 1984. *Rotary-Wing Aerodynamics*, Dover Publications, New York, Vol. II.

Studebaker, K. and Matuska, D. 1993. "Variable Diameter Tiltrotor Wind Tunnel Test Results," 49th Annual Forum of the American Helicopter Soc., St. Louis, MO, May 19–21.

Tarttelin, P. C. and Martyn, A. W. 1994. "In Flight Research with Instrumented Main and Tail Rotor Blades Using the DRA Bedford Aeromechanics Research Lynx Helicopter," AGARD CP-552.

Torok, M. S. and Ream, D. T. 1993. "Investigation of Empennage Airloads Induced by a Helicopter Main Rotor Wake," 49th Annual Forum of the American Helicopter Soc., St. Louis, MO, May 19–21.

Tung, C. and Lee, S. 1994. "Evaluation of Hover Performance Prediction Codes," 50th Annual Forum of the American Helicopter Soc., Washington DC, May 11–13.

Vuillet, A. and Morelli, F. 1986. "New Aerodynamic Design of the Fenestron for Improved Performance," 12th European Rotorcraft Forum, Garmisch-Partenkirchen, Germany, Sept. 22–25.

Vuillet, A., Allongue, M., Philippe, J. J., and Desopper, A. 1989. "Performance and Aerodynamic Development of the Super Puma Mk II Main Rotor with New SPP8 Blade Tip Design," 15th European Rotorcraft Forum, Amsterdam, The Netherlands, Sept. 12–15.

Vuillet, A. 1990. "Rotor and Blade Aerodynamic Design," AGARD R-781.

Weinstock, S. 1991. "Formulation of a Simplified Model of Rotor-Horizontal Stabilizer Interactions and Comparison with Experimental Measurements," 17th European Rotorcraft Forum, Berlin, Germany, Sept. 24–27.

Wheatley, J. B. 1935. "The Influence of Wing Setting on the Wing Load and Rotor Speed of a PCA-2 Autogiro as Determined in Flight," NACA Rep. 536.

Wiesner, W. and Kohler, G. 1973. "Design Guidelines for Tail Rotors," US Army USAAMRDL Report D210-10687-1.

Wiesner, W. and Kohler, G. 1974. "Tail Rotor Performance in Presence of Main Rotor, Ground, and Winds," *J. American Helicopter Soc.*, 19 (3), pp. 2–9.

Wilby, P. G. 1998. "Shockwaves in the Rotor World – A Personal Perspective of 30 Years of Rotor Aerodynamic Research in the UK," *The Aeronaut. J.*, 102 (1013), pp. 113–128.

Williams, R. and Montana, P. S. 1975. "A Comprehensive Plan for Drag Reduction," in *Rotorcraft Parasitic Drag*, 31st Annual Forum of the American Helicopter Soc., Washington DC, May 13–15.

Williams, R. M. 1976. "Application of Circulation Control Rotor Technology to a Stopped Rotor Aircraft Design," *Vertica*, 1, pp. 3–15.

Wilson, F. T. 1984. "Design and Testing of a Large Scale Helicopter Model in the RAE 5 Metre Pressurised Wind Tunnel," 10th European Rotorcraft Forum, The Hague, The Netherlands, Aug. 28–31.

Wilson, F. T. 1990. "Fuselage Aerodynamic Design Issues and Rotor/Fuselage Interactinal Aerodynamics – Pt. 1: Practical Design Issues," AGARD-R-781.

Wilson, J. C. 1975. "Rotorcraft Low-Speed Download Drag Definition and Its Reduction," Rotorcraft Parasite Drag. 31st Annual National Forum, American Helicopter Soc., Washington DC, May 13–15.

Wilson, J. C. and Kelley, H. L. 1983. "Measured Aerodynamic Forces on Three Typical Helicopter Tail Boom Cross Sections," *J. American Helicopter Soc.*, 28 (4), pp. 68–71.

Wilson, J. C. and Kelley, H. L. 1986. "Aerodynamic Characteristics of Several Current Helicopter Tail Boom Cross Sections Including the Effect of Spoilers," NASA TP-2506, AVSCOMTR-85-B-3.

Wilson, J. C., Kelly, H. L., and Donahue, C. C. 1988. "Development in Helicopter Tailboom Strake Application," Proc. of the RAeS Conf. on Helicopter Handling Qualities and Control, Nov. 15–17, London.

Winn, A. L. and Logan, A. H. 1990. "The MDHC NOTAR System," RAeS Conf. on Helicopter Yaw Control Concepts, London, February/March.

Wood, T. L. and Peryea, M. A. 1991. "Reduction of Tilt-Rotor Download," 47th Annual Forum of the American Helicopter Soc., Phoenix, AZ, May 6–8.

Yeager, W. T, Jr., Mantay, W. R., Singleton, J. D. Wilbur, M. L., Cramer, R. G., and Singleton, J. D. 1987. "Wind-Tunnel Evaluation of an Advanced Main-Rotor Blade Design for a Utility-Class Helicopter," NASA Technical Memorandum 89129.

Yeager, W. T, Jr., Noonan, K. W., Singleton, J. D. Wilbur, M. L., and Minick, P. H. 1997. "Performance and Vibratory Loads Data from a Wind-Tunnel Test of a Model Helicopter Main-Rotor Blade with a Paddle-Type Tip," NASA TM-4754, ARL TR-1283, ATCOM TR-97-A-006.

CHAPTER 7

Rotor Airfoil Aerodynamics

It [the Göttingen-429] is a reasonably efficient airfoil, although others give greater lift and a great many different curves [airfoil shapes] are used for designing [fixed-wing] airplanes. But, the important advantage of this particular type is that its center of lift or pressure is approximately the same at all angles which it may assume in flight. This is not true of other types of airfoil, so that center of pressure travel is a factor to be reckoned with in using them.

Juan de la Cierva (1931; in reference to the twisting moment produced on autogiro blades by a cambered airfoil.)

7.1 Introduction

The goal of this chapter is to review the aerodynamics of airfoils and to discuss their potential impact on rotor performance. An improved understanding and predictive capability of airfoil characteristics will always lead to an improved analysis capability of existing rotor designs and may ultimately lead to new rotors optimized for greater performance in both hover and high speed forward flight. The selection of airfoil sections for helicopter rotors is more difficult than for a fixed-wing aircraft because they are not point designs; that is, the angle of attack and Mach number vary continuously at all blade elements on the rotor and one airfoil section cannot meet all the various aerodynamic requirements. Overall, low pitching moment airfoil designs are essential to maintain low torsional loads on the blades and low control forces.

On early helicopters, little attention was paid to the selection of airfoil section because there were just too many other technical problems to solve. Although the NACA had developed some special helicopter airfoils in the late 1940s, it was not until the middle of the 1960s that airfoil sections specifically tailored to meet the special requirements of helicopters became more widely used by manufacturers. Since then, the major helicopter manufacturers and research organizations have developed various families of improved airfoil profiles for use on helicopter rotors. Each airfoil profile within the family will have specific aerodynamic and geometric attributes optimized for different radial positions on the blade. The construction of a blade with multiple airfoil sections along its length is made easier today because of composite materials technology, which makes the design and fabrication costs comparable to one with a single airfoil.

Historically, airfoils are obtained through an evolutionary process, where both theory and experiment go hand-in-hand to meet specific operating requirements. The tools to help design airfoils that have specific aerodynamic characteristics have been available since the 1920s. The development of the thin-airfoil theory by Munk (1922, 1924) led to an understanding of how camber affected the chordwise pressure loading. This allowed the effects of camber to be isolated from thickness, but the effects combined as required by linear superposition. The problem of defining the airfoil pressure distribution for an airfoil of arbitrary shape was tackled semianalytically by Theodorsen & Garrick (1932). The design of practical airfoil shapes was further aided by methods representing airfoil thickness, such as the conformal transformation developed by Prandtl & Tiejens (1934). This made it possible to compute pressure distributions and lift characteristics, at least for some specially shaped "Joukowski" airfoils. The aerodynamic properties of Joukowski airfoils were studied in the late 1920s at Göttingen, Germany and by the NACA – see Schrenk (1927) and Von Mises (1945). Abbott et al. (1945) developed a numerical method to predict chordwise pressure distributions and airfoil characteristics based on an extension of thin-airfoil theory, where the increment in loading distribution associated with camber could be combined with loading from thickness and angle of attack.

By the 1960s, "panel methods" coupled with boundary layer displacement corrections were available. Much of the pioneering work with panel methods was done by Hess & Smith (1967) and by Rubbert (1964). Inverse panel methods allowed airfoils to be designed to meet specific requirements. Kennedy & Marsden (1978) were one of the first to develop such methods. Generally, the airfoil designs were optimized for maximum lift. Eppler & Somers (1980) discuss an alternative method for inverse airfoil design using conformal mapping with boundary layer corrections. Hicks & McCroskey (1980) discuss the numerical optimization of a helicopter airfoil, with experimental validation. The advent of numerical methods for transonic airfoil design also meant that for the first time helicopter airfoil shapes could be more carefully designed to meet advancing blade requirements. Sloof et al. (1975) and Narramore & Yen (1982, 1997) discuss transonic airfoil design methods for helicopter rotors.

While most airfoil designs have been conducted for 2-D flows, the complicated flow near the tip of a helicopter blade demands 3-D prediction methods as well. Caradonna and coworkers (1972, 1976, 1978) were major contributors to 2-D and 3-D transonic flow prediction methods for helicopter blades using finite-difference methods. The advent of finite-difference solvers for the Euler and Navier–Stokes equations has led to increasing sophistication in airfoil design tools – see, for example, McCroskey & Baeder (1985), Malone et al. (1989), Bezard (1992), and Narramore (1994). However, the extreme operating conditions and often highly unsteady flow environment found on helicopters means that rotor airfoils must still be tested in a wind tunnel to fully assess their aerodynamic performance, because modern computational tools have not yet matured to a level where turbulent flow separation and stall effects can be predicted with acceptable accuracy.

7.2 Rotor Airfoil Requirements

A representative operating lift coefficient, C_l, versus incident Mach number, M, for a section near the tip of a helicopter rotor in forward flight is illustrated in Fig. 7.1. Superimposed on this figure is the approximate static stall boundary for the NACA 0012 airfoil. The advancing blade operates at low angles of attack but at high subsonic or transonic conditions, whereas the retreating blade operates at low Mach numbers and high lift coefficients. Overall, it will be seen that the airfoil section operates very close to the stall boundary. For this particular case, it is apparent that the rotor limits are likely to be imposed by retreating blade stall. However, the use of another airfoil shape on the rotor may reduce the critical Mach number and make the onset of compressibility effects more important in limiting rotor performance. Therefore, there has been a great deal of emphasis in rotor design on maximizing the lifting capability of rotor airfoil sections to simultaneously alleviate both compressibility effects and retreating blade stall. The normal rotor design point, which is shown in the inset of Fig. 7.1, must also allow sufficient margin from the boundary for perturbations in angle of attack and Mach number associated with maneuvering flight and turbulent air.

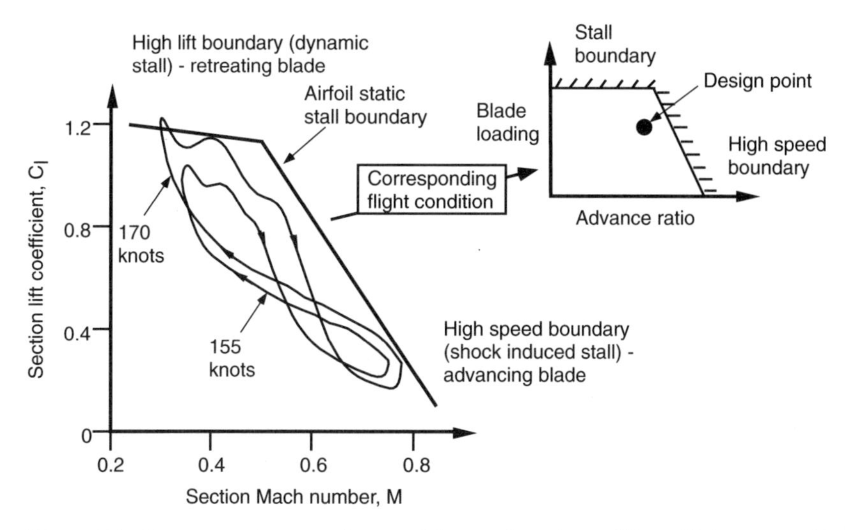

Figure 7.1 Representative operating conditions and maximum lift coefficient versus Mach number boundary for a helicopter rotor section.

The general requirements for a good helicopter rotor airfoil are:

1. A high maximum lift coefficient, $C_{l_{\max}}$. This will allow a rotor with lower solidity and lighter weight or will permit flight at high rotor thrusts and under high maneuver load factors.
2. A high drag divergence Mach number. This will permit flight at high forward speeds without prohibitive power loss or increase in noise levels.
3. A good lift-to-drag ratio over a wide range of Mach number. This will give the rotor a low profile power consumption and a low autorotative rate of descent.
4. A low pitching moment. This will help minimize blade torsion moments, minimize vibrations, and keep control loads to reasonable values.

In airfoil design, it turns out that many of these requirements are conflicting in that they cannot all be simultaneously achieved with the use of a single airfoil shape. However, much can be done to maximize one or more of the airfoil performance attributes without drastically compromising another. To do this, however, requires an understanding of the key (and interrelated) effects of factors such as airfoil shape, angle of attack, Reynolds number, and Mach number.

7.3 Reynolds Number and Mach Number

Two of the most well known parameters used in aerodynamics are the *Reynolds number* and the *Mach number*. For an airfoil, its size is described by a characteristic length based on the chord, c. When the airfoil is moving through a fluid of viscosity μ, density ρ, and sonic velocity a, and with a speed V_∞ at some relative orientation to the flow, then the method of *dimensional analysis* shows that the force on the airfoil, F, can be written in functional form as

$$\frac{F}{\rho V_\infty^2 c^2} = f\left(\frac{\rho V_\infty c}{\mu}, \frac{V_\infty}{a}\right) = f(Re, M). \tag{7.1}$$

The combinations $\rho V_\infty c/\mu$ and V_∞/a are called the Reynolds number (denoted by Re) and the Mach number (denoted by M), repectively. These parameters have both independent and interdependent influences, which complicates the understanding of the problem of finding their effects on the aerodynamic force.

7.3.1 *Reynolds Number*

The physical significance of the Reynolds number is that it represents the ratio of the inertial forces to the viscous forces in the fluid. This can be seen by writing the Reynolds number as

$$Re = \frac{\rho V_\infty c}{\mu} = \frac{\rho V_\infty c(V_\infty c)}{\mu(V_\infty c)} = \frac{\rho V_\infty^2 c^2}{\mu(V_\infty/c)c^2}. \tag{7.2}$$

On the numerator, $\rho V_\infty^2 c^2$ represents an inertial force. The coefficient of viscosity, μ, is the shear force per unit area per unit velocity gradient. The denominator, therefore, is a viscous force. For an ideal fluid, the Reynolds number is effectively infinite. However, when viscous forces are dominant, the Reynolds number is small in value.

Rotor airfoil design can only be meaningful if assessed at the operational Reynolds numbers and Mach numbers actually found on helicopter rotors. Figure 7.2 illustrates the working Reynolds number and Mach number ranges typical of helicopter rotors, both at full scale and model scale. The maximum lift coefficient, $C_{l_{\max}}$, can be used as one indicator of the significance of viscous effects. At the low end of the practical Reynolds number range for rotors (i.e., for chord Reynolds number in the range $10^5 < Re < 10^6$), most airfoils have relatively low values of $C_{l_{\max}}$. This is because the viscous forces are more dominant and the flow will separate more readily from the airfoil surface, all other factors being equal. In the range $Re = 1 \times 10^6$ to 3×10^6 the greatest changes in $C_{l_{\max}}$ generally occur, and for $Re > 4 \times 10^6$ any changes in $C_{l_{\max}}$ are found to be relatively gradual. Clearly, subscale rotor models, tail rotors, and possibly even some small main rotors can fall into the range where the airfoil characteristics can be sensitive to Reynolds number. In particular, for subscale

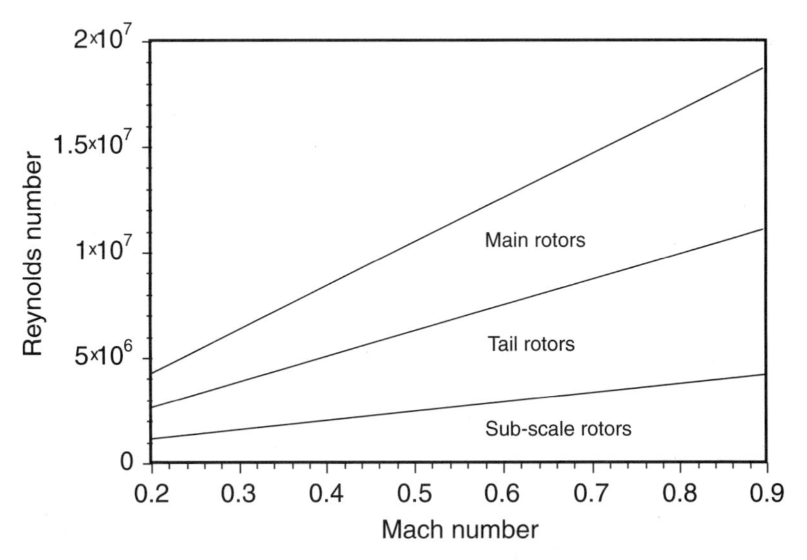

Figure 7.2 Typical ranges of Reynolds number and Mach number found on helicopter rotors.

rotors the chord Reynolds numbers on the retreating blade are typically below 10^6 and, therefore, are more susceptible to viscous effects. This is important when assessing the performance characteristics obtained from subscale rotor tests and extrapolating the results to full scale.

7.3.2 *Concept of the Boundary Layer*

At low angles of attack, the effects of viscosity are confined to a thin region near the surface of the airfoil known as the *boundary layer*. Historically, the concept of the boundary layer was first proposed by Prandtl (1928). Real fluids do not slip at a solid boundary and there will be no relative motion between the fluid and the surface of the airfoil. Therefore, there is a region close to the airfoil where the velocity rises from zero to the external flow velocity, u_e. This boundary layer region is very thin, being a small fraction of the airfoil chord. The concept is illustrated in Fig. 7.3. On an airfoil the boundary layer will vary from practically zero thickness near the leading edge to a few percent of the chord at the trailing edge.

Boundary layers are found to be of two main types: *laminar* or *turbulent*. A third type can be considered to be *transitional*. A comparison of the profile shapes of a laminar and turbulent boundary layer is also shown in Fig. 7.3. The parameter δ is the boundary layer thickness, which is defined as the value of y for which 99% of the external flow velocity is recovered (i.e., $u = 0.99u_e$). The flow in a laminar boundary layer is smooth and free of any mixing of fluid between successive layers, whereas the flow in a turbulent boundary layer is characterized by significant mixing between layers of the fluid. This produces a momentum transfer through the boundary layer, and so the distribution of velocity in a turbulent boundary layer is characterized by larger velocities closer to the airfoil surface. Also, for the same reasons, a turbulent boundary layer has a greater thickness compared with a laminar boundary layer that develops under the same pressure gradient. See also Schlichting (1979) and Young (1989).

Viscous stresses are produced whenever there is relative motion between adjacent fluid elements, and these stresses produce a resistance that tends to retard the motion of the fluid.

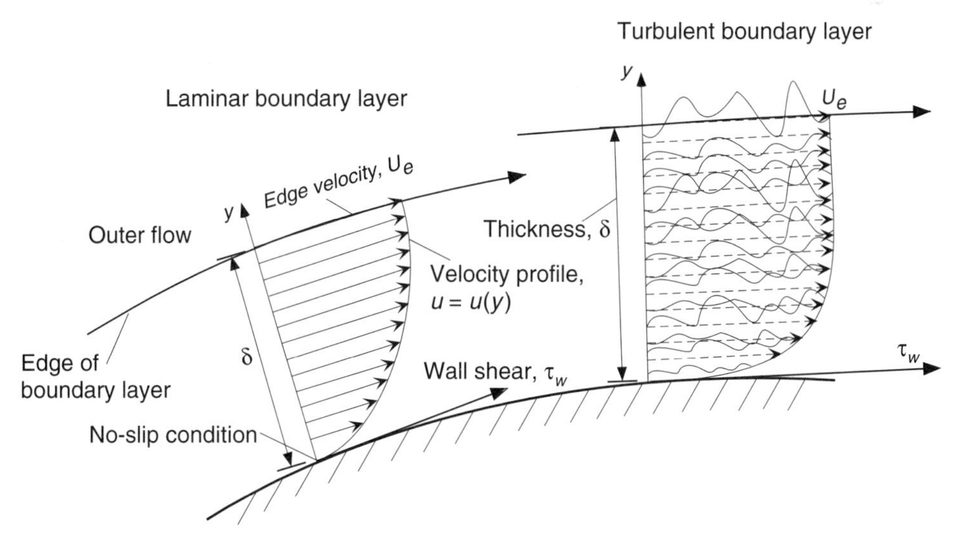

Figure 7.3 Development of a boundary layer on a solid surface.

The viscous shear stress, τ, is related to the absolute viscosity, μ, by

$$\tau = \mu \frac{\partial u}{\partial y}, \tag{7.3}$$

where $\partial u/\partial y$ is the rate at which the velocity increases perpendicular to the surface. (Note that the partial derivative is used because u can vary not only with y but also in other directions.) Thus, as implied by Fig. 7.3, the shear stress, τ_w, produced on the surface of an airfoil will be greater with a turbulent boundary layer than for a laminar one because

$$\tau_w = \mu \left(\frac{\partial u}{\partial y} \right)_{y=0}. \tag{7.4}$$

The development of a laminar boundary layer in a zero pressure gradient flow can be computed exactly. This result was first obtained by Blasius with an improved solution by Kuo (1953). In Blasius's solution the local skin friction coefficient, c_f, on one side of a flat plate is

$$c_f = \frac{\tau_w}{\frac{1}{2}\rho V_\infty^2} = 0.664\, Re_x^{-0.5}, \tag{7.5}$$

where Re_x is the Reynolds number based on the distance from the leading edge of the plate. The net shear stress drag coefficient of the plate will be

$$C_d = \frac{2}{c} \int_0^c c_f \, dx = 1.328\, Re_x^{-0.5}, \tag{7.6}$$

where c is the chord of the plate. This result is plotted in Fig. 7.4. At Reynolds numbers above about 5×10^5, the boundary layer may become turbulent. The reason for this is that at Reynolds numbers above a certain minimum value, natural flow disturbances (although often caused prematurely by surface roughness) can cause a *transition* from a laminar to a turbulent boundary layer. Helicopter airfoils will typically exhibit laminar flow only over a

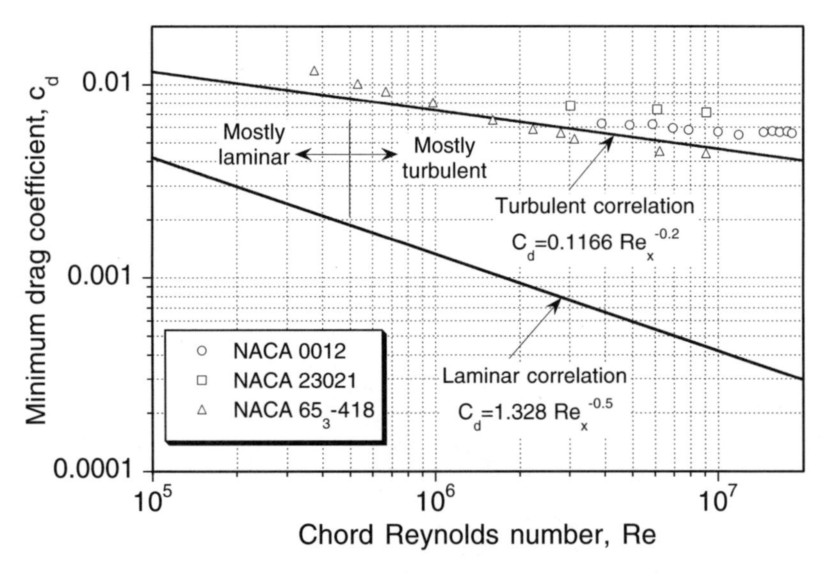

Figure 7.4 Viscous drag of a flat plate compared to minimum drag coefficients of several NACA airfoils. Data source: Abbott & von Doenhoff (1949).

few percent of the chord. For the turbulent boundary layer development on a flat plate, the skin friction coefficient c_f on one side of the plate is found to be close to

$$c_f = 0.0583\, Re_x^{-0.2}, \tag{7.7}$$

so that

$$C_d = \frac{2}{c}\int_0^c c_f\, dx = 0.1166\, Re_x^{-0.2}. \tag{7.8}$$

The validity of the latter expression is limited to a Re_x range between 10^5 and 10^9. Below $Re_x = 5 \times 10^5$ the boundary layer can normally be assumed to be laminar. The drag of a flat plate with a fully turbulent boundary layer is also plotted in Fig. 7.4 and is compared to the minimum drag coefficients of several NACA airfoils. The results suggest that the turbulent flat-plate solution is a good approximation to the viscous (shear) drag on airfoils over the practical range of Reynolds numbers to be found on helicopters.

Note also that a turbulent boundary layer will be thicker than a laminar one. If δ is defined as the value where $u = 0.99u_e$ then for a laminar boundary layer

$$\delta \simeq 5.2x\, Re^{-0.5} \tag{7.9}$$

and for a turbulent boundary layer

$$\delta \simeq 0.37x\, Re^{-0.2}. \tag{7.10}$$

Therefore, in addition to a higher viscous shear on the airfoil surface the presence of a turbulent boundary layer will result in a higher overall profile drag compared to an airfoil with a fully laminar boundary layer.

It is found that the developing boundary layer on an airfoil is sensitive to the pressure gradient. In the boundary layer, a simplified form of the Navier–Stokes equations applies. The x-component of the momentum equation becomes

$$u\frac{\partial u}{\partial x} + v\frac{\partial u}{\partial y} = -\frac{1}{\rho}\frac{\partial p}{\partial x} + \nu\frac{\partial^2 u}{\partial y^2}. \tag{7.11}$$

The pressure gradient is denoted by $\partial p/\partial x$ along the surface. When the gradient is positive or adverse, that is, one in which $\partial p/\partial x > 0$, the pressure force is in the direction that tries to decelerate the flow. The resulting force on the fluid is particularly strong near the surface of the airfoil where the velocity is low; therefore $\partial u/\partial y$ near $y = 0$ becomes smaller the longer the adverse pressure gradient persists. This effect is shown in Fig. 7.5. At some distance downstream a point is reached where

$$\left(\frac{\partial u}{\partial y}\right)_{y=0} = 0 \tag{7.12}$$

and the direction of the flow reverses near the surface. This point is called the *separation point*, because the flow breaks away and leaves the surface. Under these conditions, the concept of the boundary layer breaks down and a recirculating flow (wake) is left downstream of the separation point; downstream of separation the effects of viscosity influence an extensive region of the flow. Flow separation is one of the least understood phenomena in fluid mechanics, even under steady 2-D conditions. In unsteady (external) flows, the onset of separation is more complicated and there can be significant regions of flow reversal within the boundary layer even in the absence of separation – see McCroskey (1977).

Turbulent boundary layers are much less susceptible to separation than laminar boundary layers because of the higher mixing and interlayer momentum transfer of the fluid. At higher

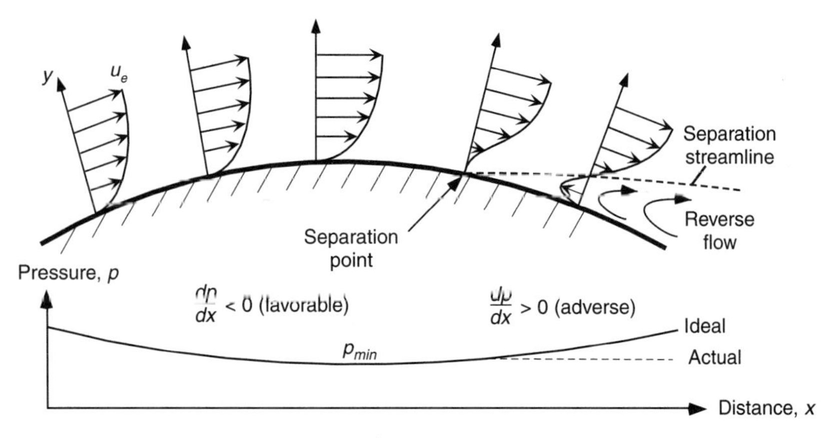

Figure 7.5 Development of the boundary layer in a simple external pressure gradient.

Reynolds numbers, the pressure rise required to separate a turbulent boundary layer may be an order of magnitude larger than that required to separate a laminar boundary layer. On rotor airfoils, which tend to have relatively sharp leading edges and peak suction pressures close to the leading edge, a steep adverse pressure gradient is found over most of the chord, and a laminar boundary layer can only exist for a very short distance from the stagnation point (typically, 2–15% chord). After laminar separation occurs, the flow will temporarily leave the airfoil surface but undergo a transition process and immediately reattach again as a turbulent boundary layer. This can leave a small region of recirculating separated flow that is known as a *laminar separation bubble*, as shown schematically in Fig. 7.6. The presence of laminar separation bubbles can be observed by a small constant pressure region in the measurement at the chordwise pressure distribution near the leading edge of the airfoil. With surface oil flow visualization, the bubble is evidenced by an accumulation of oil. The formation of laminar separation bubbles is common for the types of airfoils and Reynolds numbers found on helicopter rotors. Further details of the laminar separation bubble phenomenon are given by Horton (1967) and Liebeck (1992).

At low angles of attack, the turbulent boundary layer will generally extend all the way to the airfoil trailing edge. At higher angles of attack, the increasing intensity of the adverse pressure gradients will ultimately cause the turbulent boundary layer to begin to separate – see Thwaites (1960). Many rotor airfoils stall by the process of progressive turbulent trailing-edge separation, whereby the separation point starts at the trailing edge and moves forward on the chord with increasing angle of attack. This is shown schematically in

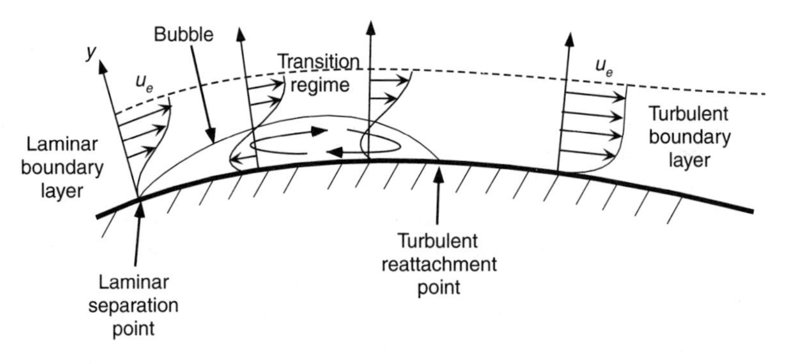

Figure 7.6 Schematic showing the structure of a laminar separation bubble.

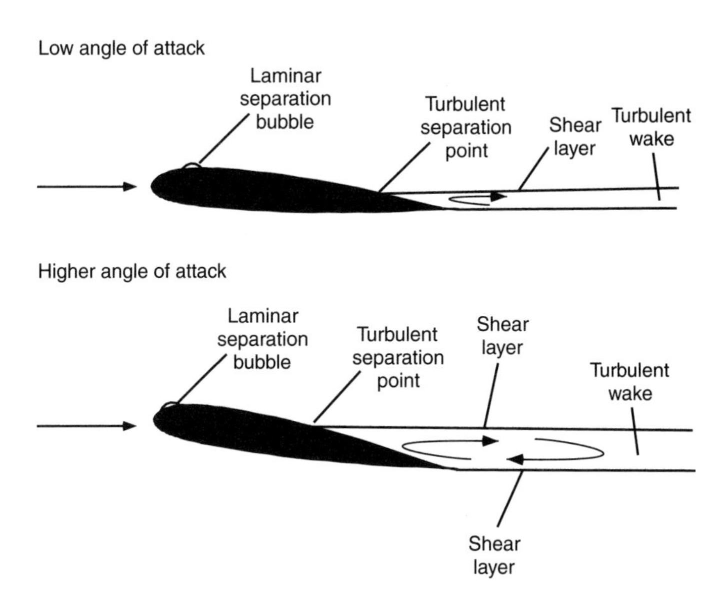

Figure 7.7 Development of trailing-edge flow separation on an airfoil.

Fig. 7.7. Progressive turbulent trailing-edge separation has a deleterious effect on the airfoil performance compared to that obtained with fully attached flow, as evident when the forces and moments are plotted versus angle of attack – see Section 7.8.

7.3.3 *Mach Number*

The ratio of the speed of the flow to that of the speed of sound is called the *Mach number* and can be interpreted as a ratio of inertia forces in the fluid to forces resulting from compressibility. The speed of sound is the speed at which pressure disturbances are propagated through the air. The speed of sound is given by

$$a = \sqrt{\gamma R T}, \tag{7.13}$$

where $\gamma = 1.4$ is the ratio of specific heats for air, R is the gas constant, and T is absolute temperature. The speed of sound is approximately 1,116.45 ft s^{-1} or 340 ms^{-1} at sea level on a standard day. For an incompressible flow, $a = \infty$ so that $M = 0$.

When the air flows past an airfoil, the local velocity at the surface outside the boundary layer may be greater or less than the free-stream velocity. The highest velocities occur near the leading edge and over the upper surface of the airfoil. If the free-stream velocity is low enough, the flow velocity remains *subsonic* everywhere. Such flows are relatively easy to analyze because the governing equations are linear. At higher free-stream velocities, however, the acceleration of the flow will eventually cause regions of supersonic flow. This is a mixed or *transonic* flow, which although predominantly subsonic, has an embedded supersonic pocket, as shown by Fig. 7.8. This problem is much more difficult to analyze because of the nonlinear nature of the governing flow equations.

The value of the free-stream Mach number, M_∞, where the flow first becomes locally sonic ($M = 1$) is called the *critical Mach number*, M_c. The value of M_c will also depend on the angle of attack of the airfoil. Further increases in M_∞ cause the extent of the supersonic region to grow, and the region becomes terminated by a *shock wave* that is initially almost perpendicular to the airfoil surface. Across the shock wave there is a rapid increase in

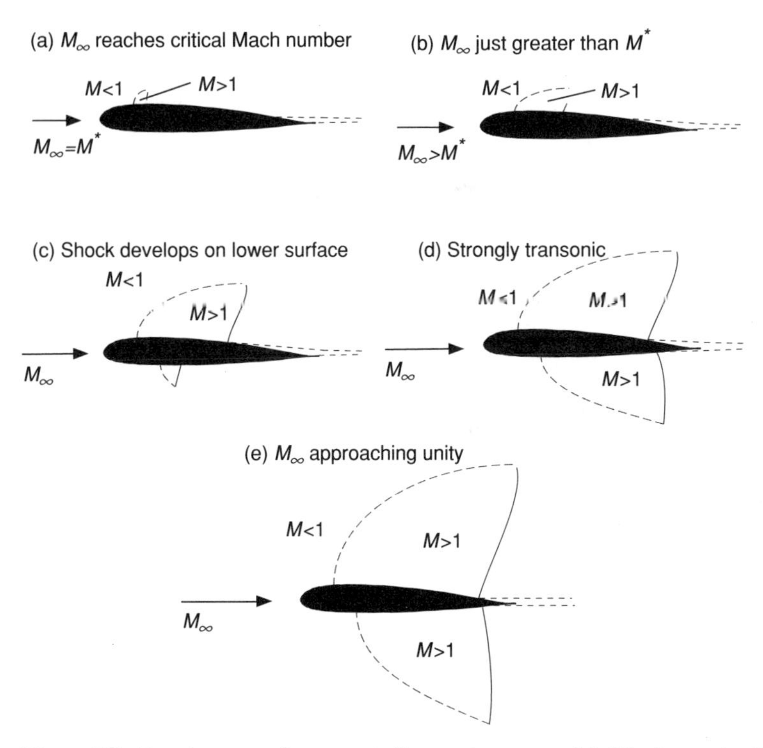

Figure 7.8 Development of supersonic flow pockets on an airfoil for increasing free-stream Mach number. (a) M_∞ = critical Mach number, M^*. (b) $M_\infty > M^*$ with developing shock wave. (c) Shock wave develops on lower surface. (d) Lower surface shock wave moves quickly toward trailing edge. (e) M_∞ approaching unity.

pressure and also an entropy change whereby energy is converted to heat, resulting in a form of drag known as *wave drag*. The interaction of the shock wave and the boundary layer results in an increase in skin friction drag. Moreover, the high adverse pressure gradients found in the vicinity of the shock wave make boundary layer thickening and an increase in pressure drag inevitable.

With increasing Mach number (or angle of attack) the shock wave will strengthen, move aft, and become more oblique to the airfoil surface. At some point, depending on the angle of attack and airfoil shape, the flow on the lower surface also becomes supersonic, resulting in a second supersonic pocket and terminating shock wave. With further increases in M_∞, the lower surface shock wave moves quickly toward the trailing edge; this is followed by the rapid aft movement of the upper surface shock wave. If at any point during this process the shock wave becomes sufficiently strong, then the high adverse pressure gradients will cause the boundary layer to separate, causing a loss of lift and an increase in drag known as *shock induced stall*. See Liepman (1946) and Pearcy (1955) for good discussions on shock wave/boundary layer interactions.

In Fig. 7.1, it has already been shown that rotor airfoils operate in a relatively diverse aerodynamic regime during each revolution of the rotor, ranging from low subsonic speeds at the root of the blade to transonic flow conditions at the tips. Compressibility effects can manifest before sonic speed is reached locally on the airfoil. Even when the free-stream Mach number is quite low, say about 0.3 on the retreating blade, the high angles of attack found under dynamic (unsteady) conditions mean that a supercritical flow region can exist near the leading edge of the airfoil. Consequently, compressibility issues need to be addressed very carefully to ensure that the selected airfoil sections will perform satisfactorily under the conditions found on a helicopter rotor.

7.4 Airfoil Shape Definition

The importance of the airfoil shape on the rotor behavior was well known to Juan de la Cierva – see de la Cierva & Rose (1931). De la Cierva originally used a symmetric airfoil section on his autogiros, but he later changed the airfoil to a highly cambered 17% thick section. While having a higher stalling angle of attack, this airfoil also had a higher pitching moment, which resulted in blade twisting and control problems and finally led to a crash of a C-30 – see Beavan & Lock (1939). This event, and the generally low torsional stiffness of early wood and fabric helicopter blades, led to the almost universal use of symmetric airfoil sections on helicopters produced prior to 1960. Although symmetric airfoils offered a reasonable overall compromise in terms of maximum lift coefficient, low pitching moments, and high drag divergence Mach numbers, they were by no means optimal for attaining maximum performance from the rotor. However, it was not until the early 1960s that a serious effort came about to improve airfoil sections for use on helicopters.

As early as 1920, a number of research institutions had begun to examine the characteristics of various airfoils and organize the results into families of airfoils, basically in an effort to determine the profile shapes that were best suited for specific applications. In the United States, NACA conducted a comprehensive and systematic study of the effect of airfoil shape on aerodynamic characteristics. Existing cambered airfoils such as the Clark-Y and Göttingen 398 sections were known from early experiments to have good aerodynamic characteristics. These airfoils were used by NACA as a basis and were found to have geometrically similar shapes when the camber was removed and the airfoils were reduced to the same thickness-to-chord ratio. NACA then followed a procedure where a given airfoil could be constructed of a *thickness shape* that was distributed around a *camber line*. This allowed the systematic construction of several families of airfoil sections. The use of linearized methods also enabled the chordwise aerodynamic loading associated with camber and thickness to be studied, allowing means of designing airfoils to meet specific purposes. The various families of airfoils developed by NACA were then tested in the wind tunnel to document the effects of varying the important geometrical parameters on the airfoil lift, drag, and pitching moment characteristics as a function of angle of attack, Reynolds number, and Mach number. Variables found to have important effects on the airfoil characteristics included the maximum camber and its distance aft of the leading edge and the leading-edge nose curvature (nose radius) of the airfoil. A summary of the results are documented in considerable detail by Abbott et al. (1945) and Abbott & von Doenhoff (1949).

Two of the most popular airfoils used on many early helicopters were the symmetric NACA 0012 and NACA 0015 sections. These airfoils were found to have low pitching moments about the 1/4-chord and good low speed as well as high speed (transonic) performance, giving a relatively high maximum lift and a relatively high drag divergence Mach number. Because these airfoils were also relatively thick, the stiffness of the blade could be maintained while keeping blade weight to a minimum. The upper and lower surfaces of the NACA four-digit symmetrical sections (or thickness envelopes) are described by the polynomial

$$\pm\frac{y_t}{c} = \bar{y}_t = 5\bar{t}[0.29690\sqrt{\bar{x}} - 0.12600\bar{x} - 0.35160\bar{x}^2 + 0.28430\bar{x}^3 - 0.10150\bar{x}^4], \tag{7.14}$$

where $t/c = \bar{t}$ = maximum thickness as a fraction of chord. For example, for the NACA

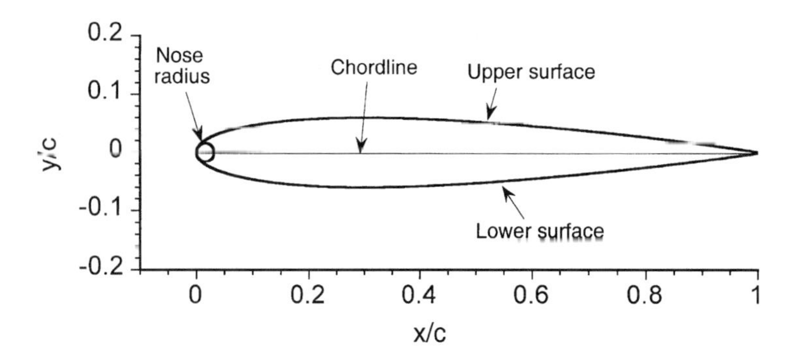

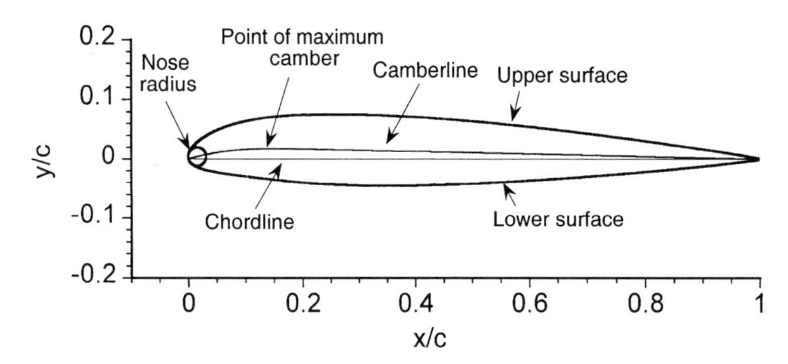

Figure 7.9 NACA airfoil construction, with inscribed nose radius. (a) NACA four-digit 0012 airfoil. (b) NACA five-digit 23012 airfoil section.

0012 airfoil $\bar{t} = 0.12$. The corresponding leading-edge radius of the airfoil is $r_t = 1.1019\bar{t}^2$. The center for this leading-edge radius is found by drawing a straight line through the end of the chord at the origin of the axes and laying off a distance along the x axis that is equal to the leading-edge radius. The nose radius is then inscribed and faired onto the upper and lower surfaces of the airfoil. The resulting shape is shown in Fig. 7.9(a). The analytic form of the construction lends itself easily to computer generation of the airfoil coordinates – see Ladson et al. (1995).

A similar approach is used for cambered airfoils, but the mean line (camber line) is now used for laying out the airfoil shape. The camberline can be specified as $y_c = y_c(x_c)$. If the slope of the camberline makes an angle θ with the chord line, then the airfoil shape is obtained by plotting the thickness distribution at right angles to the slope of the camberline. This will give the upper (x_u, y_u) and lower coordinates (x_l, y_l) of the airfoil:

$$\bar{x}_u = \bar{x} - \bar{y}_t \sin\theta \quad \text{and} \quad \bar{y}_u = \bar{y}_c + \bar{y}_t \cos\theta,$$
$$\bar{x}_l = \bar{x} + \bar{y}_t \sin\theta \quad \text{and} \quad \bar{y}_l = \bar{y}_c - \bar{y}_t \cos\theta,$$

where $\theta = \tan^{-1} dy_c/dx_c$. In effect, the shape is being defined as a tangent to a series of circles of radius y_t with centers on the camberline. For cambered airfoils, the center for the leading-edge radius is found by drawing a straight line through the end of the chord at the origin of the axes but with slope equal to the slope of the camberline at $\bar{x} = 0.005$, and

Table 7.1. *Numerical Values for Three-Digit Camberlines Used in the NACA Five-Digit Airfoil Construction. Source: Abbott & von Doenhoff (1949)*

Mean Line	p	m	k_1
210	0.05	0.0580	361.4
220	0.10	0.1260	51.64
230	0.15	0.2025	15.957
240	0.20	0.2900	6.643
250	0.25	0.3910	3.230

then laying off a distance that is equal to the leading-edge radius. Because of the form of the geometric construction of cambered airfoils, the leading-edge part of the nose radius protrudes very slightly forward of the origin at $x = 0$. As with the symmetric airfoils, the nose radius is then inscribed and faired onto the upper and lower surfaces.

Various series of mean lines were developed by the NACA, and some of the resulting sections (and derivatives) can be found on modern helicopter blades. Many mean lines, such as the three-digit mean lines, are defined by two equations derived to produce shapes having a progressively decreasing camber line slope from the leading edge to the trailing edge. The camberline slope is zero at the point of maximum camber (denoted by p) and is constant aft of point m over the trailing edge. The equations for these mean lines are

$$\bar{y_c} = \frac{1}{6}k_1[\bar{x}^3 - 3m\bar{x}^2 + m^2(3-m)\bar{x}] \quad \text{from } \bar{x} = 0 \text{ to } \bar{x} = m, \tag{7.15}$$

$$\bar{y_c} = \frac{1}{6}k_1 m^3 (1 - \bar{x}) \quad \text{from } \bar{x} = m \text{ to } \bar{x} = 1. \tag{7.16}$$

Values of p were selected to give five positions of maximum camber, and values of k_1 were selected to give a design lift coefficient of 0.3. The various values of p, m, and k_1 for the three-digit camberlines are given in Table 7.1. Figure 7.9(b) shows the graphical construction of the NACA 23012 section, which has been used as a baseline for many modern helicopter airfoil sections. It is derived from the 230 mean line plus the 0012 thickness distribution, and it has a maximum camber at 15% chord and a thickness-to-chord ratio of 12%.

Modifications to the NACA four-digit and five-digit series of airfoil sections include reflex to produce zero pitching moment [see Jacobs & Pinkerton (1935)] and changes in the nose radius and position of thickness [see Stack & von Doenhoff (1934)]. The latter sections have seen some use on helicopter rotors and are denoted by a two-digit suffix, such as the NACA 0012-64 and NACA 23012-64. The first integer after the dash indicates the relative magnitude of the nose radius – with a standard nose radius denoted by 6 and a sharp radius by 0. The second digit indicates the position of maximum thickness in tenths of chord.

An early series of special helicopter sections was designed at NACA by Tetervin (1943) and Stivers & Rice (1946). These airfoils, which have NACA 3-H-00 through NACA 10-H-00 series designators, were designed to have lower overall drag over a useful range of lift coefficients but still retain relatively low pitching moments. These airfoils are also discussed by Gessow & Myers (1952). Another set of NACA airfoils that has seen some use in helicopter applications is the six-digit series. These airfoils were designed to achieve lower drag, higher drag divergence Mach numbers, and higher values of maximum lift. The shapes of these airfoils is such that they are conducive to maintaining an extensive run of laminar flow over the leading-edge region and, thereby, lower skin friction drag, at least over a range of angle of attack that is limited to low lift coefficients. This is achieved by using camberlines that have a uniform loading from the leading edge to a distance $\bar{x} = a$, and thereafter the loading decreases linearly to zero at the trailing edge. The favorable pressure gradients tend to give the airfoils lower drag compared to other airfoils, but these characteristics are easily spoiled by surface contaminants or other transition-causing disturbances. There are many designator combinations used in the six-digit number system. For example, consider the NACA 64_3-215 $a = 0.5$ section. In this case, the number 6 denotes the airfoil series, and the 4 denotes the position of minimum pressure in tenths of chord for the basic symmetric section. The 3 denotes the range of lift coefficient in tenths above and below the design lift coefficient for which low drag may be obtained. The 2 after the dash indicates a design lift coefficient of 0.2, and 15 denotes a 15% thickness-to-chord ratio. These and many other airfoil designators are explained by Abbott & von Doenhoff (1949).

7.5 Airfoil Pressure Distributions

7.5.1 *Pressure Coefficient*

In measurements or calculations of the flow about airfoils, the surface pressure data are conventionally presented in terms of the *pressure coefficient* C_p. In incompressible flow, the definition of C_p follows from *Bernoulli's equation*,

$$p_\infty + \frac{1}{2}\rho_\infty V_\infty^2 = p + \frac{1}{2}\rho_\infty V^2, \tag{7.17}$$

where p and V are the local pressure and velocity, and p_∞ and V_∞ are the corresponding free-stream conditions at infinity (i.e., far upstream of the airfoil). Using the standard abbreviation for the dynamic pressure, $q_\infty = \frac{1}{2}\rho_\infty V_\infty^2$, we can define the pressure coefficient as

$$C_p = \frac{p - p_\infty}{q_\infty} = 1 - \left(\frac{V}{V_\infty}\right)^2. \tag{7.18}$$

For an incompressible flow, $C_p = 1$ at a stagnation point (i.e., where $V = 0$). Also, $C_p = 0$ far from the body where $V = V_\infty$.

For compressible flow, the pressure coefficient is defined in terms of the Mach number. The dynamic pressure can be expressed in terms of M_∞ as follows:

$$q_\infty = \frac{1}{2}\rho_\infty V_\infty^2 = \frac{1}{2}\frac{\gamma p_\infty}{\gamma p_\infty}\rho_\infty V_\infty^2 = \frac{\gamma}{2}p_\infty\left(\frac{\rho_\infty}{\gamma p_\infty}\right)V_\infty^2. \tag{7.19}$$

Also, because $a_\infty^2 = \gamma p_\infty/\rho_\infty$, the above equation becomes

$$q_\infty = \frac{\gamma}{2}p_\infty\frac{V_\infty^2}{a_\infty^2} = \frac{\gamma}{2}p_\infty M_\infty^2 \tag{7.20}$$

and the pressure coefficient can be written as

$$C_p = \frac{2}{\gamma M_\infty^2}\left(\frac{p}{p_\infty} - 1\right), \tag{7.21}$$

which is simply an alternative form of the pressure coefficient for compressible flow. Most airfoil pressure distributions are plotted using this definition of C_p. With this definition, we can define the critical condition when the minimum pressure reaches the sonic pressure p^*

and $M_\infty = M^*$. The critical pressure coefficient is

$$C_p^* = \frac{2}{\gamma (M^*)^2}\left(\frac{p^*}{p_\infty} - 1\right). \tag{7.22}$$

Using the energy equation applied to isentropic flow along a streamline gives

$$\frac{V_\infty^2}{2} + \frac{a_\infty^2}{\gamma - 1} = \frac{V^2}{2} + \frac{a^2}{\gamma - 1}, \tag{7.23}$$

and for the condition when $V = a = a^*$ then

$$\frac{(M^*)^2}{2} + \frac{1}{\gamma - 1} = \left(\frac{a^*}{a_\infty}\right)^2 \left(\frac{1}{2} + \frac{1}{\gamma - 1}\right) \tag{7.24}$$

or

$$\left(\frac{a^*}{a_\infty}\right)^2 = (M^*)^2 \left(\frac{\gamma - 1}{\gamma + 1}\right) + \frac{2}{\gamma + 1}. \tag{7.25}$$

But since

$$\frac{p^*}{p_\infty} = \left(\frac{a^*}{a_\infty}\right)^{2\gamma/(\gamma - 1)} \tag{7.26}$$

substituting into Eq. 7.22 gives

$$C_p^* = \frac{2}{\gamma (M^*)^2}\left[\left(\frac{\gamma - 1}{\gamma + 1}(M^*)^2 + \frac{2}{\gamma + 1}\right)^{\gamma/(\gamma - 1)} - 1\right]. \tag{7.27}$$

Note that C_p^* is the pressure coefficient at the point on the airfoil when sonic conditions are first achieved, i.e., $M_\infty = M^*$. This point is generally not known a priori and is usually predicted on the basis of the minimum pressure coefficient found from incompressible flow.

7.5.2 *Synthesis of Chordwise Pressure*

The ideas of synthesizing the distribution of chordwise pressure on the basis of contributions from camber and thickness envelopes are discussed by Abbott & von Doenhoff (1949). The velocity distribution about the airfoil, v/V_∞, can be considered to be composed of three separate and independent contributions that are combined by linear superposition. These components of loading are: 1. A distribution of velocity v_t/V_∞ corresponding to that of the basic thickness form at zero angle of attack. 2. A distribution of velocity v_c/V_∞ corresponding to the camberline when operating at its "ideal" angle of attack. The ideal angle of attack corresponds to $-\alpha_0$, where α_0 is the angle of attack for zero-lift. 3. A distribution of chordwise loading – the "additional" loading v_a/V_∞. This loading closely corresponds to the thin-airfoil result for a flat-plate at angle of attack, although Abbott & von Doenhoff give tabulated values of v_a/V_∞ for airfoils with finite thickness as obtained from other theoretical means.

The local loading is equal to the difference in velocity between the upper and lower surfaces of the airfoil. In accordance with the principles of vortex sheets and thin airfoil concepts, the velocity increment on one side of the airfoil surface is equal to the velocity decrement on the other surface. Therefore, the pressure coefficient can be obtained from

$$C_p = 1 - \left(\frac{v_t}{V_\infty} \pm \frac{v_c}{V_\infty} \pm \frac{v_a}{V_\infty}\right)^2. \tag{7.28}$$

Values for v_t and v_c can be read directly from the tables in Abbott & von Doenhoff (1949). Values of v_c will be tabulated for a specified design lift coefficient. The additional loading is a function of airfoil thickness, and is usually tabulated for a lift coefficient of 1.0. Because the results will usually be required for lift coefficients other than 1.0, they must be scaled by multiplying the values of the additional loading v_a by $f(\alpha)$ where

$$f(\alpha) = \frac{C_l - C_{l_i}}{C_{l_0}}, \tag{7.29}$$

and where C_{l_i} is the ideal lift coefficient, C_l is the desired lift coefficient, and C_{l_0} is the lift coefficient for which the values of v_a were tabulated.

The perturbation velocity distributions for a large number of camberlines and thickness envelopes are tabulated by Abbott & von Doenhoff (1949), and the method of superposition offers a simple and convenient way of constructing estimates of chordwise pressure for any derived airfoil shape. More importantly, however, the technique provides key elements in the understanding of how airfoil shape affects the aerodynamic characteristics. An example of the technique is shown in Fig. 7.10, where the velocity distribution across the chord of a NACA 0012 airfoil is plotted in terms of the constituent parts. We see that the technique gives good agreement with the measurements. However, the technique is obviously restricted to low (subsonic) speeds and low angles of attack where the assumption of linearity can be justified. (See also Question 7.4.)

7.5.3 *Measurements of Chordwise Pressure*

Representative measurements of chordwise pressure distributions about an airfoil at various angles of attack below stall in a subsonic flow are shown in Fig. 7.11. Although there is a stagnation region over the lower leading-edge region where C_p is positive, over most of the airfoil C_p is negative. Note that as the angle of attack is increased from zero, the pressure reduction on the upper surface increases both in intensity and extent. The high adverse pressure gradients downstream of the leading edge make the boundary layer thicker,

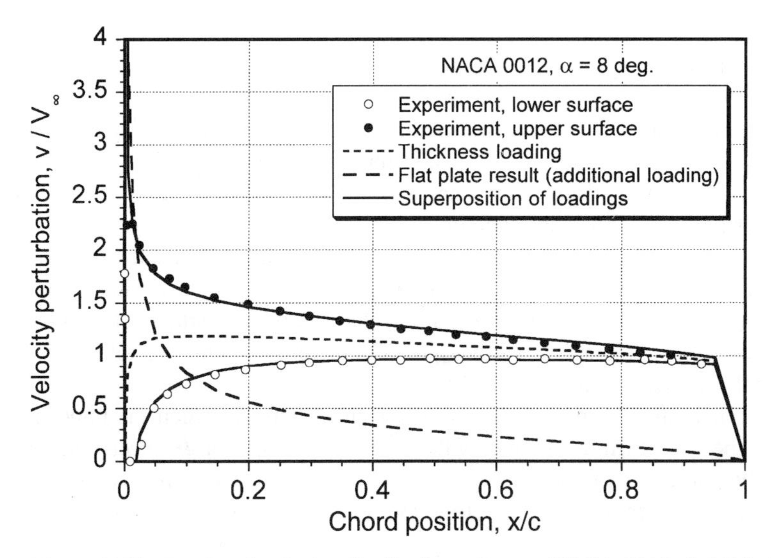

Figure 7.10 Predicted velocity distributions about a NACA 0012 airfoil in low speed (subsonic) flow, $M_\infty = 0.38$, using method of linear superposition. Data source: Bingham & Noonan (1982).

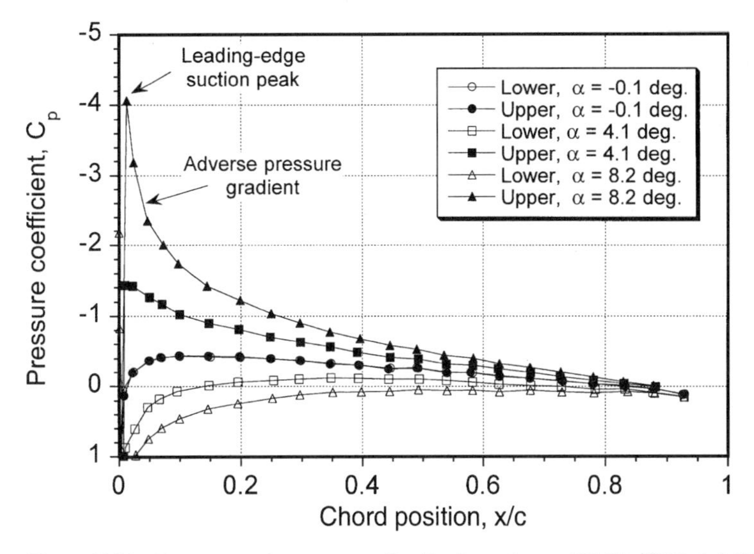

Figure 7.11 Representative pressure distributions about a NACA 0012 airfoil in low speed (subsonic) flow, $M_\infty = 0.38$. Data source: Bingham & Noonan (1982).

and ultimately when the adverse pressure gradients become too large, the boundary layer will separate from the surface causing stall.

With the formation of regions of supercritical flow, such as found toward the tip of the rotor blade, the airfoil pressure distribution changes considerably. Measurements of the pressure distribution about an airfoil in a developing transonic flow are plotted in Fig. 7.12 for a condition where the flow is just subcritical ($M_\infty = 0.68$), and also for a supercritical condition ($M_\infty = 0.77$). The large adverse pressure gradients found near the leading edge in the subsonic case now move further aft to near the mid-chord of the airfoil, which is a result of the formation of a supersonic flow region and a shock wave. The adverse pressure

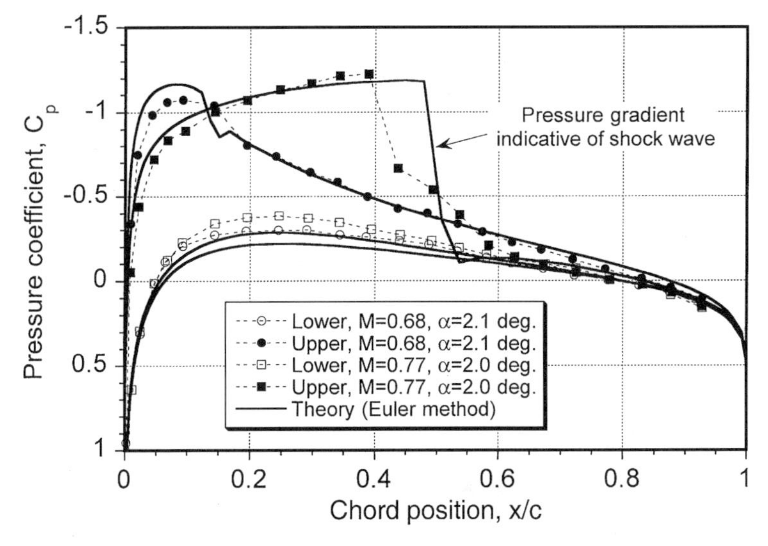

Figure 7.12 Representative chordwise pressure distribution about an airfoil in a developing transonic flow. NACA 0012, $\alpha \approx 2^\circ$, with $M_\infty = 0.68$ and $M_\infty = 0.77$. Data source: Bingham & Noonan (1982).

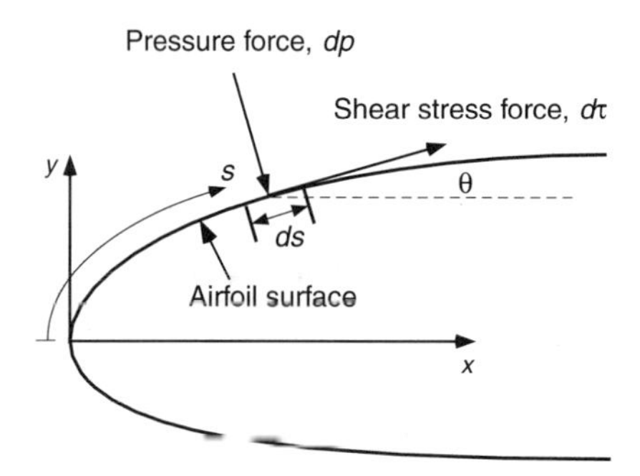

Figure 7.13 Pressure and shear forces acting on an element of the airfoil surface.

gradients near the shock wave make the boundary layer on the surface more susceptible to thickening and separation. When the shock wave becomes stronger, flow separation will occur at the foot of the shock and the airfoil will ultimately stall.

7.6 Aerodynamics of a Typical Airfoil Section

The resultant forces and moments acting on a typical section of the blade are the net result of the action of the distributed pressure and viscous shear forces, as shown schematically by Fig. 7.13. These forces and moments are obtained by integrating the local values of pressure and shear stress acting normal and parallel to the surface around the airfoil. The forces can be resolved into a wind-axis system (lift and drag) or a chord-axis system (normal force and axial force), as shown in Fig. 7.14. The lift force per unit length, L, acts normal to the velocity, V_∞, and the drag, D, is parallel to V_∞. Alternatively, this lift force can be decomposed into the sum of two other forces as shown in Fig. 7.14: the normal force, N, which acts normal to the airfoil chord, and the leading-edge suction force or axial force, A, which points upstream and acts parallel to the chord.

7.6.1 *Integration of Distributed Forces*

Surface shear contributions to the normal force and the pitching moment are small and can usually be neglected. For the axial and drag forces, the shear stress contribution has a measurable effect and should always be included. If we consider the pressure forces

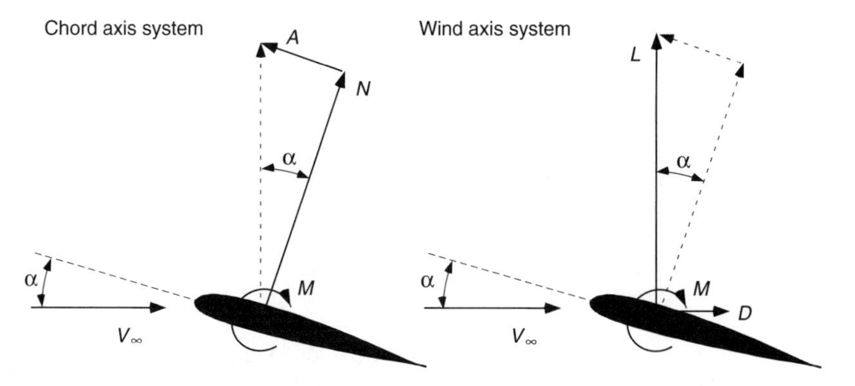

Figure 7.14 Decomposition of distributed surface pressure into resultant forces on an airfoil.

alone, then on the upper surface

$$dN_u = -p_u ds_u \cos\theta_u \quad \text{and} \quad dA_u = -p_u ds_u \sin\theta_u, \tag{7.30}$$

where θ is the local surface slope. On the lower surface

$$dN_l = p_l ds_l \cos\theta_l \quad \text{and} \quad dA_l = p_l ds_l \sin\theta_l. \tag{7.31}$$

Integrating gives the normal and leading-edge suction forces

$$N = -\int_{LE}^{TE} p_u \cos\theta ds_u + \int_{LE}^{TE} p_l \cos\theta ds_l, \tag{7.32}$$

$$A = -\int_{LE}^{TE} p_u \sin\theta ds_u + \int_{LE}^{TE} p_l \sin\theta ds_l \tag{7.33}$$

and the moment about the leading edge by

$$M_{LE} = \int_{LE}^{TE} (p_u x \cos\theta + p_l y \sin\theta) ds_u - \int_{LE}^{TE} (-p_l x \cos\theta + p_u y \sin\theta) ds_l. \tag{7.34}$$

The lift and the pressure drag are obtained by resolving the chordwise (axial) and normal forces through the angle of attack α to give

$$L = N \cos\alpha + A \sin\alpha, \tag{7.35}$$

$$D = N \sin\alpha - A \cos\alpha. \tag{7.36}$$

The corresponding force and moment coefficients are defined as

$$\begin{aligned} C_n &= \frac{N}{q_\infty c}, \quad C_a = \frac{A}{q_\infty c}, \quad C_m = \frac{M}{q_\infty c^2}, \\ C_l &= \frac{L}{q_\infty c}, \quad C_d = \frac{D}{q_\infty c}, \end{aligned} \tag{7.37}$$

where c is the airfoil reference chord. For a thin airfoil this gives the normal, suction, and moment coefficients in terms of the differential pressure coefficient ΔC_p:

$$C_n = \frac{1}{c}\int_0^c (C_{p_l} - C_{p_u}) dx = \frac{1}{c}\int_0^c \Delta C_p \, dx = \int_0^1 \Delta C_p \, d\bar{x}, \tag{7.38}$$

$$C_a = \frac{1}{c}\int_0^c \left(\frac{dy_l}{dx} C_{p_l} - \frac{dy_u}{dx} C_{p_u}\right) dx, \tag{7.39}$$

$$C_{m_{LE}} = -\frac{1}{c^2}\int_0^c (C_{p_l} - C_{p_u}) x \, dx = -\frac{1}{c^2}\int_0^c \Delta C_p x \, dx = -\int_0^1 \Delta C_p \bar{x} \, d\bar{x}. \tag{7.40}$$

The normal force coefficient can generally be used interchangeably with the lift coefficient in the low angle of attack regime. This can be seen by resolving the normal force and axial forces in the lift direction, that is,

$$C_l = C_n \cos\alpha + C_a \sin\alpha. \tag{7.41}$$

Also, for 2-D potential flow we note that $C_a = C_n \tan\alpha$ ($C_d = 0$), so that

$$C_l = C_n \cos\alpha + C_n \tan\alpha \sin\alpha = C_n \cos\alpha + O(\alpha^2) \approx C_n. \tag{7.42}$$

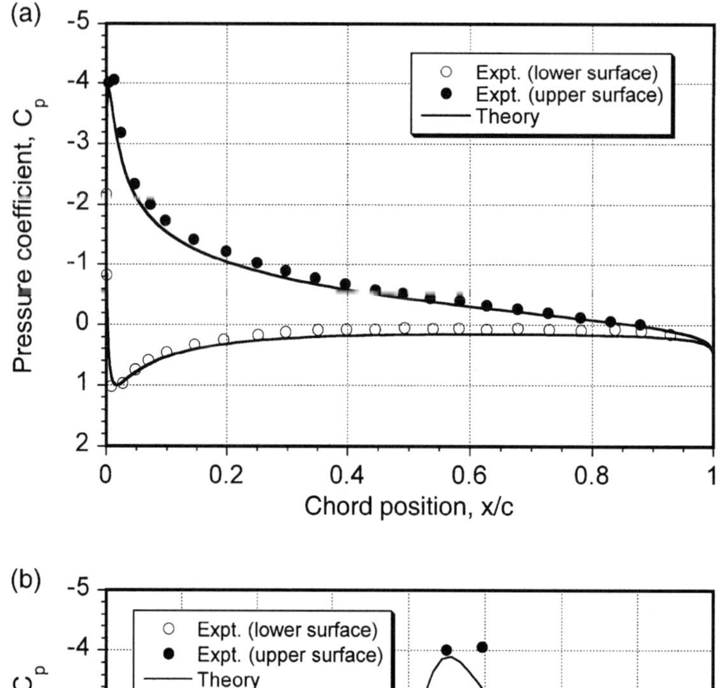

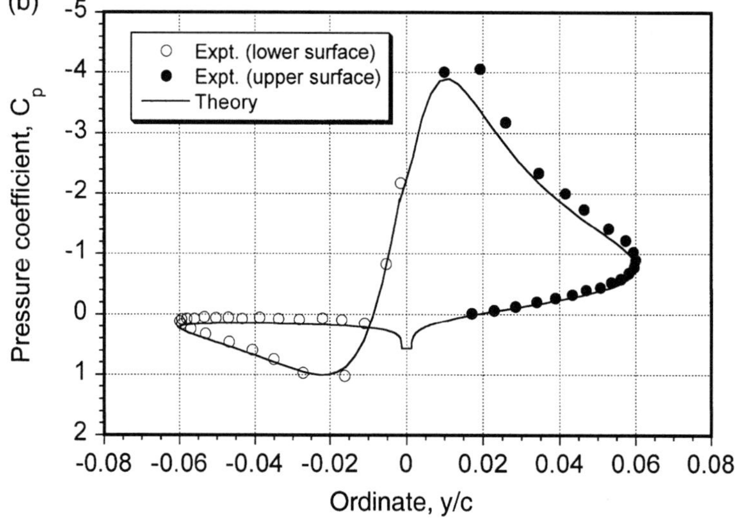

Figure 7.15 Representative airfoil pressure distributions in subsonic flow. (a) C_p versus chord dimension, x/c. (b) C_p versus ordinate, y/c. NACA 0012, $\alpha = 8.2°$. Data source: Bingham & Noonan (1982).

For low to moderate angles of attack (i.e., for angles of attack less than 15 degrees), C_l and C_n have almost the same numerical values.

7.6.2 *Pressure Integration*

The accuracy with which the values of C_n, C_a, and C_m can be measured (or computed) depends on the number and location of the pressure points over the airfoil surface. A typical subsonic pressure distribution is shown in Fig. 7.15 and is compared to theory based on a standard panel method solution. The agreement is good, and any slight differences can likely be attributed to experimental uncertainty and airfoil/wind-tunnel interference effects. For subsonic flows, the suction peak and high adverse pressure gradients occur near the leading edge, so that there must be a bias of pressure points in the leading-edge region to minimize errors in the integration process. At higher subsonic speeds and when transonic flow develops over the airfoil, the largest pressure gradients occur downstream of the leading edge, usually near any shock waves. Therefore, the pressure points must be relocated to this region to adequately resolve the pressure distribution and maintain the accuracy in the calculation of the integrated loads. While numerically this can be easily done by respecifying the locations where C_p is to be calculated, in an experiment this is not usually possible because relocating pressure instrumentation on a wing section can be expensive or impractical. Note from Fig. 7.15(b) that the integration with respect to airfoil thickness becomes particularly difficult if the suction peak and pressure gradients are not adequately resolved.

Special interpolation methods can sometimes be used to maintain accuracy in the calculation of total forces and moments from discrete pressure points, as long as this is done with caution. One way of improving the accuracy with sparse numbers of points in the leading-edge region is to apply a transformation to the measured or computed pressure distribution. One such transformation is

$$C_p^* = 2C_p\sqrt{x/c} \quad \text{and} \quad x^*/c = \sqrt{x/c}, \tag{7.43}$$

as used by St. Hillaire et al. (1979) and McCroskey et al. (1982). The transformed variables C_p^* and x^*/c are then plotted in the conventional way. This transformation has the effect of providing a better definition of the leading-edge pressure peak, as shown by Fig. 7.16, and also generating an additional point at $(x^*/c, C_p^*) = (0, 0)$. The coefficients C_n and $C_{m_{LE}}$ are given in terms of the transformed variables by

$$C_n = \int_0^1 \Delta C_p^* \, d(x^*/c) \quad \text{and} \quad C_{m_{LE}} = -\int_0^1 \Delta C_p^* (x^*/c)^2 \, d(x^*/c) \tag{7.44}$$

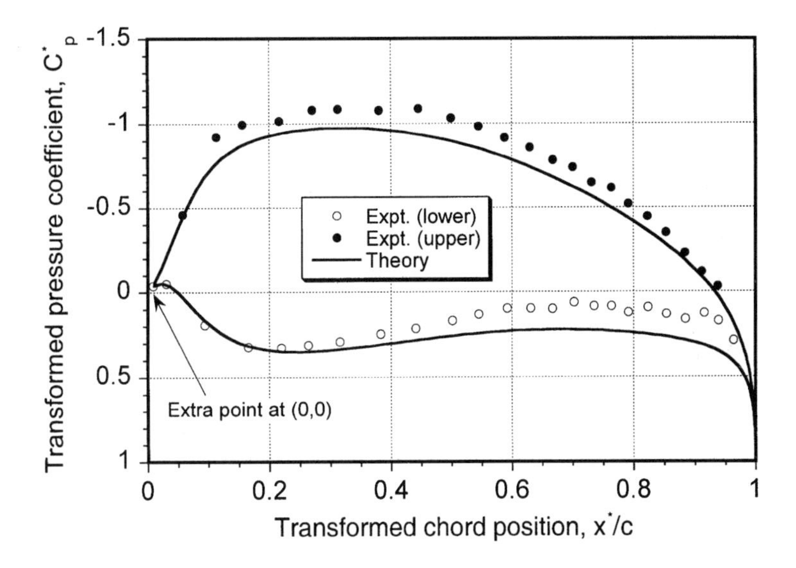

Figure 7.16 Transformation of chordwise pressure distribution to aid numerical integration. NACA 0012, $\alpha = 8.2^\circ$. Data source: Bingham & Noonan (1982).

7.6.3 *Typical Force and Moment Results*

The data plotted in Fig. 7.17 show the low speed lift force coefficient, 1/4-chord pitching moment coefficient, and drag coefficient variations as functions of angle of attack for the SC1095 and SC1095-R8 airfoils at $M = 0.3$.[1] The results are shown over an extended angle of attack that ranges from fully attached flow, where the lifting characteristics of an airfoil below stall are not substantially influenced by the presence of the boundary layer, to the fully stalled conditions where the flow has detached from the upper surface of the airfoil.

Below stall the resultant pressure forces on the airfoil are only slightly affected by thickness and camber (provided they are small) and this is usually the case for most rotor airfoils. The lift on the airfoil section is proportional to its angle of attack and the local dynamic pressure. The lift per unit span of the blade can be written as

$$L = \frac{1}{2}\rho_\infty V_\infty^2 c C_l = \frac{1}{2}\rho_\infty V_\infty^2 c C_{l_\alpha}(\alpha - \alpha_0), \tag{7.45}$$

where C_{l_α} is the lift-curve-slope and is measured in per degree or per radian angle of attack. In coefficient form

$$C_l = C_{l_\alpha}(\alpha - \alpha_0) \quad \text{in the low } \alpha \text{ range}, \tag{7.46}$$

where α_0 is the angle of attack for zero lift or zero-lift angle. The above relation comes under the category of *linearized aerodynamics*. For a real fluid, the lift varies linearly with angle of attack and is within about 10% of the above relation up to an angle of about 10 to 15 degrees, depending on the Mach number. Figure 7.17 shows that above a certain angle, the lift decreases and the pitching moment becomes increasingly nose down (negative). This is a result of the onset of flow separation from the upper surface of the airfoil.

The effects of compressibility manifest as an increase in the effective angle of attack, which increases the lift on the airfoil. In other words, the effects of increasing Mach number appear as an increase in lift-curve slope. In linearized subsonic flow, this increase in lift-curve-slope is given mathematically by the Prandtl–Glauert or Glauert rule [see Glauert (1927)] in which

$$C_{l_\alpha}(M_\infty) = \frac{2\pi}{\sqrt{1 - M_\infty^2}}, \tag{7.47}$$

where 2π is the lift-curve-slope in incompressible flow and the Glauert factor is $\sqrt{1 - M_\infty^2}$, with M_∞ as the free-stream Mach number. In practice, the theoretical value of the lift-curve-slope is replaced by $2\pi\eta_l$, where η_l can be viewed as an efficiency factor. The Glauert factor is found to give good agreement with experimental measurements of the lift-curve-slope of airfoils of thin to moderate thickness-to-chord ratio. An example is shown in Fig. 7.18 where the lift-curve-slope for several airfoils in the NACA 00-series is plotted versus Mach number. The agreement with the Glauert factor is generally good up until the critical Mach number, M^*. Beyond this, the developments of shock waves and their interactions with the boundary layer produce a reduction in the lift-curve-slope throughout the transonic range.

7.7 Pitching Moment

Because the pitching moment can be sensitive to small changes in the pressure distribution, its variation with angle of attack can be difficult to compute accurately. It may

[1] The SC1095-R8 has a modified nose shape and a slightly smaller thickness-to-chord ratio than the SC1095. For this reason, it is sometimes referred to as the SC1094-R8 – see Flemming (1982).

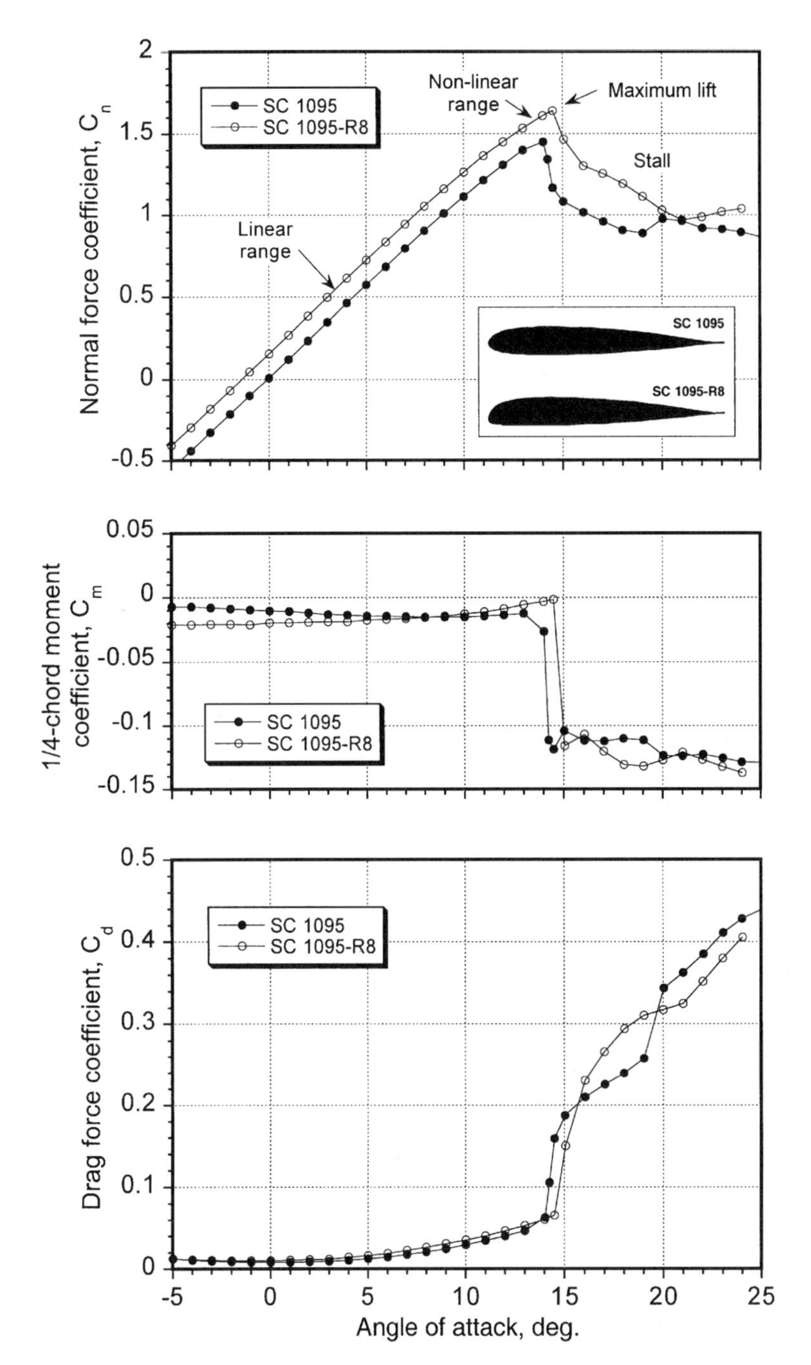

Figure 7.17 Variation of C_n, C_m, and C_d with angle of attack for the SC1095 and SC1095-R8 airfoils at $M = 0.3$ and $Re = 3 \times 10^6$. Source: Leishman (1996).

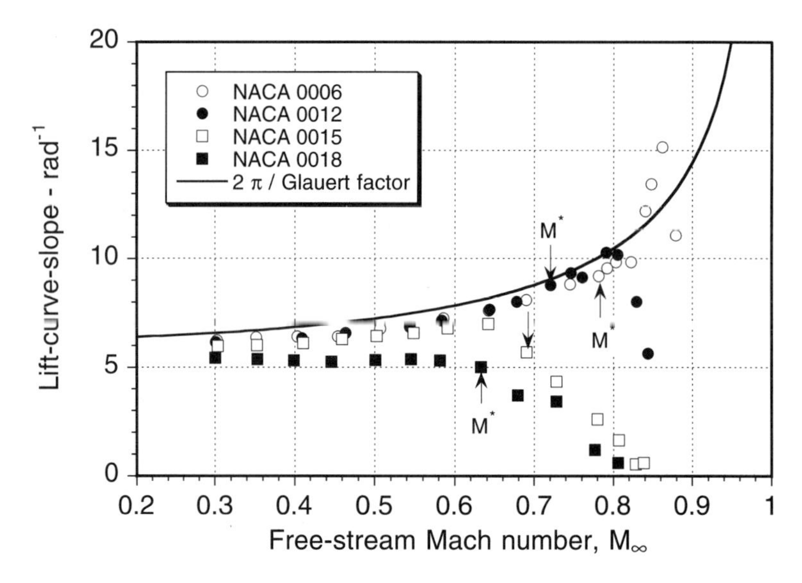

Figure 7.18 Comparison of measured lift-curve-slope for the NACA 00-series with predictions made by using Glauert factor. Data source: Riegels (1961).

be estimated experimentally by direct measurement from a balance or by the integration of chordwise pressure about some reference point, as described previously. In either case, the pitching moment coefficient depends on the reference point chosen. For helicopter work, the 1/4-chord point is normally used. Converting from one reference point to another uses the rules of statics. For example, assume the normal force and pitching moment are known about a point a distance x_a from the leading-edge of the airfoil and it is desired to find the pitching moment about another point, say at a distance x behind the leading edge (see Fig. 7.19). Taking moments about the airfoil leading edge in each case gives

$$M_x = M_a + N(x - x_a). \tag{7.48}$$

Converting to coefficient form by dividing by $\frac{1}{2}\rho_\infty V_\infty^2 c^2$ gives

$$C_{m_x} = C_{m_a} + C_n\left(\bar{x} - \bar{x}_a\right). \tag{7.49}$$

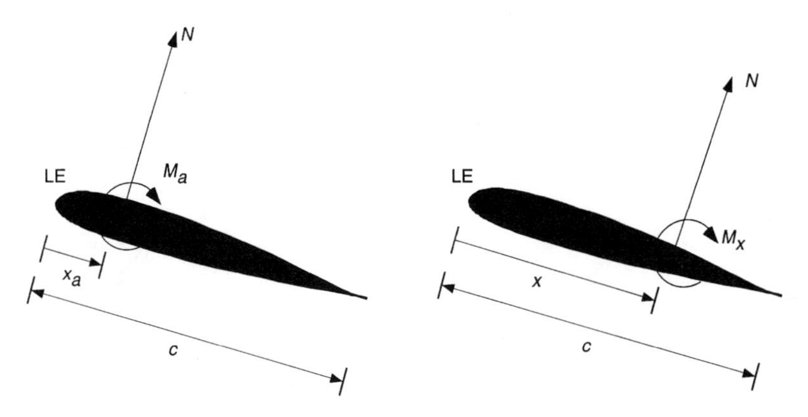

Figure 7.19 Equivalent point loadings on an airfoil.

As an example, if the known pitching moment is about the leading edge, $C_{m_{LE}}$, then $\bar{x}_a = 0$ and the above equation becomes

$$C_{m_x} = C_{m_{LE}} + \bar{x} C_n. \tag{7.50}$$

7.7.1 *Aerodynamic Center*

There is one special point on an airfoil for which it is found that C_m is constant and independent of the angle of attack. This point is called the *aerodynamic center*. For angles of attack up to a few degrees below the stalling angle it is a fixed point on the chord and is relatively close to 1/4-chord. For a flat-plate airfoil in inviscid flow, the aerodynamic center is theoretically on the 1/4-chord axis. However, the thickness of the airfoil and viscous effects as a result of the development of a boundary layer usually cause the aerodynamic center to move a few percent further forward or aft of the 1/4-chord.

The aerodynamic center on an airfoil can be found with a knowledge of the normal force coefficient and the moment coefficient about any other known point. If the aerodynamic center is assumed to lie at a distance x_{ac} behind the leading edge, then

$$C_{m_a} = C_{m_{ac}} + C_n \left(\frac{x_a}{c} - \frac{x_{ac}}{c} \right) = C_{m_{ac}} + C_n (\bar{x}_a - \bar{x}_{ac}). \tag{7.51}$$

Differentiating the above equation with respect to C_n gives

$$\frac{dC_{m_a}}{dC_n} = \frac{dC_{m_{ac}}}{dC_n} + (\bar{x}_a - \bar{x}_{ac}). \tag{7.52}$$

By definition, the aerodynamic center is that point about which the moment is independent of C_n and so the first term on the right-hand side is zero. Therefore,

$$\bar{x}_{ac} = \bar{x}_a - \frac{dC_{m_a}}{dC_n}. \tag{7.53}$$

One can obtain dC_{m_a}/dC_n using

$$\frac{dC_{m_a}}{dC_n} = \frac{dC_{m_a}}{d\alpha} \frac{d\alpha}{dC_n}, \tag{7.54}$$

or if C_{m_a} is plotted versus C_n and the slope of the resulting line found then the aerodynamic center can be determined using Eq. 7.53. Typical results are shown in Fig. 7.20. In the attached flow regime, the data lie on a straight line, the slope of which gives the offset of the aerodynamic center from the 1/4-chord axis. For the SC1095 airfoil and at this Mach number and Reynolds number the aerodynamic center is located at approximately 25.6% chord, and for the SC1095-R8 the aerodynamic center is at 24.7% chord.

The effects of compressibility tend to move the aerodynamic center further aft of the 1/4-chord, and for supersonic flow the aerodynamic center is situated near 50% chord. The measured behavior of the NACA 00-series airfoils is shown in Fig. 7.21, where it is apparent that the aerodynamic center initially tends to drift slightly forward with increasing Mach number. The explanation for this behavior lies in the effects of the pressure distribution on the developing boundary layer. Higher Mach numbers tend to form a pressure distribution that resembles that given by an increase in thickness. Thin airfoils, such as the NACA 0006, tend to have an aerodynamic center that remains close to the theoretical flat-plate value of 1/4-chord. Also, because of the higher critical Mach number of thin sections, the onset of shock waves is delayed and so aft movement of the aerodynamic center occurs at relatively higher subsonic Mach numbers. When shock waves do occur, the movement of the aerodynamic

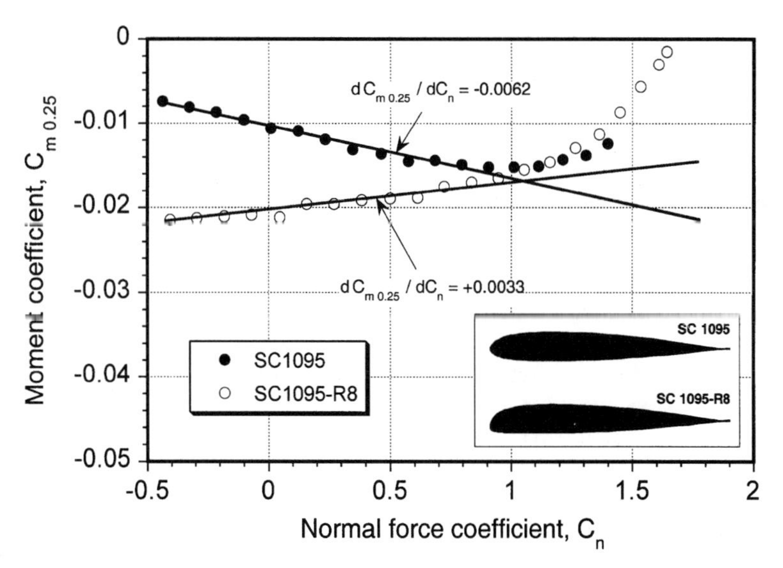

Figure 7.20 Variation of $C_{m_{0.25}}$ with C_n for SC1095 and SC1095-R8 airfoils at $M_\infty = 0.3$. Data source: Leishman (1996).

center is dictated by the relative position of the upper and lower surface shock waves. Because these positions will be a function of both airfoil shape and angle of attack, a simple generalization of the aerodynamic center position in the transonic regime is not possible.

7.7.2 *Center of Pressure*

As shown previously by means of Fig. 7.14, the net aerodynamic loads on an airfoil may be represented by a normal force, an axial (leading-edge suction or axial force), and a

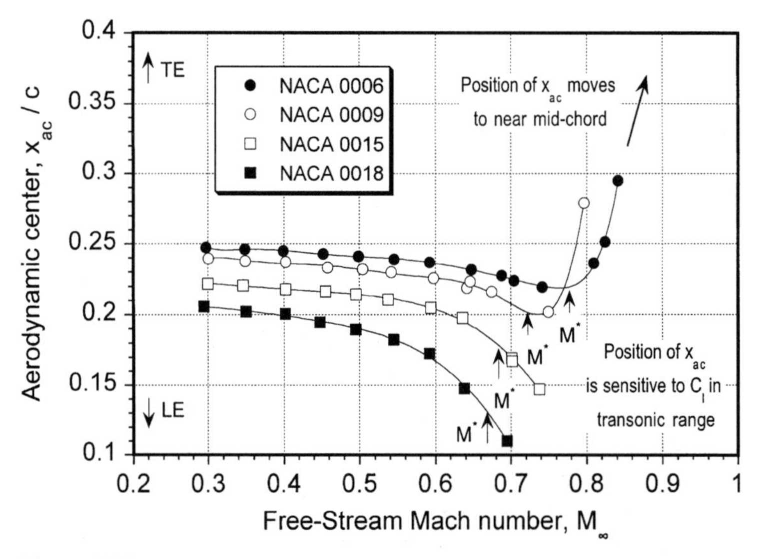

Figure 7.21 Variation of aerodynamic center with Mach number for NACA 00-series airfoil sections. Data source: Riegels (1961).

pitching moment. For each value of the normal force a single point can be determined about which the pitching moment is zero. Therefore, the loading on the airfoil can be replaced only by a normal and axial force acting at this point, which is called the *center of pressure*. The aerodynamic center is a fixed point on the airfoil at a given Mach number and below stall, but the center of pressure moves to different locations on the chord with changes in angle of attack. For example, one may find the center of pressure, x_{cp}, on an airfoil using the normal force and pitching moment about the quarter-chord. Using Fig. 7.19 with $x_a = c/4$ and $x = x_{cp}$, we take moments about the leading edge to obtain

$$M_{LE} = M_{0.25} - N \times \frac{c}{4} = -N x_{cp}, \tag{7.55}$$

and in coefficient form this becomes

$$C_{m_{0.25}} = C_n \left(\frac{1}{4} - \bar{x}_{cp} \right), \tag{7.56}$$

so that the center of pressure is given by

$$\bar{x}_{cp} = \frac{1}{4} - \frac{C_{m_{0.25}}}{C_n}. \tag{7.57}$$

The coefficient $C_{m_{0.25}}$ is generally negative, so that the center of pressure will be behind the aerodynamic center. This is shown in Fig. 7.22, where it will also be noticed that the center of pressure moves forward and approaches the aerodynamic center for higher values of C_n, but below stall. After the airfoil stalls, the center of pressure stabilizes and approaches mid-chord. This reflects the form of the pressure distribution over the upper surface of the airfoil, which is much more uniform when the flow is separated.

7.7.3 *Effect of Airfoil Shape on Moments*

The effect of the airfoil shape on the pitching moment characteristics is of utmost concern for the design of helicopter rotors. As mentioned earlier, experiences with cambered

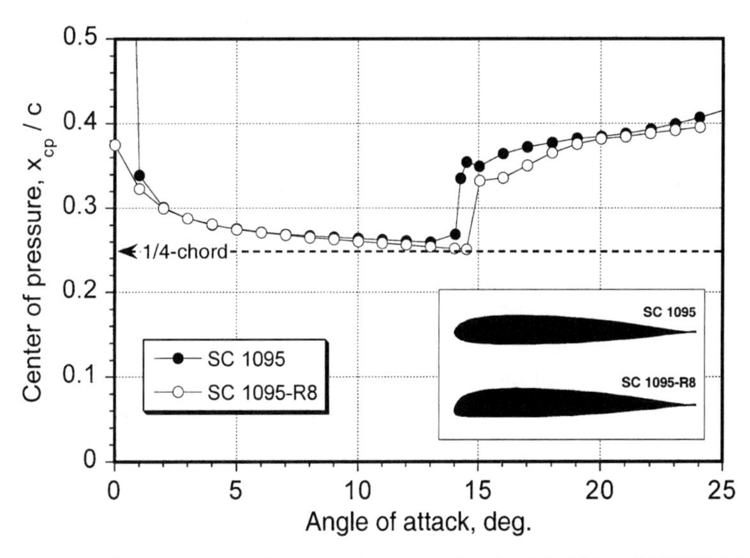

Figure 7.22 Variation of center of pressure for the SC1095 and SC1095-R8 airfoils at $M_\infty = 0.3$. Data source: Leishman (1996).

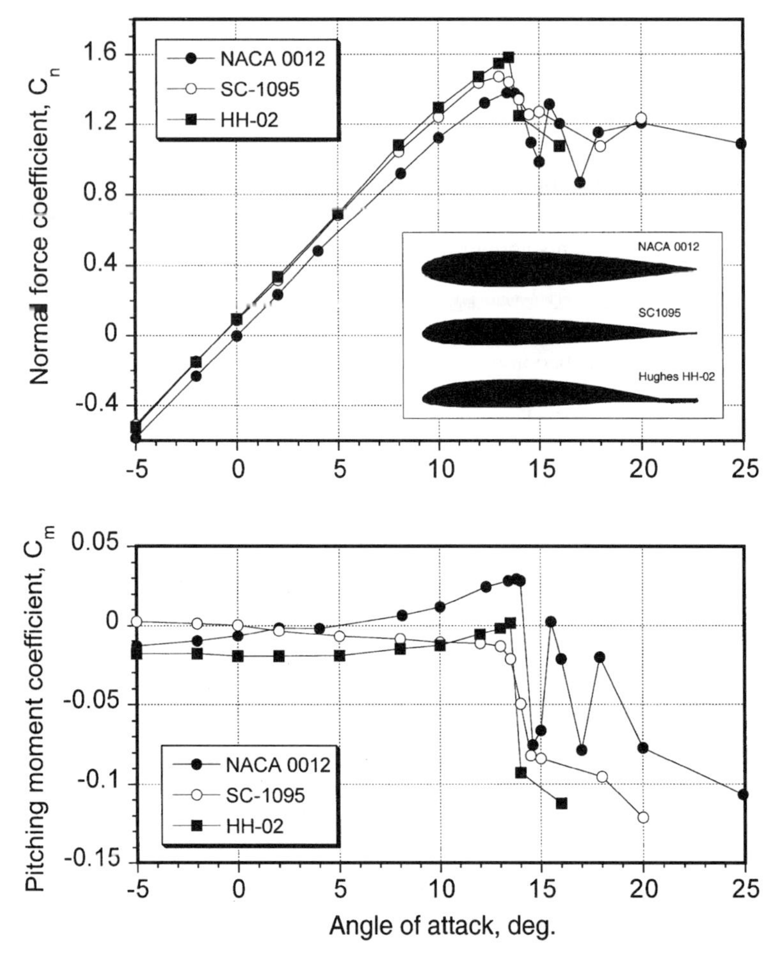

Figure 7.23 Comparison of lift and pitching moments on symmetric and cambered airfoils for $M_\infty = 0.3$. Data source: McCroskey et al. (1982).

airfoils on early autogiros resulted in significant blade twisting and high control loads. For many years, this led to the exclusive use of low pitching moment (symmetrical) sections on all helicopters. The development of torsionally stiffer composite blades and control systems has alleviated this problem somewhat, so that the better high-lift capabilities of cambered airfoils can now be utilized.

Figure 7.23 shows lift and pitching moment results for a NACA 0012 in comparison with two cambered helicopter rotor airfoils, one with a relatively small amount of camber (the SC1095) and one with a larger amount of camber (the HH-02). Both the NACA 0012 and SC1095 airfoils have relatively low zero-lift pitching moments, C_{m_0}, whereas the HH-02 has a higher pitching moment. The actual aerodynamic pitching moment that can be tolerated on a given rotor depends on the structural and dynamic characteristics of the blades, as well as the type of hub and control system. Traditionally, various individual airfoils have been developed based on the minimization of C_{m_0}. But by allowing radially varying airfoil sections on the blade, pitching moment limits placed on the airfoil section itself can be relaxed somewhat; the main requirement is that the resultant moment that is reacted by the blade and control system be minimized as much as possible. As described briefly in

Chapter 6, this is sometimes accomplished by using a reflex cambered airfoil at the inboard sections of the blade where the need for a high maximum lift coefficient is not as stringent.

The effect of camber and chordwise position of maximum camber can be studied using the thin-airfoil theory (see appendix). The NACA four-digit airfoils have a camberline that can be conveniently expressed in terms of two parabolic arcs that are tangent at the position of maximum camber. The camberline is defined by

$$\bar{y}_c = \frac{m}{p^2}(2p\bar{x} - \bar{x}^2) \quad \text{for } \bar{x} \leq p, \tag{7.58}$$

$$\bar{y}_c = \frac{m}{(1-p)^2}[(1-2p) + 2p\bar{x} - \bar{x}^2] \quad \text{for } \bar{x} \geq p, \tag{7.59}$$

where m is the maximum camber and p is the position of maximum camber, both as fractions of the chord. For example, the NACA 2412 airfoil has 2% camber at 40% chord with a thickness ratio of 12%. To use the thin-airfoil theory, the slope of the camberline is required. Differentiating the expressions for y_c gives

$$\frac{d\bar{y}_c}{d\bar{x}} = \frac{2m}{p^2}(p - \bar{x}) \quad \text{for } \bar{x} \leq p, \tag{7.60}$$

$$\frac{d\bar{y}_c}{d\bar{x}} = \frac{2m}{(1-p)^2}(p - \bar{x}) \quad \text{for } \bar{x} \geq p, \tag{7.61}$$

and converting to an angular coordinate using $\theta = \cos^{-1}(1 - 2\bar{x})$ gives

$$\frac{d\bar{y}_c}{d\bar{x}} = \frac{m}{p^2}(2p - 1 + \cos\theta) \quad \text{for } \theta \leq \theta_p, \tag{7.62}$$

$$\frac{d\bar{y}_c}{d\bar{x}} = \frac{m}{(1-p)^2}(2p - 1 + \cos\theta) \quad \text{for } \theta \geq \theta_p. \tag{7.63}$$

The thin-airfoil theory gives the pitching moment about the 1/4-chord as $C_{m_{0.25}} \equiv C_{m_0} = -\pi(A_1 - A_2)/4$, where

$$\begin{aligned} A_1 &= \frac{2}{\pi}\int_0^{\pi} \frac{d\bar{y}_c}{d\bar{x}} \cos\theta \, d\theta \\ &= \frac{2m}{\pi p^2}\left[(2p-1)\sin\theta_p + \frac{1}{4}\sin 2\theta_p + \frac{\theta_p}{2}\right] \\ &\quad - \frac{2m}{\pi(1-p)^2}\left[(2p-1)\sin\theta_p + \frac{1}{4}\sin 2\theta_p - \frac{1}{2}(\pi - \theta_p)\right], \end{aligned} \tag{7.64}$$

$$\begin{aligned} A_2 &= \frac{2}{\pi}\int_0^{\pi} \frac{d\bar{y}_c}{d\bar{x}} \cos 2\theta \, d\theta \\ &= \left(\frac{m}{\pi p^2} - \frac{m}{\pi(1-p)^2}\right)\left[(2p-1)\sin 2\theta_p + \sin\theta_p - \frac{1}{3}\sin^3\theta_p\right]. \end{aligned} \tag{7.65}$$

The results are summarized in Fig. 7.24 where the zero-lift pitching moment coefficient is plotted versus the point of maximum camber, p. Notice that forward camber has a much smaller effect on the pitching moment compared to camber applied near the trailing edge of the airfoil. Generally, the thin-airfoil theory will tend to slightly overpredict the pitching moment, but the trends are correct.

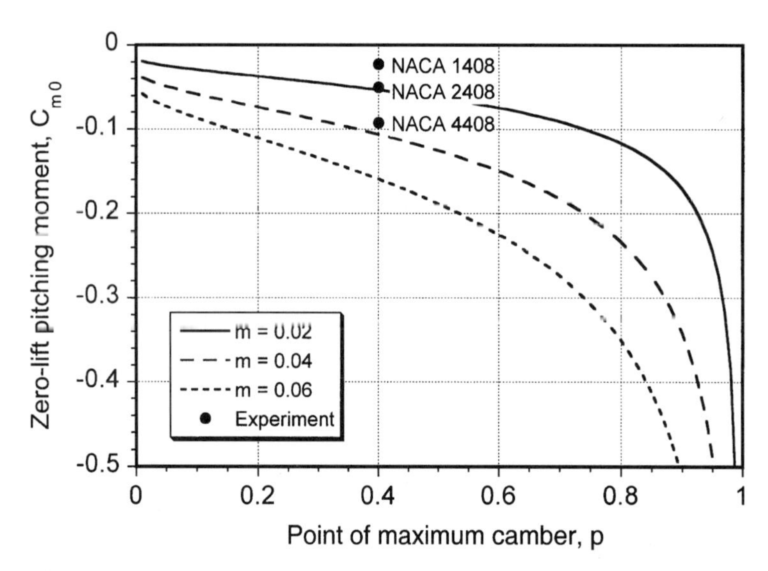

Figure 7.24 Effect of chordwise point of maximum camber on pitching moment. Data source: Abbott & von Doenhoff (1949).

7.7.4 *Use of Tabs*

On helicopter airfoils, trailing-edge tabs are frequently used to help negate the pitching moment produced by positive camber over the leading-edge region, while retaining the benefits of a high maximum lift achieved by the use of camber. An example is the HH-02 shown previously in Fig. 7.23 or the VR-7 shown in Fig. 7.25. As shown in Fig. 7.24, the mean camberline near the trailing edge significantly affects the pitching moment, and also the zero lift angle of the section is altered. On early helicopter rotors, trailing-edge tabs were used to move the aerodynamic center further aft to help control or delay the onset of torsional flutter. Furthermore, an aft aerodynamic center is desirable as less nose weight has to be installed in the blade to achieve the proper forward position of the center of gravity.

By means of the thin-airfoil theory, the contributions to the lift and moment from the tab can be obtained for a given tab size and angle of deflection. Assume that the ratio of the tab chord to the airfoil chord is given by E, as shown in Fig. 7.25, and the tab deflection angle is η. An upward deflection of a tab makes the ordinates of the mean camberline more

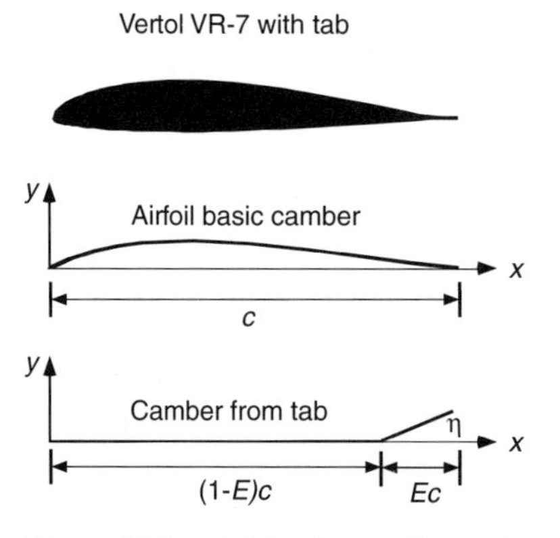

Figure 7.25 Airfoil with a trailing-edge tab and thin airfoil model.

positive in the trailing-edge region. As a consequence, the zero lift angle becomes more positive, and the lift and net pitching moment are reduced for a given angle of attack. The camberline slope is zero for $\bar{x} \leq (1 - E)$ and takes a value η for $\bar{x} > (1 - E)$. Thin-airfoil theory gives for the A_0 coefficient

$$A_0 = -\frac{1}{\pi}\int_0^{\pi} \frac{dy}{dx} d\theta = -\frac{1}{\pi}\int_{\theta_t}^{\pi} \eta \, d\theta$$
$$= \left(\frac{\theta_t}{\pi} - 1\right)\eta, \tag{7.66}$$

where θ_t is the value of θ at the tab [i.e., $\theta_t = \cos^{-1}(2E - 1)$]. In a similar way, A_1 and A_2 can be evaluated as

$$A_1 = \frac{2\eta}{\pi}\sin\theta_t, \qquad A_2 = \frac{\eta}{\pi}\sin\theta_t.$$

The incremental lift and moment from tab deflection can, therefore, be written as

$$\frac{dC_l}{d\eta} = -2(\pi - \theta_t - \sin\theta_t), \tag{7.67}$$

$$\frac{dC_{m_{0.25}}}{d\eta} = -\frac{1}{2}\sin\theta_t(\cos\theta_t - 1). \tag{7.68}$$

Figure 7.26 shows the results of tab deflection on the lift and moment as a function of the ratio of the tab to airfoil chord. Experimental results, which have been collated by Prouty (1986), correlate reasonably well with the predictions made by thin-airfoil theory. The larger tabs have the ability to produce moderately large pitching moments, but without any significant penalty in terms of maximum lift. Drag penalties associated with trailing-edge tabs (or flaps) are small provided the tab deflection remains below about 5 degrees. If the deflection angle is significant, the adverse pressure gradients found in the region where the tab is initiated will lead to boundary layer thickening and a reduction of maximum lift and increase in profile drag.

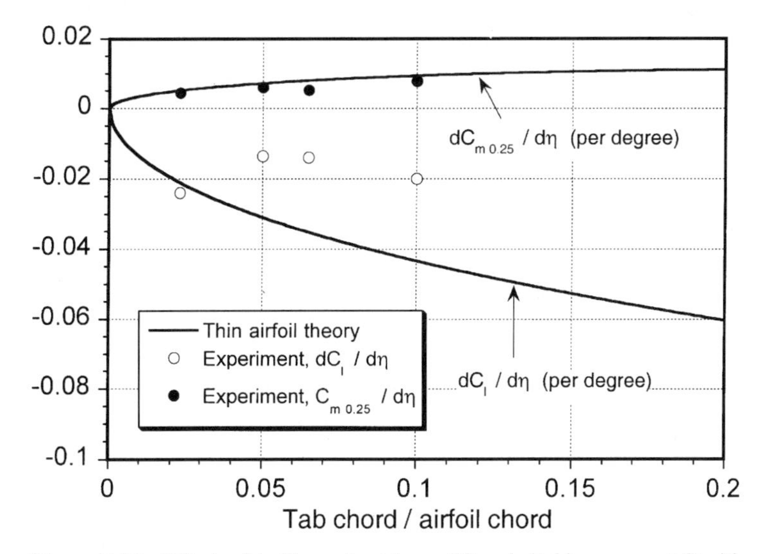

Figure 7.26 Effects of trailing-edge tabs on lift and pitching moment. Positive upward tab deflection. Data source: Collated results by Prouty (1986).

7.8 Maximum Lift and Stall Characteristics

As previously noted, at higher angles of attack the adverse pressure gradients produced on the upper surface of the airfoil result in a progressive increase in the thickness of the boundary layer and cause some deviation from the linear lift versus angle of attack behavior (see Fig. 7.7). Eventually, the flow will separate causing stall. On many airfoils, the onset of flow separation and stall occurs gradually with increasing angle of attack, but on some airfoils, particularly those with sharp leading edges, the separation may occur quite suddenly. In the stalled flow regime, the flow over the upper surface of the airfoil is characterized by a region of fairly constant static pressure. The pitching moment about the 1/4-chord is much more negative (nose-down) because with the almost constant pressure over the upper surface the center of pressure is now close to mid-chord. Less lift is generated by the airfoil because of the reduction in circulation and loss of suction near the leading edge, and the drag is greater. In addition, under these separated flow conditions, steady flow no longer prevails, with turbulence and often vortices being alternately shed from the upper surface into the wake. Measurements will generally show fairly large fluctuating forces and moments under stalled flow conditions.

One of the most important characteristics used to judge the performance of an airfoil is the maximum static lift capability. This is a quantity that is not easily predicted, even with state-of-the-art computational methods, and reliance must be placed on experimental measurements. However, even from an experimental perspective, absolute values of $C_{l_{\max}}$ are difficult to guarantee with high precision, and especially between tests performed in different wind tunnels. This is mainly because of the uncertainties associated with the testing technique, the aspect ratio of the wing or airfoil specimen, wall interference effects, and different turbulence intensities in different wind tunnels. McCroskey (1987) gives a good overview of the problem. However, it appears that if airfoil measurements are performed in a consistent manner and to uniform data accuracy standards, such as those defined by Steinle & Stanewsky (1982), the measurements can be considered reliable. However, the formation of test specific 3-D stall developments near maximum lift are difficult to avoid – see Moss & Murdin (1965). See also Prouty (1975) for a survey of 2-D rotor airfoil measurements.

Abbott & von Doenoff (1949) have documented a summary of airfoil section measurements made at Reynolds numbers of 3 to 9 million and Mach numbers up to 0.2. These Reynolds numbers are close to the range encountered by the retreating blade on a full scale helicopter rotor and provide a consistent basis from which to review the stall characteristics of airfoils, in general. The maximum lift that can be developed by an airfoil when operating at a steady angle of attack is related to the type of stall characteristic of that airfoil. At low speeds, airfoils generally fall into three static stall categories, as identified by McCullough & Gault (1953) and Gault (1957). These types are: thin-airfoil stall, leading-edge stall, and trailing-edge stall. The results show that thin-airfoil and leading-edge stalls can be fairly sensitive to changes in airfoil shape, whereas trailing-edge stall is insensitive. Most conventional rotor airfoils fall into the category of trailing-edge or leading-edge stall types at low to moderate Mach numbers. It is also common for a *mixed stall* behavior to occur on some airfoils, which is a stall characteristic that is not clearly one type or another.

The three low speed static stall characteristics can be illustrated by comparing the lift and moment results for a given family of airfoils that have different thickness ratios. For example, Fig. 7.27 shows results for the NACA 63 series airfoil section for thickness-to-chord ratios of 6%, 12%, 15%, and 21%. *Thin-airfoil stall* occurs on the NACA 63-006 airfoil. The sharp nose radius produces high adverse pressures near the leading edge, which results in separation of the laminar boundary layer at low angles of attack. Initially the flow

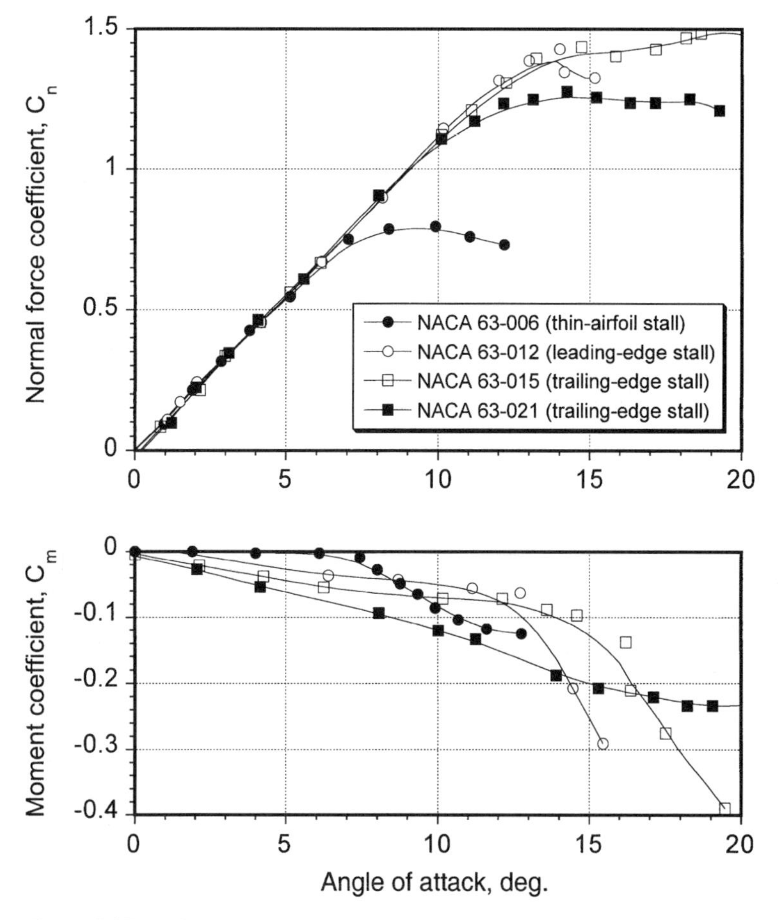

Figure 7.27 Lift and pitching moment characteristics of the NACA 63-series airfoil sections with different thickness-to-chord ratios. Data source: Abbott & von Doenhoff (1949).

reattaches as a turbulent boundary layer, but as the angle of attack is increased the turbulent reattachment point moves aft producing a *long separation bubble*. The formation of the bubble often causes a slight jog in the lift-curve-slope. Ultimately, the boundary layer fails to reattach and the airfoil stalls. Thin-airfoil stall occurs mainly at low Reynolds numbers and, as shown by Fig. 7.27, involves a fairly gentle lift stall characteristic with a break in the moment curve at low angle of attack. This is followed by a progressive increase in nose-down pitching moment as the separation bubble envelops more of the upper surface of the airfoil. Airfoils that exhibit thin-airfoil stall also tend to show considerable hysteresis in the airloads, and different results can be obtained depending on whether the angle of attack is set in the wind tunnel before turning on the flow or vice versa. The former tends to replicate flow reattachment from stalled conditions.

The NACA 63-012 airfoil exhibits what is known as *leading-edge stall.* This airfoil achieves a much higher value of maximum lift, followed by an abrupt stall leading to sharp breaks in the lift and moment curves. The leading-edge stall mechanism involves the participation of a *laminar separation bubble* in the leading-edge region, as previously discussed and shown in Fig. 7.6. Transition to a turbulent boundary layer occurs at this bubble. While the airfoil is sufficiently thin to promote laminar separation, the adverse pressure gradients are mild enough to cause the transitional turbulent boundary layer to reattach forming this bubble, which is typically 2–3% of chord and is normally situated immediately downstream of the leading-edge suction peak. As the angle of attack is increased, this bubble moves forward on the chord toward the leading edge. Eventually, the adverse pressure gradient becomes more severe, preventing the turbulent boundary layer from reattaching to the surface, and the bubble can be said to "burst." Alternatively, the bubble may remain present and the turbulent boundary layer may separate abruptly immediately downstream of the bubble. This is called leading-edge stall by the *reseparation* mechanism – see Evans & Mort (1959). As both bubble burst and reseparation result in a sudden change in the flow pattern around the airfoil, it is often difficult to discern the actual leading-edge separation mechanism. It can usually be determined with the aid of flow visualization, which can confirm the existence of a laminar separation bubble in the poststall regime. At the Reynolds numbers of practical interest on full-scale helicopters, reseparation is the normal mechanism for leading-edge stall.

The third type of static stall shown in Fig. 7.27, which is exhibited by the relatively thicker NACA 63-015 and the very thick NACA 63-021 airfoil, is called *trailing-edge stall.* This is a fairly common stall mechanism found with the higher t/c ratios and camberlines typical of some modern rotor airfoils, and it is caused by the relatively gradual movement of the turbulent separation point from the trailing edge toward the leading edge. Trailing-edge stall produces a progressive rounding of lift behavior near maximum lift with a gentle pitching moment break. Another characteristic of the trailing-edge stall mechanism is that the pitching moments exhibit a pronounced nose-up tendency just prior to maximum lift. For lower t/c ratios a high value of maximum lift is still produced on airfoils that exhibit trailing-edge stall, although generally not as high as for airfoils that exhibit leading-edge stall. With increasing t/c, the maximum lift decreases rapidly with trailing-edge separation occurring at progressively lower angles of attack. Some airfoils can experience abrupt trailing-edge stall, particularly at higher Reynolds numbers, which manifest as a rapid forward movement of the trailing-edge separation point and in the absence of supporting evidence such as pressure distributions and flow visualization can be easily confused with leading-edge stall.

In light of the foregoing, it is apparent that several potential methods can be used to increase the $C_{l_{\max}}$ of an airfoil. One method is to change the airfoil thickness, such as going from the NACA 0009 to the NACA 0012 or another in the series. Figures 7.28 and 7.29 summarize the experimental results for the NACA airfoils. There are clearly significant advantages in using some thickness to avoid the formation of a long laminar separation bubble, but there is no substantial benefits in terms of $C_{l_{\max}}$ for $t/c > 12\%$, unless thicker airfoils are needed to meet blade structural requirements. Another method is to introduce forward camber for a given airfoil thickness, such as in going from the NACA 0012 to the 23012. These results are summarized in Fig. 7.29. This latter modification gives a more gradual curvature to the airfoil at the leading edge, resulting in a less adverse pressure gradient at a given angle of attack or lift coefficient. As shown in both Figs. 7.28 and 7.29, all of the cambered airfoils exhibit higher values of $C_{l_{\max}}$ compared to the symmetric NACA 00-series.

7.8.1 *Effects of Reynolds Number*

At low Mach numbers, both the type of stall and the maximum lift coefficient are affected by Reynolds number. A basic study of the relationships was first made by Jacobs & Sherman (1937). Increasing the Reynolds number increases the inertial effects in the flow, which will dominate over the viscous effects and, all other factors being equal, will thin the

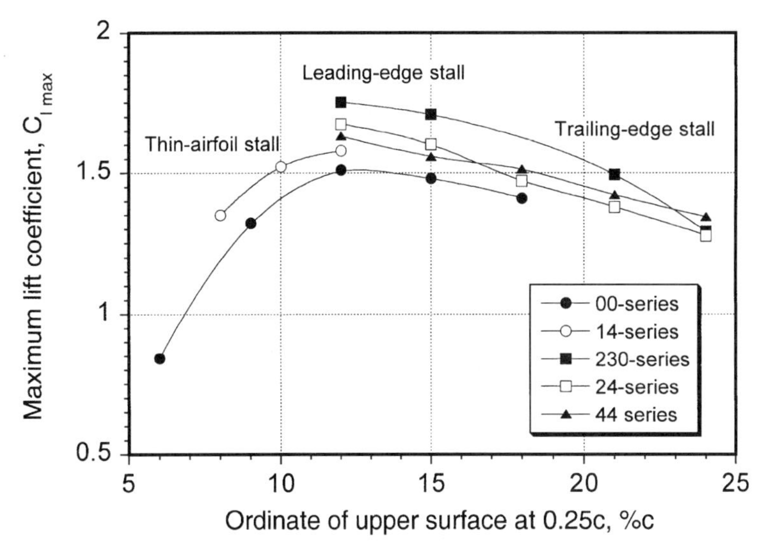

Figure 7.28 Summary of results showing combined effect of thickness and camber on $C_{l_{max}}$. Data source: Abbott & von Doenhoff (1949)

boundary layer and delay the onset of flow separation to higher values of angle of attack and lift coefficient. Figure 7.30 shows the effects of increasing Reynolds number for the NACA 63- and 64-series airfoils. The main trend is an increase in $C_{l_{max}}$, but at a very small rate after a certain value of Re is reached. The main difficulty in assessing the effects of Reynolds number is the interdependent effects of Mach number. The Mach number, M_∞, may be written in terms of the Reynolds number, Re, as

$$M_\infty = \left(\frac{\mu}{\rho\, a}\right)\frac{Re}{c}, \tag{7.69}$$

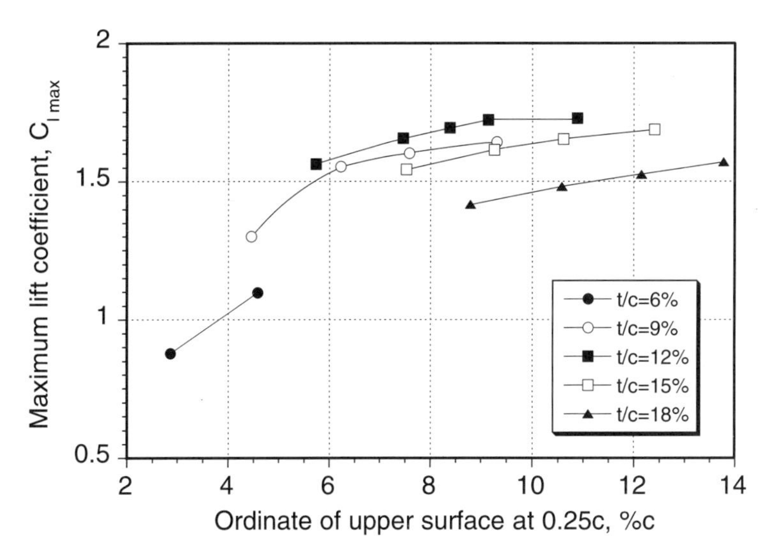

Figure 7.29 Results showing effect of nose camber on $C_{l_{max}}$. Data source: Abbott & von Doenhoff (1949).

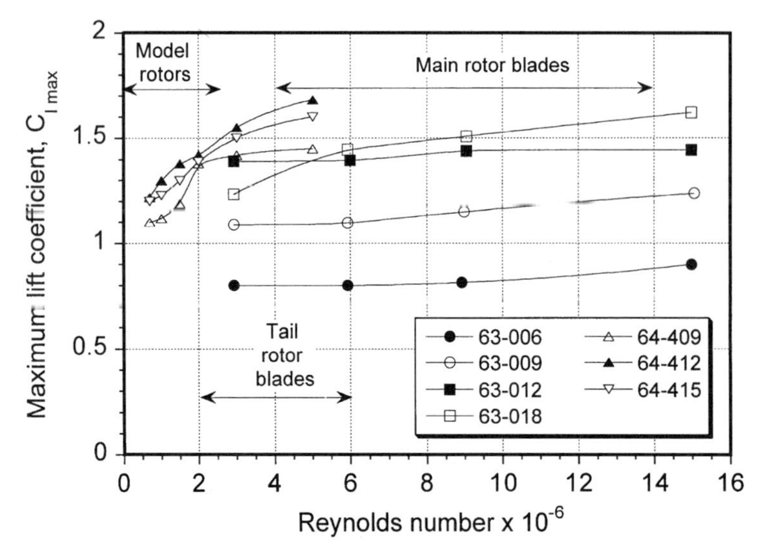

Figure 7.30 Effect of Reynolds number on the maximum lift coefficient of the NACA 63-series and 64-series airfoils. Data source: Racisz (1952).

where c is the airfoil chord. Unless special (pressurized) wind tunnels are used where the Reynolds number can be varied completely independently of Mach number, a change in free-stream Mach number will always be accompanied by a change in Reynolds number and vice versa.

Up to Mach numbers of about 0.4, the effects of varying Reynolds number on the maximum lift and stall characteristics can be significant. For example, results showing the independent Mach number and Reynolds number variation on the maximum lift of a NACA 64-210 airfoil section are given in Fig. 7.31. These data show that the Reynolds number has a strong influence on the maximum lift capability at a given Mach number, with larger

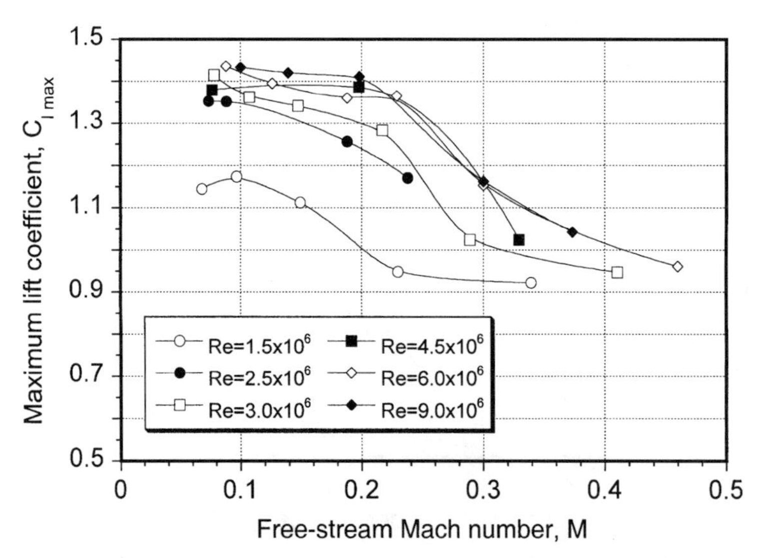

Figure 7.31 Results of independent Reynolds number and Mach number variations on the $C_{l_{max}}$ of a NACA 64-210 airfoil. Data source: Racisz (1952).

Reynolds numbers leading to higher values of $C_{l_{max}}$ for a given Mach number. At higher free-stream Mach numbers the effects of Reynolds number are secondary compared to the effects of compressibility, as confirmed by the fact that curves for each Reynolds number show a converging trend. These effects of Reynolds number should be borne in mind when analyzing results from model rotor tests, where the blade Reynolds numbers may be one quarter or less of the full-scale rotor values.

In addition to the effects of Reynolds number and airfoil shape, any roughness on the leading edge of the airfoil may also affect the stall type. Roughness causes premature transition from a laminar to a turbulent boundary layer flow, thereby increasing the effective Reynolds number. Generally, prematurely tripping the boundary layer will always make the airfoil exhibit a trailing-edge (gradual) stall, whereby the point of turbulent separation moves forward on the chord with increasing angle of attack. In wind-tunnel experiments, standard roughness is applied to the airfoils in the form of carborundum grains or other standardized approaches – see, for example, Loftin (1945) and Abbott & von Doenhoff (1949). In the case of an airfoil that initially exhibits thin-airfoil stall, leading-edge roughness will eliminate the laminar separation bubble and will give the airfoil a gradual turbulent trailing-edge stall characteristic (see Fig. 7.32). As shown by the NACA 63-006 airfoil, this is usually accompanied by a mild increase in $C_{l_{max}}$. However, the application of standard roughness to an airfoil that initially exhibits leading-edge stall will cause the boundary layer to be more susceptible to turbulent boundary layer separation at a lower angle of attack. Therefore, as shown by the 63-012 airfoil, this results in a gradual trailing-edge stall characteristic and a significant loss of maximum lift capability.

7.8.2 *Effects of Mach Number*

As described previously, airfoil characteristics at Mach numbers above about 0.3 are affected by the compressible nature of the flow. As an example, the effects of Mach number on the lift and moment characteristics of a NACA 0012 are shown in Fig. 7.33. Two effects are immediately noticeable. First, there is an increase in the lift-curve-slope with

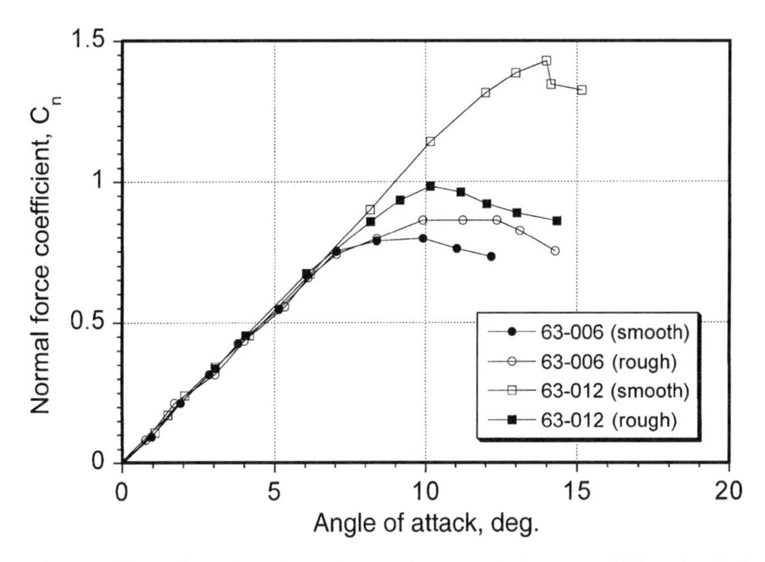

Figure 7.32 Effect of leading-edge roughness on the low-speed lift and stall characteristics of the 63-006 and 63-012 airfoils: Data source: Abbott & von Doenhoff (1949).

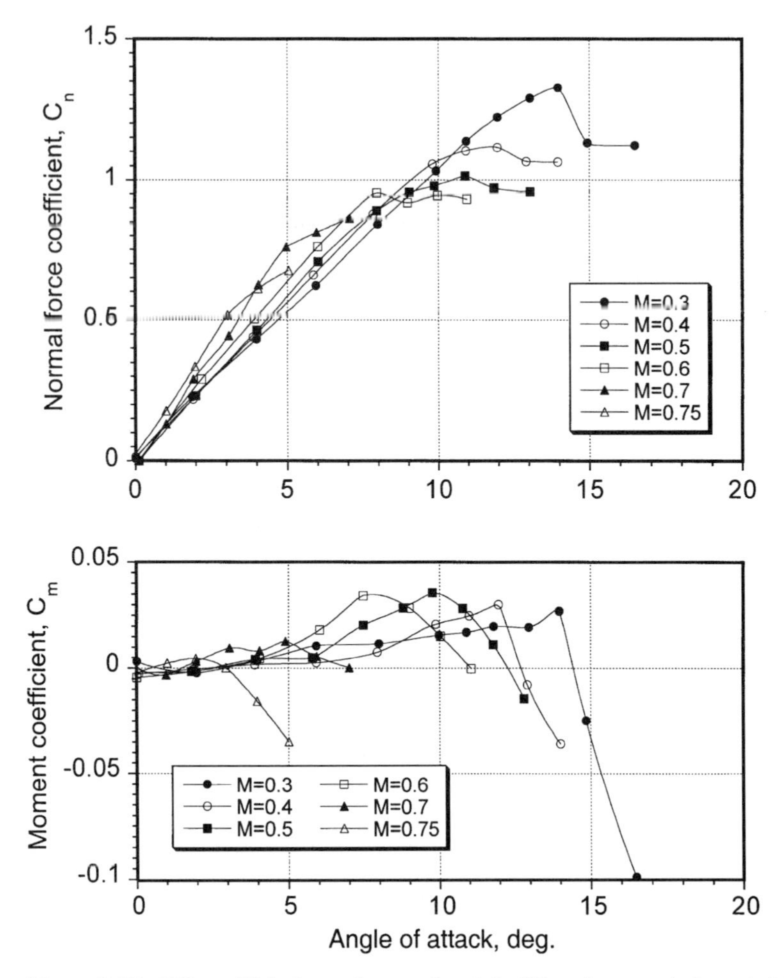

Figure 7.33 Effect of Mach number on the static lift and moment characteristics of a NACA 0012 airfoil. Data source: Wood (1979).

increasing Mach number. This is the well-known Glauert effect, as discussed previously. Physically, the effect occurs because pressure disturbances cannot propagate as far upstream in a given time as the Mach number increases. This increases the streamline curvature near the leading edge of the airfoil and manifests as an increase in the effective angle of attack. Second, the value of maximum lift coinciding with the break in the pitching moment curve tends to decrease with increasing Mach number. This is because of the onset of flow separation produced by the high adverse pressure gradients near the leading edge at low free-stream Mach numbers or near the shock wave at higher Mach numbers.

To explain the airfoil behavior at higher Mach numbers consider the presentation in Figs. 7.34 and 7.35. Figure 7.34 shows results of the upper surface pressure distribution on the NACA 0012 for a range of angles of attack at a free-stream Mach number of 0.64. As the angle of attack is increased from 0 to 2 degrees, the flow at the leading edge of the airfoil becomes mildly supercritical. Under these conditions the flow returns to subsonic conditions by passing through a shock wave before reaching the trailing edge. With increasing angle of attack, the extent of supercritical flow increases and the shock wave moves aft on the chord and strengthens. However, as the angle of attack reaches 6 degrees the shock wave becomes

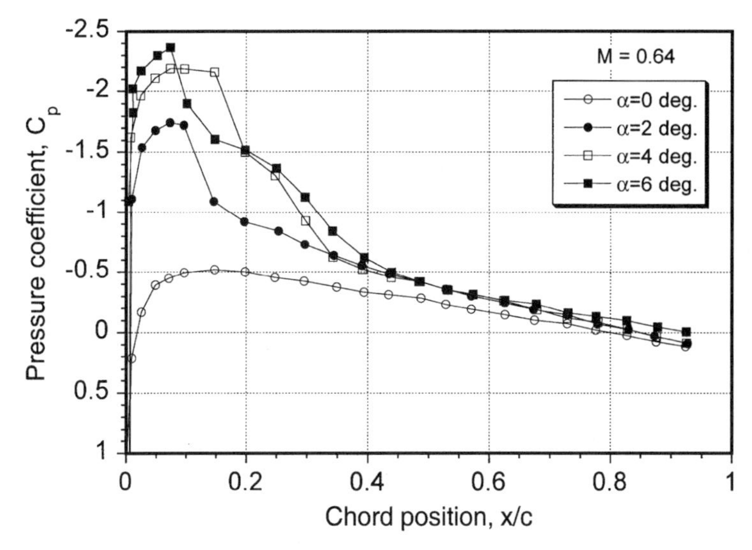

Figure 7.34 Effect of increasing angle of attack on the upper surface pressure distribution of a NACA 0012 airfoil at $M_\infty = 0.64$. Data source: Bingham & Noonan (1982).

more oblique to the airfoil surface and starts to move forward. The pressure behind the shock also decreases, as reflected in the break in the pitching moment shown in Fig. 7.33. This indicates the onset of flow separation. However, the flow reattaches again some distance downstream of the shock wave, in effect producing a turbulent separation bubble. This bubble is conceptually similar to a laminar separation bubble and contains a region of low velocity, constant pressure, recirculating flow. During this process, the lift coefficient shows a departure from the linear C_l versus α behavior but continues to increase with increasing angle of attack.

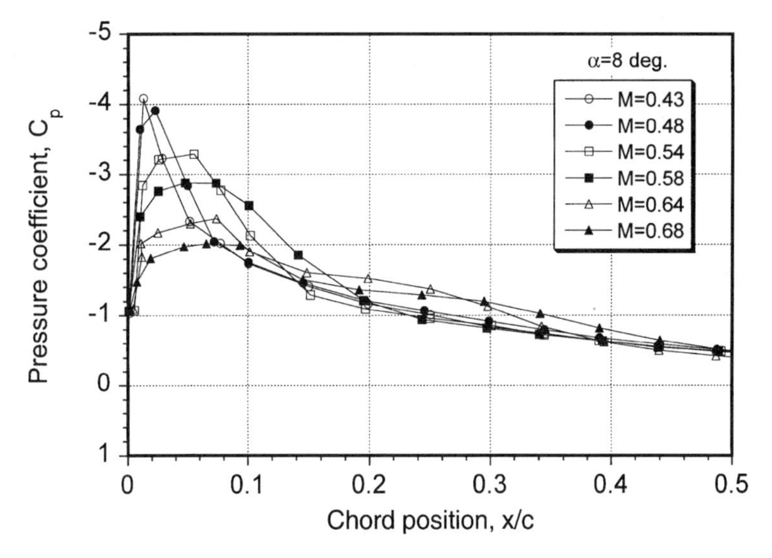

Figure 7.35 Effect of increasing free-stream Mach number on the upper surface pressure distribution of a NACA 0012 airfoil at a constant angle of attack of 8°. Data source: Bingham & Noonan (1982).

Figure 7.35 shows the upper surface pressure distribution over the leading-edge region of the NACA 0012 for a series of increasing values of free-stream Mach number when the airfoil is held at a constant angle of attack of 8 degrees. As M_∞ is increased from 0.43 to 0.48, the subsonic form of the pressure distribution changes as the flow near the leading edge becomes mildly supercritical and the steepest adverse gradient moves further aft to 20% chord. With further increases in M_∞, the shock wave strengthens and moves aft as the extent of supercritical flow increases. Eventually at $M_\infty = 0.64$, the shock wave becomes sufficiently strong to produce flow separation at the foot of the shock, which again causes the shock to move forward on the chord and become more oblique to the airfoil surface. Again, the flow reattaches forming a turbulent separation bubble. Further increases in M_∞ cause the bubble to lengthen, and eventually the boundary layer will fail to reattach, producing complete flow separation over the upper surface of the airfoil. This is called *shock induced stall.*

Figure 7.36 shows the effect of Mach number on the maximum value of lift of the airfoil for nominally attached flow. Because the lift continues to increase after the onset of flow separation, these values have been derived by determining the value of C_l corresponding to the break in the pitching moment curve (i.e., the initial onset of flow separation), which will also be coincident with an increase in drag. For lower values of M_∞ the curves are relatively flat and the airfoils retain their maximum lift, although the effects of compressibility clearly cause the values of $C_{l_{\max}}$ to decrease with increasing Mach number. For values of M_∞ greater than about 0.5, the onset of supercritical flow causes the maximum values of lift to quickly decrease for further increases in M_∞, at least for the NACA 0012 and NACA 23015 airfoils. At the lower Mach numbers, the camber of the NACA 23015 airfoil increases the value of $C_{l_{\max}}$ relative to the NACA 0012, but this advantage is lost at the higher Mach numbers because the camber and slightly greater thickness gives the NACA 23015 airfoil a lower critical Mach number. For the higher Mach numbers, the range of angles of attack over which the airfoils can operate without some flow separation is considerably reduced.

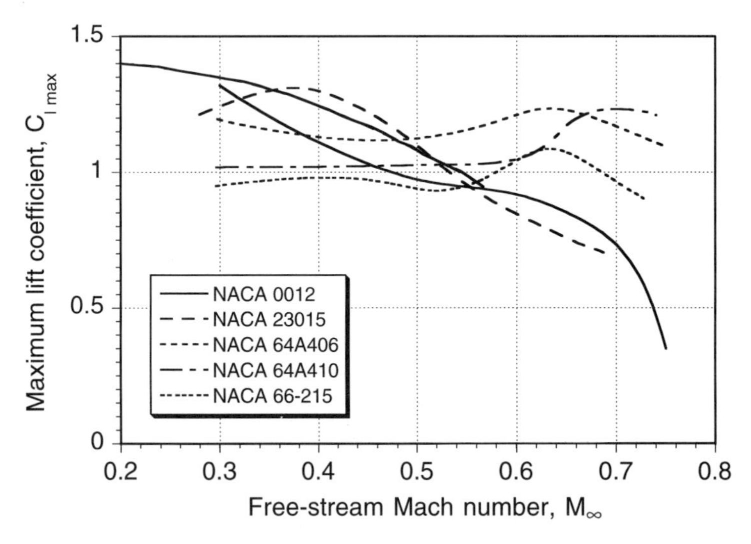

Figure 7.36 Composite results showing effect of Mach number on maximum lift coefficient coinciding with pitching moment break for several "conventional" and "supercritical-like" airfoil sections. Data sources include: Racisz (1952), Stivers (1954), Bensen et al. (1973), and Wood (1979).

The steady reduction of maximum lift coefficient with increasing Mach number shown for many airfoils is not necessarily a characteristic of all airfoils. Figure 7.36 also shows the $C_{l_{max}}$ behavior of the NACA 64-series and 66-series airfoils. While the values of $C_{l_{max}}$ attainable by this airfoil at low Mach numbers is inferior to either of the NACA 0012 or 23015 airfoils, there is a beneficial effect on $C_{l_{max}}$ at the higher Mach numbers. The NACA 66-215 and 64-series airfoils have a point of maximum camber fairly far aft, and so the airfoil exhibits a higher critical Mach number and a favorable effect on the strength of the developing shock wave. This is similar to the behavior of *supercritical* airfoils, which have been briefly discussed previously. Improved airfoils for transonic flow applications were originally devised by NASA – see Whitcomb & Clark (1965) and Whitcomb (1976). To achieve this behavior, the leading-edge geometry (thickness and camber) must be shaped to produce a longer run of supersonic flow but a weaker recovery shock. These conditions, however, can generally only be obtained for very specific combinations of angles of attack and Mach numbers. For other "off-design" operating conditions these airfoils usually suffer lift and drag penalties. Supercritical airfoils have been used for many years on jet transport aircraft, and "supercritical-like" sections are used as tip sections on some helicopters. At subsonic speeds, supercritical airfoils can exhibit higher leading-edge suction peaks and are likely to encourage stall at lower angles of attack compared to conventional airfoils and thus would not be suitable for the inboard parts of helicopter blades.

As previously discussed, the effects of compressibility cause the aerodynamic center to migrate rearward on the chord with increasing Mach number. As the flow becomes transonic, the aerodynamic center shifts more quickly rearward resulting in the *Mach tuck* phenomenon. Mach tuck manifests as a rapid increase in nose-down pitching moment for a small change in Mach number. If Mach tuck occurs on the advancing tip of a helicopter rotor, then it may produce high blade and control loads and may effectively limit the forward flight speed. For most airfoils, the Mach tuck characteristic appears at a slightly higher Mach number after drag divergence as the shock waves begin to strengthen and move more rapidly aft on the airfoil surface.

Because it is possible for the rotor to exceed the tip drag divergence Mach number under various flight regimes such as during dives, it is obviously desirable to minimize the nose-down pitching moment trend to as high a Mach number as possible. The data in Fig. 7.37, taken from Ferri (1945), compare pitching moment results at a constant $C_l = 0.1$. Figure 7.37 shows that the nose-down trend in the pitching moment is minimized for symmetric airfoils. It is found that camber applied further to the rear of the airfoil generally results in a more severe tuck Mach problem than for an airfoil with nose camber. The increased pitching moment is caused by changes in the pressure distribution that result from the development of supercritical flow and shock waves. Thus supercritical airfoil design techniques can delay the Mach tuck problem to higher free-stream Mach numbers.

The tip of the advancing blade operates at low lift but at high Mach numbers. Therefore, from a power consumption point of view the drag characteristics of the airfoil at high Mach numbers are also important. Figure 7.38 shows the drag characteristics of several airfoils as a function of M_∞. In the low Mach number range, the drag stays nominally constant or may decrease slightly because of an interdependent effect of increasing Reynolds number. It is only when the critical Mach number is reached that the drag shows a more rapid rise. This drag rise occurs at lower Mach numbers as the thickness of the airfoil is increased. The same effect is obtained by increasing the angle of attack. The effect of camber also decreases the critical Mach number and so reduces the Mach number at which the drag rise occurs compared to symmetric airfoils of the same thickness.

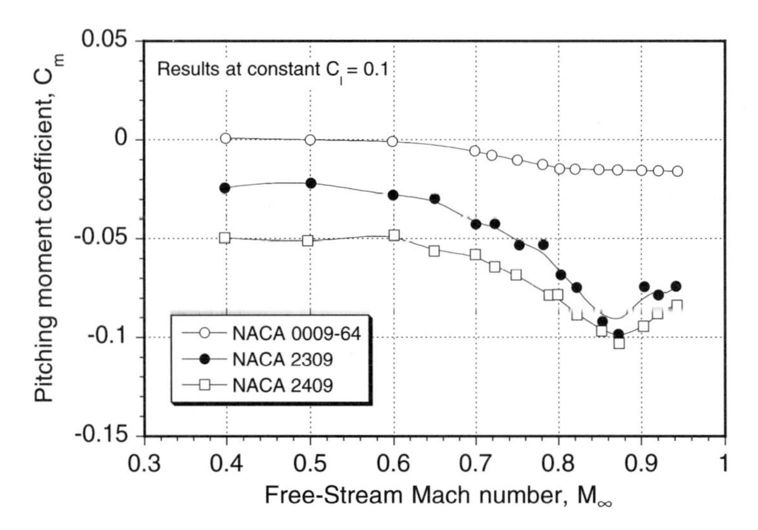

Figure 7.37 Effect increasing free-stream Mach number on the pitching moment at constant lift coefficient for several NACA series airfoils. Data source: Ferri (1945).

The value of M_∞ at which the drag coefficient increases significantly is known as the *drag divergence Mach number*, M_{dd}. Physically, drag divergence occurs because of an entropy loss through the shock wave coupled with additional pressure drag as a result of shock induced separation. Nitzberg & Crandall (1949) discuss the phenomenon in detail. The value of M_{dd} is often defined as the Mach number for which $dC_d/dM_\infty > 0.1$ or, alternatively, the Mach number for which the drag coefficient is twice its incompressible value at the same C_l or α. The drag divergence Mach numbers of airfoils at zero lift are plotted in Fig. 7.39 as a function of thickness ratio. It can be readily observed that thinner airfoils have a much higher drag divergence Mach number and will be a natural choice for the blade tip region.

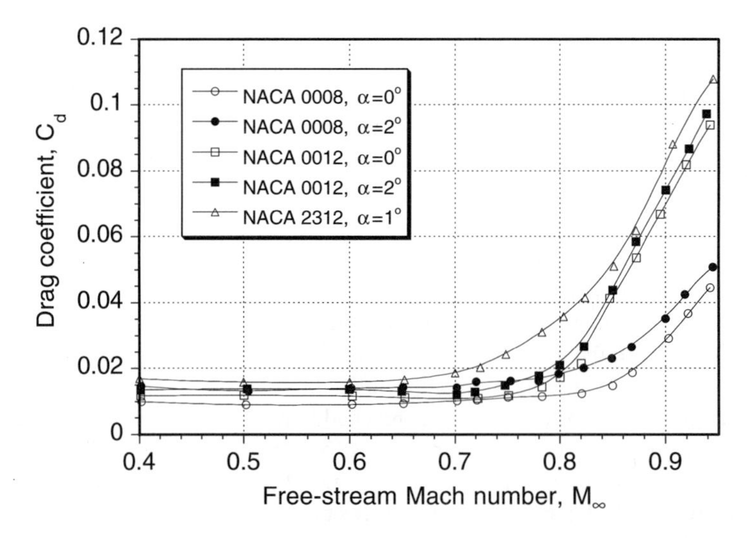

Figure 7.38 Effect of increasing free-stream Mach number on the drag of several NACA series airfoils. Data source: Ferri (1945).

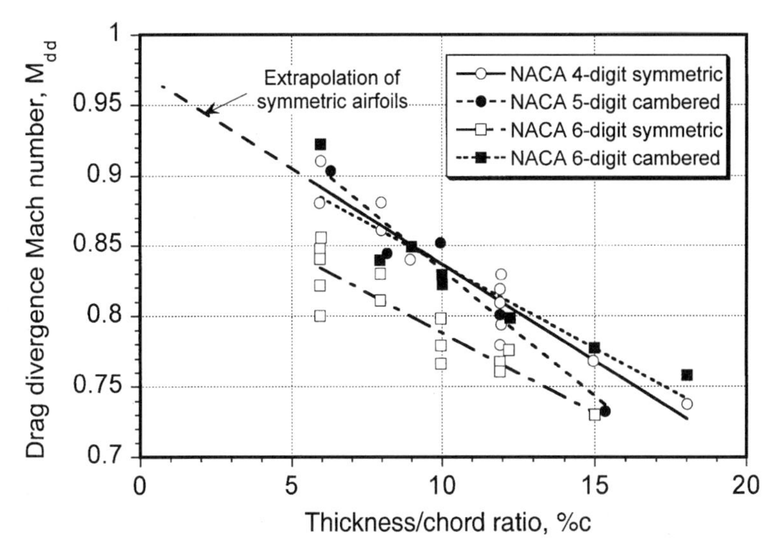

Figure 7.39 Drag divergence Mach numbers of several NACA series airfoils at zero lift. Data source: Collated information by Prouty (1986).

7.9 Advanced Rotor Airfoil Design

Several research programs have been undertaken with a view of improving helicopter performance by careful design and optimization of rotor airfoils. With appropriate design of both the airfoil sections and the blade geometry itself, conventional helicopters can now operate at flight speeds approaching 200 kts without being limited by significant stall or compressibility effects. To achieve these speeds by expansion of the flight boundary, it is necessary to vary the airfoil section along the blade to give optimal performance for the extreme operating regimes encountered at that blade station.

A program of airfoil development has been conducted by many of the major helicopter manufacturers and government research organizations. As previously discussed, the NACA 0012 airfoil represents a good compromise between high maximum lift, low pitching moment, and high drag divergence Mach number performance for an uncambered airfoil. As shown in Fig. 7.39, reducing the thickness ratio of the airfoil gives a marked improvement to the high Mach number performance. This allows a higher forward flight speed to be obtained with the rotor prior to the onset of increased rotor torque demands. Alternatively, for a given operational forward speed the rotor tip speed can be increased without incurring penalties of drag divergence and flow separation, and so rotor solidity could be reduced, thereby saving weight. However, reducing blade thickness limits the maximum lift capability of the airfoil at low Mach numbers and can seriously impact the retreating blade performance of the rotor. Therefore, generally the airfoil section thickness must be maintained to give a compromised high Mach number performance, while the high-lift performance is much improved by adding leading-edge camber. As shown previously in Figs. 7.28 and 7.29, the maximum lift capability of an airfoil improves rapidly with the addition of nose camber.

Unfortunately, the addition of camber also affects the shock strength on the airfoil lower surface at higher Mach numbers, causing a reduction in the drag divergence Mach number. Nevertheless, careful addition of leading-edge camber can restore the $C_{l_{\max}}$ performance back to the levels of at least the NACA 0012, while still retaining the advantages of a higher

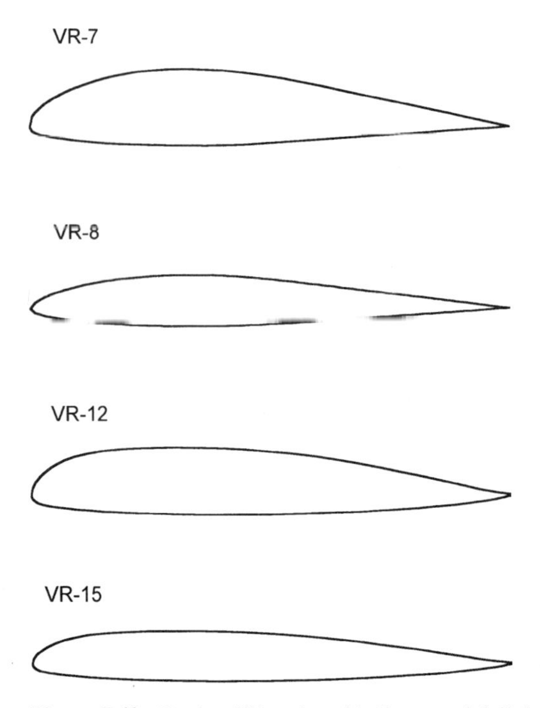

Figure 7.40 Boeing VR series of helicopter airfoil development.

drag divergence Mach number. Further improvements in high-lift capability can only be achieved by extending the camber further to the rear of the airfoil, but as shown previously in Fig. 7.24, this is at the expense of some increase in pitching moment. On the retreating side of the rotor disk, the high pitching moments that are associated with cambered airfoils can normally be tolerated as the dynamic pressure is relatively low. Yet, on the advancing side, the dynamic pressure is large and so the blade moments can be significant enough to manifest in high control loads, and possibly result in flight envelope restrictions. The final design of the airfoil section is usually a compromise in this regard.

The evolution of the Boeing (Vertol) VR airfoil series is shown in Fig. 7.40 – see Dadone & Fusushima (1975) and Dadone (1978, 1982, 1987). The latest airfoil sections, the VR-12 and the VR-15, represent the best compromise in terms of maximum lift capability at the lower Mach numbers typical of the retreating blade while maximizing the drag divergence Mach number and meeting hover requirements and control load limitations. These sections were designed with the aid of numerical methods using a potential flow/boundary layer interaction analysis and a viscous transonic analysis. These analyses were previously validated against experimental measurements on other airfoil shapes so that they could be used with confidence in the design process.

The ONERA has conducted a systematic development of helicopter airfoil sections [see Thilbert & Gallot (1977)]. These airfoil shapes are designated as the OA-family, for which a selection is shown in Fig. 7.41. The philosophy behind the design of these airfoil shapes also follows that of high $C_{l_{\max}}$ capability at low Mach numbers and a high drag divergence Mach number. The OA-206 is an example of a thin supercritical-like section, which exhibits a high drag divergence Mach number and gives potentially large improvement in advancing blade performance. The OA-209 is an example of an airfoil that is a compromise between advancing and retreating blade requirements, with gains in $C_{l_{\max}}$ relative to the NACA 0012 at low Mach numbers with some increase in the drag divergence Mach number. Recall that a high $C_{l_{\max}}$ capability is only required on the outboard sections of the blade (between

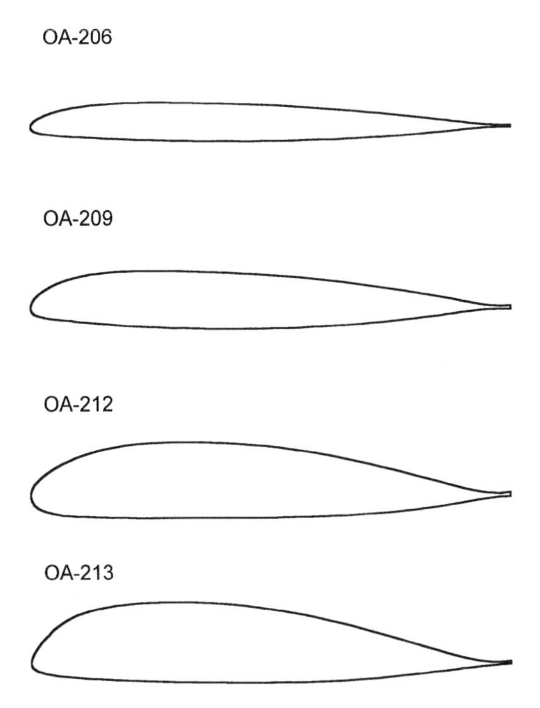

Figure 7.41 The ONERA OA airfoil series of helicopter airfoils.

50–80% of radius) and maximum lifting performance is relaxed further inboard. Thus, some reflex camber can be used on the inboard sections to generate lower pitching moments. The OA-212 is an example of such a reflexed section.

The Royal Aircraft Establishment and Westland Helicopters have conducted a systematic development of helicopter sections since the late 1960s. A good review of this work is given by Wilby (1979, 1998). The first airfoil in the series, the RAE (NPL) 9615, used nose camber to give a moderate increase in $C_{l_{max}}$ compared to the NACA 0012, with a small increase in drag divergence Mach number. Both improvements were made with only a small increase in pitching moment. Later airfoils that were developed included the 12% thick RAE 9645, which has more aft camber and gives a 30% increase in $C_{l_{max}}$ relative to the NACA 0012. The RAE 9648 is a 12% thick reflexed airfoil, which gives a significant nose-up pitching moment while retaining most of the advantages of the RAE 9645 (see also Fig. 6.16). The RAE 9634 airfoil is a thinner 8% thick section, which is designed to minimize transonic flow effects by giving a higher drag divergence Mach number and to delay the nose-down pitching moment trend to as high a Mach number as possible.

A series of high-lift low pitching moment airfoils have been devised by the US Army and NASA for helicopter applications. These are designated as the NASA RC-series – see Bingham (1975) and Bingham & Noonan (1982). The RC(3) airfoil families use a careful combination of nose camber, trailing-edge reflex camber, and a supercritical type thickness distribution to extract the highest static $C_{l_{max}}$ from the airfoil whilst retaining a very low pitching moment and a high drag divergence Mach number. The NASA RC(4) and RC(5) series were designed by Noonan (1990) for high maximum lift coefficients and are suitable for the inboard part of the blade. The RC(5) family has a lower thickness than the RC(4) family forward of the point of maximum thickness. The RC(6) series is described by Noonan (1991) and is a development of the RC(3) series designed for application at the tip of the blade. Unlike the VR, OA and RAE airfoils, thus far, the NASA RC-series airfoils have not been used on production helicopters.

7.10 Representing Static Airfoil Characteristics

The details of the airfoil pressure distribution and the variation with Mach number are important for airfoil design and airfoil selection purposes. However, in helicopter rotor performance and airloads analysis the details of the pressure distributions are usually not required because these are too expensive and time consuming to compute on a routine basis. More often it is required to mathematically model the airfoil characteristics in terms of lift, pitching moment, and drag coefficients as functions of angle of attack. These models can then be incorporated into a blade element model, such as those discussed in Chapter 3. Computational tools have not yet evolved to the point where the performance of the airfoil at high angles of attack near stall and at high subsonic and transonic Mach numbers near drag divergence can be obtained with confidence. Therefore, reliance is placed on experimental measurements. Experimental characteristics of the airfoil are usually obtained at discrete values of angle of attack from measurements in 2-D wind-tunnel tests.

7.10.1 *Linear Aerodynamics*

In the low angle of attack regime, and at subsonic Mach numbers, the airloads can be adequately modeled using the equations

$$C_n \approx C_l = c_0 + c_1\alpha, \tag{7.70}$$

$$C_m = m_0 + m_1\alpha, \tag{7.71}$$

$$C_d = d_0 + d_1\alpha + d_2\alpha^2, \tag{7.72}$$

where c_0, c_1, m_0, m_1, d_0, d_1, and d_2 are empirically derived coefficients obtained through curve fitting to the airfoil measurements. It is very important to recognize that these polynomials can only be used to represent airfoil characteristics below stall; they are totally invalid with significant amounts of flow separation or in the stalled flow regime.

The aerodynamic significance of the constants $c_0, c_1, m_0, m_1, d_0, d_1$, and d_2 are not always recognized but can be readily established. Usually it is desirable to express the airloads in terms of well-known (measurable) aerodynamic parameters, such as lift-curve-slope, zero lift angle, etc. The usual way of representing the lift is by using the equation

$$C_l = C_{l_\alpha}(\alpha - \alpha_0), \tag{7.73}$$

where C_{l_α} is the lift-curve-slope and α_0 is the zero lift angle of attack. Expanding gives

$$C_l = C_{l_\alpha}\alpha - C_{l_\alpha}\alpha_0 = c_0 + c_1\alpha. \tag{7.74}$$

Therefore, $c_0 = -C_{l_\alpha}\alpha_0$ and $c_1 = C_{l_\alpha}$. The pitching moment about the 1/4-chord is represented by

$$C_m = C_{m_0} + \left(\frac{1}{4} - \frac{x_{ac}}{c}\right) C_l, \tag{7.75}$$

where C_{m_0} is the zero lift moment and x_{ac} is the position of the aerodynamic center from the leading edge. Substituting for C_l and expanding gives

$$C_m = C_{m_0} + \left(\frac{1}{4} - \frac{x_{ac}}{c}\right) C_{l_\alpha}\alpha - \left(\frac{1}{4} - \frac{x_{ac}}{c}\right) C_{l_\alpha}\alpha_0, \tag{7.76}$$

which is in the form of

$$C_m = m_0 + m_1\alpha. \tag{7.77}$$

Therefore, the coefficient m_0 includes a combination of the usual aerodynamic parameters C_{m_0}, C_{l_α}, as well as x_{ac}. The coefficient m_1 is a term that represents the offset of the aerodynamic center from the 1/4-chord axis. The pressure drag can be obtained by resolving the components of the normal force and axial force through the angle of attack using

$$C_{d_p} = C_n \sin\alpha - C_a \cos\alpha. \tag{7.78}$$

The total drag (profile drag) is obtained by adding the contribution due to viscous shear to the pressure drag. (This is roughly C_{d_0} at low angles of attack.) Therefore, the total drag can be written as

$$C_d = C_{d_0} + C_n \sin\alpha - C_a \cos\alpha. \tag{7.79}$$

For real (viscous) flows, the axial force coefficient is no longer given by $C_a = C_n \tan\alpha$ because the effects of viscosity limit the leading-edge suction to some value that is less than the inviscid value (see also Sections 7.6.1 and 8.15.7). This effect can be modeled by using a leading-edge suction recovery factor, η_a, such that C_a is now given by

$$C_a = \eta_a C_n \tan\alpha. \tag{7.80}$$

Typically, η_a is close to but less than one. Thus, the pressure drag becomes

$$C_d = C_{d_0} + C_n \sin\alpha - \eta_a C_n \tan\alpha \cos\alpha \tag{7.81}$$

$$= C_{d_0} + C_n (1 - \eta_a) \sin\alpha. \tag{7.82}$$

For small angles, $\sin\alpha \approx \alpha$ and $C_n \approx C_l = C_{l_\alpha}\alpha$ so that the total drag is

$$\begin{aligned} C_d &= C_{d_0} + C_{l_\alpha}(1 - \eta_a)\alpha^2 \\ &= C_{d_0} + C_{l_\alpha}(1 - \eta_a)(\alpha - \alpha_0)^2 \\ &= C_{d_0} + C_{l_\alpha}(1 - \eta_a)\alpha^2 - 2C_{l_\alpha}(1 - \eta_a)\alpha_0\alpha + C_{l_\alpha}(1 - \eta_a)\alpha_0^2, \end{aligned} \tag{7.83}$$

which is of the form

$$C_d = d_0 + d_1\alpha + d_2\alpha^2. \tag{7.84}$$

This equation, like the previous equations for the lift and moment, is only valid in the regimes where the flow is fully attached. Note that $d_1 = 0$ for a symmetric airfoil ($\alpha_0 = 0$).

7.10.2 *Nonlinear Aerodynamics*

In the high angle of attack regime, separation and stall occur and the airloads become highly nonlinear functions of angle of attack. Therefore, the representation of the airfoil characteristics in this regime becomes more difficult. Usually this cannot be easily accomplished by using simple polynomial curve fits to the test data. However, mathematically representing the nonlinear airfoil characteristics by means of analytic functions or equations can still be accomplished in several different ways.

7.10.3 *Table Look-Up*

One common way of representing nonlinear airfoil characteristics is to store the measured airfoil data as tables. Generally, one table must be provided for each Mach number. A computer program can be written to manipulate these data and to extract interpolated values of C_l (or C_n), C_m, and C_d for any specified angle of attack and Mach number. This is a procedure used by large comprehensive rotor analyses, and it is also used for flight

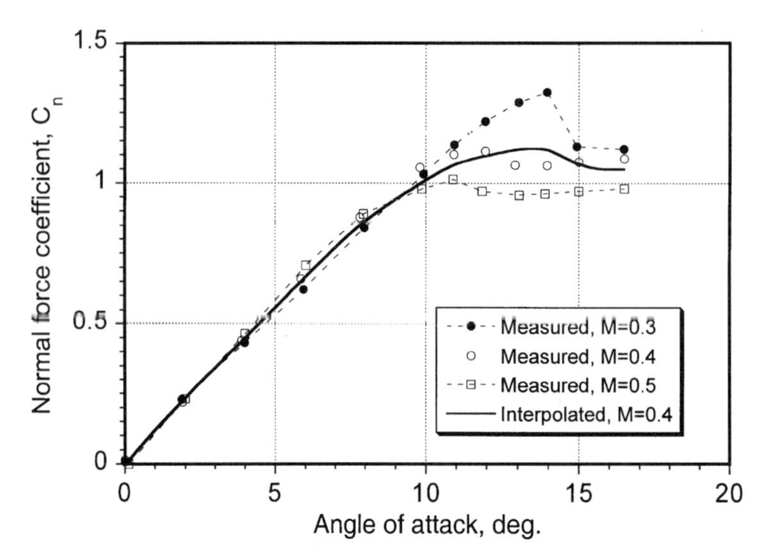

Figure 7.42 Interpolated static lift versus angle of attack characteristics using table look-up.

simulation work. The simulation program is usually executed using an array processor, which can perform table look-up and interpolation exceedingly quickly.

Representative experimental results for C_l versus α for a rotor airfoil at Mach numbers of 0.3, 0.4, and 0.5 are shown in Fig. 7.42. The results for the intermediate Mach number of 0.4 were found by linear interpolation between the results at $M_\infty = 0.3$ and $M_\infty = 0.5$. It will be seen that, in the low α range, the results are accurate, but for conditions near stall less reliable results may be produced. If the measurements are spaced at sufficiently small increments in Mach number, then the resulting interpolated data using this kind of scheme are normally viable.

7.10.4 *Direct Curve Fitting*

Curve fitting the airfoil characteristics using higher-order polynomials is another possible way of representing the nonlinear airfoil characteristics in the high angle of attack regime. For example, the lift may be represented using

$$C_l = \sum_{n=1}^{N} a_n \alpha^n. \tag{7.85}$$

Generally, as low an order of polynomial as possible would be used. Typically, $N = 3$ gives a reasonable approximation. Alternatively, a ratio of polynomials of the form

$$C_l = \frac{\displaystyle\sum_{n=1}^{N} a_n \alpha^{2n-1}}{1 + \displaystyle\sum_{m=1}^{M} b_m \alpha^{2m}} \tag{7.86}$$

will often prove acceptable. In each case, the coefficients in the series are obtained in a least-squares sense. As shown by Fig. 7.43, comparisons of direct curve fitting techniques with test data are reasonably adequate but tend to underpredict the magnitude of the airloads near maximum lift and also change the nature of the stall characteristic. This is typical

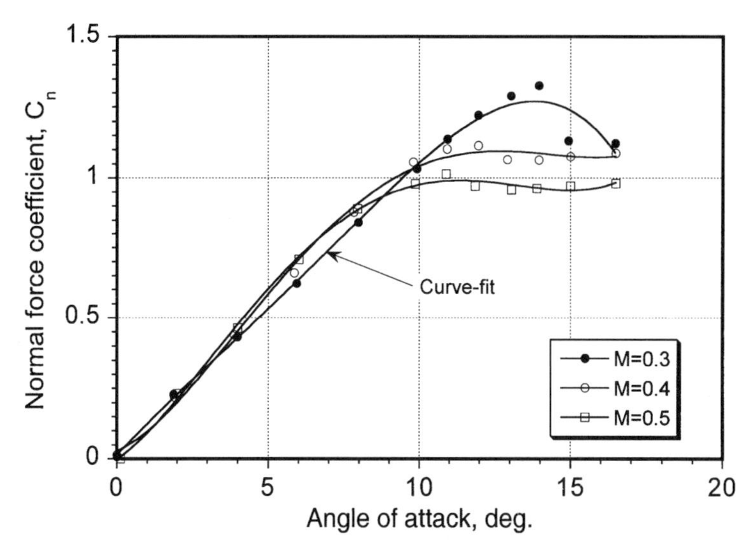

Figure 7.43 Curve-fitting static lift versus angle of attack characteristics using a higher-order polynomial in α.

whenever the airloads change rapidly with respect to angle of attack. However, polynomial representations generally do not work well over an extended range of angle of attack.

Alternatively, the representation of the airfoil characteristics can be broken into smaller ranges of angle of attack. For example, one curve can be used below stall, and another curve(s) in the poststall regime, while requiring that the curves be piecewise continuous. This method has some difficulties, especially when interpolating the curve fit coefficients and break points between Mach numbers. Also, the conditional branching in the computer program consumes more time, which may not be acceptable in many applications.

7.10.5 *Beddoes Method*

Another method, originally devised by Beddoes (1983), uses the Kirchhoff/Helmholtz solution for the lift on a flat plate with a fixed separation point to model the nonlinear airfoil characteristics. In the Kirchhoff/Helmholtz model, the lift or normal force coefficient on the airfoil, C_n, is approximated by the equation

$$C_n = 2\pi \left(\frac{1 + \sqrt{f}}{2} \right)^2 \alpha, \tag{7.87}$$

where 2π is the normal force-curve-slope for incompressible flow, f is the trailing-edge separation point (nondimensionalized with respect to chord), and α is the angle of attack – see Thwaites (1960). Thus, if the separation point can be determined, it is easy to compute the normal force coefficient. The expression in Eq. 7.87 may also be extended to encompass compressible flows where 2π is replaced by the force-curve-slope at the appropriate Reynolds number and Mach number [i.e., $C_{n_\alpha}(Re, M_\infty)$] and α is measured relative to the zero lift angle, that is,

$$C_n = C_{n_\alpha}(Re, M_\infty) \left(\frac{1 + \sqrt{f}}{2} \right)^2 (\alpha - \alpha_0). \tag{7.88}$$

To practically implement this procedure, the relationship between the separation point, f, and the angle of attack, α, must be obtained. These data are not generally known; however, an "effective" trailing-edge separation point, f, can then be deduced from the experimental measurements of the static C_n variation with α by rearranging Eq. 7.88 to solve directly for f, that is,

$$f = \left(2\sqrt{\frac{C_n}{C_{n_\alpha}(\alpha - \alpha_0)}} - 1 \right)^2. \tag{7.89}$$

The resulting curves are given by Beddoes (1983) and have been recalculated in Fig. 7.44 for the NACA 0012 airfoil over a range of Mach numbers.

After the values of f have been found, the lift at any angle of attack can be found by interpolating the values of f and finding the corresponding nonlinear value of C_n using Eq. 7.88. Beddoes (1983) shows that the f versus α curves all have a characteristic shape

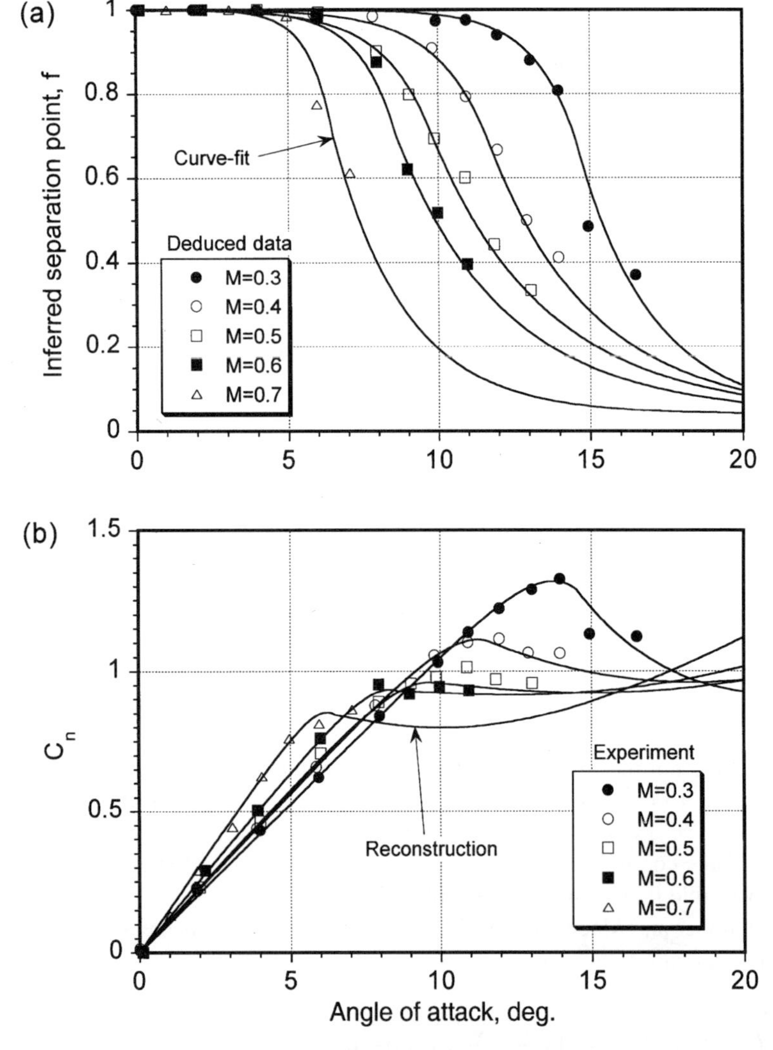

Figure 7.44 Reconstructions of static airfoil characteristics using the Beddoes method. (a) Position of effective separation point. (b) Reconstruction of the nonlinear lift.

for all Mach numbers, which turns out to be extremely convenient because the variations can then be generalized empirically in a fairly simple manner using the curve fits

$$f = \begin{cases} 1 - 0.3\exp\left(\dfrac{(\alpha - \alpha_0) - \alpha_1}{S_1}\right) & \text{if } \alpha \leq \alpha_1, \\ 0.04 + 0.66\exp\left(\dfrac{\alpha_1 - (\alpha - \alpha_0)}{S_2}\right) & \text{if } \alpha > \alpha_1. \end{cases}$$

The coefficients S_1 and S_2 define the static stall characteristic, that is, whether the stall occurs progressively or abruptly. The value of α_1 defines the break point corresponding to $f = 0.7$. This point is defined only as a matter of convenience; however, the value of α when $f \approx 0.7$ closely corresponds to the static stall angle of attack. The coefficients S_1, and S_2 and α_1 are easily determined for different Mach numbers from the static lift data. Their values are read from a data table [see Beddoes (1983)], with values for intermediate Mach numbers being performed using interpolation, as required.

The pitching moment as a function of angle of attack can also be established from Kirchhoff/Helmholtz theory. However, in practice the resulting equation is found inadequate and Beddoes has suggested several empirical relations. From the airfoil static data, the center of pressure at any angle of attack may be determined from the ratio C_m/C_n (allowing for the zero lift moment C_{m_0}). The variation can be plotted versus the corresponding value of the separation point and curve fitted using a low-order polynomial. One suitable curve fit is to use the form

$$\frac{C_{m_{0.25}}}{C_n} = k_0 + k_1(1 - f) + k_2 \sin(2\pi f^m), \tag{7.90}$$

where $k_0 = (0.25 - x_{ac})$ is the aerodynamic center offset from the 1/4-chord. The constant k_1 gives the direct effect on the center of pressure due to the growth of the separated flow region, and the constant k_2 helps describe the shape of the moment break at stall. Again, the values of k_0, k_1, k_2, and m can be obtained using a least-squares fit.

7.10.6 *High Angle of Attack Range*

An understanding of the aerodynamic behavior of airfoils in the high angle of attack regime is important for an understanding of effects produced in the reverse flow regime on the rotor. In the reverse flow region, the angle of attack increases through stall beyond 90° as the direction of the relative flow vector now changes from the trailing edge toward the leading edge of the airfoil. Although the velocities and dynamic pressure are low in this region, the high drag coefficients associated with flow separation may result in a higher overall parasitic power penalty.

The aerodynamics of airfoils at extreme angles of attack beyond the normal static stall angle and in reversed flow have been measured by Lock & Townend (1925), Anderson (1931), Naumann (1942), Critzos et al. (1955), and Leishman (1996). All measurements are at low Mach numbers. Typical results are reproduced in Fig. 7.45, which shows the lift, pitching moment, and drag characteristics as a function of angle of attack. Although the airfoil shape makes some difference to the nature of the stall characteristic at positive and negative angles of attack and in reverse flow, when the flow becomes fully separated the results become mostly independent of airfoil shape and are close to the values for a flat plate.

In the high angle of attack range, the quasi-steady lift coefficient can be modeled using

$$C_l = A \sin 2(\alpha - \alpha_0), \tag{7.91}$$

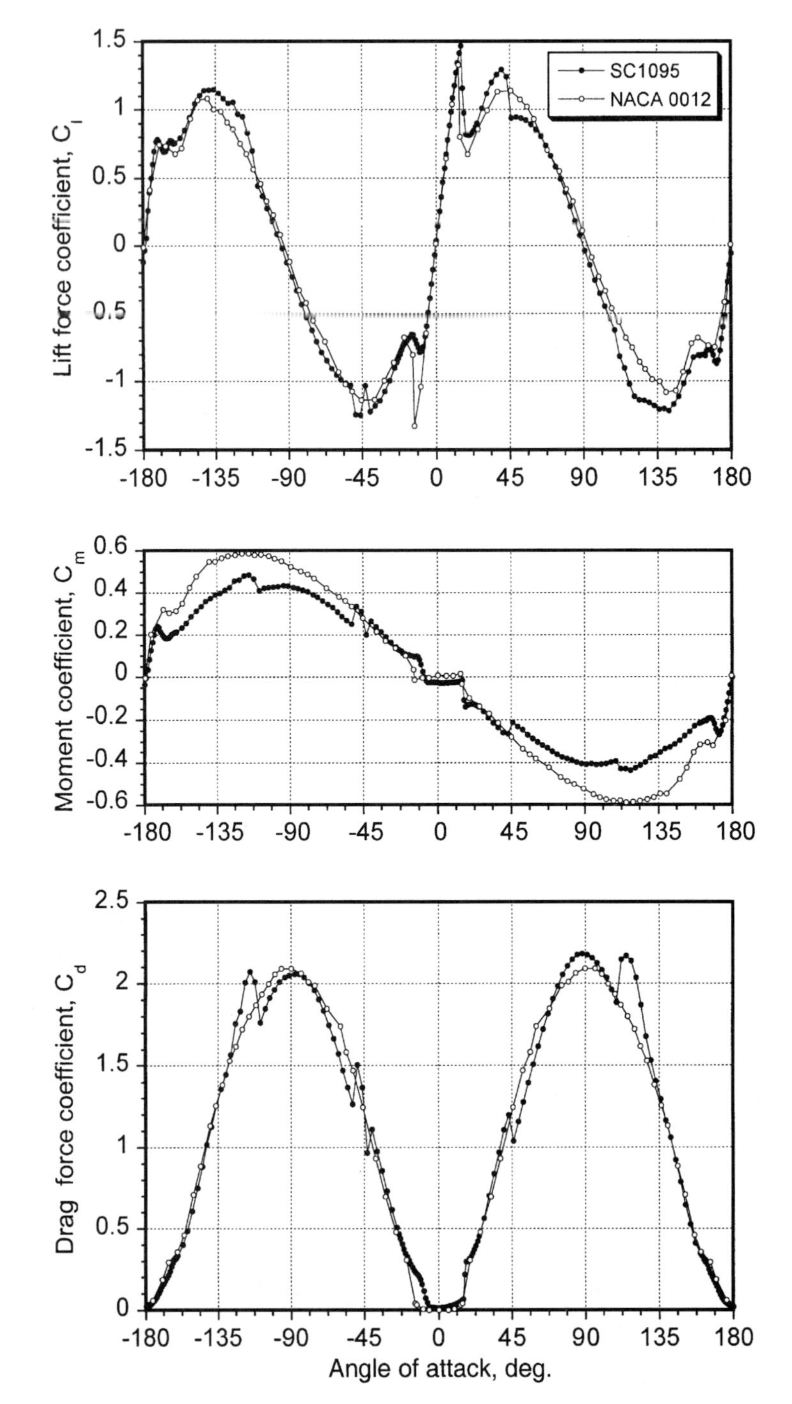

Figure 7.45 Measurements of lift, pitching moment, and drag for a complete 360° range of angle of attack at low Mach number. Data source: Critzos et al. (1955) and Leishman (1997).

where $A = 1.1$ for the NACA 0012 and $A = 1.25$ for the SC1095 when determined in a least-squares sense, with α being measured in degrees. An average value of 1.175 would seem reasonable for use with an arbitrary airfoil where the force coefficients are not known. For either airfoil, the center of pressure in the poststall regime moves close to mid-chord when $\alpha = 90°$ (i.e., $\bar{x}_{cp} = 0.5$). For the ranges $20° \leq (\alpha - \alpha_0) \leq 160°$ and $-160° \leq (\alpha - \alpha_0) \leq -20°$ the center of pressure varies approximately linearly with α and can be represented by

$$\bar{x}_{cp} = 0.33 + 0.0019(\alpha - \alpha_0) \quad \text{for } 20° \leq (\alpha - \alpha_0) \leq 160°, \tag{7.92}$$

$$\bar{x}_{cp} = 0.33 - 0.0019(\alpha - \alpha_0) \quad \text{for } -160° \leq (\alpha - \alpha_0) \leq -20°. \tag{7.93}$$

Alternatively, the pitching moment in the poststall regime can be modeled directly using

$$C_m = B \sin(\alpha - \alpha_0) + C \sin 2(\alpha - \alpha_0), \tag{7.94}$$

where $B = -0.5$ and $C = 0.11$ are the averages for the two airfoils, which would seem appropriate choices for an arbitrary airfoil. Finally, the drag can be modeled using

$$C_d = D + E \cos 2(\alpha - \alpha_0), \tag{7.95}$$

with $D = 1.135$ and $E = -1.05$ for both airfoils.

While the forgoing results are based on the assumption that the flow in the reverse flow region is quasi-steady, these estimates of the airloads may not necessarily be representative of the actual flow state on the rotor. This is because the reverse flow region is actually dynamic (unsteady) in nature, and the flow may not have time to adjust to the quasi-steady values. For a typical helicopter it can be shown that the aerodynamic time scales are between 2 and 6 semi-chords of airfoil travel through the flow in the reverse flow region. Based on known results from unsteady airfoil measurements (see Chapter 8), these time scales are too short to allow the flow to readjust, even if the angles of attack were to become low enough to promote flow reattachment with reverse flow. For this reason, it is more accurate to calculate the airloads in the reverse flow region by means of a *dynamic stall* model (see Chapter 9). However, in the absence of such a model or other information it would be sufficient to assume that $C_l = 0$, $\bar{x}_{cp} = 0.5$, and $C_d = 2.05$ in the reverse flow region when using the blade element method. (Note: C_d for a flat plate ≈ 2 at $\alpha = 90°$.)

7.11 Chapter Review

This chapter has described some basic aerodynamic characteristics of airfoil sections and has provided a basis for assessing the potential impact of airfoil design and selection on rotor performance. Methods of geometrically defining airfoils have been described following the NACA approach of combining camberlines and thickness envelopes. This is justified because many of the airfoils used on current helicopter rotors have their origin in the NACA sections. Also, the ideas of combining loading distributions associated with thickness and camber allow the primary effects of geometric shape on the chordwise loading and overall aerodynamic characteristics to be assessed. However, the primary discussion in this chapter has been centered around published measurements of static airfoil characteristics. These results, while from more than one primary source, allow a good basis from which to compare airfoil behavior and for the most part are considered relatively unbiased by the wind-tunnel test facility or by testing techniques. For these latter reasons, it has not been considered fruitful to dwell on the relative merits of airfoils designed by the various competing helicopter manufacturers.

Airfoils designed for helicopter applications have traditionally been obtained through a long evolutionary process, in which various levels of theory and experimental measurements have been meshed together in the pursuit of airfoil shapes with higher values of maximum lift, better lift-to-drag ratios, lower pitching moments, and higher drag divergence Mach numbers. It has been shown that, in general, these requirements are conflicting, making the design of general purpose rotor airfoils extremely challenging. Instead, various families of airfoils have been developed and optimized to meet the specific needs of different parts of the rotor blade. For example, airfoils with high camber and moderate thickness, which give high values of maximum lift, are used between 60 and 85% of blade radius. Much thinner airfoils, perhaps even those with supercritical-like shapes, give relatively high drag divergence Mach numbers and have been designed for the tip region of the blade ($>85\%R$). The use of different airfoils along the blade is made easier today because of composite manufacturing capability, which involves little additional design and production costs over a blade made with a single airfoil section.

The principles of integrating surface pressure and shear stress distributions to obtain lift, drag, and pitching moment coefficients have been reviewed. Typical airfoil characteristics have been discussed, along with the limits of conventional linearized theories in predicting this behavior. Because of the importance of low pitching moments in the design of helicopter airfoils, the ideas of aerodynamic center and center of pressure have been reviewed. The influence of Reynolds number and Mach number on airfoil characteristics have been highlighted. Although these parameters have both dependent and interdependent effects on airfoil behavior, it has been possible to isolate the basic effects and assess their significance on maximum lift and other important airfoil characteristics. Because the design of airfoils with high values of maximum lift can result in smaller and lighter rotors with lower solidity, the geometric shape of the airfoil and other factors affecting maximum lift have been discussed in detail.

While an improved understanding of airfoil characteristics will usually lead to an improved analysis capability of existing rotor designs, and may lead to new rotors optimized for greater performance in both hover and high speed forward flight, the performance of the rotor cannot be completely parameterized on the basis of static (steady) considerations alone. Therefore, in the next two chapters, the important role of unsteady aerodynamics on the problem of airfoil behavior will be discussed.

7.12 Questions

7.1. Write a computer program to plot the geometry of the NACA four-digit airfoil sections. Plot the shape of the NACA 0018 airfoil section. Using an enlarged plot, clearly show the location of the nose radius and how it is faired into the lower and upper surfaces.

7.2. Use the 230 camberline to construct the coordinates of the NACA 23018 airfoil section. Clearly show the construction of the nose radius relative to the camberline and in relation to the lower and upper surfaces of the airfoil, and show how the nose radius protrudes slightly forward of $x = 0$.

7.3. Construct the family of airfoils: 22015, 23015, and 25015. By means of thin-airfoil theory, compute the zero-lift angle of attack and pitching moment coefficient for this series of airfoils. Comment on your results.

7.4. Consider the family of airfoils: NACA 24006, 24012, and 24018. Use linear superposition principles and the tabulated results in Abbott & von Doenhoff to show the effects of thickness on the distribution of chordwise pressure. Compare the results at the same mean lift coefficient of 1.0. Comment on your results.

7.5. The upper and lower surface pressure coefficients on a particular airfoil at a particular angle of attack are approximated by the functions

$$C_{p_u}(\bar{x}) = -(100\bar{x}^{3/4}\, e^{-(30\bar{x}-1)} + 0.75\bar{x}^{1/7}(1-\bar{x})),$$
$$C_{p_l}(\bar{x}) = 10\bar{x} \quad \text{for } 0 \le \bar{x} \le 0.1,$$
$$C_{p_l}(\bar{x}) = 1.11(1-\bar{x}) \quad \text{for } 0.1 \le \bar{x} \le 1.$$

The airfoil geometry is given by the functions

$$y(\bar{x}) = \pm 0.06\sqrt{\left(1 - \tfrac{1}{9}(8\bar{x}-3)^2\right)} \quad \text{for } 0 \le \bar{x} \le 3/5,$$
$$y(\bar{x}) = \pm 0.12(1-\bar{x}) \quad \text{for } 3/5 \le \bar{x} \le 1.$$

Plot the pressure distribution with respect to chord and with respect to thickness. Estimate numerically, and for different numbers of points: (a) the normal force coefficient, (b) the axial force (leading-edge suction) coefficient, (c) the pitching moment coefficient about the leading-edge, (c) the pitching moment coefficient about the 0.27-chord.

7.6. Test results for a new helicopter airfoil section show that the normal force coefficient and the pitching moment coefficient about the 1/4-chord vary with the angle of attack as given in the table below for Mach numbers of 0.3, 0.5, and 0.7. For each Mach number, find the lift-curve-slope in the low angle of attack range, the zero-lift angle of attack, the aerodynamic center, and the zero-lift moment coefficient. Comment on your results.

$M_\infty = 0.3$				$M_\infty = 0.5$				$M_\infty = 0.7$			
α	C_n	α	C_m	α	C_n	α	C_m	α	C_n	α	C_m
−10.0	−0.690	−16.0	0.199	−10.0	−0.690	−16.0	0.199	−10.0	−0.730	−16.0	0.199
−8.0	−0.610	−16.0	0.198	−8.0	−0.600	−16.0	0.184	−8.0	−0.640	−16.0	0.130
−6.0	−0.530	−8.0	−0.012	−6.0	−0.510	−8.0	−0.0104	−6.0	−0.720	−8.0	−0.0068
−4.0	−0.312	−7.0	−0.011	−4.0	−0.350	−7.0	−0.0114	−4.0	−0.420	−7.0	−0.0252
−3.0	−0.201	−5.0	−0.0091	−3.0	−0.229	−5.0	−0.0100	−3.0	−0.270	−5.0	−0.0335
−2.0	−0.093	−3.0	−0.0081	−2.0	−0.108	−3.0	−0.0095	−2.0	−0.120	−3.0	−0.0243
−1.5	−0.039	−2.5	−0.0071	−1.5	−0.047	−2.5	−0.0091	−1.5	−0.045	−2.5	−0.0243
−1.0	0.058	−1.0	−0.0060	−1.0	0.014	−1.0	−0.0092	−1.0	0.030	−1.0	−0.0151
0.0	0.125	0.0	−0.0050	0.0	0.135	0.0	−0.0071	0.0	0.180	0.0	−0.0140
1.0	0.232	2.0	−0.0040	1.0	0.254	2.0	−0.0040	1.0	0.333	2.0	−0.0050
2.0	0.339	4.0	−0.0010	2.0	0.374	4.0	−0.0009	2.0	0.486	4.0	−0.0093
3.0	0.447	6.5	0.0010	3.0	0.493	6.5	0.0042	3.0	0.639	6.5	−0.0323
3.5	0.500	12.0	0.0058	3.5	0.553	12.0	−0.0100	3.5	0.708	12.0	−0.121
4.0	0.554	15.5	0.0045	4.0	0.612	15.5	−0.0489	4.0	0.770	15.5	−0.0122
6.0	0.768	16.0	−0.0610	6.0	0.851	16.0	−0.0610	6.0	0.875	16.0	−0.0610
8.0	0.983			8.0	1.060			8.0	0.857		
10.0	1.180			10.0	1.210			10.0	0.903		
12.0	1.360			12.0	1.220			12.0	0.963		
14.0	1.540			14.0	1.250			14.0	1.022		
16.0	1.650			16.0	1.207			16.0	1.081		
17.0	1.653			17.0	1.189			17.0	1.105		

7.7. An 11 cm chord model of the airfoil in the previous question is to be tested in a 30 cm-by-25 cm high speed pressurized wind-tunnel. The airfoil completely spans the larger dimension of the working section and can be pivoted about the 1/4-axis by an actuator system. The wind-tunnel operates at 5 atmospheres and the temperature in the working section is 15°. *Estimate* the design loads for the model support structure and actuator system. Comment on your results.

7.8. When testing a new airfoil section for helicopter applications, the lift variation at $M_\infty = 0.4$ shows an abrupt break at an angle of attack of 13.5 degrees. What static stall mechanism may be displayed by this airfoil? Describe the various steps that would be taken to confirm the stall type.

7.9. The drag on an airfoil (at low angles of attack) can be estimated by measuring the streamwise velocity in the wake behind the airfoil in a plane normal to the chord. The resulting distribution of velocity can be approximated by the equation

$$\frac{v(y)}{V_\infty} = \left[1 - \frac{1}{2}\cos\left(\frac{\pi y}{2W}\right)\right],$$

where $2W$ is the total width of the wake normal to the airfoil chord. Using the "momentum deficiency" approach (application of momentum conservation equation in integral form to a control volume surrounding the airfoil and its wake), find an expression for the drag coefficient C_d in terms of W.

Bibliography

Abbott, I. H., von Doenhoff, A. E., and Stivers, L. S., Jr. 1945. "Summary of Airfoil Data," NACA Rep. No. 824.

Abbott, I. H. and von Doenhoff, A. E. 1949. *Theory of Wing Sections, Including a Summary of Airfoil Data,* McGraw-Hill Book Co., Inc., NewYork. (Also, Dover Publications, New York, 1959.)

Anderson, R. F. 1931. "The Aerodynamic Characteristics of Six Commonly Used Airfoils over a Large Range of Positive and Negative Angles of Attack," NACA TN 397.

Beavan, J. A. and Lock, C. N. H. 1939. "The Effect of Blade Twist on the Characteristics of the C.30 Autogyro," ARC R & M 1727.

Beddoes, T. S. 1983. "Representation of Airfoil Behavior," *Vertica*, 7 (2), pp. 183–197.

Benson, R. G., Dadone, L. U., Gormont, R. E., and Kohler, G. 1973. "Influence of Airfoils on Stall Flutter Boundaries of Articulated Helicopter Rotors." *J. American Helicopter Soc.*, 18 (1), pp. 36–46.

Bezard, H. 1992. "Rotor Blade Design by Numerical Optimization and Unsteady Calculations," 48th Annual Forum of the American Helicopter Soc., Washington DC, June 3–5.

Bingham, G. J. 1975. "An Analytical Evaluation of Airfoil Sections for Helicopter Rotor Applications," NASA TN D-7796.

Bingham, G. J. and Noonan, K. W. 1982. "Two-Dimensional Aerodynamic Characteristics of Three Rotorcraft Airfoils at Mach Numbers from 0.35 to 0.90," NASA TP-2000, AVRADCOM TR-82-B-2.

Caradonna, F. X. and Isom, M. P. 1972. "Subsonic and Transonic Potential Flow over Helicopter Rotor Blades," *AIAA J.*, 10 (12), pp. 1606–1612.

Caradonna, F. X. and Isom, M. P. 1976. "Numerical Calculation of Unsteady Transonic Potential Flow over Helicopter Rotor Blades," *AIAA J.*, 14 (4), pp. 482–487.

Caradonna, F. X. and Philippe, J. J. 1978. "The Flow over a Helicopter Blade Tip in the Transonic Regime," *Vertica*, 2 (1), pp. 43–60.

Critzos, C. C., Heyson, H. H., and Boswinkle, R. W. 1955. "Aerodynamic Characteristics of NACA 0012 Airfoil Section at Angles of Attack from 0° to 180°," NACA TN 3361.

Dadone, L. U. and Fusushima, T. 1975. "A Review of Design Objectives for Advanced Helicopter Rotor Airfoils," American Helicopter Soc. Northeast Region Symp., Hartford, CT.

Dadone, L. U. 1978. "Rotor Airfoil Optimization: An Understanding of the Physical Limits," 34th Annual Forum of the American Helicopter Soc., Washington DC, May 15–17.

Dadone, L. U. 1982. "The Role of Analysis in the Aerodynamic Design of Advanced Rotors," AGARD CP 334.

Dadone, L. U. 1987. "Future Directions in Helicopter Rotor Development," American Helicopter Soc. Aeroacoustics Specialists' Meeting, Arlington, TX, Feb. 25–27.

de la Cierva, J. and Rose, D. 1931. *Wings of Tomorrow: The Story of the Autogiro*, Brewer, Warren & Putnam, New York.

Eppler, R. and Somers, D. M. 1980. "A Computer Program for the Design and Analysis of Low-Speed Airfoils, Including Transition," NASA TM-80210 (Supplement NASA TM-81862).

Evans, W. T. and Mort, K. W. 1959. "Analysis of Computed Flow Parameters for a Set of Sudden Stalls in Low Speed Two-Dimensional Flow," NASA TN D-85.

Ferri, A. 1945. "Completed Tabulation in the United States of Tests of 24 Airfoils at High Mach Numbers," NACA WR L-143. Also ACR No. L5E21.

Flemming, R. J. 1982. "An Experimental Evaluation of Advanced Rotorcraft Airfoils in the NASA Ames Eleven-Foot Wind Tunnel," NASA CR-166587.

Gault, D. E. 1957. "A Correlation of Low Speed Airfoil Section Stalling Characteristics with Reynolds Number and Airfoil Geometry," NACA TN 3963.

Gessow, A. and Myers, G. C. 1952. *Aerodynamics of the Helicopter*, Macmillan Co., New York (republished by Frederick Ungar Publishing, New York, 1967), pp. 244–249.

Glauert, H, 1927. "The Effect of Compressibility on the Lift of an Aerofoil," ARC R & M 1095.

Hess, J. L. and Smith, A. M. O. 1967. "Calculation of Potential Flow about Arbitrary Bodies," *Progress in Aeronaut. Sci.*, 8 (1), Pergamon Press, pp. 1–138.

Hicks, R. M. and McCroskey, W. J. 1980. "An Experimental Evaluation of a Helicopter Rotor Section Designed by Numerical Optimization," NASA TM 78622.

Horton, H. P. 1967. "A Semi-Empirical Theory for the Growth and Bursting of Laminar Separation Bubbles," ARC CP 1073.

Jacobs, E. N. and Pinkerton, R. M. 1935. "Tests in the Variable-Density Wind Tunnel of Related Airfoils Having the Maximum Camber Unusually Far Forward," NACA Report 537.

Jacobs, E. N. and Sherman, A. 1937. "Airfoil Section Characteristics as Affected by Variations of the Reynolds Number," NACA TR 586.

Kennedy, J. L. and Marsden, D. J. 1978. "A Potential Flow Design Method for Multi-Component Airfoil Sections," *J. of Aircraft*, 15 (1), pp. 47–52.

Kuo, Y. H. 1953. "On the Flow of an Incompressible Viscous Fluid Past a Flat Plate at Moderate Reynolds Numbers," *J. of Math. & Phys.*, 32, pp. 83–100.

Ladson, C. L., Brooks, C. W., Hill, A. S., and Sproles, D. W. 1995. "Computer Program to Obtain Ordinates for NACA Airfoils," NASA Technical Memorandum 4741.

Leishman, J. G. 1996. "Experimental Investigation into the Aerodynamic Characteristics of Helicopter Rotor Airfoils with Ballistic Damage," ARL-CR-295.

Liebeck, R. H. 1992. "Laminar Separation Bubbles and Airfoil Design at Low Reynolds Numbers," AIAA Paper 92-2735-CP.

Liepman, H. W. 1946. "The Interaction between Boundary Layer and Shock Waves in Transonic Flow," *J. Aeronaut. Sci.*, 13 (912), pp. 623–637.

Lock, C. N. H. and Townend, H. C. H. 1925. "Lift and Drag of Two Aerofoils Measured over 360-Degree Range of Incidence," ARC R & M 958.

Loftin, L. K. 1945. "Effects of Specific Types of Surface Roughness on Boundary-Layer Transition," NACA ACR L5J29a (Wartime Report No. L-48).

Malone, J. B., Narramore, J. C., and Sankar, L. N. 1989. "An Efficient Airfoil Design Method Using the Navier–Stokes Equations," AGARD FPD Specialists' Meeting on Computational Methods for Aerodynamic Design (Inverse) and Optimization, Loen, Norway, May 22–23.

McCroskey, W. J. 1977. "Some Current Research in Unsteady Fluid Dynamics," *J. of Fluids Eng.*, 99 (1), pp. 8–39.

McCroskey, W. J., McAlister, K. W., Carr, L. W., and Pucci, S. L. 1982. "An Experimental Study of Dynamic Stall on Advanced Airfoil Sections," Vols. 1,2 & 3 NASA TM-84245.

McCroskey, W. J. and Baeder, J. D. 1985. "Some Recent Advances in Computational Aerodynamics for Helicopter Applications," NASA TM-86777.

McCroskey, W. J. 1987. "A Critical Assessment of Wind Tunnel Results for the NACA 0012 Airfoil," AGARD Fluid Dynamics Panel Symp. on Aerodynamic Data Accuracy and Quality: Requirements and Capabilities on Wind Tunnel Testing, Naples, Italy, 28 Sept.–2 Oct. See also NASA Technical Memorandum 100019.

McCullough, G. B. and Gault, D. E. 1953. "Examples of Three Types of Stall," NACA TN 2502.

Moss, G. F. and Murdin, P. M. 1965. "Two Dimensional Low-Speed Tunnel Tests on the NACA 0012 Section Including Measurements Made during Pitching Oscillations at the Stall," ARC C & P 1145.

Munk, M. M. 1922. "General Theory of Thin Wing Sections," NACA TR 142.

Munk, M. M. 1924. "Elements of the Wing Section Theory and of the Wing Theory," NACA TR 191.

Narramore, J. C. and Yen, J. G. 1982. "A New Transonic Airfoil Design Method and Its Application to Helicopter Rotor Airfoil Design," 38th Annual Forum of the American Helicopter Soc., Anaheim, CA, May 4–7.

Narramore, J. C. 1994. "Computational Fluid Dynamics Development and Validation at Bell Helicopter," AGARD CP-552.

Narramore, J. C. and Yen, J. G. 1997. "Computing Nonlinear Airfoil Characteristics at High Mach Numbers," American Helicopter Soc. Technical Specialists' Meeting on Rotorcraft Acoustics and Aerodynamics, Williamsburg, Oct. 28–30.

Naumann, A. 1942. "Pressure Distribution on Wings in Reversed Flow," NACA TM 1011.

Nitzberg, G. E. and Crandall, S. 1949. "A Study of Flow Changes Associated with Airfoil Section Drag Rise at Supercritical Speeds," NACA TN 1813.

Noonan, K. W. 1990. "Aerodynamic Characteristics of Two Rotorcraft Airfoils Designed for Application to the Inboard Region of a Main Rotor Blade," NASA TP-3009, AVSCOM TR-90-B-005.

Noonan, K. W. 1991. "Aerodynamic Characteristics of a Rotorcraft Airfoil Designed for the Tip Region of a Main Rotor Blade," NASA Technical Memorandum 4264, AVSCOM Technical Report 91-B-003.

Pearcy, H. H. 1955. "Some Effects of Shock-Induced Separation of Turbulent Boundary Layer in Transonic Flow Past Aerofoils," ARC R & M 3108.

Prandtl, L. 1928. "Motion of Fluids with Very Little Viscosity," NACA TN 452. (Originally presented to the 3rd Int. Mathematical Congress, Heidelberg, Germany, 1904.)

Prandtl, L. and Tietjens, O. G. 1934. *Fundamentals of Hydro- and Aeromechanics*, United Engineering Trustees, Inc. Published by Dover, New York, 1957.

Prouty, R. W. 1975. "A State-of-the-Art Survey of Two-Dimensional Airfoil Data." *J. American Helicopter Soc.*, 20 (4), 14–25.

Prouty, R. W. 1986. *Helicopter Performance, Stability, and Control*, PWS Engineering Publishing, Boston, MA. Chapter 6.

Racisz, S. F. 1952. "Effects of Independent Variations of Mach Number and Reynolds Number on the Maximum Lift Coefficients of Four NACA 6-Series Airfoil Sections," NACA TN 2824.

Reigels, F. W. 1961. *Aerofoil Sections – Results from Wind-Tunnel Investigations and Theoretical Foundations*, translated from the German by D. G. Randall, Butterworths, London.

Rubbert, P. E. 1964. "Theoretical Characteristics of Arbitrary Wings by a Nonplanar Vortex Lattice Method," Boeing Co. D6-9244.

Schlichting, H. 1979. *Boundary Layer Theory*, 7th Ed., McGraw-Hill, New York.

Schrenk, O. 1927. "Systematic Investigation of Joukowski Wing Sections," NACA TM 422.

Sloof, J. W., Wortmann, F. X. and Duhon, J. M. 1975. "The Development of Transonic Airfoils for Helicopters," 31st Annual National Forum of the American Helicopter Soc., Washington DC, May 13–15.

Stack, J. and von Doenhoff, A. E. 1934. "Tests of 16 Related Airfoils at High Speeds," NACA Report 492.

Steinle, F. and Stanewsky, E. 1982. "Wind Tunnel Flow Quality and Data Accuracy Requirements," AGARD Advisory Report 184.

Stivers, L. S., Jr. and Rice, F. J., Jr. 1946. "Aerodynamic Characteristics of Four NACA Airfoil Sections Designed for Helicopter Rotor Blades," NACA RB L5K02.

Stivers, L. S., Jr. 1954. "Effects of Subsonic Mach Number on the Forces and Pressure Distributions on Four NACA 64A-Series Airfoil Sections at Angles of Attack as High as 28°," NACA TN 3162.

St. Hilaire, A. O., Carta, F. O., Fink, M. R., and Jepson, W. D. 1979. "The Influence of Sweep on the Aerodynamic Loading of a NACA 0012 Airfoil," Vol. 1 – Technical Report, NASA CR 3092.

Tetervin, N. 1943. "Tests in the NACA Two-Dimensional Low-Turbulence Tunnel of Airfoil Sections Designed to Have Small Pitching Moments and High Lift-Drag Ratios," NACA CB 3I13.

Theodorsen, T. and Garrick, I. E. 1932. "General Potential Theory of Arbitrary Wing Sections," NACA TR 452.

Thibert, J. J. and Gallot, J. 1977. "A New Airfoil Family for Rotor Blades," 3rd European Rotorcraft and Powered Lift Aircraft Forum, Germany, Sept.

Thwaites, B. 1960. *Incompressible Aerodynamics*, Oxford University Press, Oxford, England.

Von Mises, R. 1945. *Theory of Flight*, McGraw-Hill, New York. Republished 1959 by Dover Publications, Inc., New York.

Whitcomb, R. T. and Clark, L. R. 1965. "An Airfoil Shape for Efficient Flight at Supercritical Mach Numbers," NASA TM X-1109.

Whitcomb, R. T. 1976. "Advanced Transonic Aerodynamic Technology," NASA CP 2001 *Advances in Engineering Science*, Vol. 4.

Wilby, P. G. 1979. "The Aerodynamic Characteristics of Some New RAE Blade Sections and Their Potential Influence on Rotor Performance," 5th European Rotorcraft Forum, Amsterdam, The Netherlands, Sept. 4–7.

Wilby, P. G. 1998. "Shockwaves in the Rotor World – A Personal Perspective of 30 Years of Rotor Aerodynamic Research in the UK," *The Aeronaut. J.*, 102 (1013), pp. 113–128.

Wood, M. E. 1979. "Results from Oscillatory Pitch Tests on the NACA 0012 Blade Section," Aircraft Research Association, Bedford, UK, ARA Memo 220.

Young, A. D. 1989. *Boundary Layers*, AIAA Education Series, Washington DC.

CHAPTER 8

Unsteady Aerodynamics

The addition of the dimension "time" to steady aerodynamics has far-reaching effects, both practial and theoretical. There is the practical necessity for coping with many important problems involving nonsteady phenomena such as flutter, buffeting, transient flows, gusts, dynamic response in flight, maneuvers, and stability. Apart from the many applications, theoretical nonsteady aerodynamics embraces and sheds light on the realm of steady aerodynamics and introduces interesting new methods.

I. E. Garrick (1957)

8.1 Introduction

The confidence levels in the design of new helicopters are greatly improved by the ability to predict accurately the aerodynamic behavior of the rotor system at all corners of the operational flight envelope. One difficulty and uncertainty in this process is to fully account for *unsteady aerodynamic effects*, especially in high speed forward flight and during maneuvers. In the previous chapter, the quasi-steady aerodynamic characteristics of rotor airfoils have been discussed in detail. Yet, these attributes alone are not necessarily the best indicator as to whether a given airfoil will operate successfully in the rotor environment or will meet the demands required of any given rotor design. The next level of consideration is to examine unsteady aerodynamic effects and to assess their impact on the prediction of the airloads and performance of helicopter rotors.

In the context of rotor airloads prediction, the mathematical modeling of unsteady airfoil behavior is one of formidable complexity. While the absence of significant flow separation reduces somewhat the complexity of the problem, a complete understanding of unsteady airfoil behavior even in attached flow has not yet been obtained. The additional problem of dynamic flow separation[1] is still the subject of ongoing research, and completely satisfactory models of the problem have not yet been developed. This is particularly true for the rotor case, where the rotor blades encounter a broad spectrum of unsteady effects from a number of different sources. The most obvious are the excursions in angle of attack resulting from blade flapping and pitch control inputs. The additional effects of the rotor wake, with its embedded concentrated tip vortices, lead to regions of the rotor disk that can experience large perturbations in angle of attack over very short time scales. The problems are compounded by the three-dimensional effects found at the blade tips, which can be locally transonic during forward flight. Therefore, the problems of defining accurately the unsteady aerodynamic flow field on the rotor is quite formidable.

The principle focus of the present chapter is to describe the key physical features and techniques for modeling the unsteady aerodynamic effects found on airfoils operating under nominally attached flow conditions away from stall. The essential physics of nonsteady airfoil problems can be observed from simplified two-dimensional (2-D) experiments, and interpretations of the behavior can be supported by theoretical or numerical models. The

[1] This is called dynamic stall and is discussed in detail in Chapter 9.

"classical" unsteady aerodynamic theories describing the observed behavior have formed the basis for many types of rotor analyses. The tools for the analysis of 2-D, incompressible, unsteady aerodynamic problems were laid down by 1940, with the extension to compressible flows complete by 1950. The most authoritative source documenting these theories is Bisplinghoff et al. (1955). Lomax (1968) and Lomax et al. (1952) have provided a basis from which to develop linearized unsteady aerodynamic models applicable to compressible flows. The mathematical elegance and computational simplicity of these linearized approaches are attractive to the rotor analyst. Although there have been a plethora of "new" unsteady aerodynamic theories developed over the years for helicopter applications, most have their origin in the classical theories. Also, while the classical theories assume linearity in the airloads, the assumption of linearity can probably be justified for many of the problems encountered on the rotor, in practice. The proof of this latter statement is not always easy to justify, mostly because of other uncertainties in the problem such as those resulting from the rotor wake. The advent of nonlinear methods based on finite-difference approximations to the Euler and Navier–Stokes equations has provided new results that help the rotor analyst define the limits of the simpler, linear models and may give guidance in developing improved and more practical unsteady aerodynamic models for future use in blade airloads prediction, aeroelastic analysis, and rotor design.

8.2 Sources of Unsteady Aerodynamic Loading

Figure 8.1 summarizes the various sources of unsteady forcing effects that may affect the blade airloads. The angle of attack environment of a typical blade element is the resultant of a combination of forcing from collective and cyclic blade pitch, twist angle, elastic torsion, blade flapping velocity, and elastic bending. The induced downwash effects from the trailed wake system and the locally high velocity field perturbations produced by discrete tip vortices are also of primary importance, and their effects on the airloads must be fully considered.

At the blade element level, the various effects described in Fig. 8.1 can be decomposed into perturbations to the local angle of attack and velocity field, as shown in Fig. 8.2. At low angles of attack with fully attached flow, the various sources of unsteady effects manifest primarily as moderate amplitude and phase variations relative to the quasi-steady airloads. However, at higher angles of attack when time-dependent flow separation from the airfoil may be involved, a phenomenon that has become known as *dynamic stall* may occur. This phenomenon is manifest by large overshoots in the values of the lift, drag, and pitching moment relative to the quasi-steady stall values. Dynamic stall is also accompanied by much larger phase variations in the unsteady airloads as a result of significant hysteresis in the flow developments; that is, the values of the airloads at the same angle of attack may be very different depending on whether the flow is separating, fully stalled, or reattaching. As will be discussed in Chapter 9, the amplitude and phase effects produced by the stalled airloads can lead to various aeroelastic problems on the rotor that may seriously limit its performance compared to that assumed or predicted on the basis of quasi-steady flow assumptions.

8.3 Blade Wake

The wake from the rotating blade comprises, in part, a vortical shear layer or vortex sheet, with a concentrated vortex formed at the blade tip. The vortex sheet is comprised of vorticity with vectors aligned mainly normal to and parallel to the trailing edge of the

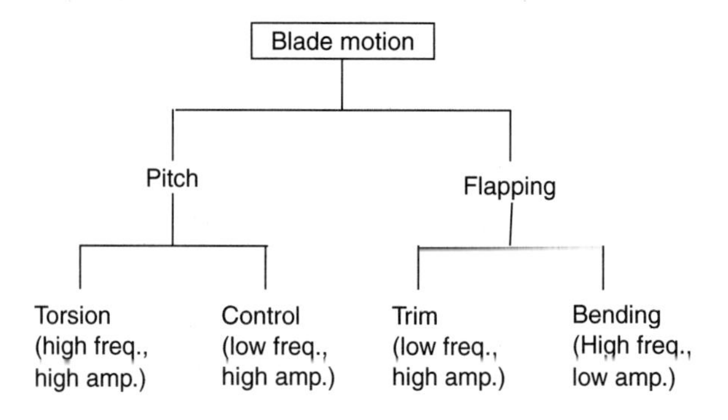

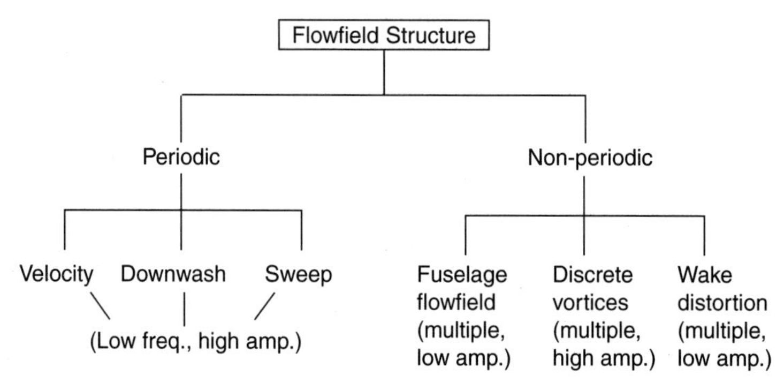

Figure 8.1 Possible sources of unsteady aerodynamic loading on a helicopter rotor. Adapted from Beddoes (1980).

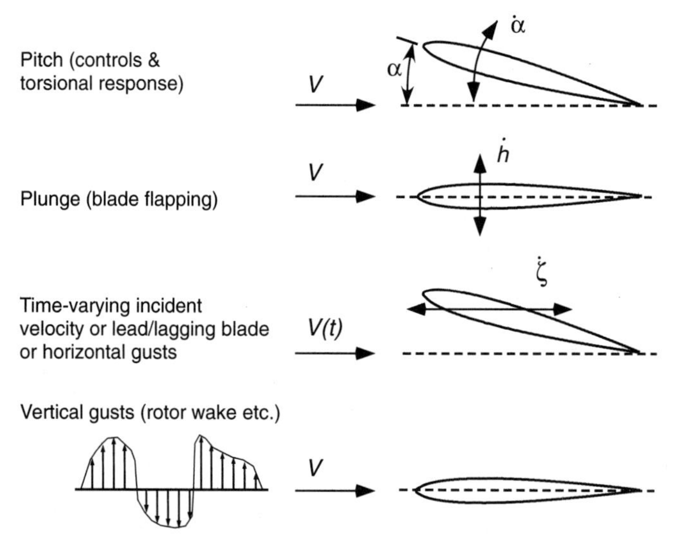

Figure 8.2 Decomposition of unsteady aerodynamic forcing terms at the blade element level.

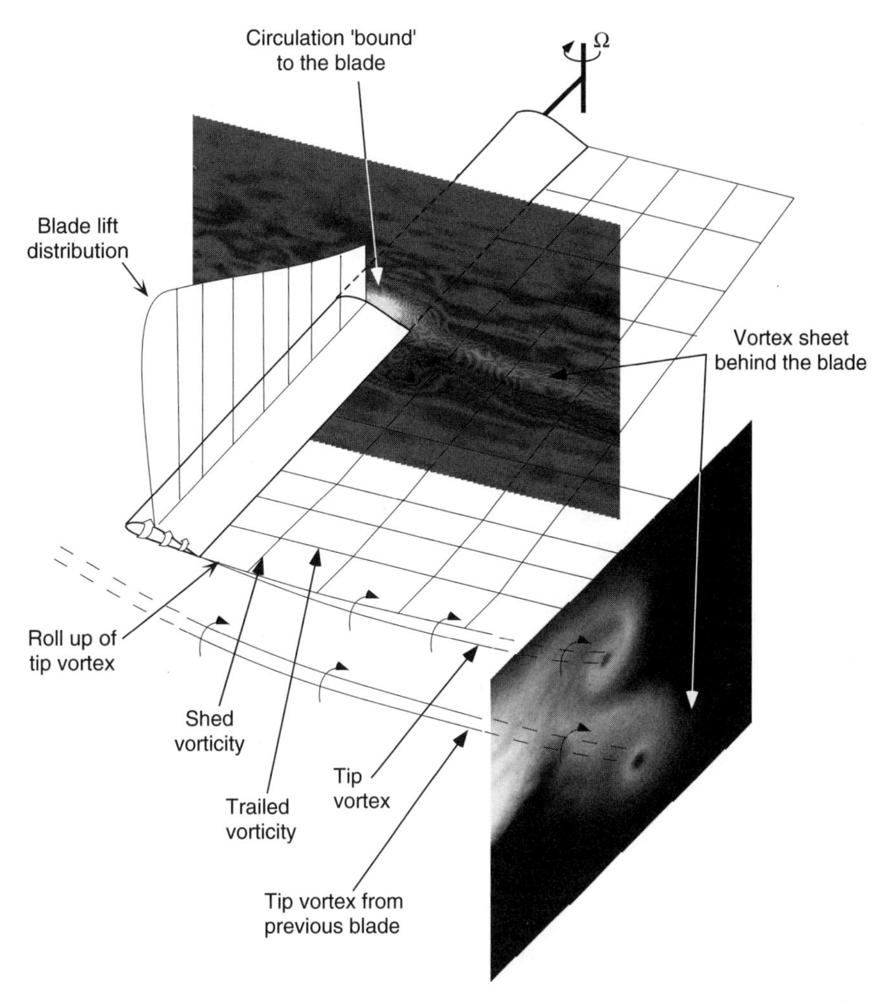

Figure 8.3 Schematic showing the wake and tip vortex rollup behind a single blade and the interaction with the vortex from another blade.

blade, as shown in Fig. 8.3, which is a schematic reconstruction of the wake behind a rotor blade using both flow field measurements and flow visualization. The strength of the former component (the trailed vorticity) is related to the spanwise gradient of lift (circulation, Γ) on the blade, that is, $\partial\Gamma/\partial r$, whereas the latter component (the shed vorticity) is related to the time rate of change of lift on the blade, that is, $\partial\Gamma/\partial t$. Experiments with rotors have shown that the blade tip vortices are almost fully rolled up within only a few degrees of blade rotation. Because the aerodynamic loading on the blades is biased toward their tips, the tip vortex is of high overall strength with significant induced velocities.

While the fundamental process of the blade wake and tip vortex formation is similar to that of a fixed wing, it will be appreciated that a complication with rotors is that the wakes and tip vortices from other blades can lie close to the plane of blade rotation and can have large induced effects on the blade lift distribution. Because of these mutual effects, the problem of calculating the detailed airloads over the rotor is quite formidable. Miller (1964) has examined the higher harmonics of the rotor loading in forward flight and has concluded that the effects of the trailed wake (tip vortices) are generally more important than the shed wake; only for very low advance ratios or for hover do the effects of the "returning" shed wake or the shed wake from other blades seem to be important. However, the effects of the "near" shed wake (that is, the wake vorticity immediately behind each blade) on the blade from which it was generated were found to be important under all flight regimes. These observations suggest that the overall aerodynamic environment at the blade (specifically, the angle of attack) is determined mainly by the trailed wake (tip vortex) system. In general, *unsteady* aerodynamic effects are relatively local and are a consequence of the time history of the vorticity contained in the shed wake immediately behind each blade. Such observations often permit simplified forms of mathematical analysis to be pursued, without substantial loss of accuracy in predicting the airloads on the rotor.

8.4 Reduced Frequency and Reduced Time

One important parameter used in the description of unsteady aerodynamics and unsteady airfoil behavior is the *reduced frequency*. This parameter is used to characterize the *degree of unsteadiness* of the problem. It can be shown that the reduced frequency appears when nondimensionalizing the Navier–Stokes equations. Alternatively, it can be shown using dimensional analysis that the resultant force F on an airfoil of chord c oscillating at angular frequency ω in a flow of velocity V can be written in functional form as

$$\frac{F}{\rho V^2 c^2} = f\left(\frac{\rho V c}{\mu}, \frac{V}{a}, \frac{\omega c}{V}\right) = f\,(Re, M, k), \tag{8.1}$$

see Question 8.1. As noted previously in Chapter 7, the resultant force, F, depends on the Reynolds number, Re, and the Mach number, M, but now the reduced frequency, k, of the flow is a third parameter to be considered. The reduced frequency is normally defined in terms of the airfoil semi-chord, $b = c/2$, so that

$$k = \frac{\omega b}{V} = \frac{\omega c}{2V}. \tag{8.2}$$

For $k = 0$, the flow is steady. For $0 \leq k \leq 0.05$, the flow can be considered quasi-steady; that is, unsteady effects are generally small, and for some problems they may be neglected completely. Usually flows with characteristic reduced frequencies of 0.05 and above are considered unsteady, and the unsteady terms in the governing equations cannot be routinely neglected. Problems that have characteristic reduced frequencies of 0.2 and above are considered highly unsteady, and the unsteady terms, such as those associated with acceleration effects, will begin to dominate the behavior of the airloads.

For a helicopter rotor in forward flight the reduced frequency at any blade element is an ambiguous parameter because the local sectional velocity (which appears in the denominator of the reduced frequency expression) is constantly changing. However, a first-order approximation for k can give useful information about the degree of unsteadiness found on a rotor and the necessity of modeling unsteady aerodynamic effects in any form of analysis. Consider first the unsteady effects induced by rigid blade flapping, for which the first flap frequency is about 1.05Ω for an articulated rotor. Then the reduced frequency, k_{75}, at the 75% radius location ($r = 0.75$) will be

$$k_{75} = \frac{1.05\Omega c}{2 \times 0.75\Omega R} = \frac{2.8c}{R}, \tag{8.3}$$

assuming for simplicity that the local velocity at the blade element is just the rotational velocity, $r\Omega R$. For a typical helicopter rotor with a blade aspect ratio $R/c \approx 10$, then

$k_{75} \approx 0.07$, which is in the unsteady range. Also, because the reduced frequencies increase further inboard on the blade owing to the lower values of local sectional velocity, k may become relatively large. Consider further the first elastic torsion mode, which is typically about 3–4Ω. In this case, at the tip the reduced frequency associated with airloads generated by torsional displacements is in excess of 0.2. At these reduced frequencies, there is a significant amplitude and phasing introduced into the airloads by the effects of the unsteady aerodynamics, and the modeling of unsteady aerodynamic effects is critical if erroneous predictions of the airloads are to be avoided.

It should be appreciated that the above calculations are only very approximate and serve only to illustrate the *potential* significance of unsteady effects and the need to model such effects in predicting the airloads in rotor problems. For quantification of more transient problems, the concept of a single reduced frequency in terms of characterizing the degree of unsteadiness of the problem begins to lose its significance. Under these circumstances it is normal to work with *reduced time*, s, where

$$s = \frac{1}{b}\int_0^t V\,dt = \frac{2}{c}\int_0^t V\,dt, \tag{8.4}$$

which represents the relative distance traveled by the airfoil through the flow in terms of airfoil semi-chords during a time interval t. It has been found useful to characterize many of the events occurring in unsteady aerodynamics, such as dynamic stall or blade encounters with blade tip vortices, in terms of a reduced time parameter.

8.5 Unsteady Attached Flow

A prerequisite in any unsteady aerodynamic theory is the ability to model accurately the unsteady airloads at the blade element under attached flow conditions. In the first instance, the most elementary level of approximation is to consider incompressible, 2-D flow. This avoids the need to model the wake from other blades (a problem considered in detail in Chapter 10) and allows convenient analytical and semi-analytical mathematical solutions to be incorporated into the rotor analysis. However, the helicopter rotor analyst is still faced with several compromises. First, the assumptions and limitations of any model must be properly assessed and understood. For example, neglecting the compressibility of the flow is not readily justified for rotor problems. This justification requires that not only must the local free stream Mach number be low, but the frequency of the source of unsteady effects must be small compared to the sonic velocity, that is, the product $\omega c/a \ll 1$, where a is the speed of sound. This means that the characteristic reduced frequency must also be small. The reduced frequency can be written as $k = \omega c/2Ma$, so that $Mk \ll 1$ to justify the assumption of incompressible flow. Second, any model must be written in a mathematical form that can be coupled into the structural dynamic model of the rotor system. For example, in some cases it may be desirable to write the aerodynamic model at each blade element in the form of ordinary differential equations. Third, because the blade element unsteady aerodynamic model is contained within radial and azimuthal integration loops, computational time considerations are important, and this alone can limit the allowable level of sophistication possible with any mathematical model.

The most fundamental approach to the modeling of unsteady aerodynamic effects is through an extension of steady, 2-D thin-airfoil theory. This gives a good level of analysis of the problem and provides considerable insight into the physics responsible for the unsteady behavior. Results for unsteady airfoil problems have been formulated in both the time domain and the frequency domain, primarily by Wagner (1925), Theodorsen (1935),

Küssner (1935), and von Kármán & Sears (1938). These solutions all have the same root in unsteady thin-airfoil theory and give exact analytic (closed-form) solutions for the pressure distribution (hence, the forces and moments) for different forcing conditions, (i.e., for perturbations in angle of attack or an imposed nonuniform vertical distribution of chordwise velocity). While these methods are valid for 2-D and incompressible flows, and were primarily intended for fixed-wing aeroelastic applications, they have also formed the foundation for several extensions to subsonic compressible flow and also to specific types of rotating-wing problems. For example, one extension of Theodorsen's theory is attributed to Loewy (1957), which is a solution that approximates the effects of the shed wake vorticity below the rotor, as laid down by the blade and by other blades.

The unsteady compressible (subsonic) thin-airfoil problem has also received considerable attention – see, for example, Lomax (1968) and Lomax et al. (1952). Even though in some cases the local flow may have an incident Mach number that may be low, the product Mk must still be $\ll 1$ if incompressibility is to be justified. Because the governing equation in a compressible flow is the hyperbolic wave equation compared to the elliptic nature of Laplace's equation for incompressible flow [see Karamcheti (1966)], unsteady aerodynamic theories cannot be obtained in a corresponding exact, convenient analytical form. There are, however, some limited exact solutions and numerical solutions available. These can be used to great advantage in the development of semi-analytic or semi-empirical methods for unsteady subsonic compressible flows, which are formulated in the spirit of the classical incompressible theories but are still computationally practical enough to be included within helicopter rotor analyses.

8.6 Quasi-Steady Thin-Airfoil Theory

The unsteady airfoil problem can be tackled initially using the classical, incompressible, steady, thin-airfoil theory. This is equivalent to setting all the unsteady terms in the governing equations equal to zero and will be called the *quasi-steady* problem. The unsteady motion of the airfoil produces a normal perturbation velocity across the chord, for which a solution for the vortex sheet strength on the airfoil, γ_b, can be found to maintain flow tangency on the chordline. An angle of attack, α, or plunge velocity, $\dot{h}$, produces a uniform velocity perturbation normal to the chord, as shown in Fig. 8.4. For an angle of attack perturbation, $w(x) = V\alpha =$ constant. Similarly, for a steady plunge velocity, $w(x) = -\dot{h} =$ constant. The pitch-rate term produces a linear variation in normal perturbation velocity. For a pitch rate imposed about an axis at a semi-chords from the mid-chord, then $w(x) = \dot{\alpha}(x - ab)$, so that the induced camber is a parabolic arc, as also shown in Fig. 8.4.

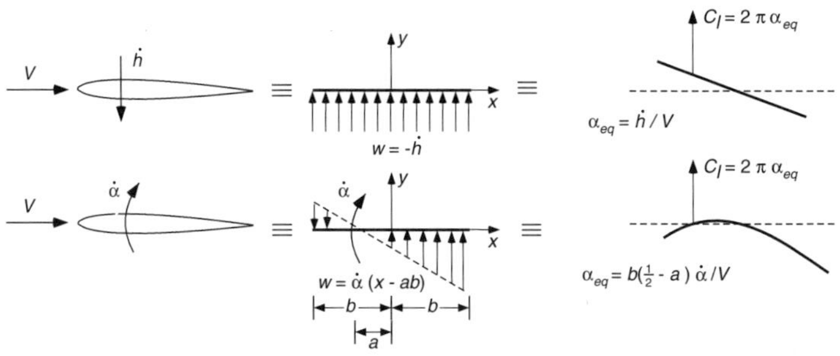

Figure 8.4 Normal velocity perturbation and effective induced camber for plunge velocity and pitch rate about an axis located at 1/4-chord.

Table 8.1. *Coefficients for the Quasi-Steady Airfoil Problem Using Thin-Airfoil Theory*

Coefficient	A_0	A_1	A_2
α	α	0	0
$\dot{h}$	$\dot{h}/V$	0	0
$\dot{\alpha}$	$-\dot{\alpha}ab/V$	$\dot{\alpha}b/V$	0

The quasi-steady contribution to the airloads follows directly from thin-airfoil theory (see appendix) using the following solutions for the Fourier harmonics:

$$A_0 = \alpha - \frac{1}{\pi}\int_0^{\pi} \frac{w}{V} d\theta \quad \text{and} \quad A_n = \frac{2}{\pi}\int_0^{\pi} \frac{w}{V} \cos n\theta \, d\theta, \tag{8.5}$$

where $x = -b\cos\theta$ based on a coordinate system at mid-chord (see Fig. 8.4). The lift and moment are given by

$$C_l = 2\pi\left(A_0 + \frac{A_1}{2}\right) \quad \text{and} \quad C_{m_{0.25}} = -\frac{\pi}{4}\left(A_1 - A_2\right). \tag{8.6}$$

The results for this problem are summarized in Table 8.1. The quasi-steady lift and moment are

$$C_l = 2\pi\left[\alpha + \frac{\dot{h}}{V} + b\left(\frac{1}{2} - a\right)\frac{\dot{\alpha}}{V}\right], \tag{8.7}$$

$$C_{m_{0.25}} = -\frac{\pi}{4}\frac{\dot{\alpha}b}{V}, \tag{8.8}$$

$$C_{m_{0.5}} = \left[-\frac{\pi}{4} + \frac{1}{2}\left(\frac{1}{2} - a\right)\right]\frac{\dot{\alpha}b}{V}. \tag{8.9}$$

For a pitching axis at the 1/4-chord ($a = -1/2$), the term inside the square brackets in Eq. 8.7 will be seen to be the effective angle of attack at the 3/4-chord. For this reason, the 3/4-chord point is sometimes called the rear neutral point. Also, note that the pitching moment about the 1/4-chord resulting from the pitch rate contribution is independent of the pitch axis location, a. (See also Question 8.2.)

8.7 Theodorsen's Theory

Theodeorsen's theory, which is widely used by fixed-wing analysts, forms one root for many of the unsteady aerodynamic solution methods used for helicopter analysis. The problem of finding the airloads on an oscillating airfoil was first tackled by Glauert (1929) but was properly solved by Theodorsen (1935). Theodorsen's approach gives a solution to the unsteady airloads on a 2-D harmonically oscillated airfoil in inviscid, incompressible flow, and subject to small disturbance assumptions. The basic model is shown in Fig. 8.5. Both the airfoil and its shed wake are represented by a vortex sheet, with the shed wake extending as a planar surface from the trailing edge downstream to infinity. The shed wake comprises countercirculation that is shed at the airfoil trailing edge and is convected downstream at the free-stream velocity. The assumption of a planar wake is justified if the angle of attack disturbances remain small. As with the standard thin-airfoil theory, the bound vorticity, γ_b,

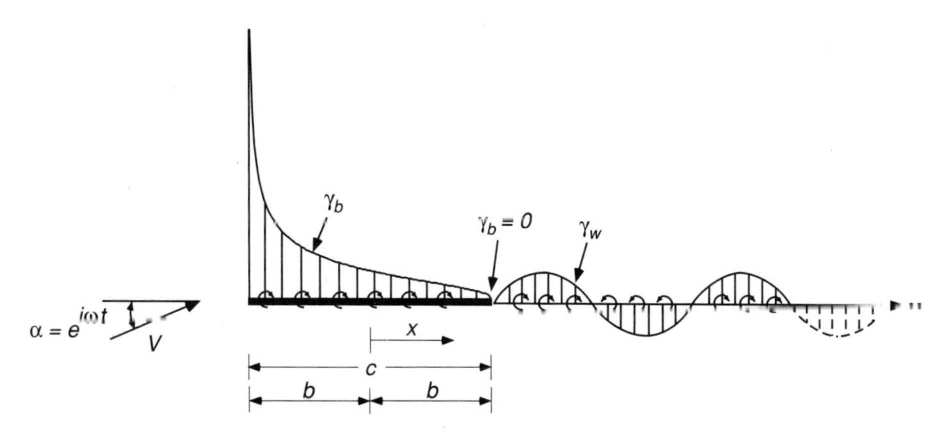

Figure 8.5 Mathematical model of a harmonically oscillated thin airfoil used by Theodorsen.

can sustain a pressure difference and, therefore, a lift force. The wake vorticity, γ_w, however, must be force free with zero net pressure jump over the sheet.

Theodorsen's problem was to obtain the solution for the loading, γ_b, on the airfoil surface under harmonic forcing conditions. The governing integral equation is

$$w(x, t) = \frac{1}{2\pi} \int_0^c \frac{\gamma_b(x, t)}{(x - x_0)} dx + \frac{1}{2\pi} \int_c^\infty \frac{\gamma_w(x, t)}{(x - x_0)} dx, \tag{8.10}$$

where w is the downwash on the airfoil surface. This equation must be solved subject to invoking the Kutta hypothesis at the trailing edge (i.e., $\gamma_b(c, t) = 0$). There is also a connection to be drawn between the change in circulation about the airfoil and the circulation shed into the wake. Conservation of circulation requires that

$$\gamma_w(c, t)\, dx = -\frac{d\Gamma(t)}{dt} dt. \tag{8.11}$$

Assuming that the shed vortices are convected at the free-stream velocity, V, this gives $dx = V dt$ and so

$$V \gamma_w(c, t) = -\frac{d\Gamma(t)}{dt}, \tag{8.12}$$

where the airfoil circulation $\Gamma(t)$ is given by

$$\Gamma(t) = \int_0^c \gamma_b(x, t)\, dx. \tag{8.13}$$

This wake vorticity changes the downwash velocity over the airfoil and, therefore, the loads on the airfoil are also affected. So long as the circulation about the airfoil is changing with respect to time, circulation is continuously shed into the wake and will continuously affect the aerodynamic loads on the airfoil. In the limit as the forcing becomes zero, the shed wake vorticity cast off the trailing edge of the airfoil becomes zero, and the remaining circulation in the wake convects downstream to infinity. In this case, the problem is modeled by the standard quasi-steady thin-airfoil theory.

The problem posed above is certainly not trivial, but for simple harmonic motion the solution is given by Theodorsen (1935) in a simple form to represent a transfer function between the forcing (angle of attack) and the aerodynamic response (pressure distribution, lift, and moment). Theodorsen's approach is summarized by Bisplinghoff et al. (1955) and also by another approach attributed to Schwarz. See also Bramwell (1976) and Johnson (1980) for a good exposition of the theory. For a general motion, where an airfoil of chord $c = 2b$ is undergoing a combination of pitching (α, $\dot{\alpha}$) and plunging (h) motion in a flow of steady velocity V, Theodorsen gives for the lift

$$L = \pi\rho V^2 b\left[\frac{b}{V^2}\ddot{h} + \frac{b}{V}\dot{\alpha} - \frac{b^2}{V^2}a\ddot{\alpha}\right] + 2\pi\rho V^2 b\left[\frac{\dot{h}}{V} + \alpha + \frac{b\dot{\alpha}}{V}\left(\frac{1}{2} - a\right)\right]C(k), \tag{8.14}$$

where a is the pitch axis location relative to the mid-chord of the airfoil and is measured in terms of semi-chords. The corresponding moment about mid-chord is

$$M = -\rho b^2\left(\pi\left[\frac{1}{2} - a\right]Vb\dot{\alpha} + \pi b^2\left[\frac{1}{8} + a^2\right]\ddot{\alpha} - a\pi b\ddot{h}\right) + 2\rho V b^2\pi\left(a + \frac{1}{2}\right)\left(V\alpha + \dot{h} + b\left[\frac{1}{2} - a\right]\dot{\alpha}\right)C(k). \tag{8.15}$$

The first set of terms in Eqs. 8.14 and 8.15 result from flow acceleration effects (i.e., a *non-circulatory* or *apparent mass* effect). The second terms arise from the creation of circulation about the airfoil (i.e., a *circulatory* effect). The circulatory term $C(k) = F(k) + iG(k)$ is a complex valued function known as *Theodorsen's function*, which accounts for the effects of the shed wake on the unsteady airloads.

The noncirculatory or apparent mass terms arise from the $\partial\phi/\partial t$ term contained in the unsteady Bernoulli equation (Kelvin's equation) [see Karamcheti (1966)] and account for the pressure forces required to accelerate the fluid in the vicinity of the airfoil. For example, for a thin airfoil of chord $c = 2b$ moving normal to its surface at velocity $w(t)$, the noncirculatory fluid force, F^{nc}, acting on the surface is

$$F^{nc} = -M_a\frac{dw}{dt}. \tag{8.16}$$

The term M_a is known as the apparent mass and in this case is given by

$$M_a = \pi\rho b^2 = \pi\rho\frac{c^2}{4}, \tag{8.17}$$

which can be identified with the $\ddot{h}$ term in Eq. 8.14. Other apparent mass forces and moments will be produced for different types of forcing – see Bisplinghoff et al. (1955) for a further discussion.

In coefficient form, Theodorsen's results in Eqs. 8.14 & 8.15 can be expressed as

$$C_l = \pi b\left[\frac{\dot{\alpha}}{V} + \frac{\ddot{h}}{V^2} - \frac{ba\ddot{\alpha}}{V^2}\right] + 2\pi C(k)\left[\frac{\dot{h}}{V} + \alpha + b\left(\frac{1}{2} - a\right)\frac{\dot{\alpha}}{V}\right], \tag{8.18}$$

$$C_m = \frac{\pi}{2}\left[\frac{ba\ddot{h}}{V^2} - \frac{b^2}{V^2}\left(\frac{1}{8} + a^2\right)\ddot{\alpha}\right] + \pi\left(a + \frac{1}{2}\right) \times \left[\frac{\dot{h}}{V} + \alpha + b\left(\frac{1}{2} - a\right)\frac{\dot{\alpha}}{V}\right]C(k) - \frac{\pi}{2}\left[\left(\frac{1}{2} - a\right)\frac{b\dot{\alpha}}{V}\right]. \tag{8.19}$$

The last term in the pitching moment expression is a circulatory term, albeit a quasi-steady term not associated with the $C(k)$ function. Furthermore, the effects of an aerodynamic center offset from the 1/4-chord should be added to the pitching moment by multiplying the circulatory value of the lift by the offset distance – see Chapter 7.

Theodorsen's function is expressed in terms of Hankel functions, H, with the reduced frequency k as the argument, where

$$C(k) = F(k) + iG(k) = \frac{H_1^{(2)}(k)}{H_1^{(2)}(k) + iH_0^{(2)}(k)}. \tag{8.20}$$

The Hankel function is defined as $H_\nu^{(2)} = J_\nu - iY_\nu$, with J_ν and Y_ν being Bessel functions of the first and second kind, respectively. Implicitly recognizing that each Bessel function has an argument k, then the real or in-phase ($\Re$) part and imaginary or out-of-phase part ($\Im$) can be written as

$$\Re C(k) = F = \frac{J_1(J_1 + Y_0) + Y_1(Y_1 - J_0)}{(J_1 + Y_0)^2 + (J_0 - Y_1)^2},$$

and

$$\Im C(k) = G = -\frac{Y_1 Y_0 + J_1 J_0}{(J_1 + Y_0)^2 + (J_0 - Y_1)^2}.$$

Theodorsen's function is plotted in Fig. 8.6. The amplitude and phase of Theodorsen's function are given by

$$|C(k)| = \sqrt{F^2 + G^2} \quad \text{and} \quad \phi = \tan^{-1}\left(\frac{G}{F}\right) \tag{8.21}$$

repectively.

It will be appreciated from Fig. 8.6 that Theodorsen's function serves to introduce an amplitude reduction and phase lag effect on the *circulatory* part of the lift response compared to the result obtained under quasi-steady conditions. The basic effect can be seen if a pure

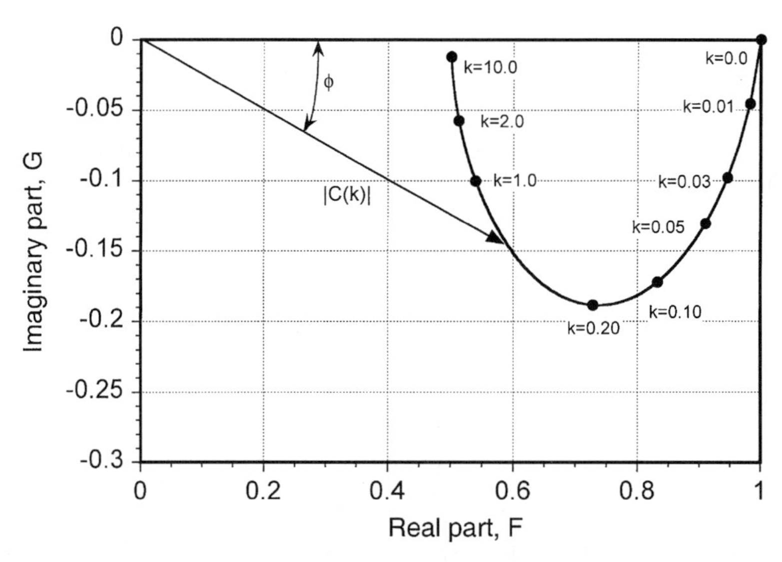

Figure 8.6 Theodorsen's function plotted as real and imaginary parts.

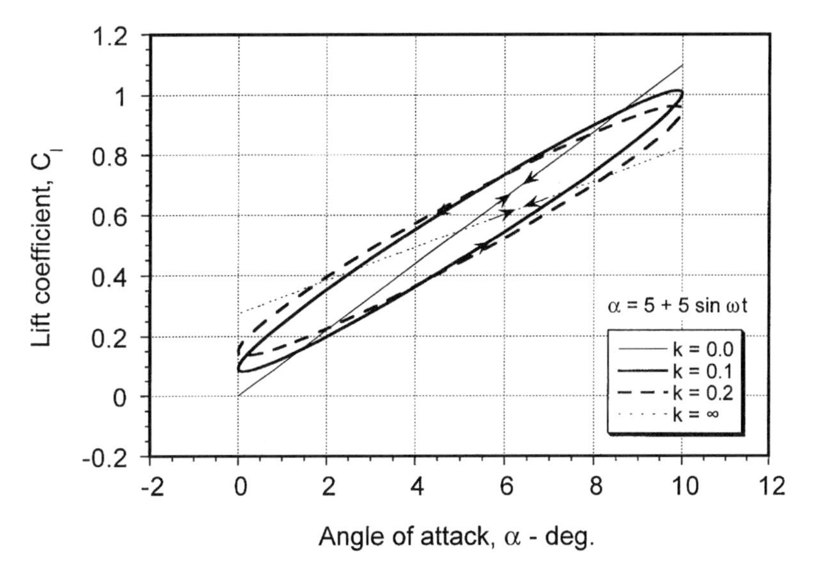

Figure 8.7 The effects of Theodorsen's function on the circulatory part of the unsteady lift response for a sinusoidal variation in angle of attack.

oscillatory variation in angle of attack is considered, that is, $\alpha = \bar{\alpha} \sin \omega t$.[2] In this case the circulatory part of the airfoil lift coefficient is given by

$$C_l = 2\pi\bar{\alpha}\; C(k) = 2\pi\bar{\alpha}\; [F(k) + iG(k)]\,. \tag{8.22}$$

Representative results from the Theodorsen model are shown in Fig. 8.7. For $k = 0$ the steady-state lift behavior is obtained, that is, C_l is linearly proportional to α. However, as k is increased, the lift plots develop into hysteresis loops, and these loops rotate such that the amplitude of the lift response (half of the peak-to-peak value) decreases with increasing reduced frequency. These loops are circumvented in a counterclockwise direction such that the lift is lower than the steady value when α is increasing with time and higher than the steady value when α is decreasing with time, (i.e., there is a phase lag). Note that for infinite reduced frequency the circulatory part of the lift amplitude is half that at $k = 0$ and there is no phase lag angle.

8.7.1 *Pure Angle of Attack Oscillation*

It is now possible to consider the effects of both the circulatory and the non-circulatory contributions to the unsteady lift. Consider first a pure harmonic variation in α, that is, $\alpha = \bar{\alpha}e^{i\omega t}$. Substituting into the expression for the lift given by Eq. 8.14 yields

$$L = 2\pi\rho V^2 b\left[C(k) + i\pi\frac{\omega b}{V}\right]\bar{\alpha}e^{i\omega t}. \tag{8.23}$$

In terms of the lift coefficient, the result is

$$C_l = \frac{L}{\rho V^2 b} = [2\pi(F + iG) + i\pi k]\bar{\alpha}e^{i\omega t}. \tag{8.24}$$

[2] Note that this does not represent an oscillating airfoil in angle of attack because the pitch-rate terms are not included.

The complete term inside the square brackets can be considered as the lift transfer function. The first term inside the brackets is the circulatory term, and the second term is the apparent mass contribution. Note that the apparent mass contribution is proportional to the reduced frequency and leads the forcing by a phase angle of $\pi/2$. If the result is normalized by $2\pi|\bar{\alpha}|$ then

$$\frac{|C_l|}{2\pi|\bar{\alpha}|} = (F + iG) + i\frac{k}{2}. \tag{8.25}$$

The equivalent result for the pitching moment about mid-chord is

$$\frac{|C_m|}{|\bar{\alpha}|} = -i\frac{\pi k}{2}. \tag{8.26}$$

The results for the lift are shown in Fig. 8.8, where the significance of the apparent mass contribution to both the amplitude and phase can be appreciated. At lower values of

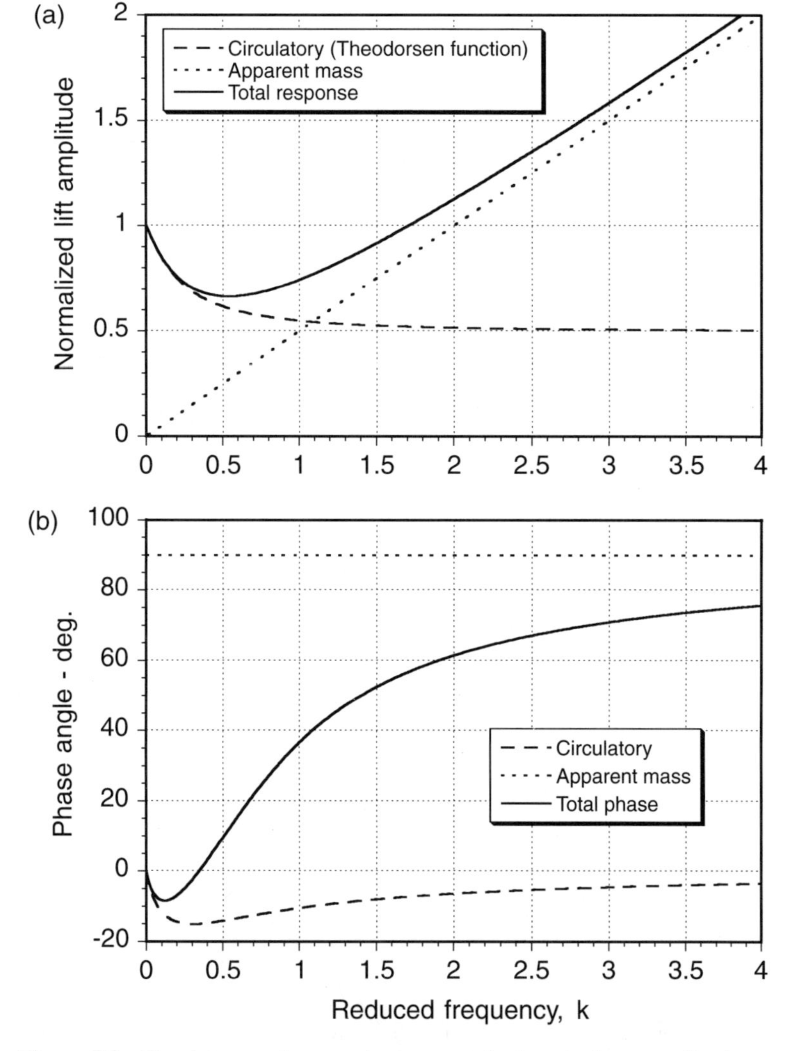

Figure 8.8 Circulatory and apparent mass contributions to the normalized unsteady lift response for a pure sinusoidal angle of attack oscillation. (a) Lift amplitude. (b) Phase of lift.

reduced frequency (say, $k < 0.1$) the noncirculatory or apparent mass forces are small, and the circulatory terms dominate the solution. At higher values of reduced frequency, the apparent mass forces clearly dominate. By setting $C(k) = 1$, that is, $F = 1$ and $G = 0$, in Eq. 8.25, the effects of the shed wake are removed and the quasi-steady result is obtained.[3]

8.7.2 *Pure Plunging Oscillation*

Consider now a harmonic plunging motion, such as would be contributed by blade flapping. Here, the forcing is $h = \bar{h}e^{i\omega t}$ so that $\dot{h} = i\omega\bar{h}e^{i\omega t}$ and $\ddot{h} = -\omega^2\bar{h}e^{i\omega t}$. Substituting into the expression for the lift given by Eq. 8.14 and solving for the lift coefficient gives

$$C_l = [2\pi k(iF - G) - \pi k^2]\frac{\bar{h}}{b}e^{i\omega t}. \tag{8.27}$$

Again, the complete term inside the square brackets can be considered as the lift transfer function. The first term inside the brackets is the circulatory term, and the second term is the apparent mass contribution. Note that for this problem the circulatory part of the lift response leads the forcing displacement h by a phase angle of $\pi/2$. Also, the apparent mass force leads the circulatory part of the response by a phase angle of $\pi/2$, or the forcing by a phase angle of π. The corresponding pitching moment about mid-chord for this case is

$$C_m = \left(\frac{\pi}{4}\right) k^2 \frac{\bar{h}}{b}e^{i\omega t}. \tag{8.28}$$

A comparison of Theodorsen's result with experimental data for an airfoil oscillating in plunge is shown in Fig. 8.9. The results are plotted as the first harmonic normalized amplitude of the lift and pitching moment about the 1/4-chord and their corresponding phase angles as functions of reduced frequency. The experimental results are taken from Halfman (1951), where measurements were made on an oscillating NACA 0012 airfoil. These tests were conducted for Mach numbers less than 0.1 and a Reynolds number of $\approx 10^6$. It is significant that a relatively high reduced frequency of 0.4 was attained in the experiment. Figure 8.9 shows that that there is fairly good agreement between Theodorsen's theory and Halfman's measurements. Note the sign of the lift phase angle, which changes from a lag (less than $-90°$) to a lead (greater than $-90°$) at higher reduced frequencies as the noncirculatory effects become more dominant. Note also that the phase of the pitching moment shows a significant digression from the results obtained with the baseline Theodorsen theory. However, this behavior can be explained by means of an aerodynamic center location that is not at the 1/4-chord. The experimental results suggest an aerodynamic center for this airfoil and test conditions that is located at 23.5% chord.

8.7.3 *Pitch Oscillations*

For harmonic pitch oscillations, additional terms involving $\dot{\alpha}$ appear in the equations for the aerodynamic response. The forcing is now given by $\alpha = \bar{\alpha}e^{i\omega t}$, and the pitch

[3] Note that the apparent mass contributions to the forces and moments, which are proportional to the instantaneous motion, are often included as part of the quasi-steady result. If the apparent mass terms are neglected, then the standard quasi-steady thin-airfoil result is obtained.

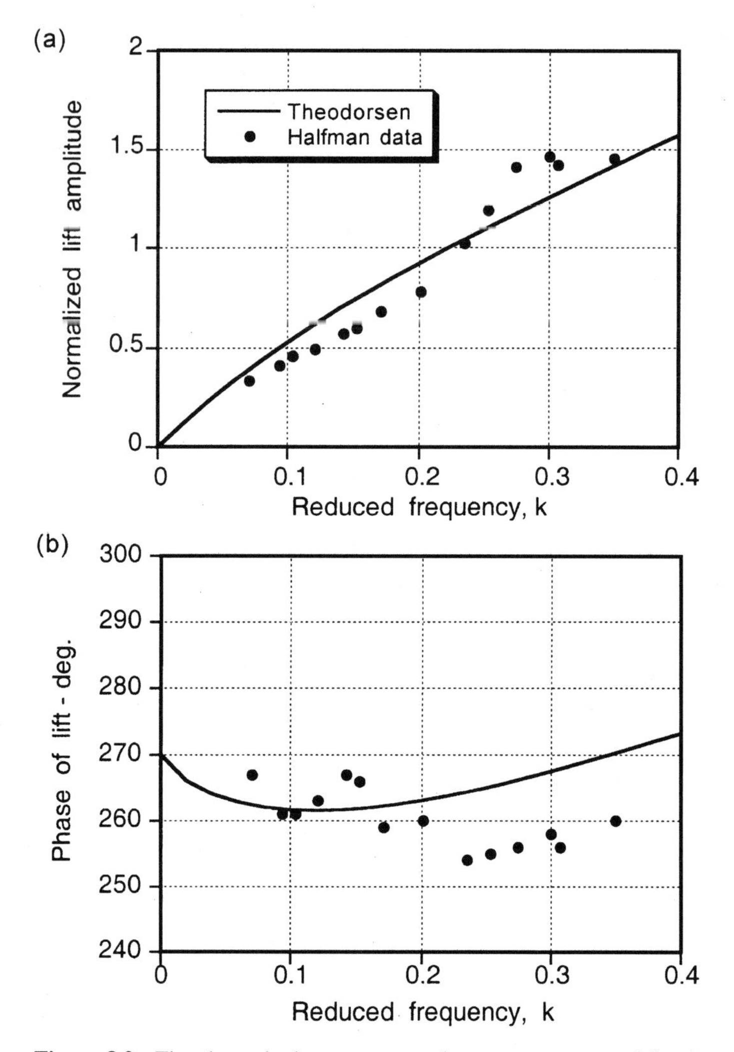

Figure 8.9 Theodorsen's theory compared to measurements of first harmonic unsteady lift and pitching moment for an airfoil oscillating in plunge. (a) Lift amplitude. (b) Phase of lift. (c) Pitching moment amplitude about 1/4-chord. (d) Phase of pitching moment.

rate by $\dot{\alpha} = i\omega\bar{\alpha}e^{i\omega t}$. In this case, the lift coefficient for pitching about 1/4-chord is

$$C_l = 2\pi\left(F[1+ik] + G[i-k]\right)\bar{\alpha}e^{i\omega t} + \pi k\left(i - \frac{k}{2}\right)\bar{\alpha}e^{i\omega t}. \tag{8.29}$$

The corresponding pitching moment about mid-chord is

$$C_m = \frac{\pi}{2}k\left(\frac{3}{8}k - i\right)e^{i\omega t}. \tag{8.30}$$

The results from this solution are compared with experimental measurements in Fig. 8.10, again in terms of the normalized lift and moment amplitude about the 1/4-chord and their corresponding phase angles versus reduced frequency. In addition to the low Mach number results of Halfman (1951), this figure shows data measured by Rainey (1957), which are for

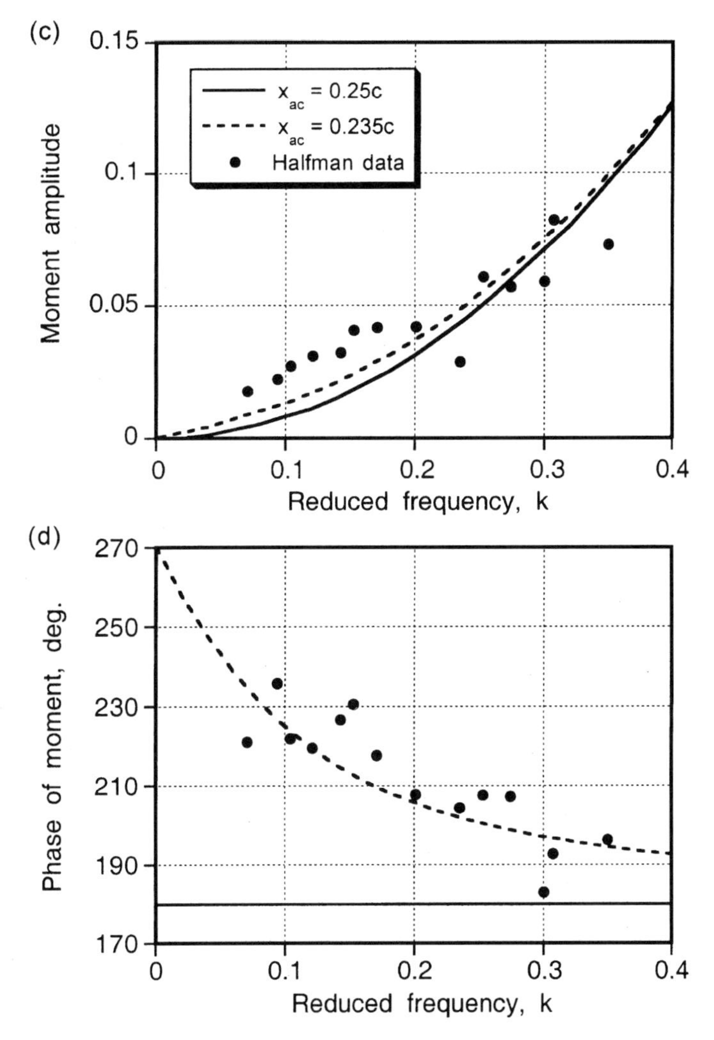

Figure 8.9 (*Continued*)

a Mach number of about 0.3 and a Reynolds number of 5.3×10^6. It is significant in the latter case that a reduced frequency of up to 0.6 was obtained in the experiment, which gives a good opportunity to examine the applicability of the linearized incompressible aerodynamic theory at higher reduced frequencies.

As shown in Fig. 8.10, the lift amplitude initially decreases with increasing k because of the effects of the shed wake. The amplitude begins to increase again for $k > 0.5$ as the apparent mass forces begin to dominate the airloads. This is also shown by the phase angle, which exhibits an increasing lead for $k > 0.3$. For the lift amplitude, Theodorsen's theory compares well with Rainey's results. The agreement with Halfman's results are not quite as good for the lift amplitude but are better in phase. The amplitude of the 1/4-chord pitching moment increases quickly with increasing k, with the agreement of theory and experiment being excellent. Again, the phase of the pitching moment response shows a behavior that can be explained in terms of shift of the aerodynamic center from the 1/4-chord. As for the plunging case, Halfman's data suggest an aerodynamic center at 23.5% chord. Rainey's data, however, suggest that the aerodynamic center is further back at 28% chord. Overall,

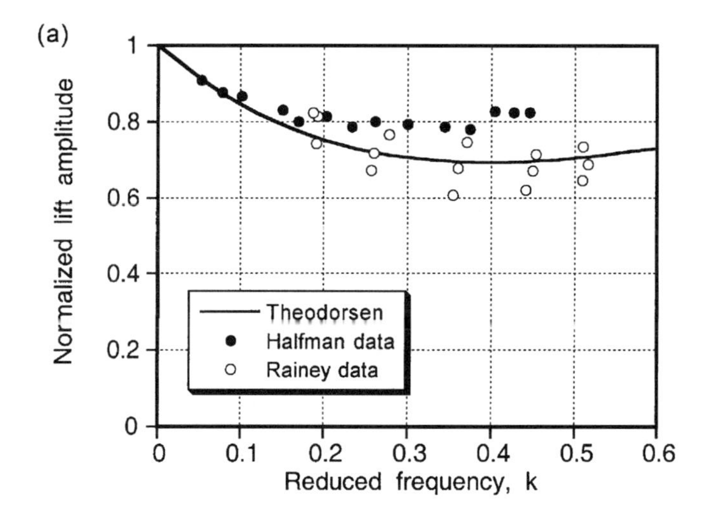

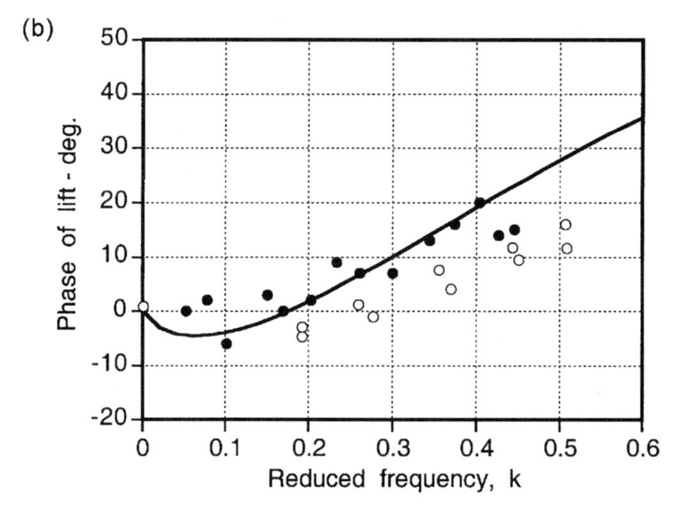

Figure 8.10 Theodorsen's theory compared to measurements of first harmonic unsteady lift and pitching moment for an airfoil oscillating in pitch. (a) Lift amplitude. (b) Phase of lift. (c) Pitching moment amplitude about 1/4-chord. (d) Phase of pitching moment.

the correlation obtained between Theordorsen's theory and experimental measurements for airfoils oscillating in pitch and plunge is quite good, giving considerable support to the validity of Theodorsen's theory, at least for low Mach numbers and up to moderate values of reduced frequency.

8.8 The Returning Wake: Loewy's Problem

Theodorsen's theory has considered an isolated 2-D thin airfoil with the wake convected downstream to infinity. For rotorcraft work, this is perhaps a questionable assumption because the rotor blade sections may encounter the wake vorticity from previous blades as well as the returning wake from the blade in question. This fact was acknowledged by Loewy (1957) and by Jones (1958) who set up a model of a 2-D blade section with a returning shed wake, as shown in Fig. 8.11.

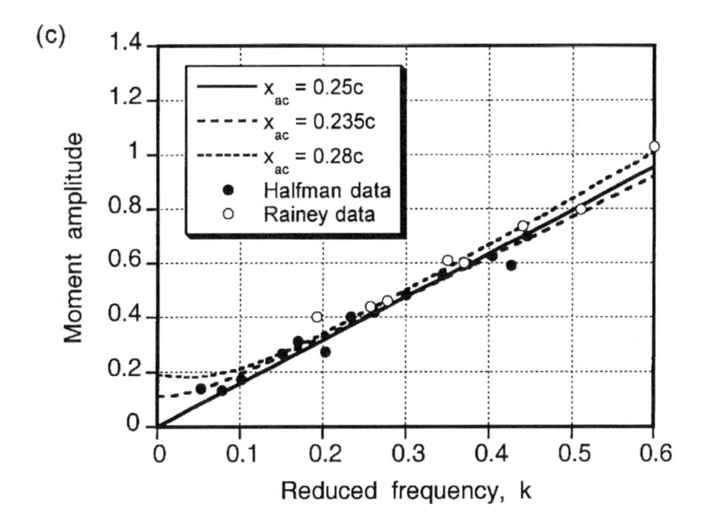

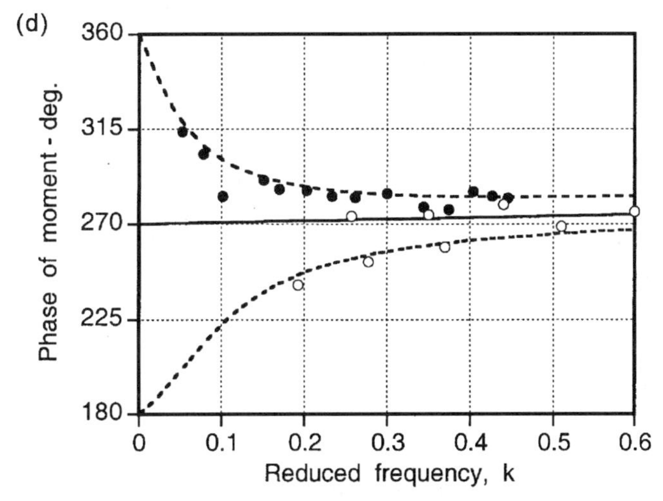

Figure 8.10 *(Continued)*

This returning wake can be modeled as a series of planar 2-D vortex sheets, just as in Theodorsen's method, but with a vertical separation h that depends on the mean induced velocity through the rotor disk and the number of rotor blades. Loewy (1957) has shown that in this case the lift on the blade section can be expressed by replacing Theodorsen's function by

$$C'\left(k, \frac{\omega}{\Omega}, h\right) = \frac{H_1^{(2)}(k) + 2J_1(k)W}{H_1^2(k) + i H_0^{(2)}(k) + 2[J_1(k) + i J_0(k)]W}, \tag{8.31}$$

where $C'(k)$ is known as the Loewy function, with argument of reduced frequency k. For a single blade, the complex valued W function is given by

$$W\left(\frac{kh}{b}, \frac{\omega}{\Omega}\right) = \left(e^{kh/b}e^{i2\pi(\omega/\Omega)} - 1\right)^{-1}. \tag{8.32}$$

If $\omega/\Omega =$ an integer, then all the shed wake effects are in phase. Note from Eqs. 8.31 and 8.32 that as $h \to \infty$ then $W \to 0$ and $C'(k) \to C(k)$, and Loewy's function approaches

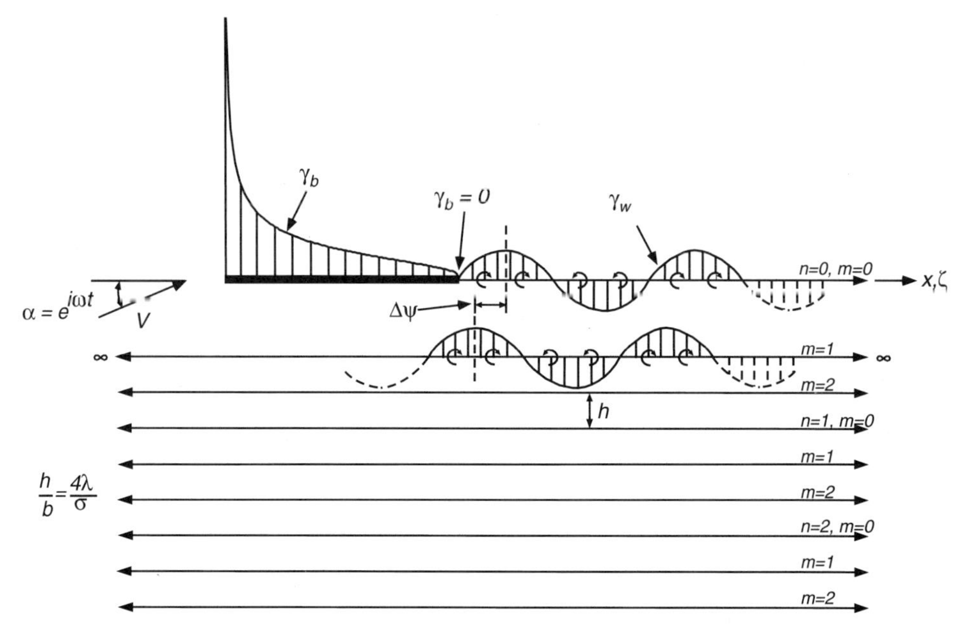

Figure 8.11 Schematic of Loewy's problem showing the returning nature of the shed wake below a rotor.

Theodorsen's result, as it should. For a rotor with N_b blades, the W function is modified to read

$$W\left(\frac{kh}{b}, \frac{\omega}{\Omega}, \Delta\psi, N_b\right) = \left(e^{kh/b} e^{i2\pi(\omega/N_b\Omega)} e^{i\Delta\psi\omega/\Omega} - 1\right)^{-1}, \tag{8.33}$$

where the parameter $\omega/N_b\Omega$ now controls the wake phasing.

The wake spacing ratio h/b can be determined from the spacing of the helical vortex sheets that are laid down below the rotor. If an average induced velocity $v_i = \lambda\Omega R$ is assumed, then during a single rotor revolution the shed wake generated by a single blade will be at a distance $v_i(2\pi/\Omega) = h$ below the rotor. For multiple blades, the spacing will be $v_i(2\pi)/(\Omega N_b)$, that is,

$$\frac{h}{b} = \frac{\lambda\Omega R 2\pi}{\Omega N_b b} = \frac{4\lambda}{\sigma} \tag{8.34}$$

with σ as the rotor solidity. For $\Delta\psi = 0$, which means that the only phase shift in the wake vorticity results from the spacing between the blades, then

$$W\left(\frac{kh}{b}, \frac{\omega}{\Omega}, 0, N_b\right) = \left(e^{kh/b} e^{i2\pi(\omega/N_b\Omega)} - 1\right)^{-1}. \tag{8.35}$$

Representative results from Loewy's theory for a one-bladed rotor are shown in Fig. 8.12, where we see that the main consequence of including the shed vorticity below the blade is that it serves to amplify or attenuate the unsteady lift response, depending on the reduced frequency, wake spacing, and wake phase. The most important effects are for lower reduced frequencies, with oscillations at the harmonics of the rotor rotational frequency. Using typical helicopter values of $\lambda \approx 0.05$ and $\sigma = 0.1$ gives $h/b \approx 2$. Therefore, for a helicopter the Loewy function predicts that $|C'| \approx 0.5$ over most of the reduced frequency range

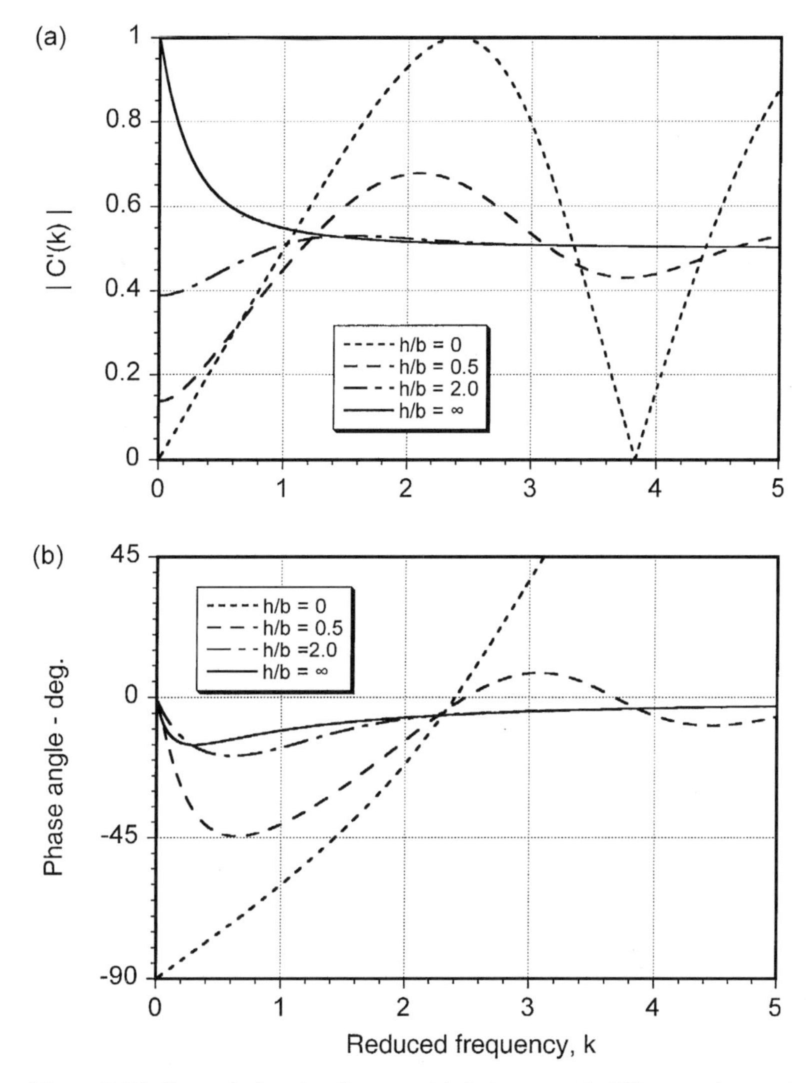

Figure 8.12 Loewy's function for a one-bladed rotor with different wake spacings and $\omega/\Omega =$ integer. (a) Lift amplitude. (b) Phase of lift.

and so will be an important effect. This can lead to lower damping of blade flapping and flapwise elastic bending modes and will increase the vibratory response of the blade to harmonic airloads. However, it is only for very low advance ratios or for hover that the Loewy effect seems important. Daughaday et al. (1959) have conducted indirect validation of the Loewy effect from measured blade flap bending moments and transient flapping decay data. The reduction in flap damping at frequencies that were multiples of the rotor rotational frequency was verified. Other verifications of the Loewy effect and the implications for rotor aeroelasticity have been made by Ham et al. (1958), Silverai & Brooks (1959), and Anderson & Watts (1975). Hammond & Pierce (1972) have considered a development of the Loewy problem for subsonic compressible flow. Whereas the effects of compressibility are small for ω/Ω near unity, larger differences are noted for $\omega/\Omega < 1$. No equivalent analytic theory to the Loewy problem is available for forward flight; the only option is to solve the problem numerically through discrete vortex tracking, a problem discussed in Chapter 10.

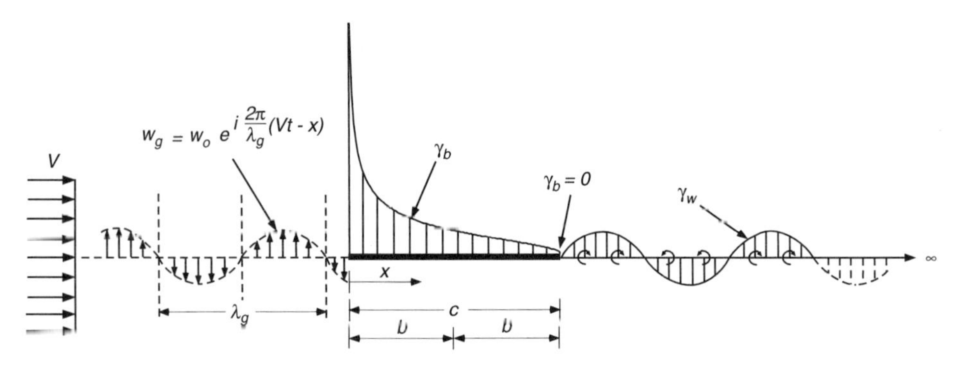

Figure 8.13 Model of a thin airfoil encountering a sinusoidal vertical gust (Sears's problem).

8.9 The Sinusoidal Gust: Sears's Problem

Von Kármán & Sears (1938) analyzed the problem of a thin airfoil moving through a sinusoidal vertical gust field. This is also a frequency domain solution. The gust can be considered as an upwash velocity that is uniformly convected by the free stream, as shown in Fig. 8.13. The forcing function in this case is

$$\begin{aligned} w_g(x,t) &= \sin\left(\omega_g t - \frac{\omega_g x}{V}\right) \\ &= \sin\omega_g t\ \cos\left(\frac{\omega_g x}{V}\right) - \cos\omega_g t\ \sin\left(\frac{\omega_g x}{V}\right), \end{aligned} \tag{8.36}$$

where ω_g is the gust frequency. There are two cases of interest. First, if the gust is referenced to the airfoil leading edge, then $x = 0$ and so Eq. 8.36 simply becomes $w_g(t) = \sin\omega_g t$. Second, if the gust is referenced to the mid-chord, then $x = b = c/2$ and the forcing becomes $w_g(t) = \cos k_g \sin\omega_g t - \sin k_g \cos\omega_g t$, which is equivalent to a phase shift (see also Question 8.6). The mid-chord was the reference point used in the original work of von Kármán & Sears (1938). In this case, the final result for the lift can be written as

$$C_l(t) = 2\pi\left(\frac{w_0}{V}\right)S(k_g), \tag{8.37}$$

where $S(k_g)$ is known as Sears's function. The gust encounter frequency is given by

$$k_g = \frac{2\pi V}{\lambda_g}, \tag{8.38}$$

where λ_g is the wavelength of the gust (see Fig. 8.13). Sears's function can also be computed exactly in terms of Bessel functions and is given by

$$S(k_g) = (J_0(k_g) - i J_1(k_g))\, C(k_g) + i J_1(k_g) \tag{8.39}$$

or in terms real and imaginary parts as

$$\begin{aligned} \Re S(k_g) &= F(k_g)\, J_0\,(k_g) + G(k_g)\, J_1(k_g), \\ \Im S(k_g) &= G(k_g)\, J_0\,(k_g) - F(k_g)\, J_1\,(k_g) + J_1(k_g). \end{aligned}$$

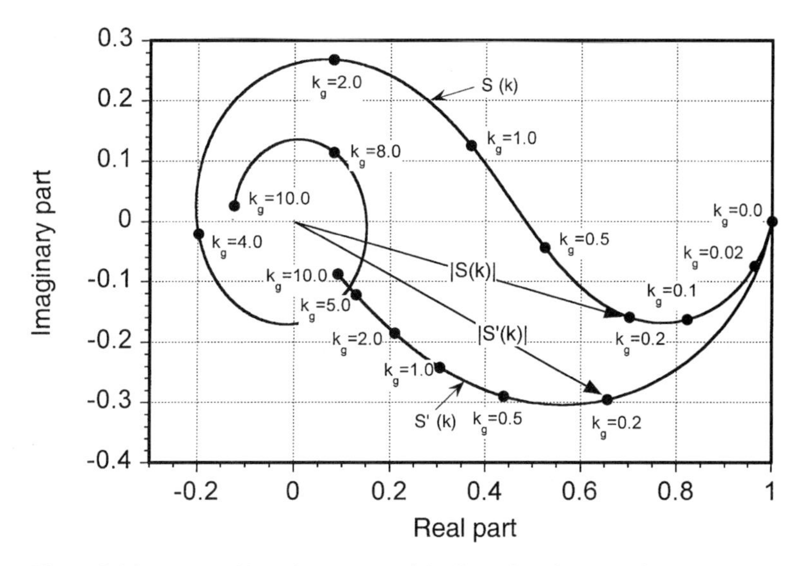

Figure 8.14 Real and imaginary parts of the Sears function, as referenced to the midchord and also to the leading edge of the airfoil.

If the gust is referenced to the leading edge of the airfoil, the result must be transformed as described previously. This function will be called S' and can be written as

$$\Re S'(k_g) = \Re S \cos k_g + \Im S \sin k_g,$$

$$\Im S'(k_g) = -\Re S \sin k_g + \Im S \cos k_g,$$

which is equivalent to a frequency dependent phase shift. The two results are plotted in Fig. 8.14. Note that the peculiar spiral shape of the S transfer function arises only when the gust front is referenced to the mid-chord of the airfoil. If the gust response is computed relative to the leading edge, then the S' transfer function is obtained. In application the gust front reference point is frequently confused in the published literature. While the differences are small at low reduced frequencies, the errors will be significant for $k_g > 0.2$.

The Sears function and the Theodorsen function are compared in Fig. 8.15 in terms of amplitude and phase angle as a function of reduced frequency. At low reduced frequencies the functions converge, but for $k > 0.1$ the differences become increasingly large. Note that as $k \to \infty$ then $|C(k)| \to 1/2$, and the corresponding phase angle $\to 0$. For the Sears function, the asymptotic behavior is $|S(k_g) \propto 1/\sqrt{2\pi k_g}$. When referenced to the mid-chord, then phase angle is proportional to $k_g - \pi/4$, or $-\pi/4$ if the leading edge is used as the reference point.

8.10 Indicial Response: Wagner's Problem

Theodorsen's lift deficiency approach has found use in many problems in both fixed-wing and helicopter aeroelasticity. However, for a rotor analysis Theodorsen's theory is somewhat less useful because in a rotor environment the non-steady value of V means that argument k (the reduced frequency) is, strictly speaking, an ambiguous parameter. Therefore, a theory formulated in the time domain is more general and is usually more useful. Wagner (1925) has obtained a solution for the indicial lift on a thin airfoil undergoing

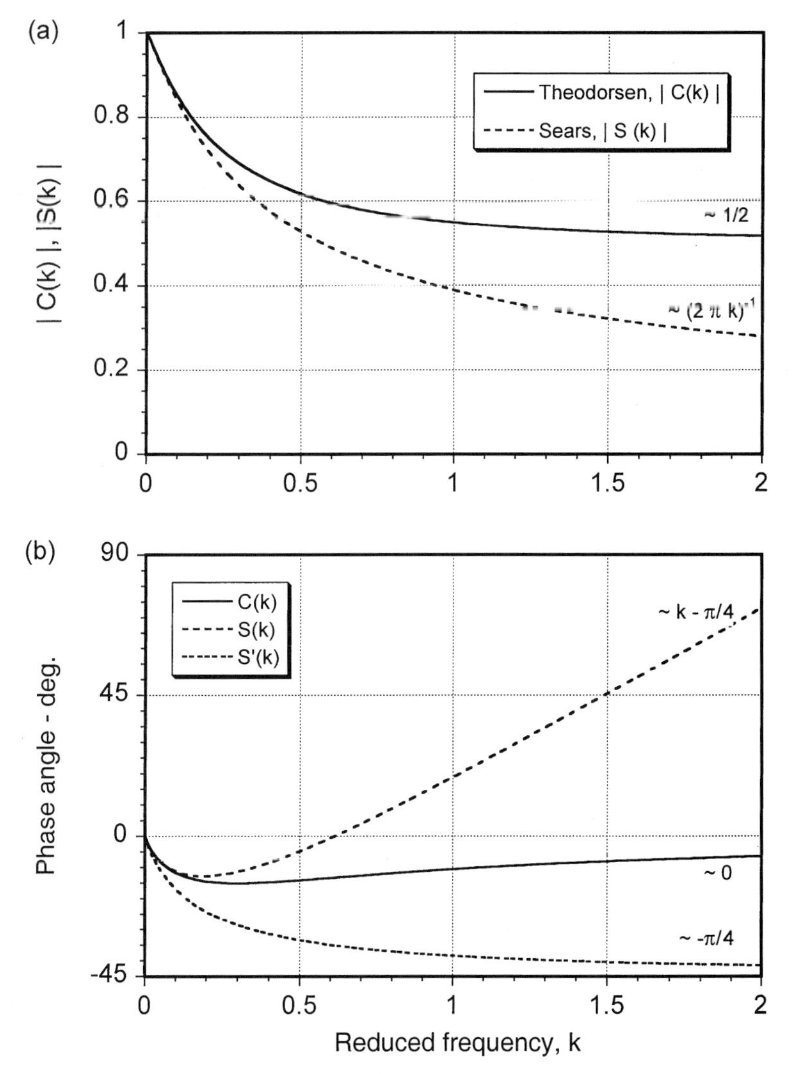

Figure 8.15 The Theodorsen and Sears functions in terms of amplitude and phase angle as a function of reduced frequency. (a) Amplitude. (b) Phase.

a step change in angle of attack in incompressible flow.[4] The transient chordwise pressure loading is given by

$$\frac{\Delta C_p(\bar{x}, s)}{\alpha} = \frac{4}{V}\delta(t)\sqrt{(1-\bar{x})\bar{x}} + 4\phi(s)\sqrt{\frac{1-\bar{x}}{\bar{x}}}, \tag{8.40}$$

where $\phi(s)$ is called Wagner's function and, by analogy with the Theodorsen function, accounts for the effects of the shed wake. As defined previously, the variable s represents the distance traveled by the airfoil in semi-chords. The first term in Eq. 8.40 is the apparent mass contribution, which for a step input appears as a Dirac-delta function. The corresponding

[4] By definition, an indicial function is the response to a disturbance that is applied instantaneously at time zero and held constant thereafter, that is, a disturbance given by a step function. In this case, $w = 0$ for $t < 0$, and $w = V\alpha$ for $t > 0$.

result for an indicial change in pitch rate about the leading edge is given by

$$\frac{\Delta C_p(\bar{x},t)}{q} = \frac{\delta(t)}{V}(1+2\bar{x})\sqrt{(1-\bar{x})\bar{x}}$$

$$+(3\phi(s)-1)\sqrt{\frac{1-\bar{x}}{\bar{x}}}+4\sqrt{(1-\bar{x})\bar{x}}. \tag{8.41}$$

Again, the first term is an apparent mass term, with the second term being circulatory and affected by the shed wake. The third term is a quasi-steady term, with an analogous term also appearing in Theodorsen's result.

Wagner's function, $\phi(s)$, is known exactly [see, for example, Lomax (1968)] and is plotted in Fig. 8.16. Note that the apparent mass loading is responsible for the initial infinite pulse in the lift as $s = 0$. Thereafter, the function builds asymptotically from one half to a final value of unity as $s \to \infty$. In Wagner's problem, the aerodynamic center is at mid-chord at $s = 0$ and moves immediately to the $1/4$-chord for $s > 0$. The resulting variation in the lift coefficient for a step change in angle of attack, α, can be written as

$$C_l(t) = \frac{\pi c}{2V}\delta(t) + 2\pi\alpha\phi(s), \tag{8.42}$$

where $2\pi\alpha$ is the steady-state lift coefficient, as given by steady thin-airfoil theory.

For rotor analyses, the indicial lift response makes a useful starting point in the development of a general time-domain unsteady aerodynamic theory. If the indicial response is known, then the unsteady loads to arbitrary changes in angle of attack can be obtained through the superposition of indicial aerodynamic responses using the Duhamel integral. Consider a general system in response to a general forcing function $f(t)$, $t \geq 0$. If the indicial response ϕ of the system is known, then the output $y(t)$ of the system can be written in terms of Duhamel's integral as

$$y(t) = f(0)\,\phi(t) + \int_0^t \frac{df}{dt}\,\phi(t-\sigma)\,d\sigma. \tag{8.43}$$

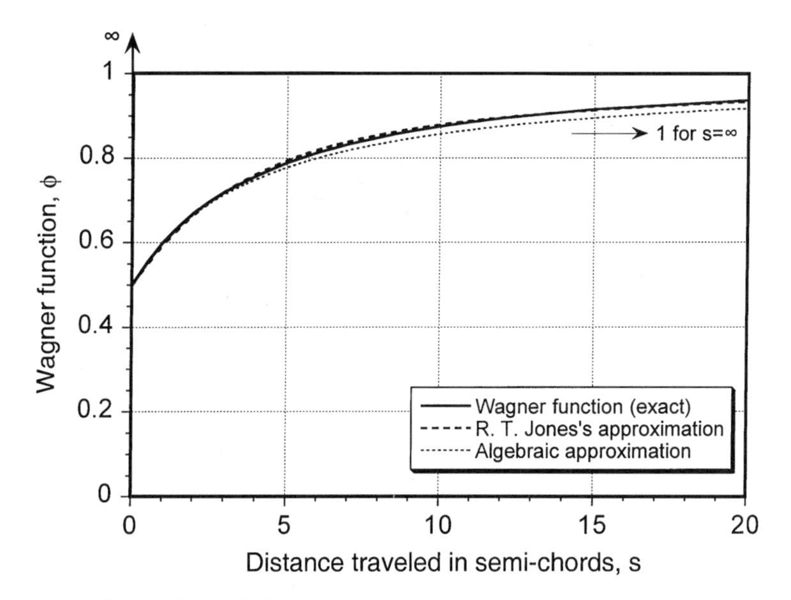

Figure 8.16 Wagner's function for a step change in angle of attack.

By analogy with Eq. 8.43, the circulatory part of the lift coefficient, C_l^c, in response to an arbitrary variation in angle of attack can now be written in terms of the Wagner function as

$$C_l^c(t) = 2\pi\left(\alpha(0)\phi(s) + \int_0^s \frac{d\alpha(\sigma)}{dt}\phi(s-\sigma)\,d\sigma\right) = 2\pi\alpha_e(t), \tag{8.44}$$

where α_e simply represents an effective angle of attack and contains within it all of the time-history effects on the lift because of the shed wake. Note that if V = constant, then $s = 2Vt/c$. In addition, the appropriate apparent mass terms must be added to get the total lift. For incompressible flow the apparent mass terms are proportional to the instantaneous motion, and so they all appear outside the Duhamel integral.

The Duhamel integral in Eq. 8.44 can be solved analytically or numerically. Analytical solutions are mostly restricted to simple forcing functions, and numerical methods must be employed in the general case. The main difficulty in solving Duhamel's integral is, however, with the Wagner function itself. Although the Wagner function is known exactly, its evaluation is not in a convenient analytic form. Therefore, it is usually replaced by a simple exponential or algebraic approximation. When this is done, a whole series of practical numerical tools for computing the unsteady aerodynamics can be unleashed. One approximation to the Wagner function, attributed to R. T. Jones (1938, 1940), is written as a two-term exponential series with four coefficients, that is,

$$\phi(s) \approx 1.0 - 0.165e^{-0.0455s} - 0.335e^{-0.3s}, \tag{8.45}$$

as shown in Fig. 8.16. This approximation is found to agree with the exact solution to an accuracy that is within 1%. Another approximation to the Wagner function is attributed to W. P. Jones (1945); here

$$\phi(s) \approx 1.0 - 0.165e^{-0.041s} - 0.335e^{-0.32s}. \tag{8.46}$$

In each case it will be noted that $A_1 + A_2 = 0.5$, according to Wagner's exact result.

The main advantage of the exponential approximation is that it has a simple Laplace transform. While the exponential behavior of the Wagner function is not an exact representation of the physical behavior, it is usually sufficiently accurate for practical calculations.[5] An alternative algebraic approximation to the Wagner function suggested by Garrick (1938) is

$$\phi(s) \approx \frac{s+2}{s+4}, \tag{8.47}$$

which, although not as accurate as the exponential approximation except in the limit as $s \to \infty$, it agrees with both the exact solution and the exponential approximation to within 2%.

8.11 The Sharp-Edged Gust: Küssner's Problem

In a rotor flow field, the rotor wake (and specifically the tip vortices) produces a highly nonuniform induced velocity across the plane of the disk. Therefore, a typical blade element encounters a nonuniform vertical upwash/downwash field as it rotates. It is, therefore, important to distinguish properly the effects on the airloads arising from angle of attack changes from blade motion (in effect, a plunging and pitching motion at the blade element)

[5] For some applications the exponential approximation to the indicial response may not be considered adequate. This is usually because the rate of approach to the asymptotic value is not as correct for the exponential approximation compared to the exact behavior. This effect, however, is more of academic interest rather than of any practical importance.

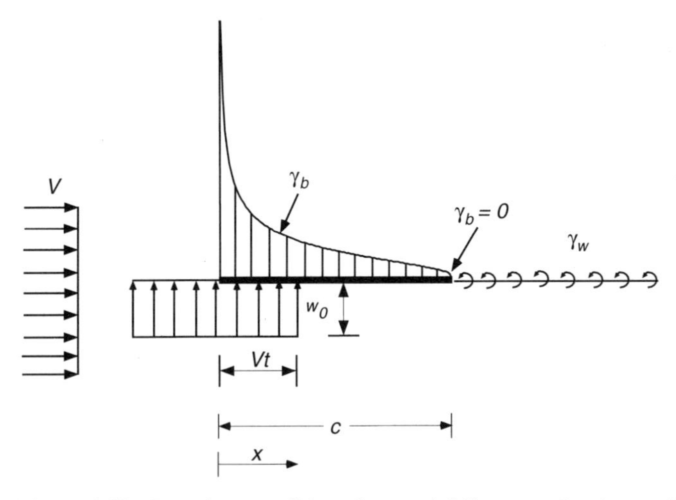

Figure 8.17 Boundary conditions for an airfoil penetrating into a sharp-edged vertical gust.

from the effects resulting from the rotor wake induced velocity field (in effect, a vertical gust velocity normal to the blade element). This distinction has important effects on the airloads and should not be overlooked in the mathematical modeling of helicopter problems.

The problem of finding the transient lift response on a thin airfoil entering a sharp-edged vertical gust (that is, a vertical upwash velocity) was first tackled by Küssner (1935) and properly solved by von Kármán and Sears (1938). In this problem, the upwash velocity, w_g, is defined relative to an axis at the leading edge as

$$w_g = \begin{cases} 0 & \text{if} \quad \bar{x} > Vt/c \\ w_0 & \text{if} \quad \bar{x} \leq Vt/c \end{cases} \tag{8.48}$$

as shown in Fig. 8.17. Recall that, in Wagner's problem, the angle of attack changes instantaneously over the whole chord at $s = 0$. In Küssner's problem, however, the quasi-steady angle of attack changes progressively as the airfoil penetrates into the gust front. At $s = 2$, the airfoil becomes fully immersed in the gust. The resulting variation in the lift coefficient can be written in a similar way to Wagner's solution such that

$$C_l(t) = 2\pi \left(\frac{w_0}{V} \right) \psi(s), \tag{8.49}$$

where $\psi(s)$ is known as Küssner's function and is plotted in Fig. 8.18. Compared to the Wagner function, it will be seen that the Küssner function builds from an initial value of zero and asymptotes to unity for $s \rightarrow \infty$. Küssner's function is also known exactly, albeit not in a convenient analytic form. Von Kármán & Sears (1938) also show that the aerodynamic center always acts at the 1/4-chord of the airfoil for all s. This is perhaps a surprising result, but it has been verified experimentally.

The Küssner function can be used with the Duhamel superposition integral to find the lift response to an arbitrary vertical upwash field, where the lift coefficient can be obtained using

$$C_l(t) = \frac{2\pi}{V} \left(w_g(0)\psi(s) + \int_0^s \frac{dw_g(\sigma)}{dt} \psi(s - \sigma)\, d\sigma \right). \tag{8.50}$$

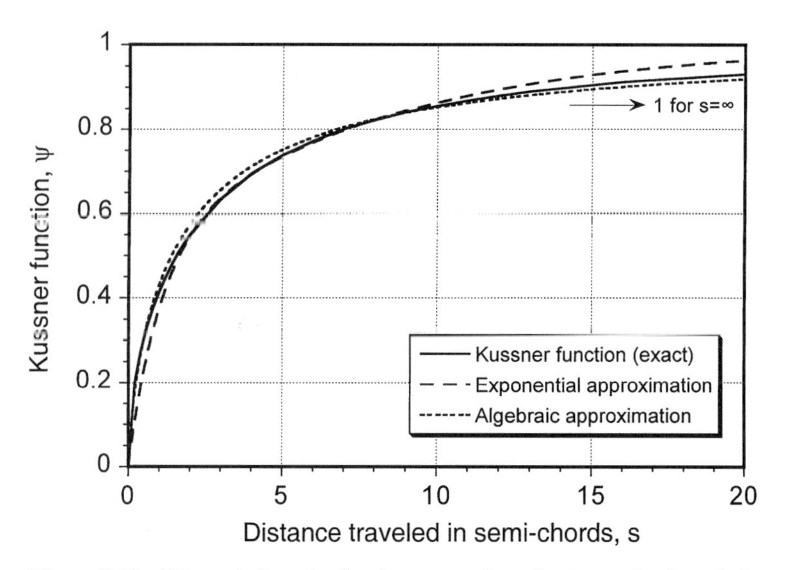

Figure 8.18 Küssner's function for the penetration of a sharp-edged vertical gust.

Note that this equation is analogous to the result for the lift resulting from arbitrary variations in angle of attack, as given previously, but with a different indicial function being used. To enable practical calculations using Duhamel superposition, the Küssner function, like the Wagner function, is usually replaced by an exponential approximation. One approximation is given by Sears & Sparks (1941) as

$$\psi(s) \approx 1 - 0.5e^{-0.13s} - 0.5e^{-1.0s}, \tag{8.51}$$

which is shown in Fig. 8.18. Alternatively, an algebraic approximation that is often used for the Küssner function is

$$\psi(s) \approx \frac{s^2 + s}{s^2 + 2.82s + 0.80}. \tag{8.52}$$

See Bisplinghoff et al. (1955). Neither approximation represents the correct vertical tangent of the ψ curve at $s = 0$, but this is of no practical significance.

8.12 Traveling Sharp-Edged Gust: Miles's Problem

Results for the lift on a thin 2-D airfoil encountering traveling (in-plane convecting) vertical sharp-edge gusts in incompressible flow have been obtained by Miles (1956) in terms of a gust speed ratio, λ:

$$\lambda = \frac{V}{(V + V_g)} \tag{8.53}$$

where V is the velocity of the airfoil and V_g is the in-plane component of the gust convection velocity relative to the airfoil. The normal assumption made in most rotor aerodynamic analyses is that the wake (tip vortices and corresponding induced velocity field) are stationary (nonconvecting) with respect to the rotor (i.e., rigid wake assumption), so that $\lambda = 1$ at all blade elements over the rotor disk. However, the self-induced velocities generated by the vortex wake system results in a continuously changing and nonuniform convection of

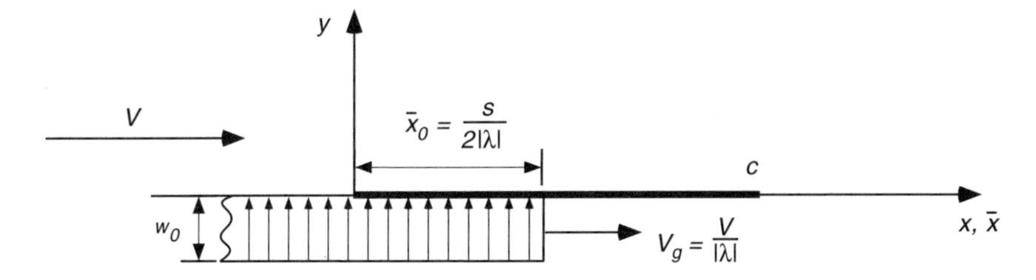

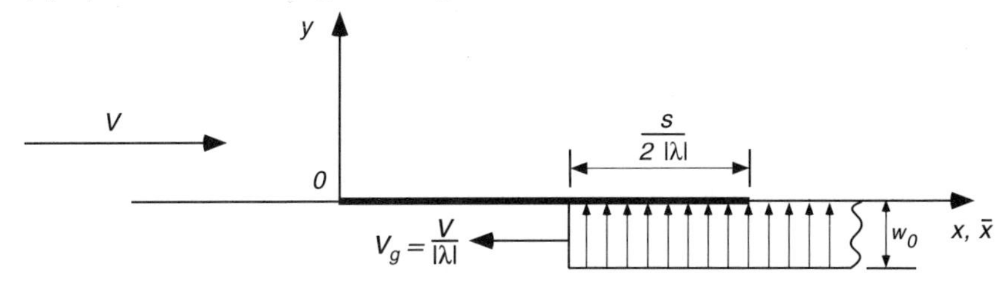

Figure 8.19 Boundary conditions for an airfoil encountering a convecting sharp-edged vertical gust.

the induced velocity field with respect to the rotor, and this may produce values of λ less than or greater than unity.

Miles (1956) showed that as the propagation speed, V_g, of the traveling gust front increased from zero to ∞ (λ decreases from 1 to 0), the solution for the unsteady lift changes from the Küssner result to the Wagner result, with a variety of intermediate transitional results being obtained. Miles's results were later generalized by Drischler & Diederich (1957) who obtained continuous semi-analytical forms for both the lift and pitching moment gust functions. The problem has also been solved by Leishman (1997) using the reverse flow theorems. All approaches make use of either algebraic or exponential approximations to the Wagner function to facilitate numerical solutions.

Consider a 2-D airfoil traveling with velocity V and subject to a vertical sharp-edged gust field of magnitude w_0 convecting with velocity $V_g = (\lambda^{-1} - 1)V$, as shown in Fig. 8.19. Note that when the gust field is stationary, $\lambda = 1$, and when the field is traveling toward the airfoil at infinite speed, $\lambda = 0$. For the sharp-edged gust, the primary boundary condition is that the downwash, w, is zero on the part of the airfoil that has not reached the gust front. This means that for a downstream traveling gust referenced to the leading edge

$$w_g = \begin{cases} 0 & \text{if } \bar{x} > \bar{x}_0 = V\lambda^{-1}\, t/c = s/2\lambda, \\ w_0 & \text{if } \bar{x} < \bar{x}_0 = V\lambda^{-1}\, t/c = s/2\lambda, \end{cases} \tag{8.54}$$

and for an upstream traveling gust

$$w_g = \begin{cases} 0 & \text{if } \bar{x} < \bar{x}_0 = 1 - V|\lambda|^{-1}\, t/c = 1 - s/2|\lambda|, \\ w_0 & \text{if } \bar{x} > \bar{x}_0 = 1 - V|\lambda|^{-1}\, t/c = 1 - s/2|\lambda|. \end{cases} \tag{8.55}$$

The problem is shown schematically in Fig. 8.19. In either case, it will be seen that, like the Küssner problem, the quasi-steady angle of attack on the airfoil changes progressively as a function of time as the airfoil penetrates into the gust front. For a stationary gust $\lambda = 1$, and under incompressible flow assumptions, this is equivalent to solving Küssner's problem. For $\lambda = 0$, this is equivalent to Wagner's problem.

One approach to solving the convecting gust problem involves using the reciprocal or reverse flow theorems – see Flax (1952, 1953), Brown (1949), Jones (1951), and Heaslet & Spreiter (1952). The main utility of reverse flow theorems is that they build from known solutions for airfoil flows and obviate the need to start each new problem from first principles. They are ideally suited to solving various indicial problems, both analytically and numerically. The reverse flow theorems have been used by Leishman (1994) and Hariharan & Leishman (1996) to calculate the indicial responses of airfoils with flaps. General forms of the aerodynamic reverse flow theorems have been established by Heaslet & Spreiter (1952). The first theorem states that: "The lift in steady or indicial motion of one airfoil having arbitrary twist and camber is equal to the integral over the planform of the product of the local angle of attack and the loading per unit angle of attack at the corresponding points on a second flat-plate airfoil of identical planform but moving in the reverse direction." In application, consider two airfoils, one moving in a forward direction and the other in a reverse direction. The first airfoil (the unknown problem) has an arbitrary angle of attack distribution $\alpha_1(x_1)$, which could be produced by a vertical gust field (or a flap). The second airfoil is a flat plate at constant angle of attack, $\alpha_2 =$ constant, which is assumed to have a known aerodynamic loading over the chord. The boundary conditions are

$$\alpha_1 = \alpha_1(x_1), \qquad \alpha_2 = \text{const.} \tag{8.56}$$

The first reverse flow theorem gives the result that

$$\alpha_2 C_{l_1} = \int_1 \alpha_2 \Delta C_{p_1}\, dx_1 = \int_2 \alpha_1 \Delta C_{p_2}\, dx_2. \tag{8.57}$$

In other words, the lift coefficient on the first airfoil can be found from the loading on the second airfoil by integrating the known solution and the local chordwise angle of attack using

$$C_{l_1} = \int_2 \alpha_1 \left(\frac{\Delta C_{p_2}}{\alpha_2} \right) dx_2. \tag{8.58}$$

This relatively simple but powerful technique allows some remarkable simplifications in solving both steady and transient airfoil problems. The utility of the theorems, however, extends only to the integrated forces and moments on the airfoil and not to pressure distributions, which may be required for some problems.

For $M = 0$, the chordwise pressure loading for an indicial change in angle of attack is given by Eq. 8.40. By using the reverse flow theorems, the time-varying (indicial) lift on the airfoil for a traveling sharp-edged vertical gust can be obtained by integration of this known solution over the appropriate part of the airfoil affected by the gust front, but when the airfoil is moving in the reverse direction. For the downstream traveling vertical gust, this is equivalent to integrating the known flat-plate loading from the trailing edge to the leading edge of the gust front at $\bar{x}_0$ (see Fig. 8.19). For the upstream traveling vertical gust, the known loading must be integrated from the leading edge of the airfoil up to $\bar{x}_0$. It will be immediately apparent that different results, both quasi-steady and unsteady, will be produced for downstream versus upstream traveling vertical gusts.

For incompressible flows, the noncirculatory part of the unsteady lift can be written in terms of the instantaneous upwash over the airfoil. For a traveling sharp-edged vertical gust, results can be obtained analytically by integrating the first term of Eq. 8.40 with the boundary conditions given in Eqs. 8.54 and 8.55. For a downstream traveling gust the noncirculatory lift can be shown to be

$$\frac{C_l^{nc}(t)}{(w_0/V)} = \frac{1}{2}\frac{\partial}{\partial t}\left(\frac{\sin 2\theta_0}{2} - \theta_0 + \pi\right), \tag{8.59}$$

where $\theta_0 = \cos^{-1}(1 - 2\bar{x}_0)$ so that $\theta_0 = 0$ at the time when the gust front is at the airfoil leading edge and $\theta_0 = \pi$ at the trailing edge. For the upstream traveling sharp-edged vertical gust, the corresponding result for the noncirculatory lift is

$$\frac{C_l^{nc}(t)}{(w_0/V)} = \frac{1}{2}\frac{\partial}{\partial t}\left(\theta_0 - \frac{\sin 2\theta_0}{2}\right). \tag{8.60}$$

Equations 8.59 and 8.60 can be evaluated numerically at discrete values of time as the gust front proceeds over the airfoil, with the time derivatives being evaluated by means of finite differences.

Unlike the apparent mass terms, the circulatory parts of the unsteady lift depend on the prior time history of the gust field, and so the lift must be obtained by Duhamel superposition. For a downstream traveling gust, the quasi-steady part of the circulatory lift can be obtained analytically by integration of the second term in Eq. 8.40 with the application of the appropriate boundary conditions. For the downstream traveling sharp-edged gust, it can be shown (see Question 8.12) that the quasi-steady circulatory lift is

$$\frac{C_l^{qs}}{(w_0/V)} = 2(\pi - \theta_0 - \sin\theta_0) \tag{8.61}$$

or, in terms of equivalent angle of attack,

$$\frac{C_l^{qs}}{(w_0/V)} = 2\pi\alpha^{qs} = 2\pi\left(1 - \frac{\theta_0}{\pi} - \frac{\sin\theta_0}{\pi}\right). \tag{8.62}$$

For the upstream traveling sharp-edged gust, the equivalent quasi-steady angle of attack is

$$\alpha^{qs} = \left(\frac{\theta_0}{\pi} + \frac{\sin\theta_0}{\pi}\right)\left(\frac{w_0}{V}\right). \tag{8.63}$$

The net unsteady circulatory lift is then determined numerically by Duhamel superposition with the instantaneous or quasi-steady equivalent angle of attack and the Wagner function using Eq. 8.44. The calculation of the corresponding unsteady pitching moments proceeds by a similar process. This problem, however, is somewhat more difficult to solve because a second reverse flow theorem and the chordwise loading as a result of pitch rate must be used in addition to the angle of attack result – see Leishman (1997).

Results for the unsteady lift and pitching moment for downstream traveling sharp-edged gusts are shown in Fig. 8.20. For $\lambda = 0\,(V_g = \pm\infty)$, the results lead to the Wagner function. For $\lambda = 1$ $(V_g = 0)$, the results reduce to the Küssner function. For intermediate values of λ, note that a different series of results are obtained as the gust propagation speed increases from zero ($\lambda = 1$). The noncirculatory term is responsible for the very large peaks in the lift produced as λ decreases. The lift reaches a maximum at the point when the airfoil is about half way into the gust. It can be seen that the magnitudes of these peaks are often larger than the steady-state lift coefficient of 2π per radian angle of attack. For vertical gusts that move with the wing at velocities less than 0 ($\lambda > 1$), the noncirculatory part of the lift is small and the circulatory lift grows only very slowly with time. The corresponding pitching moment shows a change in the sign of the center of pressure for λ greater or less than one. For the stationary gust ($\lambda = 1$), the center of pressure remains at the 1/4-chord throughout

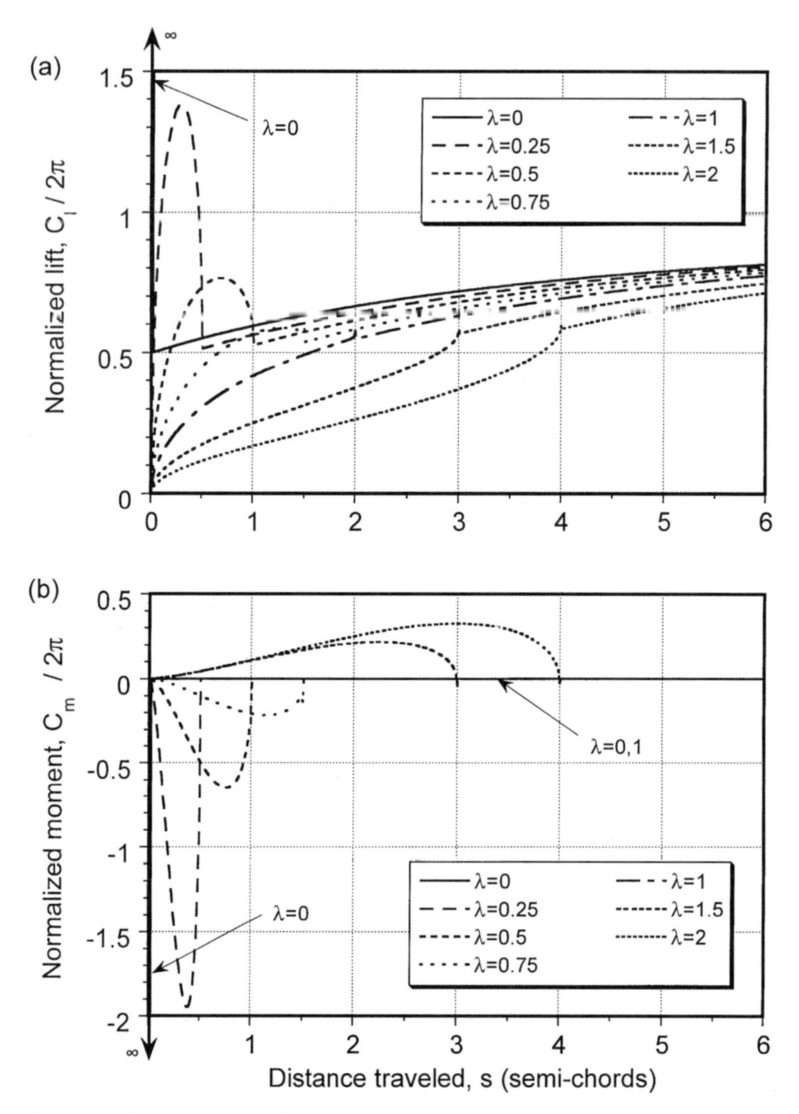

Figure 8.20 Airloads produced for a series of downstream moving sharp-edged vertical gusts. (a) Lift. (b) Pitching moment about 1/4-chord.

the motion, a result proved analytically by von Kármán & Sears (1938). As the gust speed approaches infinity, the peak in the pitching moment approaches $-\pi\delta(t)$ with the center of pressure moving to mid-chord. For receding vertical gusts, the center of pressure moves in front of the 1/4-chord.

Results for upstream traveling vertical gusts are shown in Fig. 8.21. Again, for quickly traveling gusts the results approach the Wagner function. For progressively slower gusts, large peaks in the lift and pitching moment appear as a result of the noncirculatory contributions to the airfoil loading. Note that the noncirculatory terms are the same for any value of $|\lambda|$ but that the total transient value of the lift is higher than for a downstream traveling gust. The reasons for this will be apparent from a comparison of Eqs. 8.62 and 8.63, which simply prove that a gust affecting the trailing edge of the airfoil will have a larger effect on the circulatory lift than a gust affecting the same percentage of the leading edge. For the

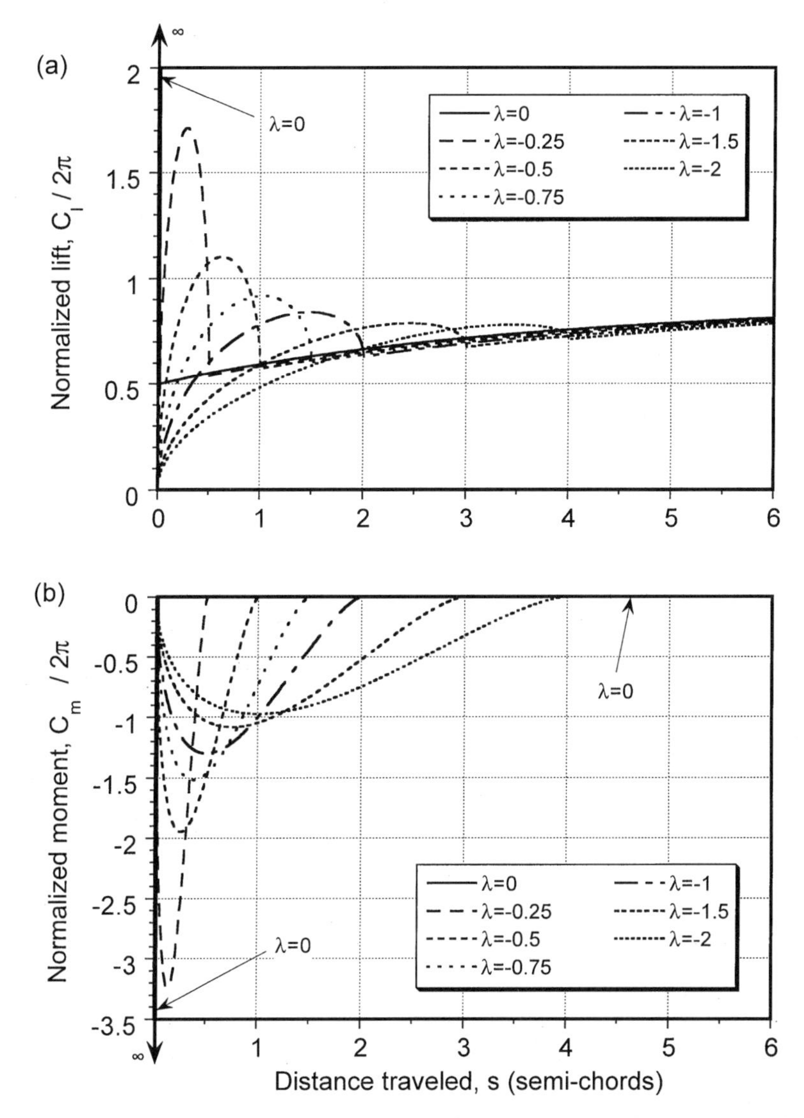

Figure 8.21 Airloads produced for a series of upstream moving sharp-edged vertical gusts. (a) Lift. (b) Pitching moment about 1/4-chord.

same reasons, a trailing edge flap deflection is more effective in producing a change in lift than a leading-edge flap.

8.13 Time-Varying Incident Velocity

The previously described analyses have considered the local free-stream velocity to be constant. However, in forward flight a typical rotor blade section will encounter a time-varying incident velocity because

$$V = U_T(y, \psi) = \Omega y + \mu \Omega R \sin \psi. \tag{8.64}$$

Under these conditions, there are additional unsteady aerodynamic effects to be considered. These effects include both noncirculatory and circulatory contributions. With a time-varying

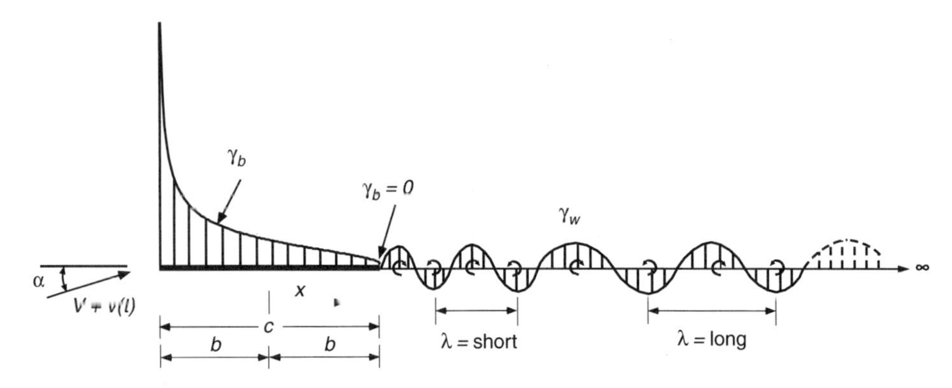

Figure 8.22 A time-varying free-stream velocity results in a nonuniform convection speed of the shed wake.

incident flow velocity, the shed wake behind the airfoil is convected at a nonuniform velocity (see Fig. 8.22), and this causes several complications in the mathematical treatment of the induced velocity effects. Isaacs (1945, 1946) derived a closed-form solution for the additional unsteady aerodynamic effects of a harmonically varying free-stream velocity. Greenberg (1947) published a similar theory, but he made a high frequency assumption about the shed wake behind the airfoil to obtain a solution in terms of the Theodorsen function alone. An approximate theory for this problem was also developed by Kottapalli (1985), in which small "lead-lag" airfoil oscillation amplitudes with respect to the mean velocity were assumed. Johnson (1980) has also discussed the general problem of a varying velocity on unsteady airloads. Using the same assumptions made by Isaacs, complete expressions are given for the lift and pitching moment on an airfoil executing harmonic plunge and pitch motion about an arbitrary pitch axis. Essentially, Johnson's conclusion is that the approximation using the Theodorsen function with the local reduced frequency is appropriate for flow oscillation amplitudes of up to 70% of the mean velocity. For small flow oscillation amplitudes, the Theodorsen function calculated with reduced frequency argument based on the mean velocity will be accurate enough, which effectively means neglecting the unsteady free-stream effects.

The theoretical problem of a time-varying free-stream velocity on the unsteady aerodynamic response has also been examined by van der Wall & Leishman (1994), both from the classical frequency-domain approach and also from the time domain. This work reemphasizes that one of the significant effects of an oscillating incident flow velocity is a nonuniform convection velocity of the shed wake vorticity behind the airfoil. Van der Wall & Leishman (1994) compare and contrast five theories representing the effect of the unsteady incident velocity variations: Isaacs's theory, Greenberg's theory, Theodorsen's theory combined with unsteady incident velocity, Kottapalli's theory, and Duhamel superposition with the Wagner function. They found that, strictly speaking, all of the theories represent the case of a fore–aft moving airfoil instead of an unsteady incident velocity; this latter case should be more correctly viewed as a propagating horizontal gust field. As described by Eq. 8.64, a helicopter rotor blade section in forward flight encounters both unsteady sectional velocity (the superposition of rotation and forward flight velocity components), but in addition there is a fore–aft motion (through the lead-lag degree of freedom). However, it was found that in the range of reduced frequencies encountered by a helicopter blade, the results obtained from all the various theories are essentially equivalent. Therefore, the interpretation of an unsteady free stream as equivalent to fore – aft motion of the airfoil can be viewed as a good approximation for the helicopter case. All of the theories cited above lead to the same noncirculatory expressions, and all were found to reduce to Theodorsen's theory (or equivalent) when the varying part of the incident flow oscillation amplitude becomes zero.

As mentioned previously, in the time domain the effect of a time-varying velocity in an incompressible flow can be modeled using Duhamel superposition with the Wagner function. Again, the appropriate apparent mass forces must be included in the solution, even though they will be small for the frequencies and amplitudes of the fluctuating velocity typically found on a helicopter. In this case

$$C_l(t) = \frac{\pi b}{V^2}\left(\ddot{h} + \frac{d(V\alpha)}{dt} - ab\ddot{\alpha}\right) + \frac{2\pi}{V}\left(V\alpha(0)\,\phi(s) + \int_0^s \frac{d(V\alpha)(\sigma)}{dt}\phi(s-\sigma)\,d\sigma\right). \tag{8.65}$$

The distance traveled through the flow is now proportional to the area under the velocity versus time curve:

$$s = \frac{2}{c}\int_0^t V dt. \tag{8.66}$$

Van der Wall & Leishman compare results obtained with this time-domain method with Isaac's frequency domain solution, where the agreement is essentially exact.

Results calculated using Eq. 8.65 are shown in Fig. 8.23 for the lift on an airfoil at a constant angle of attack, α_0, experiencing a harmonic oscillation of velocity of the form $V(t) = V_\infty(1 + \lambda \sin \omega t)$ at a reduced frequency, k, of 0.2 with $\lambda = 0, 0.2, 0.4, 0.6, 0.8$. The Duhamel integral was solved by means of a finite-difference approximation using the numerical scheme, discussed in Section 8.14.1. The lift in Fig. 8.23 is shown as a fraction of the quasi-steady lift, $(C_l)_{qs} = 2\pi\alpha_0$. Also shown in this figure are results computed by means of a finite-difference method based on solutions to the Euler equations, which are

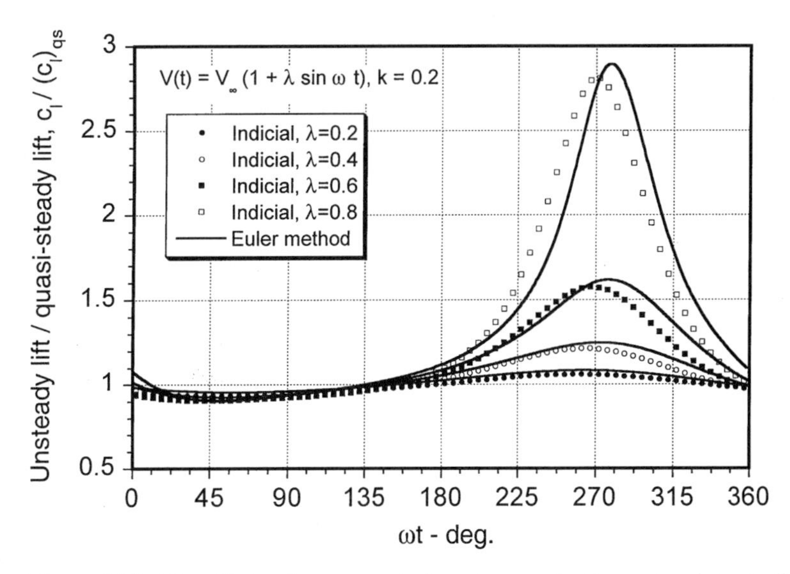

Figure 8.23 The effects of a nonsteady free-stream velocity on the unsteady lift for an airfoil with a constant angle of attack.

taken from van der Wall & Leishman (1994). It will be seen from Fig. 8.23 that at a given time the primary effect of unsteadiness is either an increase or attenuation of the unsteady lift compared to the quasi-steady result, which is a result already noted for oscillations in angle of attack. For $0^\circ < \omega t < 90^\circ$, the local velocity (and thus the product $V\alpha$) is increasing. Here, the shed circulation has the opposite sign to the airfoil circulation and so the unsteady lift is less than the quasi-steady value. The higher local velocity here means that the shed wake is further away from the airfoil in a given time, and this effect contributes to a more rapid buildup in the circulatory lift compared to the case where the wake convection speed would be assumed uniform. For $90^\circ \leq \omega t < 270^\circ$, the local velocity is decreasing. Here, the shed circulation has the same sign as the airfoil circulation, which tends to increase the lift over the quasi-steady value. Furthermore, for $180^\circ \leq \omega t < 360^\circ$ a velocity lower than the mean incident velocity means that the shed wake is closer to the airfoil; this then enhances the basic unsteady effect and the lift quickly becomes much greater than the quasi-steady values. Finally for $270^\circ \leq \omega t < 360^\circ$, the local velocity is increasing again. With the change in sign of the wake circulation and the increasing wake convection velocity, the lift drops quickly to become close to the quasi-steady value again. Note that in Fig. 8.23 there is good agreement between the incompressible theory and the Euler method[6], despite the slight phase differences. Other theoretical work on modeling unsteady free-stream velocity effects is given by Ashley et al. (1952) and Friedmann & Yuan (1977).

8.14 Indicial Response Method

As described previously, if the indicial aerodynamic response(s) can be determined, then these solutions form a powerful means of finding the unsteady aerodynamic forces and moments in the time domain as a result of arbitrary variations in angle of attack and/or inflow velocity by using Duhamel superposition. The effects of a nonconstant incident flow velocity may also be handled by means of this approach. There are two main difficulties with this method. The first is to determine the indicial response functions themselves. While they are known exactly for incompressible flow (e.g. Wagner and Küssner functions), they are not known exactly for subsonic compressible flow. Nevertheless, there are various techniques that can be used to find and approximate forms of the indicial response from aerodynamic transfer functions obtained from experimental measurements. Second, numerical methods must be devised to enable the superposition process to be conducted accurately and efficiently.

The time-varying value of the lift coefficient, $C_l(s)$, can be expressed as a function of angle of attack, $\alpha(s)$, in terms of the Duhamel integral as

$$C_l(s) = C_{l_\alpha}\left[\alpha(s_0)\,\phi(s) + \int_{s_0}^{s} \frac{d\alpha}{ds}(\sigma)\,\phi(s-\sigma)\,d\sigma\right] = C_{l_\alpha}\alpha_e(s), \tag{8.67}$$

where $\phi(s)$ is indicial response to a unit step input α and C_{l_α} is the lift-curve-slope ($= 2\pi$/radian for incompressible flow). If the integral is evaluated, then the term $\alpha_e(s)$ can be viewed as an effective angle of attack in that it contains all the time-history information.[7] Equation 8.67 is usually solved numerically for discrete values of time. For a discretely

[6] The Euler finite–difference method is computationally expensive, being approximately five orders of magnitude greater than the cost of the solution obtained using Duhamel superposition.

[7] Note that the result in Eq. 8.67 can also be extended to finite wings, where ϕ represents the indicial aerodynamic response of the entire wing as opposed to a single 2-D section – see Jones (1940) and Lomax (1968).

sampled system at times $s = s_0, \sigma_i \ldots \sigma_2, \sigma_1, s$, then $\alpha_e(s)$ can be written using Eq. 8.67 as

$$\begin{aligned}
\alpha_e(s) &= \alpha(s_0)\,\phi(s) + \sum_{i=1}^{\infty} \frac{d\alpha}{ds}(\sigma_i)\,\phi(s-\sigma_i)\Delta\sigma_i \\
&= \alpha(s_0)\,\phi(s) + \alpha'(\sigma_1)\,\phi(s-\sigma_1)\Delta\sigma_1 + \alpha'(\sigma_2)\,\phi(s-\sigma_2)\Delta\sigma_2 + \cdots \\
&\quad + \alpha'(\sigma_i)\,\phi(s-\sigma_i)\Delta\sigma_i + \cdots,
\end{aligned} \tag{8.68}$$

with the summation extending over all inputs that have acted up to the instant s. Therefore, the result for $\alpha_e(s)$ requires the storage of $\alpha'(s), \alpha'(\sigma_1), \ldots$ at all previous time steps and the repeated reevaluation of the indicial function for each $s - \sigma_i$ at each new time step. Obviously, in most cases information at a large number of previous time steps must be retained. Also, the indicial response function is not always known in a convenient simple analytic form, such as with the Wagner and Küssner functions, and a relatively large number of numerical operations must be performed. Fortunately, there are alternative approaches to the problem.

8.14.1 *Recurrence Solution to the Duhamel Integral*

If a general two-term exponentially growing indicial function is assumed such that

$$\phi(s) = 1 - A_1 e^{-b_1 s} - A_2 e^{-b_2 s} \tag{8.69}$$

then the Duhamel integral in Eq. 8.67 can be written as

$$\begin{aligned}
\alpha_e(s) &= \alpha(s_0)\,\phi(s) + \int_{s_0}^{s} \frac{d\alpha}{ds}(\sigma)\,\phi(s-\sigma)\,d\sigma \\
&= \alpha(s_0)(1 - A_1 e^{-b_1 s} - A_2 e^{-b_2 s}) \\
&\quad + \int_{s_0}^{s} \frac{d\alpha}{ds}(\sigma)(1 - A_1 e^{-b_1(s-\sigma)} - A_2 e^{-b_2(s-\sigma)})d\sigma \\
&= \alpha(s_0) - A_1\alpha(s_0)e^{-b_1 s} - A_2\alpha(s_0)e^{-b_2 s} + \int_{s_0}^{s} d\alpha(s) \\
&\quad - A_1 \int_{s_0}^{s} \frac{d\alpha}{ds}(\sigma)e^{-b_1(s-\sigma)}d\sigma - A_2 \int_{s_0}^{s} \frac{d\alpha}{ds}(\sigma)e^{-b_2(s-\sigma)}\,d\sigma.
\end{aligned} \tag{8.70}$$

Note that the terms $A_1\alpha(s_0)e^{-b_1 s}$ and $A_2\alpha(s_0)e^{-b_2 s}$ containing the initial value of α are short-term transients and can be neglected. Therefore, the Duhamel integral can be rewritten as

$$\alpha_e(s) = \alpha(s) - X(s) - Y(s), \tag{8.71}$$

which is in the notation used by Beddoes (1976, 1984), where the X and Y terms are given by

$$X(s) = A_1 \int_{s_0}^{s} \frac{d\alpha}{ds}(\sigma)e^{-b_1(s-\sigma)}\,d\sigma, \tag{8.72}$$

$$Y(s) = A_2 \int_{s_0}^{s} \frac{d\alpha}{ds}(\sigma)e^{-b_2(s-\sigma)}\,d\sigma. \tag{8.73}$$

Assuming a continuously sampled system with time step Δs (which may be nonuniform) and that $s_0 = 0$, then we have at the next time step $s + \Delta s$:

$$X(s+\Delta s) = A_1 \int_{0}^{s+\Delta s} \frac{d\alpha}{ds}(\sigma)e^{-b_1(s+\Delta s-\sigma)}\,d\sigma. \tag{8.74}$$

Splitting the integral into two parts gives

$$X(s+\Delta s) = A_1 e^{-b_1 \Delta s} \int_0^s \frac{d\alpha}{ds}(\sigma) e^{-b_1(s-\sigma)}\, d\sigma + A_1 \int_s^{s+\Delta s} \frac{d\alpha}{ds}(\sigma) e^{-b_1(s+\Delta s-\sigma)}\, d\sigma \tag{8.75}$$

$$= X(s) e^{-b_1 \Delta s} + A_1 \int_s^{s+\Delta s} \frac{d\alpha}{ds}(\sigma) e^{-b_1(s+\Delta s-\sigma)}\, d\sigma$$

$$= X(s) e^{-b_1 \Delta s} + I \tag{8.76}$$

Note that this new value, $X(s+\Delta s)$, is a one-step recursive formula in terms of the previous value, $X(s)$, and a new increment, I, over the new period. Consider now the evaluation of the I term:

$$I = A_1 \int_s^{s+\Delta s} \frac{d\alpha}{ds}(\sigma) e^{-b_1(s+\Delta s-\sigma)}\, d\sigma$$

$$= A_1 e^{-b_1(s+\Delta s)} \int_s^{s+\Delta s} \frac{d\alpha}{ds}(\sigma) e^{b_1 \sigma}\, d\sigma$$

$$= A_1 e^{-b_1(s+\Delta s)} \int_s^{s+\Delta s} \frac{d\alpha}{ds}(\sigma) f(\sigma)\, d\sigma \tag{8.77}$$

with $f(\sigma) = e^{b_1 \sigma}$. At this point, several simplifying assumptions can be made. Introducing a simple backward-difference approximation for $d\alpha/ds$ at time $s + \Delta s$ gives

$$\left.\frac{d\alpha}{ds}\right|_{s+\Delta s} = \frac{\alpha(s+\Delta s) - \alpha(s)}{\Delta s} = \frac{\Delta \alpha_{s+\Delta s}}{\Delta s}, \tag{8.78}$$

which has an error of order $\alpha''(s+\Delta s)\Delta s$. Alternatively, one could use

$$\left.\frac{d\alpha}{ds}\right|_{s+\Delta s} = \frac{3\alpha(s+\Delta s) - 4\alpha(s) + \alpha(s-\Delta s)}{2\Delta s}, \tag{8.79}$$

which has an error of order $\alpha'''(s+\Delta s)(\Delta s)^2$, although this scheme requires the storage of α at two previous time steps. The remaining part of the integral involving $f(\sigma)$ can be evaluated exactly and I becomes

$$I = A_1 \left(\frac{\Delta \alpha_{s+\Delta s}}{\Delta s}\right) \left(\frac{1 - e^{-b_1 \Delta s}}{b_1}\right) \tag{8.80}$$

when using Eq. 8.78. Now, if $b_1 \Delta s$ is small so that $b_1^2 (\Delta s)^2$ and higher powers can be neglected, then

$$\frac{1 - e^{-b_1 \Delta s}}{b_1} \approx \Delta s. \tag{8.81}$$

Therefore, we obtain the relatively simple result that

$$I = A_1 \left(\frac{\Delta \alpha_{s+\Delta s}}{\Delta s}\right) \Delta s = A_1 \Delta \alpha_{s+\Delta s}. \tag{8.82}$$

It will be seen that this latter result is equivalent to setting $f(\sigma) = \text{constant} = e^{b_1(s+\Delta s)}$ over the sample period (i.e., using the rectangle rule of integration) and has a local error of order

$(\Delta s)^2$. When Eq. 8.82 is introduced into Eq. 8.76, this gives the recurrence formula

$$X(s + \Delta s) = X(s)e^{-b_1 \Delta s} + A_1 \Delta \alpha_{s+\Delta s} \tag{8.83}$$

or

$$X(s) = X(s - \Delta s)e^{-b_1 \Delta s} + A_1 \Delta \alpha_s. \tag{8.84}$$

Proceeding in a similar fashion for the Y term in Eq. 8.73 and using Eq. 8.70 gives

$$\begin{aligned} \alpha_e(s) &= \alpha(0) + \int_0^s d\alpha(s) - X(s) - Y(s) \\ &= \alpha(s) - X(s) - Y(s), \end{aligned} \tag{8.85}$$

where the $X(s)$ and $Y(s)$ terms are given by the one-step recursive formulas that will be denoted by Algorithm D-1:

$$X(s) = X(s - \Delta s)e^{-b_1 \Delta s} + A_1 \Delta \alpha_s, \tag{8.86}$$

$$Y(s) = Y(s - \Delta s)e^{-b_2 \Delta s} + A_2 \Delta \alpha_s. \tag{8.87}$$

Note that recursive functions X and Y contain all the time-history information of the unsteady aerodynamics and are simply updated once at each time step, thereby providing numerically efficient solutions to the unsteady lift for arbitrary variations in α. Obviously, the above results can be extended to other modes of forcing and to any number of exponential terms that may be used to represent the indicial function. The error in this algorithm results mostly from the approximation used in Eq. 8.81, and it can be shown that the relative error in the integral is

$$\epsilon = 2 - \frac{b_1 \Delta s}{1 - e^{-b_1 \Delta s}} - \frac{b_2 \Delta s}{1 - e^{-b_2 \Delta s}}. \tag{8.88}$$

Generally, to obtain errors of less than 5%, each of the products $b_1 \Delta s$ and $b_2 \Delta s$ must be less than 0.05. This requires a relatively small time step and may not always be practical in many helicopter rotor problems.

To minimize errors associated with larger time steps, various alternative sets of recursive formulas can be obtained. If Δs (or the products $b_1 \Delta s$ or $b_2 \Delta s$) is large, another approximation based on the midpoint rule can be used. In this case let $f(\sigma) \approx f(s + \Delta s/2)$, and thus

$$\begin{aligned} I &= A_1 e^{-b_1(s+\Delta s)} \left(\frac{\Delta \alpha_{s+\Delta s}}{\Delta s} \right) \int_s^{t+\Delta s} f(\sigma)\, d\sigma \\ &= A_1 e^{-b_1(s+\Delta s)} \left(\frac{\Delta \alpha_{s+\Delta s}}{\Delta s} \right) e^{b_1(s+\Delta s/2)}\, \Delta s \\ &= A_1 \Delta \alpha_{s+\Delta s}\, e^{-b_1 \Delta s/2}. \end{aligned} \tag{8.89}$$

This gives a method that will be denoted as Algorithm D-2:

$$X(s) = X(s - \Delta s)e^{-b_1 \Delta s} + A_1 \Delta \alpha_s\, e^{-b_1 \Delta s/2}, \tag{8.90}$$

$$Y(s) = Y(s - \Delta s)e^{-b_2 \Delta s} + A_2 \Delta \alpha_s\, e^{-b_2 \Delta s/2}. \tag{8.91}$$

This recurrence algorithm was first used by Beddoes (1984). This method has a local error of order $(\Delta s)^3$, and the relative error in the integral can be shown to be

$$\epsilon = 2 - \frac{b_1 \Delta s\, e^{-b_1 \Delta s/2}}{1 - e^{-b_1 \Delta s}} - \frac{b_2 \Delta s\, e^{-b_2 \Delta s/2}}{1 - e^{-b_2 \Delta s}}. \tag{8.92}$$

In this case, Eq. 8.92 suggests that errors of less than 1% from an exact solution will be obtained if both $b_1 \Delta s$ and $b_2 \Delta s$ are less than 0.25, which is a much more practically realizable option in a rotor analysis.

While the above mentioned methods have seen some previous use in comprehensive rotor analyses, other methods based on the trapezoidal rule or Simpson's rule can also be used to evaluate I. Using Simpson's rule, then

$$
\begin{aligned}
I &= A_1 e^{-b_1(s+\Delta s)} \left(\frac{\Delta \alpha_{s+\Delta s}}{\Delta s} \right) \int_s^{s+\Delta s} f(\sigma)\, d\sigma \\
&= A_1 e^{-b_1(s+\Delta s)} \left(\frac{\Delta \alpha_{s+\Delta s}}{\Delta s} \right) \left(\frac{e^{b_1 s} + 4e^{b_1(s+\Delta s/2)} + e^{b_1(s+\Delta s)}}{6} \right) \Delta s \\
&= \frac{A_1}{6} \Delta \alpha_{s+\Delta s} (1 + 4e^{-b_1 \Delta s/2} + e^{-b_1(s+\Delta s)}),
\end{aligned} \tag{8.93}
$$

which has a local error of order $(\Delta s)^5$, but the overall error in the integration process is still of order $(\Delta s)^2$. Therefore, the Algorithm D-3 is obtained where

$$
X(s) = X(s - \Delta s) e^{-b_1 \Delta s} + \frac{A_1}{6} \Delta \alpha_s \left(1 + 4e^{-b_1 \Delta s/2} + e^{-b_1 \Delta s}\right), \tag{8.94}
$$

$$
Y(s) = Y(s - \Delta s) e^{-b_2 \Delta s} + \frac{A_2}{6} \Delta \alpha_s \left(1 + 4e^{-b_2 \Delta s/2} + e^{-b_2 \Delta s}\right). \tag{8.95}
$$

In this case, the relative error becomes

$$
\epsilon = 2 - \frac{b_1 \Delta s \left(1 + 4e^{-b_1 \Delta s/2} + e^{-b_1 \Delta s}\right)}{6(1 - e^{-b_1 \Delta s})} - \frac{b_2 \Delta s \left(1 + 4e^{-b_2 \Delta s/2} + e^{-b_2 \Delta s}\right)}{6(1 - e^{-b_2 \Delta s})}. \tag{8.96}
$$

Generally, when using Algorithm D-3, errors of less than 0.05% from an exact solution will be obtained if both $b_1 \Delta t$ and $b_2 \Delta s$ are less than 0.5. Therefore, despite some minor additional computational overhead, it will normally be the preferred method if Δs must be chosen to take larger values. (See Question 8.10 for a solution to this problem using the trapezoidal rule.)

8.14.2 *State-Space Solution*

For some applications, it is more convenient if the unsteady aerodynamic model is written in the form of differential equations (i.e., in state-space or state-variable form). The differential equations describing the unsteady aerodynamics can then be directly appended to the structural dynamic equations governing the airfoil or blade motion. One advantage of this approach is that the stability of an aeroelastic problem for a wing section or rotor blade can then be obtained by formulating it as an eigenvalue problem or using a Floquet stability analysis. Alternatively, the problems can be studied by time integration of the governing equations using standard numerical algorithms.

One of the most fundamental concepts associated with the description of any dynamic system, aerodynamic or otherwise, is the *state* of the system. The state describes the internal behavior of that system and is simply the information required at a given instant in time to allow the determination of the outputs from the system given future inputs. In other words, the state of the system determines its present condition; it is represented by a set of appropriately chosen variables describing the internal mechanics of the system. These variables are called the state variables and define an n-dimensional vector space $\mathbf{x}$ called the state-space in which one coordinate is defined by each of the state variables $x_1, x_2, \ldots, x_n$. A general nth-order differential system with m inputs and p outputs may be represented by n first-order differential equations

$$\dot{\mathbf{x}} = \mathbf{A}\mathbf{x} + \mathbf{B}\mathbf{u}, \tag{8.97}$$

with the output equations

$$\mathbf{y} = \mathbf{C}\mathbf{x} + \mathbf{D}\mathbf{u}, \tag{8.98}$$

where $\dot{x} = dx/dt$; $\mathbf{u} = u_i$, $i = 1, 2, \ldots, m$ are the system inputs, and the $\mathbf{y} = y_i$, $i = 1, 2, \ldots, p$ are the system outputs; $\mathbf{x} = x_i$, $i = 1, 2, \ldots, n$ are the states of the system. See Franklin et al. (1994) for more information about this approach.

The state equations describing the behavior of a 2-D unsteady airfoil can be obtained through direct application of Laplace transform methods to the indicial response. Consider the indicial lift response, ϕ, which is to be approximated by the exponential function given previously by Eq. 8.69. This function can be written in the time domain as

$$\phi(t) = 1.0 - A_1 e^{-b_1\left(\frac{2V}{c}\right)t} - A_2 e^{-b_2\left(\frac{2V}{c}\right)t}. \tag{8.99}$$

Initially, let $\phi(0) = 0 = 1 - A_1 - A_2$. Then the corresponding impulse response, $h(t)$, is given by

$$h(t) = A_1 b_1 \left(\frac{2V}{c}\right) e^{-b_1\left(\frac{2V}{c}\right)t} + A_2 b_2 \left(\frac{2V}{c}\right) e^{-b_2\left(\frac{2V}{c}\right)t}. \tag{8.100}$$

The Laplace transform of the impulse response is

$$\mathcal{L}[h(t)] = \frac{A_1 b_1 \left(\frac{2V}{c}\right)}{p + b_1 \left(\frac{2V}{c}\right)} + \frac{A_2 b_2 \left(\frac{2V}{c}\right)}{p + b_2 \left(\frac{2V}{c}\right)}, \tag{8.101}$$

which can be rearranged to yield the transfer functions as the Padé approximant

$$\mathcal{L}[h(t)] = \frac{(A_1 b_1 + A_2 b_2)\left(\frac{2V}{c}\right) p + (b_1 b_2)\left(\frac{2V}{c}\right)^2}{p^2 + (b_1 + b_2)\left(\frac{2V}{c}\right) p + (b_1 b_2)\left(\frac{2V}{c}\right)^2}. \tag{8.102}$$

From this transfer function, the lift response to an input $\alpha(t)$ can be directly written in state-space form as

$$\begin{Bmatrix} \dot{x}_1 \\ \dot{x}_2 \end{Bmatrix} = \left(\frac{2V}{c}\right) \begin{bmatrix} -b_1 & 0 \\ 0 & -b_2 \end{bmatrix} \begin{Bmatrix} x_1 \\ x_2 \end{Bmatrix} + \begin{Bmatrix} 1 \\ 1 \end{Bmatrix} \alpha(t), \tag{8.103}$$

and the output equation for the lift coefficient is

$$C_l(t) = C_{l_\alpha} \left(\frac{2V}{c}\right) [\, A_1 b_1 \quad A_2 b_2 \,] \begin{Bmatrix} x_1 \\ x_2 \end{Bmatrix}, \tag{8.104}$$

with C_{l_α} being recognized as the lift-curve-slope of the airfoil. These equations are in the form of Eqs. 8.97 and 8.98, where in this case the matrix $\mathbf{D}$ is equal to zero. An alternative realization of the system is the form

$$\begin{Bmatrix} \dot{x}_1 \\ \dot{x}_2 \end{Bmatrix} = \begin{bmatrix} 0 & 1 \\ -b_1 b_2 \left(\frac{2V}{c}\right)^2 & -(b_1 + b_2)\left(\frac{2V}{c}\right) \end{bmatrix} \begin{Bmatrix} x_1 \\ x_2 \end{Bmatrix} + \begin{Bmatrix} 0 \\ 1 \end{Bmatrix} \alpha(t), \quad (8.105)$$

and the corresponding equation for the lift coefficient is

$$C_l(t) = C_{l_\alpha} \left[b_1 b_2 \left(\frac{2V}{c}\right)^2 \quad (A_1 b_1 + A_2 b_2)\left(\frac{2V}{c}\right) \right] \begin{Bmatrix} x_1 \\ x_2 \end{Bmatrix}. \quad (8.106)$$

For a nonzero initial condition, such as when using the Wagner function approximation, the form of these equations is only slightly different. In this case

$$\begin{Bmatrix} \dot{x}_1 \\ \dot{x}_2 \end{Bmatrix} = \begin{bmatrix} 0 & 1 \\ -b_1 b_2 \left(\frac{2V}{c}\right)^2 & -(b_1 + b_2)\left(\frac{2V}{c}\right) \end{bmatrix} \begin{Bmatrix} x_1 \\ x_2 \end{Bmatrix} + \begin{Bmatrix} 0 \\ 1 \end{Bmatrix} \alpha(t), \quad (8.107)$$

and the equation for the lift coefficient is

$$C_l(t) = C_{l_\alpha} \left[(A_1 + A_2) b_1 b_2 \left(\frac{2V}{c}\right)^2 \; (A_1 b_1 + A_2 b_2)\left(\frac{2V}{c}\right) \right] \begin{Bmatrix} x_1 \\ x_2 \end{Bmatrix} + C_{l_\alpha}(1 - A_1 - A_2)\,\alpha(t). \quad (8.108)$$

For example, if R. T. Jones's approximation to the Wagner function is used (Eq. 8.45), then after substituting the numerical values of the coefficients we get

$$\begin{Bmatrix} \dot{x}_1 \\ \dot{x}_2 \end{Bmatrix} = \begin{bmatrix} 0 & 1 \\ -0.01375 \left(\frac{2V}{c}\right)^2 & -0.3455\left(\frac{2V}{c}\right) \end{bmatrix} \begin{Bmatrix} x_1 \\ x_2 \end{Bmatrix} + \begin{Bmatrix} 0 \\ 1 \end{Bmatrix} \alpha(t), \quad (8.109)$$

with

$$C_l(t) = 2\pi \left[0.006825 \left(\frac{2V}{c}\right)^2 \; 0.10805\left(\frac{2V}{c}\right) \right] \begin{Bmatrix} x_1 \\ x_2 \end{Bmatrix} + \frac{\pi}{2}\,\alpha(t), \quad (8.110)$$

where $C_{l_\alpha} = 2\pi$ The extra term $\frac{\pi}{2}\alpha$ on the right hand side of the above equation arises because of the non-zero initial conditions for the Wagner function, that is, $\phi(0) = 1 - A_1 - A_2 = 1/2$. Conversely, if we apply a unit step input to the above state-space equations and set the initial states to be zero (i.e., $\alpha(t) = 1$ for $t \geq 0$ and $x_1(0) = x_2(0) = 0$), the resulting response is exactly Jones's approximation to the Wagner function (see Question 8.11). In fact, it is clear that the Theodorsen solution, the Wagner solution with Duhamel superposition, and the above state-space model are simply different mathematical realizations of the same aerodynamic system. Dinyavari & Friedmann (1986) and Leishman & Nguyen (1990) describe state-space unsteady aerodynamic models of 2-D airfoil sections using Jones's approximation to the Wagner function for the incompressible flow case. Friedmann (1985) describes a more general, but still approximate, state-space realization of Loewy's function. This is done by representing the exact lift transfer function as a higher-order Padé approximation – see also Vepa (1976).

8.15 Subsonic Compressible Flow

All rotor problems involve compressibility to some degree. However, for subsonic compressible flow there is no convenient analytic equivalent to Theodorsen's theory, nor to the Wagner or Küssner functions. This is because of the nature of the governing equation, which for the subsonic case is the hyperbolic wave equation versus the elliptic Laplace's equation for incompressible flow [Karamcheti (1966)]. Therefore, alternative numerical means of finding the indicial response and sharp-edge gust functions must be derived. Because a time-domain representation of the unsteady aerodynamics is sought, it is convenient to address the subsonic problem starting from the indicial response. Although this is by no means the only basis from which the problem can be tackled, the indicial approach gives considerable physical insight into the problem.

For a compressible flow, the initial loading on the airfoil after an indicial (step) input comprises a pressure wave system with a compression wave on one surface of the airfoil and an expansion wave on the other. This initial pressure loading on the surface can be computed directly using piston theory, which is a local wave equation solution for the unsteady airloads [see Lomax (1968)]. The piston theory gives a result valid for any Mach number, M, but only at the instant after the perturbation has been applied (i.e., at $t = 0$ or $s = 0$). If we consider a small element of the airfoil surface subject to a change in normal velocity Δw, then piston theory gives the difference in pressure across the surface as $\Delta p = 2\rho a \Delta w$, so that

$$\Delta C_p(x, t = 0) = \frac{2\rho a \Delta w(x)}{\frac{1}{2}\rho V^2} = \left(\frac{4}{M}\right)\frac{\Delta w(x)}{V}. \tag{8.111}$$

Consider now a thin airfoil undergoing a simultaneous plunging and pitching about the 1/4-chord. As described previously by means of Fig. 8.4, the normal velocity on the airfoil is composed of two primary modes: a first uniform perturbation in w from the pure angle of attack contribution with another uniform perturbation resulting from plunge velocity, $\dot{h}/V$, and a second perturbation mode from the pure pitch rate of the airfoil q. By convention it will be assumed that the pitching motion takes place about the 1/4-chord, although the result can be generalized to any pitch axis. For a step change in each mode, applying piston theory and integrating across the chord gives the normal force coefficients[8] at $t = 0$ as

$$\Delta C_{n_\alpha}(0) = \frac{4}{M}\Delta\alpha \quad \text{and} \quad \Delta C_{n_q}(0) = \frac{1}{M}\Delta q \tag{8.112}$$

and the initial moment coefficients about the 1/4-chord as

$$\Delta C_{m_\alpha}(0) = -\frac{1}{M}\Delta\alpha \quad \text{and} \quad \Delta C_{m_q}(0) = -\frac{7}{12M}\Delta q. \tag{8.113}$$

For subsequent time, pressure waves from the airfoil propagate at the local speed of sound and, in the absence of any other forcing, the loading will decay rapidly with time from these initial "noncirculatory" values. For a compressible flow the noncirculatory terms do not appear as an infinite pulse at $s = 0$ as for incompressible flow but are finite in magnitude and decay more slowly from their initial values. Therefore, it will be appreciated that noncirculatory terms can no longer be assumed proportional to the *instantaneous* blade motion (i.e., $\alpha, \dot{\alpha}, \dot{h}$) as they are for the incompressible case and, like the circulatory terms, they must also depend on the time history of the forcing. Physically this is because for an incompressible flow pressure waves are propagated at infinite velocity, whereas for a real flow the disturbances propagate at the speed of sound.

[8] Note that in the thin-airfoil solution and also because low angles of attack are assumed, the normal force and lift force are usually used synonymously.

Such a transient behavior is difficult to compute in subsonic flow, but some analytic solutions can be obtained for limited values of time after the step input has been applied. Using an analogy of unsteady 2-D subsonic flow with steady supersonic flow, solutions to the wave equation have been evaluated exactly by Lomax et al. (1952) to obtain the chordwise pressure loading on the airfoil in the short period in the range $0 \leq s \leq 2M/(M+1)$. For a unit step change in angle of attack

$$\frac{\Delta C_p^{\alpha}(x, \hat{t})}{\alpha} = \Re \left\{ \frac{8}{\pi(1+M)} \sqrt{\frac{\hat{t} - x'}{M\hat{t} + x'}} + \frac{4}{\pi M} \left[\cos^{-1}\left(\frac{\hat{t}(1+M) - 2(c - x')}{\hat{t}(1-M)} \right) - \cos^{-1}\left(\frac{2x' - \hat{t}(1-M)}{\hat{t}(1+M)} \right) \right] \right\}, \tag{8.114}$$

where the domain is $x' = x - M\hat{t}$. This equation is valid for the early period $0 \leq \hat{t} \leq c/(1+M)$. The symbol $\Re$ refers to the real part, where the real parts of the arc cosine of numbers greater than 1 and less than 1 are 0 and π, respectively. Results are shown in Fig. 8.24 for $M = 0.5$. Note the nature of the chordwise pressure loading as the upstream and downstream moving waves pass over the chord of the airfoil. Even after a very short time, the growth in circulation has been established, but the final value steady state is only obtained after a relatively long time. Stahara & Spreiter (1976) and Singh & Baeder (1997a) show numerical solutions to the same problem using finite-difference methods.

By integrating the exact chordwise pressure loading solution in Eq. 8.114, the lift (normal force) and moment during the short period $0 \leq s \leq 2M/(1+M)$ can be determined analytically. The lift coefficient resulting from a step change in angle of attack is

$$\frac{C_{n_\alpha}(s)}{\alpha} = \frac{4}{M}\left[1 - \frac{1-M}{2M} s\right] \tag{8.115}$$

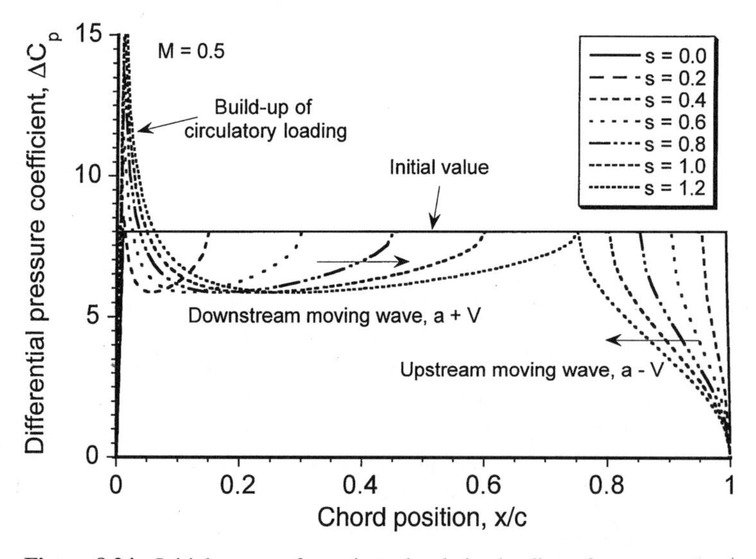

Figure 8.24 Initial stages of transient chordwise loading after a step change in angle of attack in subsonic flow at $M = 0.5$, showing the decay of the noncirculatory pressure loading and buildup of circulatory loading.

and the pitching moment coefficient about the 1/4-chord is

$$\frac{C_{m_\alpha}(s)}{\alpha} = -\frac{1}{M}\left[1 - \frac{1-M}{2M}s + \frac{M-2}{4M}s^2\right]. \tag{8.116}$$

Similarly, by using the corresponding chordwise loading for pitch rate, it is found that

$$\frac{C_{n_q}(s)}{q} = \frac{1}{M}\left[1 - \frac{1-M}{2M}s + \left(1 - \frac{M}{2}\right)\frac{s^2}{2M}\right] \tag{8.117}$$

and

$$\frac{C_{m_q}(s)}{q} = \frac{1}{M}\left[-\frac{7}{12} + \frac{5(1-M)}{8M}s - \frac{1-M^2}{8M^2}s^2 + \frac{(1-M)^3 + 4M}{64M^2}s^3\right]. \tag{8.118}$$

8.15.1 *Approximations to the Indicial Response*

The exact results obtained above are only valid for short values of time but they are useful because they provide guidance in developing approximations for the indicial functions that are in a more convenient analytic form to derive recurrence solutions to the Duhamel integral or for state-space realizations. In the general case, the indicial normal force and 1/4-chord pitching moment response to a step change in angle of attack, α, and a step change in nondimensional pitch rate about the 1/4-chord, q $(= \dot{\alpha}c/V)$, can be represented by the equations

$$\frac{C_{n_\alpha}(s)}{\alpha} = \frac{4}{M}\phi_\alpha^{nc}(s, M) + \frac{2\pi}{\beta}\phi_\alpha^{c}(s, M), \tag{8.119}$$

$$\frac{C_{m_\alpha}(s)}{\alpha} = -\frac{1}{M}\phi_{\alpha_m}^{nc}(s, M) + \frac{2\pi}{\beta}\phi_\alpha^{c}(s, M)\,(0.25 - x_{ac}), \tag{8.120}$$

$$\frac{C_{n_q}(s)}{q} = \frac{1}{M}\phi_q^{nc}(s, M) + \frac{\pi}{\beta}\phi_q^{c}(s, M), \tag{8.121}$$

$$\frac{C_{m_q}(s)}{q} = -\frac{7}{12M}\phi_{q_m}^{nc}(s, M) - \frac{\pi}{8\beta}\phi_{q_m}^{c}(s, M), \tag{8.122}$$

where the superscript $(.)^{nc}$ refers to the assumed noncirculatory part of the response, and the superscript $(.)^{c}$ refers to the assumed circulatory part. While it must be remembered that in the subsonic case this decomposition of the total loading is only an idealization, it is convenient for engineering purposes to handle the problem this way. The β term is the Glauert compressibility factor for linearized subsonic flow (i.e., $\beta = \sqrt{1-M^2}$). In the above equations, the initial values of the indicial response functions are given directly by piston theory as

$$\frac{C_{n_\alpha}(s=0)}{\alpha} = \frac{4}{M} \quad \text{and} \quad \frac{C_{m_\alpha}(s=0)}{\alpha} = -\frac{1}{M},$$

$$\frac{C_{n_q}(s=0)}{q} = \frac{1}{M} \quad \text{and} \quad \frac{C_{m_q}(s=0)}{q} = -\frac{7}{12M}.$$

The final values are given by the steady linearized subsonic theory as

$$\frac{C_{n_\alpha}(s=\infty)}{\alpha} = \frac{2\pi}{\beta} \quad \text{and} \quad \frac{C_{m_\alpha}(s=\infty)}{\alpha} = \frac{2\pi}{\beta}\,(0.25 - x_{ac}),$$

$$\frac{C_{n_q}(s=\infty)}{q} = \frac{\pi}{\beta} \quad \text{and} \quad \frac{C_{m_q}(s=\infty)}{q} = -\frac{\pi}{8\beta}.$$

It should be noted that in practical applications the linearized value of the lift-curve-slope, $2\pi/\beta$, can be replaced by the experimental value for a given airfoil at the appropriate Reynolds number and Mach number,[9] which is generally denoted by C_{n_α}. Additionally, the second term in Eq. 8.120 represents the contribution to the pitching moment resulting from a Mach number dependent offset of the aerodynamic center from the airfoil 1/4-chord axis, an effect previously described for the incompressible flow case in Section 8.7. As shown in Chapter 7, the values of the aerodynamic center, x_{ac}, for a given airfoil can be obtained from static airfoil measurements at the appropriate Mach number (e.g. Fig. 7.21). Also note that the second term of Eq. 8.122 represents the induced camber pitching moment resulting from pitch-rate motion, as given by the quasi-steady thin-airfoil result described earlier in Section 8.6. Therefore, it will be appreciated that the various indicial response functions denoted by ϕ represent the time-dependent behavior of the lift and pitching moment between $s = 0$ and $s = \infty$.

8.15.2 *Indicial Lift from Angle of Attack*

The lift or normal force response to a unit step change in angle of attack, α, can be idealized as the sum of a decaying noncirculatory part, $C^{nc}_{n_\alpha}$, and a growing circulatory part, $C^{c}_{n_\alpha}$, that is,

$$C_{n_\alpha}(s) = C^{nc}_{n_\alpha}(s, M) + C^{c}_{n_\alpha}(s, M) \tag{8.123}$$

or in terms of the indicial functions

$$C_{n_\alpha}(s) = \frac{4}{M}\phi^{nc}_{\alpha}(s, M) + \frac{2\pi}{\beta}\phi^{c}_{\alpha}(s, M). \tag{8.124}$$

Using reciprocal relations, the noncirculatory component of the loading on the airfoil can be theoretically extracted from the total lift response. This was first done by Mazelsky (1952a). The results can be closely approximated as an exponential decaying function, and so the noncirculatory part of lift for a unit step change in α may be written as

$$C^{nc}_{n_\alpha}(s, M) = \frac{4}{M}\phi^{nc}_{\alpha}(s, M) = \frac{4}{M}\exp\left(\frac{-s}{T'_\alpha}\right), \tag{8.125}$$

where $T'_\alpha = T'_\alpha(M) > 0$ is a Mach number dependent time constant still to be defined. For the growing (circulatory) part of the total indicial response, Mazelski (1952a,b) appears to have been one of the first investigators to use an exponential approximation of the form

$$\phi^{c}(s, M) = 1 - \sum_{n=1}^{N} A_n e^{-b_n s}; \quad \sum_{n=1}^{N} A_n = 1, b_n > 0 \tag{8.126}$$

for compressible flow, where the coefficients A_n and b_n will vary as a function of Mach number. These coefficients were obtained indirectly by Mazelsky by relating back into the time domain from numerical results (transfer functions) obtained for oscillating airfoils in the frequency domain. A similar approach has also been suggested by Dowell (1980) to obtain approximations to the indicial response for incompressible flow using Theodorsen's exact result, and for compressible flows by using transfer functions numerically computed using finite-difference solutions to the unsteady flow problem. In general, direct indicial

[9] The effects of Reynolds number and Mach number are generally implied, and the functional dependency on Re and M will be omitted for brevity.

type calculations are rare in the published literature, but direct indicial and vertical sharp-edged gust solutions have been performed by McCroskey & Goorjan (1983) using various small-disturbance, full-potential, and Euler finite-difference solvers. A series of more elaborate direct indicial calculations have recently been performed by Parameswaran & Baeder (1997) who have computed indicial angle of attack, pitch rate, and sharp-edged gust results using an Euler/Navier–Stokes finite-difference method. These solutions provide good check cases for indicial problems that cannot be simulated by experimentation and are also not amenable to exact analytical treatment. However, these solutions are only available at relatively high computational cost, and even then they are still subject to certain approximations and limitations.

While Mazelsky's exponential approximation in Eq. 8.126 may be acceptable for applications in fixed-wing analyses, it is not entirely convenient for a helicopter rotor analysis. This is because each blade station encounters a different local Mach number as a function of both blade radial location and azimuth angle. Therefore, repeated interpolation of the A_n and b_n coefficients will be required between successive Mach numbers to find the locally effective indicial function. Although relatively simple in concept, there is relatively large computational overhead associated with this process. In addition, it must be recognized that when superposition is applied to find the unsteady lift, each exponential term in the series in Eq. 8.191 contributes an additional state or deficiency function. To this end, Beddoes (1984) and Leishman (1987a, 1993) have assumed that the circulatory lift can be expressed in terms of a two-term growing exponential function, but one that scales directly with Mach number alone. The lift function $\phi_\alpha^c(s, M)$ takes the general form

$$\phi_\alpha^c(s, M) = 1 - A_1 e^{-b_1 \beta^2 s} - A_2 e^{-b_2 \beta^2 s}; \; A_1 + A_2 = 1, b_1, b_2 > 0, \tag{8.127}$$

where $\beta = \sqrt{1 - M^2}$, and the A and b coefficients are fixed and independent of Mach number.[10] The use of the β^2 term in the indicial function reflects the effects of compressibility (Mach number) on the growth of the circulatory part of the lift (through the effects of the shed wake). This manifests as larger lags in the growth of lift at higher Mach numbers. Such a behavior is well known from both a theoretical standpoint [see Osborne (1973)] as well as from experimental observations with oscillating airfoils. Furthermore, in a practical sense, not only is this simple compressibility scaling approach a computationally efficient way of accounting for compressibility effects in the wake, but it is also more accurate than repeated linear interpolation of the coefficients between discrete Mach numbers.

The coefficients A_1, A_2, b_1, and b_2 must be assumed as initially unknown, and although they could be derived in a number of ways, they can be reliably determined by relating back from frequency domain results using experimental measurements for oscillating airfoils in subsonic flow. It is possible to reduce the number of coefficients implicit in the indicial representation in Eq. 8.124 from five to four by obtaining an expression for the noncirculatory time constant, T'_α, in terms of the coefficients A_1, A_2, b_1, and b_2. Differentiating the approximation in Eq. 8.124 (using Eqs. 8.125 and 8.127) and the exact result in Eq. 8.115, and equating the gradients at $s = 0$, gives the time constant in the s domain as

$$T'_\alpha(M) = \frac{4M}{2(1 - M) + 2\pi\beta M^2 (A_1 b_1 + A_2 b_2)}. \tag{8.128}$$

[10] This equation can be shown valid up to at least the critical Mach number of the airfoil, beyond which nonlinear effects do not allow such simple generalizations because of the development of transonic flow.

Using the results $T_\alpha = (c/2V)T'_\alpha$ and $M = V/a$ gives the time constant in the t domain as

$$T_\alpha = [(1 - M) + \pi\beta M^2(A_1 b_1 + A_2 b_2)]^{-1}\frac{c}{a} = \kappa_\alpha T_i. \tag{8.129}$$

The advantage of taking this approach is that, regardless of the values of the coefficients A_1, A_2, b_1, and b_2, the noncirculatory constant, T_α, will always be adjusted to give the correct initial value and slope of the total indicial response as given by exact linear theory in Eq. 8.112. The continuity between the initial impulsive and succeeding circulatory loading is then obtained using linear superposition, as given by Eq. 8.124.

8.15.3 *Indicial Lift from Pitch Rate*

The indicial lift response to a step change in pitch rate q about the 1/4-chord can also be written as the sum of a noncirculatory part, $C^{nc}_{n_q}$ and a circulatory part, $C^c_{n_q}$, that is,

$$C_{n_q}(s) = C^{nc}_{n_q}(s, M) + C^c_{n_q}(s, M) \tag{8.130}$$

or in terms of the indicial functions

$$C_{n_q}(s) = \frac{1}{M}\phi^{nc}_q(s, M) + \frac{\pi}{\beta}\phi^c_q(s, M). \tag{8.131}$$

Numerous references have shown that, for incompressible flow, the chordwise pressure variation on the airfoil is the same as the thin airfoil loading but is independent of the mode of motion. Therefore, an angular velocity about some point can be considered equivalent to an angular velocity about some other point plus an angle of attack. In particular, the indicial lift for a pitch rate about the 3/4-chord position in incompressible flow is an impulse at $s = 0$ and is exactly zero thereafter. Subsequently, it follows that for incompressible flow $\phi^c_\alpha = \phi^c_q$. In linearized subsonic flow, the circulatory lift lag still remains an intrinsic function of the fluid itself. The chordwise pressure variation remains the same as the steady thin-airfoil loading and is unaffected by the mode of forcing or pitch axis location. Therefore, on a thin airfoil in subsonic flow, the lift always acts at the 1/4-chord point. In view of the foregoing, it is valid to assume $\phi^c_\alpha(s, M) = \phi^c_q(s, M)$ for subsonic compressible flow as well as incompressible flow without any loss of rigor. The circulatory part of the lift from pitch rate about the 1/4-chord, therefore, can be written as

$$C^c_{n_q}(s) = \frac{\pi}{\beta}[1 - A_1 e^{-b_1\beta^2 s} - A_2 e^{-b_2\beta^2 s}]. \tag{8.132}$$

The noncirculatory function can be assumed to be of the form

$$C^{nc}_{n_q}(s) = \frac{1}{M}\phi^{nc}_q(s) = \frac{1}{M}\exp\left(\frac{-s}{T'_q}\right). \tag{8.133}$$

Using Eqs. 8.132 and 8.133 and following the same procedure as before where the slopes of the approximating indicial function and the exact result (in this case, Eq. 8.117) are matched at $s = 0$ gives the time constant as

$$T_q = \left(\frac{c}{2V}\right)T'_q = [(1 - M) + 2\pi\beta M^2(A_1 b_1 + A_2 b_2)]^{-1}\frac{c}{a} = \kappa_q T_i. \tag{8.134}$$

8.15.4 Determination of Indicial Function Coefficients

The coefficients A_1, A_2, b_1, and b_2 are used to define the intermediate behavior of the indicial lift approximations. Using experimental results for the total lift and moment response resulting from a prescribed harmonic forcing, such as oscillations in pitch or plunge, we can relate these data back to empirically determine the coefficients of the indicial functions. Experimental data are available from a number of sources that comprise the unsteady aerodynamic lift and moment response from pitch and plunge oscillations performed under nominally attached flow conditions, that is, in the region where linearized aerodynamics are appropriate. It is essential that the data selected be for attached flow conditions as the presence of nonlinearities due to separation effects introduces further complications in the validation of the indicial responses. In Beddoes (1984) and Leishman (1993), data were taken mainly from the results of Liiva et al. (1968), Wood (1979), and Davis & Malcolm (1980).

Because the indicial functions have been completely defined as exponential functions of time, they may be easily manipulated using Laplace transforms. If we consider just the circulatory part of the lift from angle of attack given by Eqs. 8.124 and 8.127, the Laplace transform of this part of the indicial response function yields

$$C_n^c(p) = C_{n_\alpha}(M)\left[\frac{1}{p} - \frac{A_1 T_1}{1+T_1 p} - \frac{A_2 T_2}{1+T_2 p}\right], \tag{8.135}$$

where

$$T_1 = \frac{c}{2Vb_1\beta^2} \quad \text{and} \quad T_2 = \frac{c}{2Vb_2\beta^2}.$$

The Laplace transform of the forcing function (which in this case is a step change in angle of attack) is given by $\alpha(p) = 1/p$. The circulatory lift transfer function can be simplified to

$$\frac{C_n^c(p)}{\alpha(p)} = C_{n_\alpha}(M)\left[(1 - A_1 - A_2) + \frac{A_1}{1+T_1 p} + \frac{A_2}{1+T_2 p}\right]. \tag{8.136}$$

Because the initial value of the circulatory indicial response is zero, then $(1 - A_1 - A_2) = 0$ and the lift transfer function simplifies further to

$$\frac{C_n^c(p)}{\alpha(p)} = C_{n_\alpha}(M)\left[\frac{A_1}{1+T_1 p} + \frac{A_2}{1+T_2 p}\right]. \tag{8.137}$$

For the noncirculatory lift from angle of attack, the Laplace transform of the indicial response yields the transfer function

$$\frac{C_n^{nc}(p)}{\alpha(p)} = \frac{4}{M}\left(\frac{T'_\alpha p}{1+T'_\alpha p}\right), \tag{8.138}$$

and the transfer functions for the other lift components can be derived in a similar way. Therefore, for any forcing function for which a Laplace transform may be easily derived (such as a sinusoid), the lift and pitching moment response may be derived in explicit form using inverse Laplace transforms in terms of the assumed form of the indicial response function (see Questions 8.4–8.7). For example, the lift response to a harmonic pitch oscillation with a reduced frequency k about the airfoil 1/4-chord axis may be obtained after some manipulation as

$$\Re\, C_{n_\alpha}(k, M) = C_{n_\alpha}\left[\frac{A_1 b_1^2 \beta^4}{b_1^2\beta^4 + k^2} + \frac{A_2 b_2^2 \beta^4}{b_2^2\beta^4 + k^2}\right] + \frac{4}{M}\left[\frac{4\kappa_\alpha^2 M^2 k^2}{1 + 4\kappa_\alpha^2 M^2 k^2}\right], \tag{8.139}$$

$$\Im\, C_{n_\alpha}(k, M) = -C_{n_\alpha}\left[\frac{A_1 b_1 \beta^2 k}{b_1^2\beta^4 + k^2} + \frac{A_2 b_2 \beta^2 k}{b_2^2\beta^4 + k^2}\right] + \frac{4}{M}\left[\frac{2\kappa_\alpha M k}{1 + 4\kappa_\alpha M^2 k^2}\right], \tag{8.140}$$

$$\Re\, C_{n_q}(k, M) = C_{n_\alpha}\left[\frac{A_1 b_1 \beta^2 k^2}{b_1^2\beta^4 + k^2} + \frac{A_2 b_2 \beta^2 k^2}{b_2^2\beta^4 + k^2}\right] - \frac{1}{M}\left[\frac{4\kappa_q M^2 k^2}{1 + 4\kappa_q^2 M^2 k^2}\right], \tag{8.141}$$

$$\Im\, C_{n_q}(k, M) = C_{n_\alpha}\left[\frac{A_1 \beta^4 k}{b_1^2\beta^4 + k^2} + \frac{A_2 \beta^4 k}{b_2^2\beta^4 + k^2}\right] + \frac{1}{M}\left[\frac{8\kappa_q^2 M^2 k^3}{1 + 4\kappa_q^2 M^2 k^2}\right], \tag{8.142}$$

where $\Re$ and $\Im$ denote the real and imaginary parts of the response respectively. The resultant frequency response can be obtained by summing the various contributions to the real and imaginary components. A similar approach can be adopted to find the response from harmonic plunge oscillations, which of course, are useful because they do not include any pitch rate (q) terms in the response.

A vector $\mathbf{x}$ can now be defined that consists of the coefficients A_1, A_2, b_1, and b_2 used in the indicial lift function:

$$\mathbf{x}^T = \{A_1\ A_2;\ b_1\ b_2;\ \delta_\alpha, \delta_q\}. \tag{8.143}$$

The additional factors δ_α and δ_q are empirical values chosen to modify the initial value of the indicial response functions to account for the possiblities of finite-span effects present in the airfoil measurements. It has been shown, however, that their values are always close to unity, suggesting that the measurements selected are indeed representative of 2-D unsteady flow. The vector in Eq. 8.143 must be chosen to minimize the difference between the explicit solution based on the assumed indicial response approximations and the reference solutions (in this case, experimental results) for the unsteady lift in the frequency domain. If the real and imaginary parts, $F_m(M_i)$ and $G_m(M_i)$, respectively, are known at up to M values of reduced frequency and at each of I values of Mach number, an objective function $\bar{J}(\mathbf{x})$ can be defined as

$$\bar{J} = \sum_{i=1}^{I} J(\mathbf{x}, M_i), \tag{8.144}$$

where

$$J(\mathbf{x}, M_i) = \sum_{m=1}^{M} [F_m(M_i) - \Re C_n(\mathbf{x}, M_i, k_m)]^2 + [G_m(M_i) - \Im C_n(\mathbf{x}, M_i, k_m)]^2 . \tag{8.145}$$

The minimum of the objective function $\bar{J}$ in the parameter space $\mathbf{x}$ will give the best approximation to the known (reference) lift transfer function. Therefore, the objective function minimization algorithm is basically a nonlinear programming problem of minimizing $\bar{J}(\mathbf{x})$ subject to the constraints

$$A_n, b_n > 0, n = 1, 2 \quad \text{and} \quad 0 \le \delta_\alpha, \delta_q \le 1 \quad \text{and} \quad \sum_{n=1}^{N} A_n = 1. \tag{8.146}$$

The equality constraint may be replaced by a penalty function. Note that although Table 8.2 shows that the different sets of experimental data result in slightly different values of the

Table 8.2. *Coefficients of Indicial Lift Approximation Deduced from Oscillating Airfoil Experiments*

Data Source	A_1	A_2	b_1	b_2	δ_α	δ_q
Boeing Data	0.636	0.364	0.339	0.249	0.77	0.70
ARA Data	0.625	0.375	0.310	0.312	1.00	1.00
NASA Data	0.482	0.518	0.684	0.235	0.72	0.70
All Data	0.918	0.082	0.366	0.102	0.85	0.73

coefficients A_1, A_2, b_1, and b_2, the differences are of limited practical significance; the quality of the derived indicial functions is as good as the quality of the experimental data.

8.15.5 *Indicial Pitching Moment from Angle of Attack*

A similar process to that described previously can be used to find approximations for the other indicial response functions. The indicial pitching moment response about the 1/4-chord resulting from a step change in angle of attack, α, can also be written as the sum of a noncirculatory part, $C^{nc}_{m_\alpha}$, and a circulatory part, $C^{c}_{m_\alpha}$, that is,

$$C_{m_\alpha}(s) = C^{nc}_{m_\alpha}(s, M) + C^{c}_{m_\alpha}(s, M) \tag{8.147}$$

or in terms of the indicial functions

$$C_{m_\alpha}(s) = -\frac{1}{M}\phi^{nc}_{\alpha_m}(s, M) + \frac{2\pi}{\beta}(0.25 - x_{ac})\,\phi^{c}_{\alpha}(s, M). \tag{8.148}$$

For the noncirculatory part, a convenient general expression for the indicial function is of the form

$$\phi^{nc}_{\alpha_m}(s) = A_3 \exp\left(\frac{-s}{b_3 T'_{\alpha_m}}\right) + A_4 \exp\left(\frac{-s}{b_4 T'_{\alpha_m}}\right). \tag{8.149}$$

One approximation to this function uses the values $A_3 = 1.5$, $A_4 = -0.5$, $b_3 = 0.25$, and $b_4 = 0.1$. Following a similar approach to that used above for the lift, but now using Eq. 8.148, gives the noncirculatory time constant as

$$T_{\alpha_m} = \left(\frac{c}{2V}\right) T'_{\alpha_m} = \left[\frac{A_3 b_4 + A_4 b_3}{b_3 b_4 (1 - M)}\right] \frac{c}{a} = \kappa_{\alpha_m} T_i. \tag{8.150}$$

8.15.6 *Indicial Pitching Moment from Pitch Rate*

Finally, for the indicial moment response about the 1/4-chord resulting from a step change in pitch rate, q, about the 1/4-chord, this can be written as the sum of a noncirculatory part, $C^{nc}_{m_q}$, and a circulatory part, C^{c}_{mq}, that is,

$$C_{n_q}(s) = C^{nc}_{m_q}(s, M) + C^{c}_{m_q}(s, M) \tag{8.151}$$

or in terms of the indicial functions

$$C_{m_q}(s) = \frac{-7}{12M}\phi^{nc}_{q_m}(s, M) + \frac{-\pi}{8\beta}\phi^{c}_{q_m}(s, M). \tag{8.152}$$

The circulatory part is assumed to be of the form

$$C^{c}_{m_q}(s) = \frac{-\pi}{8\beta}\phi^{c}_{q_m} = \frac{-\pi}{8\beta}\left[1 - A_5 \exp(-b_5 \beta^2 s)\right], \tag{8.153}$$

where $A_5 = 1.0$ and $b_5 = 5.0$ are found to be satisfactory. For the noncirculatory part

$$C^{nc}_{m_q}(s) = \frac{-7}{12M}\phi^{nc}_{q_m} = \frac{-7}{12M}\exp\left(\frac{-s}{T'_{m_q}}\right). \tag{8.154}$$

Following the same procedure as before, but using Eq. 8.118, leads to the noncirculatory time constant

$$T_{q_m} = \left(\frac{c}{2V}\right)T'_{q_m} = \left[\frac{7}{15(1-M) + 3\pi\beta M^2 A_5 b_5}\right]\frac{c}{a} = \kappa_{q_m} T_i. \tag{8.155}$$

Thus, the noncirculatory time constants T_α, T_{α_m}, T_q, and T_{q_m} used in the indicial response functions are now defined in terms of the other coefficients. These functions are only weakly dependent on Mach number in the low subsonic flow regime; however, their values increase rapidly as $M = 1$ is approached. The four basic indicial functions are plotted in Figs. 8.25

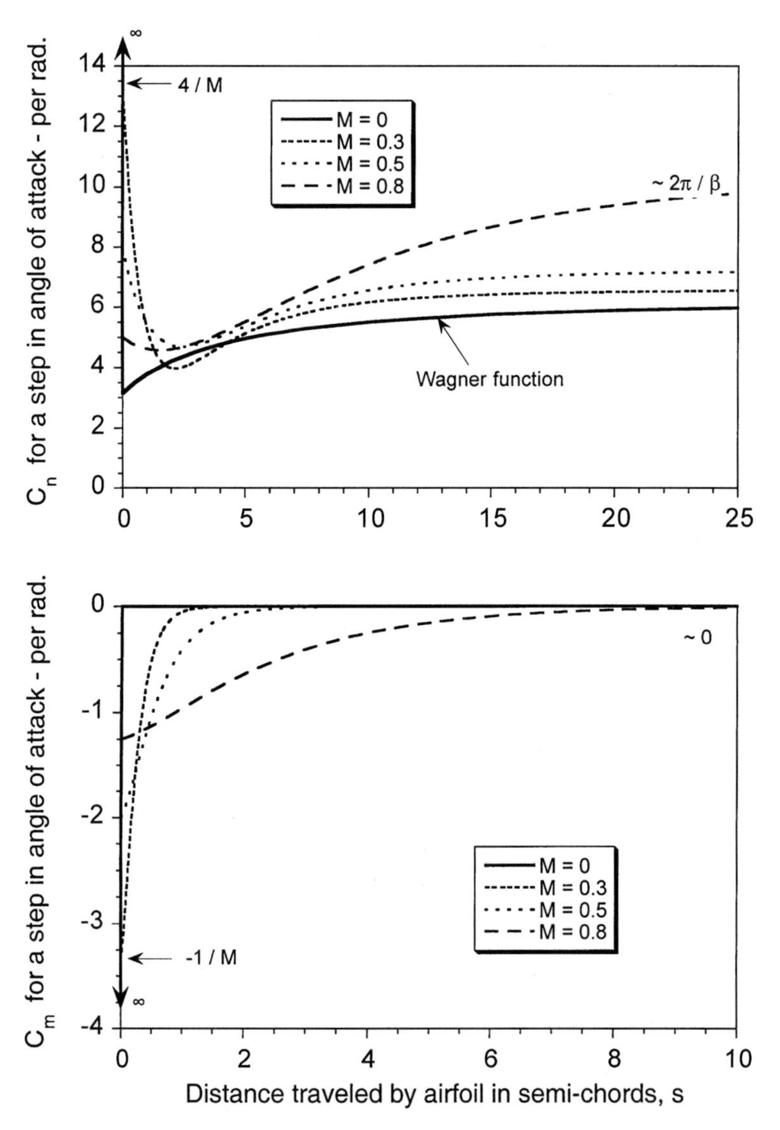

Figure 8.25 Indicial lift and pitching moment resulting from a step change in angle of attack.

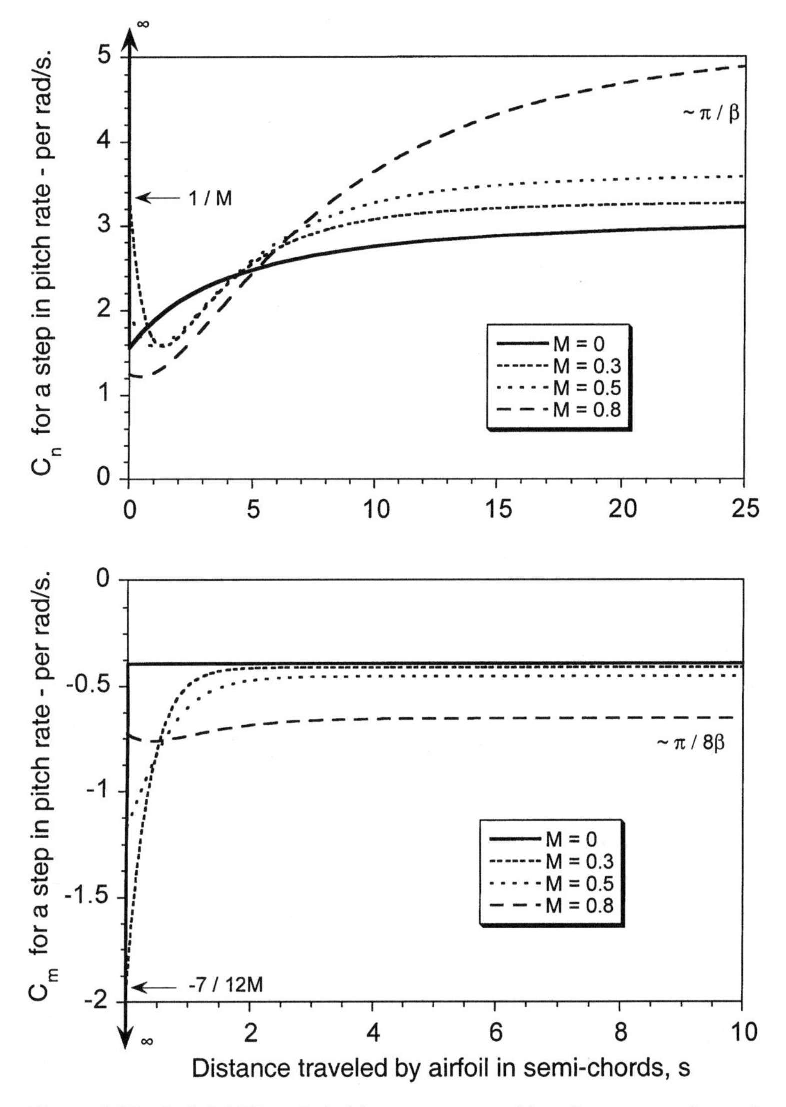

Figure 8.26 Indicial lift and pitching moment resulting from a step change in pitch rate about 1/4–chord.

and 8.26 for several Mach numbers to illustrate their behavior. The considerable influence of Mach number compared to the incompressible values should be noted. In application, these subsonic indicial aerodynamic functions are manipulated in exactly the same way as that shown previously for the incompressible case when using the exponential approximation to the Wagner or Küssner functions. For example, in the case of the subsonic lift to an arbitrary variation in α, three recurrence equations or three states instead of two must now be used, with both the circulatory and the noncirculatory terms having a time-history effect.

8.15.7 *Unsteady Axial Force and Drag*

Helicopter rotor blades have a much lower stiffness and effective damping than fixed-wing aircraft for the in-plane (lead-lag) degree of freedom. Whereas the flap and torsion degrees of freedom are primarily influenced by the lift and pitching moment respectively, the lead-lag degree-of-freedom is influenced by the drag. Furthermore, the blade lead-lag motion may couple with the flap or torsion degrees of freedom and may lead to an aeroelastic instability of the blades. As shown in Chapter 4, these coupling effects result from both the Coriolis forces and the aerodynamic loads. Therefore, for a comprehensive unsteady aerodynamic model of the rotor system it is necessary to include aerodynamic loads for all three degrees of freedom.

For steady flow conditions, the pressure drag coefficient C_{d_p} may be computed by resolving the normal force and axial force (or leading-edge suction force) coefficients through the angle of attack α using

$$C_{d_p} = C_n \sin\alpha - C_a \cos\alpha, \tag{8.156}$$

as shown in Section 7.10.1. For steady flow the normal force coefficient is given in terms of the force-curve-slope, $C_{n_\alpha}(M)$, at a given Mach number, M, and the angle of attack, α, as

$$C_n = C_{n_\alpha}(M)\,\alpha. \tag{8.157}$$

The corresponding leading-edge suction force coefficient for these conditions is given by

$$C_a = C_{n_\alpha}(M)\,\alpha\,\tan\alpha, \tag{8.158}$$

where α is measured in radians. Now, it is well known that for steady potential flow the pressure drag, C_{d_p}, is identically zero (d'Alembert's paradox), that is,

$$C_{d_p} = C_n \sin\alpha - C_a \cos\alpha = C_{n_\alpha}(M)\,\alpha \sin\alpha - C_{n_\alpha}(M)\,\alpha \tan\alpha \cos\alpha = 0. \tag{8.159}$$

However, for a real flow there is a net pressure drag on the airfoil because of viscous effects on the chordwise pressure distribution. As shown in Section 7.10.1, the inability of the airfoil to attain 100% leading-edge suction is modeled using the recovery factor η_a such that

$$C_a = \eta_a\, C_{n_\alpha}(M)\;\alpha\;\tan\alpha, \tag{8.160}$$

where the value of η_a may be adjusted empirically as necessary to give the best fit with the static chord force and/or drag test data. Typically, the value of η_a is approximately 0.95.

The viscous (shear stress) drag is represented by the term C_{d_0} and is also a function of Mach number. However, its value is nominally constant for the angle of attack range below stall. Thus, the total drag in steady flow is given by the sum of the pressure and viscous shear components as

$$C_d = C_{d_0} + C_{d_p} = C_{d_0} + C_n \sin\alpha - C_a \cos\alpha. \tag{8.161}$$

The unsteady axial force coefficient, $C_a(t)$, depends only on the circulatory component of the loading. This is easily proved by considering the chordwise form of the circulatory and noncirculatory pressure loadings. The circulatory form is given by the standard thin-airfoil result, namely

$$\Delta C_p^c(\bar{x}, t) = 4\left(\alpha_e + \frac{\dot{\alpha}_e c}{2V}\right)\sqrt{\frac{1-\bar{x}}{\bar{x}}}. \tag{8.162}$$

The important point here is that the circulatory form has a leading-edge pressure singularity, and this form of pressure distribution is obtained no matter what the value of the effective angle of attack α_e. The noncirculatory form (at time zero) is given by the piston theory result

$$\Delta C_p^{nc}(\bar{x}, t = 0) = \frac{4}{M}\alpha. \tag{8.163}$$

While this initial loading changes with time as pressure waves propagate from the airfoil, no leading-edge singularity exists for any time. The general expression for the leading-edge suction force is

$$C_a = \frac{\pi}{8} \lim_{\bar{x} \to 0} \left\{ \Delta C_p^2 \bar{x} \right\}, \tag{8.164}$$

and inserting the circulatory loading gives

$$C_a = \frac{\pi}{8} \lim_{\bar{x} \to 0} \left\{ 16 \left(\alpha_e + \frac{\dot{\alpha}_e c}{2V} \right)^2 \left(1 - \frac{\bar{x}}{c} \right) \right\} = 2\pi \left(\alpha_e + \frac{\dot{\alpha}_e c}{2V} \right)^2. \tag{8.165}$$

Thus, as a consequence of the above the unsteady axial force may be obtained directly from the steady result by replacing α by the instantaneously effective angle of attack $\alpha_e + \dot{\alpha}_e c/2V$, that is,

$$C_a(t) = \eta_a \, C_n(t) \tan\left(\alpha_e + \frac{\dot{\alpha}_e c}{2V} \right) \approx \eta_a \, C_{n_\alpha}(M) \left(\alpha_e + \frac{\dot{\alpha}_e c}{2V} \right)^2. \tag{8.166}$$

Note that the noncirculatory loading does not appear in this expression. The unsteady pressure drag (if this is required) is obtained by resolving the total normal force and axial force coefficients through the angle of attack α to obtain

$$C_d(t) = C_n(t) \sin\alpha - C_a(t) \cos\alpha + C_{d_0}. \tag{8.167}$$

See also Leishman (1987b) for further discussion on unsteady drag.

8.15.8 *State-Space Aerodynamic Model for Compressible Flow*

By suitably generalizing the indicial response in terms of exponential functions and Mach number as shown previously, the corresponding state-space realization may be obtained for each component of the loading in a subsonic compressible flow – see Leishman & Nguyen (1990). Firstly, consider the normal force response to continuous forcing in terms of angle of attack. From Section 8.15.2, the circulatory normal force response to a variation in angle of attack can be written in state-space form as

$$\begin{Bmatrix} \dot{x}_1 \\ \dot{x}_2 \end{Bmatrix} = \left(\frac{2V}{c} \right) \beta^2 \begin{bmatrix} -b_1 & 0 \\ 0 & -b_2 \end{bmatrix} \begin{Bmatrix} x_1 \\ x_2 \end{Bmatrix} + \begin{Bmatrix} 1 \\ 1 \end{Bmatrix} \alpha_{3/4}(t), \tag{8.168}$$

with the output equation for the normal force coefficient given by

$$C_n^c(t) = \frac{2\pi}{\beta} \left(\frac{2V}{c} \right) \beta^2 \left[A_1 b_1 \;\; A_2 b_2 \right] \begin{Bmatrix} x_1 \\ x_2 \end{Bmatrix}, \tag{8.169}$$

where $2\pi/\beta$ is the lift-curve-slope for linearized compressible flow and $\alpha_{3/4}$ is the angle of attack at the 3/4-chord, that is,

$$\alpha_{3/4}(t) = \alpha(t) + \frac{q(t)}{2}.$$

Similarly, the noncirculatory normal force from angle of attack can be written in the state-space representation as

$$\dot{x}_3 = \alpha(t) - \frac{1}{\kappa_\alpha T_i} x_3 = \alpha(t) + a_{33} x_3, \tag{8.170}$$

with the output equation for the normal force coefficient given by

$$C_{n_\alpha}^{nc}(t) = \frac{4}{M} \dot{x}_3. \tag{8.171}$$

The remaining state equations for the pitching moment and pitch rate terms can be derived in a similar way using the other indicial response approximation discussed previously in Sections 8.15.3, 8.15.5 and 8.15.6. The individual components of aerodynamic loading are then linearly combined to obtain the overall aerodynamic response. For example, the total normal force coefficient is given by

$$C_n(t) = C_n^c(t) + C_{n_\alpha}^{nc}(t) + C_{n_q}^{nc}(t), \tag{8.172}$$

and a similar equation holds for the pitching moment about the 1/4-chord. Thus, the overall unsteady aerodynamic response can be described in terms of a two-input, two-output system where the inputs are the airfoil angle of attack and pitch rate and the outputs are the unsteady normal force (lift) and pitching moment. It can be shown that by rearranging the state equations, they can be represented in the general form

$$\begin{Bmatrix} \dot{x}_1 \\ \dot{x}_2 \\ \dot{x}_3 \\ \dot{x}_4 \\ \dot{x}_5 \\ \dot{x}_6 \\ \dot{x}_7 \\ \dot{x}_8 \end{Bmatrix} = \begin{bmatrix} a_{11} & 0 & 0 & 0 & 0 & 0 & 0 & 0 \\ 0 & a_{22} & 0 & 0 & 0 & 0 & 0 & 0 \\ 0 & 0 & a_{33} & 0 & 0 & 0 & 0 & 0 \\ 0 & 0 & 0 & a_{44} & 0 & 0 & 0 & 0 \\ 0 & 0 & 0 & 0 & a_{55} & 0 & 0 & 0 \\ 0 & 0 & 0 & 0 & 0 & a_{66} & 0 & 0 \\ 0 & 0 & 0 & 0 & 0 & 0 & a_{77} & 0 \\ 0 & 0 & 0 & 0 & 0 & 0 & 0 & a_{88} \end{bmatrix} \begin{Bmatrix} x_1 \\ x_2 \\ x_3 \\ x_4 \\ x_5 \\ x_6 \\ x_7 \\ x_8 \end{Bmatrix} + \begin{bmatrix} 1 & 0.5 \\ 1 & 0.5 \\ 1 & 0 \\ 0 & 1 \\ 1 & 0 \\ 1 & 0 \\ 0 & 1 \\ 0 & 1 \end{bmatrix} \begin{Bmatrix} \alpha \\ q \end{Bmatrix},$$

$$\begin{Bmatrix} C_n \\ C_m \end{Bmatrix} = \begin{bmatrix} c_{11} & c_{12} & c_{13} & c_{14} & 0 & 0 & 0 & 0 \\ c_{21} & c_{22} & 0 & 0 & c_{25} & c_{26} & c_{27} & c_{28} \end{bmatrix} \begin{Bmatrix} x_1 \\ x_2 \\ x_3 \\ x_4 \\ x_5 \\ x_6 \\ x_7 \\ x_8 \end{Bmatrix} + \begin{bmatrix} \frac{4}{M} & \frac{1}{M} \\ \frac{-1}{M} & \frac{-7}{12} \end{bmatrix} \begin{Bmatrix} \alpha \\ q \end{Bmatrix},$$

which is the form of Eqs. 8.97 and 8.98. The total aerodynamic lift and pitching moment response to an arbitrary time history of α and q can be obtained from the above state equations by integrating numerically using a standard ordinary differential equation solver.

The unsteady axial force (in-plane force) and pressure drag on the airfoil may also be obtained in terms of the state variables. From the output equation, the effective angle of attack of the airfoil, α_e, because of the shed wake (circulatory) terms can be written in terms of the states x_1 and x_2 as

$$\alpha_e(t) = \beta^2 \left(\frac{2V}{c} \right) (A_1 b_1 x_1 + A_2 b_2 x_2). \tag{8.173}$$

Therefore, as shown in Section 8.15.7, the corresponding axial force, C_a, is given in terms of α_e as

$$C_a(t) = \frac{2\pi}{\beta} \alpha_e^2(t), \tag{8.174}$$

which involves a bilinear combination of the states x_1 and x_2. Thus, as a by-product of the above system representation for the unsteady lift, the necessary information may be extracted from the system at a given instant of time to obtain the unsteady axial force component. Finally, the instantaneous pressure drag can be obtained by resolving the components of the normal force and axial forces through the geometric angle of attack α using Eq. 8.167.

8.16 Comparison with Experiment

Whereas the formulation of the subsonic unsteady aerodynamic model in Section 8.15 has been derived, in part, by using experimental measurements in the frequency domain, the validity of the model must be reconfirmed by comparing with experimental results of unsteady forces and moments as functions of time or angle of attack. The unsteady airloads in response to an arbitrary time history of α and q can be obtained by using the previously derived indicial functions with Duhamel superposition in the form of the recurrence equations or by integration of the state-space equations. Results will now be shown for both a pure plunging oscillation and also for a pitching oscillation; the absence of pitch rate in the plunge oscillations makes it possible to isolate the effects of the pitch rate in the unsteady aerodynamic response. For the plunging case, an 'equivalent' quasi-steady angle of attack can be defined using

$$\alpha_{eq}(t) = \tan^{-1}\left(\frac{\dot{h}}{V}\right) \approx \frac{\dot{h}}{V}. \tag{8.175}$$

Note that, unlike the pitch angle (angle of attack), the equivalent angle of attack in Eq. 8.175 is not a directly measurable quantity because it depends on the velocity V.

Representative variations in normal (lift) force and pitching moment are shown in Fig. 8.27 for harmonic plunge forcing and in Fig. 8.28 for harmonic pitch forcing. In each case, the results obtained from the subsonic unsteady aerodynamic model are compared to experimental measurements made by Liiva et al. (1968). The pitch and plunge data have been selected for approximately the same reduced frequency and mean angle of attack, although the "equivalent" amplitude of forcing is slightly different. The results show the expected characteristic elliptical shaped normal force and pitching moment loops symptomatic of attached flow. It is clear that in both cases there is a reduction of the unsteady "lift slopes" compared to the steady case, but with a larger phase lag (width of the hysteresis loop) for the plunging case. The width (amplitude) of the moment loop for the pitching oscillation is much larger than that for the plunge oscillation because of the pitch rate contributions.[11] For each mode of forcing, it is apparent that the major axis of the unsteady pitching moment loop is closely aligned with the steady moment slope. This positive slope is a result of the aerodynamic center being slightly forward of the $1/4$-chord measurement axis.

Further results showing quality of the predictions of the unsteady lift in compressible flow and the improvement that can be obtained over Theodorsen's theory are given in Fig. 8.29 for pitching oscillations. To summarize as many conditions as possible, the results are shown in terms of lift and moment amplitude and corresponding phase angle versus reduced frequency. To put the test data on a common basis, the amplitudes of the response have been expressed as ratios by normalizing with respect to the measured static lift-curve-slope. Plunge oscillation data are more scarce, but as mentioned previously, they are particularly useful because of the absence of loading contributions from pitch rate (q) terms. It is clear from the results obtained in Fig. 8.29 that there is a good correlation between

[11] Note that there is some offset in the C_m curves, which has been attributed to thermal drift of the pressure transducers used in the experiment.

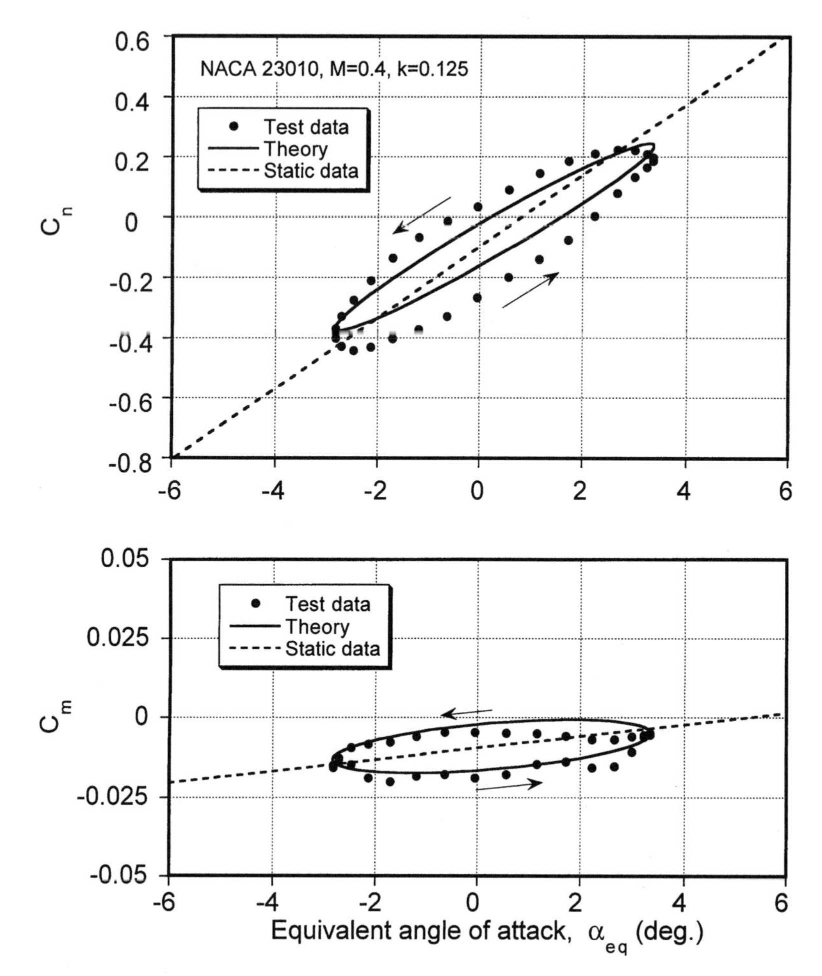

Figure 8.27 Comparison of theory and experiment for the lift and pitching moment coefficients in fully attached flow under oscillatory plunge forcing conditions, at $M = 0.4$.

the predicted and measured lift response, and this provides considerable support for using this type of modeling over the Mach number and reduced frequency range appropriate for helicopter applications. The compressible flow theory compares much more favorably with test data than Theodorsen's theory alone, particularly for the phase. The pitching moment is, however, somewhat more difficult to predict without the use of additional empirical results. Note from Fig. 8.29(d) that there is a gradual divergence of the phase angle as the reduced frequency tends to zero. As described previously in section 8.7, this is attributable to an offset of the aerodynamic center from the 1/4-chord moment reference point, and it emphasizes the need to carefully represent the changing aerodynamic center with Mach number in a rotor simulation if accurate predictions of the unsteady aerodynamic effects are to be obtained.

8.17 Nonuniform Vertical Velocity Field

In the rotor plane, there are a large number of vortical disturbances that lie in proximity to the blades. This is especially significant on the advancing and retreating sides of

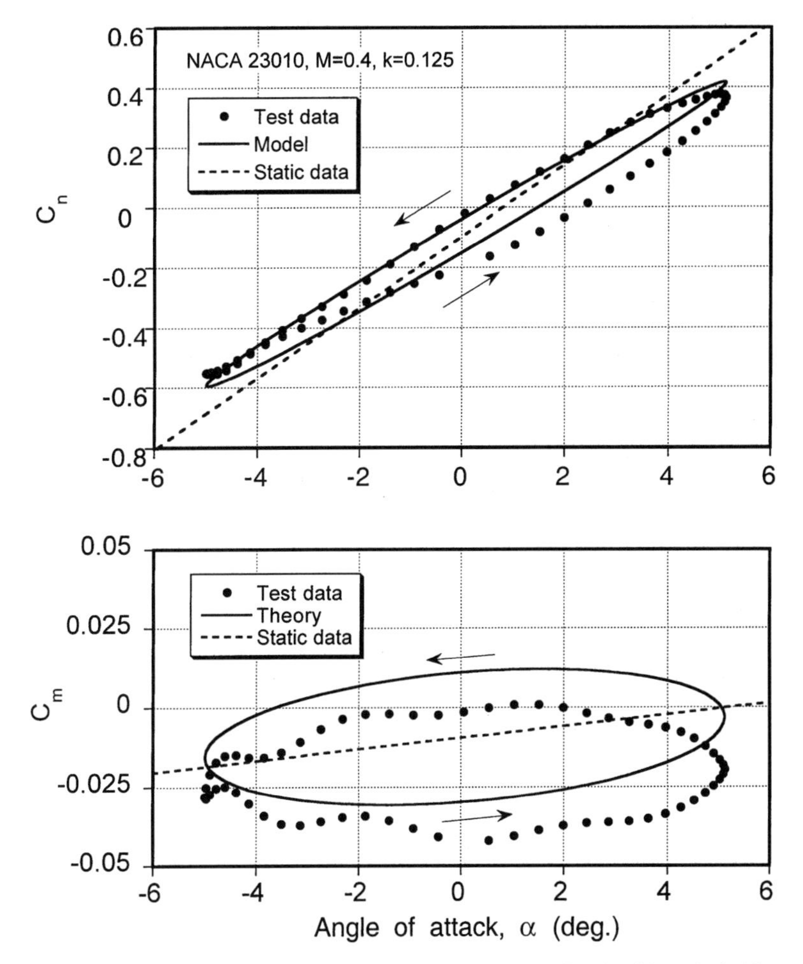

Figure 8.28 Comparison of theory and measurements for the lift and pitching moment coefficients in fully attached flow under oscillatory pitch forcing conditions, at $M = 0.4$.

the rotor where the blades may interact with tip vortices.[12] The unsteady forces produced on a rotor blade arise primarily because of the vertical velocity between the wake disturbance (gust field) and the airfoil surface. In linear theory, this is treated as an imposed unsteady upwash field, which must be used to satisfy the boundary conditions of flow tangency on the airfoil surface. As described previously, within the assumptions of linear theory, incompressible solutions for the sinusoidal vertical gust problem have been solved exactly by Sears (1940), and solutions for the sharp-edge vertical gust problem have been found by Küssner (1935), von Kármán & Sears (1938), and Miles (1956).

8.17.1 *Exact Subsonic Linear Theory*

For the subsonic compressible flow case, the problem of finding the sharp-edge vertical gust response, $\psi(s, M)$, was considered by Lomax (1953) using a similar approach

[12] This is called the blade–vortex interaction or BVI problem. Accurate predictions of BVI airloadings and the related rotor noise propagation are becoming increasingly important design issues to meet more stringent civilian and military noise requirements.

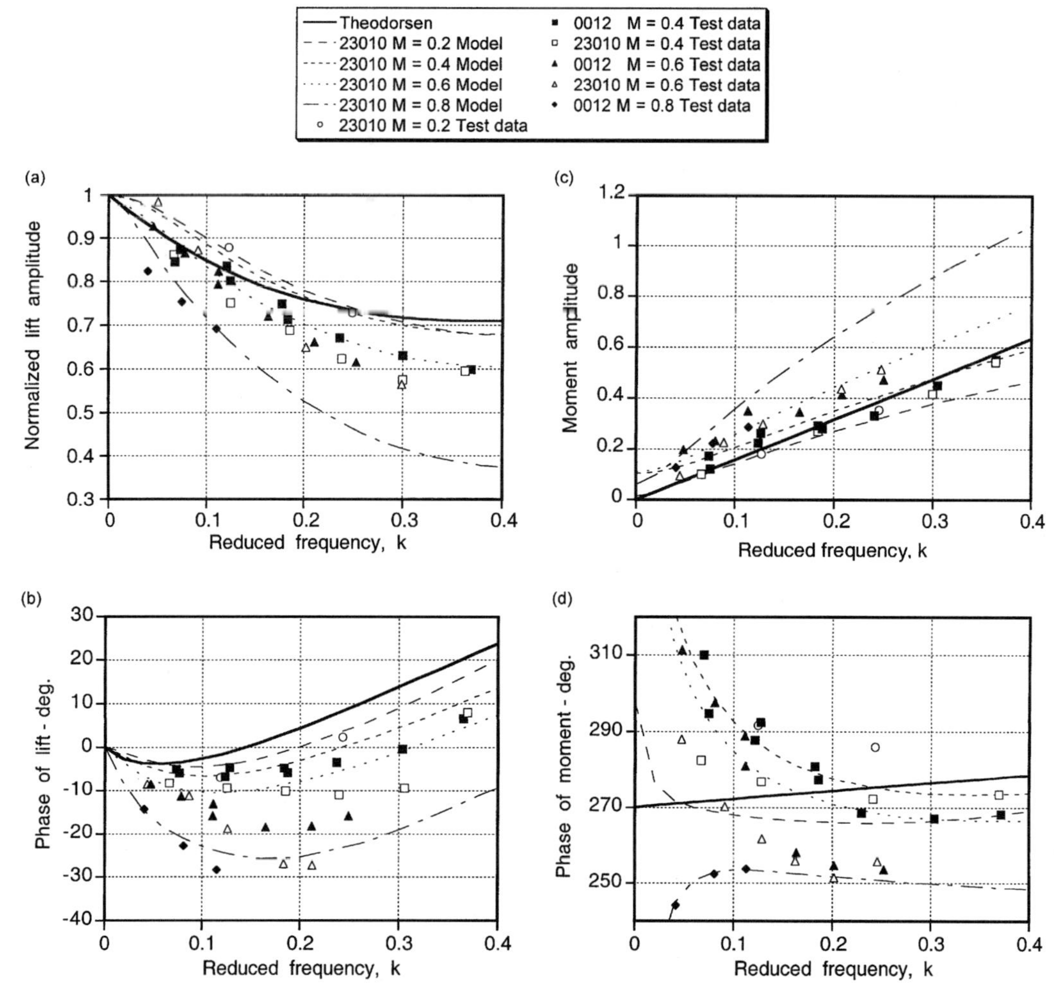

Figure 8.29 Comparison of the compressible unsteady airfoil theory with pitch oscillation test data at various Mach numbers. (a) Lift amplitude. (b) Phase of lift. (c) Pitching moment amplitude. (d) Phase of pitching moment.

to that described earlier in Section 8.15 to derive the indicial responses from changes in airfoil angle of attack and pitch rate. The subsonic vertical gust result was also obtained by Heaslet & Sprieter (1952) by means of reciprocal relations. The actual calculations are fairly involved, but Lomax (1953) has shown that exact analytical expressions for the airfoil pressure distribution can be found for a limited period of time after the vertical gust entry. For the period $0 \leq s \leq 2M/(1+M)$ the lift coefficient is given by

$$\Delta C_n^g(s) = \frac{2s}{\sqrt{M}}\left(\frac{\Delta w_g}{V}\right). \tag{8.176}$$

The corresponding pitching moment coefficient about the airfoil 1/4-chord for the period $0 \leq s \leq 2M/(1+M)$ can also be found, giving

$$\Delta C_m^g(s) = \frac{8s - \left[\dfrac{(M+1)}{M}\right]s^2}{4\sqrt{M}}\left(\frac{\Delta w_g}{V}\right). \tag{8.177}$$

Although these results are valid for less than one-chord length of airfoil travel for all subsonic Mach numbers, these analytic solutions are exact within the underlying assumptions of linearized, subsonic, unsteady thin-airfoil theory.

One interesting result from Eq. 8.176 is that increasing Mach number decreases the *initial* rate of lift production for a given distance traveled during the vertical gust penetration, perhaps not an intuitive result. However, a similar result has been shown in Section 8.15 for the indicial angle of attack case, where there is an increasing lag in the development of the circulatory lift for higher subsonic Mach numbers – see also Bisplinghoff et al. (1955). Using Eq. 8.176 it can be shown that the lift builds very rapidly during the vertical gust penetration, reaching close to one third of its final value ($2\pi/\beta$) shortly after the airfoil becomes fully immersed in the vertical gust (that is, when $s = 2$).

The position of the center of pressure during the vertical gust penetration is also of some interest. Helicopter rotor blades tend to be relatively compliant in torsion compared to fixed-wing aircraft, and so the variation in airfoil pitching moment about the elastic axis can be very important. For the period $0 \leq s \leq 2M/(1+M)$, the center of pressure can be computed from Eqs. 8.176 and 8.177 giving

$$\bar{x}_{cp}(s, M) = \frac{s}{8}\left[\frac{M+1}{M}\right], \tag{8.178}$$

which shows that $\bar{x}_{cp}$ moves quickly aft to the 1/4-chord location within the short period $s = 2M/(1+M)$. Therefore, for most practical purposes it is sufficient to assume that the aerodynamic center remains at the 1/4-chord throughout the vertical gust penetration. This result is also consistent with the incompressible solution of von Kármán & Sears (1938). For later values of time up to $s = 4M/(1-M^2)$, solutions for the airfoil pressure distribution during the vertical gust penetration take a more complicated form, and the determination of the lift and moment is only possible by means of numerical methods. For $s > 4M/(1-M^2)$ no exact solutions to the sharp-edge vertical gust problem are possible in subsonic flow by means of the linear theory; consequently, other and usually more approximate methods must be adopted.

8.17.2 *Approximations to the Sharp-Edged Gust Functions*

Mazelsky (1952a) and Mazelsky & Drischler (1952) have obtained exponential approximations to the stationary sharp-edged vertical gust functions at various Mach numbers by means of reciprocal theorems in conjunction with numerical solutions for airloads computed in the frequency domain. A suitable functional approximation for the sharp-edged vertical gust is of the form

$$\psi(s, M) \approx 1 - \sum_{i=1}^{N} G_i e^{-g_i s}, \tag{8.179}$$

where the G_i and g_i coefficients will all be Mach number dependent, and with $\sum_{i=1}^{N} G_i = 1$ and $g_i, i = 1, \ldots, N > 0$. The corresponding lift is given by

$$\Delta C_n^g(t, M) = \frac{2\pi}{\beta}\psi(s, M)\left(\frac{\Delta w_g}{V}\right) = \frac{2\pi}{\beta}\left(1 - \sum_{i=1}^{N} G_i e^{-g_i s}\right)\left(\frac{\Delta w_g}{V}\right). \tag{8.180}$$

It has been shown previously in Section 8.15.2 that the circulatory part of the total lift from a step change in angle of attack in subsonic compressible flow can be approximated by a two-term exponential function, and for all subsonic Mach numbers the results are related through a characteristic time that can be scaled in terms of Mach number alone. For later values of time the sharp-edge vertical gust and indicial angle of attack functions must approach each other; thus it is reasonable to also assume a similar behavior for the vertical gust function so it can be approximated by

$$\psi(s, M) \approx 1 - \sum_{i=1}^{N} G_i e^{-g_i \beta^2 s}, \tag{8.181}$$

where $1 - \sum_{i=1}^{N} G_i = 0$, but now the Gs and gs are fixed and considered independent of Mach number. Like the indicial angle of attack functions given in Section 8.15.2, such an approximation can be assumed to be valid up to at least the critical Mach number of the airfoil, beyond which nonlinear effects associated with transonic flow do not allow for such relatively simple generalizations.

Like the indicial lift for angle of attack, a solution for the coefficients in Eq. 8.181 can be formulated as a least-squares optimization problem, with several imposed constraints. In this case, however, there are no equivalent experimental results for airfoils in sinusoidal gust fields. Therefore, direct time-domain results obtained using analytic and finite-difference solutions can be used as a reference to find the gust function approximations in exponential form. To obtain an approximation to the exact linear theory, one constraint can be imposed by matching the time rate of change of the exact solution (Eq. 8.176) and approximate solution (Eq 8.180) at $s = 0$. This helps constrain the solution to ensure that the exact result will always be closely obtained in the initial stages. This part of the response is particularly important for transient aerodynamic phenomena such as BVI. Differentiating Eqs. 8.176 and 8.180 with respect to s and equating their gradients at $s = 0$ leads to a definition for the first constraint, namely

$$\sum_{i=1}^{N} G_i g_i = \frac{1}{\pi\sqrt{M}\beta} = \frac{1}{\pi\sqrt{M}\sqrt{1 - M^2}} = \text{constant}. \tag{8.182}$$

This cannot be obtained over the entire subsonic Mach number regime; however, an evaluation of right-hand side of Eq. 8.182 shows that it is numerically close to 0.6 over the practical range $0.2 \leq M \leq 0.8$. As $M \to 0$, the slope tends to infinity at $s = 0$, which is consistent with the exact solution given by von Kármán & Sears (1938). In the latter case, with any common type of exponential approximation to the exact incompressible solution [see, for example, Bisplinghoff et al. (1955)] the gradients cannot be matched as $s \to 0$. In addition, a constraint is imposed for the initial conditions, namely $\sum_{i=1}^{N} G_i - 1 = 0$, and also $G_i,\ g_i,\ > 0, i = 1, 2, \ldots, N$. Also, as $s \to \infty$, the airloads approach the value given by the usual steady-state subsonic linearized airfoil theory (i.e., $C_n^g(s = \infty, M) = 2\pi/\beta$). In a parallel way to that described previously for the indicial angle of attack case, a $2N$-dimensional vector of unknown coefficients can now be defined as

$$\mathbf{x}^T = \{G_1\ G_2\ \ldots\ G_N;\ g_1\ g_2\ \ldots\ g_N\} \tag{8.183}$$

The vector in Eq. 8.183 can be chosen to minimize the differences between the approximating exponential vertical gust function and the exact or any reference solutions over the domain of s and M.

Exact results for the vertical gust response in subsonic flow can be computed using the solutions given by Lomax (1968) and Lomax et al. (1952) but only up to $s = 4M/(1 - M^2)$. For the higher subsonic Mach numbers this corresponds to finding an exact solution up to

Table 8.3. *Summary of Sharp-Edged Vertical Gust Function Coefficients*

	G_1	G_2	G_3	g_1	g_2	g_3
$\psi(s)$*	0.5	0.5	–	0.130	1.0	–
$\psi(s, M = 0.5)$**	0.390	0.407	0.203	0.0716	0.374	2.165
$\psi(s, M)$ (linear)	0.527	0.473	–	0.100	1.367	–
$\psi(s, M = 0.5)$ (linear)	0.527	0.473	–	0.075	1.025	–
$\psi(s, M)$ (CFD)	0.670	0.330	–	0.1753	1.637	–

* From Bisplinghoff et al. (1955).
** From Mazelsky & Drischler (1952).

about 10 semi-chord lengths of airfoil travel. Other results for the sharp-edge gust problem have been made available by Singh & Baeder (1997b), who have used finite-difference (CFD) solutions to the Euler equations. Note that the exponential approximations to the gust response, which are shown in Fig. 8.30, match the exact solutions and the CFD solutions almost precisely. It is apparent that although the final values increase with increasing Mach number, the initial growth in lift is less.

The resulting coefficients for the indicial gust functions are given in Table 8.3. Results for the $N = 3$ case at $M = 0.5$ as given by Mazelsky & Drischler (1952) are also tabulated. The coefficients of the generalized subsonic vertical gust function as $M \to 0$ are close to those given by R. T. Jones (1938, 1940) for the incompressible case (the $N = 2$ exponential approximation to the Küssner function) and confirms that the results for the subsonic case are closely approximated by scaling the g coefficients by β^2; that is, the aerodynamic vertical gust responses are related in subsonic flow, albeit approximately, through a characteristic time.

8.17.3 *Response to an Arbitrary Vertical Gust*

If the subsonic sharp-edged gust functions are approximated in exponential form, the same techniques used previously to find the the total unsteady lift to an arbitrary vertical gust field can be applied. Within the assumptions of the linear theory, a general stationary vertical gust field, $w_g(x, t)$, can be decomposed into a series of sharp-edged vertical gusts of small magnitude. When the approximation to the aerodynamic response for sharp-edged vertical gust is found, then the response to an arbitrary vertical gust field can be found using linear superposition by means of Duhamel's integral. The response to a continuous vertical gust field may be written analytically as

$$\Delta C_n^g(s) = \frac{2\pi}{\beta}\left[\frac{1}{V}\int_0^s \frac{dw_g}{dt}\psi(s - \sigma, M)\, d\sigma\right]. \tag{8.184}$$

As described previously, the Duhamel superposition can be performed numerically in various ways, including the state-space (continuous time) form or the one-step recursive formulation (discrete time) form. In the latter case, a finite-difference approximation to the Duhamel integral leads to a solution for the lift that may be constructed from an accumulating series of small vertical gust inputs using

$$\Delta C_n^g(t) = \frac{2\pi}{\beta}\frac{1}{V}\left[w_g(s) - Z_1(s) - Z_2(s)\right], \tag{8.185}$$

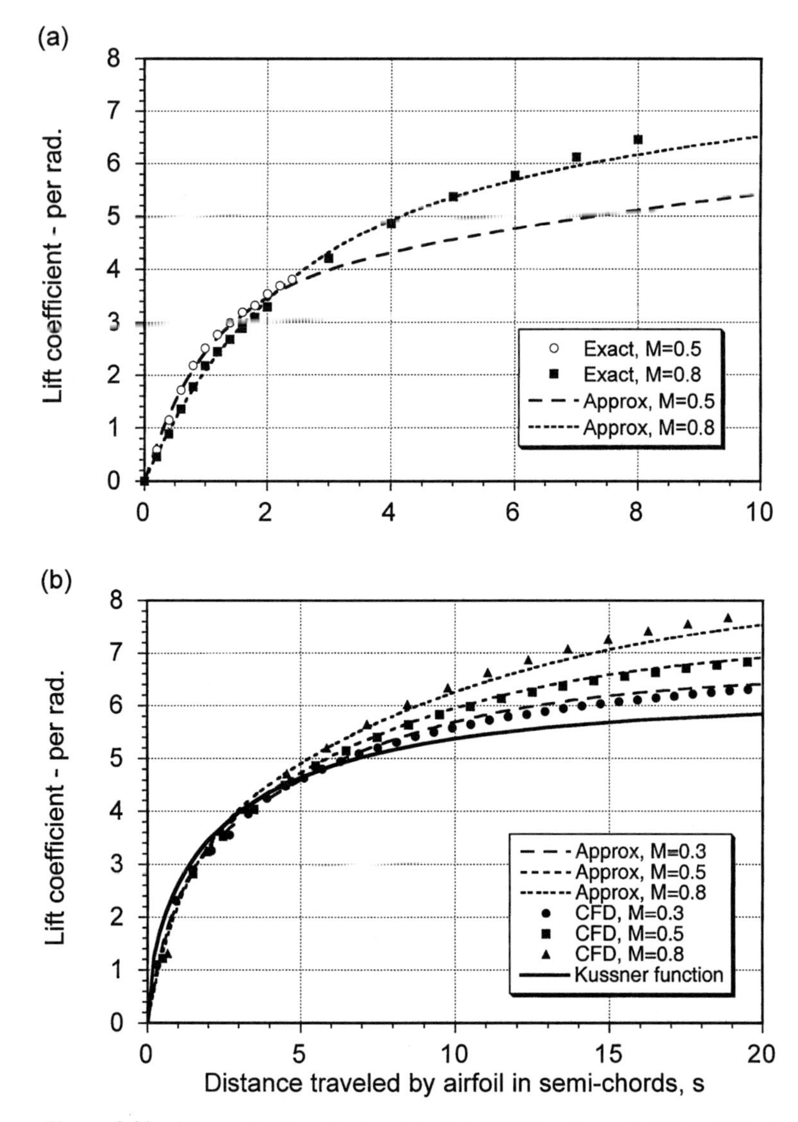

Figure 8.30 Sharp-edged vertical gust indicial lift functions at different Mach numbers. (a) Short values of time. (b) Longer values of time.

where the terms Z_1 and Z_2 are the deficiency functions. In this case, using Algorithm D-1 given previously in Section 8.14.1, the deficiency functions are given by the one-step recursive formulas

$$Z_1(s) = Z_1(s - \Delta s)E_1 + G_1\left[w_g(s) - w_g(s - \Delta s)\right], \tag{8.186}$$

$$Z_2(s) = Z_2(s - \Delta s)E_2 + G_2\left[w_g(s) - w_g(s - \Delta s)\right], \tag{8.187}$$

where $E_1 = \exp(-g_1\beta^2\Delta s)$ and $E_2 = \exp(-g_2\beta^2\Delta s)$.

By applying Laplace transforms to the exponential approximation to the sharp-edged vertical gust function in Eq. 8.181, the lift transfer function relating the output (the lift) to the input (the vertical gust field) can be obtained. From the transfer function, the alternative state-space form of the equations can be written as

$$\begin{Bmatrix} \dot{z}_1(t) \\ \dot{z}_2(t) \end{Bmatrix} = \begin{bmatrix} 0 & 1 \\ -g_1 g_2 \left(\frac{2V}{c}\right)^2 \beta^4 & -(g_1 + g_2)\left(\frac{2V}{c}\right)\beta^2 \end{bmatrix} \begin{Bmatrix} z_1(t) \\ z_2(t) \end{Bmatrix} + \begin{Bmatrix} 0 \\ 1 \end{Bmatrix} \frac{\Delta w_g(t)}{V}, \tag{8.188}$$

with the corresponding output equation for the total normal force (lift) coefficient for the arbitrary vertical gust field as

$$\Delta C_n^g(t) = \frac{2\pi}{\beta} \left[(g_1 g_2)\left(\frac{2V}{c}\right)^2 \beta^4 \quad (G_1 g_1 + G_2 g_2)\left(\frac{2V}{c}\right)\beta^2 \right] \begin{Bmatrix} z_1(t) \\ z_2(t) \end{Bmatrix}. \tag{8.189}$$

These equations can then be solved using a standard ordinary differential equation solver for any arbitrarily imposed vertical gust field.

8.17.4 *Blade Vortex Interaction (BVI) Problem*

Blade vortex interaction (BVI) is a practical example of an intense vertical velocity field with high velocity gradients. The 2-D BVI problem, which is shown schematically in Fig. 8.31, has been widely addressed in the literature – see, for example, McCroskey & Goorjian (1983) and Singh & Baeder (1997b). While passing an airfoil at a predetermined distance, a convecting vortex of positive circulation produces a downwash velocity while upstream of the blade (airfoil), and this changes to an upwash as it moves downstream. This situation leads to a rapidly and continuously changing angle of attack, resulting in highly unsteady aerodynamic loads. The acoustic pressure (or noise) propagated to an observer from such a blade–vortex encounter is related, as given by the compact source limit, to the time rate of change of the lift.

Representative numerical calculations will now be shown for the unsteady loads on a NACA 0012 airfoil interacting with a convecting vortex of nondimensional strength $\hat{\Gamma} = \Gamma/(Vc) = 0.2$. The airfoil was set to zero angle of attack. The interacting vortex was assumed to have a Lamb-like normal velocity distribution given by

$$w_g(x, y) = -\frac{\hat{\Gamma}}{2\pi}\frac{(x - x_v)}{r^2}\left[1 - \exp\left(\frac{-r^2}{r_c^2}\right)\right], \tag{8.190}$$

with $r^2 = (x - x_v)^2 + (y - y_v)^2$, where x_v, y_v refer to the position of the vortex relative to a coordinate axis at the leading edge of the airfoil. The reciprocal influence of the airfoil on the vortex convection velocity and trajectory was neglected, and we will assume that the vortex remains undisturbed from the path $y_v = y_0 = -0.26$. As shown by Srinivasian & McCroskey (1987), the effect of vortex trajectory distortion appears to be of secondary importance to the overall airloads except for the transonic case when the vortex may pass

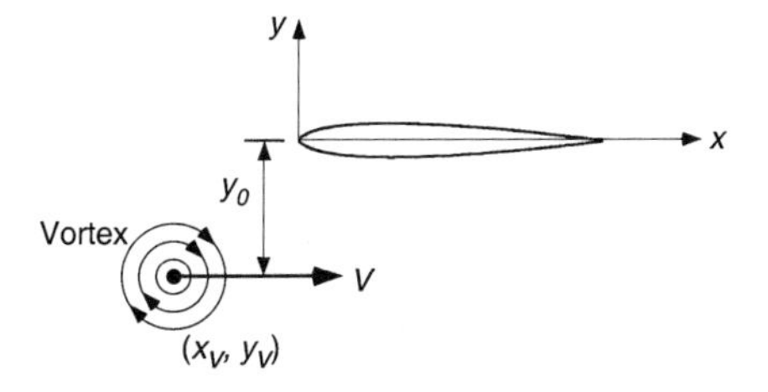

Figure 8.31 Two-dimensional model of the blade vortex interaction (BVI) problem.

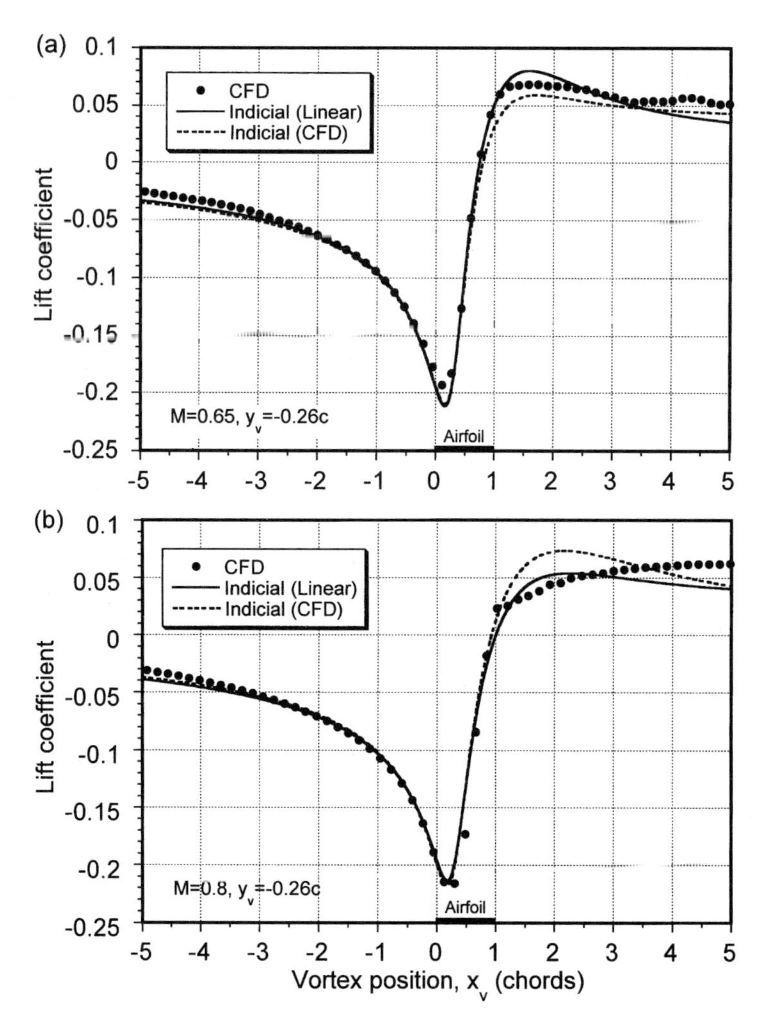

Figure 8.32 Comparison of indicial method with Euler finite-difference result for the lift during a 2-D vortex–airfoil interaction. $\hat{\Gamma} = 0.2$, and $y_0 = -0.26$. (a) $M = 0.65$ (subsonic). (b) $M = 0.8$ (weakly transonic).

close to a shock wave. A viscous core of dimensions $r_c = 0.05c$ was used for the calculations, although the interaction between the airfoil and the vortex is sufficiently spaced in this case that the core radius does not play a role.

Results for Mach numbers between 0.65 and 0.8 are shown in Fig. 8.32. The results are compared to solutions obtained using a finite-difference solution based on the Euler equations. (The cost of the latter is at least four orders of magnitude greater than the indicial method.) Note the good agreement between the indicial approach and the finite-difference solution, the results essentially confirming the validity of linear theory for this problem. Even for the higher Mach number of 0.8, where some nonlinearities because of the transonic nature of the flow might be expected, the agreement is good, although there is a somewhat larger lift overshoot downstream of the airfoil trailing edge compared to that predicted by the Euler method. The results in Fig. 8.32 also indicate that the peak-to-peak value of the lift response becomes attenuated with increasing Mach number, which is exactly the opposite to the result obtained using quasi-steady, subsonic, unsteady airfoil theory or incompressible unsteady airfoil theory. Also, it is apparent that the effects of increasing Mach number introduces a significantly larger phase lag in the lift response, which obviously becomes a significant consideration when accurate noise predictions are an issue (see also Question 8.13).

8.17.5 *Convecting Vertical Gusts in Subsonic Flow*

In the subsonic case, the time-varying lift and moment during the penetration of a convecting sharp-edged gust can only be found by numerical means. The problem, however, can be simplified if it is approached by using the reverse flow theorems, as discussed previously in Section 8.12 – see also Leishman (1997). Results for the lift on the airfoil penetrating a convecting sharp-edged gust at a Mach number of 0.5 are shown in Fig. 8.33, for various gust speed ratios. Results at other Mach numbers are qualitatively similar. Note that the effect of increasing gust convection velocity is to increase the rate of buildup of lift, analogous to the incompressible case shown previously in Figs. 8.20 and 8.21. For the subsonic case the center of pressure is always initially forward of the 1/4-chord as the airfoil penetrates into the gust front, but it moves back quickly again after the airfoil becomes fully immersed in the gust. After only a short time, the aerodynamic center can be considered to act at the 1/4-chord.

Lift and pitching moment results for convecting sharp-edged gusts in subsonic flow have also been computed by Singh & Baeder (1997b) using an Euler finite-difference method. Representative results for the lift are shown in Fig. 8.33 for a Mach number of 0.5 and for several gust speed ratios. Note that in the early period where $s < 2M/(1+M)$, or 0.67 semi-chords at this Mach number, the lift varies linearly with time as predicted by the exact subsonic linear theory, the rate of growth increasing with increasing gust convection velocity. The comparisons are excellent and lend significant credibility to the Euler results, which can provide valuable solutions for later values of time where exact analytic solutions are not possible. Like the incompressible results, the Euler results predict an initial lift overshoot for $s > 2M/(1 + M)$ that reaches a peak when the airfoil is about halfway into the gust.

To use the indicial method to examine arbitary convecting vertical gust problems, the convecting sharp-edged gust solutions must be approximated by exponential functions. One suitable exponential approximation to the lift produced on an airfoil encountering a convecting sharp-edged gust is

$$\frac{C_l(s)}{w_0/V} = C_{l_\alpha}\left(1 - \sum_{i=1}^{N} G_i e^{-g_i s}\right) + G_{(N+1)} e^{-g_{(N+1)} s} - G_{(N+1)} e^{-g_{(N+2)} s}, \qquad (8.191)$$

where all the coefficients will, in general, be Mach number dependent. To satisfy the initial conditions at $s = 0$ then $\sum_{i=1}^{N} G_i = 1$. Also, $g_i > 0$ for $i = 1, \ldots, N+2$. The transient shown in the lift response at small values of time for fast traveling gusts is represented by the second two terms in Eq. 8.191, where the coefficient G_{N+1} and the differences in the values of the time constants g_{N+1} and g_{N+2} will affect the size and width of this transient. Physically, this transient is a result of the accumulation of pressure waves. In the limit when $\lambda \to 0$, the magnitude of the transient approaches the piston theory value of $4/M$. Results of this procedure are shown in Fig. 8.33 for several gust speed ratios at a Mach number of 0.5. It is seen that while an exact fit to the initial transient at smaller values of λ cannot be obtained, an acceptable level of accuracy is possible for values of λ that are not too far from unity.

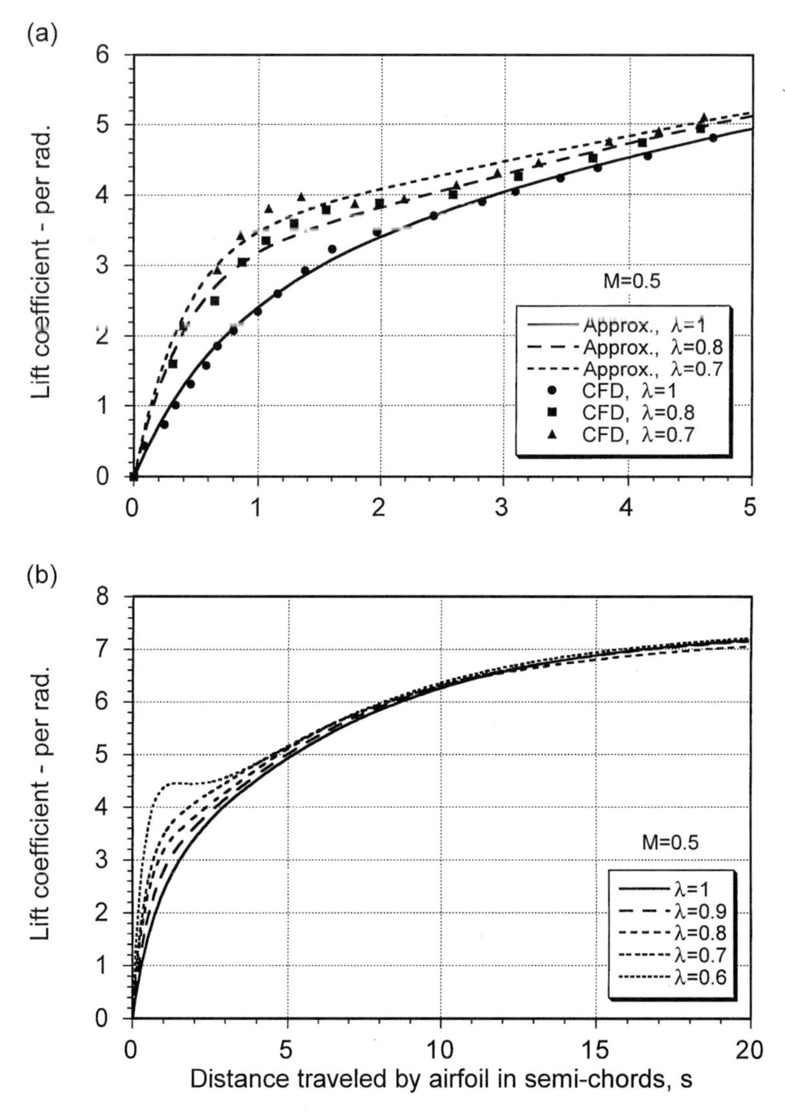

Figure 8.33 Convecting sharp-edged vertical gust indicial lift functions for different gust speed ratios at $M = 0.5$. (a) Short values of time. (b) Longer values of time.

The problem of convecting vortices on the unsteady lift and noise generation has been addressed by Singh & Baeder (1997b) and Leishman (1997). Results for the unsteady lift on the airfoil for the same conditions of Fig. 8.32 are shown in Fig. 8.34 for several gust speed ratios. The results are all referenced to the $\lambda = 1$ case, so that for downstream traveling vortices the BVI encounter occurs progressively earlier in time (or distance). We see that an increase in vortex convection speed (decrease in λ) progressively increases the peak-to-peak value of the unsteady lift, but more importantly, increases the time rate of change of lift. This will be reflected in the acoustics, where the BVI sound pulse will increase significantly in magnitude even for values of λ not too much lower than unity. Figure 8.34 also shows results for the BVI problem as computed directly using the Euler finite-difference method. The correlation of the indicial results with the Euler solution is excellent. Overall, these

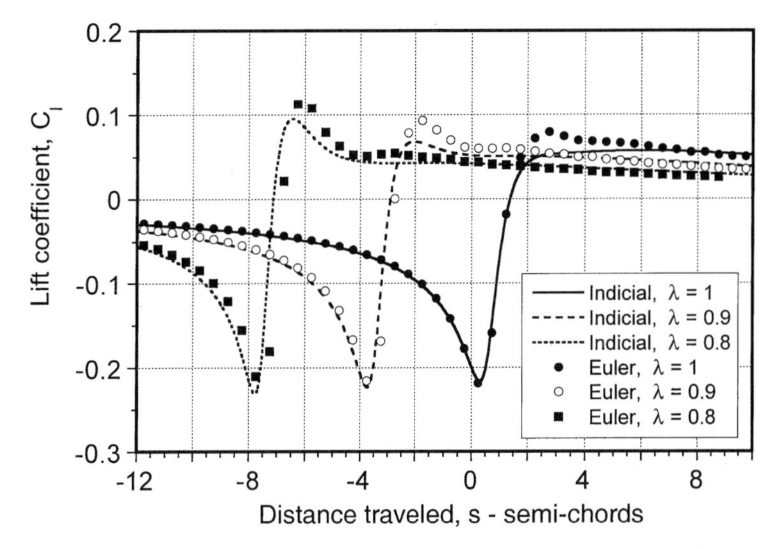

Figure 8.34 Indicial and Euler finite-difference predictions of unsteady lift for downstream convecting vortices. $\Gamma / V_\infty c = 0.2$, $h = -0.25c$, and $M = 0.5$.

results indicate that the gust speed ratio will be a necessary parameter to account for in helicopter blade airloads and aeroacoustics analyses.

8.18 Dynamic Inflow

Whereas the previous theories have dealt with the unsteady aerodynamics at the blade element level, in many cases it is desirable to directly account for unsteady effects on net rotor forces and moments. While the localized unsteady aerodynamic effects are clearly complicated, for some problems it is useful to seek a more global approach to modeling their effects, which is called "dynamic inflow." Consider, for example, a flight dynamics analysis. During a transient maneuver, the unsteady effects will cause the inflow through the rotor to be different from that produced under steady level flight conditions. This will change the rotor forces and moments, and so the rotor flapping response will be affected. See also Curtiss & Shupe (1971). The essence of these ideas can be attributed to Sissingh (1952), but many variations of the dynamic inflow theory and applications thereof have appeared in the literature. Crews et al. (1973), Ormiston (1976), Peters (1974), and Curtiss (1986) have put forth variations of the dynamic inflow theory and have showed that the effects can have an influence on rotor response and various problems in rotor dynamics and aeroelasticity. The most popular model of dynamic inflow is that of Pitt & Peters (1983), which has seen several applications in the literature – see Gaonkar & Peters (1986a,b) and Peters & HaQuang (1988). The ideas can also be extended to represent unsteady effects on 2-D airfoils – see Peters & He (1995).

Momentum theory can be used to relate the aerodynamic forces and moments on the rotor to the inflow across the disk. The thrust, T, is given by

$$T = 2\rho \int_0^R \int_0^{2\pi} v^2 r \, d\psi \, dr, \tag{8.192}$$

where v is the inflow velocity. Similarly, the pitching moment (positive nose-up) will be given by

$$M_y = -2\rho \int_0^R \int_0^{2\pi} v^2 r^2 \cos\psi \, d\psi \, dr \tag{8.193}$$

and the rolling moment (positive roll to starboard) is given by

$$M_x = -2\rho \int_0^R \int_0^{2\pi} v^2 r^2 \sin\psi \, d\psi \, dr. \tag{8.194}$$

A key assumption is the form of inflow distribution over the rotor disk. As shown in Chapter 3, a popular assumption is to use a linear distribution of the form

$$v = v(r, \psi) = v_0 + v_c r \cos\psi + v_s r \sin\psi, \tag{8.195}$$

where v_0, v_c, and v_s are the uniform, longitudinal, and lateral contributions to the inflow, respectively. This allows a coupling of the rotor thrust and moments to the induced velocity field. This can be written in matrix form as

$$\begin{Bmatrix} v_0 \\ v_c \\ v_s \end{Bmatrix} = [L] \begin{Bmatrix} C_T \\ C_{M_y} \\ C_{M_x} \end{Bmatrix}_{\text{aero}}, \tag{8.196}$$

where $[L]$ is a coupling or "gain" matrix. The subscript "aero" implies that only aerodynamic contributions are to be included; rotor inertial terms are omitted. In a local momentum analysis of the problem, the $[L]$ matrix is a diagonal matrix, where v_0 is related to C_T, v_c is related to C_{M_y}, and v_s is related to C_{M_x}. These relations, however, may be coupled and the matrix will be fully populated. In forward flight, the coefficients in the gain matrix also become functions of the wake skew angle. Gaonkar & Peters (1986a,b) review the development of the gain matrix, although it appears that more experimental verification may be required to validate the assumptions made.

As explained earlier, for two-dimensional airfoils unsteady aerodynamic effects manifest as a lag in the build-up of the lift in response to a change in angle of attack. In the case of a rotor, there are similar effects where there is a lag in the build-up of the rotor thrust and inflow in response to a change in collective pitch. These ideas were first studied by Carpenter & Fridovitch (1953), where additional 'apparent mass' terms that are attributed to the inertia of the fluid surrounding the rotor were included in the modeling of the problem. In the case of a hovering rotor, the thrust on the rotor can now be written as the modified simple momentum equation

$$T = 0.637\rho \frac{4}{3}\pi R^2 \dot{v}_i + 2\rho A v_i^2, \tag{8.197}$$

where the first term on the right-hand side of this equation is an apparent mass contribution. In coefficient form this equation can be written as

$$C_T = \frac{0.637(4/3)}{\Omega}\dot{\lambda}_i + 2\lambda_i^2 = \frac{0.849}{\Omega}\dot{\lambda}_i + 2\lambda_i^2. \tag{8.198}$$

This latter equation can be linearized about a mean operating state using $\lambda_i = \bar{\lambda}_i + \delta\lambda_i$ and $C_T = \bar{C}_T + \delta C_T$, which after simplification and neglecting terms of $O(\delta\lambda_i)^2$ gives

$$\left(\frac{0.849}{4\Omega\bar{\lambda}_i}\right)\delta\dot{\lambda}_i + \delta\lambda_i = \left(\frac{1}{4\bar{\lambda}_i}\right)\delta C_T. \tag{8.199}$$

This is a simple first-order differential equation that relates the change in inflow, $\delta\lambda_i$, to the change in rotor thrust, δC_T. The time constant of this dynamic system is

$$\tau_\lambda = \frac{0.849}{4\bar{\lambda}_i \Omega}. \tag{8.200}$$

For a typical helicopter rotor where $\bar{\lambda}_i \approx 0.05$ and $\Omega \approx 40$ rad/s, τ_λ will be of the order of 0.1 seconds. The dynamic adjustments that take place in the time averaged inflow are, therefore, relatively rapid in real-time. However, it will be noted that a time lag of 0.1 seconds corresponds to over half a rotor revolution, so at the local (blade element) level the local adjustments to the flow do, in fact, occur over relatively long aerodynamic time scales.

In the general case, such ideas can be introduced through the use of a time constant matrix $[\tau] = [L][M]$ where $[M]$ is a matrix of unsteady terms. The dynamic inflow model is now written as

$$[\tau]\left\{\begin{array}{c}\dot{v}_0\\ \dot{v}_c\\ \dot{v}_s\end{array}\right\} + \left\{\begin{array}{c}v_0\\ v_c\\ v_s\end{array}\right\} = [L]\left\{\begin{array}{c}C_T\\ C_{M_y}\\ C_{M_x}\end{array}\right\}_{\text{aero}}, \tag{8.201}$$

or

$$[M]\left\{\begin{array}{c}\dot{v}_0\\ \dot{v}_c\\ \dot{v}_s\end{array}\right\} + [L]^{-1}\left\{\begin{array}{c}v_0\\ v_c\\ v_s\end{array}\right\} = \left\{\begin{array}{c}C_T\\ C_{M_y}\\ C_{M_x}\end{array}\right\}_{\text{aero}}. \tag{8.202}$$

These ordinary differential equations can be used to relate the unsteady inflow to the rotor thrust and moments. The time constants are also functions of the wake skew angle, which are derived from vortex theory. In the complete problem, the equations of motion of the blades must be introduced using a blade element approach. This provides the complete relationship among the blade loads, the inflow, and rotor response.

Many refinements of the basic dynamic inflow approach can be found in the literature. The coupling and time-constant matrices have been derived from a number of other theories, including actuator disk and vortex theories. For extensive information on dynamic inflow modeling, see Johnson (1980). The most recent work on dynamic inflow has been put forth by Peters and colleagues – see Peters et al. (1987), Peters & HaQuang (1988), Peters & He (1989,1995), and Peters et al. (1995). Their work also encompasses the classical theories of Theodorsen and Loewy. Dynamic inflow models have found utility for various problems in rotor aeroelasticity, as well as for helicopter flight dynamics – see Chen (1989) and Padfield (1996).

8.19 Chapter Review

The focus of the present chapter has been to describe the key physical features of the unsteady aerodynamic effects found on airfoils operating under nominally attached flow conditions and away from stall. The contributions of circulatory and noncirculatory effects have been described, and their effects on the airloads have been explained through the use of "classical" unsteady aerodynamic theories. These theories have their origin in thin-airfoil theory, with allowance for a shed vortical wake of nonzero strength downstream of the airfoil. Of primary significance is that unsteady effects manifest as phase differences between the forcing function and the aerodynamic response, and these are functions of the reduced frequency, the Mach number, and the mode of forcing. While most of the classical unsteady aerodynamic theories are elegant in mathematical form, they are restricted to fully incompressible flows. This is an assumption that is hard to justify for helicopter problems, where both the local Mach numbers and effective reduced frequencies are generally high enough to render incompressible flow assumptions invalid, at least under the strictest terms. However, sometimes even some allowance for unsteady effects provides a better predictive capability than if quasi-steady flow alone is assumed.

For subsonic flows no convenient exact analytic solutions are available for unsteady airfoil problems, at least not over the entire time domain, and numerical solutions must be sought. However, the extension of the classical incompressible methods to subsonic compressible flows can be approached using many of the same fundamental principles as for incompressible flow, albeit with certain levels of approximation. It has been shown that compressibility effects generally manifest as increased phase lags between the forcing function and the unsteady aerodynamic response. For some transient problems, such as blade vortex interactions, the treatment of compressibility proves essential if the correct amplitude and phasing of the aerodynamic loads are to be obtained. Wherever possible, validation of the various methods have been conducted with experimental results. Unfortunately, many of the problems of interest are difficult to simulate experimentally, and recourse to indirect validation has been the only choice. However, the recent advent of nonlinear methods based on finite-difference approximations to the Euler or Navier–Stokes equations has provided a new standard that continuously helps define the limits of applicability of the classical theories.

Dynamic inflow represents another formulation of the unsteady aerodynamics problem, where the effects can be included in a form of differential equations to represent the relationship between the time-dependent inflow velocities and the rotor forces and moments. While perhaps offering less flexibility in the ability to represent some aspects of the unsteady airfoil problem, this approach is attractive for many problems in rotor analysis, particularly those in flight dynamics and aeroelasticity.

8.20 Questions

8.1. Show using dimensional analysis that the resultant force F on an airfoil of chord c oscillating at angular frequency ω in a flow of velocity V, density ρ, sonic velocity a, and viscosity μ, can be written in functional form as

$$\frac{F}{\rho V^2 c^2} = f\left(\frac{\rho V c}{\mu}, \frac{V}{a}, \frac{\omega c}{V}\right) = f\,(Re, M, k).$$

8.2. Use the quasi-steady thin-airfoil theory to compute the instantaneous differential pressure distribution on an airfoil for a series of nondimensional pitch rates, $q = \dot{\alpha}c/V$ about an axis at 1/4-chord. Plot and compare your results at the same value of lift coefficient. Determine and plot the chordwise pressure gradient, again at the same value of lift coefficient. Comment on results in regard to the likely effect on the boundary layer development.

8.3. According to Theodorsen's theory, how does the unsteady motion of an airfoil affect the lift response? Briefly, explain why frequency domain methods such as Theodorsen's theory cannot be easily applied to a rotor in forward flight.

8.4. If the Wagner function is assumed to be represented (in general) as the exponential series

$$\phi(s) = 1 - \sum_{j=1}^{N} A_j e^{-b_j s}$$

then show, by the application of Laplace transforms, that it is possible to write an approximation to the Theodorsen function in terms of real and imaginary parts, i.e., $C(k) = F(k) + iG(k)$, as

$$F(k) \approx \frac{1}{2} + \sum_{j=1}^{N} \frac{A_j b_j^2}{\left(b_j^2 + k^2\right)},$$

$$G(k) \approx -\sum_{j=1}^{N} \frac{A_j b_j k}{\left(b_j^2 + k^2\right)}.$$

Compare the approximation derived above using the $N = 2$ result for ϕ given by R. T. Jones with Theodorsen's exact solution. Comment on your result.

8.5. If the gust response function for the penetration of a stationary sharp-edged gust is assumed to be represented by the equation

$$\psi(s) \approx 1 - \sum_{j=1}^{N} G_j e^{-g_j s}$$

show that for a pure first sinusoidal gust referenced to the leading-edge of the airfoil the resulting real and imaginary parts of lift response can be written as

$$\Re\, C_l^g(k_g) = 2\pi \left[\sum_{j=1}^{N} \frac{G_j g_j^2}{g_j^2 + k_g^2} \right]$$

and

$$\Im\, C_l^g(k) = -2\pi \left[\sum_{j=1}^{N} \frac{G_j g_j k_g}{g_j^2 + k_g^2} \right],$$

where k_g is the gust reduced frequency. Use an exponential approximation to the Küssner function to estimate the Sears's function. Comment on your results.

8.6. Show, by a suitable transformation, that when the sinusoidal gust in the previous problem is referenced to the mid-chord then

$$\Re\, C_n^g(k_g) = 2\pi \left[\cos k_g \sum_{j=1}^{N} \frac{G_j g_j^2}{g_j^2 + k_g^2} - \sin k_g \sum_{j=1}^{N} \frac{G_j g_j k_g}{g_j^2 + k_g^2} \right],$$

$$\Im\, C_n^g(k_g) = -2\pi \left[\cos k_g \sum_{j=1}^{N} \frac{G_j g_j k_g}{g_j^2 + k_g^2} + \sin k_g \sum_{j=1}^{N} \frac{G_j g_j^2 k_g}{g_j^2 + k_g^2} \right].$$

8.7. If the sharp-edged gust function in subsonic flow is assumed to be represented by the equation

$$\psi(s) \approx 1 - \sum_{j=1}^{N} G_j e^{-g_j \beta^2 s}$$

show that for a pure first harmonic sinusoidal gust referenced to the leading-edge of the airfoil the resulting real and imaginary parts of lift response can be written as

$$\Re\, C_l^g(k_g, M) = \frac{2\pi}{\beta} \left[\sum_{j=1}^{N} \frac{G_j g_j^2 \beta^4}{g_j^2 \beta^4 + k_g^2} \right]$$

and

$$\Im\, C_l^g(k_g, M) = -\frac{2\pi}{\beta} \left[\sum_{j=1}^{N} \frac{G_j g_j \beta^2 k_g}{g_j^2 \beta^4 + k_g^2} \right],$$

where k_g is the gust reduced frequency.

8.8. Program a solution to the Loewy function. Show the results of the lift transfer function and corresponding phase angle for a case where ω/Ω = integer, integer $+1/4$, integer $+1/2$, and integer $+3/4$. Use a primary wake spacing parameter $h/b = 4.0$. Discuss your results.

8.9. Program the numerical solution to the Duhamel superposition problem with the $N = 2$ exponential approximation to the Wagner function using the one-step recurrence equations. Use at least two of the numerical methods given in the text. Consider a sinusoidal forcing at various reduced frequencies. Compare the exact results for the time-history of the circulatory lift as obtained using Theodorsen's theory with the numerical results obtained using the recurrence equations. Also, compare the results from the recurrence solutions to the frequency domain approximation to Theodorsen's theory using the Wagner function coefficients. Use several values of the time step, Δs (or points per cycle), to illustrate your answer.

8.10. Starting from a general indicial function in terms of a two term exponentially growing function, use the trapezoidal rule of integration to numerically approximate the Duhamel integral to find the solution to an arbitrary input in the form of one-step recurrence equations.

8.11. For some applications, it is convenient to write the unsteady aerodynamics of a two-dimensional airfoil in the form of differential equations, i.e., in state-space form. Program a numerical solution for the unsteady lift in incompressible flow due to changes in angle of attack using the appropriate state-space equations. Use a convenient ordinary differential equation solver of your choice. Compute and plot the lift coefficient for an input comprising: (a) a sinusoidal oscillation, (b) a step input.

8.12. Use the state-space model derived in the previous question to numerically calculate the circulation and the total lift (circulatory lift plus apparent mass lift) on a flat plate airfoil penetrating a stationary (non-convecting) sharp-edged gust. Use thin-airfoil theory and the first reverse flow theorem to calculate the quasi-steady angle of attack during the gust penetration.

8.13. Use Duhamel superposition with an exponential approximation to the sharp-edged gust to calculate numerical results for the unsteady lift on a flat plate airfoil interacting with a convecting Lamb-like desingularized vortex of nondimensional strength $\hat{\Gamma} = \Gamma/Vc = 0.2$ and core size $r_0 = 0.05$, which convects at a distance $y_v = 0.2$ chords below the airfoil. The airfoil is set to zero angle of attack. Plot your results as a function of vortex position, x_v, for several values of free-stream Mach number. Compare your results with the incompressible ($M = 0$) case using the Küssner function. Find the time-derivative of the lift, which is an approximation to the acoustic pressure generated by the interaction. Discuss your results.

Bibliography

Anderson, W. D. and Watts, G. A. 1975. "Rotor Blade Wake Flutter – A Comparison of Theory and Experiment," 31st Annual National Forum of the American Helicopter Soc., Washington DC, May 13–15.

Ashley, H., Dugundji, J., and Nielson, D. O. 1952. "Two Methods for Predicting Air Loads on a Wing in Accelerated Motion," *J. Aeronaut. Sci.*, 19 (8), pp. 543–552.

Beddoes, T. S. 1976. "A Synthesis of Unsteady Aerodynamic Effects Including Stall Hysteresis," *Vertica*, 1 (2), pp. 113–123.

Beddoes, T. S. 1980. "Unsteady Flows Associated with Helicopter Rotors," AGARD Report 679.

Beddoes, T. S. 1984. "Practical Computation of Unsteady Lift," *Vertica*, 8 (1), pp. 55–71.

Bisplinghoff, R. L., Ashley, H., and Halfman, R. L. 1955. *Aeroelasticity*, Addison-Wesley Publishing Co., Reading, MA.

Bramwell, A. R. S. 1976. *Helicopter Dynamics*, Edward Arnold, London, Chapter 8.

Brown, C. E. 1949. "The Reversibility Theorem for Thin Airfoils in Subsonic and Supersonic Flow," NACA TN 1944.

Carpenter, P. J. and Friedovitch, B. 1953. "The Effect of a Rapid Blade Pitch Increase on the Thrust and Induced Velocity of a Full-Scale Helicopter Rotor," NACA TN 3044.

Chen, R. T. 1989. "A Survey of Nonuniform Inflow Models for Rotorcraft Flight Dynamics and Control Applications," 15th European Rotorcraft Forum, Amsterdam, The Netherlands, Sept. 12–15.

Crews, S. T., Hohenemeser, K. H., and Ormiston, R. A. 1973. "An Unsteady Wake Model for a Hingeless Rotor," *J. of Aircraft*, 10 (12), pp. 758–759.

Curtiss, H. C. amd Shupe, N. K. 1971. "A Stability and Control Theory for Hingeless Rotors," 26th Annual National Forum of the American Helicopter Soc., Washington DC, May.

Curtiss, H. C. 1986. "Stability and Control Modeling," 12th European Rotorcraft Forum, Garmisch-Partenkirchen, Germany, Sept. 22–25.

Daughaday, H., Du Walt, F., and Gates, C. 1959. "Investigation of Helicopter Blade Flutter and Load Amplification Problems," *J. American Helicopter Soc.*, 2 (3), pp. 27–46.

Davis, S. S. and Malcolm, G. N. 1980. "Experimental Unsteady Aerodynamics of Conventional and Supercritical Airfoils," NASA TN 81221.

Dinyavari, M. A. H. and Friedmann, P. P. 1986. "Application of Time-Domain Unsteady Aerodynamics to Rotary-Wing Aeroelasticity," *AIAA J.*, 24 (9), pp. 1424–1432.

Dowell, E. H. 1980. "A Simple Method of Converting Frequency Domain Aerodynamics to the Time Domain," NASA TM-81844.

Drischler, J. A. and Diederich, F. W. 1957. "Lift and Moment Responses to Penetration of Sharp-Edged Traveling Vertical Gusts, with Application to Penetration of Weak Blast Waves," NACA TN 3956.

Flax, A. H. 1952. "General Reverse Flow and Variational Theorems in Lifting-Surface Theory," *J. Aeronaut. Sci.*, 19 (6), pp. 361–374.

Flax, A. H. 1953. "Reverse-Flow and Variational Theorems for Lifting Surfaces in Nonstationary Compressible Flow," *J. Aeronaut. Sci.*, 19 (2), pp. 120–126.

Franklin, G. F., Powell, J. D., and Abbas, E.-N. 1994. *Feedback Control of Dynamic Systems*, 3rd ed., Addison-Wesley Publishing Co., Reading, MA, Chapters 2 & 7.

Friedmann, P. and Yuan, C. 1977. "Effect of Modified Aerodynamic Strip Theories on Rotor Blade Aeroelastic Stability," *AIAA J.*, 15 (7), pp. 932–940.

Friedmann, P. P. 1985. "A New Look at Arbitrary Motion Unsteady Aerodynamics and Its Application to Rotary-Wing Aeroelasticity," 2nd Int. Symp. on Aeroelasticity and Structural Dynamics, Aachen, Germany, April 1–3.

Gaonkar, G. H. and Peters, D. A. 1986a. "Effectiveness of Current Dynamic Inflow Models in Hover and Forward Flight," *J. American Helicopter Soc.*, 31 (2), pp. 47–57.

Gaonkar, G. H. and Peters. D. A. 1986b. "Review of Dynamic Inflow Modeling for Rotorcraft Flight Dynamics," AIAA Paper 86-0845-CP.

Garrick, I. E. 1938. "On Some Reciprocal Relations in the Theory of Non-Stationary Flows," NACA Report 629.

Garrick, I. E. 1957. "Nonsteady Wing Characteristics," Section F, pp. 658, Aerodynamic Components Aircraft at High Speeds, Vol. VII of High Speed Aerodynamics and Jet Propulsion, ed. A. F. Donovan & H. R. Lawrence, Princeton University Press, Princeton, N. J.

Glauert, H. 1929. "The Force and Moment on an Oscillating Airfoil," ARC R & M 1242.

Greenberg, J. M. 1947. "Airfoil in Sinusoidal Motion in a Pulsating Stream," NACA TN 1326.

Halfman, R. 1951. "Experimental Aerodynamic Derivatives of a Sinusoidally Oscillating Airfoil in Two-Dimensional Flow," NACA TN 2465.

Ham, N. D., Moser, H. H., and Zvara, J. 1958. "Investigation of Rotor Response to Vibratory Aerodynamic Inputs – Experimental Results and Correlation with Theory," Wright Air Development Center, WADC TR 58–87, Part 1.

Hammond, C. E. and Pierce, G. A. 1972. "A Compressible Unsteady Aerodynamic Theory for Helicopter Rotors," AGARD-CP-111.

Hariharan, N. and Leishman, J. G. 1996. "Unsteady Aerodynamics of a Flapped Airfoil in Subsonic Flow by Indicial Concepts," *J. of Aircraft*, 33 (5), pp. 855–868.

Heaslet, M.A. and Sprieter, J.R. 1952. "Reciprocity Relations in Aerodynamics," NACA Report 1119.

Isaacs, R. 1945. "Airfoil Theory for Flows of Variable Velocity," *J. Aeronaut. Sci.*, 12 (1), pp. 113–117.

Isaacs, R. 1946. "Airfoil Theory for Rotary Wing Aircraft," *J. Aeronaut. Sci.*, 13 (4), pp. 218–220.

Johnson, W. 1980. *Helicopter Theory*, Princeton University Press, Princeton, NJ, Chapter 10.

Jones, J. P. 1958. "The Influence of the Wake on the Flutter and Vibration of Rotor Blades," *The Aeronaut. Quart.*, 9 (3), pp. 258–286.

Jones, R. T. 1938. "Operational Treatment of the Non-Uniform Lift Theory in Airplane Dynamics," NASA Technical Note 667.

Jones, R. T. 1940. "The Unsteady Lift of a Wing of Finite Aspect Ratio," NACA Report 681.

Jones, R. T. 1951. "The Minimum Drag of Thin Wings in Frictionless Flow," *J. Aeronaut. Sci.*, 18 (2), pp. 75–81.

Jones, W. P. 1945. "Aerodynamic Forces on Wings in Nonuniform Motion," ARC R & M 2117.

Karamcheti, K. 1966. *Principles of Ideal-Fluid Aerodynamics*, John Wiley & Sons, Inc., New York.

Kottappalli, S. B. R. 1985. "Unsteady Aerodynamics of Oscillating Airfoils with Inplane Motions, *J. American Helicopter Soc.* 30 (1), pp. 62–63.

Küssner, H. G. 1935. "Zusammenfassender Bericht über den instationären Auftreib von Flügeln," *Luftfahrtforschung*, 13 (12), p. 410.

Leishman, J. G. 1987a. "Validation of Approximate Indicial Aerodynamic Functions for Two-Dimensional Subsonic Flow," *J. of Aircraft*, 25 (10), pp. 914–922.

Leishman, J. G. 1987b. "A Two-Dimensional Model for Airfoil Unsteady Drag below Stall," *J. of Aircraft*, 25 (7), pp. 665–666.

Leishman, J. G. and Nguyen, K. Q. 1990. "A State-Space Representation of Unsteady Aerodynamic Behavior," *AIAA J.*, 28 (5), pp. 836–845.

Leishman, J. G. 1993. "Indicial Lift Approximations for Two–Dimensional Subsonic Flow as Obtained from Oscillatory Measurements," *J. of Aircraft*, 30 (3), pp. 340–351.

Leishman, J. G. 1994. "Unsteady Lift of a Flapped Airfoil by Indicial Concepts," *J. of Aircraft*, 31 (2), pp. 288–297.

Leishman, J. G. 1997. "Unsteady Aerodynamics of Airfoils Encountering Traveling Gusts and Vortices," *J. of Aircraft*, 34 (6), pp. 719–729.

Liiva, J., Davenport, F. J., Gray, L., and Walton, I. C. 1968. "Two-Dimensional Tests of Airfoils Oscillating Near Stall," USAAVLABS TR 68–13, Vols I & II.

Loewy, R. G. 1957. "A Two-Dimensional Approximation to the Unsteady Aerodynamics of Rotary Wings," *J. American Helicopter Soc.*, 24 (2), pp. 81–92.

Lomax, H. 1953. "Lift Developed on Unrestrained Rectangular Wings Entering Gusts and Subsonic and Supersonic Speeds," NACA Technical Note 2925.

Lomax, H. 1968. "Indicial Aerodynamics," Chapter 6, AGARD Manual on Aeroelasticity.

Lomax, H., Heaslet, M. A., Fuller, F. B., and Sluder, L. 1952. "Two and Three Dimensional Unsteady Lift Problems in High Speed Flight," NACA Report 1077.

Mazelsky, B. 1952a. "Determination of Indicial Lift and Moment of a Two-Dimensional Pitching Airfoil at Subsonic Mach Numbers from Oscillatory Coefficients with Numerical Calculations for a Mach Number of 0.7," NACA TN 2613.

Mazelsky, B. 1952b. "On the Noncirculatory Flow about a Two-Dimensional Airfoil at Subsonic Speeds," *J. Aeronaut. Sci.*, 19 (12), pp. 848–849.

Mazelsky, B. and Drischler, J. A. 1952. "Numerical Determination of Indicial Lift and Moment Functions of a Two-Dimensional Sinking and Pitching Airfoil at Mach Numbers 0.5 and 0.6," NACA TN 2739.

McCroskey, W. J. and Goorjian, P. M., 1983. "Interactions of Airfoils with Vertical Gusts and Concentrated Vortices in Unsteady Transonic Flow," AIAA Paper 83–1691.

Miles, J. W. 1956. "The Aerodynamic Force on an Airfoil in a Moving Gust," *J. Aeronaut. Sci.*, 23 (11), pp. 1044–1050.

Miller, R. H. 1964. "Unsteady Air Loads on Helicopter Rotor Blades," *J. Royal Aeronaut. Soc.*, 68 (640), pp. 217–229.

Ormiston, R. A. 1976. "Application of Simplified Inflow Models to Rotorcraft Dynamic Analysis," *J. American Helicopter Soc.*, 21 (3), pp. 34–37.

Osborne, C. 1973. "Unsteady Thin Airfoil Theory in Subsonic Flow," *AIAA J.*, 11 (2), pp. 205–209.

Padfield, G. D. 1996. *Helicopter Flight Dynamics: The Theory and Applications of Flying Qualities and Simulation Modeling*, Blackwell Science, Ltd., Oxford, and AIAA Education Series, Washington DC.

Parameswaran, V. and Baeder, J. D. 1997. "Indicial Aerodynamics in Compressible Flow – Direct Calculations," *J. of Aircraft*, 34 (1), pp. 131–133.

Peters, D. A. 1974. "Hingeless Rotor Frequency Response with Unsteady Aerodynamics," AHS/NASA Specialists' Meeting on Rotorcraft Dynamics, NASA SP–362.

Peters, D. A., Boyd, D. D., and He, C. J. 1987. "Finite-State Induced-Flow Model for Rotors in Hover and Forward Flight," 43rd Annual Forum of the American Helicopter Soc., St. Louis, MO, May 18–20.

Peters, D. A. and HaQuang, N. 1988. "Dynamic Inflow for Practical Applications," *J. American Helicopter Soc.*, 33 (4), pp. 64–68.

Peters, D. A. and He, C. J. 1989. "Correlation of Measured Induced Velocities with a Finite-State Wake Model," 45th Annual Forum of the American Helicopter Soc., Boston, MA, May 22–24.

Peters, D. A., Karunamoorthy, S., and Cae, W.-M. 1995. "Finite State Induced Flow Models Part I: Two Dimensional Thin Airfoil," *J. of Aircraft*, 32 (2), pp. 313–322.

Peters, D. A. and He, C. J. 1995. "Finite State Induced Flow Models Part II: Three-Dimensional Rotor Disk," *J. of Aircraft*, 32 (2), pp. 323–333.

Pitt, D. M. and Peters, D. A. 1983. "Rotor Dynamic Inflow Derivatives and Time Constants from Various Inflow Models," 9th European Rotorcraft Forum, Stresa, Italy, Sept. 13–15.

Rainey, A. G. 1957. "Measurement of Aerodynamic Forces for Various Mean Angles of Attack on an Airfoil Oscillating in Pitch and on Two Finite-Span Wings Oscillating in Bending with Emphasis on Damping in the Stall," NACA Report 1305.

Sears, W. R. 1940. "Operational Methods in the Theory of Airfoils in Nonuniform Motion," *J. Franklin Inst.*, 230, pp. 95–111.

Sears, W. R. and Sparks, B. O. 1941. "On the Reaction of an Elastic Wing to Vertical Gusts," *J. Aeronaut. Sci.*, 9 (2), pp. 64–51.

Singh, R. and Baeder, J. D. 1997a. "The Direct Calculation of Indicial Lift Response of a Wing Using Computational Fluid Dynamics," *J. of Aircraft*, 35 (4), pp. 465–471.

Singh, R. and Baeder, J. D. 1997b. "Generalized Moving Gust Response Using CFD with Application to Airfoil–Vortex Interaction," 15th AIAA Applied Aerodynamics Conf., Atlanta, GA, June.

Silviera, M. A. and Brooks, G. W. 1959. "Dynamic-Model Investigation of the Damping of Flapwise Bending Modes of Two-Blade Rotors in Hovering and a Comparison with Quasistatic and Unsteady Aerodynamic Theories," NACA TN D-175.

Sissingh, G. J. 1952. "The Effect of Induced Velocity Variation on Helicopter Rotor Damping in Pitch and Roll," ARC Technical Report G P No. 101.

Srinivasan, G.R. and McCroskey, W.J. 1987. "Numerical Simulations of Unsteady Airfoil Interactions," *Vertica*, 11 (1/2), pp. 3–28.

Stahara, S. S., Spreiter, J. R., 1976. "Research on Unsteady Transonic Flow Theory," NEAR TR-107, Jan. For Office of Naval Research, Arlington, VA.

Theodorsen, T. 1935. "General Theory of Aerodynamic Instability and the Mechanism of Flutter," NACA Report 496.

van der Wall, B. and Leishman, J. G. 1994. "The Influence of Variable Flow Velocity on Unsteady Airfoil Behavior," *J. American Helicopter Soc.*, 39 (4), pp. 288–297.

Vepa, R. 1976. "On the Use of Padé Approximants to Represent Unsteady Aerodynamic Loads for Arbitrary Small Motions of Wings," AIAA Paper 76–17, AIAA Aerospace Sci. Meeting and Exhibit, Jan.

von Kármán, Th. and Sears, W. R. 1938. "Airfoil Theory for Non-Uniform Motion," *J. Aeronaut. Sci.*, 5 (10), pp. 379–390.

Wagner, H. 1925. "Über die Entstehung des dynamischen Auftriebes von Tragflügeln," *Zeitschrift für Angewandte Mathematik und Mechanick*, 5 (1), p. 17.

Wood, M. E. 1979. "Results from Oscillatory Pitch Tests on the NACA 0012 Blade Section," ARA Memo 220, Aircraft Research Association, Bedford, UK.

CHAPTER 9

Dynamic Stall

> Fortunately, engineers and technologists do not wait until everything is completely understood before building and trying new devices. Even so, an improved understanding of fundamental unsteady fluid flow processes can serve to stimulate new innovations, as well as improvements in the performance, reliability, and costs of many existing machines. Therefore, research in unsteady fluid dynamics seems assured a lively future in modern industrial societies.
>
> William J. McCroskey (1975)

9.1 Introduction

The phenomenon of dynamic stall has long been known to be a factor that limits helicopter rotor performance at high forward flight speeds or in high "g" maneuvers because of the onset of large torsional airloads and vibrations on the blades – see, for example, Tarzanin (1972), McCroskey & Fisher (1972), and McHugh (1978). Whereas for a fixed-wing aircraft, stall occurs at low flight speeds, stall on a helicopter rotor will occur at relatively high airspeeds as the advancing and retreating blades begin to operate close to the limits where the flow can feasibly remain attached to the airfoil surfaces. In this regard, the advancing blade operates at low angles of attack but close to its shock-induced separation boundary. The retreating blade operates at much lower Mach numbers but encounters very high angles of attack close to stall. Because of the time-varying blade element angle of attack resulting from blade flapping, cyclic pitch inputs, and wake inflow, separation and stall occur on a rotor in a very dynamic or time-dependent manner. The stall phenomenon is, therefore, referred to as "dynamic stall." Despite the fact that the static stall characteristics of airfoils were discussed extensively in Chapter 7, the problem of flow separation and stall must now be reassessed from a dynamic perspective.

Following the general definition given by McCroskey and coworkers (1976, 1982), dynamic stall will occur on any airfoil or other lifting surface when it is subjected to time-dependent pitching, plunging or vertical translation, or other type of non-steady motion that takes the effective angle of attack above its normal static stall angle. Under these circumstances, the physics of flow separation and the development of stall have been shown to be fundamentally different from the stall mechanism exhibited by the same airfoil under static (quasi-steady) conditions (i.e., where $k = 0$). Dynamic stall is, in part, distinguished by a delay in the onset of flow separation to a higher angle of attack than would occur statically. This initial delay in stall onset is obviously advantageous as far as the performance and operational flight envelope of a helicopter rotor is concerned. However, when dynamic separation does occur, it is found to be characterized by the shedding of a concentrated vortical disturbance from the leading-edge region of the airfoil. As long as this vortex disturbance stays over the airfoil upper surface, it acts to enhance the lift being produced. Yet, the vortex flow pattern is not stable, and the vortex is quickly swept over the chord of the blade by the oncoming flow. This produces a rapid aft movement of the center of pressure, which results in large nose-down pitching moments on the airfoil section and an increase in torsional loads on the blades. This is the main adverse characteristic of dynamic stall that concerns the rotor analyst.

The consideration of dynamic stall in the rotor design process will more accurately define the operational and performance boundaries of the helicopter. Generally the rotor will be first designed so that the onset of high blade loads, aeroelastic problems, or limits in overall performance are not limiting factors on the basis of linear and nonlinear quasi-steady aerodynamic assumptions. Nonlinearities in the airloads associated with dynamic stall can introduce effects that give rise to dangerously high blade stresses, vibrations and control loads. One such nonlinear phenomena is called *stall flutter*.[1] Because of the significant hysteresis in the airloads as functions of angle of attack that takes place during dynamic stall, and also because of the possibilities of lower aerodynamic damping, an otherwise stable elastic blade mode can become unstable if flow separation is present. Therefore, the onset of dynamic stall generally defines the overall lifting and propulsive performance limits of a helicopter rotor.

The accurate prediction of the combination of angle of attack and Mach number that will produce dynamic stall onset, as well as the prediction of the subsequent effects of dynamic stall on rotor loads and performance, is not an easy task. The phenomenon of dynamic stall is not fully understood and is still the subject of much research on both experimental and numerical fronts – see the review article by Carr (1988). The very large number of publications on the phenomenon (see also the list of references for this chapter) illustrates the importance of dynamic stall in the more complete aerodynamic and aeroelastic analysis of the helicopter rotor and the difficulties in both measuring and predicting the phenomenon. The complicated nonlinear physics of dynamic stall means that the behavior can only be completely modeled by means of numerical solutions to the Navier–Stokes equations (using computational fluid dynamics or CFD). This, like other CFD problems that involve unsteady, compressible, separated flows, the solution to dynamic stall problems is a formidable task that is not yet practical. However, since the early 1990s, the rapid increase in computer resources has enabled considerable progress to be made in modeling dynamic stall by means of first-principle based CFD approaches – see, for example, Srinivasan et al. (1993), Ekaterinaris et al. (1994), and Landgrebe (1994). CFD methods will eventually prevail, but for the most part these methods are currently impractical for routine use as part of a rotor design process. For engineering analyses, the modeling of dynamic stall also remains a particularly challenging problem. To this end, a large number of semi-empirical models have been developed for use in rotor design. A brief discussion of some of these methods will be described in this chapter, along with a demonstration of their general capabilities in predicting dynamic stall induced airloads. Although giving good results, these models are not strictly predictive tools and can really only be used confidently for conditions that are bounded by validation with experimental data.

9.2 Flow Topology of Dynamic Stall

The effects of unsteady motion on unsteady airfoil behavior and dynamic flow separation have been recognized for many years, mainly through studies of oscillating airfoils in wind-tunnel experiments. As mentioned previously, for an increasing angle of attack it has been observed that the flow remains attached to the upper surface of an airfoil at angles of attack that are much higher than could be attained quasi-statically, giving a corresponding increase in maximum lift. Kramer (1932) was one of the first investigators to observe the phenomenon. The delay in the onset of flow separation under unsteady

[1] This is different from classical flutter, which involves fully attached flow.

conditions is a result of three primary unsteady phenomena. First, during the conditions where the angle of attack is increasing with respect to time, the unsteadiness of the flow resulting from circulation that is shed into the wake at the trailing edge of the airfoil causes a reduction in the lift and adverse pressure gradients compared to the steady case at the same angle of attack. This "classical" effect has been described previously in Chapter 8. Second, by virtue of a kinematic induced camber effect, which has also been described in Chapter 8, a positive pitch rate further decreases the leading-edge pressure and pressure gradients for a given value of lift. This can be considered a quasi-steady effect. Ericsson (1967), Carta (1971), Johnson & Ham (1972), Ericsson & Redding (1972), McCroskey (1973), and Beddoes (1978) have given a good summary of these basic effects from the perspective of unsteady airfoil theory. Third, in response to the external pressure gradients, there are also additional unsteady effects that occur within the boundary layer, including the existence of flow reversals in the absence of any significant separation – see McAlister & Carr (1979). These unsteady boundary layer effects have been quantitatively examined by Scruggs et al. (1974), Telionis (1975), and McCroskey (1975). Although the behavior of unsteady turbulent boundary layers is still not fully understood, the onset of separation on airfoils is generally found to be delayed by unsteady effects such as those associated with increasing pitch rate. Coupled with the aforementioned pressure gradient reductions, the resulting lag in the formation of boundary layer separation causes the onset of dynamic stall to be averted to a significantly higher angle of attack than would be obtained under quasi-steady conditions.

Ultimately, however, with increasing angle of attack, the high adverse pressure gradients that build up near the leading edge under dynamic conditions cause flow separation to occur there. Experimental evidence suggests the formation of a shear layer that forms just downstream of the leading edge, which quickly rolls up and forms a vortical disturbance. This feature is now known to be a very characteristic aspect of dynamic stall and is shown in the flow visualization images in Fig. 9.1. Not long after it is formed, this vortical disturbance leaves the leading-edge region and begins to convect over the upper surface of the airfoil. This induces a pressure wave that sustains lift and produces airloads well in excess of those obtained under steady conditions. A qualitative understanding of this vortex shedding phenomenon was first given by Ham (1968) and McCroskey (1972a,b), and is reviewed by Beddoes (1979). A great number of subsequent experimental studies have provided a

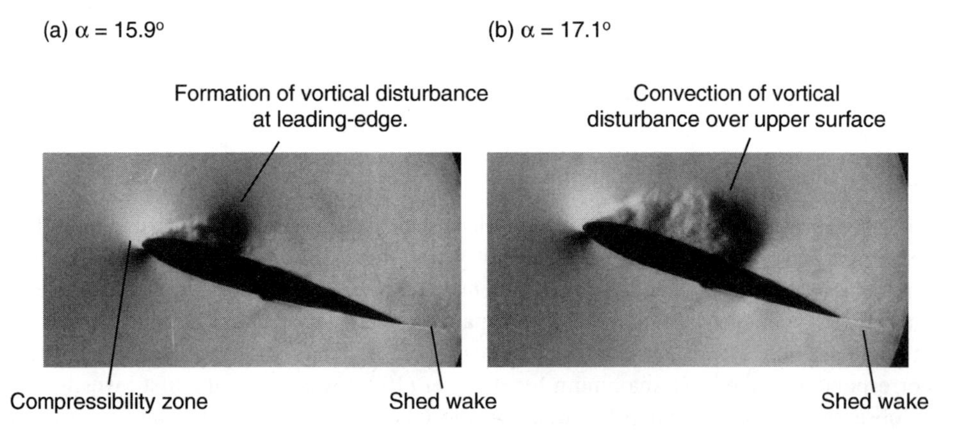

Figure 9.1 Visualization of dynamic stall using schlieren. Source: Chandrasekhara & Carr (1990) and courtesy of M. S. Chandrasekhara.

much more comprehensive physical understanding of the factors that determine the onset of dynamic stall, including the important influence of compressibility – see Beddoes (1978, 1983), Lorber (1992), and Chandrasekhara & Carr (1990, 1994). However, there have been fewer experimental studies of dynamic stall at the combinations of Reynolds numbers and Mach numbers that would be useful to the helicopter analyst. Fortunately, several studies have been commissioned to study the effects of compressibility on the quantitative effects of dynamic stall for airfoils operating at or near to full-scale rotor Reynolds numbers – see, for example, Liiva et al. (1968) and Wood (1979). Generally, the results have shown that the qualitative features of the dynamic stall process remain similar over a fairly wide range of Mach numbers and also under different types of forcing conditions (i.e., for pitching oscillations, plunging oscillations, and ramp or constant angular rate motions). Yet, the quantitative behavior of the airloads show subtle variations with Mach number, especially for different airfoil shapes. It is these more subtle aspects of the dynamic stall problem that make its accurate prediction difficult for the helicopter rotor analyst.

The various stages of the dynamic stall process are summarized schematically in Fig. 9.2. Stage 1 represents the delay in the onset of separation in response to reduction in adverse pressure gradients produced by the kinematics of pitch rate (induced camber), the influence of the shed wake, and the unsteady boundary layer response. Stage 2 of the dynamic stall process involves flow separation and the formation of a vortex disturbance that is cast off from the leading-edge region of the airfoil. This vortex disturbance provides additional lift on the airfoil so long as it stays over the upper surface. In some cases, primarily at low Mach numbers, the additional "lift overshoots" produced by this process may be between 50 and 100% of the static value of maximum lift. The effective lift-curve-slope may also increase during this process. These, often surprisingly large increments in lift, are also accompanied by significant increases in nose-down pitching moment, which results from an aft moving center of pressure as the vortex disturbance is swept downstream across the chord. The speed at which the vortex convects downstream has been documented to be between one third and one half of the free-stream velocity – see Beddoes (1976) and Galbraith et al. (1986). It will also be seen from Fig. 9.2, that the sudden break in the lift coefficient at the start of Stage 3 occurs at a higher angle of attack than that for the divergence in the pitching moment; that is, the moment break (moment stall) occurs at the onset of vortex shedding (start of Stage 2), whereas the lift break (lift stall) occurs when the vortex passes into the wake (end of Stage 2 and start of Stage 3).

After the vortex disturbance passes the trailing edge of the airfoil and becomes entrained into the turbulent wake downstream of the airfoil, the flow on the upper surface progresses to a state of full separation. This is referred to as Stage 4 of the dynamic stall process; it is accompanied by a sudden loss of lift, a peak in the pressure drag, and a maximum in nose-down pitching moment. In this flow state, the airloads are approximately the same as those found under steady conditions at the same angle of attack. Flow reattachment can take place if and when the angle of attack of the airfoil becomes low enough again. However, there is generally a significant lag in this process. First, there is a general lag in the reorganization of the flow from the fully separated state until it becomes amenable to reattachment. Second, there is a lag because of the reverse kinematic "induced camber" effect on the leading-edge pressure gradient by the negative pitch rate. Therefore, full flow reattachment may not be obtained until the airfoil is well below its normal static stall angle, as denoted by Stage 5 in Fig. 9.2. In this particular case, it is apparent that the angle of attack falls to as low as 5° before the flow can be considered as fully attached. Because of these lags in the development of the various flow states, a large amount of hysteresis is present in all three components of the airloads. These hysteresis effects are the source of

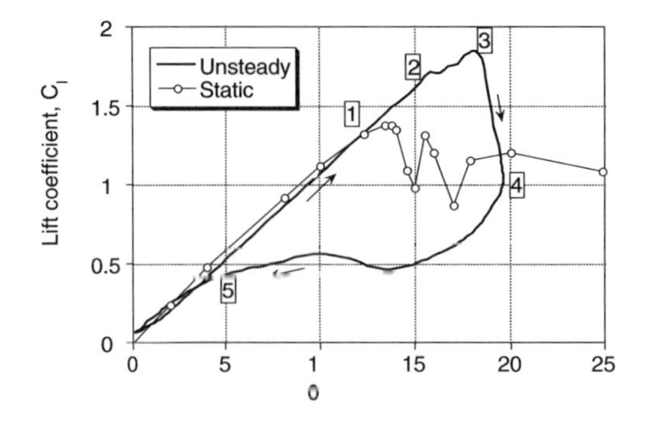

Stage 1: Airfoil exceeds static stall angle, flow reversals take place in boundary layer.

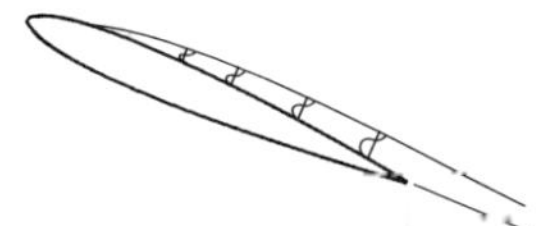

Stage 2: Flow separation at the leading-edge, formation of a 'spilled' vortex. Moment stall.

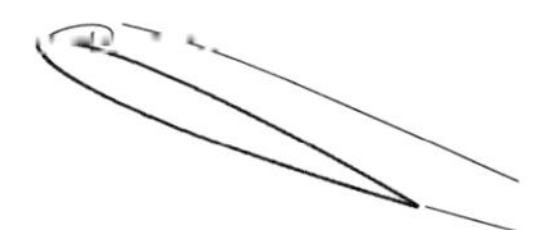

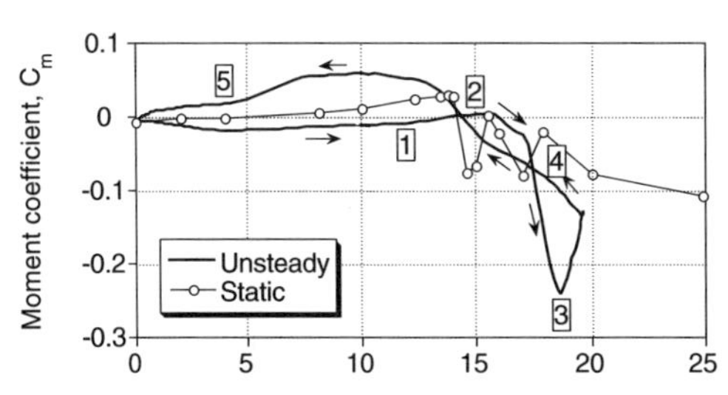

Stage 2-3: Vortex convects over chord, induces extra lift and aft center of pressure movement.

Stage 3-4: Lift stall. After vortex reaches trailing-edge, flow progresses to a state of full separation.

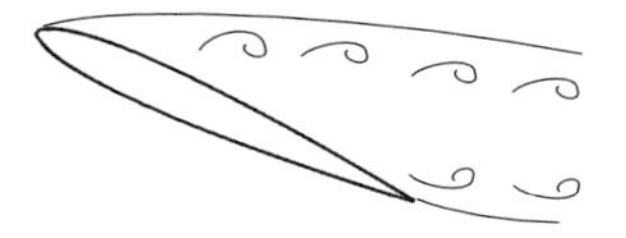

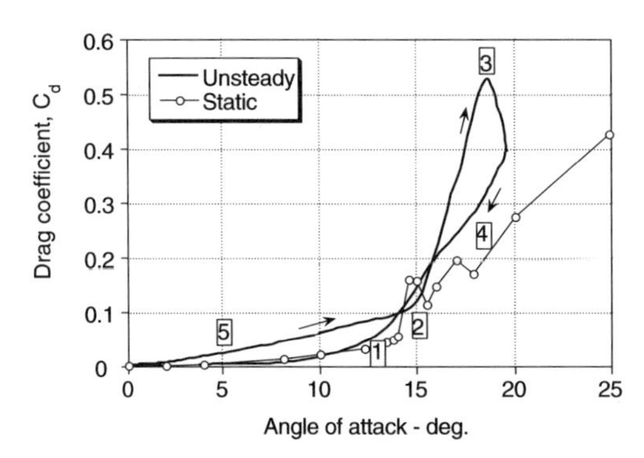

Stage 5: When angle of attack becomes low enough, flow reattaches front to back.

Figure 9.2 Schematic showing the essential flow topology and the unsteady airloads during the dynamic stall of an oscillating 2-D airfoil. Adapted from Carr et al. (1977) and McCroskey et al. (1982).

reduced aerodynamic damping, which, as mentioned previously, can potentially lead to a variety of aeroelastic problems on the rotor.

9.3 Dynamic Stall in the Rotor Environment

While much of what is known about dynamic stall has been obtained from idealized experiments on 2-D airfoils in wind tunnels, it is important to recognize that when dynamic stall occurs on the rotor, it has a more three-dimensional character than described above and may simultaneously occur over several radial and azimuthal sectors of the rotor disk. Continued advances in miniature pressure transducer technology and high speed data acquisition and telemetry systems have enabled a more detailed understanding of dynamic stall as it manifests on helicopters during actual flight. Issacs & Harrison (1989) and

Bousman (1998) provide good in-flight documentation of the dynamic stall phenomenon. The results shown in Fig. 9.3 are adapted from Bousman (1998) and show the time histories of the lift and pitching moment at various radial stations on the blade of a UH-60 helicopter during a pull-up maneuver at $\mu \approx 0.3$ and $C_T/\sigma \approx 0.17$. The results are presented in terms of the nondimensional quantities M^2C_n and M^2C_m, because these quantities give a better quantitative measure of the local airloads produced on the rotor.

Using the unsteady chordwise pressures as an indicator, Bousman (1998) has identified three locations on the rotor disk for this flight condition that show the lift overshoots and large nose-down pitching moments that are characteristic features of dynamic stall. On Fig. 9.3 these are marked by points M (moment stall) and by points L (lift stall). Remember that dynamic lift stall always occurs after moment stall. At these particular flight conditions, it is apparent that dynamic stall encompasses relatively large areas of the rotor disk. In particular, note that the occurrence of dynamic stall causes large transients in the pitching moments, especially between $r = 0.77$ and $r = 0.92$ in the first quadrant of the disk, and also on the retreating side near $\psi = 270°$. Operating the rotor at thrusts or airspeeds beyond this flight condition will result in high structural loads and stresses that can quickly exceed the fatigue or endurance limits of the rotor and/or control system (see also Fig. 6.6). Even though the rotor is usually able to operate with some amount of stall, the very rapid growth in the blade torsion and other structural loads because of dynamic stall is normally a limiting factor in the overall operational flight envelope of helicopters – see also Benson et al. (1973) and Stepniewski & Keys (1984).

9.4 Effects of Forcing Conditions on Dynamic Stall

Dynamic stall has been extensively studied experimentally, mostly using oscillating 2-D airfoils in wind-tunnel experiments. This simulates the quasi-periodic first harmonic angle of attack variations that are found on helicopter rotors during forward flight. The majority of the documented experimental results are for airfoils oscillating in pitch, but there are some limited results available for plunging oscillations (vertical translation or so-called heaving), as well as for constant angular rate (ramp) type motion. The former plunging experiments are useful in that pitch rate effects (i.e., $\dot{\alpha}$ effects) can be isolated from the problem. The latter ramp tests are useful in that acceleration effects (i.e., $\ddot{\alpha}$ effects) can be eliminated, somewhat simplifying the understanding of the various effects produced by the airfoil kinematics on the unsteady airloads. However, both pitch and ramp experiments are difficult to perform, especially for the Mach numbers and effective reduced frequencies required to validate any mathematical model of dynamic stall.

A few examples showing the variation of the unsteady airloads with variations in the oscillatory forcing parameters will now be described. With variations in parameters such as the amplitude of the angle of attack oscillation, the mean angle of attack, and the reduced frequency, the various dynamic stall events such as separation onset, leading-edge vortex shedding, and flow reattachment all shift around the cycle to different angles of attack. This results in significant quantitative changes in the airloads. Varying one parameter, while holding the others nominally constant, can help to provide a better understanding of the physics of dynamic stall. To this end, Figs. 9.4 through 9.6 summarize the effect on the lift and pitching moment for increasing mean angle of attack, increasing reduced frequency, and increasing Mach number, respectively. The results are for a NACA 0012 airfoil taken from Wood (1979), where the airloads were obtained by integrating static pressures measured using miniature pressure transducers distributed around a section of a 2-D wing.

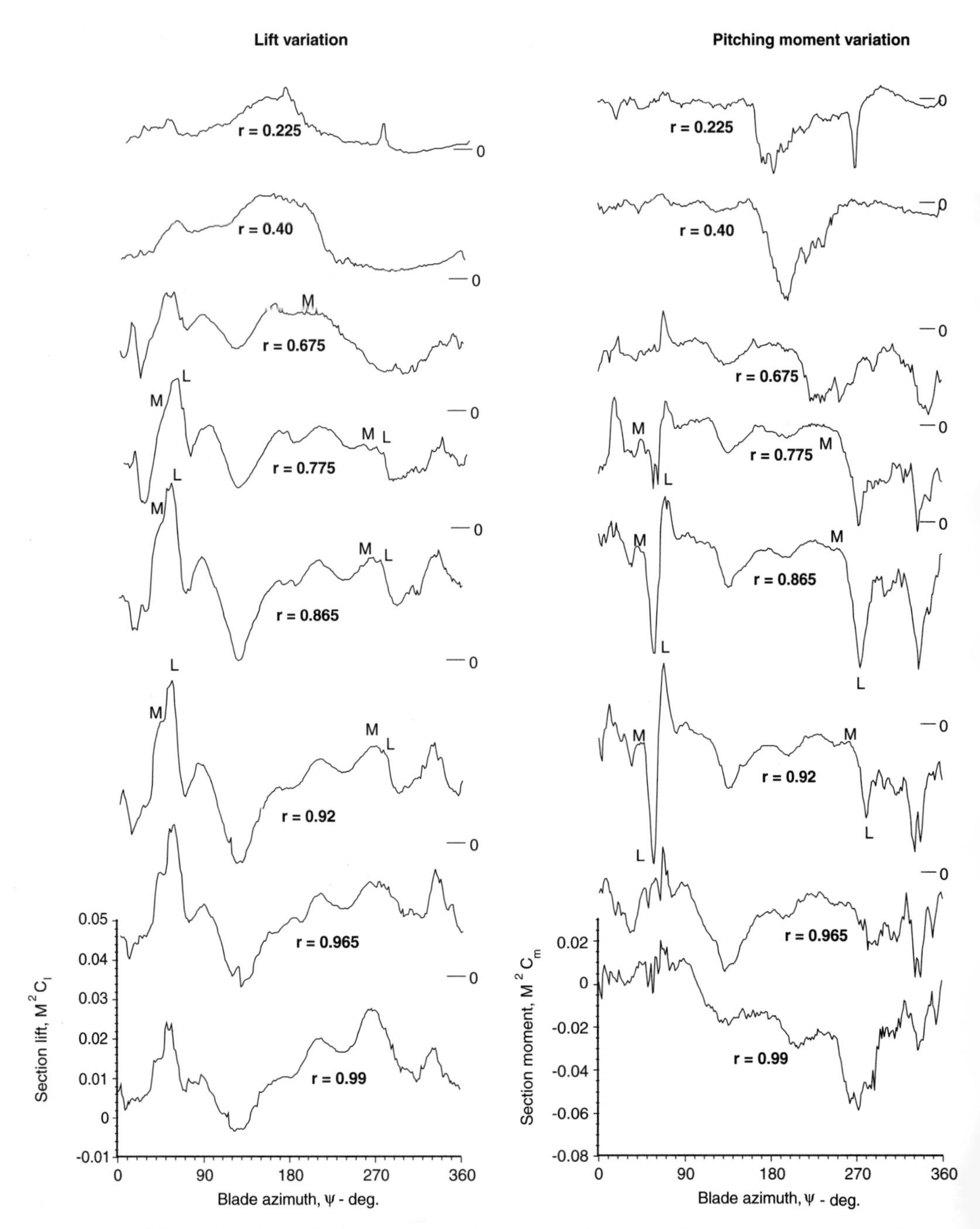

Figure 9.3 In-flight measurements of sectional lift and pitching moment over the blade of a UH-60 operating at high thrust in forward flight indicating the occurrence of dynamic stall. M = moment stall; L = lift stall. Data source: Bousman (1998).

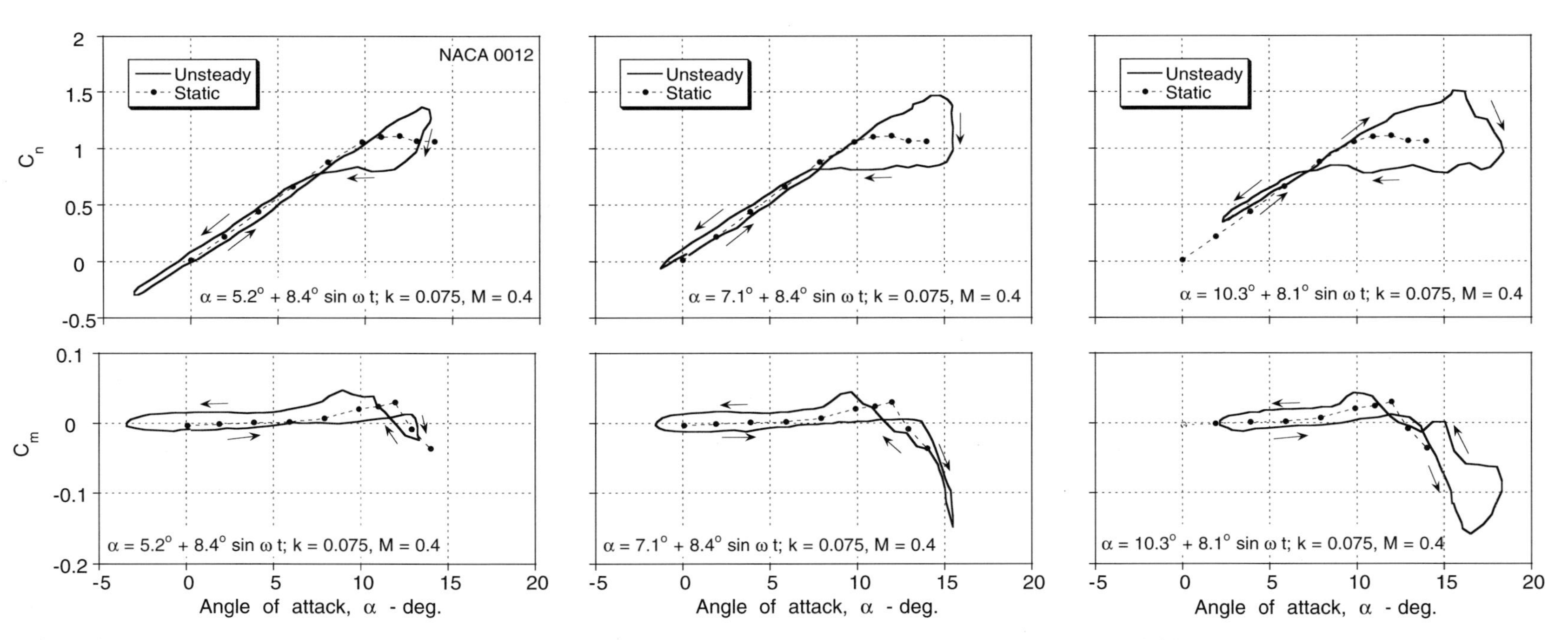

Figure 9.4 Effects of increasing mean angle of attack on the unsteady lift and pitching moment of an oscillating NACA 0012 airfoil, at $M = 0.4$, with $k = 0.075$.

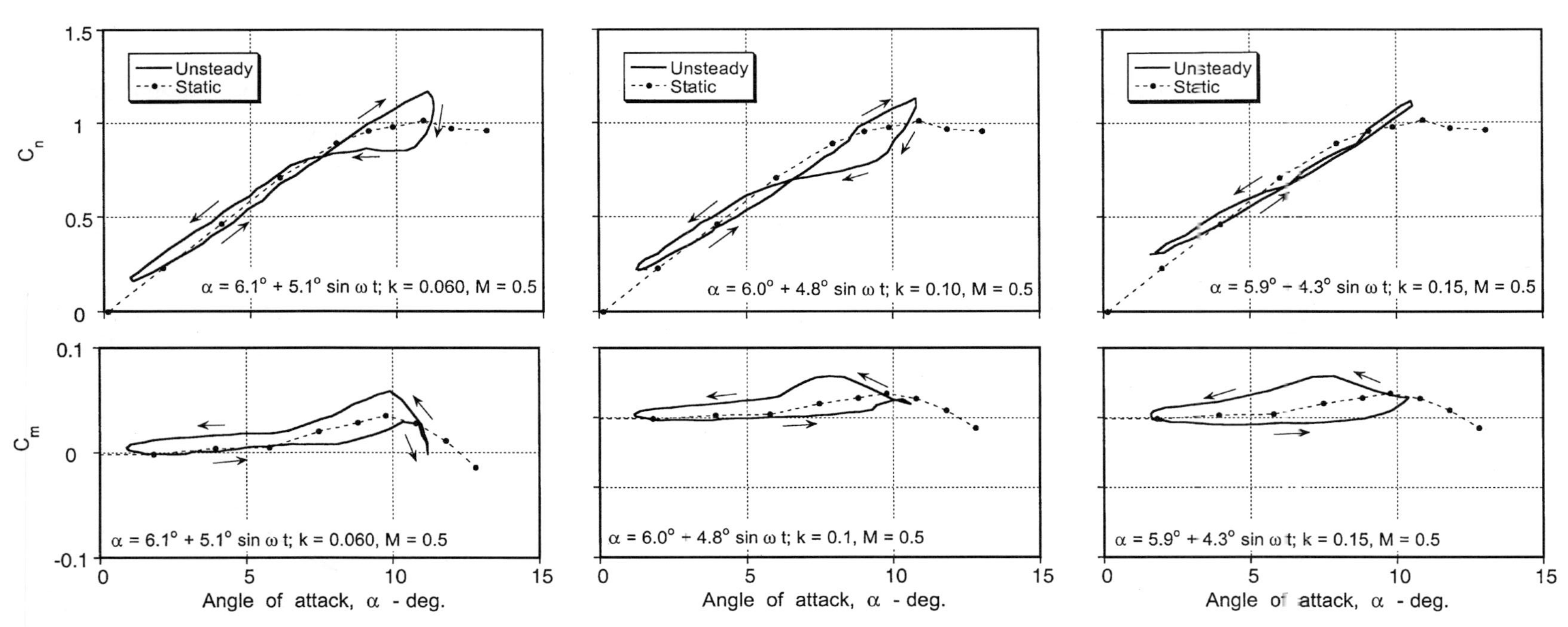

Figure 9.5 Effects of reduced frequency on the unsteady lift and pitching moment of an oscillating NACA 0012 airfoil, with $\alpha \approx 6^\circ + 5^\circ \sin \omega t$, at $M = 0.5$.

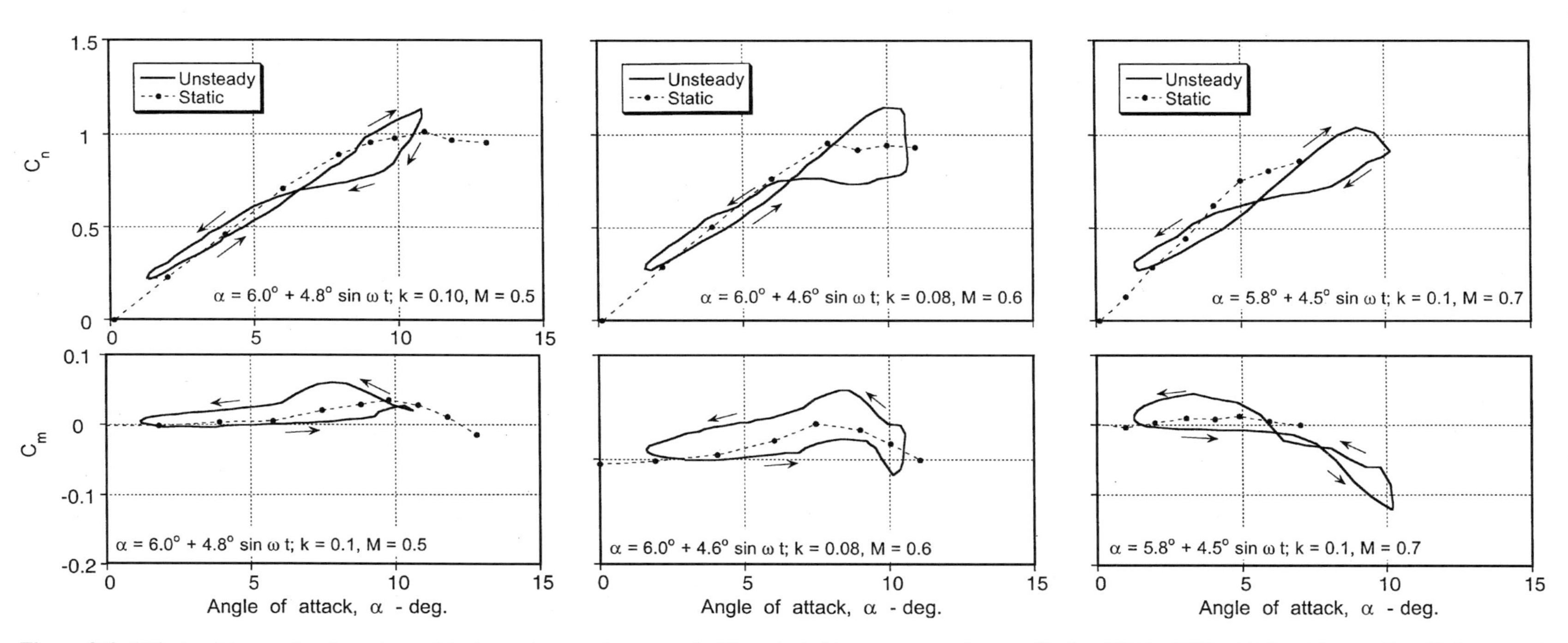

Figure 9.6 Effects of increasing free-stream Mach number on the unsteady lift and pitching moment of an oscillating NACA 0012 airfoil, with $\alpha \approx 6^\circ + 5^\circ \sin \omega t$, and $k \approx 0.1$.

Figure 9.4 shows the effects of increasing mean angle of attack, while holding the amplitude of oscillation at approximately 8.4°, and the reduced frequency at approximately 0.075. These results show the effects on the airloads as the flow state progresses from nominally attached conditions at the lowest mean angle, through "light" dynamic stall, and into "deep" dynamic stall for the highest mean angle. These "light" and "deep" dynamic stall descriptors were suggested by McCroskey et al. (1976, 1980) and Carr et al. (1977, 1978). For the lowest mean angle of attack of 5.2°, Fig. 9.4 shows a typical case of stall onset. This is when the combination of forcing conditions is just sufficient to cause some minor flow separation from the airfoil. Before this occurs, however, there is a delay in the onset of lift stall to a higher angle of attack and to a higher value of lift. The airloads then show some small deviations from the attached flow behavior (which would be almost pure elliptical loops) near the maximum angle of attack in the cycle as a result of the limited flow separation. As the angle of attack is reduced on the downstroke of the cycle the flow reattaches, as indicated by the return to the nominally elliptical shapes of the loop, but it will be noted that the angle of attack at which reattachment occurs is significantly below the static stall angle.

The second case shown in Fig. 9.4 is for a mean angle of attack of 7.1°, which is high enough to cause stronger leading-edge vortex shedding and the creation of airloads that are more typical of "light" dynamic stall. In this case the maximum dynamic lift coefficient was about 0.4 higher than the static value, but the large nose-down pitching moment is of more significance. Also, because of the clockwise loop that is now introduced into the pitching moment curve (the curve now looks like a figure eight), this forcing condition represents a situation with significantly reduced torsional aerodynamic damping.[2] The third case shown in Fig. 9.4 is for a mean angle of attack of 10.3°, which is high enough such that the behavior would now be characterized as "strong" dynamic stall. Leading-edge vortex shedding, again, contributes significantly to increased values of lift but gives a particularly large increase in the nose-down pitching moment. Because of the higher mean angle of attack, now a larger part of the oscillation cycle involves partly or fully separated flow, and so there are larger hysteresis effects. There is now another large counter clockwise loop introduced into the moment curve, which gives a return to high torsional damping. Flow reattachment is delayed to a fairly low angle of attack during the downstroke motion. This indicates the relatively long time scales required for the flow to reorganize after strong dynamic stall has occurred and to allow conditions conducive for flow reattachment.

Figure 9.5 shows the powerful effect of reduced frequency on the lift and pitching moment responses. In the first case, results are shown for a relatively low reduced frequency of 0.06, and for an angle of attack history that is just sufficient to produce stall onset or light dynamic stall. The other two cases are for the same nominal angle of attack forcing, but the reduced frequency has been increased to 0.1 and then to 0.15. Of particular note in these cases is that vortex shedding is delayed with increasing reduced frequency until it finally occurs at the maximum angle of attack achieved in the cycle. This delay in stall onset is, in part, because of the so-called kinematic induced camber effect that is associated with pitch rate. This effect progressively alleviates the leading-edge pressure gradients for a given value of lift and thus delays the onset of flow separation to a higher angle of attack. At the same time, increasing the reduced frequency also delays the onset of flow reattachment, if flow separation occurs. In the third case, it is apparent that a high enough reduced frequency can be attained to prevent flow separation from being initiated at any point in the cycle. If the

[2] When negative aerodynamic torsional damping occurs, there is a possibility of aeroelastic problems on the rotor, including stall flutter. This effect is discussed quantitatively later in this chapter.

mean angle of attack were to be increased further, then a higher reduced frequency would be required to prevent flow separation.

Finally, Fig. 9.6 illustrates the effect on dynamic stall by increasing the Mach number from 0.4 to 0.7, but under the same nominal forcing conditions, that is, for the same approximate angle of attack schedule and reduced frequency. It has been shown previously in Chapter 7 that the effects of compressibility manifest as a lower angle for attack for static stall onset. These effects are carried forth into the dynamic regime, where for the same forcing the degree of stall penetration and amount of hysteresis in the airloads are found to increase with increasing Mach number. Recall that dynamic stall onset is indicated by the break in the pitching moment. In the third case shown in Fig. 9.6, which is for a Mach number of 0.7, the dynamic stall onset involves the participation of a shock wave. This shock wave introduces a more complicated behavior in the center of pressure movement during the flow separation and reattachment process. Beddoes (1983) and Chandrasekhara & Carr (1990, 1994) give a detailed discussion of the role of shock waves in the dynamic stall process.

9.5 Modeling of Dynamic Stall

Mathematical models that attempt to predict the effects of dynamic stall currently range from relatively simple empirical or semi-empirical models to sophisticated computational fluid dynamics (CFD) methods. Because dynamic stall is characterized by large recirculating separated flow regimes, a proper CFD simulation can only be achieved by using the full Navier–Stokes equations with a suitable turbulence model. CFD methods have now begun to show some promise in predicting 2-D and 3-D dynamic stall events – for example, see Srinivasan et al. (1993) and Ekaterinaris et al. (1994). However, the quantitative predictions of the airloads are not yet satisfactory in the stalled regime and during flow reattachment, and especially not at the Reynolds numbers and Mach numbers relevant to helicopter rotors. In this regard, the accurate prediction of the laminar to turbulent boundary layer transition is a key issue. Furthermore, the computational resources for these CFD solutions are prohibitive other than for use as research tools, and for the foreseeable future more approximate models of dynamic stall will have to be used in rotor design work.

Some of the mathematical models of dynamic stall in current use are a form of resynthesis of the measured unsteady airloads, which are based on results from 2-D oscillating airfoils in wind-tunnel experiments. Other so-called semi-empirical models of dynamic stall contain simplified representations of the essential physics using sets of linear and nonlinear equations for the lift, drag, and pitching moment. The nonlinear equations may have many empirical coefficients, which must be deduced from unsteady airfoil measurements. However, the root of these models is usually based on classical unsteady thin-airfoil theory, as discussed in Chapter 8. The development of the nonlinear part of such models are more subjective and require skillful interpretation of experimental data. As a result, most of these models remain in a perpetual state of flux as the level of detail is refined and/or more experimental data become available for formulation and/or correlation purposes.

While semi-empirical models are usually adequate for most rotor design purposes, they usually lack rigor and generality when applied to different airfoils and at different Mach numbers for which 2-D experimental measurements may not be available. Another major problem with some of these types of models is that a significant number of empirical coefficients must be derived. Generally, a set of coefficients for the model must be derived for each and every airfoil, and also over the appropriate range of Mach numbers, assuming measurements available. In cases where experimental measurements are not available, the models cannot be used with the same confidence levels to predict the nonlinear airloads. Other common limitations with these models include the accuracy with which the stall onset can be predicted, that is, the prediction of the combination of unsteady angle of attack and Mach number that produce the onset of dynamic leading-edge flow separation. In these cases, computer coding of the model must be done with extreme care to ensure that logic or conditional branching in the algorithm does not cause nonphysical transients in the predictions of the unsteady airloads, especially if large time (azimuth) steps are involved. This undesirable behavior may produce erroneous predictions of stall and stall flutter, which would be considered unacceptable for rotor design purposes. Therefore, it is important for the analyst to build up a confidence level with any model selected for the design process. While a review of the literature will show a large number of experiments on dynamic stall that could be used for such purposes, the problem is usually that the full range of Reynolds numbers, Mach numbers, reduced frequencies, and, to some extent, airfoil shapes cannot be studied in the same test facility or wind tunnel. Therefore, the same problems and uncertainties in data quality that were discussed in Chapter 8 in regard to comparing static airfoil characteristics also apply in the dynamic case.

9.5.1 *Engineering Models of Dynamic Stall*

Some typical engineering models that have been used (or are currently being used) for modeling dynamic stall and that may be suitable for rotor airloads predictions and rotor design analyses are:

1. UTRC α, A, B Method: This is a resynthesis method, with the approach being described by Carta et al. (1970) and Bielawa (1975). The basis of this method is that in attached flow the airloads can be expressed in terms of the forcing parameters α, $A = \dot{\alpha}c/2V$, and $B = \ddot{\alpha}c^2/4V^2$. In an attempt to isolate the dynamic contributions to the airloads, the static coefficients are subtracted from the total airloads. By appropriate cross-plotting and interpolating for given instantaneous values of these parameters, the contributions to the dynamic airloads can be reconstructed and added to the static values. The method has met with some success, but large data tables must be generated for each airfoil and for each Mach number. A development of the model that obviates the need for large tables is given by Bielawa (1975).
2. Boeing-Vertol "gamma" Function Method: This model was initially developed by Gross & Harris (1969) and Gormont (1973). In the 'gamma' function method, the influence of airfoil motion is determined by computing an effective angle of attack. First, the influence of plunging and pitching effects in attached flow is obtained by applying a "correction" to the angle of attack derived from Theodorsen's theory at the appropriate reduced frequency of the forcing. From this, a second correction is applied from the instantaneous value of the "gamma" function. This gamma function is obtained empirically as a function of Mach number from 2-D oscillating airfoil tests on the appropriate airfoil. The corrected angle of attack is then used to obtain values of the airloads from the static force and moment curves. This has the effect of delaying the onset of stall to higher angles of attack with increasing pitch rate, a result observed experimentally. The pitching moment is obtained from an empirically determined center-of-pressure function. Good predictions of the unsteady airloads are possible with this method [see Harris et al. (1970)], but the quantitative accuracy even for 2-D airfoils is deficient for conditions of stall

onset or light dynamic stall. This model has been used in a large comprehensive rotor analysis for the prediction of helicopter rotor airloads – see Gormont (1973).

3. Time-Delay Method: Beddoes (1976, 1978) has developed a semi-empirical model for dynamic stall. Unlike the previously described resynthesis methods, the philosophy behind this method is to try to model, albeit still in a very simplified manner, the basic physics of the dynamic stall process itself. The method is based in the time domain. The behavior of the airloads in attached flow is obtained from Duhamel superposition using the Wagner indicial response function. Corrections are applied to this function to account for the effects of compressibility. Although scaling the Wagner function in this way is not rigorous, the results obtained approximately replicate the increased lag in the unsteady loads resulting from compressibility effects. The key feature of this model is the use of two nondimensional time delays (based on the semi-chords of airfoil travel). These time delays represent periods of nondimensional time that exist between identifiable dynamic stall flow states. The first time delay represents the delay in the onset of separation after the static stall angle has been exceeded and the time required for the initial separation to develop. The second time delay represents the time during which the leading-edge vortex shedding process occurs. These time delays have been obtained from an analysis of many airfoil tests over a relatively wide range of Mach number. Similar studies have been done by Galbraith et al. (1986) at lower Mach numbers. The results from the time-delay model have been shown to give good predictions of the unsteady airloads on 2-D airfoils. The method also requires relatively few empirical constants.
4. Gangwani's Method: Gangwani (1982, 1984) has developed a synthesized airfoil method for the prediction of dynamic stall. This method is also based in the time domain. To model the airloads in attached flow, a "Mach-scaled" Wagner function with a finite-difference approximation to the Duhamel superposition integral is used, very much in the same manner as for the Beddoes time-delay model. In the nonlinear part of the model, a series of equations with several empirical coefficients are used to represent the forces and moments produced by the various dynamic stall events. These equations are based on the determination of "delayed" angles of attack, with the coefficients being derived from steady and unsteady airfoil data. Although a disadvantage with the method is the relatively large number of equations and empirical coefficients, nearly all of which are derived from oscillating airfoil data, credible predictions of the unsteady airloads have been demonstrated. The capabilities of this model have been independently evaluated by Reddy & Kaza (1997). One of the main difficulties with this method seems to be in predicting flow reattachment after dynamic stall.
5. Johnson's Method: Johnson (1969, 1974) has developed a relatively simple representation for incorporating dynamic stall effects on the sectional airloads. The experimental data of Ham & Garelick (1968) were used to develop the model in a form that could be used to correct the static stall lift and pitching moments as functions of pitch rate. Stall onset was represented by defining vortex shedding to occur just above the static stall angle of attack. It was assumed that vortex shedding produced increments in lift and nose-down moment that increased linearly to a peak value over a finite time, followed by a decay back to the static loads. Reasonable predictions of the unsteady lift seem possible with this method, although the pitching moment predictions appear less pleasing.
6. Leishman–Beddoes Method: Leishman & Beddoes (1986, 1989) have developed a model capable of representing the unsteady lift, pitching moment, and drag

characteristics of an airfoil undergoing dynamic stall. The model has been developed to overcome certain shortcomings of other unsteady aerodynamic models that were available up to about 1980 for use in rotor design and aeroelasticity analysis. The emphasis in this model is on a more complete physical representation of the overall unsteady aerodynamic problem, while still keeping the complexity of the analysis down to minimize computational overheads. The model was initially developed by Beddoes (1983) and Leishman & Beddoes (1986, 1989), with various refinements by Leishman (1989) and Tyler & Leishman (1992). Extensive validation of the model has been conducted with experimental measurements.

This model consists of three distinct subsystems: 1. an attached flow model for the unsteady (linear) airloads, 2. a separated flow model for the nonlinear airloads, and 3. a dynamic stall model for the leading-edge vortex induced airloads. An important feature is that more rigorous representations of compressibility effects are included in the attached flow part of the model. These are represented using the compressible indicial response functions (see Section 8.15), along with linear superposition in the form of more accurate finite-difference approximations to the Duhamel's integral. The treatment of nonlinear aerodynamic effects associated with separated flows are derived from Kirchhoff/Helmholtz theory, which can be used to relate the airfoil lift to the angle of attack and the trailing-edge separation point, a technique discussed previously in Section 7.10.5. In application, the experimental static lift stall characteristics are used with the Kirchhoff/Helmholtz model to define an effective separation point variation that can then be generalized empirically and used to accurately reconstruct the nonlinear airloads for any angle of attack. To represent the effects of dynamic stall, a third subsystem emulates the dynamic effects of the accretion of vorticity into a concentrated leading-edge vortex, the passage of this vortex across the upper surface of the airfoil, and its eventual convection downstream. The dynamic stall process begins when an equivalent leading-edge pressure parameter reaches a Mach number dependent critical value indicative of leading-edge or shock-induced separation. To simulate the effects of the complex viscous dominated changes in the flow topology during dynamic stall, the various time constants that describe the behavior of this third subsystem, and also of the trailing-edge separation point subsystem, are modified in a logically determined sequence. One significant advantage of this method is that it uses relatively few empirical coefficients, with all but four being derived from static airfoil data.

7. ONERA Method: This model describes the unsteady airfoil behavior in both attached flow and during dynamic stall using a set of nonlinear differential equations. The model was first described by Tran & Pitot (1981), Tran & Falchero (1981), and McAlister et al. (1984), with various modifications by Peters (1985). A version of the ONERA model has been evaluated by Reddy & Kaza (1987). A later version of the model (the ONERA Edlin model) is documented by Pitot (1989). The coefficients in the equations of this model are determined by parameter identification from experimental measurements on oscillating airfoils. As in several other models including the Leishman–Beddoes model, the airloads are expressed as a sum of two components: a component associated with the linear (attached flow) behavior and an increment that represents a deviation from the linear value resulting from stall. The model requires 22 empirical coefficients. The later ONERA BH model, as documented by Truong (1993, 1996), requires 18 coefficients and also adapts the Kirchhoff/Helmholtz trailing-edge separation scheme from the Leishman–Beddoes model.

Generally, reasonable predictions of the unsteady airloads are obtained with the ONERA model – see also Tan & Carr (1996) and Nguyen & Johnson (1998). However, like many of the models, the predictions are deficient for flow reattachment after dynamic stall. Based on the experimental studies of Green & Galbraith (1995), the modeling of dynamic flow reattachment requires as much care as for modeling dynamic flow separation, especially if accurate predictions of aerodynamic torsional damping are an objective. One advantage of the ONERA method is that because all of the equations for the lift, drag, and pitching moment are written as differential equations, they are in a form that can be immediately useful for various types of aeroelastic analysis. However, the Leishman–Beddoes model has also been cast into differential equation form – see Leishman & Nugyen (1990) and Leishman & Crouse (1989). Furthermore, it would seem that the structure of some of the other semi-empirical dynamic stall models may also lend themselves to be cast into the same mathematical form, if required by the analyst.

9.5.2 *Capabilities of 2-D Dynamic Stall Modeling*

Based on the foregoing, there are several dynamic stall models available to the helicopter rotor analyst, all of which represent, to an engineering level of approximation, the forces and moments produced on an airfoil during dynamic stall. Reddy & Kaza (1987) have compared and contrasted several of these models, from which the general quantitative capabilities and deficiencies were independently documented. Used intelligently, it would seem that most of these models are adequate for use in a wide variety of rotor analyses. Much comes down to the confidence levels that can be established through correlation studies, both at a 2-D level as well as inside the rotor environment. Tan & Carr (1996) have presented a summary of results for a number of currently used semi-empirical dynamic stall models, as well as first-principles based CFD approaches to the dynamic stall problem. Results have been compared for both an oscillating 2-D airfoil and an idealized 3-D problem in the form of an oscillating cantilevered finite wing (see Section 9.9). Again, it seems that if the semi-empirical models are used intelligently, very credible predictions of the integrated airloads can be obtained when compared to experimental measurements.

In the following discussion, predictions made by the Leishman & Beddoes (1986, 1989) model will be used to show the general capabilities of the semi-empirical dynamic stall models. While it should be appreciated that some models might produce different or better results, the idea here is simply to illustrate the overall levels of performance that could be expected from these types of engineering models.

It is highly desirable to expose the models to different types of forcing, that is, not solely for oscillatory pitching as might have been used for formulation purposes. As mentioned previously, a large proportion of the aerodynamic changes in angle of attack that take place in the rotor environment come from blade flapping, which is equivalent to a plunging or heaving type of forcing at the blade element level. One underlying assumption that seems to be made in nearly all the various unsteady aerodynamic models described above is that the effects of blade motions and wake inflow variations can be adequately represented by a lumped "equivalent" angle of attack. This proves an adequate assumption for fully attached flows, but Fukushima & Dadone (1977) and Ericsson & Reding (1983) have postulated that more fundamental differences may exist in the dynamic stall airloads when different modes of motion are imposed (i.e., pitching versus plunging). This perhaps raises some questions about the general capabilities of the various dynamic stall methods to predict accurately the unsteady airloads for completely arbitrary variations in angle of attack, pitch rate, etc. However, while many 2-D oscillating airfoil experiments have been conducted to study dynamic stall, only a few experimental facilities can simulate both pitching and plunging oscillations or other combined motions so that this can be verified. This is mainly because plunge experiments are mechanically difficult to perform in a wind-tunnel environment, especially over the wide range of amplitudes and reduced frequencies that would be necessary to validate any model. Liiva et al. (1968) and Carta (1979) have measured oscillatory airloads under oscillatory pitch and plunging conditions. However, only the results obtained by Liiva et al. (1968) are for a range of Reynolds numbers and Mach numbers that are representative of those found on a helicopter rotor.

Figure 9.7 shows the normal force and pitching moment responses obtained for oscillatory pitching and plunging into dynamic stall. Note that the general features of dynamic stall are evident here, with both results showing a qualitative similarity for both types of forcing. The hysteresis loops predicted by the dynamic stall model are in good agreement with the experimental measurements – the only significant discrepancy is during the flow reattachment process. In both cases, stall onset is clearly postponed well below the static stall angle of attack, followed by leading-edge vortex shedding. Both conditions suggest that an organized vortex is formed and transported downstream over the chord, although for the plunge case, the degree of stall penetration is slightly less.

Figure 9.8 shows dynamic stall loops for pitching and plunging oscillations performed at a higher mean angle of attack. These particular results are for conditions such that the minimum angle of attack is sufficiently high that the flow can only reattach to the airfoil surface during the latter part of the downstroke of the motion. The agreement of the model with test data is again very good, for both pitch and plunge oscillations. The pitching moment response shows that stall onset is predicted a little early for pitch and later for plunge; in a practical sense these differences are small, however. The maximum nose-down pitching moment for the pitching airfoil is overpredicted, but it is predicted better for the plunging airfoil.

Carta (1979) has concluded that dynamic stall occurred on the airfoils during certain pitch oscillation cases but not in the corresponding plunge cases, even though the same equivalent angle of attack history was imposed. It appears that this behavior can, in part, be traced to the (inviscid) pressure distribution at the leading edge of the airfoil. For pitch oscillations the favorable "induced camber" effect discussed previously means that, in principle, the leading edge pressure conditions that delimit attached flow should be met at a lower equivalent angle of attack than for the plunge oscillations. Tyler & Leishman (1992) have confirmed that the stall onset behavior is related to the additional effect of pitch rate contribution to the unsteady airloads during pitch oscillations. Therefore, for "equivalent" conditions the degree of stall penetration and lift and moment hysteresis should be somewhat greater for an equivalent pitch oscillation, confirming Carta's observation.

Ericsson & Reding (1983, 1984) have put forward an alternative theory, postulating that the differences in the airloads seen between dynamic stall in oscillatory pitching and plunging is a result of two viscous phenomena: the "spilled" leading-edge vortex but also another effect called the "leading-edge jet effect." While the leading-edge vortex shedding phenomenon has been well documented experimentally, the role of the leading-edge jet effect in the dynamic stall problem is much less clear. Ericson & Reding suggest that this "leading-edge jet effect" helps to delay the onset of leading-edge separation on a pitching airfoil by producing a fuller boundary layer profile. The final result is, however, opposite to that inferred from both Liiva's data and Carta's data where it can be seen that for nominally "equivalent" pitch/plunge forcing a pitching airfoil will generally stall at a lower equivalent angle of attack than for the plunging airfoil.

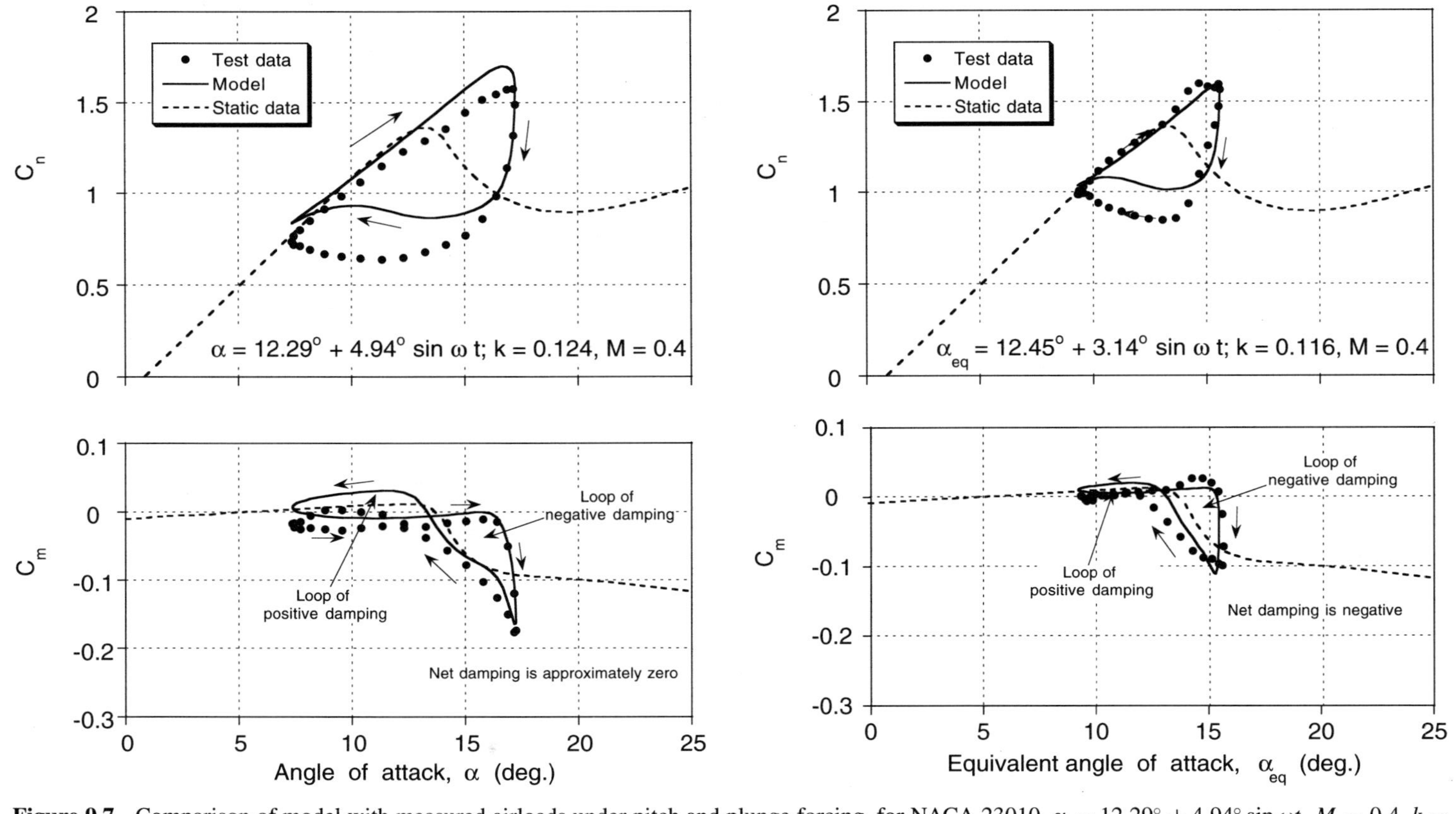

Figure 9.7 Comparison of model with measured airloads under pitch and plunge forcing, for NACA 23010, $\alpha = 12.29^\circ + 4.94^\circ \sin \omega t$, $M = 0.4$, $k = 0.124$; $\alpha_{eq} = 12.45^\circ + 3.14^\circ \sin \omega t$, $M = 0.4$, $k = 0.116$.

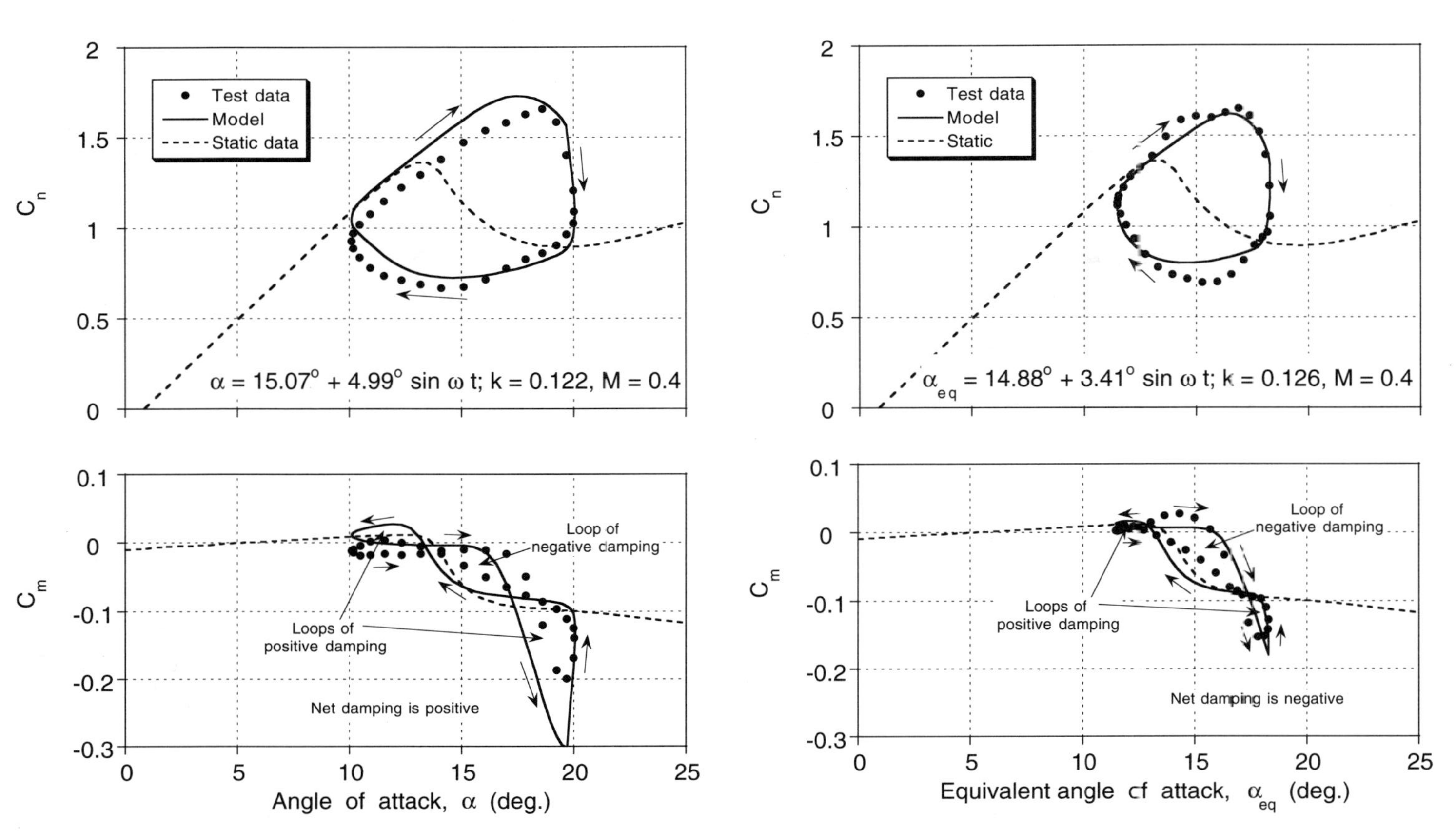

Figure 9.8 Comparison of model with measured airloads under pitch and plunge forcing, for NACA 23010. $\alpha = 15.07^\circ + 4.99^\circ \sin \omega t$, $M = 0.4$, $k = 0.122$; $\alpha = 14.88^\circ + 3.41^\circ \sin \omega t$, $M = 0.4$, $k = 0.126$.

Overall, it can be seen from the results shown in Figs. 9.7 and 9.8 that the dynamic stall model provides fairly satisfactory predictions of the unsteady airloads, for both pitching and plunging oscillations. Again, these results can only be considered as representative, and other semi-empirical models may provide better results. None of the current models, however, can be considered as the last word on the problem, and it is likely that improved semi-empirical models of dynamic stall will continue to be developed for helicopter rotor applications.

9.6 Torsional Damping

As mentioned previously, the highly nonlinear airloads obtained by operating the rotor in proximity to retreating blade stall can introduce aeroelastic stability problems. One problem is called *stall flutter*, and this occurs when negative aerodynamic torsional damping begins to convert an otherwise stable aeroelastic blade mode into a divergent or high amplitude limit cycle oscillation. A torsional damping factor, $D.F.$ or C_W can be defined by the line integral

$$D.F. = C_W = \oint C_m(\alpha) d\alpha, \tag{9.1}$$

which is *positive* when it corresponds to a *counterclockwise* loop in the C_m versus α curve – see Carta (1967). If the torsional damping is negative, this would tend to promote an aeroelastic divergence or flutter.

The progressive change from positive to reduced or negative torsional damping (for a given amplitude and reduced frequency of oscillation) is apparent as the mean angle of attack is increased, as previously illustrated by the results shown in Fig. 9.4. It has been shown that the onset of light dynamic stall introduces a second clockwise loop of *negative* damping so that the moment curve now looks like a figure eight. This second loop then grows in size with increasing mean angle of attack. Further penetration into dynamic stall produces a moment break that is now early enough in the oscillation cycle that the peak nose-down moment occurs while the angle of attack is still increasing. This introduces another loop in the counterclockwise (positive) sense. By this mechanism, more positive torsional damping is restored when deep stall penetration occurs. Increasing the reduced frequency can act to delay the onset of stall to a higher angle of attack and can suppress the amount of flow separation, thereby reducing the negative damping, as shown by Fig. 9.5.

The aerodynamic torsional damping of an airfoil in the dynamic stall regime is a very difficult quantity to predict. While the various semi-empirical dynamic stall models appear to give good predictions of the net forces and pitching moments, they do not always give good estimates of torsional aerodynamic damping. Because damping is an integrated quantity, the combination of conditions that determine dynamic stall onset and phasing of the center-of-pressure movement during vortex shedding must be determined very accurately. Measurements of the overall torsional damping trend versus mean angle of attack, α_0, is shown in Fig. 9.9, for both pitch and plunge forcing. The damping is normalized by the theoretical damping as given by the classical incompressible flow unsteady aerodynamic theory. Using the results in Section 8.7, then for pitching oscillations

$$C_m = \frac{-1}{2V^2}\left(\pi V b\dot{\alpha} + \frac{3}{8}\pi b^2 \ddot{\alpha}\right) \tag{9.2}$$

$$= \frac{\pi}{2}k\left(\frac{3}{8}k - i\right)\bar{\alpha}e^{i\omega t} \tag{9.3}$$

$$= \bar{C}_m \sin(\omega t + \phi), \tag{9.4}$$

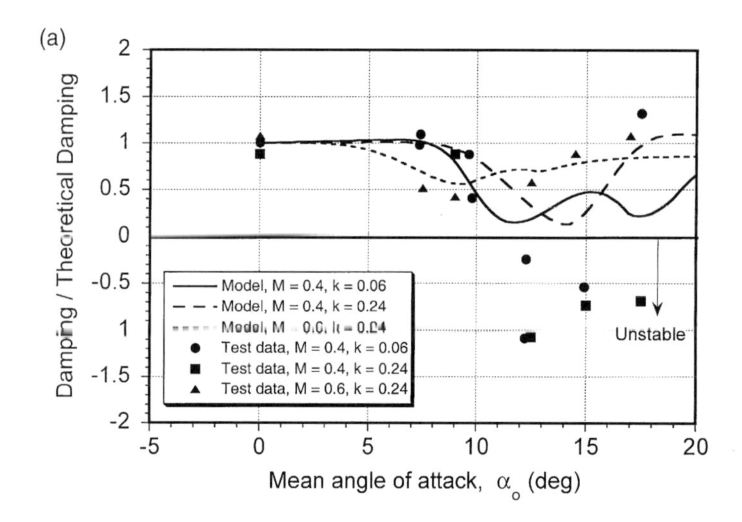

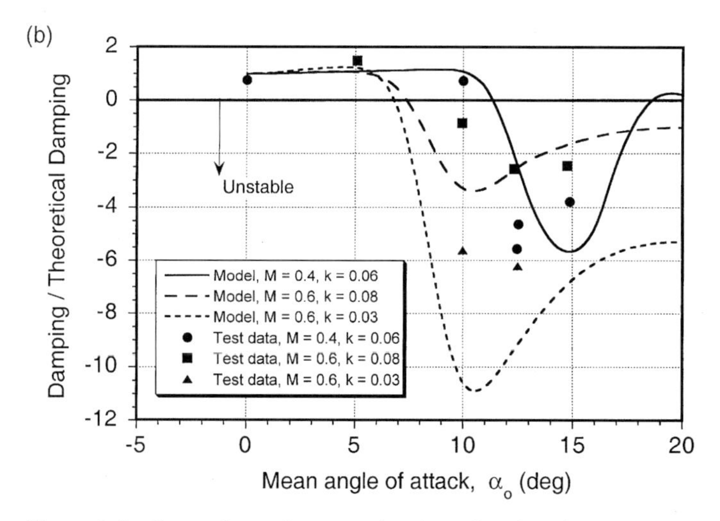

Figure 9.9 Comparison of measured and predicted torsional aerodynamic damping for (a) pitch oscillations and (b) plunge oscillations.

where

$$\bar{C}_m = \frac{\pi}{2} k\bar{\alpha}\sqrt{1 + \frac{9}{64}k^2} \quad \text{and} \quad \phi = \sin^{-1}\left(-\frac{8}{3k}\right). \tag{9.5}$$

Substituting this expression into Eq. 9.1 gives

$$C_W = \pi k\bar{\alpha}\sqrt{1 + \frac{9}{64}k^2}. \tag{9.6}$$

By a similar process, the theoretical torsional damping for plunging motion is found to be

$$C_W = \frac{\pi}{2} k\bar{\alpha}_{\text{eq}}. \tag{9.7}$$

where $\bar{\alpha}_{\text{eq}}$ is given by Eq. 8.175. These results show that in fully attached flow the torsional damping for an airfoil undergoing pitching motion is approximately twice that for equivalent plunging motion. This is because of the effects of pitch rate contributions to the noncirculatory moment, which causes the area contained within the pitching moment loop to be greater for a pitching airfoil than for a plunging airfoil under the same equivalent angle of attack forcing conditions. This result is also clearly seen in the experimental measurements on oscillating airfoils, as shown previously in Chapter 8 (Figs. 8.27 and 8.28).

The explanation for the observed behavior of the torsional damping can be deduced from Fig. 9.4, where it has been shown that for higher mean angles of attack the pitching moment develops into two loops at dynamic stall onset, with a decrease in torsional damping. With further increase in mean angle of attack, the second loop increases in size and the net damping during the cycle rapidly decreases. Figure 9.9 shows that for both the pitch and plunge cases very low or negative damping is obtained at dynamic stall onset. At some point, just after stall onset, the mean angle of attack is sufficient such that significant negative damping reoccurs for the pitch oscillation. Although stall onset occurs at a lower equivalent angle of attack for an airfoil oscillating in pitch, the inherently lower damping for the plunging airfoil in attached flow means that the damping may well become negative at a lower mean angle of attack for conditions of oscillatory plunging. As the mean angle of attack is increased further, the damping increases again and ultimately becomes positive when fully separated flow conditions exist. See Carta (1967) for a detailed discussion of torsional damping under dynamic stall conditions.

9.7 Effects of Sweep Angle on Dynamic Stall

The local sweep or yaw angle at a blade element on a helicopter rotor in forward flight can be significant. The radial component of the velocity relative to the leading edge of the blade is the source of this sweep angle, as shown in Fig. 9.10. The sweep angle, Λ, is defined in terms of the normal and radial velocity components U_T and U_R, respectively, by

$$\Lambda(r, \psi) = \tan^{-1}\left(\frac{\mu\cos\psi}{r + \mu\sin\psi}\right) = \tan^{-1}\left(\frac{U_R}{U_T}\right). \tag{9.8}$$

Examples of the iso-sweep angle distribution over the rotor disk are shown in Fig. 9.11 for advance ratios of 0.05 and 0.3. At the higher advance ratios, the sweep angles can exceed 30°

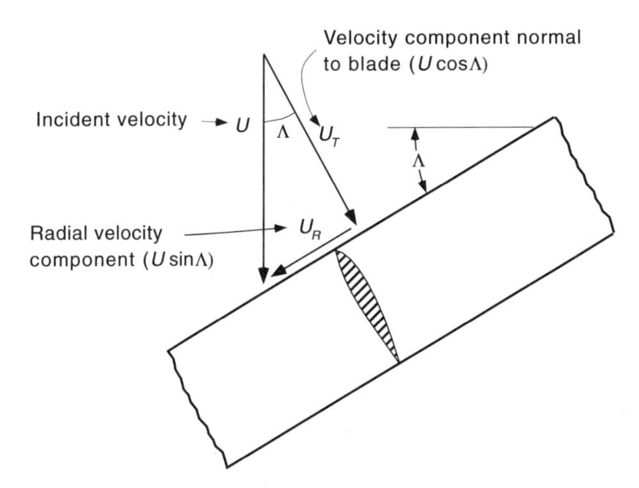

Figure 9.10 Velocity components relative to the blade element and definition of sweep angle.

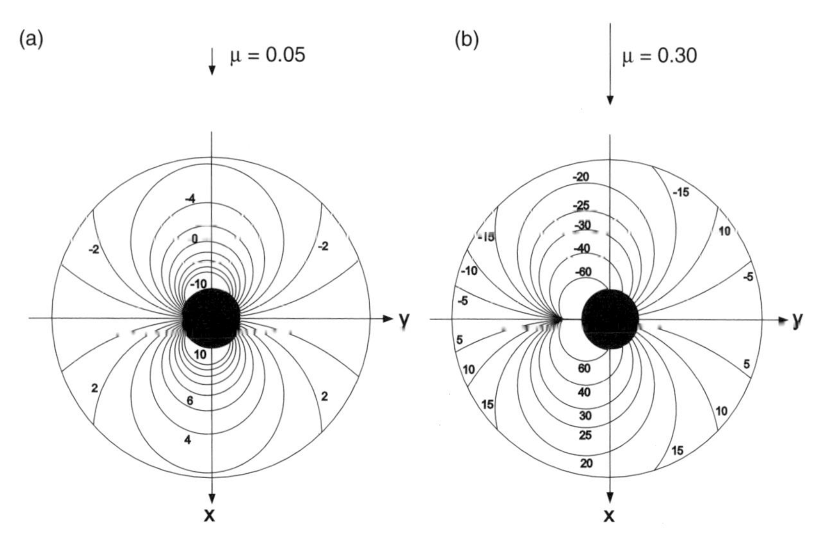

Figure 9.11 Iso-sweep angles over the rotor disk in forward flight. (a) $\mu = 0.05$. (b) $\mu = 0.30$. Angles are in degrees.

over some parts of the disk. Overall they clearly become significant enough that their effects would need to be assessed experimentally. Any effects on the aerodynamics so produced will also need to be represented and properly integrated within a model of dynamic stall.

In the classical blade element theory, one usually neglects the effect of sweep on the lift and pitching moments. This is in accordance with the independence principle of sweep – see Jones & Cohen (1957). However, when an airfoil is operated at high angles of attack near stall this may not be a valid assumption. For example, the effect of sweep angle on the *static lift* characteristics of a swept 2-D airfoil is shown in Fig. 9.12. The results are based on the measurements of Purser & Spearman (1951) and are presented in the conventional blade element format – that is, in terms of angle of attack and velocities normal to the leading edge of the airfoil section. When the data are presented this way, the results for the lift, pitching moment, and drag show good correlation and confirm that, at least in the attached flow regime, the independence principle is a valid assumption. However, note that in the high angle of attack region a much higher lift coefficient is obtained for the larger sweep angles. This is because of favorable effects on the spanwise development of the boundary layer, which tend to delay the onset of flow separation on the wing to a higher angle of attack – see also Dwyer & McCroskey (1971). Similar results have been found experimentally by St. Hillaire et al. (1979) and St. Hillaire & Carta (1979, 1983a,b).

The upshot of these observations is that if these sweep effects are carried forth into the rotor regime, then they will tend to delay the onset of stall on a rotor to higher values of thrust. Various earlier studies of the problem, including those by Harris (1966) and Gormont (1973), suggest that improvements in rotor thrust prediction can be obtained by including a model that accounts for a delayed stall angle of attack and higher maximum lift as a function of sweep angle. However, it must be remembered that when stall occurs on the rotor in forward flight, the stall is actually dynamic in nature. The fundamental question is whether the increase in static lift shown in Fig. 9.12 is also carried forth into the dynamic stall regime.

To assess this problem, a comprehensive series of experiments has been conducted by St. Hillaire & Carta (1979, 1983a) where a NACA 0012 airfoil at a constant sweep

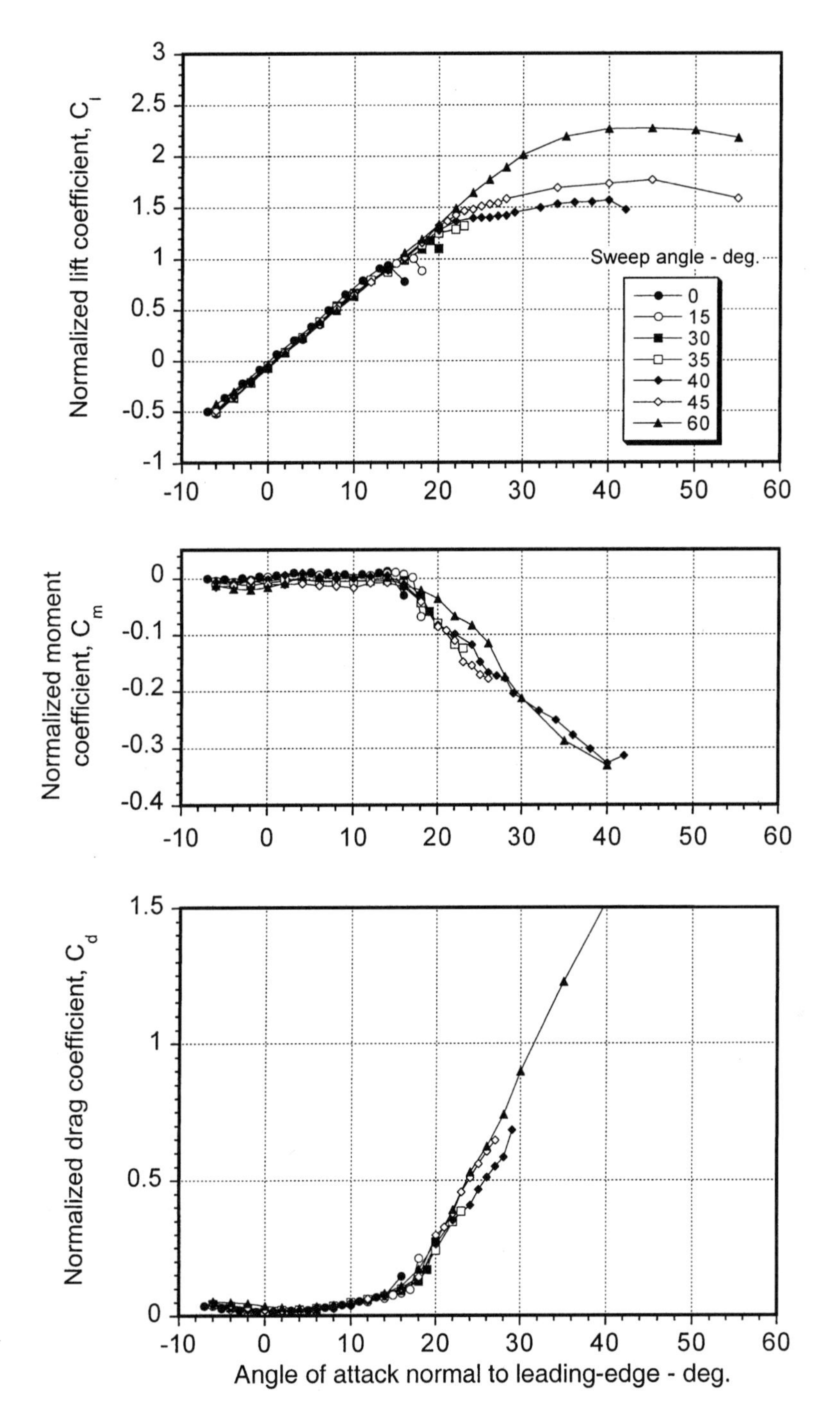

Figure 9.12 Effect of sweep angle on the measured static lift characteristics of an airfoil. The results are presented in the conventional blade element convention.

angle of 30° was used. These oscillating airfoil tests were performed at Mach numbers of 0.3 and 0.4 and Reynolds numbers that are nominally full rotor scale. The experimental airloadings were determined by the integration of sectional pressures measured by pressure transducers distributed about a section at the midspan of the model. Leishman (1989) has also conducted an analysis of these data. It appears that for fully attached unsteady flows the independence principle also applies. Any unsteady effects associated with sweep are small in the attached flow regime – probably smaller than uncertainties associated with the measurements themselves. However, in the dynamic stall regime, there are other characteristics of swept flow that are worthy of consideration.

Figure 9.13 shows a typical behavior of the lift and pitching moment in the dynamic stall regime for a pitch oscillation. A feature of these dynamic stall results is that compared with the static case where approximately a 20% higher maximum C_n was attained for $\Lambda = 30°$, the unsteady case shows a delay in dynamic lift stall to a higher angle of attack, but not to a significantly higher maximum value of lift. Also, for $\Lambda = 30°$ somewhat narrower lift hysteresis loops are produced, and the mean value of lift is somewhat higher. Thus, it appears that the effects of sweep on the rotor may serve to provide an overall increase in rotor thrust compared to predictions obtained when sweep effects are not included. It

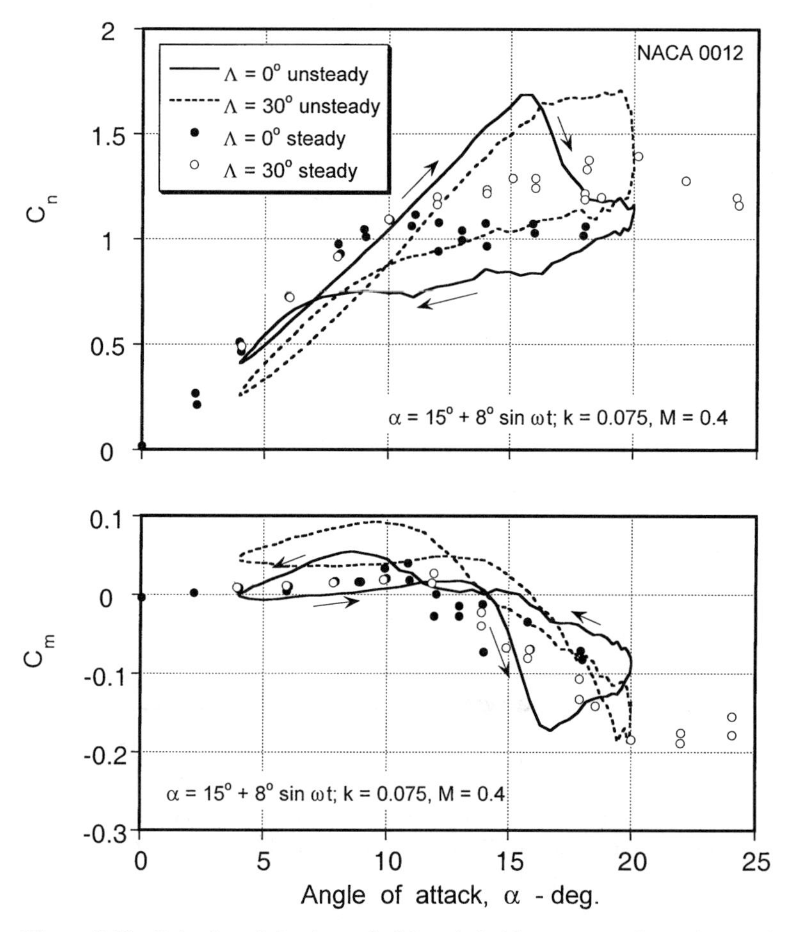

Figure 9.13 Behavior of the dynamic lift and pitching moment for a pitch oscillation in swept and unswept flow.

should be noted that the expected increase in thrust is not because of higher mean values of C_n, but simply because the mean value of the lift is higher. The divergence in the pitching moment (moment stall) occurs at the same nominal value of angle of attack for both the $\Lambda = 0°$ and $\Lambda = 30°$ cases. However, the slope of the moment curve during the next part of the cycle is clearly less for the $\Lambda = 30°$ case. Also, the minimum pitching moment is reached at a higher angle of attack. This suggests that the delay in dynamic lift stall to a higher angle of attack in swept flow is due, in part, to a lower velocity at which the shed leading-edge vortex is convected over the chord. The comprehensive analysis performed on the airfoil pressure time histories by St. Hillaire & Carta (1983a) also supports this observation. In light of these results, it would appear that from a modeling perspective simple corrections to the stall angle of attack based on steady flow observations, such as suggested by Harris (1966), may give the desired effect on rotor performance predictions, but for the wrong underlying reasons. This problem, like several others in rotating-wing aerodynamics, illustrates the difficulties in simply extrapolating observations of quasi-2-D steady airfoil behavior to the complicated flow environment found on rotors.

9.8 Effect of Airfoil Shape on Dynamic Stall

McCroskey et al. (1980, 1982) have studied systematically the effects of airfoil shape on the dynamic stall characteristics of several airfoils. Some of these airfoils are unsuitable for use on helicopters, but the results serve to bracket the effect of airfoil shape on the problem, albeit only at relatively low Mach numbers. Wilby (1984, 1996, 1998) reports another study of airfoil shape on the dynamic stall problem and over a much wider range of Mach numbers, including transonic flow. This author presents a tantalizing glimpse of a wealth of information on the dynamic stall problem; however, it is unfortunate that very little of these data have been formally published in the open literature. The datum airfoil used by McCroskey et al. (1980, 1982) was the ubiquitous NACA 0012, and results are summarized here for two other airfoils that are representative of modern rotor airfoils used on current production helicopters, namely the HH-02 airfoil (used on the AH-64 Apache) and the SC1095 airfoil (used on the UH-60 Blackhawk). These are both cambered airfoils with approximately 9.5% thickness to chord ratios and can be considered representative of modern helicopter airfoil designs. The HH-02 has considerably more leading-edge camber than the SC1095, and has the distinction of a large trailing-edge tab (Fig. 7.23).

The results of airfoil shape on the dynamic stall airloads are summarized in Fig. 9.14, which shows the normal force and pitching moment for stall onset conditions through deep dynamic stall. In each case, the results are compared to the static airloads for that airfoil. The test data shown are for a free-stream Mach number of 0.3, which is the highest Mach number that was tested. At the lowest mean angle of attack of 5°, the maximum angle of attack becomes just large enough to initiate some minor leading-edge separation, as evidenced by the distortion in the nominally elliptical hysteresis loops near the maximum angle of attack. All three airfoils exhibit a significant increase in maximum lift over the static values. It is clear, however, that the HH-02 and SC1095 airfoils maintain attached flow to slightly higher angles of attack, with correspondingly higher values of C_n, and, thereby, exhibit a slightly superior lifting performance to the NACA 0012. This is consistent with the static behavior of these airfoils (see also Chapter 7 and Fig. 7.23). Therefore, it can be concluded that airfoils designed for high static lift capability should also exhibit a higher angle of attack capability before stall when operated under dynamic conditions.

The forces and pitching moments for a case of moderately strong dynamic stall, which occurs for a mean angle of attack of 10°, are also shown in Fig. 9.14. Under these conditions,

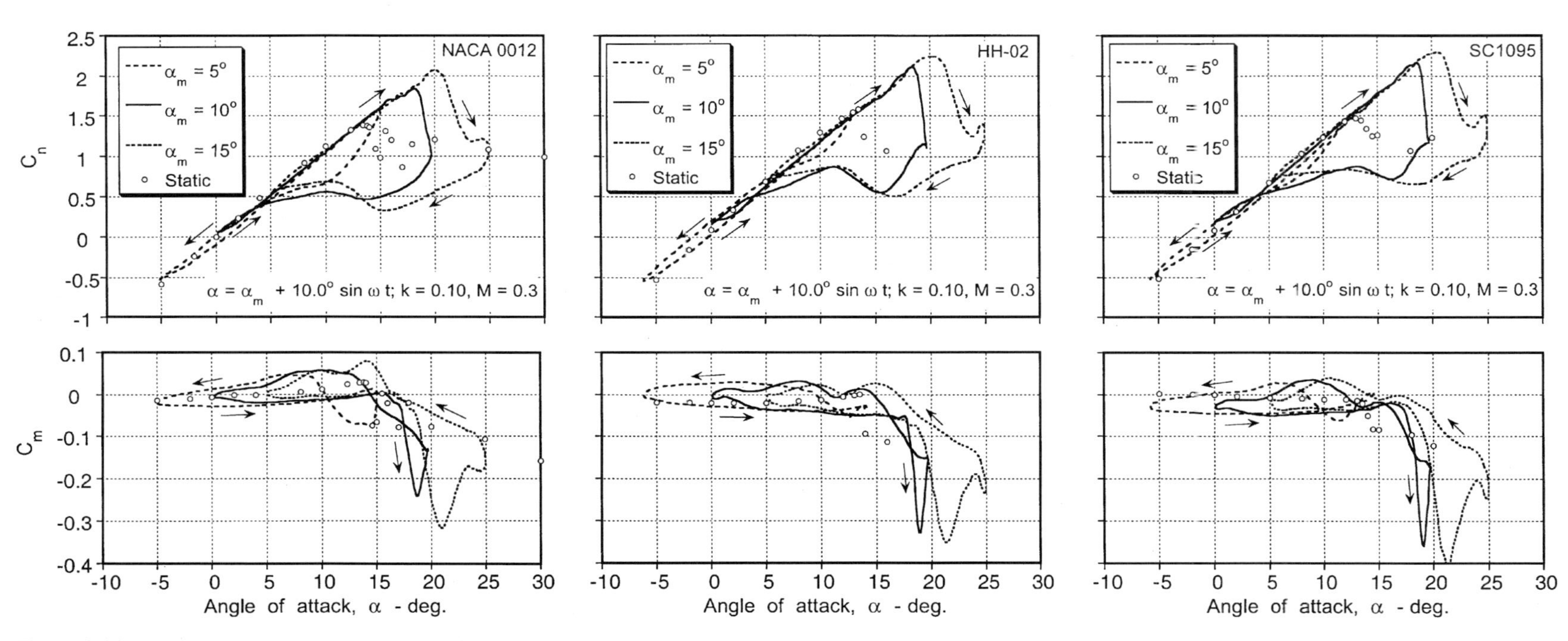

Figure 9.14 Unsteady airloads on NACA 0012, HH-02, and SC1095 airfoil sections.

relatively strong leading-edge vortex shedding is initiated and the characteristic lift overshoots and strong nose-down pitching moment behavior of dynamic stall are intensified. Considerable hysteresis in the lift and moment behavior is also present for these conditions. All three airfoils exhibit a qualitatively similar type of behavior, although it is apparent that there are measurable quantitative differences. The NACA 0012 exhibits moment stall at a lower angle of attack to either the HH-02 or the SC1095 airfoils, although the lift stall occurs at approximately the same angle of attack for all three airfoils. Both the NACA 0012 and the SC1095 airfoils exhibit a well-rounded moment break at the onset of dynamic stall in comparison to the HH-02, which has a very abrupt moment break. This suggests that some trailing-edge separation is still present on the NACA 0012 and SC1095 prior to the onset of leading-edge separation and dynamic stall. This also suggests that to some extent the static stall behavior of the airfoil is actually carried over into the dynamic stall regime. Both the HH-02 and SC1095 airfoils exhibit a slightly greater maximum dynamic lift over the NACA 0012. Again, these static lift gains appear to be carried over somewhat into the dynamic regime. However, the NACA 0012 clearly exhibits a smaller peak value of (nose-down) pitching moment compared to the other two airfoils. This suggests a weaker shed leading-edge vortex for the NACA 0012 airfoil.

The traditional approach in designing helicopter rotor airfoils is to maximize the quasi-steady lift and minimize the pitching moment. Little emphasis is usually placed on the consequences of the unsteady behavior of the airfoil. Figure 9.14 reinforces this point, where the oscillatory lift and moment for a mean angle of attack of 15° indicate strong leading-edge vortex shedding, producing significant increments in normal force and pitching moment. As for a mean angle of attack of 10°, all three airfoils exhibit a qualitatively similar type of dynamic stall behavior, with both the HH-02 and the SC1095 exhibiting increased values of maximum dynamic lift over the NACA 0012. (It should also be noted that for each airfoil there is evidence of secondary vortex shedding near the maximum angles of attack, which manifest as smaller secondary peaks in the normal force and pitching moment). It is significant that while under static conditions the SC1095 exhibits a gain in maximum C_n of about 0.1 over the HH-02 airfoil, under these particular dynamic conditions there is almost no difference in maximum C_n between these two airfoils. This indicates that whereas the maximum lift coefficient may be a useful measure of airfoil performance under static conditions, this does not necessarily appear to be an indication of the dynamic lift capability of the airfoil. Also, although the HH-02 and SC1095 give approximately the same value of maximum dynamic lift, the maximum nose-down pitching moment is clearly greater for the SC1095. This is despite the fact that the HH-02 has a higher zero-lift pitching moment under quasi-steady conditions. Therefore, like the lift coefficient, the design of airfoils for low static pitching moments does not necessarily guarantee that low dynamic pitching moments will also be produced.

9.9 Three-Dimensional Effects on Dynamic Stall

As discussed earlier in this chapter, in the rotor environment the problem of dynamic stall must be considered as fully three dimensional. Stall will occur over different parts of the blade and at different blade azimuth angles, and needless to say, the resulting flow can be very complicated. Apart from the three-dimensional effects associated with swept flows, the problem of three-dimensional unsteady separating flows are still poorly understood. As mentioned earlier, the various mathematical models used for dynamic stall prediction are still heavily empirical, with reliance being placed almost exclusively on oscillating 2-D airfoils for formulation and validation purposes. It is only recently that attempts to validate these models has been made for 3-D dynamic stall problems, albeit these are still idealized problems compared to those found in the rotor environment.

The problem of three-dimensional dynamic stall has been studied experimentally by Lorber et al. (1991), Lorber (1992), and Pizialli (1994). Both experiments have used a cantilevered semi-span wing that was oscillated in angle of attack through stall and, like the 2-D tests, was designed to simulate the variations in angle of attack encountered by a helicopter rotor blade. Miniature pressure transducers were distributed at various stations along the wing, and the measured pressures have been integrated to determine sectional lift, drag, and moment. While Pizialli's results are limited to a Mach number of 0.3, Lorber's results cover a wider range of Mach number from 0.2 to 0.6 and also for wing sweep angles of 0, 15, and 30 degrees.

Representative results from the experiments of Lorber et al. (1991) are shown in Figs. 9.15 and 9.16. These results document the unsteady lift and moment at five spanwise stations from near the midspan out toward the tip of the wing. Results are shown for an oscillation

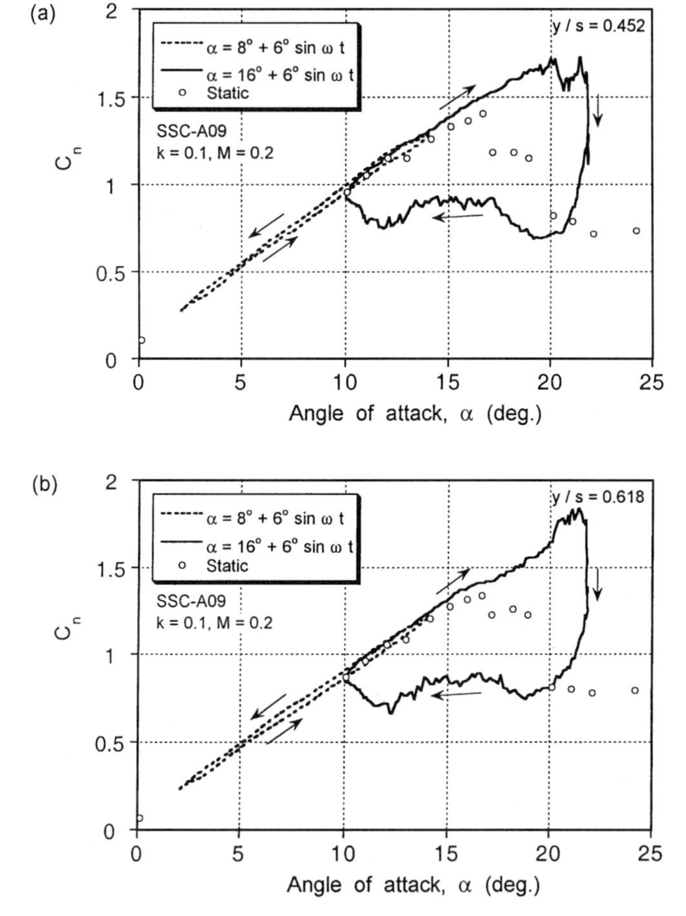

Figure 9.15 Measurements of unsteady lift on a cantilevered wing undergoing oscillations in angle of attack for nominally attached flow and into dynamic stall. $M = 0.2$.

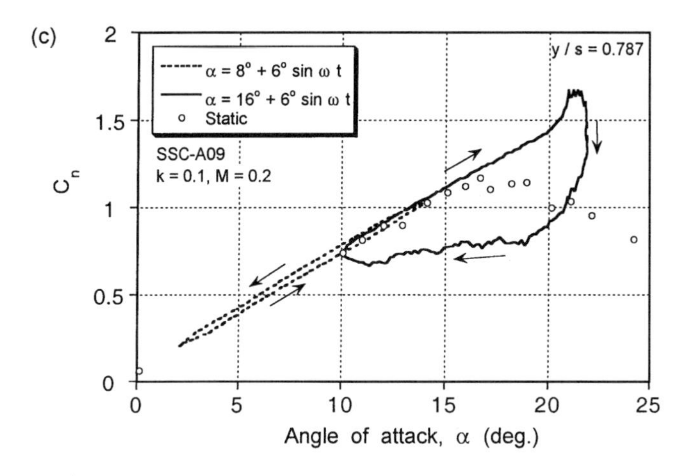

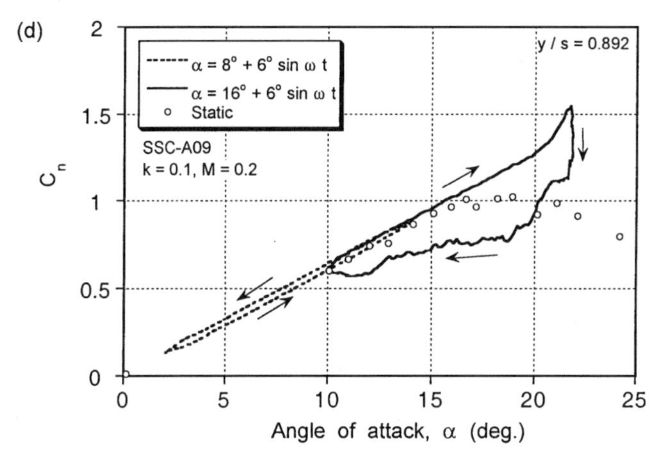

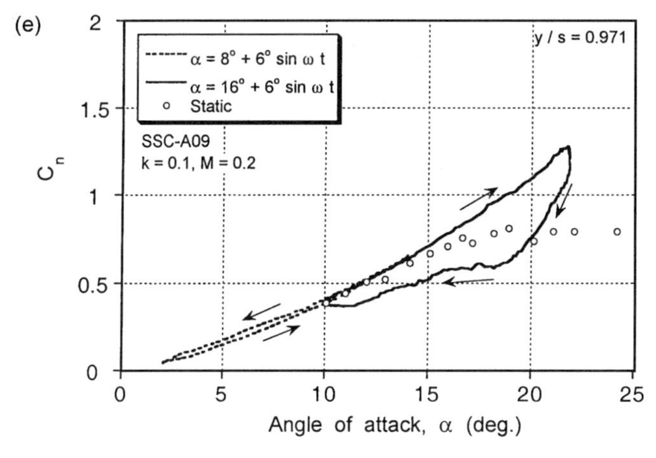

Figure 9.15 *(continued)*

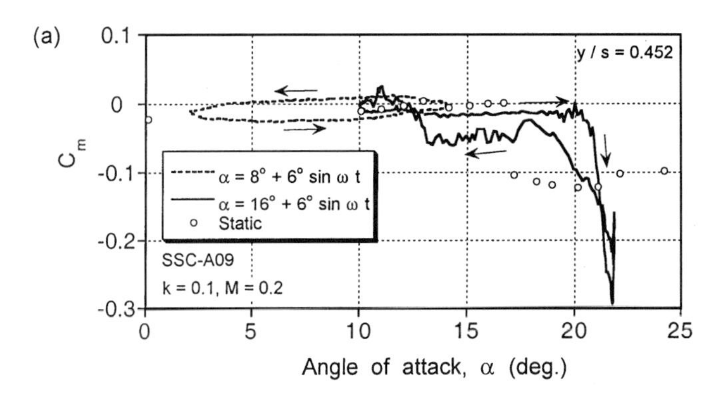

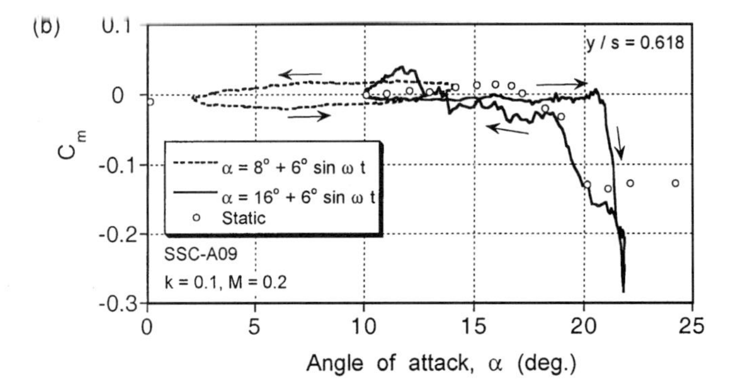

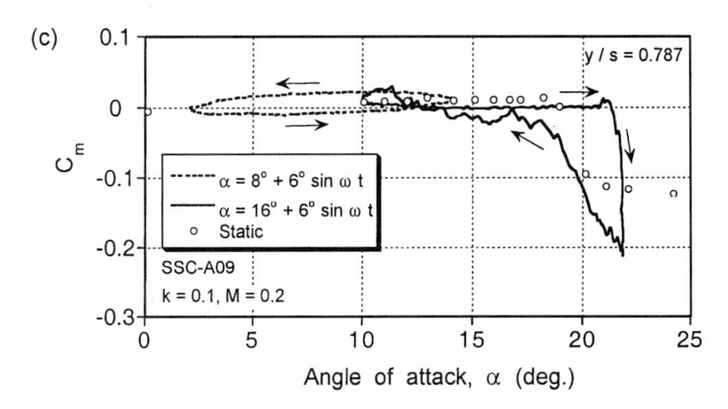

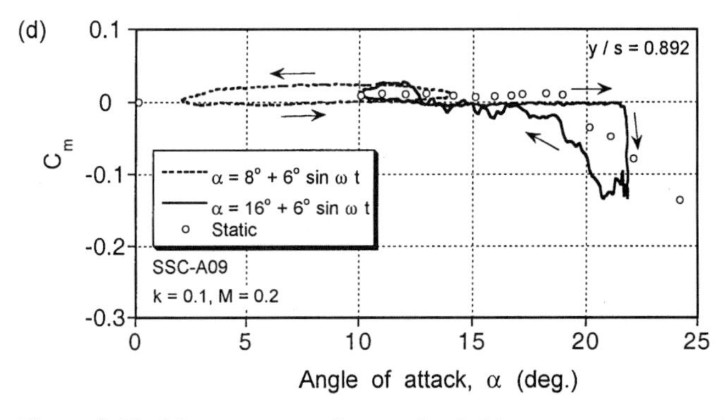

Figure 9.16 Measurements of unsteady pitching moment on a cantilevered wing undergoing oscillations in angle of attack for nominally attached flow and into dynamic stall. $M = 0.2$.

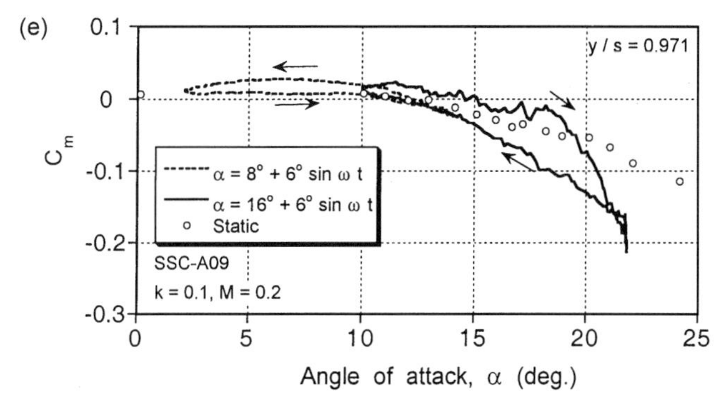

Figure 9.16 *(continued)*

below stall and for a typical case of dynamic stall at the same reduced frequency. The static measurements at each station of the wing are also shown for reference. For oscillations below stall, characteristic elliptical hysteresis loops are formed with no particularly unusual behavior compared to that expected from 2-D considerations. Note, however, the gradual reduction in the average lift-curve-slope when moving outboard toward the tip, although this is a quasi-steady effect and would be predicted by any finite-wing model. The only exception is at the outermost wing station, where both the steady and unsteady lift and moment hysteresis loops show a different characteristic, with a more nonlinear quasi-steady behavior. This is the influence of the tip vortex, which lies over the tip region and provides an element of steady "vortex" lift. Other than for this one section of the wing, unsteady thin-airfoil theory, with the quasi-steady induced effects accounted for by a lifting-line or other finite-wing method, will provide a good approximation to the unsteady airloads – see Tan & Carr (1996).

The dynamic stall characteristics on 3-D finite wings are noted to be qualitatively similar to those found on oscillating 2-D airfoils. Figures 9.15 and 9.16 show that at the four inner stations, the airloads exhibit the lift overshoots, large nose-down pitching moments, and hysteresis effects that are characteristic of 2-D dynamic stall. Note that when moving outboard from the innermost station the degree of dynamic stall penetration is reduced. Again, this is mainly a steady effect associated with the reduction in the effective quasi-steady angle of attack because of the induced effects from the tip vortex. In fact, the results appear qualitatively similar to those obtained with a reduction in the mean angle of attack in the 2-D case. While lift and moment overshoots are also obtained at the outermost station, the results are less transient suggesting that leading-edge vortex shedding does not occur. In fact, it would appear that the tip vortex alone dominates the flow field here and the angle of attack never becomes large enough to permit stall to occur.

Overall, the results indicate that the oscillating finite-wing problem can still be considered nominally two dimensional as far as dynamic stall is concerned, and 2-D models with appropriate allowance for the induced angles of attack for finite spans should give at least engineering levels of predictive capability. This has been confirmed by results shown by Tan & Carr (1996), who compare results of the Piziali experiments with a number of the better known semi-empirical dynamic stall models, some of which have been described previously in Section 9.5.1. However, this conclusion is based only on a limited validation study, and much further work still needs to be done to analyze the problem of 3-D dynamic stall if predictions of helicopter airloads are to be improved.

9.10 Time-Varying Velocity Effects

As previously discussed in Chapter 8, in forward flight a blade element will encounter a time-varying incident velocity and there will be additional unsteady aerodynamic effects to be considered. In nominally attached flow, these effects include more complicated circulatory contributions resulting from the nonuniform shed wake convection velocity, also with additional noncirculatory contributions. The problem of dynamic stall under these conditions is not completely understood but has been studied experimentally on 2-D oscillating airfoils by Pierce et al. (1978a,b), Maresca et al. (1981), and Favier et al. (1988). A time-varying onset velocity was obtained in the experiment of Pierce et al. by using choking of the upstream flow by means of rotating vanes and in the experiment of Maresca et al. and Favier et al. by means of fore-and-aft movement of the airfoil. In both experiments the angle of attack and free-stream velocity were varied harmonically with different relative phase angles.

Pierce et al. (1978a,b) have measured only the airfoil pitching moment, with a view to understanding the possible effects of varying free-stream velocity on the torsional aerodynamic damping at dynamic stall onset. The measurements in fully attached flow were found to be in good agreement with the "classical" unsteady thin-airfoil models, as discussed in Chapter 8. In the vicinity of stall, reduced aerodynamic damping was observed, although, as shown previously, this is a characteristic found with steady onset flows and is related to the phasing of the dynamic stall events with respect to the forcing function (α, $\dot{\alpha}$ etc.).

In the other experiments by Maresca and Favier, measurements of the lift, drag, pitching moment, and chordwise pressure distribution have suggested some considerable influence of the varying free-stream velocity on the dynamic stall process. Depending on the phasing of the velocity, variation with respect to the angle of attack variation, initiation of leading-edge vortex shedding and the chordwise convection of this vortex appear to be different. Favier et al. (1988) measured a phase lead of the unsteady lift response for conditions with constant free-stream velocity, but these seem to be at variance both with other measurements for the same problem and also with theory. In this regard, the various problems associated with subscale Reynolds number simulation of the problem cannot be overlooked. The issue of time-varying incident flow velocity, unfortunately, has not yet been studied using the various mathematical models of dynamic stall, and it would seem to be an ideal problem whose investigation is overdue. It would also seem that because of the difficulties in conducting experiments of this problem, it would form a good challenge for the various first-principles based CFD approaches to dynamic stall modeling currently under development.

9.11 Prediction of In-Flight Airloads

While it has been shown previously that the problem of dynamic stall can be conveniently dissected, analyzed, and modeled into more manageable subproblems, it is only when the elements of the submodels are combined into the full rotor simulation, including the rotor wake (inflow) and elastic blade dynamics, that the true benefits of any improved modeling can be realized. Many authors have shown the effects of including representations of dynamic stall for rotor loads and aeroelastic predictions – see, for example, Tarzanin (1972), Gormont (1973), Carlson et al. (1974), Yen & Yuce (1992), Johnson (1969, 1974), and Nguyen & Johnson (1998). The results are, however, somewhat mixed. Generally, by including unsteady aerodynamics and dynamic stall models, better predictions of the phasing of the unsteady airloads with respect to blade aziumuth are obtained, along with correspondingly better predictions of overall rotor performance at the extremes of the flight

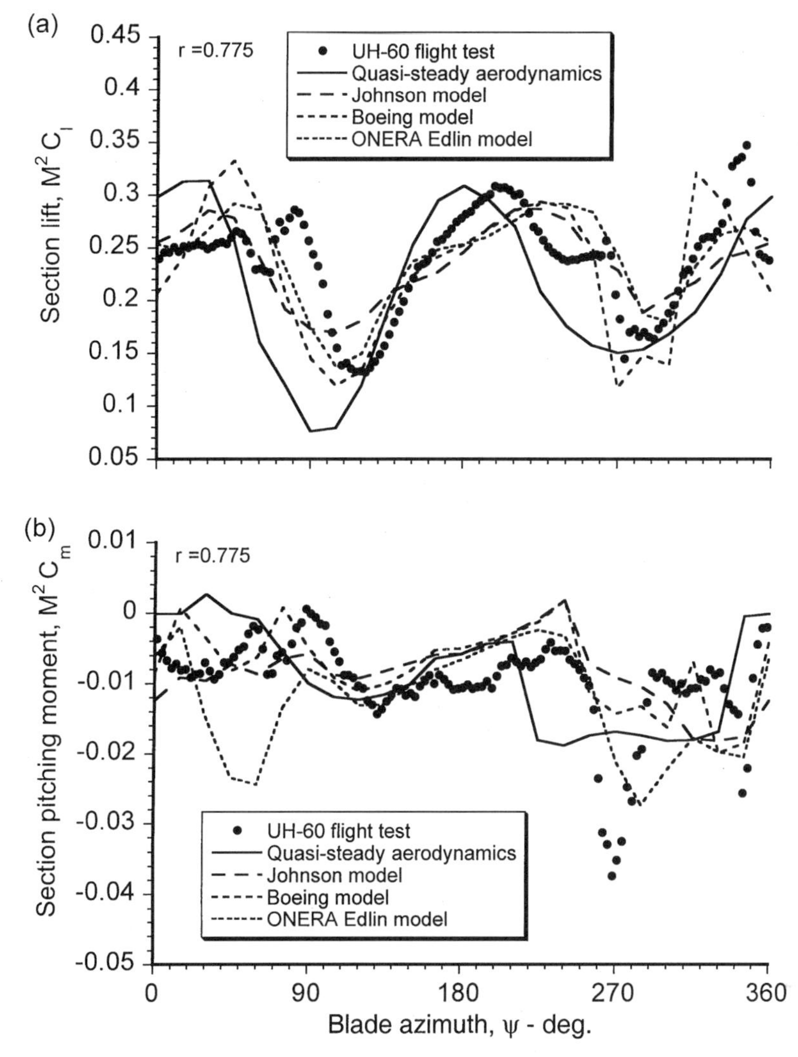

Figure 9.17 Predictions of in-flight dynamic stall airloads using semi-empirical stall models.

envelope. However, the mixed results indicate that although it may be possible to reproduce the unsteady airloads on a 2-D section, the prediction of in-flight airloads are much more difficult.

A representative example is given in Fig. 9.17, which shows the lift (as M^2C_l) and the pitching moment (as M^2C_m) on a section of a rotor blade for a case where dynamic stall is present. The flight test results are taken from Bousman (1998), and the calculated results are taken from Nguyen & Johnson (1998). The unsteady airloads are computed with different dynamic stall models, while keeping fixed all other elements of the rotor model. As seen from Fig. 9.17, the agreement of the predictions with the measurements is indeed very mixed. The improvement over the use of quasi-steady aerodynamics is, however, clearly better, particularly in regard to the phase of the lift predictions. However, the predictions of the section pitching moment are probably less than what would be considered acceptable. For example, one of the models predicts large nose-down moments associated with dynamic stall at points over the disk where there is clearly no stall indicated in the flight test measurements. In other cases, stall is not predicted in the regions where flight tests clearly indicate otherwise.

The fault, however, lies not only with the stall models. The main difficulty in the rotor simulation is the proper calculation of the combination of angles of attack and Mach number that will delimit attached flow, and these calculations must properly include models of the rotor wake and blade motion in a fully coupled sense. Because of the high angle of attack gradients that exist over the disk resulting from the wake inflow, the time steps (level of discretization) used in the rotor analysis can determine whether or not dynamic stall is initiated at all. In other words, it is possible to completely miss the stall event because the time step is too large to resolve the phenomenon. Furthermore, vortex shedding during dynamic stall occurs over relatively short time scales, generally of the order of 8 semi-chords of airfoil travel – see Galbraith et al. (1986). Therefore, if translated into an azimuth step for the rotor calculation, the azimuth step must be of the order of 2–5 degrees to accurately represent dynamic stall. In the calculations shown in Fig. 9.17, a rotor azimuth step of 15° was used, which is probably too large to capture the correct details of stall onset and subsequent aft center-of-pressure movement resulting from dynamic stall. Therefore, this level of predictive capability is of great concern to the rotor analyst and shows that even if the ability to model 2-D (or even 3-D) unsteady airloads has been realized, the prediction of dynamic stall airloads in the rotor environment is still a problem that is at the limit of current modeling capabilities. Equivalent arguments apply for phenomena such as blade vortex interaction (BVI), which occur over time scales of the order of 5 semi-chords of airfoil travel. In this case, the time (azimuth) step must be of the order of one degree or less to resolve the phenomena.

9.12 Chapter Review

The phenomenon of dynamic stall has been shown to be an important consideration in helicopter design because it ultimately limits main rotor performance. It has been shown that dynamic stall is characterized by a favorable delay in onset of flow separation to higher angles of attack. This is followed by the less favorable phenomenon of leading-edge vortex shedding. As long as this vortex stays over the airfoil, it acts to enhance the lift being produced. However, as this vortex is swept over the blade chord, the aft-moving center of pressure induces large nose-down pitching moments. These moments can manifest as high torsional airloads on the blades and may also induce aeroelastic problems such as stall flutter. Although dynamic stall can occur at various regions over the rotor disk, depending on the flight condition, it usually occurs on the retreating blade during high speed forward flight or during maneuvers such as tight turns or pull-ups. Therefore, the consideration of dynamic stall phenomenon represents a necessary refinement in the rotor design process and will more accurately define the operational and performance boundaries of the helicopter.

While the prediction of the conditions for dynamic stall onset and the subsequent effects of dynamic stall clearly forms an essential part of any rotor design process, it has been shown that it is not a problem that is yet fully understood, nor is it easily predicted. For engineering analyses, the modeling of dynamic stall still remains a particularly challenging problem. This is mainly because of the need to balance physical accuracy with computational efficiency and/or the need to formulate a model of dynamic stall in a particular mathematical form. To this end, a number of semi-empirical models have been developed for use in rotor design work, most of which have their roots in classical unsteady thin-airfoil theory. A brief discussion of these semi-empirical methods has been presented, along with a demonstration of their general capabilities in predicting the unsteady airloads during dynamic stall. Generally, predictions are good when measurements are available for validation or empirical refinement of the model, but their capabilities for general airfoil shapes and for completely arbitrary variations of angle of attack and Mach number are less certain. Future research will almost certainly devise more capable and better validated engineering models of the dynamic stall problem until such time that CFD methods capable of modeling three-dimensional, compressible, unsteady, separated flows become more practical to use on a routine basis.

Other factors that may influence the phenomenon of dynamic stall on a rotor have also been discussed. These include compressibility effects, airfoil shape, sweep angle, unsteady onset velocity effects, and three-dimensional effects. It appears that whereas the qualitative characteristics of dynamic stall are similar at all Mach numbers, there are subtle quantitative differences in the unsteady airloads that may be difficult to represent accurately with the context of engineering models. The effect of airfoil shape on the problem of dynamic stall is still not fully understood. It appears that airfoils designed for high values of static maximum lift will also show high values of lift in the dynamic case. However, it also appears that the values of maximum dynamic lift are less sensitive to airfoil shape than in the static case. Similar arguments apply for the pitching moment, and airfoils designed for low static pitching moments may not necessarily exhibit lower dynamic moments. This issue perhaps opens up fruitful avenues of research, where new rotor airfoils might be designed to meet dynamic lift and pitching moment requirements. The effects of sweep angle on the dynamic stall process have been shown to be significant, and sweep angle will be a necessary parameter to include in the modeling process. Three-dimensional effects associated with dynamic stall are clearly not fully understood. However, at least on the basis of idealized 3-D dynamic stall experiments on finite wings, the physics of the problem are similar to the 2-D case when corrected for the additional quasi-steady induced effects associated with finite span.

There remain many uncertainties in the prediction (and possibly control) of dynamic stall and also in the proper validation of predictions with measured airloads on the rotor. This is because of the need to also define accurately the blade motions and elastic deformation of the rotor blades, as well as the aerodynamic environment on the rotor in terms of induced angle of attack and induced velocity field. In regard to the latter, the inflow models discussed in Chapters 3 and 8 often prove inadequate because of the strong local induced velocity variations produced by discrete tip vortices in the rotor wake. This latter problem is considered next in Chapter 10. The problem of modeling the rotor wake, however, is just as formidable as the dynamic stall problem, perhaps even more so. The overall level of rotor analysis capability is only as strong as the weakest link in the chain. Until it becomes possible to model all aerodynamic aspects of the rotor problem, and improve upon the existing deficiencies and uncertainties that are present in the predictive models, helicopter design cycle times will continue to be long and costs will remain high.

9.13 Questions

9.1. For a 2-D airfoil operating under dynamic stall conditions, explain the difference between "lift-stall" and "moment-stall." Show some typical lift and moment curves for an airfoil oscillating in angle of attack. Your discussion should include a description of the basic flow physics as well as the integrated forces and moments.

9.2. The phenomenon of dynamic stall is often studied by analyzing measurements made on oscillating airfoils. The dynamic stall vortex convection speed is to be estimated from a series of such experiments where the lift and pitching moment are available as functions of angle of attack. Outline the approach you would take to determine the average vortex convection speed (in semi-chords of airfoil travel).

9.3. The phenomenon of dynamic stall can be characterized by time-scales defined in terms of semi-chord lengths of airfoil travel. At low Mach numbers, the time-scales associated with vortex shedding are of the order of 8 semi-chords. Estimate the angular resolution (azimuthal time-step) that would be required to properly resolve the dynamic stall phenomenon in a typical helicopter rotor analysis.

9.4. The phenomenon of dynamic stall is often associated with reduced or negative torsional aerodynamic damping. Explain what this means and why it occurs, and describe the possible consequences on a helicopter rotor.

Bibliography

Beddoes, T. S. 1976. "A Synthesis of Unsteady Aerodynamic Effects Including Stall Hysteresis," *Vertica*, 1, pp. 113–123.

Beddoes, T. S. 1978. "Onset of Leading Edge Separation Effects under Dynamic Conditions and Low Mach Number," 34th Annual Forum of the American Helicopter Soc., Washington DC, May 15–17.

Beddoes, T. S. 1979. "A Qualitative Discussion of Dynamic Stall," AGARD Report 679.

Beddoes, T. S. 1983. "Representation of Airfoil Behavior," *Vertica*, 7 (2), pp. 183–197.

Benson, R. G., Dadone, L. U., Gormont, R. E., and Kohler, G. R. 1973. "Influence of Airfoils on Stall Flutter Boundaries of Articulated Helicopter Rotors," *J. American Helicopter Soc.*, 18 (1), pp. 36–46.

Bielawa, R. L. 1975. "Synthesized Unsteady Airfoil Data with Applications to Stall Flutter Calculations," 31st Annual Forum of the American Helicopter Soc., Washington DC, May 13–15.

Bousman, W. G. 1998. "A Qualitative Examination of Dynamic Stall from Flight Test Data," *J. American Helicopter Soc.*, 43 (4), pp. 279–295.

Carlson, R. G., Blackwell, R. H, Commerford, G. L., and Mirick, P. H. 1974. "Dynamic Stall Modeling and Correlation with Experimental Data on Airfoils and Rotors," NASA SP-352.

Carr, L. W., McAlister, K. W., and McCroskey, W. J. 1977. "Analysis of the Development of Dynamic Stall Based on Oscillating Airfoil Measurements," NASA TN D-8382.

Carr, L. W., McAlister, K. W., and McCroskey, W. J. 1978. "Dynamic Stall Experiments on the NACA 0012 Airfoil," NASA TP-1100.

Carr, L. W. 1988. "Progress in Analysis and Prediction of Dynamic Stall," *J. of Aircraft*, 25 (1), pp. 6–17.

Carta, F. O. 1967. "An Analysis of the Stall Flutter Instability of Helicopter Rotor Blades," *J. American Helicopter Soc.*, 12 (4), pp. 1–18.

Carta, F. O., Casellini, L. M., Arcidiacono, P. J., and Elman, H. L. 1970. "Analytical Study of Helicopter Rotor Stall Flutter," 26th Annual Forum of the American Helicopter Soc., Washington DC, June.

Carta, F. O. 1971. "Effect of Unsteady Pressure Gradient Reduction on Dynamic Stall Delay," *J. of Aircraft*, 8 (10), pp. 839–840.

Carta, F. O. 1979. "A Comparison of the Pitching and Plunging Response of an Oscillating Airfoil," NASA CR-3172.

Chandrasekhara, M. S. and Carr, L. W. 1990. "Flow Visualization Studies of the Mach Number Effects on the Dynamic Stall of Oscillating Airfoils," *J. of Aircraft*, 27 (6), pp. 516–522.

Chandrasekhara, M. S. and Carr, L. W. 1994. "Compressibility Effects on Dynamic Stall of Oscillating Airfoils," AGARD CP-552.

Dwyer, H. A. and McCroskey, W. J. 1971. "Crossflow and Unsteady Boundary-Layer Effects on Rotating Blades," *AIAA J.*, 9 (8), pp. 1498–1505.

Ekaterinaris, J. A., Srinivasan, G. R., and McCroskey, W. J. 1994. "Present Capabilities of Predicting Two-Dimensional Dynamic Stall," AGARD CP-552.

Ericsson, L. E. 1967. "Comments on Unsteady Airfoil Stall," *J. of Aircraft*, 4 (5), pp. 478–480.

Ericsson, L. E. and Reding, J. P. 1972. "Dynamic Stall of Helicopter Blades," *J. American Helicopter Soc.*, 17 (1), Jan., pp. 10–19.

Ericsson, L. E. and Reding, J. P. 1983. "The Difference between the Effects of Pitch and Plunge on Dynamic Airfoil Stall," 9th European Rotorcraft Forum, Stesa, Italy, Sept. 13–15.

Ericsson, L. E. and Reding, J. P. 1984. "Unsteady Flow Concepts for Dynamic Stall Analysis," *J. of Aircraft*, 21 (8), pp. 601–606.

Favier, D, Agnes, A, Barbi, C., and Maresca, C. 1988. "Combined Translation/Pitch Motion: A New Airfoil Dynamic Stall Simulation," *J. of Aircraft*, 25 (9), Sept., pp. 805–814.

Fukushima, T. and Dadone, L. U. 1977. "Comparison of Dynamic Stall Phenomena for Pitching and Vertical Translation Motions," NASA CR 2793.

Galbraith, R. A. McD. Niven, A. J., and Seto, L. Y. 1986. "On the Duration of Low Speed Dynamic Stall," Paper ICAS-86-2.4.3. Proc. of the International Committee of the Aeronautical Sciences.

Gangwani, S. T. 1982. "Prediction of Dynamic Stall and Unsteady Airloads on Rotor Blades," *J. American Helicopter Soc.*, 27 (4), pp. 57–64.

Gangwani, S. T. 1984. "Synthesized Airfoil Data Method for Prediction of Dynamic Stall and Unsteady Airloads," *Vertica*, 8 (2), pp. 93–118.

Gormont, R. E. 1973. "A Mathematical Model of Unsteady Aerodynamics and Radial Flow for Application to Helicopter Rotors," USAAVLABS TR 72-67.

Green, R. B. and Galbraith, R. A. M. 1995. "Dynamic Recovery to Fully Attached Aerofoil Flow from Deep Stall," *AIAA J.*, 33 (8), pp. 1433–1440.

Gross, D. W. and Harris, F. D. 1969. "Prediction of In-Flight Stalled Airloads from Oscillating Airfoil Data," 25th Annual Forum of the American Helicopter Soc., Washington DC, May 14–16.

Ham, N. D. and Garelick, M. S. 1968. "Dynamic Stall Considerations in Helicopter Rotors," *J. American Helicopter Soc.*, 13 (2), pp. 49–55.

Ham, N. D. 1968. "Aerodynamic Loading on a 2-D Airfoil During Dynamic Stall," *AIAA J.*, 6 (10), pp. 1927–1934.

Harris, F. D. 1966. "Preliminary Study of Radial Flow Effects on Rotor Blades," *J. American Helicopter Soc.*, 11 (3), pp. 1–21.

Harris, F. D., Tarzanin, F. J., and Fisher, R. K. 1970. "Rotor High Speed Performance, Theory vs Test," *J. American Helicopter Soc.*, 15 (3), pp. 35–44.

Isaacs, N. C. G. and Harrison, R. J. 1989. "Identification of Retreating Blade Stall Mechanisms Using Flight Test Pressure Measurements," 45th Annual Forum of the American Helicopter Soc., Boston, MA, May 22–24.

Johnson, W. 1969. "The Effect of Dynamic Stall on the Response and Airloading of Helicopter Rotor Blades," *J. American Helicopter Soc.*, 14 (2), pp. 68–77.

Johnson, W. and Ham, N. D. 1972. "On the Mechanism of Dynamic Stall," *J. American Helicopter Soc.*, 17 (4), 36–45.

Johnson, W. 1974. "Comparison of Three Methods for Calculation of Helicopter Rotor Blade Loading and Stresses Due to Stall," NASA TN D-7833.

Jones, R. T. and Cohen, D. 1957. "Aerodynamics of Wings at High Speeds," Vol. VII of *High Speed Aerodynamics and Jet Propulsion*, Section A, Chapter 1, pp. 36–48, Aerodynamic Components of Aircraft at High Speeds, A. F. Donovan & H. R. Lawrence (eds.), Princeton University Press, Princeton, NJ.

Kramer, M. 1932. "Increase in the Maximum Lift of an Airfoil Due to a Sudden Increase in Its Effective Angle of Attack Resulting from a Gust," NACA Technical Memorandum 687.

Landgrebe, A. J. 1994. "New Directions in Rotorcraft Computational Aerodynamics Research in the U.S.," AGARD CP-552.

Leishman, J. G. and Beddoes, T. S. 1986. "A Generalized Method for Unsteady Airfoil Behavior and Dynamic Stall Using the Indicial Method," 42nd Annual Forum of the American Helicopter Soc., Washington DC, June.

Leishman, J. G. and Beddoes, T. S. 1989. "A Semi-Empirical Model for Dynamic Stall," *J. American Helicopter Soc.*, 34 (3), pp. 3–17.

Leishman, J. G. 1989. "Modeling Sweep Effects on Dynamic Stall," *J. American Helicopter Soc.*, 34 (3), pp. 18–29.

Leishman, J. G. and Crouse, G. L. 1989. "State-Space Model for Unsteady Airfoil Behavior and Dynamic Stall," AIAA Paper 89-1219.

Leishman, J. G. and Nugyen, K. Q. 1990. "A State-Space Representation of Unsteady Aerodynamic Behavior," *AIAA J.*, 28 (5), pp. 836–845.

Liiva, J., Davenport, F. J., Gray, L., and Walton, I. C. 1968. "2-D Tests of Airfoils Oscillating Near Stall," USAAVLABS TR 68-13, Vols. I & II.

Lorber, P. F., Covino, A. F., and Carta, F. O. 1991. "Dynamic Stall Experiments on a Swept Three-Dimensional Wing in Compressible Flow," AIAA Paper 91-1795.

Lorber, P. F. 1992. "Compressibility Effects on the Dynamic Stall of a Three-Dimensional Wing," AIAA Paper 92-0191.

Maresca, C. A., Favier, D. J., and Rebont, J. M. 1981. "Unsteady Aerodynamics of an Aerofoil at High Angles of Incidence Performing Various Linear Oscillation in a Uniform Stream," *J. American Helicopter Soc.*, 26 (2), pp. 40–45.

McAlister, K. W. and Carr, L. W. 1979. "Water Tunnel Visualizations of Dynamic Stall," *J. of Fluids Engineering*, 101, Sept., pp. 376–380.

McAlister, K. W., Lambert, O., and Pitot, D. 1984, "Application of the ONERA Model of Dynamic Stall," NASA Technical Paper 2399, AVSCOM Technical Report 84-A-3.

McCroskey, W. J. 1972a. "Dynamic Stall of Airfoils and Helicopter Rotors," AGARD Report No. 595.

McCroskey, W. J. 1972b. "Recent Developments in Rotor Blade Stall," AGARD CP 111.

McCroskey, W. J. and Fisher, R. K., Jr. 1972. "Detailed Aerodynamic Measurements on a Model Rotor in the Blade Stall Regime," *J. American Helicopter Soc.*, 17 (1), pp. 20–30.

McCroskey, W. J. 1973. "Inviscid Flowfield of an Unsteady Airfoil," *AIAA J.*, 11 (8), pp. 1130–1136.

McCroskey, W. J. 1975. "Some Current Research in Unsteady Fluid Dynamics," ASME Freeman Scholar Lecture, *J. of Fluids Engineering*, Transactions of the ASME, pp. 8–38.

McCroskey, W. J., Carr, L. W., and McAlister, K. W. 1976. "Dynamic Stall Experiments on Oscillating Airfoils," *AIAA J.*, 14 (1), pp. 57–63.

McCroskey, W. J., McAlister, K. W., Carr, L. W., Pucci, S. K., Lambert, O., and Indergand, R. F. 1980. "Dynamic Stall on Advanced Airfoil Sections," 36th Annual Forum of the American Helicopter Soc., Washington DC, May 13–15.

McCroskey, W. J., McAlister, K. W., Carr, L. W., and Pucci, S. L. 1982. "An Experimental Study of Dynamic Stall on Advanced Airfoil Sections," Vols. 1, 2 & 3 NASA TM-84245.

McHugh, F. 1978. "What are the Lift and Propulsive Limits at High Speed for the Conventional Rotor?," 34th Annual Forum of the American Helicopter Soc., Washington DC, May.

Nguyen, K. Q. and Johnson, W. 1998. "Evaluation of Dynamic Stall Models with UH-60 Airloads Flight Test Data," 45th Annual Forum of the American Helicopter Soc., Washington DC, May 20–22.

Peters, D. A. 1985. "Toward a Unified Lift Model for Use in Rotor Blade Stability Analysis," *J. American Helicopter Soc.*, 30 (3), pp. 32–42.

Pierce, G. A., Kunz, D. L., and Malone, J. B. 1978a. "The Effect of Varying Freestream Velocity on Airfoil Dynamic Stall Characteristics," *J. American Helicopter Soc.*, 23 (2), pp. 27–33.

Pierce, G. A., Kunz, D. L., and Malone, J. B. 1978b. "2-D Dynamic Stall as Simulated in a Varying Free-Stream," Final Report under Grant NGR 11-002-185, NASA Langley Research Center.

Pitot, D. 1989. "Differential Equation Modeling of Dynamic Stall," La Reserche Aerospatiale, No. 1989-6.

Piziali, R. A. 1994. "2-D and 3-D Oscillating Wing Aerodynamics for a Range of Angles of Attack Including Stall," NASA Technical Memorandum 4632, USAATCOM Technical Report 94-A-011.

Purser, P. E. and Spearman, M. L. 1951. "Wind Tunnel Tests at Low Speeds of Swept and Yawed Wings Having Various Planforms," NACA TN 2445.

Reddy, T. S. R. and Kaza, K. R. V. 1987. "A Comparative Study of Some Dynamic Stall Models," NASA TM-88917.

Scruggs, R. M., Nash, J. F., and Singleton, R. E. 1974. "Analysis of Flow-Reversal Delay for a Pitching Foil," Paper 74-183, 12th AIAA Aerospace Sci. Meeting, Washington DC, Jan. 30–Feb. 1.

Srinivasan, G. R., Ekaterinas, J. A., and McCroskey, W. J. 1993. "Dynamic Stall of an Oscillating Wing Part1: Evaluation of Turbulence Models," Paper 93-3403, AIAA 11th Applied Aerodynamics Conf., Monterey, CA, Aug. 9–11.

Stepniewski, W. Z. and Keys, C. N. 1984. *Rotary-Wing Aerodynamics*, Dover Publications, New York, pp. 103–116.

St. Hillaire, A. O., Carta, F. O., Fink, M. R., and Jepson, W. D. 1979. "The Influence of Sweep on the Aerodynamic Loading of a NACA 0012 Airfoil," Vol. 1 – Technical Report, NASA CR 3092.

St. Hillaire, A. O. and Carta, F. O. 1979. "The Influence of Sweep on the Aerodynamic Loading of a NACA 0012 Airfoil," Vol. 2 – Data Report, NASA CR 145350.

St. Hillaire, A. O. and Carta, F. O. 1983a. "Analysis of Unswept and Swept Wing Chordwise Pressure Data from an Oscillating NACA 0012 Airfoil Experiment," Vol. 1 – Technical Report, NASA CR 3567.

St. Hillaire, A. O. and Carta, F. O. 1983b. "Analysis of Unswept and Swept Wing Chordwise Pressure Data from an Oscillating NACA 0012 Airfoil Experiment," Vol. 2 – Data Report, NASA CR 165927.

Tan, C. M. and Carr, L. W. 1996. "The AFDD Int. Dynamic Stall Workshop on Correlation of Dynamic Stall Models with 3-D Dynamic Stall Data," NASA TM-110375, USAATCOM TR-96-A-009.

Tarzanin, F. J., Jr. 1972. "Prediction of Control Loads Due to Blade Stall," *J. American Helicopter Soc.*, 17 (2), pp. 33–46.

Telionis, D. P. 1975. "Calculations of Time-Dependent Boundary Layers," Symp. on Unsteady Aerodynamics, Kinney, R. B., ed., University of Arizona, Tucson, pp. 155–190.

Tran, C. T. and Pitot, D. 1981. "Semi-Empirical Model for the Dynamic Stall of Airfoils in View of the Application to the Calculation of the Responses of a Helicopter Blade in Forward Flight," *Vertica*, 5 (1), pp. 35–53.

Tran, C. T. and Falchero, D. 1981. "Application of the ONERA Dynamic Stall Model to a Helicopter Rotor Blade in Forward Flight," 7th European Rotorcraft Forum, Garmisch-Partenkirchen, Sept. 22–25.

Truong, V. K. 1993. "A 2-D Dynamic Stall Model Based on a Hopf Bifurcation," 19th European Rotorcraft Forum, Cernobbio, Italy, Sept. 14–16.

Truong, V. K. 1996. "Prediction of Helicopter Rotor Airloads Based on Physical Modeling of 3-D Unsteady Aerodynamics," 22nd European Rotorcraft Forum, Brighton, UK, Sept. 17–19.

Tyler, J. C. and Leishman, J. G. 1992. "An Analysis of Pitch and Plunge Effects on Unsteady Airfoil Behavior," *J. American Helicopter Soc*, 37 (3), pp. 69–82.

Wilby, P. G. 1984. "An Experimental Investigation of the Influence of a Range of Aerofoil Design Features on Dynamic Stall Onset," 10th European Rotorcraft Forum, The Hague, The Netherlands, Aug. 28–31.

Wilby, P. G. 1996. "The Development of Rotor Airfoil Testing in the UK – The Creation of a Capability to Exploit a Design Opportunity," 22nd European Rotorcraft Forum, Brighton, UK, Sept. 16–18.

Wilby, P. G. 1998. "Shockwaves in the Rotor World – A Personal Perspective of 30 Years of Rotor Aerodynamic Research in the UK," *The Aeronaut. J.*, 102 (1013), pp. 113–128.

Wood, M. E. 1979. "Results from Oscillatory Pitch Tests on the NACA 0012 Blade Section," Aircraft Research Association, Bedford, UK, ARA Memo 220.

Yen, J. G. and Yuce, M. 1992. "Correlation of Pitch-Link Loads in Deep Stall on Bearingless Rotors," *J. American Helicopter Soc.*, 37 (4), pp. 4–15.

CHAPTER 10

Rotor Wakes and Tip Vortices

We have built new rotor test stands, new blades, with emphasis on heavily instrumented blades, new measurement techniques – it is mind boggling to me to see how well, and in how much detail, we can measure the flow field of a rotor. Modeling the flow field of a rotor is a terribly difficult problem. The airplane people have it easy compared to us. Capturing the details of the rotor wake and of what happens in the vicinity of the blade is very difficult.

Leone U. Dadone (1995)

10.1 Introduction

The significant physical features of helicopter rotor wakes, and some of the more advanced mathematical tools for modeling the wake, are discussed in this chapter. A helicopter rotor wake is dominated by strong vortices that are trailed from the tips of each blade. The nature of the rotor wake, in terms of its geometry, strength and the aerodynamic effects produced on the blades, depends principally on the operating state and flight condition of the helicopter. In hover, the tip vortices follow nominally helical trajectories below the rotor. This is the simplest operating state to understand. During forward flight, the rotor wake is skewed back behind the rotor by the oncoming flow, and a series of more complex interlocking, but nominally epicycloidal, vortex trajectories are produced. Under these conditions, the increased mutual proximity of many of the vortex filaments results in stronger vortex–vortex interactions and complicated distortions to the evolving wake topology. At the lateral edges of the wake, the individual vortex filaments are found to roll up into a pair of merging vortex bundles, somewhat like those that would trail from the tips of a low aspect ratio fixed wing.

The highly three-dimensional nature of a helicopter rotor wake, as well as the sensitivity of the wake to the geometric and operational parameters of the helicopter, means that the details of the wake are difficult to study experimentally, as well as to compute by means of mathematical models. Recent advances in experimental techniques have been substantial and now allow measurements to be made with a fidelity that was impossible only a few years ago. However, there are many physical phenomena involving the formation and evolution of blade tip vortices and rotor wakes that are still not well understood, and it is here that future research on helicopter aerodynamics must be focused. Landgrebe (1988) and McCroskey (1995) review the state-of-the-art capabilities in modeling helicopter rotor wakes, while Leishman & Bagai (1998) give an overview of key characteristic physical features of rotor wakes and some experimental challenges involved in their measurement.

10.2 Flow Visualization Techniques

Flow visualization using subscale rotor models has been the primary method used to study the physics of helicopter rotor wakes. Some of the earliest studies were by Taylor (1950) and Dingeldein (1954), who used balsa dust to seed the rotor wake. A more popular method is to use a dense white smoke, which is entrained into the wake rendering it visible when illuminated by a suitable strobed lighting source. Density gradient flow visualization methods, such as strobed (phase-resolved) shadowgraphy and schlieren (and to a lesser extent, interferometry), have also been used to study helicopter rotor wakes. The phenomena that have been studied include: tip vortex formation, blade–vortex interactions, vortex–airframe surface interactions, main rotor wake–tail rotor interactions, ground interference, multi rotor flows, and the wake roll-up in forward flight.

10.2.1 *Smoke Flow Visualization*

Smoke flow visualization of subscale helicopter rotor wakes has been performed by Gray (1956), Piziali & Trenka (1970), Landgrebe (1971, 1972), Brand et al. (1988), Mercker & Pengel (1992), Ghee & Elliott (1995), and by many other investigators. A dense white smoke is entrained into the rotor wake and blade tip vortices. When stroboscopically illuminated, the smoke particles reflect light, allowing a photograph of the flow structure to be recorded.[1] The tip vortices appear as circular regions devoid of smoke, which is a result of centrifugal forces produced on the smoke particles near the vortex cores (see also Section 10.3). Ideally, to give accurate spatial information, the wake must be illuminated using a thin light sheet, preferably using a laser. In another form of smoke flow visualization, smoke is ejected from the blade tip directly into the tip vortices. This renders the vortices visible as 3-D tubular trails with central voids. Such methods have been used to visualize the tip vortices in hover [see Gray (1956)] and in forward flight [see Müller (1990a)]. In a water tunnel, dye ejected from the blade tips gives similar results – see Werlé and Armand (1969). Projected smoke filament techniques, such as those of Steinhoff (1985) and Müller (1990b, 1994), have also received some attention.

While the application of smoke visualization techniques has had good success for subscale rotor models, experiments on full scale helicopter rotors are rare. A general limitation of the smoke flow technique, however, is that smoke particles are quickly dispersed, and lower particle concentrations make the tip vortices harder to visualize. However, by using smoke "bombs" attached to the blade tips, the method has been used to visualize part of the wake structure of a CH-46 tandem rotor helicopter - see Spencer (1969) and Sternfeld & Schairer (1969). Using the same technique, the wake generated by a coaxial Ka-34 helicopter has been documented by Akimov et al. (1994).

10.2.2 *Density Gradient Methods*

A fundamental requirement for the successful application of density gradient methods of flow visualization such as schlieren or shadowgraphy is that the flow contain regions with significant density inhomogeneities. Because the refractive index of air is directly proportional to its density, planes with density variations in the flow field cause incident light rays normal to these planes to be refracted. Using a cutoff or other device, the angular deflection of the light rays gives a schlieren effect. If the rays are directly cast onto a projection screen, this will result in a shadowgram or shadowgraph. While shock waves and acoustic waves are readily visualized with density gradient techniques, it is not always appreciated that the flow near rotor tip vortices will also produce large enough compressibility effects. However, to achieve such effects with subscale rotors, the tip speeds must be at or close to

[1] While high speed black & white film is traditionally used, the use of a high resolution CCD video camera is a common modern format.

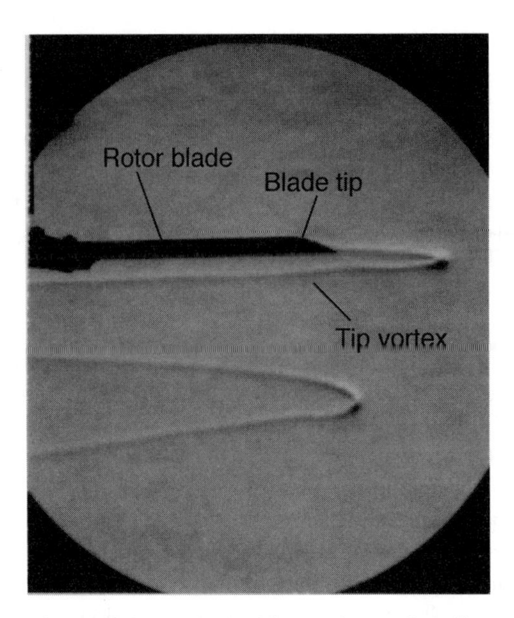

Figure 10.1 Strobed schlieren of part of a helicopter rotor wake formed during simulated hovering flight. Source: Tangler (1977) and courtesy of James Tangler.

full scale, and the rotor must be operated at relatively high thrust coefficients to generate strong vortices. General overviews of density gradient techniques of flow visualization are given by Holder & North (1963) and Merzkirch (1981).

Walters & Skujins (1972), Tangler et al. (1973), Tangler (1977), and Moedersheim et al. (1994) have used continuous strobed schlieren to observe the tip vortices generated by subscale rotors. An example of Tangler's work is shown in Fig. 10.1, where the tip vortices can be clearly seen. Tangler (1977) also shows that various 3-D acoustic wave phenomena generated by the rotor can be rendered visible with schlieren. A limitation with schlieren, however, is the small field of view because of the need to use high quality mirrors. Direct shadowgraphy, in comparison, uses no mirrors and so allows a much larger field and angle of view. However, because of the relatively weak density gradients found in rotor vortices, a shadowgraph screen must be made of a high efficiency retroreflector material giving sufficient contrast to allow the image to be recorded on photographic film – see Moedersheim et al. (1994) and Winburn et al. (1996). Parthasarthy et al. (1985), Norman & Light (1987), Light et al. (1990), Bagai & Leishman (1992a,b), Swanson (1993), and Lorber et al. (1994) have used strobed shadowgraphy to study tip vortex formation, blade vortex interactions, and rotor wake–airframe interaction phenomena. By taking advantage of the axial symmetry of the wake in hover, shadowgraphy can also allow the quantitative displacements of the tip vortices to be obtained as a function of wake age – see Norman & Light (1987) and Bagai & Leishman (1992b).

Figure 10.2 shows an example of a strobed shadowgraph detailing the flow immediately surrounding a tip vortex. This image was captured using a modified wide field-of-view shadowgraph system, incorporating a beam splitter arrangement – see Bagai & Leishman (1992a). This simple adaptation of the basic wide-field shadowgraph technique allows on-axis viewing and maximizes the efficiency of the retroreflective projection screen – see Winburn et al. (1996). While the resulting image is, in fact, a 2-D rendering of a curved 3-D vortex filament, the light rays at the right of the image are almost parallel to the vortex axis.

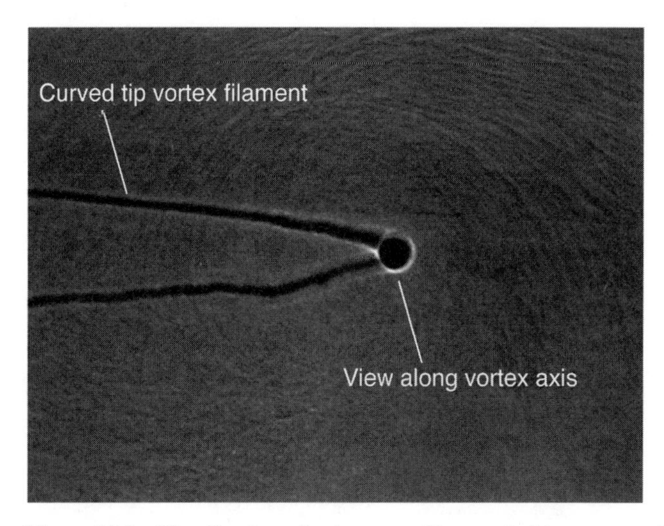

Figure 10.2 Visualization of a tip vortex filament inside a rotor wake by means of wide-field shadowgraphy. Source: University of Maryland.

A dark circular nucleus surrounded by a circular ring or "halo" is formed at the center of the image, where the light rays are refracted away from the vortex axis. The dimensions of this nucleus can be related to the density gradients in the flow, and under certain justifiable assumptions those can be related to the velocity field in the vortex – see Parthasarthy et al. (1985), Norman & Light (1987), and Bagai & Leishman (1994). Outside the immediate core region, the striations present in the shadowgraph image indicate significant turbulence in the flow surrounding the vortices.

10.2.3 *Natural Condensation Effects*

Evidence of the rotor wake can sometimes be seen through natural condensation of water vapor inside the blade tip vortices. Published photographs of the phenomena are relatively rare, but examples are given by Felker et al. (1986), Campbell & Chambers (1994), and McVeigh et al. (1997). One example is shown in Fig. 10.3, with another being shown previously in Fig. 2.2. The results obtained often appear similar to smoke flow visualization, exhibiting characteristic tubular trails with large central voids marking the positions of the vortices. These voids are caused by centripetal accelerations on the vapor particles as well as by the thermodynamic aspects of the problem. Although condensation flow visualization has been achieved in a wind-tunnel environment [Dadone (1970)] it is usually only outdoors that the correct combination of atmospheric conditions exist, that is, when the air temperature and dew point spread are small. Even then, however, a challenge is to have the right lighting conditions and background contrast to allow a good photographic exposure.

10.3 Characteristics of the Rotor Wake in Hover

In hovering flight a helicopter rotor wake is radially axisymmetric (at least in principle) and somewhat easier to study by means of flow visualization because only one view at successive blade positions (azimuth angles) is necessary to obtain a complete three-dimensional understanding of the wake topology. Figure 10.4 shows representative flow visualization images in the wake of a two-bladed rotor operating in hover. In this case, a

Figure 10.3 Natural condensation trails in the tip vortices generated by a tilt-rotor aircraft. Source: Courtesy of US Navy, Patuxent NAS.

fine mist of submicron atomized oil particles was illuminated with a thin laser light sheet that was positioned in a radial plane extending through the rotational axis of the rotor. The sheet was pulsed at a frequency of one flash per rotor revolution, so as to create an instantaneous illumination of the wake – see Martin et al. (1999) for further details. The results in Fig. 10.4 are for two blade azimuth positions.

Note the existence of two major flow features in Fig. 10.4. First, the blade tip vortex cores are identified by the dark seed voids. Wherever the local velocities are high enough

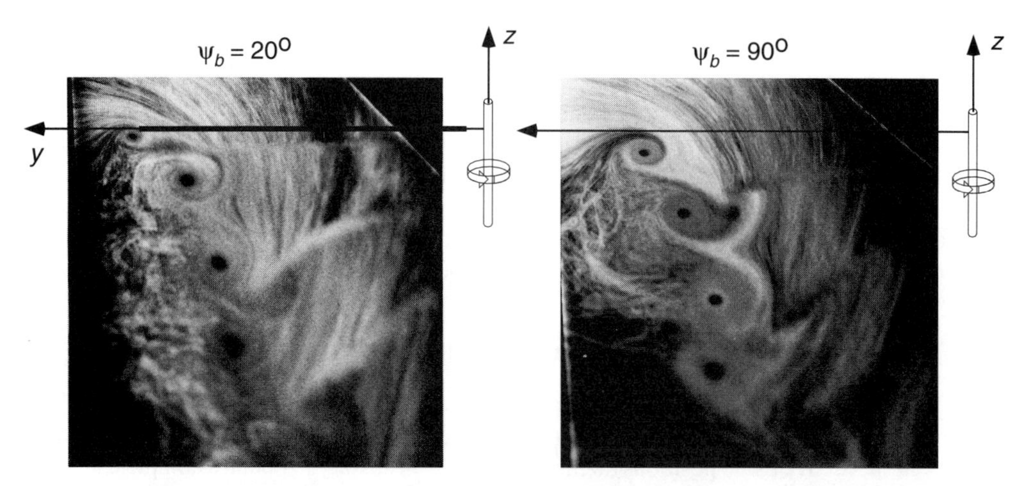

Figure 10.4 Laser light sheet flow visualization images of the rotor wake structure during simulated hovering flight. Source: University of Maryland.

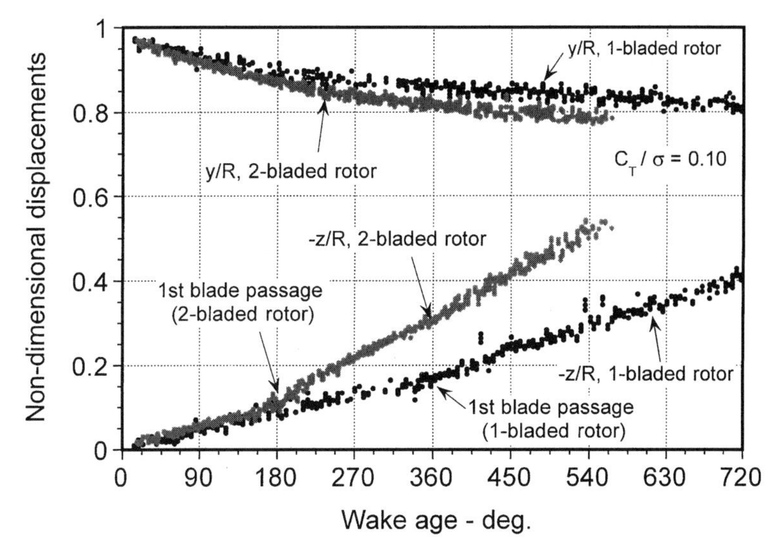

Figure 10.5 Tip vortex displacements of one- and two-bladed rotors operating in hover. $C_T/\sigma \approx 0.1$. Data source: Leishman (1998) and the University of Maryland.

to cause centrifugal forces on the seed particles, they will spiral radially outward. The particles will reach a radial equilibrium only when the centrifugal and pressure forces are in balance. The resulting voids can be larger than the actual viscous core size of the tip vortex – see Leishman (1996). Second, there is a shear layer trailed behind the blade, which is evidenced in Fig. 10.4 by a discontinuity in the streaklines. This shear layer is formed by merging of the boundary layers from the upper and lower surfaces of the blade, which contain both negative and positive vorticity. This shear layer is often referred to as a vortex sheet.[2] The sheet has a strength that is related to the gradient of lift (circulation) over the blade. Initially, the vortex sheet is seen to extend over the entire blade span. Thereafter, both the sheet and the tip vortex are convected below the rotor disk. The significant inward (radial) contraction of the tip vortices below the rotor is clearly evident. Both the tip vortex and the inner sheet are typically found to be visible for about two rotor revolutions, after which details of the individual flow structures become harder to discern. This is because of diffusion of the seed particles, as well as because the wake further downstream becomes aperiodic as it transitions into a turbulent jet.

The flow visualization images such as those shown in Fig. 10.4 can also be used to derive quantitative information by digitizing the locations of the seed void and the shear layer associated with the vortex sheet. Figure 10.5 shows the axial (z/R) and radial (y/R) displacements of the tip vortices as generated by one- and two-bladed hovering rotors operating at the same nominal blade loading coefficients; the total thrust of the two-bladed rotor was approximately twice that of the one-bladed rotor. The characteristics of the tip vortex geometry shown here are representative of the results that would be obtained with any lightly loaded hovering rotor. Up to the first blade passage, which occurs at a wake age of $\psi_w = 360/N_b$ degrees (where N_b is the number of blades), the tip vortices generated by either of the rotors convect axially only relatively slowly. However, the effective axial (slipstream) velocity (gradient of displacement curve) is noted to increase abruptly after the first blade passage. Here, the tip vortex lies close to and radially inboard of the

[2] Technically, a vortex sheet is only obtained when the vorticity lies in a surface of zero thickness.

following blade and so is subjected to an increase in downwash velocity from both the blade and its associated tip vortex. During this process, the radial position of the tip vortices contracts progressively. While the higher thrust of the two-bladed rotor leads to a slightly more rapid wake contraction, the asymptotic values are approximately the same with $y/R \approx 0.78$.

The interdependence of the vortex sheet and the tip vortex requires further discussion. The process is shown quantitatively in Fig. 10.6, where the displacements of the sheet have been digitized from flow visualization images such as those of Fig. 10.4. The vortex

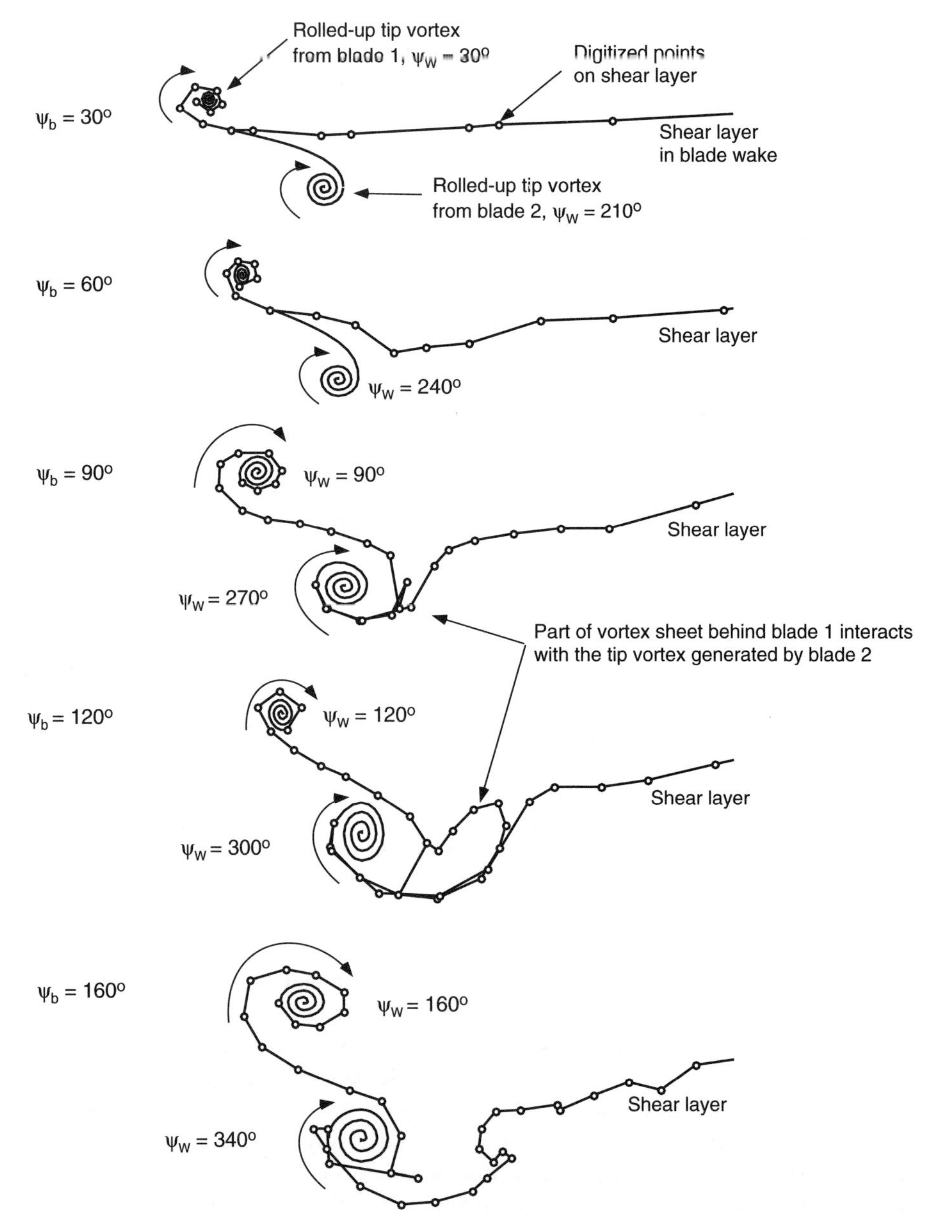

Figure 10.6 Schematic summarizing observed locations of the shear layer (vortex sheet) behind a rotor blade and its interaction with a tip vortex created by another blade.

sheet is trailed along the entire length of the blade, with a concentrated tip vortex at the outboard edge. Both the tip vortex and the sheet then convect axially downward below the rotor. The visualization images suggest that the distribution of induced velocity through the rotor is highest near the blade tip and lowest near the root. Therefore, the sheet convects more rapidly below the tip region of the blade and so becomes more progressively inclined to the rotor plane. The high swirl velocities induced by the tip vortex cause a further distortion to the sheet, and it is apparent that the outboard edge of the sheet begins to interact with the tip vortex generated by the other blade. This interaction between elements of the wakes generated by different blades is fundamentally different from that found on fixed (nonrotating) wings and illustrates another level of complexity in the understanding and modeling of helicopter rotor wakes. Free-vortex methods generally capture the effects of the vortex roll-up relatively well. However, modern CFD methods, such as those discussed by Raddatz & Pahlke (1994) and Tang & Baeder (1999), have not yet allowed the details of the vortex sheet and the tip vortex to be fully resolved because of rapid numerical diffusion associated with the finite-difference schemes.

10.4 Characteristics of the Rotor Wake in Forward Flight

As in hover, a helicopter rotor wake in forward flight is found to be dominated by the blade tip vortices. However, because of the free-stream (edgewise) component of velocity at the rotor plane in forward flight, the wake is now convected behind as well as below the rotor and it takes on a more complicated (nonaxisymmetical) form. The rotor wake geometry in forward flight is found to be sensitive to the rotor thrust, advance ratio, tip path plane (TPP) angle of attack, the presence of other rotors (such as a tail rotor, or another main rotor as in tandem or coaxial configurations), and rotor–airframe interference effects. While much is now known about the general features of rotor wakes in forward flight, much more is still to be learned about the details of the flow, especially before mathematical models of the rotor wake can be adequately validated.

There are a variety of techniques that can be used to visualize rotor wakes in forward flight, and some of these have been discussed previously in reference to hovering wake studies. Other techniques have included cavitation from the blade tips in a water tunnel, such as used by Larin (1973, 1974). Lehman (1968) and Landgrebe & Bellinger (1971) have used bubbles to trace out the tip vortices trailed from a rotor in a towing tank. Jenks et al. (1987) have used stratified layers of dye in a towing tank to observe some aspects of the wake roll-up. In a wind tunnel, smoke injection from the blade tips can provide good evidence of the epicycloidal tip vortex trajectories – see Müller (1990a,b). Laser sheet smoke flow visualization, such as used by Ghee & Elliott (1995), is considered one of the more accurate ways of documenting the spatial locations of the tip vortices.

An example documenting the general features of a helicopter rotor wake in forward flight is shown in Fig. 10.7, where smoke trailed from the blade tips was used to mark the vortex locations. It will be apparent from the top view (x–y plane) that the tip vortices are initially laid down as a series of interlocking epicycloids. Mutual interactions between the individual filaments result in some distortion of the vortex positions, but mostly in the z direction normal to the plane of the rotor (see side view). This vertical distortion of the wake is particularly strong at low advance ratios, where the tip vortices are closest together. Note from Fig. 10.7(a) that along the lateral edges of the wake the tip vortices begin to roll up into "vortex bundles" or "super vortices," as mentioned previously. Note also that the effects of the tail rotor can be seen as an expanding turbulent wake embedded inside the main rotor wake.

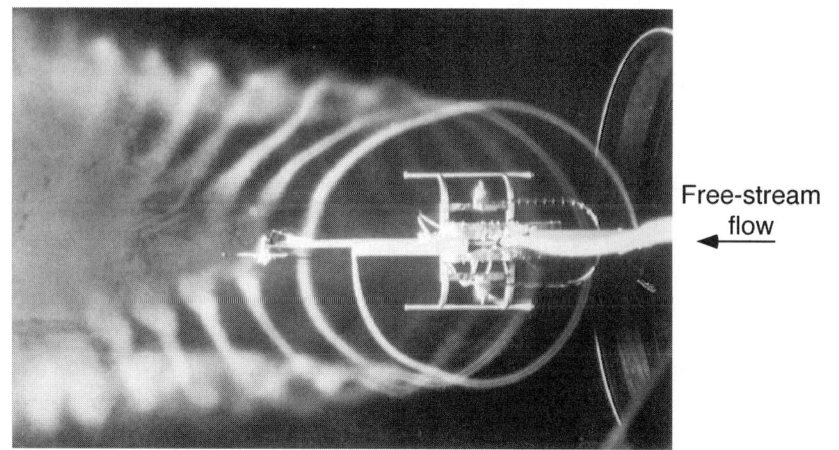

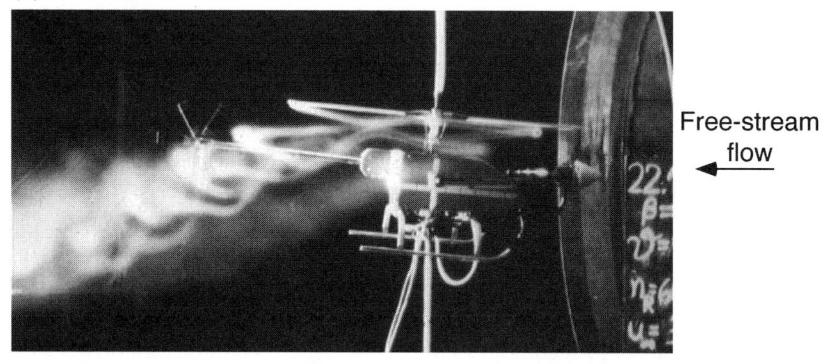

Figure 10.7 Identification of the tip vortex locations in a two-bladed rotor wake during operation in forward flight by ejecting smoke out of the blade tips. (a) Plan view (x–y plane). (b) Side view (x–z plane). Free-stream flow is from right to left. (Courtesy of Reinert Müller.)

10.4.1 *Wake Boundaries*

An understanding of the position of the rotor wake boundaries in forward flight can give much useful information to the rotor analyst. Leishman & Bagai (1991) and Bagai & Leishman (1992b) have examined the positions of the tip vortices generated by an isolated rotor and also a rotor with an airframe. The flow visualization results were obtained using shadowgraphy. The measured wake boundaries are plotted in Fig. 10.8 for hover and in forward flight at three advance ratios. Also shown are the positions of the wake that would be predicted on the basis of momentum theory considerations alone (i.e., a rigid wake; see Section 10.7.5). Clearly, the correlation with the measured displacements is poor, and the results confirm what can already be deduced from Fig. 10.7, that there are considerable mutually induced effects between vortex filaments inside the rotor wake at lower advance ratios. This leads to a powerful longitudinal (and also a lateral) asymmetry of the wake that cannot be predicted on the basis of momentum theory alone. Note from Fig. 10.8(a) that the vortices at the leading edge of the disk are initially convected above the blades and the

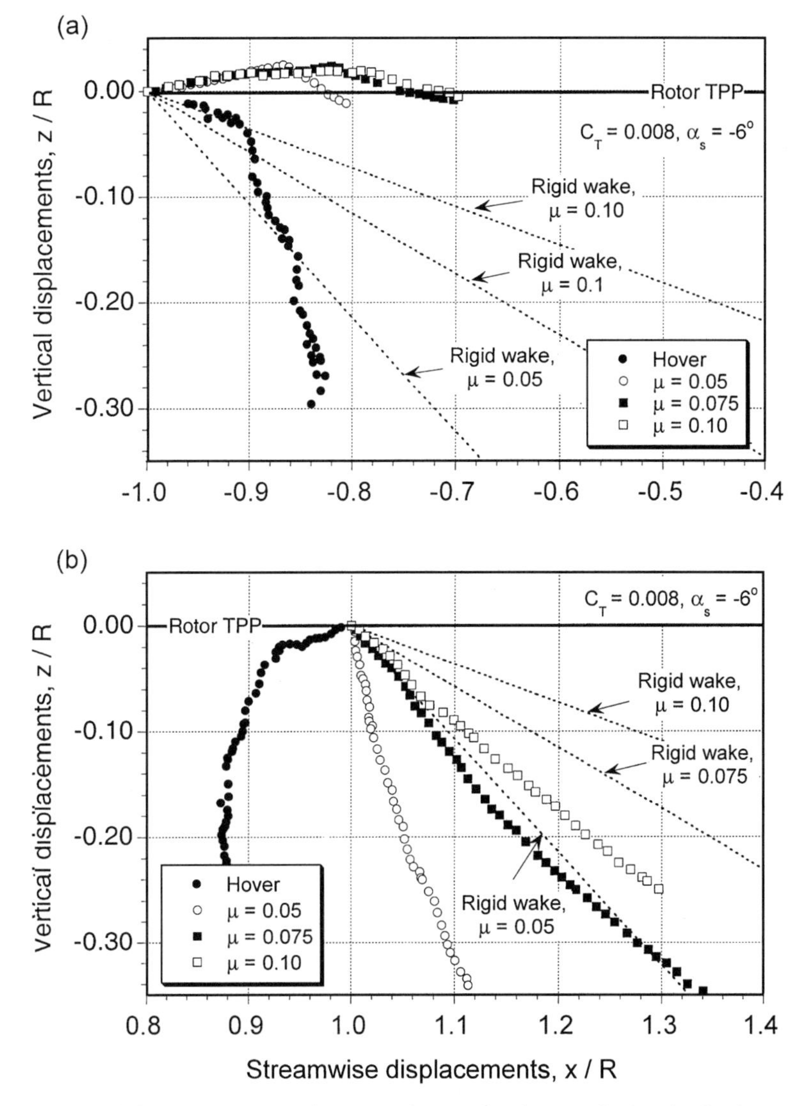

Figure 10.8 Tip vortex displacements in an x–z plane at the longitudinal centerline of an isolated rotor. Four-bladed rotor, $C_T = 0.008$, forward shaft tilt angle of 3° in forward flight. (a) Front of rotor disk. (b) Rear of rotor disk. Data source: Leishman & Bagai (1991).

rotor TPP within the first 90° of wake age, compared to the rear of the disk [Fig. 10.8(b)] where the wake vortices are convected quickly away from the rotor. This is because of a small region of upwash velocity at the leading edge of the disk and a strong longitudinal inflow gradient, which has its source in the skewness of the wake (see also the discussion in Chapter 3).

10.4.2 *Blade–Vortex Interactions (BVIs)*

It will be apparent that one of the distinctive features of rotor wakes in forward flight is the preponderance of potentially close interactions of blades and tip vortices, which

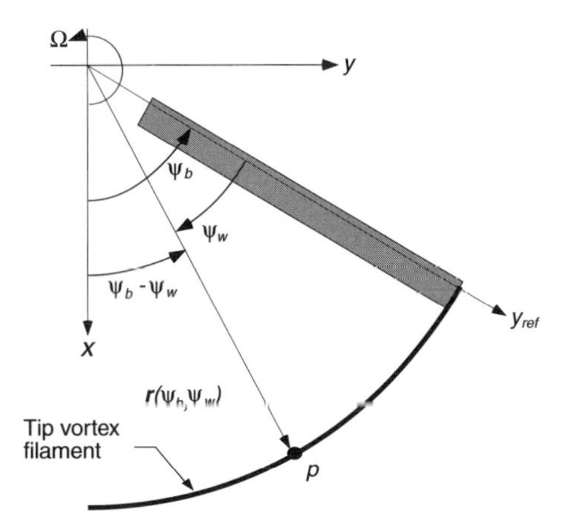

Figure 10.9 Definition of the position of an element in the vortex wake.

are called blade–vortex interactions or BVIs. When viewed from above (i.e., Fig. 10.7(a)) the trajectories formed by the tip vortices trace out closely epicycloidal forms. If the wake is assumed to be undistorted in the x–y (TPP) plane (a justifiable assumption near the rotor disk based on the example shown in Fig. 10.7), then the tip vortex trajectories are described geometrically by the parametric equations

$$\frac{x_{\text{tip}}}{R} = x'_{\text{tip}} = \cos(\psi_b - \psi_w) + \mu\psi_w, \tag{10.1}$$

$$\frac{y_{\text{tip}}}{R} = y'_{\text{tip}} = \sin(\psi_b - \psi_w), \tag{10.2}$$

where ψ_b is the position of the blade when the vortex was formed, and ψ_w is the age of the vortex element relative to that blade (see Fig. 10.9). It has been assumed that no wake contraction occurs in the radial dimension.

Examples of the wake topology when viewed from above the rotor are shown in Fig. 10.10 for three advance ratios, with results for a two-bladed rotor being shown for simplicity. It will be apparent that BVIs can occur at many different locations over the rotor disk, and also with different orientations between the blade axis and the vortex axis. If BVI occurs on the advancing side of the rotor disk, the blade and vortex axis can be nearly parallel. It has been described previously in Chapter 8 that these conditions promote locally high unsteady airloads on the blades, which can also be accompanied by significant obtrusive noise with strongly focused directivity – see Schmitz (1991).

For a rotor with N_b blades, and assuming an undistorted wake, the locus of all the potential BVIs can be determined if the equations

$$r\cos\left(\psi_b - \frac{2\pi(i-1)}{N_b}\right) = \cos(\psi_b - \psi_w) + \mu\psi_w, \tag{10.3}$$

$$r\sin\left(\psi_b - \frac{2\pi(i-1)}{N_b}\right) = \sin(\psi_b - \psi_w) \tag{10.4}$$

are simultaneously satisfied for r (on the blade) and ψ_b, where $i = 1, 2, \ldots, N_b$. (See also Questions 10.3 and 10.4.) Figure 10.11 shows these locations for a two-bladed rotor ($N_b = 2$) operating in forward flight at four advance ratios. Again, it should be appreciated that this calculation assumes that all the vortices lie in the TPP. For any one flight condition they do not, but these figures give an indication of all the possible intersection locations.

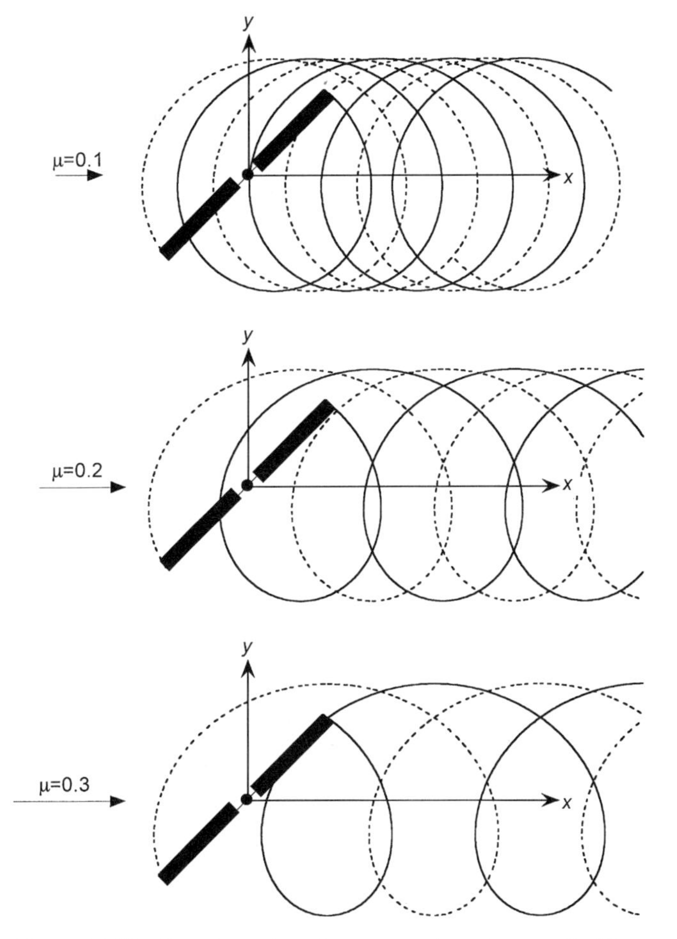

Figure 10.10 Plan view of the tip vortex trajectories as trailed from a two-bladed rotor in forward flight at different advance ratios.

Note that the largest number of potential BVIs occur in low speed forward flight. As forward speed increases, the number is reduced considerably. However, even though the number of interactions may decrease, it is their intensity (and also the trace Mach number of the interaction) that determines the potential noise and directivity of noise associated with a BVI – see Lowson (1996) and Leishman (1999).

Despite its importance to helicopter noise and vibration, clear experimental studies of the parallel BVI problem are relatively rare. Good, albeit idealized, quantitative measurements of blade vortex interactions and collisions are described by Horner et al. (1994) and Kitaplioglu & Caradonna (1994). Contributions to visualizing the problem have been achieved with the use of smoke [see Mercker & Pengal (1992) and Lorber et al. (1994)] and using strobed shadowgraphy [see Swanson (1993)].

The other type of BVI occurs when the blade and vortex axes are almost perpendicular, which is mainly over the front and rear of the disk. While the former type tends to produce the largest unsteady airloads and noise generation, the latter tends to result in more highly three-dimensional airloads and broadband noise – see Schmitz (1991). Figure 10.12 shows a sequence of flow visualization images obtained with wide-field shadowgraphy, which detail the blades encountering a perpendicular type of BVI over the leading edge of the rotor disk

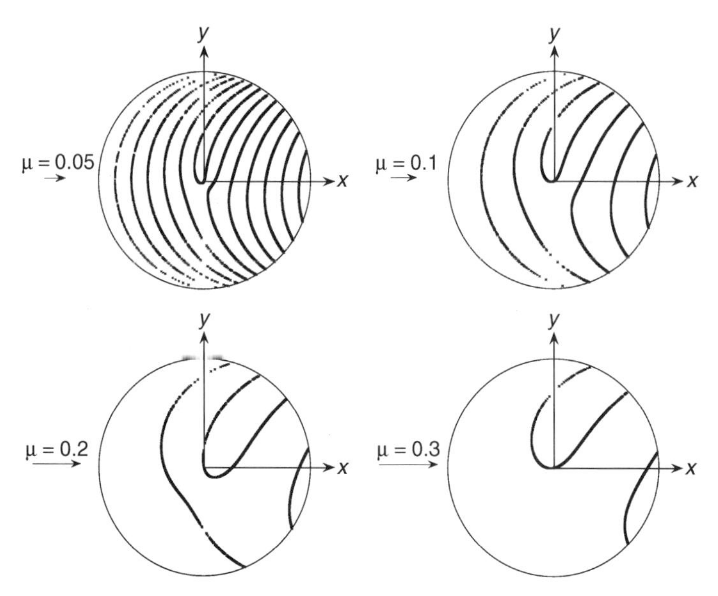

Figure 10.11 Locus of all possible BVIs over the rotor disk for a two-bladed rotor in forward flight at different advance ratios.

during low speed forward flight. We see that as the older wake vortices move downstream (to the left) they move up and over the top of the following blades (see also Fig. 10.11). A disturbance is produced on the older vortices as following blades pass underneath. After the blade passes, the tip vortices return to their original undisturbed shape. For progressively greater wake ages, the vortices are convected further downstream and the blade intersects

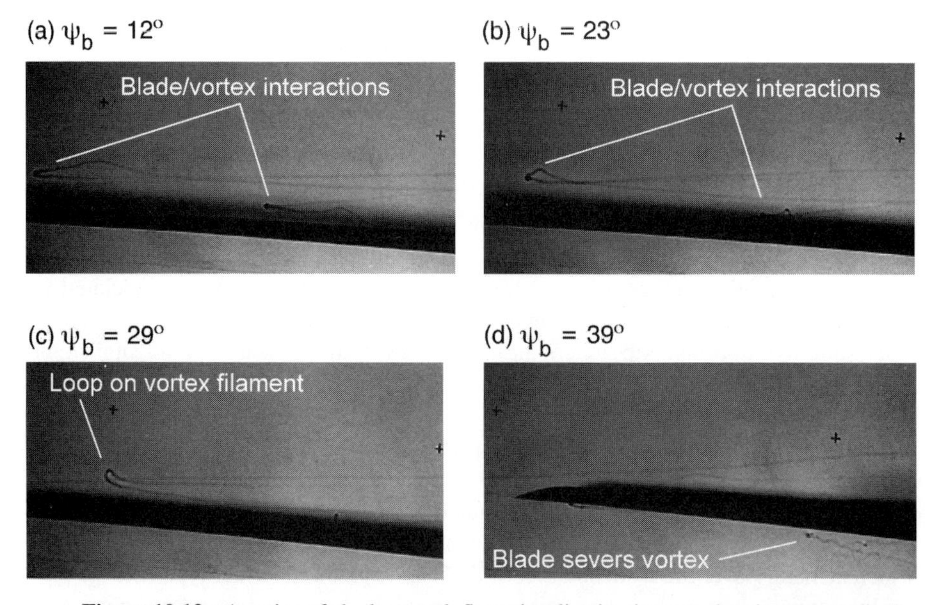

Figure 10.12 A series of shadowgraph flow visualization images showing perpendicular BVIs over the front of the rotor disk in forward flight. (a) $\psi_b = 12°$. (b) $\psi_b = 23°$. (c) $\psi_b = 29°$. (d) $\psi_b = 39°$. Free-stream flow is from left to right. Source: University of Maryland.

and severs the vortex filament. This can result in an instability or transition bifurcation characterized by a large growth in the viscous core dimensions. This is known as "vortex bursting," although the proper explanation of this phenomenon remains controversial. However, an increase in the effective core size and/or diffusion rate of the tip vortex produced by bursting is inevitable, and will affect the intensity of other potential interactions with blades. This is one type of rotor wake problem that is poorly understood and difficult to model.

10.5 Other Characteristics of Rotor Wakes

10.5.1 *Periodicity*

In most cases, the rotor wake is deterministic and the tip vortices generated by each blade will follow smooth curved and almost helical or epicycloidal paths. Also, under ideal circumstances, their spatial locations relative to the rotor will be periodic at the rotor rotational frequency. However, various types of aperiodic behavior of rotor wakes have been noticed in experiments. Aperiodicity can be defined as the random variations in the spatial locations of the vortex filaments from a mean position at a given wake age. In fixed-wing terminology, this phenomenon is referred to as vortex wandering or meandering – see Devenport et al. (1996) and Leishman (1998). If aperiodicity occurs above some threshold, then measurements based on the assumption of a periodic flow will be biased because the small random displacements of the vortices essentially average out the flow field properties at a fixed measurement point.

For helicopter rotors, the available evidence suggests that wake aperiodicity is a characteristic that is, in part, related to the nature of the rotor operating state; in hover the phenomenon is more likely to show than in forward flight. For example, measurements of the tip vortex locations in hover that are reconstructed on the basis of a series of images of the wake made at different wake ages may take the appearance of significant scatter, or even two possible geometries. Such a behavior can be traced to aperiodic flow effects and has been observed in several rotor experiments including those of Landgrebe (1972), Norman & Light (1987), Bagai & Leishman (1992b), and Leishman (1998). Most of the available experimental evidence with rotors shows that aperiodicity is pronounced only at older wake ages (older than one or two complete rotor revolutions), at low thrust coefficients (where the slipstream convection velocities are low and the tip vortices remain close to the rotor plane), or after the first blade passage. At a minimum, appropriate allowance can be made when quoting measurement uncertainties and when comparing with computations of the wake topology. From a purely scientific perspective, the challenge is to understand whether aperiodicity is always an inherent physical characteristic of rotor wakes or if it also arises from small external flow disturbances or flow recirculation in the test facility. However, the fact that they occur in laboratory experiments with rotors also suggests that they would be present in the turbulent air associated with the flight of an actual helicopter.

10.5.2 *Vortex Perturbations and Instabilities*

Besides the more random types of aperiodicity found in rotor wakes, other deterministic types of perturbations and instabilities of the tip vortices have been observed. One common type is the smooth sinuous wave type observed by Landgrebe (1971), Sullivan (1973), and others and analyzed theoretically by Levy & Forsdyke (1928), Widnall (1972), and Gupta & Loewy (1973, 1974). An example of a short sinuous wave is shown in Fig. 10.13(a). These types of perturbations can become pronounced in amplitude at older

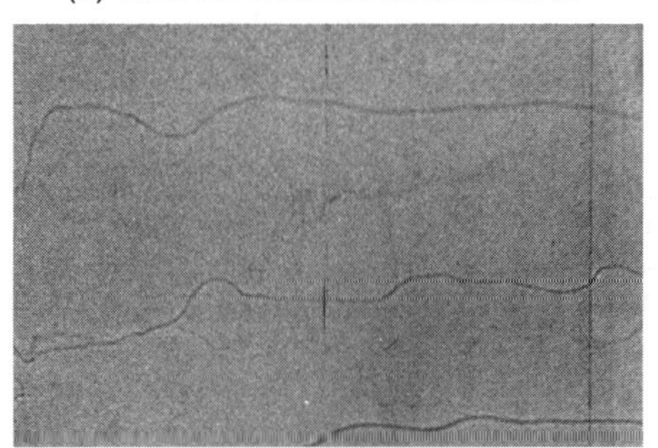

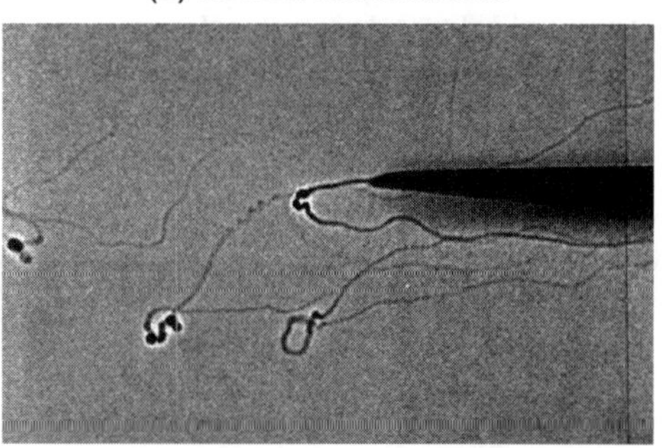

Figure 10.13 Shadowgraphs showing two types of disturbances seen on rotor tip vortices. (a) The smooth or sinuous form. (b) The "cork-screw" or helical form. Source: University of Maryland.

wake ages. Long wave disturbances or instabilities can sometimes result in pairing and looping of adjacent vortex filaments and can often be a source of some aperiodicity in the flow – see Tangler et al. (1973) and analysis by Gupta & Loewy (1973, 1974). In some cases, vortex pairing may lead to a complicated aperiodic wake formation but at a subharmonic of the rotor frequency – see Leishman & Bagai (1998). Based on various experimental observations, the onset of vortex disturbances and wake instabilities is affected by the number of blades, rotor thrust, and overall operating conditions. Forward flight experiments with helicopter rotors in wind tunnels have shown that the tip vortices appear generally free from the regular sinuous perturbations so often noted in hovering rotor wakes, although such effects cannot be discounted unilaterally.

Another type of perturbation sometimes found in rotor wakes, both in hover and forward flight, is referred to as the "cork-screw" or helical type, with an example being shown in Fig. 10.13(b). Here, the vortex filament tightly twists around on itself forming a very pronounced helix. These helical type perturbations appear to be common in the wakes of highly loaded propellers or tilt-rotors [see Norman & Light (1987)] rather than helicopter rotors, but they have been observed on both. In some cases, the disturbance travels along the vortex filament. In other cases, it is damped out, and the vortex returns to its regular (periodic) form. Occasionally, the disturbance may cause the vortex to become unstable, and it may break down or burst. An example of this latter phenomenon is shown in Fig. 10.14, where vortex bursting originates from the formation of a helical disturbance formed on the vortex just downstream of the blade.

10.6 Detailed Structure of the Tip Vortices

The roll-up of the tip vortex in terms of its strength, velocity distribution, and location defines the initial conditions for the subsequent behavior of the rotor wake. Tip vortex formation is a complex problem involving high velocities with shear, flow separation, pressure equalization, and turbulence production. On most helicopters, which will have rectangular or mildly tapered blade tips, experimental evidence shows that a single vortex is fully formed at the trailing edge of the blade tip, as seen in Fig. 10.15, which is a shadowgraph of the flow near the tip of the blade. The tip region is enveloped with a region of high vorticity, which rolls up quickly into a dominant vortex. Because rotor blades have very large pressure differences over their tip region, the resulting tip vortices have high circulation, high swirl velocities, and relatively small viscous cores.

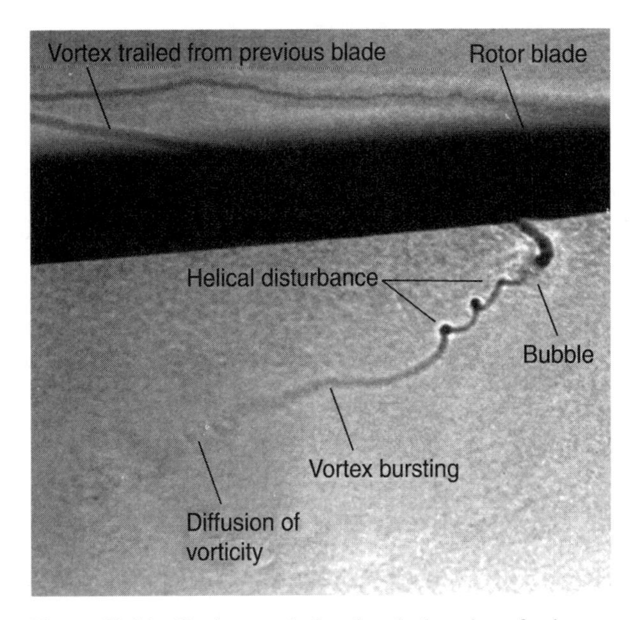

Figure 10.14 Shadowgraph showing the bursting of a tip vortex after an encounter with a blade. View from behind the blade. Source: University of Maryland.

Most vortex wake models used for rotor loads, performance, and acoustics will have some kind of semi-empirical representation of the tip vortex characteristics. Because tip vortex properties are still relatively undocumented for helicopter rotors, the required parameters to formulate a suitable model of the vortex are often interpreted or extrapolated from those measured in fixed-wing studies – see Rorke et al. (1972) and Rorke & Moffitt (1977). However, because of the sustained proximity of the blades to the tip vortices and the mutual interactions between the vortices, the validity of simply extrapolating fixed-wing results to the rotor case is still the subject of ongoing research. Another complication is that the blade tip shape is known to affect the strength and location of the blade tip vortex as it is trailed off into the wake, and these effects are even less poorly understood. For some tip shapes,

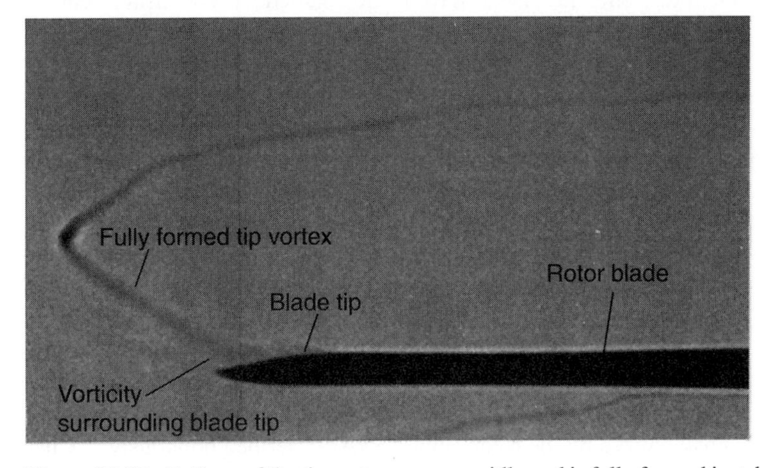

Figure 10.15 Roll-up of the tip vortex occurs rapidly and is fully formed just downstream of the trailing edge of the blade. Source: University of Maryland.

multiple vortices may be produced, although one of these is usually stronger and tends to dominate the flow.

Even for rectangular tips, the overall roll-up of the tip vortex in terms of its strength and initial location behind the blade is found to be difficult to predict. Based on classical centroid of vorticity approaches, which are used in some forms of rotor analysis, the computed vortex release locations generally tend to be predicted radially much further inward toward the hub than are observed from flow visualization experiments. Using fundamental vortex dynamics, Rule & Bliss (1998) have highlighted the complexity of the modeling problem. Modern first-principles finite-difference methods, such as those discussed by McCroskey (1995) and Tang & Baeder (1999), have achieved better success in predicting the point of vortex origination. However, for many methods, numerical diffusion and dispersion tends to produce significant errors in the subsequent vortex behavior, overpredicting the growth of the viscous core to the point that these methods cannot yet be used to confidently model the rotor wake and induced velocity field.

10.6.1 *Velocity Field*

Understanding the velocity field near the tip vortex, the vortex strength (circulation), viscous core radius, as well as how these properties change as the vortex ages is still a goal of ongoing research. The earliest reported measurements of helicopter rotor tip vortices were performed using hot-wire anemometry (HWA) – see Simons et al. (1966), Cook (1972), Caradonna & Tung (1981), and Tung et al. (1983). However, HWA is limited by spatial resolution, and probe proximity concerns make measurements difficult at early wake ages near the blade tips. Laser Doppler velocimetry (LDV) is an attractive alternative because of the nonintrusive nature of the measurements (apart from seeding) and can alleviate many limitations posed by HWA, such as physical size of the probe relative to the vortex dimensions, probe interference effects, and wire attrition. However, among the main constraints of LDV are optical access, the need to uniformly seed the flow, the need for good periodicity of the flow to allow statistical phase averaging, and a requirement for coincident measurements on all three components of velocity.

Scully & Sullivan (1972), Sullivan (1973), and Landgrebe & Johnson (1974) have reported some of the first uses of LDV systems to study rotor flows. Thomson et al. (1988) and Mahalingam & Komerath (1998) have made detailed LDV measurements of rotor tip vortices using a 1-D LDV system. A 2-D LDV system was developed by Biggers & Orloff (1975) to investigate some aspects of the wake generated by a two-bladed rotor – see also Biggers et al. (1977a,b). Hoad (1983) has developed a 2-D LDV system with large stand-off distances – see also Elliott et al. (1988) and Althoff et al. (1988). However, 1-D and 2-D LDV systems tend to suffer from reduced spatial resolution because of the elongated measurement volumes, especially with large stand-off distances, and are not always well suited for measuring the small spatial scales and steep velocity gradients found inside vortices. Three-dimensional phase-resolved LDV measurements in rotor wakes, however, have not been possible until recently – see Seelhorst et al. (1994, 1996), McAlister et al. (1995), McAlister (1996), Leishman et al. (1995), and Han et al. (1997).

Representative LDV measurements of the tip vortex induced velocity field are shown in Fig. 10.16 for several wake ages (ψ_w). The tangential component of velocity has been nondimensionalized by the rotor tip speed, and the distance from the rotational axis has been nondimensionalized by the rotor radius, R. As described previously in Chapter 2, the blade tip vortices lie on the boundary of a contracting jetlike wake, and the measurements shown in Fig. 10.16 were made on this wake boundary. Inside the wake boundary, note that the slipstream velocity is approximately constant. For increasing ψ_w, the tip vortex

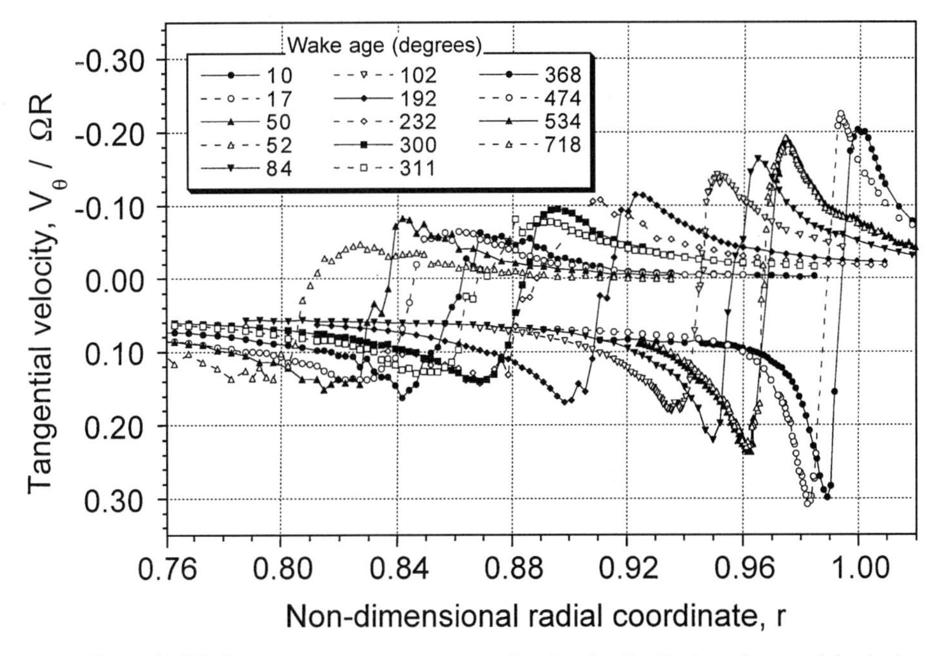

Figure 10.16 Representative measurements showing the distribution of tangential velocity surrounding a tip vortex at various wake ages. Source: Leishman et al. (1995) and the University of Maryland.

convects downward below the rotor and radially inward, and so the velocity signatures shown in Fig. 10.16 move to the left on the graphs. The results show the basic characteristics of viscous diffusion: Vorticity contained in the core diffuses radially outward with time, thereby reducing the peak swirl velocities. The distance between the two velocity peaks shown in Fig. 10.16 can be considered the viscous core diameter.

10.6.2 *Models of the Vortex*

With this general picture of the tip vortex induced velocity field in mind, a model of the tip vortex can now be hypothesized. Tip vortex models used in rotor wake simulations are typically specified in terms of a 2-D tangential (swirl) velocity profile. The other velocity components (the axial and radial) are small and are usually neglected. This, however, may not be a justifiable assumption at young wake ages. A schematic of a typical distribution of swirl velocity is shown in Fig. 10.17. The inner part of the vortex (the core region) almost rotates as a solid body, whereas the outer part behaves almost as a potential flow. The core

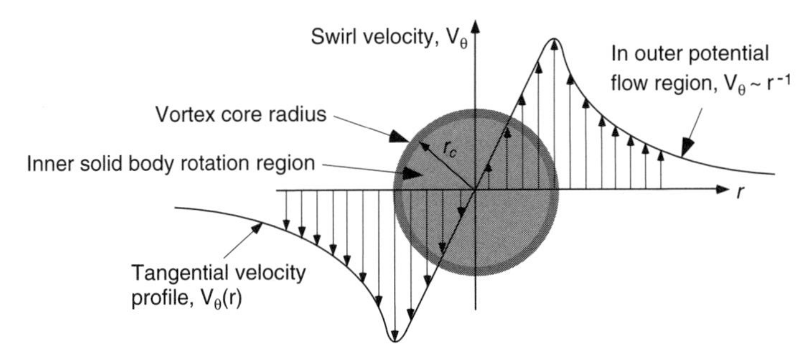

Figure 10.17 Idealization of the tangential (swirl) velocity inside a tip vortex.

radius, r_c, is defined as the radial location where V_θ is a maximum (i.e., at the core radius $\bar{r} = r/r_c = 1$). This boundary demarcates the inner (almost purely rotational) flow field from the outer (potential) flow.

The simplest model of a vortex with a finite core is the Rankine vortex, where the swirl velocity is given by

$$V_\theta(\bar{r}) = \begin{cases} \left(\dfrac{\Gamma_v}{2\pi r_c}\right)\bar{r}, & 0 \le \bar{r} \le 1, \\ \left(\dfrac{\Gamma_v}{2\pi r_c}\right)\dfrac{1}{\bar{r}}, & \bar{r} > 1. \end{cases} \tag{10.5}$$

The core is modeled as a solid body rotation, and the velocity outside decreases hyperbolically with distance as in a potential vortex. Although the Rankine vortex exhibits the key features of a viscous vortex, the swirl velocity, the vorticity, and the circulation of this model produce unrealistically higher values than are observed in practice. An alternative vortex model is given by Oseen (1912) and Lamb (1932), who solved a simplified form of the Navier–Stokes equations. The Oseen–Lamb model for the swirl velocity is

$$V_\theta(\bar{r}) = \frac{\Gamma_v}{2\pi r_c \bar{r}}(1 - e^{-\alpha \bar{r}^2}), \tag{10.6}$$

where $\alpha = 1.25643$. This solution is strictly valid only for a single viscous vortex in an unbounded incompressible flow. Newman (1959) has also derived exponential solutions for the three components of velocity in the vortex core based on a simplified Navier–Stokes formulation. The result for the swirl velocity is the same as that for the Oseen–Lamb model, but Newman shows that the axial velocity in the vortex is

$$V_z = -\frac{A}{z} e^{-\alpha \bar{r}^2}, \tag{10.7}$$

where A is a constant that can be related to the drag on the generating lifting surface. These exponential velocity profiles have been shown to give good correlations with tip vortex measurements generated by fixed wings – see, for example, Dosanjh et al. (1962).

A more general series of desingularized velocity profiles for columnar vortices with continuous distributions of the flow quantities is given by Vatistas et al. (1991). The tangential velocity in the vortex can be expressed as

$$V_\theta(r) = \left(\frac{\Gamma_v}{2\pi r_c}\right) \frac{\bar{r}}{(1 + \bar{r}^{2n})^{\frac{1}{n}}}, \tag{10.8}$$

where n is an integer. Using Eq. 10.8, the swirl velocity profiles for two specific vortex models can be rewritten in terms of the nondimensional radius, $\bar{r}$. For the $n = 1$ vortex, which is commonly known in helicopter work as the Scully (1975) vortex,[3] then

$$V_\theta(\bar{r}) = \left(\frac{\Gamma_v}{2\pi r_c}\right) \frac{\bar{r}}{(1 + \bar{r}^2)}, \tag{10.9}$$

and for the $n = 2$ vortex

$$V_\theta(\bar{r}) = \left(\frac{\Gamma_v}{2\pi r_c}\right) \frac{\bar{r}}{\sqrt{1 + \bar{r}^4}}. \tag{10.10}$$

[3] Note that if $\bar{r}$ is small, then Eq. 10.6 can be written as $V_\theta(\bar{r}) = (\frac{\Gamma}{2\pi r_c})(\frac{\bar{r}}{1+\bar{r}^2})$, which is in the same form as Eq. 10.9.

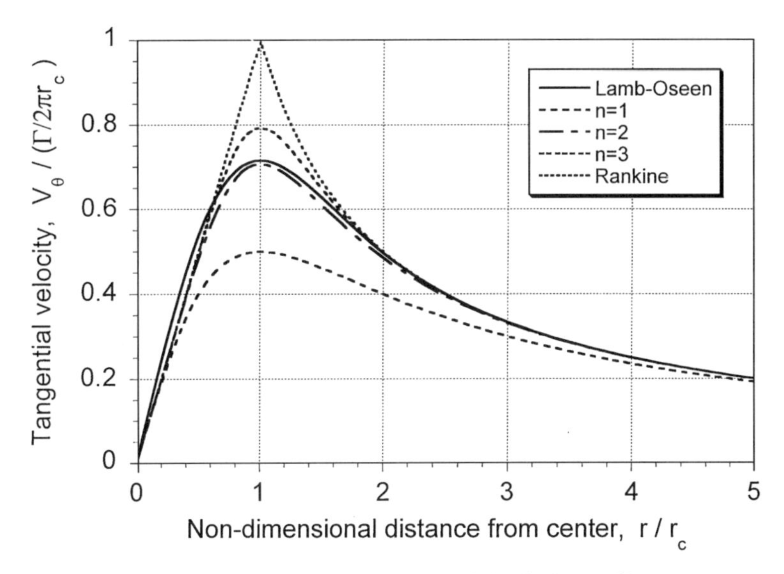

Figure 10.18 Summary of 2-D vortex tangential velocity profiles.

Note from Eq. 10.8 that as $n \to \infty$, the Rankine vortex in Eq. 10.5 is obtained. A summary of the various tangential velocity models is plotted in Fig. 10.18.

Because of the assumption of columnar vortex, the axial and radial components derived by Vatistas et al. (1991) do not apply to trailed vortices. However, a general result for the axial (and radial) components of the velocity for trailed vortices can also be deduced starting from the assumption of Eq. 10.8 and the appropriate form of the governing Navier–Stokes equations applied to a trailing vortex. In the first instance, it is reasonable to assume that the induced velocities are dependent only on r and z (i.e., the vortex is assumed to be axisymmetric). The governing Navier–Stokes equations in cylindrical coordinates for incompressible flow are written as

continuity:

$$\frac{1}{r}\frac{\partial}{\partial r}(r V_r) + \frac{\partial V_z}{\partial z} = 0; \tag{10.11}$$

r momentum:

$$V_r \frac{\partial V_r}{\partial r} + V_\infty \frac{\partial V_r}{\partial z} + V_z \frac{\partial V_r}{\partial z} - \frac{V_\theta^2}{r} = -\frac{1}{\rho}\frac{\partial p}{\partial r} + \nu\left[\nabla^2 V_r - \frac{V_r}{r^2}\right], \tag{10.12}$$

θ momentum:

$$V_r \frac{\partial V_\theta}{\partial r} + V_\infty \frac{\partial V_\theta}{\partial z} + V_z \frac{\partial V_\theta}{\partial z} + \frac{V_r V_\theta}{r} = \nu\left[\nabla^2 V_\theta - \frac{V_\theta}{r^2}\right], \tag{10.13}$$

z momentum:

$$V_r \frac{\partial V_z}{\partial r} + V_\infty \frac{\partial V_z}{\partial z} + V_z \frac{\partial V_z}{\partial z} = -\frac{1}{\rho}\frac{\partial p}{\partial z} + \nu\left[\nabla^2 V_z\right]. \tag{10.14}$$

These equations can now be nondimensionalized, with velocities being nondimensionalized by blade tip speed, distances by blade chord, and pressure by the tip dynamic pressure. An ordering scheme can then be employed to compare the magnitudes of the various nondimensional terms. The swirl and axial velocities induced by the tip vortex are known from experiments to be small compared to the free-stream velocity; that is, $\bar{V}_\theta$ and $\bar{V}_z$ are $O(\epsilon)$, where the overbar denotes a nondimensional quantity. Typically the tip vortex core radius is found to be of the order of the airfoil thickness. The tip vortex can be observed, and induces significant velocities, for several chord lengths downstream of the generating surface. This implies that if $\bar{z}$ is $O(1)$ then $\bar{r}$ is $O(\epsilon)$. The continuity equation implies that the radial velocity, $\bar{V}_r$, is much smaller and of order $O(\epsilon^2)$. This is also consistent with the observation that the radial velocities induced by the tip vortices are small – see Dosanjh et al. (1962). The radial momentum equation (Eq. 10.12) indicates that the pressure, $\bar{p}$, is $O(\epsilon^2)$. Typically, for a helicopter rotor, the chord Reynolds number at the blade tip is approximately 10^7. Even in wind-tunnel experiments with scaled models, the Reynolds numbers are always greater than 10^4. Therefore, Re can be assumed to be $O(1/\epsilon^4)$.

The conservation laws can now be rewritten with the order of magnitudes of each term being indicated below each term. Note that all the variables are now nondimensional, but the overbar has been omitted for simplicity.

continuity:

$$\underset{\epsilon}{\frac{1}{r}\frac{\partial}{\partial r}(rV_r)} + \underset{\epsilon}{\frac{\partial V_z}{\partial z}} = 0, \tag{10.15}$$

r momentum:

$$\underset{\epsilon^3}{V_r\frac{\partial V_r}{\partial r}} + \underset{\epsilon^2}{\frac{\partial V_r}{\partial z}} + \underset{\epsilon^3}{V_z\frac{\partial V_r}{\partial z}} - \underset{\epsilon}{\frac{V_\theta^2}{r}} = -\underset{\epsilon}{\frac{\partial p}{\partial r}} + \underset{\epsilon^4}{\frac{1}{Re}\left[\nabla^2 V_r - \frac{V_r}{r^2}\right]}, \tag{10.16}$$

θ momentum:

$$\underset{\epsilon^2}{V_r\frac{\partial V_\theta}{\partial r}} + \underset{\epsilon}{\frac{\partial V_\theta}{\partial z}} + \underset{\epsilon^2}{V_z\frac{\partial V_\theta}{\partial z}} + \underset{\epsilon^2}{\frac{V_r V_\theta}{r}} = \underset{\epsilon^3}{\frac{1}{Re}\left[\nabla^2 V_\theta - \frac{V_\theta}{r^2}\right]}, \tag{10.17}$$

z momentum:

$$\underset{\epsilon^2}{V_r\frac{\partial V_z}{\partial r}} + \underset{\epsilon}{\frac{\partial V_z}{\partial z}} + \underset{\epsilon^2}{V_z\frac{\partial V_z}{\partial z}} = -\underset{\epsilon^2}{\frac{\partial p}{\partial z}} + \underset{\epsilon^3}{\frac{1}{Re}\left[\nabla^2 V_z\right]}. \tag{10.18}$$

It is further assumed that because ϵ is small, higher order terms in ϵ can be neglected. Therefore, the governing equations simplify to

$$\frac{1}{r}\frac{\partial}{\partial r}(rV_r) + \frac{\partial V_z}{\partial z} = 0, \tag{10.19}$$

$$\frac{\partial p}{\partial r} - \frac{V_\theta^2}{r} = 0 \qquad [O(\epsilon^2)], \tag{10.20}$$

$$\frac{\partial V_\theta}{\partial z} = 0 \qquad [O(\epsilon^2)], \tag{10.21}$$

$$\frac{\partial V_z}{\partial z} = 0 \qquad [O(\epsilon^2)]. \tag{10.22}$$

It should be noted that the θ- and z-momentum equations indicate that the swirl and axial velocities have small gradients along the z axis and are not independent of z. This is consistent with various experimental observations that viscous diffusion is a gradual process and, therefore, the velocity gradients along the time like z axis are small. It is also observed that the peak axial velocities decrease rapidly at early wake ages, whereas the peak swirl velocities decrease more gradually ($\sim\sqrt{z}$) (i.e., the swirl velocity gradients are small compared to the axial velocity gradients). Using this information, the θ- and z-momentum equations can be written as

$$\frac{\partial V_\theta}{\partial z} = 0 \; [O(\epsilon^3)] \quad \text{and} \quad \frac{\partial V_z}{\partial z} = 0 \quad [O(\epsilon^2)], \tag{10.23}$$

and the $O(\epsilon^2)$ terms in the θ-momentum equation cancel each other, that is,

$$V_r \frac{\partial V_\theta}{\partial r} + V_z \frac{\partial V_\theta}{\partial z} + \frac{V_r V_\theta}{r} = 0. \tag{10.24}$$

The required boundary conditions given by Newman (1959) can be obtained by assuming that the vortex is generated as a "free" vortex, that is, a potential vortex of strength Γ_v at $z = 0$, and it diffuses until at large distances the vortex induced velocities become zero. These boundary conditions can be formalized as

1. At $z = 0$, $V_\theta = \Gamma_v / 2\pi r$ and V_z, $V_r = 0$. Note that there is a singularity at $r = 0$.
2. For $z > 0$, $V_\theta, V_z, V_r \to 0$ for large r.
3. As $z \to \infty$, $V_\theta, V_z, V_r \to 0$ for all r.

The governing equations have now been reduced to only two equations (Eqs. 10.19 and 10.24) but are in terms of three components of velocity, for which there can exist no unique solution. However, with the assumption of the Vatistas swirl velocity model in Eq. 10.8, the axial and radial velocities can be found using Eqs. 10.19 and 10.24. These components are found to be

$$V_z = -\frac{A}{z}\left\{1 - \frac{r^2}{\left(r_c^{2n} + r^{2n}\right)^{1/n}}\right\} \quad \text{and}$$
$$V_r = -\frac{Ar}{2z^2}\left\{1 - \frac{r^2}{\left(r_c^{2n} + r^{2n}\right)^{1/n}}\right\}, \tag{10.25}$$

where A is a constant.

The results from the above analysis of the vortex problem have been compared with some of the experimental measurements shown previously in Fig. 10.16. The time-averaged convection velocity of the tip vortex through the rotor wake has been removed, which puts the measurements into a frame of reference moving with the vortex core. Figure 10.19(a) shows the measured swirl velocities nondimensionalized with respect to the peak velocity and the radial distance with the estimated core radius. The measurements show a good correlation with the Oseen–Lamb (Eq. 10.6) and $n = 2$ (Eq. 10.10) swirl velocity models, and they also confirm the self-similar nature of the velocity field surrounding the tip vortex. Some deviations may be attributed to the presence of extraneous wake vorticity from the inboard part of the blade and other blades that are present in the measurement grid. The commonly used $n = 1$ or Scully (1975) vortex model (Eq. 10.9) seems to be a less accurate descriptor of the velocity field, at least in this case. Figure 10.19(b) shows corresponding comparisons for the nondimensional axial velocities. Again, the Lamb–Oseen and $n = 2$ models are found to give good agreement with the measured velocities. Like the swirl velocity, the

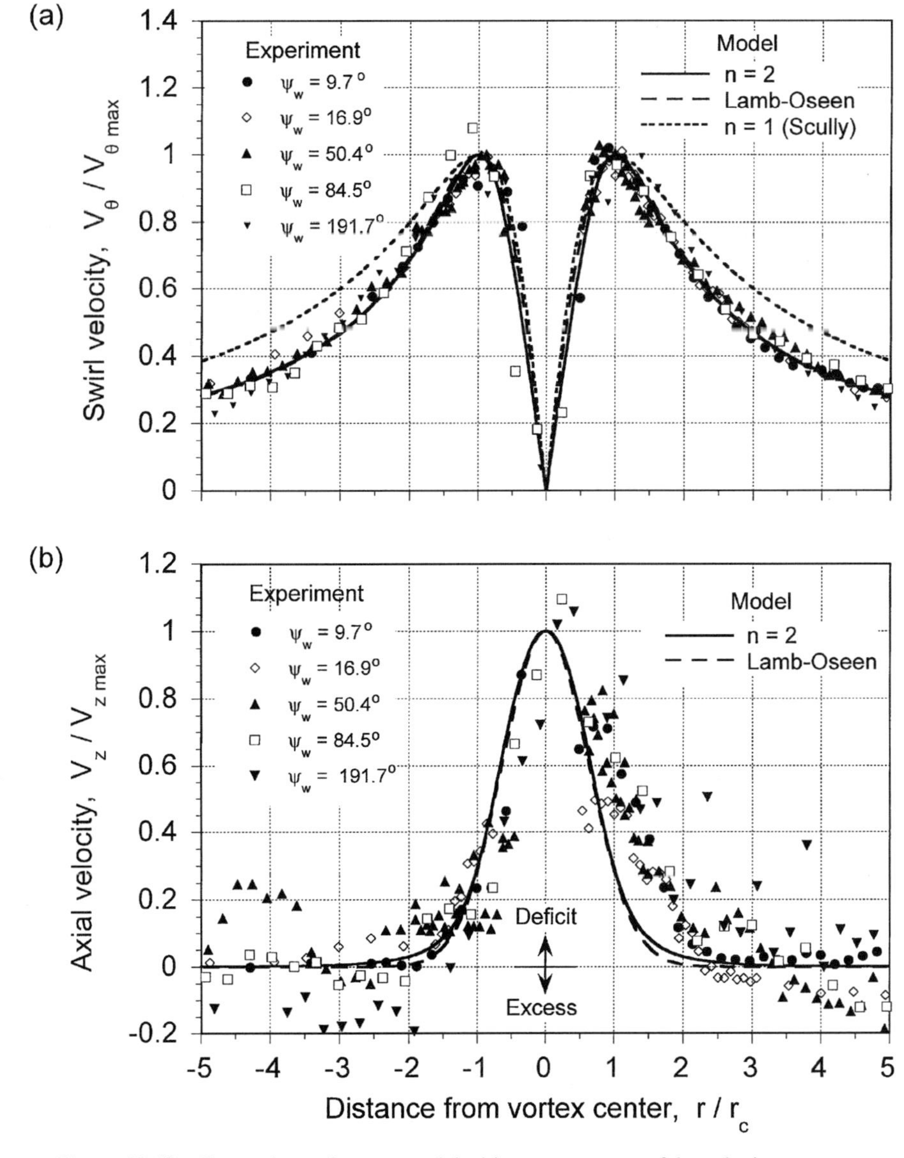

Figure 10.19 Comparison of vortex model with measurements of the velocity components inside the vortex. (a) Tangential (swirl) velocity. (b) Axial velocity.

axial velocity profiles also exhibit a self-similarity, though perhaps not as strongly. The fluctuations in the measured axial velocities are probably a result of small asymmetries in the tip vortices. However, the generally good agreement with the measurements suggests the validity of the assumptions inherent in this very basic analysis of the trailed vortex problem.

10.6.3 *Vorticity Diffusion Effects and Vortex Core Growth*

The vortex core dimension is an important parameter that can be used to help define the structure and evolution of tip vortices. Half of the distance between the two velocity

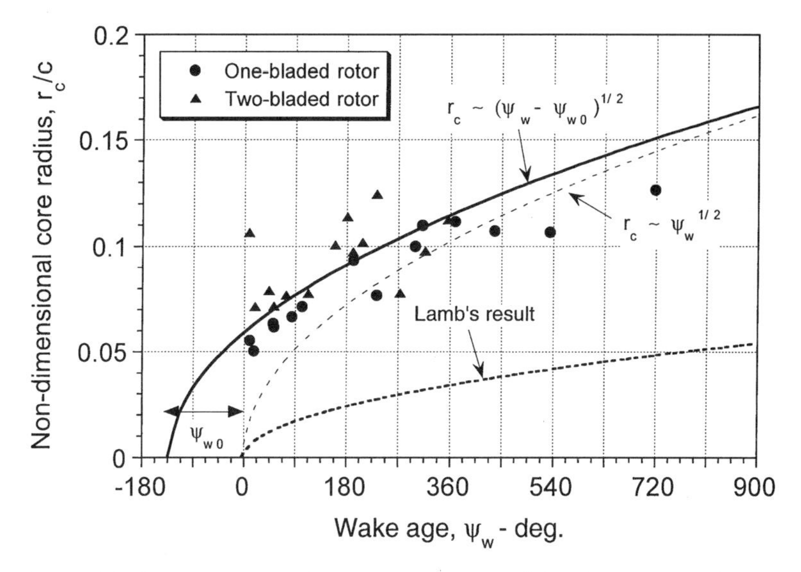

Figure 10.20 Example showing measured growth of the viscous core radius versus wake age. Rectangular tip rotor. $C_T/\sigma \approx 0.1$.

peaks shown in Fig. 10.16 can be considered as the average (nominal) core radius. This quantity has been estimated from the results shown in Fig. 10.16 and is plotted as a fraction of blade chord, c, in Fig. 10.20 as a function of wake age, ψ_w. It is apparent that there is a fairly rapid but asymptotic-like growth in the core, at least up to the first blade passage at $\psi_w = 2\pi/N_b$. The blade passage event temporarily alters the growth trend, and a slight decrease in core size is observed. This is more evident in the case of the two-bladed rotor because the tip vortex does not have time to convect through the flow and is nearly twice as close to the following blade. After the blade passage, the core continues to grow, although the growth rate appears to be slower.

A simple quantitative estimate of the growth in core radius with time can be based on Lamb's result for laminar flows – see Lamb (1932), Chigier & Corsiglia (1971), and Ogawa (1993). This result shows that without external velocity gradients the core radius varies with the square root of age according to $r_c(t) = \sqrt{4\alpha\nu t}$ (see Question 10.6). In practice, as a result of turbulence generation, the actual diffusion of vorticity contained in the vortex is known to be much quicker than molecular diffusion would suggest. It can be expected that the tip vortex initially exhibits a predominantly laminar behavior; thereafter the vortex progressively becomes turbulent. The time scales over which this degeneration to turbulence occurs must be related to the vortex Reynolds number, $\Gamma_v/\nu = Re_v$, among other factors. The rate of development of the turbulent structures near the core will alter the effective diffusion rate of the vorticity contained inside the vortex.

These effects, albeit very complicated on a fundamental level, can be incorporated into a model core growth equation using an average "turbulent" viscosity coefficient [see Dosanjh et al. (1962), Squire (1965), and Ogawa (1993)], where now $r_c(t) = \sqrt{4\alpha\delta\nu t}$. The coefficient δ takes into account the average increased rate of vorticity diffusion as a result of turbulence generation and is a coefficient that must be deduced from vortex core structure measurements. Squire (1965) has postulated that the effective (turbulent) viscosity coefficient, δ, should be a function of the circulation strength (circulation) of the trailed tip vortex, Γ_v. Because, in practice, the strength (circulation) of the tip vortex is found to remain nominally constant with time [see Dosanjh et al. (1962), Mahalingam & Komerath (1998), and

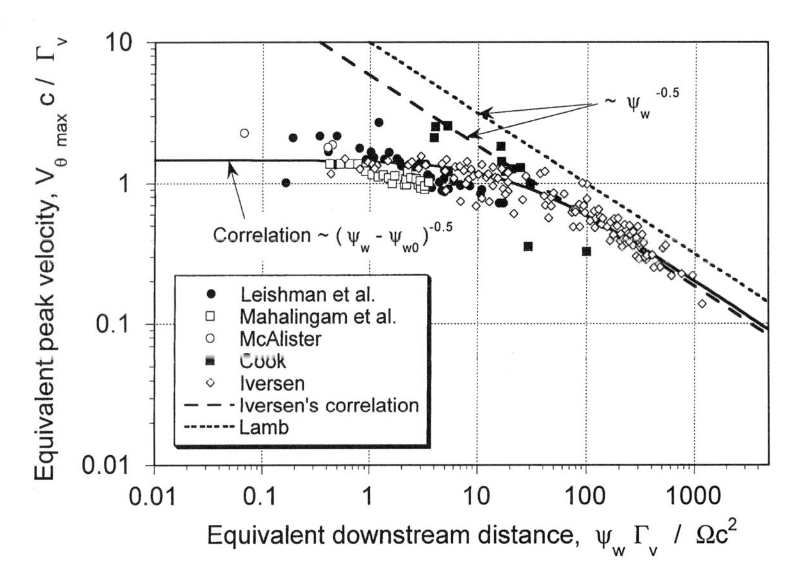

Figure 10.21 General correlation of rotor tip vortex measurements in terms of nondimensional maximum tangential velocities and nondimensional downstream distance.

Bhagwat & Leishman (1998)], it is reasonable to assume a constant value of δ. However, this hypothesis can only be verified through experimental correlation. In addition, because it is known that the vortex is already in some stage of decay immediately after its formation, it can be hypothesized that the growth curve can be originated at a virtual time, say t_0, so that $r_c(t) = \sqrt{4\alpha\delta\nu(t - t_0)}$ – see Squire (1965).

10.6.4 *Correlation of Rotor Tip Vortex Data*

Figure 10.21 shows results derived from the tip vortex measurements in terms of an equivalent maximum nondimensional tangential velocity as a function of equivalent nondimensional downstream distance from the blade tip. Following an approach similar to Iversen (1976), we can write the nondimensional velocity and distance (wake age) as

$$\bar{V}_{\theta_{\max}} = \left(\frac{V_{\theta_{\max}}}{\Omega R}\right)\left(\frac{\Omega R c}{\Gamma_v}\right) = \frac{V_{\theta_{\max}} c}{\Gamma_v}, \tag{10.26}$$

$$\bar{d} = \left(\frac{\psi_w R}{c}\right)\left(\frac{\Gamma_v}{\Omega R c}\right) = \frac{\psi_w \Gamma_v}{\Omega c^2}. \tag{10.27}$$

Figure 10.21 shows tip vortex measurements for small-scale helicopter rotors that have been published by Leishman et al. (1995), Mahalingam & Komerath (1998), and McAlister et al. (1995). Also shown are measurements by Cook (1972) using a full-scale rotor. When the results are plotted in this manner, the data emphasize a strong correlation. The decreasing trend in the core swirl velocity, which is also observed in the fixed-wing case [see Iversen (1976)] confirms that the tip vortices diffuse logarithmically with increasing wake age. The measured data show the trend $\bar{V}_{\theta_{\max}}(\bar{d}+2.197)^{\frac{1}{2}} = 2.782$. Iversen reports a correlation region after a nondimensional distance of 50, where the data show the trend $\bar{V}_{\theta_{\max}}(\bar{d})^{1/2} = 5.8$. This decreasing trend in the peak swirl velocity results from viscous diffusion. For constant net vortex circulation, the nondimensional peak velocity is inversely proportional to the core radius, and the distance $\bar{d}$ is proportional to the wake age (i.e., $\bar{V}_{\theta_{\max}} \propto r_c^{-1}$ and $\bar{d} \propto \psi_w$).

This is consistent with the experimental measurements shown in Fig. 10.20, where the core radius initially increases as the square root of wake age (i.e., $r_c \propto \sqrt{\psi_w}$). The theoretical Lamb result given previously for the viscous core growth can be written in terms of vortex age as

$$r_c = \sqrt{4\alpha\nu\left(\frac{\psi_w}{\Omega}\right)}. \tag{10.28}$$

Substituting $V_{\theta_{\max}} = \Gamma_v(1 - e^{-\alpha})/2\pi r_c$, and using Eqs. 10.26 and 10.27, gives

$$\bar{V}_{\theta_{\max}}(\bar{d})^{\frac{1}{2}} = \frac{1 - e^{-\alpha}}{4\pi}\sqrt{\frac{1}{\alpha}\left(\frac{\Gamma_v}{\nu}\right)}. \tag{10.29}$$

As shown previously, Lamb's result can be modified empirically to include an average effective (turbulent) viscosity coefficient, δ; that is, Eq. 10.29 can now be written as

$$\bar{V}_{\theta_{\max}}(\bar{d})^{\frac{1}{2}} = \frac{1 - e^{-\alpha}}{4\pi}\sqrt{\frac{1}{\alpha}\left(\frac{\Gamma_v}{\delta\nu}\right)}. \tag{10.30}$$

Comparing this with Iversen's results in Fig. 10.21, we see that Iversen's correlation is equivalent to including an effective turbulent viscosity coefficient that is proportional to the vortex Reynolds number, Γ_v/ν. Therefore, the peak velocity trend shown in Fig. 10.21 corresponds to the core growth trend

$$r_c(\psi_w) = \sqrt{4\alpha\delta\nu\left(\frac{(\psi_w - \psi_{w_0})}{\Omega}\right)} \equiv \sqrt{r_0^2 + \frac{4\alpha\delta\nu\psi_w}{\Omega}}. \tag{10.31}$$

The ordinate shift, ψ_{w_0}, results in an effective nonzero core radius, r_0, at the tip vortex origin where $\psi_w = 0^\circ$ and, therefore, a finite velocity at $\bar{d} = 0$. However, in the other two models (Lamb and Squire) the swirl velocity is singular at the origin of the tip vortex and unrealistically high at small distances (wake ages). At large downstream distances (wake ages), all three curves show the same qualitative trend, that is, the velocity is inversely related to $\sqrt{t}$, or equivalently, the core radius increases in proportion to $\sqrt{t}$.

The viscous core growth trend as a function of wake age as derived from the above correlation trend is plotted in Fig. 10.20, along with the Lamb (laminar) result and a Lamb-type trend with a higher (constant) effective turbulent viscosity. Again, the differences between the two curves are obvious at early equivalent wake ages. At later wake ages, the two curves have a qualitatively similar behavior. At large wake ages, they will be almost coincident, that is,

$$r_c \propto \sqrt{\frac{\psi_w}{\Omega}} \quad \text{for } \psi_w \gg \psi_{w_0}. \tag{10.32}$$

10.7 Vortex Models of the Rotor Wake

In vortex wake models, the convection of the tip vortices (and other concentrated vorticity) is explicitly tracked through the flow field relative to the rotor. Vortex diffusion and convection are, however, treated separately. The underlying principle is that of Helmholtz's law, also called the vorticity transport theorem. Typically, the vortical wake is modeled by

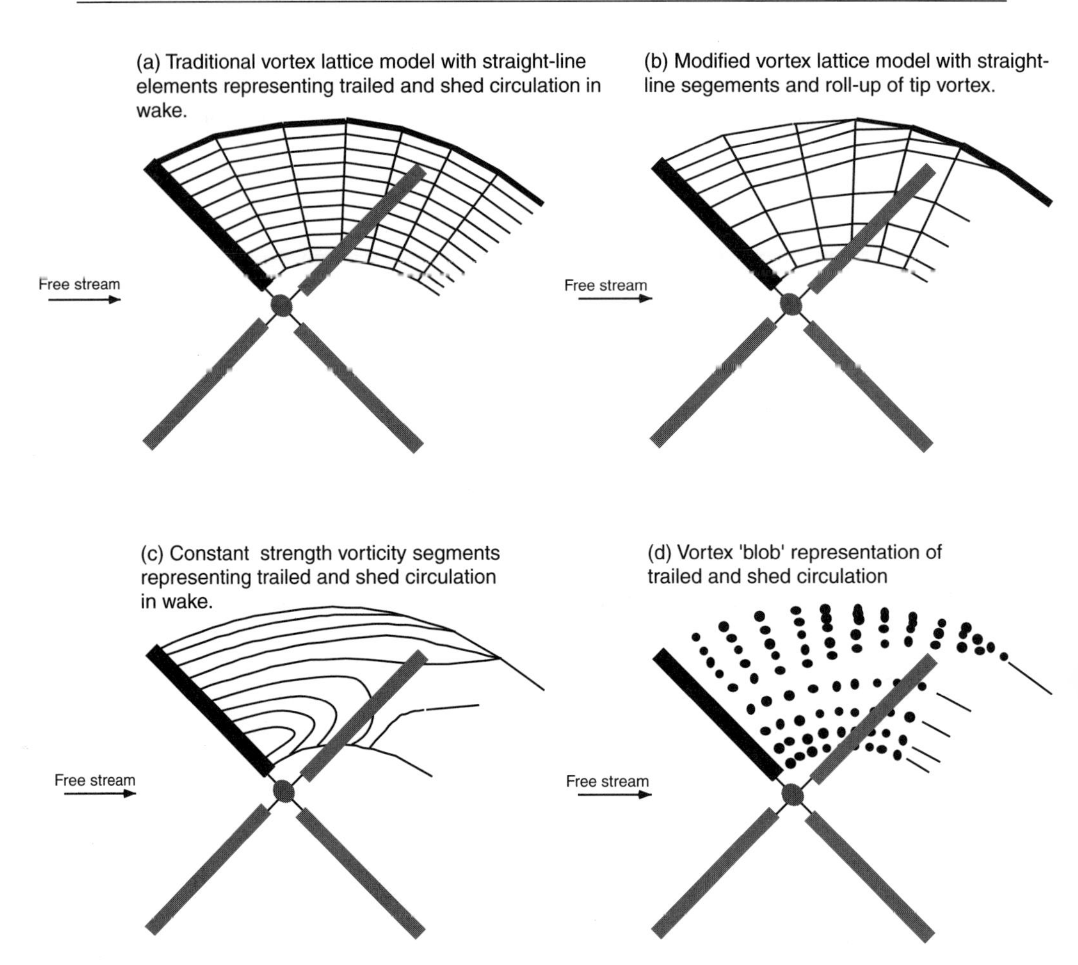

Figure 10.22 Representation of the vortex wake trailed behind a rotor blade. Note: The wake from only one blade is shown for clarity. (a) Straight-line filaments. (b) Straight filaments with a prescribed roll-up model. (c) Constant vorticity curved vortex filaments. (d) Vortex "blobs".

vortex lines[4] that are discretized into the form of a regular lattice with straight elements [see Clark & Leiper (1970), Sadler (1971a,b), and Scully (1975)], continuous curved vortex lines [see Quackenbush et al. (1988)], or vortex "blobs" [see Lee & Na (1995, 1998)]. Schematics of these representations are shown in Fig. 10.22, where the wake from only one blade is shown for clarity. For most applications, the vortex wake model will be coupled to either a lifting-line or lifting-surface representation of the rotor blades. This defines the initial strengths and locations of the wake vortices relative to the rotor and forms boundary conditions for the remainder of the "far" vortex wake.

The main advantage of vortex methods, whether the vortex be comprised of straight line filaments, curved filaments, or blobs, is that once the positions and strengths of the wake vorticity are obtained, the induced velocity field at the rotor (or elsewhere, as required) can be computed by the use of the Biot–Savart law, followed by numerical integration

[4] A vortex line is simply a curve in the fluid that is tangent to the local vorticity vector; it should not be confused with a line vortex.

over all the filaments. The process lends itself readily to the computer. However, the main disadvantage of vortex methods is one of relatively high computational expense. While the evaluation of the Biot–Savart integral is trivial for a single vortex element, it is the very large number of elements required to model the rotor wake that determines the expense because a typical vortex filament may be discretized into several hundred individual segments. For many applications in rotor analysis, it is the relatively high cost of vortex methods that limits their routine use. These methods are, however, still many orders of magnitude less expensive than finite-difference solutions based on the Euler or Navier–Stokes equations [see McCroskey (1995)].

The principal differences among all of the various discretized vortex wake models that have been developed for helicopter applications are the assumptions and methods employed for the solution of the vorticity transport equations. In prescribed vortex wake models, the difficulties inherent in trying to explicitly solve these equations directly are avoided by either specifying the locations of the wake filaments directly or by assuming an approximate velocity field near the rotor. Usually with the latter method, a very simple velocity field can be prescribed that allows an analytical solution for the wake geometry. Alternatively, when specifying the wake geometry directly, approximate relations can be derived from a combination of momentum theory and flow visualization experiments. In so-called free vortex methods, the solution for the wake is obtained from first principles. This form of solution recognizes that because parts of the vortices in the wake may interact with themselves (a self-induced effect) or with other vortices (a mutually induced effect) or with the blades, the overall behavior of the wake cannot easily be generalized or prescribed.

10.7.1 *Biot–Savart Law*

Fundamental to all vortex models is the requirement to compute the induced velocity at a point contributed by a vortex filament in the wake. This can be calculated through the application of the Biot–Savart law – see Batchelor (1967). The induced velocity $d\vec{v}$ at any point, P, a distance $\vec{r}$ from the segment $d\vec{l}$ of a vortex filament of strength Γ_v can be written in general form as

$$d\vec{v} = \frac{\Gamma_v}{4\pi}\frac{d\vec{l} \times \vec{r}}{|r|^3}. \tag{10.33}$$

Although the Biot–Savart result is singular at $r = 0$, this can be rectified by using vortex models with finite cores or by using other cutoff methods (see Section 10.6.2). Curved wake filaments would normally be discretized into a number of collocation points that are connected by straight-line vortex segments (see Fig. 10.23) because the velocity induced by a straight-line vortex segment is readily integrable. Errors associated with discretization into straight elements are usually small[5] because the radius of curvature of the rotor wake is generally large, except perhaps in the roll-up regions at the lateral edges of the wake. The total velocity at P is then obtained by integration along the lengths of the vortex filaments.

To help reduce the number of numerical operations associated with vortex methods, the complexity of the real rotor wake can be reduced into various simplified forms. A common approximation is to treat only the trailed vortex filaments, that is, the filaments that are initially trailed perpendicular to the blade span (see Fig. 8.3). These contributions to the

[5] Discretization of curved vortex filaments into straight lines will result in an induced velocity calculation that is, at best, second-order accurate. For some free vortex wake schemes, this can result in round-off errors that may be a source of numerical problems.

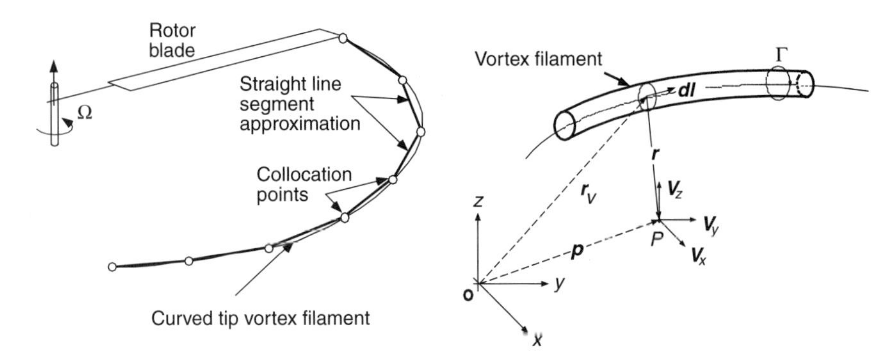

Figure 10.23 Evaluation of the induced velocity from an element of general curved vortex filament using the Biot–Savart law.

vorticity field arise because of the spanwise gradients in the distribution of circulation loading on the blades. The time-dependent aerodynamic loading results in shed wake vorticity, and to save tracking all the shed elements in the wake this effect can be approximated by means of one of the unsteady aerodynamic models considered in Chapter 8. Even with the trailed wake system alone, the problem of calculating the induced velocity field is still one of high numerical cost. However, experiments have shown the dominant structures in the rotor wake to be the tip vortices, and so a common level of modeling approximation is to consider just these tip vortices alone. Correlation studies with experimental measurements of the rotor inflow and blade loads have shown that this is a good level of approximation and does not sacrifice much physical accuracy in the problem, especially in forward flight, for a substantial reduction in computational effort.

10.7.2 *Blade Model*

The rotor blade itself is typically modeled as a series of 2-D blade elements, in the manner introduced in Chapter 3. If the induced velocities from the vortical wake can be calculated, then the induced angle of attack can be found, the lift on the elements determined, and the blade lift can be obtained by radial integration. However, to properly use vortex wake models, the loading on the blade must be properly coupled to the circulation that is trailed (and shed) into the wake. For this reason, lifting-line or reduced-order lifting-surface blade models have found use in rotor analyses. In principle, these models are similar to those used for fixed-wing analyses [see, for example, Bertin & Smith (1989)], but they must be formulated carefully for use in rotating-wing applications. Many different formulations can be found in the literature. Although the basic principles are the same, the numerical formulation can vary substantially.

A generic formulation of the problem will be described. The blade is modeled as a vortex of strength $\Gamma_b(y)$ that is "bound" to the aerodynamic center (1/4-chord). Then, using the Kutta–Joukowski theorem, the lift per unit span, dL, is given by

$$dL(y) = \rho V(y)\Gamma_b(y)\,dy = \frac{1}{2}\rho V(y)^2 c C_l(y)\,dy$$
$$= \rho V(y)^2 \pi \alpha_e(y) c\,dy, \tag{10.34}$$

where α_e is the "effective" angle of attack of that section and 2π is the lift-curve-slope of a 2-D thin airfoil. In a practical sense, this value can be replaced by the lift-curve-slope of

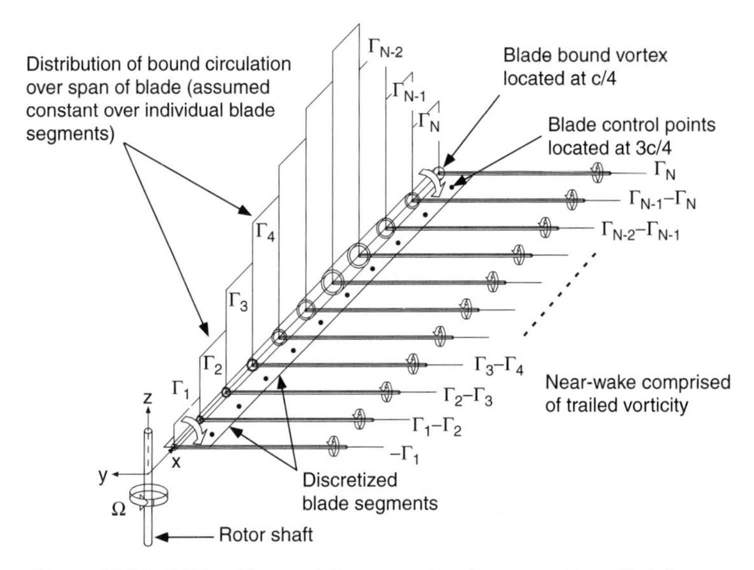

Figure 10.24 Lifting-line model representing the near wake trailed from a rotor blade.

the airfoil section at the appropriate combination of Mach number and Reynolds number. Rearranging the terms gives

$$V(y)\,\alpha_e(y) = \frac{\Gamma_b(y)}{2\pi(c/2)}. \tag{10.35}$$

For a 2-D airfoil the bound vortex is a infinite line vortex and must have a strength that is defined on the basis of the downwash at a distance $c/2$ from the $1/4$-chord, that is, the no-flow penetration condition is imposed at the $3/4$-chord. This is consistent with the unsteady thin-airfoil model of the problem described in Chapter 8. To extend this concept to three dimensions, the previous equation can also be written in terms of an influence coefficient, I_b, with

$$[I_b]_{i,j}\Gamma_{b_j} = V\alpha_i, \quad \text{where } I_b = \text{diag}[(\pi c)^{-1}]. \tag{10.36}$$

For a blade with N elements, as shown by Fig. 10.24, a control point is placed at the $3/4$-chord of each element. The objective is to solve numerically for the distribution of "bound" circulation over the blade, Γ_{b_j}, which simultaneously satisfies the no-flow penetration condition at all of the control points on the blade. Three-dimensional effects are included in the influence coefficient such that the velocity induced at the ith control point is given by $[I_{b_{ij}}]\Gamma_{b_j}$. If there are N elements on the blade, then for the ith element

$$\sum_{j=1}^{N} [I_b + I_{nw}]_{i,j}\,\Gamma_{b_j} = V\alpha_{e_i} = V\,(\theta_i - \phi_i)\,, \tag{10.37}$$

where I_{nw} is the influence of the "near-wake" vortex filaments that must trail from the edges of the blade element. These are usually modeled by a series of horseshoe vortices or vortex arcs. On the right-hand-side of the equation, ϕ_i contains the induced contributions from the remainder of the vortical wake (the "far wake") and must exclude all of the induced effects from the parts of the wake that are being defined as the near wake. Applying this equation at all control points results in a linear set of simultaneous equations of the form

$$[I_b + I_{nw}]\{\Gamma_b\} = \{V(\theta - \phi)\}, \tag{10.38}$$

which can be solved for Γ_b by standard methods for systems of simultaneous linear equations. Both direct and iterative methods can be used, although the direct method is usually preferable because of its superior numerical behavior. Note that, in general, both I_b and I_{nw} are fully populated matrices.

From the resulting circulation loading on the blade, the strengths of the vortices that trail from the edges of the blade elements (Fig. 10.24) provide a boundary condition for the far vortex wake. As mentioned previously, in many applications, only the tip vortex is modeled. In this case, all of the circulation outboard of the maximum bound circulation on the blade can be assumed to roll up into the tip vortex and thereby defines its strength. This strength is equivalent to the maximum bound circulation on the blade (see Question 10.7), although the actual strength of the tip vortex may also be adjusted empirically based on measurements, if these are available.

10.7.3 *Governing Equations for the Vortex Wake*

A general equation describing the positions of the vortex filaments in the rotor wake can be derived from Helmholz's law (the vorticity transport theorem) by assuming that for each point in the flow the local vorticity is convected at the local velocity, V_{loc}. Viscous diffusion of the vortex filaments is modeled separately, as described previously in Section 10.6.3. In the discretized problem, this convection criteria is applied to all of the collocation points that have been specified along the lengths of all the vortex lines (see Fig. 10.23). If we consider a single element of a trailed vortex filament, the fundamental equation describing the transport of the filament is

$$\frac{d\vec{r}}{dt} = \vec{V}_{\text{loc}}(\vec{r}, t), \tag{10.39}$$

where $\vec{r} = \vec{r}(\psi_w, \psi_b)$ is the position vector of a point on the filament at a time or wake age ψ_w that was trailed from the blade when it was at an azimuth angle ψ_b. The derivative on the left-hand side of this equation can be expanded to give

$$\frac{d\vec{r}(\psi_b, \psi_w)}{dt} = \frac{\partial \vec{r}}{\partial \psi_w}\left(\frac{\partial \psi_w}{\partial t}\right) + \frac{\partial \vec{r}}{\partial \psi_b}\left(\frac{\partial \psi_b}{\partial t}\right). \tag{10.40}$$

This equation can be simplified by noting that the time derivatives are

$$\frac{\partial \psi_w}{\partial t} = \frac{\partial \psi_b}{\partial t} = \Omega. \tag{10.41}$$

Therefore, the fundamental equation describing the position vector of the vortex filament is

$$\frac{\partial \vec{r}}{\partial \psi_w} + \frac{\partial \vec{r}}{\partial \psi_b} = \frac{1}{\Omega}\vec{V}_{\text{loc}}(\vec{r}, t). \tag{10.42}$$

This is a first-order, quasi-linear, hyperbolic, partial differential equation (PDE). The homogeneous portion of the equation (the left-hand side) is the linear wave equation. However, the right-hand side is, in general, a nonlinear function because of the induced effects resulting from the complete wake geometry. By assuming that every vortex filament is convected through the flow field at the local velocity, a PDE governing the geometry of a single element of the vortex filament can be written as

$$\frac{\partial \vec{r}(\psi_b, \psi_w)}{\partial \psi_b} + \frac{\partial \vec{r}(\psi_b, \psi_w)}{\partial \psi_w} = \frac{\vec{V}_\infty}{\Omega} + \frac{1}{\Omega}\sum_{j=1}^{N_v} \vec{V}_{\text{ind}}[\vec{r}(\psi_b, \psi_w), \vec{r}(\psi_j, \psi_w)]. \tag{10.43}$$

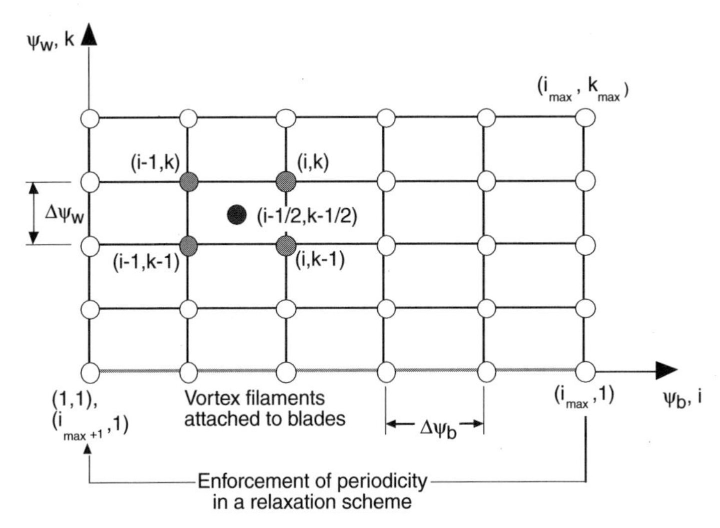

Figure 10.25 Discretized computational domain for free-vortex wake problem.

Note that the summation is carried out over the total number of trailed vortex filaments, N_v, that contribute to the induced velocity field at any given point.

To solve Eq. 10.43 using a numerical scheme, it must be spatially and temporally discretized. A computational domain can be defined as a discretized grid in time (ψ_b) and space (ψ_w), as shown in Fig. 10.25. Based on such discretizations, the partial differentials in the governing PDE can be approximated using several different types of finite-difference schemes. For example, Crouse & Leishman (1993) have used a "three-point central difference" approximation of the left-hand-side derivatives of Eq. 10.43 about point $(i - \frac{1}{2}, k - \frac{1}{2})$ using points $(i - 1, k - 1)$ and (i, k), as shown in Fig. 10.25. Bagai & Leishman (1995a,b, 1996) have used a "five-point central difference" scheme, where the derivatives were evaluated at the point $(i - \frac{1}{2}, k - \frac{1}{2})$ about points (i, k), $(i - 1, k)$, $(i, k - 1)$, and $(i - 1, k - 1)$ in the discretized computational domain, as also shown in Fig. 10.25.

The induced velocity, $\vec{V}_{\text{ind}}$, in Eq. 10.43 can be determined using the Biot–Savart law (Eq. 10.23) with

$$\vec{V}_{\text{ind}}[\vec{r}(\psi_b, \psi_w), \vec{r}(\psi_j, \psi_w)] \equiv \frac{1}{4\pi} \int \frac{\Gamma_v \, d\vec{\psi}_{w_j} \times \left(\vec{r}(\psi_b, \psi_w) - \vec{r}(\psi_j, \psi_w)\right)}{|\vec{r}(\psi_b, \psi_w) - \vec{r}(\psi_j, \psi_w)|^3}, \tag{10.44}$$

where $\vec{r}(\psi_b, \psi_w)$ is the point in the flow field influenced by the jth vortex at location $\vec{r}(\psi_j, \psi_w)$ that has strength Γ_v. The geometry of each element of a discretized vortex filament is governed by one of these equations, and the velocity term on the right-hand side couples the PDEs together. This means that the equations for all of the filaments must be solved simultaneously, and this is the primary expense of free vortex models (see Section 10.7.6).

Note that the singular nature of the Biot–Savart law can be avoided by replacing the induced velocity field of the individual vortices with a model with a desingularized core, as discussed previously in Section 10.6.2. Remember that this may also include a core growth (vorticity diffusion) model. The results for the wake geometry and induced velocity field thus obtained are probably qualitatively accurate no matter what vortex model is used. Yet, the quantitative errors associated with the wake model can only be determined through careful correlation studies with experiments documenting both the wake geometry and induced velocity field, and caution should always be employed.

10.7.4 *Prescribed Wake Models for Hovering Flight*

For hovering flight, generalized prescribed vortex wake models have been developed to enable predictions of the inflow through the disk, but without the expense and uncertainties associated with explicitly calculating the force-free positions of the wake. These models prescribe the locations of the rotor tip vortices (and sometimes also the inner vortex sheet) as functions of wake age ψ_w on the basis of experimental observations. For hovering flight, generalized prescribed vortex wake models have been developed by Landgrebe (1969, 1971, 1972), Gilmore & Gartshore (1972), Kocureck & Tangler (1976), and Kocureck & Berkowitz (1982).

Landgrebe's Model

Landgrebe (1971, 1972) studied experimentally about seventy rotor configurations with different combinations of number of blades, rotor solidity, blade twist, and blade aspect ratio. On the basis of these experiments, Landgrebe's model describes the tip vortex geometry by the equations

$$\frac{z_{\text{tip}}}{R} = \begin{cases} k_1 \psi_w & \text{for } 0 \leq \psi_w \leq 2\pi/N_b, \\ \left(\dfrac{z_{\text{tip}}}{R}\right)_{\psi_w = 2\pi/N_b} + k_2\left(\psi_w - \dfrac{2\pi}{N_b}\right) & \text{for } \psi_w \geq 2\pi/N_b \end{cases} \tag{10.45}$$

and

$$\frac{y_{\text{tip}}}{R} = r_{\text{tip}} = A + (1 - A)\exp(-\Upsilon \psi_w). \tag{10.46}$$

Note that the vertical displacements are linear with respect to wake age (i.e., they convect axially at a constant velocity). At the first blade passage when $\psi_w = 2\pi/N_b$, it has been shown previously in Fig. 10.5 that there is a sudden change in the axial convection velocity. This is reflected in the change of the coefficient from k_1 to k_2 in the equations describing the axial displacements. The axial settling rates are modeled by the empirical equations

$$k_1 = -0.25(C_T/\sigma + 0.001\theta_{tw}), \tag{10.47}$$

$$k_2 = -(1.41 + 0.0141\theta_{tw})\sqrt{C_T/2} \approx -(1 + 0.01\theta_{tw})\sqrt{C_T}, \tag{10.48}$$

where the blade twist θ_{tw} is measured in units of degrees. The radial contraction of the wake is smooth and asymptotic. The empirically derived coefficients for the radial contraction are given by $A = 0.78$ and $\Upsilon = 0.145 + 27C_T$. Note that whereas the theoretical contraction ratio of a rotor wake in hover is 0.707, in practice the contraction ratio is found consistently closer to 0.78. The vortex sheet trailed by the inner parts of the blade vary linearly with r. The outer end of the sheet ($r = 1$) is represented by

$$\left(\frac{z}{R}\right)_{r=1} = \begin{cases} K_{1,r=1}\psi_w & \text{for } 0 \leq \psi_w \leq 2\pi/N_b, \\ K_{1,r=1}\left(\dfrac{2\pi}{N_b}\right) + K_{2,r=1}\left(\psi_w - \dfrac{2\pi}{N_b}\right) & \text{for } \psi_w \geq 2\pi/N_b, \end{cases} \tag{10.49}$$

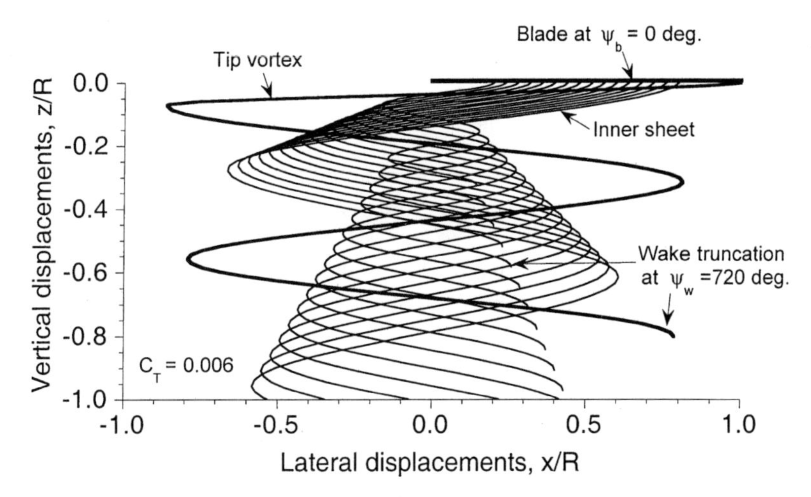

Figure 10.26 Representative prescribed hovering wake based on Landgrebe's model. $N_b = 2$; $C_T = 0.006$; $\theta_{tw} = 0$.

whereas the inner end of the sheet is represented by

$$\left(\frac{z}{R}\right)_{r=0} = \begin{cases} 0 & \text{for } 0 \leq \psi_w \leq 2\pi/N_b, \\ K_{2,r=0}\left(\psi_w - \dfrac{\pi}{2}\right) & \text{for } \psi_w \geq \pi/2. \end{cases} \tag{10.50}$$

The locations of intermediate parts of the sheet are determined by linear interpolation. The empirical coefficients of the sheet coordinates are described by

$$K_{1,r=1} = -2.2\sqrt{C_T/2}, \tag{10.51}$$

$$K_{2,r=1} = -2.7\sqrt{C_T/2}, \tag{10.52}$$

$$K_{1,r=0} = \left[\frac{\theta_{tw}}{128}(0.45\theta_{tw} + 18)\right]\sqrt{\frac{C_T}{2}}. \tag{10.53}$$

A representative hover wake geometry based on Landgrebe's model is shown in Fig. 10.26. The wake for two revolutions of the rotor are shown to convey the relative differences in the convection rates of the sheet versus that of the tip vortex. Note that the outer edge of the inner sheet convects axially downward at a rate that is approximately twice that of the tip vortex. In the hover state, it is found that the inclusion of the vortex sheet is usually necessary to enable good predictions of inflow and rotor performance.

Kocurek & Tangler's Model

Kocurek & Tangler (1976) have derived a prescribed wake model similar to that of Landgrebe using the same set of generalized equations for the tip vortex trajectories, but with different coefficients based on another series of subscale rotor experiments. This model attempts to include the number of blades as well as the blade lift distribution (through the twist rate) in the modeling of the axial settling rates. The generalized equation for k_1 is

$$k_1 = B + C\left(\frac{C_T{}^m}{N_b^n}\right), \tag{10.54}$$

where $B = -0.000729\,\theta_{tw}$, $C = -2.30.206\theta_{tw}$, $m = 1.0 - 0.25e^{0.040\theta_{tw}}$, and $n = 0.5-0.0172\theta_{tw}$. The equation for k_2 is

$$k_2 = -(C_T - C_{T_0})^{1/2}, \quad \text{where} \quad C_{T_0} = b^n(-B/C)^{1/m}. \tag{10.55}$$

The radial contraction rate parameter, Υ, is given by $\Upsilon = 4.0\sqrt{C_T}$, with $A = 0.78$, as in Landgrebe's model (Eq. 10.46). No inner vortex sheet is used in the Kocurek & Tangler model.

An example showing the expected quality of the predicted wake geometry and hover performance that can be made using a prescribed wake model is shown in Figs. 10.27 and 10.28. It is found that in this case the prescribed wake tends to slightly overpredict both the axial displacements and the radial contraction of the tip vortices. Results from a free-vortex wake model (see Section 10.7.6) are also shown in Figs. 10.27 and 10.28. Note that the

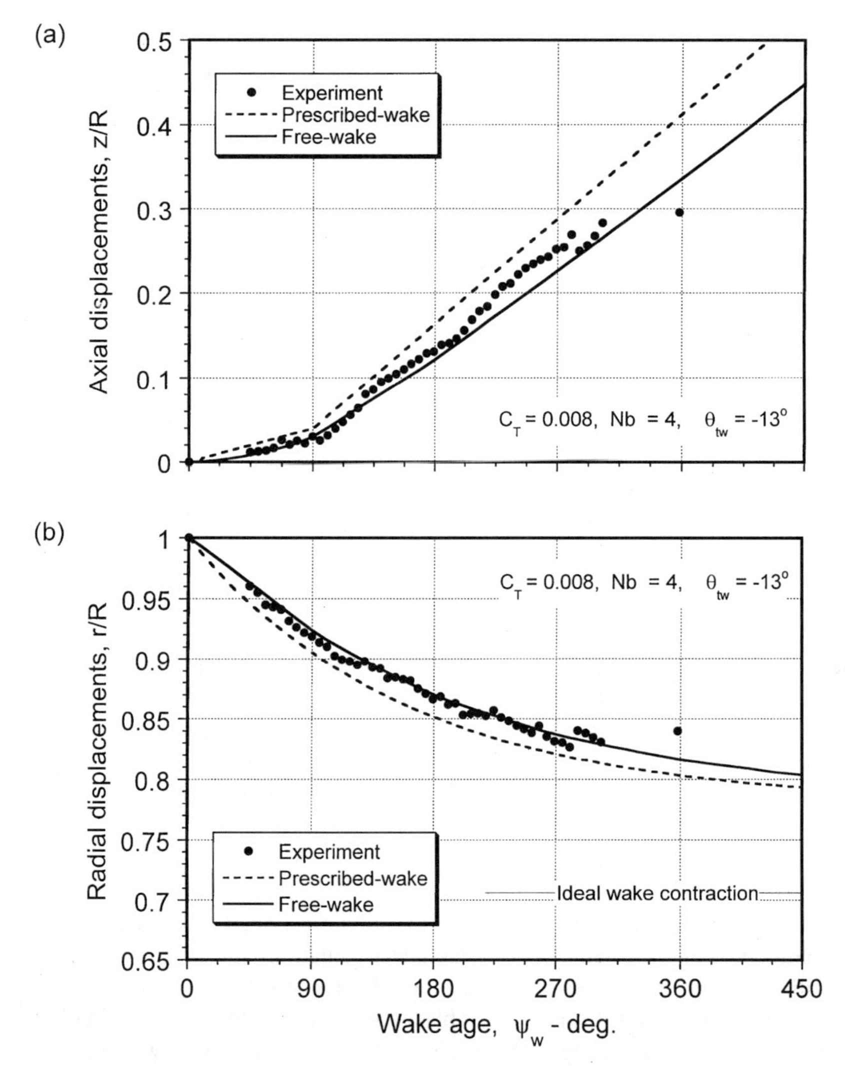

Figure 10.27 Comparison of prescribed and free-vortex wake models with experimental measurements of the tip vortex locations in hover. (a) Axial displacements. (b) Radial displacements. Four-bladed rotor; $C_T = 0.008$.

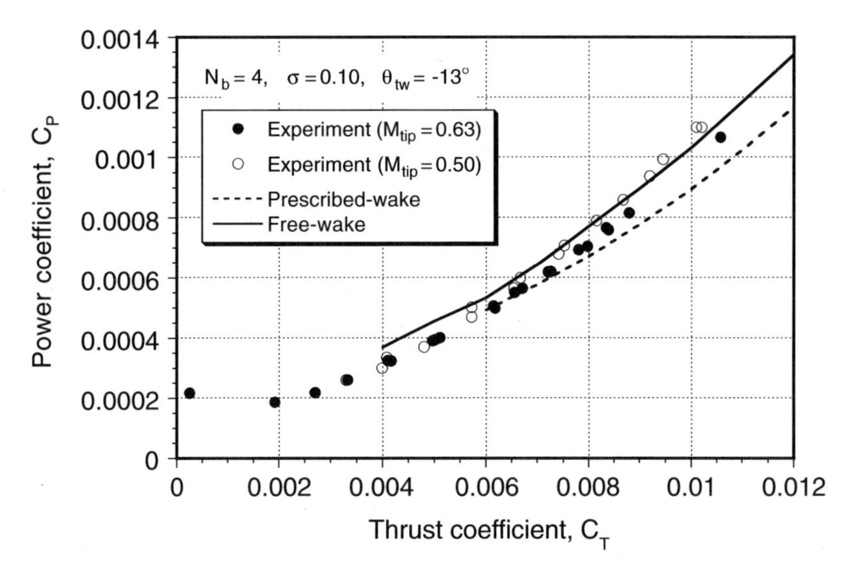

Figure 10.28 Predictions of hovering rotor performance using prescribed and free-vortex wake models. Four-bladed rotor.

free-vortex wake model gives only slightly better predictions of both the wake geometry and the hovering performance, albeit at a considerably greater computational expense.

10.7.5 *Prescribed Vortex Wake Models for Forward Flight*

Vortex Ring Model

One simple way to approximate the trailed wake vorticity from the rotor is to use a series of stacked vortex rings or a vortex tube – see Coleman et al. (1945), Castles & De Leeuw (1954), and Young (1974). The advantage of using vortex rings is that an exact (analytic) solution for the induced velocity can be obtained. In the vortex ring (tube) model, one ring represents the trailed wake system generated by one blade during one rotor revolution. The positioning of the rings is defined on the basis of simple momentum theory. The induced velocity from one desingularized vortex ring of strength Γ_v can be written as

$$V_z(r, z) = \frac{\Gamma_v}{4\pi} \int_{0+\delta\theta}^{2\pi-\delta\theta} \frac{z^2 + (R - r\cos\theta)^2}{\sqrt{r_c^2 + \left(\sqrt{z^2 + (R - r\cos\theta)^2}\right)^2}} \times \frac{R(R - r\cos\theta)\, d\theta}{(R^2 - 2Rr\cos\theta + z^2 + r^2)^{3/2}} \tag{10.56}$$

[see, for example, Lewis (1991)]. This equation can be integrated analytically using elliptic integrals of the first and second type. Alternatively, it can be integrated using Gaussian quadrature (see Question 10.1). Note that for points not on the vortex segment itself, the cutoff distance, $\delta\theta$, equals zero. However, for points on the ring vortex a logarithmic singularity occurs, and so $\delta\theta$ must be nonzero. Although the value of $\delta\theta$ can be approximated in terms of the vortex core radius and an exponential function of the kinetic energy in the vortex core [see Widnall (1972)] in a practical sense a cutoff distance of $\delta\theta \sim 10^{-4}$ can be assumed.

From this basic result, the net induced velocity at any point in the flow field can be obtained by summing the effects of all the rings. It has been found from the vortex ring (tube) model that the longitudinal distribution of inflow in the rotor TPP is approximately linear, which is in agreement with experiments – see Coleman et al. (1945). Also, it has been found that the longitudinal coefficient k_x in the linear inflow model can be approximated using $k_x = \tan(\chi/2)$, which is a result discussed previously in Section 3.4.2.

Rigid or Undistorted Wake

The rigid or undistorted vortex wake model is one in which the trailed vortices are represented by skewed helical filaments. The position of the vortex filaments is defined geometrically based on the flight conditions and momentum theory considerations. There are no self- or mutual interactions between vortex filaments. If one assumes that the induced velocity field in the wake is uniform (i.e., the induced velocity does not vary with time or location) then an exact analytical solution may be obtained for Eq. 10.42 as a special case. Making this assumption is equivalent to assuming a uniform streamwise velocity and a mean inflow, $\lambda_i =$ constant throughout the flow, that is,

$$\vec{V} = \Omega R \mu \hat{i} + \Omega R(\lambda_i)\hat{k}. \tag{10.57}$$

Under these assumptions, the PDE governing the wake geometry relative to the TPP (Eq. 10.42) can be written as

$$\frac{\partial \vec{r}}{\partial \psi_b} + \frac{\partial \vec{r}}{\partial \psi_w} = \mu \hat{i} + \lambda_i \hat{k}, \tag{10.58}$$

where ψ_w is the wake age and ψ_b is the azimuth angle of the blade at which the vortex filament was generated, also recognizing that the position vector r has now been nondimensionalized by R. Equation 10.58 is a linear, first-order, hyperbolic, PDE. The two required boundary conditions in the ψ_b and ψ_w directions are given by

$$\psi_b : \vec{r}(\psi_b, \psi_w) = \vec{r}(\psi_b + 2\pi, \psi_w), \tag{10.59}$$

$$\psi_w : \vec{r}(\psi_b, 0) = r_v \cos\psi_b \hat{i} + r_v \sin\psi_b \hat{j}, \tag{10.60}$$

where r_v is the radial release point of the trailed vortex filament from the blade ($r_v = 1$ at the tip). The analytical solution to Eq. 10.58 can be obtained using separation of variables – see Hildebrand (1976). The solution of the PDE is defined by the characteristic curve of the equation, which can be determined from the intersection of the two surfaces found by integrating the following ordinary differential equations:

$$\frac{d\psi_b}{1} = \frac{d\psi_w}{1}, \tag{10.61}$$

$$\frac{d\psi_w}{1} = \frac{d\vec{r}}{\vec{V}/\Omega R}. \tag{10.62}$$

From Eqs. 10.61 and 10.62, the two surfaces are given by $\psi_b = \psi_w + c_1$ and $\vec{r} = \vec{V}\psi_w/\Omega R + c_2$, respectively, where c_1 and c_2 are constants of integration. From these solutions, $c_1 = \psi_b - \psi_w$ and $c_2 = \vec{r} - \vec{V}\psi_w/\Omega$. The intersection of the two surfaces results in an equation relating the two constants such that $c_2 = f(c_1) \Rightarrow \vec{r} - \vec{V}\psi_w/\Omega R = f(\psi_b - \psi_w)$. Invoking the boundary condition at $\psi_w = 0$ results in $\vec{r} = r_v \cos\psi_b \hat{i} + r_v \sin\psi_b \hat{j}$, and so the exact solution can be written as

$$\vec{r}(\psi_b, \psi_w) = (\mu\psi_w + r_v \cos(\psi_b - \psi_w))\hat{i} + r_v \sin(\psi_b - \psi_w)\hat{j} + \lambda_i \psi_w \hat{k}. \tag{10.63}$$

Note that the periodic boundary condition in the ψ_b direction is also satisfied because $\cos(2\pi + \psi_b) = \cos\psi_b$ and $\sin(2\pi + \psi_b) = \sin\psi_b$. Equation 10.63 defines a skewed, undistorted helix, which is known as the rigid or undistorted vortex wake topology as introduced in Section 10.4.2. If only tip vortices are assumed ($r_v = 1$), then the solution for the tip vortex geometry relative to the rotor TPP can be described by the simple parametric equations

$$\frac{x_{\text{tip}}}{R} = x'_{\text{tip}} = \cos(\psi_b - \psi_w) + \mu\psi_w, \tag{10.64}$$

$$\frac{y_{\text{tip}}}{R} = y'_{\text{tip}} = \sin(\psi_b - \psi_w), \tag{10.65}$$

$$\frac{z_{\text{tip}}}{R} = z'_{\text{tip}} = \lambda_i\psi_w = -\mu\psi_w \tan\chi_{\text{TPP}}. \tag{10.66}$$

The wake skew angle χ_{TPP} is given by

$$\chi_{\text{TPP}} = \tan^{-1}\left(-\frac{\lambda_i}{\mu}\right), \tag{10.67}$$

where

$$\lambda_i = \mu\tan\alpha_{\text{TPP}} + \frac{\kappa C_T}{2\sqrt{\mu^2 + \lambda_i^2}} \tag{10.68}$$

and where κ is an induced power factor. Note that there is no contraction assumed in the rigid vortex wake. However, if it is found necessary to model this effect, the tip vortices can be assumed to originate at a point just inboard of the tip, say $r_v = 0.97$. Also, note that with the rigid wake equations, as $\mu \to 0$ the wake geometry reduces to a helix, but as described previously this is an unsatisfactory model of the wake in hover.

A typical rigid wake geometry is plotted in Fig. 10.29 and is compared to predictions made by a free-vortex wake solution (see Section 10.7.6), which is a better physical model of the problem. Note that the predictions are substantially different. Nevertheless, the simplicity of the rigid wake model will give the primary effects of the skewness of the wake on the inflow distribution over the rotor disk and may be attractive for applications where the full details of the rotor wake are not required, such as for integrated performance predictions. (See also the results of the rotor wake boundary shown previously in Fig. 10.8.)

Modifications to Rigid Vortex Wake Models

Any number of assumptions about the induced velocity field near the rotor can be made to specify the right-hand side of Eq. 10.42 and decouple the equations. Other types of prescribed wake models for use in forward flight have also been derived on the basis of experimental observations. The advantage of these models is that, for some small additional computational effort, much better estimates of the rotor wake geometry can be obtained compared to a rigid wake. It has been observed from flow visualization experiments that the longitudinal and lateral distortions of the wake are relatively small compared to the vertical distortions. This means that the self-induced velocities in a plane parallel to the rotor plane are small; so for all the prescribed wake models it is still justified to use the undistorted tip vortex equations given previously by Eqs. 10.64 and 10.65. The differences in the various prescribed wake models lie in the prescription of the vertical displacements of the tip vortices. Generally, these are defined using Eq. 10.66 as a basis, but with empirical or semi-empirical weighting functions being used to produce distortions.

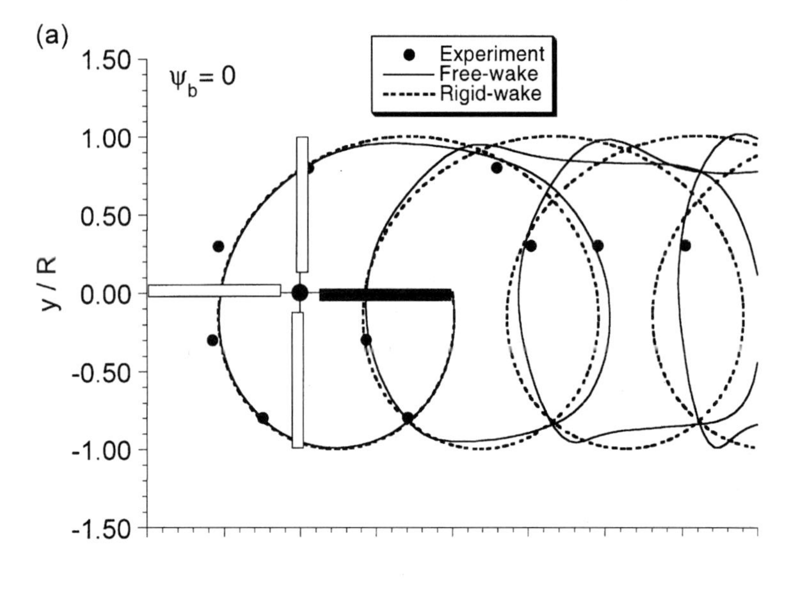

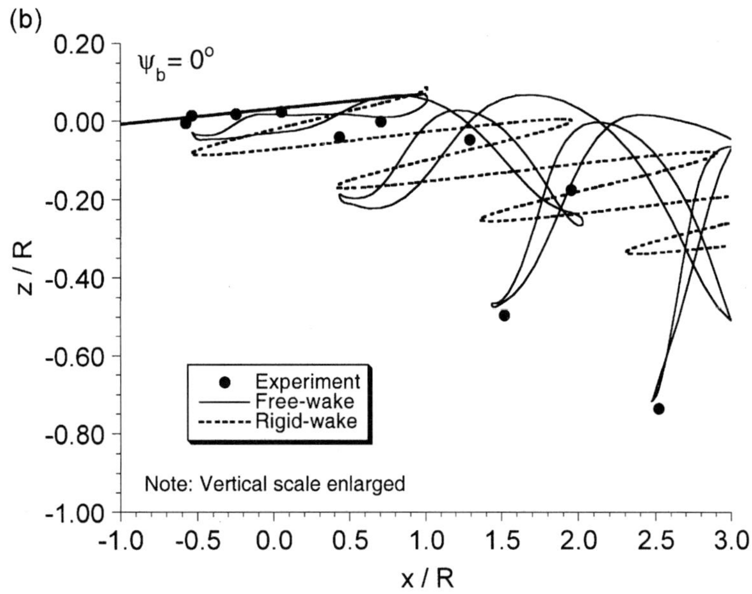

Figure 10.29 Representative rigid wake geometry in forward flight compared to free-vortex wake model and measurements. (a) Plan (top) view. (b) Side view, looking from retreating side. Four-bladed rotor; $C_T = 0.008$; $\mu = 0.15$; $\alpha_s = -3^\circ$. (Note: Results for a four-bladed rotor have been computed, with the results for one blade being shown for clarity.)

UTRC Generalized Wake Model

One such prescribed wake approach, used by Egolf & Landgrebe (1983), is called the Generalized Wake Model. The vertical displacements of the tip vortices are written as

$$\frac{z_{\text{tip}}}{R} = z'_{\text{tip}} = -\lambda_i \psi_w - EG, \tag{10.69}$$

where, like the rigid wake, λ_i is based on momentum considerations. The second term is simply a modification to the undistorted rigid wake. The E term is called an envelope function and is effectively an amplitude term, and the G term is a shape or geometric function. The vertical displacement envelope function is defined by

$$E = \begin{cases} A_0 \psi_w \exp(A_1 \psi_w) & \text{for } \psi_w \leq 4\pi, \\ M\psi_w + B & \text{for } \psi_w > 4\pi, \end{cases}$$

where the coefficients A_0, A_1, M, and B are functions of the number of blades, the advance ratio, and the rotor thrust for a given rotor. The corresponding shape function is given by the equations

$$G = \begin{cases} \sum_{n=0}^{N} \{g'_{nc} \cos n(\psi_w - \psi_b) + g'_{ns} \sin n(\psi_w - \psi_b)\} & \text{for } 0 \leq \psi_w \leq 2\pi, \\ \sum_{n=0}^{N} \{g''_{nc} \cos n(\psi_w - \psi_b) + g''_{ns} \sin n(\psi_w - \psi_b)\} & \text{for } \psi_w > 2\pi, \end{cases}$$

where again the coefficients g'_{nc}, g'_{ns}, g''_{nc}, and g''_{ns} are functions of the number of blades and advance ratio for a given rotor. Usually N takes a value of 12.

Like the hover prescribed wake models, the coefficients of the forward flight model have been deduced from experimental studies of the wake geometry. However, because forward flight tests are more difficult and time consuming to undertake, there are less data available. Nevertheless, as shown by Egolf & Landgrebe (1983), the agreement of this prescribed wake model with measured wake geometries has been found to be good. As a consequence, there is usually a significant improvement in the prediction of blade airloads compared to those obtained using an undistorted rigid wake.

Beddoes's Generalized Wake Model

Another simple but effective prescribed wake model has been developed by Beddoes (1985). In this model, it is also recognized that axial (vertical) trajectories of the tip vortices can be estimated if an assumption for the inflow distribution across the disk is made. Beddoes suggests the model

$$\lambda_i = \lambda_0(1 + Ex' - E|(y')^3|), \tag{10.70}$$

deducing this form from the measurements of Heyson & Katsoff (1957). Again, the evaluation of λ_i is based on momentum considerations. Beddoes suggests that $E = \chi$, where χ is the wake skew angle, although this is generally too large and $E = \chi/2$ has been found to give better agreement with experimental and free-vortex wake results.

Although with these assumptions the solution for the wake can be obtained by directly solving Eq. 10.42, this still involves significant expense. Instead, Beddoes (1985) specifies the equations for the wake. At the rear of the disk, in the region occupied by the wake,

$$\lambda_i = 2\lambda_0(1 - E|(y'_{\text{tip}})^3|), \tag{10.71}$$

so that the vertical displacement of the tip vortex is given by

$$z'_{\text{tip}} = -\mu_z \psi_w + \int_0^{\psi_w} \lambda \, d\psi_b. \tag{10.72}$$

Now, if $x'_{\text{tip}} < \cos(\psi_b - \psi_w)$ then the element on the vortex filament has not convected beyond the rotor disk and

$$\int_0^{\psi_w} \lambda \, d\psi_b = -\lambda_0\{1 + E[\cos(\psi_b - \psi_w) + 0.5\mu\psi_w - |(y'_{\text{tip}})^3|]\}\psi_w. \tag{10.73}$$

If $\cos(\psi_b - \psi_w) > 0$ then the vortex element has always been downstream of the disk and

$$\int_0^{\psi_w} \lambda \, d\psi_b = -\lambda_0[1 - E|(y'_{\text{tip}})^3|]\psi_w. \tag{10.74}$$

If neither of the preceding conditions are met, then the element has spent part of the time within the disk and the remainder downstream of the disk and

$$\int_0^{\psi_w} \lambda \, d\psi_b = -2\lambda_0 x'_{\text{tip}} \frac{[1 - E|(y'_{\text{tip}})^3|]}{\mu_x}. \tag{10.75}$$

On the basis of these equations, the three components of the tip vortex location with respect to the rotor can be calculated.

Figure 10.30 shows a comparison of the vertical components of the tip vortices in forward flight at $\mu = 0.15$ as obtained from the Beddoes prescribed-wake model and also from a free-vortex wake model. Because the differences between the prescribed wake and the free-vortex wake are now much smaller compared to the rigid wake, for clarity the results in Fig. 10.30 are plotted in terms of vertical displacement versus wake age relative to the blade of origination. The agreement is good, and bearing in mind the relative simplicity of the prescribed-wake solution compared to the free-vortex wake (at least two orders of magnitude less expensive to compute), it is certainly a modeling option to consider in any form of rotor analysis. After the wake geometry is known, the induced velocity field can be computed through the application of the Biot–Savart law to the discretized vortex filaments. Because this incurs the highest cost of any vortex wake model, Beddoes (1985) suggests an alternative and highly simplified approach of approximating the induced velocities from some elements of the wake by using elongated straight line segments, which retains the simplicity and low cost of semi-analytic models such as when using vortex rings.

10.7.6 *Free-Vortex Wake Analyses*

General Concepts

Free-vortex wake models solve for the vortex strengths and rotor wake geometry, and in principle they do not require experimental data for formulation purposes, apart from the various assumptions that must be made about the real viscous nature of the vortices themselves, as discussed previously in Section 10.6.2. The position vectors of the individual wake filaments are now part of the solution process. Starting from some initial condition, which may be an undistorted rigid wake or a prescribed wake, the right-hand side of the equations must be computed through the repeated application of the Biot–Savart law with a desingularized core, which will likely include a core growth (vorticity diffusion) model such as discussed in Section 10.6.4. The solution for the wake geometry then proceeds until convergence, which usually means that a periodic solution is obtained. Pioneering work on free-vortex modeling of rotor wakes was performed by Landgrebe (1969, 1971, 1972), Clark & Leiper (1970), Sadler (1971a,b), and Scully (1975), and many

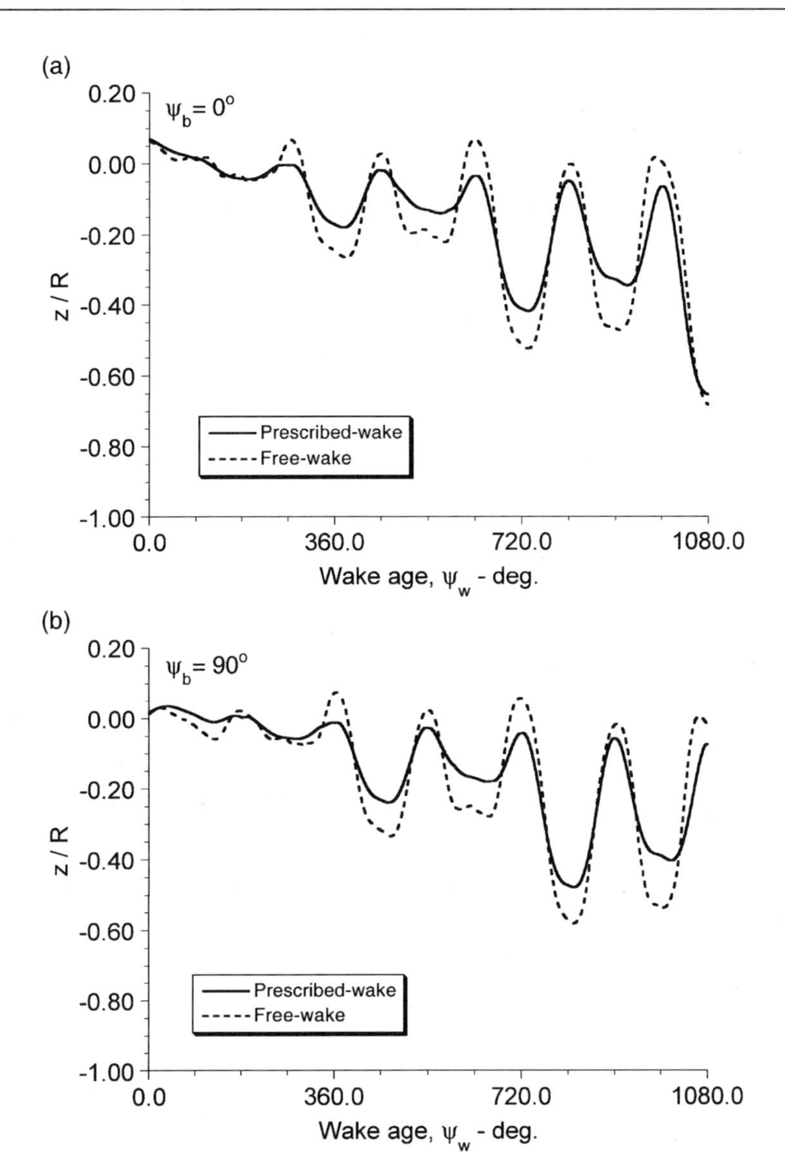

Figure 10.30 Comparison of Beddoes's prescribed wake model with predictions made by a free-vortex wake. (a) $\psi_b = 0°$. (b) $\psi_b = 90°$. (c) $\psi_b = 180°$. (d) $\psi_b = 270°$. Four-bladed rotor; $\mu = 0.15$; $C_T = 0.008$; $\alpha_s = -3°$.

variations and developments of free-vortex wake models have subsequently followed. The results obtained with free-wake models are probably qualitatively correct no matter what type of numerical scheme and vortex model are used. However, the quantitative predictions made using these methods are much less absolute, and their true capabilities can only be determined through careful and systematic correlation studies with experimental results.

In the first instance, consider the rolled up tip vortices generated by each blade. For a rotor with N_b blades there will be an equal number of intertwining tip vortices (see Fig. 10.31).

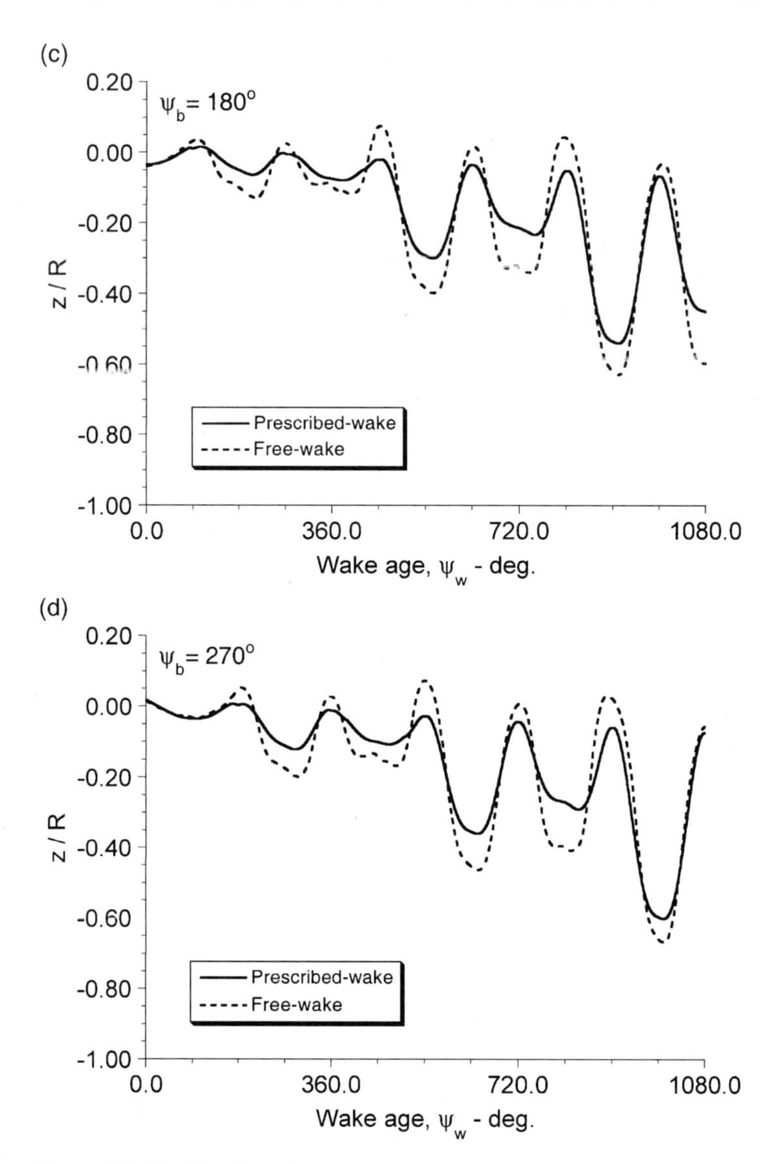

Figure 10.30 (*Continued*)

Generally, the filaments will have unequal circulation strengths along their length, which can be related to the lift (circulation) on the blade at the time of their formation. This means that both trailed and shed elements must be included in the solution to satisfy conservation of circulation. However, because this increases dramatically the number of free elements and the numerical costs, a tip vortex of average strength related to the rotor thrust (blade loading) is often assumed (see Chapter 3, Section 3.3.5). A series of collocation points can be specified on the trailed vortex filaments, and these points are numerically convected through the flow field at the local velocity in accordance with the transport equation defined previously in Eq. 10.42.

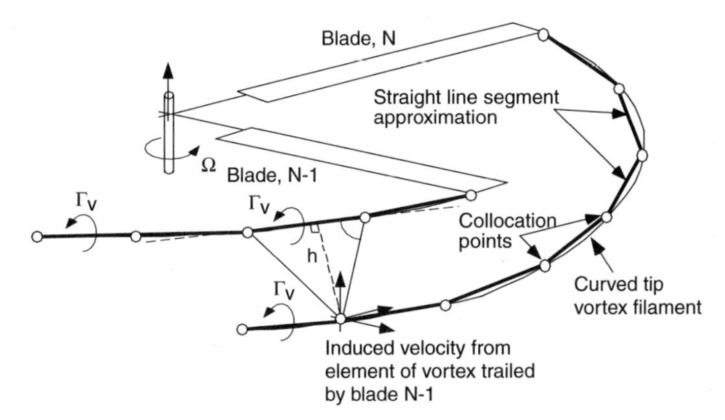

Figure 10.31 Idealization of trailed tip vortices into segments connected by collocation points for free-vortex wake modeling.

General Solution Methodologies

Free-wake models can be divided into two general types of solution methodologies, depending on the specified boundary and initial conditions: 1. time-stepping or time-marching methods and 2. relaxation or iterative methods. In the time-stepping method, the solution can be developed by an impulsive start of the rotor with no initial wake. The boundary condition specifies that each trailed vortex filament be attached to the blade at its point of origin, $\psi_w = 0$. Alternatively, a prescribed wake geometry can be specified as an initial condition. In either case, the solution is stepped in the ψ_b (azimuthal) direction using a time integration scheme, and for a steady-state flight condition, a converged solution is obtained when the transients introduced by the initial condition die out and a periodic solution is achieved. The relaxation method, in contrast, specifies that the trailed vortex elements be attached to the blades as an initial condition, whereas periodicity is enforced as the boundary condition. The solution is stepped in the ψ_w direction in an iterative manner, and convergence is obtained once the wake vortex geometry no longer distorts between successive iterations. Relaxation schemes have the advantage that they are numerically more efficient because periodicity of the wake solution can be imposed as a boundary condition. Also, relaxation schemes are generally free of the numerical instabilities that are produced in time-marching schemes because of round-off errors.

Different numerical schemes (i.e., explicit, implicit, or hybrid) can also be used within both the time-stepping and relaxation approaches. Several rotor wake analyses use explicit type methods, which are simple in concept but are particularly prone to various numerical problems associated with round-off errors. Consider, as an example, a representative PDE describing the convection of the vortex filaments such as

$$\frac{\partial r}{\partial t} + \frac{\partial r}{\partial x} = v(r). \tag{10.76}$$

Spatial discretization would result in reducing this partial differential equation into a system of simultaneous ordinary differential equations that can be written as

$$\frac{dr}{dt} = Ar + v(r), \tag{10.77}$$

where A is the space discretization matrix. An explicit method for this equation would use a finite-difference scheme such as

$$r^{n+1} = r^n + \Delta t[Ar^n + v(r^n)]. \tag{10.78}$$

Note that the updated value of r at time $n+1$ is based on values from the previous time step alone. Although this is one of the simplest methods, the truncation error in the solution for r is of order $(\Delta t)^2$ per time step, and so it can grow rapidly if the time step is too large. The high computational costs of most blade element based rotor analyses often require relatively large time (blade azimuth or $\Delta\psi_b$) steps with the wake solution, making the assessment of truncation errors an important consideration.

Implicit methods, in contrast, are free from many of these numerical problems. An implicit difference scheme to solve the example equation in Eq. 10.76 is

$$r^{n+1} = r^n + \Delta t[Ar^{n+1} + v(r^{n+1})]. \tag{10.79}$$

In this case, the right-hand side also requires the evaluation of terms at the $(n+1)$th step, information that has not yet been determined. If the velocity term, v, is a simple function of r, then the equation can be easily simplified and solved as a set of simultaneous equations. For more complex velocity functions, however, v must first be linearized using a Taylor expansion, resulting in a semi-implicit scheme. This approach was applied to the free-vortex wake problem by Miller & Bliss (1993). While numerical errors are much smaller with this technique, it becomes computationally very expensive for routine use because it requires the simultaneous solution of a large set of linear equations at each time step.

A multi step or predictor–corrector method may also be used to solve the wake equations. In this approach, an explicit method is used to generate an approximate value for the current step

$$\tilde{r}^{n+1} = r^n + \Delta t[Ar^n + v(r^n)], \tag{10.80}$$

and this approximate, or predicted value, $\tilde{r}$, is then used to generate the final approximation to r^{n+1} using a corrector step such as

$$r^{n+1} = r^n + \Delta t[A\tilde{r}^{n+1} + v(\tilde{r}^{n+1})]. \tag{10.81}$$

Predictor–corrector methods have a truncation error per step of order $(\Delta t)^3$ and exhibit considerably better stability characteristics than explicit methods alone. They also require much less computation than implicit or semi-implicit schemes. However, this scheme requires two velocity field calculations per time step compared to one for an explicit method (see Question 10.8).

Bagai & Leishman (1995a,b, 1996) have developed a free-vortex wake analysis based on a predictor–corrector scheme in a relaxation formulation, but with two modifications. First, the explicit predictor step is modified into a pseudo-implicit equation

$$\tilde{r}^{n+1} = r^n + \Delta t[A\tilde{r}^{n+1} + v(r^n)]. \tag{10.82}$$

While the forcing (velocity) function is calculated explicitly from the previous time step, the position vectors at the current time step also appear on the right-hand side ($\tilde{r}^{n+1}$). The above predictor equation, therefore, is no longer fully explicit. Second, the corrector step incorporates an averaging scheme whereby an average of the velocity function from the previous time step and the predicted value are used to update the corrected position vectors. This helps to improve the stability characteristics of the scheme. Once again, the position vectors at the current time step also appear on the right-hand side, also making the corrector step a pseudo-implicit equation, that is,

$$r^{n+1} = r^n + \Delta t\left[Ar^{n+1} + \frac{1}{2}(v(\tilde{r}^{n+1}) + v(r^n))\right]. \tag{10.83}$$

Cost of Free-Wake Solution

To understand the potential cost of a free-vortex wake solution, it is possible to estimate the number of Biot–Savart-like velocity evaluations required for a typical free-vortex wake solution. Let $N_{\psi_b} = 360°/\Delta\psi$ be the number of discrete azimuthal grid points. Likewise, for a vortex filament that is n rotor revolutions old, the number of free collocation points (= number of free vortex elements on that filament) is $N_{\psi_w} = n\ 360°/\Delta\psi_w$. Therefore, a total of $N^2_{\psi_w}$ Biot–Savart evaluations must be performed for each free-vortex filament at each of N_{ψ_b} locations to account for the total self-induced velocities at each collocation point from every other vortex element. If the wake is modeled using only a single free-vortex filament from each blade tip, this results in $N_{\psi_b} N^2_{\psi_b}$ velocity field evaluations to define the rotor wake at all blade azimuth angles. For a rotor with N_b blades, a further $N_b(N_b - 1)N_{\psi_b}N^2_{\psi_w}$ evaluations must be performed to account for mutually induced effects. Therefore, the total number of Biot–Savart evaluations required per free-vortex wake computation, N_E, is given by the equation

$$N_E = (1 + N_b(N_b - 1))N_{\psi_b}N^2_{\psi_w}. \tag{10.84}$$

For equal step sizes, $\Delta\psi_b = \Delta\psi_w \Rightarrow N_{\psi_w} = nN_{\psi_b}$, and so the total number of evaluations becomes

$$N_E = (1 + N_b(N_b - 1))nN^3_{\psi_b}. \tag{10.85}$$

For a typical four-bladed helicopter rotor modeled with three revolutions of free-tip vortices and equal discretization step sizes of 10 degrees, Eq. 10.85 suggests that the Biot–Savart integral must be evaluated over 1.8×10^6 times to cover the entire computational domain just once. Doubling the resolution (i.e., using step sizes half the original size, such that $\Delta\psi_b = 5°$) requires eight times that number, or over 14.5×10^6 velocity evaluations. For additional free vortex filaments, the mutual interactions among all the additional free vortices must also be computed leading to a very substantial computational effort.

Various ideas have been used to decrease these computational times and make free-vortex methods more attractive for routine use. Often, methods can be used to differentiate between vortex elements in the "near field" and "far field," the latter of which have a smaller influence and can be either excluded from the calculation altogether or can be included by lumping together the induced effects of several "far" elements. Miller & Bliss (1993) employ an analytical and numerical matching technique, in which a "near field" solution for the vortex near the core is matched to a "far field" solution. Substantial reductions in computational cost are possible. An extension of this approach is discussed by Miller (1993). Bagai & Leishman (1998) have approached the problem through adaptive refinement of the finite-difference grid used to solve the governing vortex transport equations, with interpolation of known information onto intermediate points in the wake. Reductions in computational effort of over one order of magnitude are possible. Sarpkaya (1989) discusses other methods that can potentially be used to reduce the computational time of vortex methods.

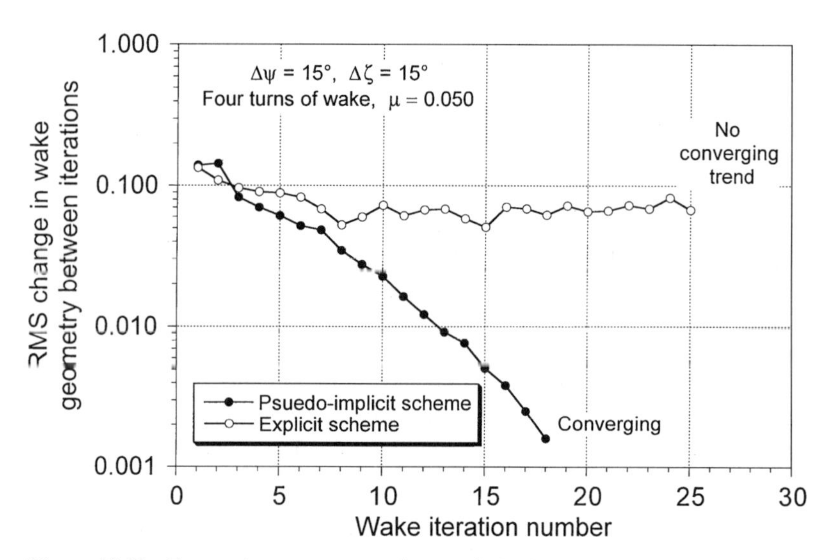

Figure 10.32 Free-wake convergence characteristics for explicit and psuedo-implicit free-vortex wake schemes. Four turns of free-vortex wake, with $\mu = 0.05$, $C_T = 0.008$, and $\alpha_s = -6°$.

Numerical Characteristics of Free-Wake Models

The advantage of using implicit or pseudo-implicit free-vortex wake methods is that many of the numerical and round-off error problems associated with explicit methods can be avoided. Typical convergence characteristics of an explicit and a pseudo-implicit free-vortex wake scheme are shown in Fig. 10.32 for a relatively low advance ratio of 0.05, where the self- and mutually induced effects from the tip vortices are strong. In each case, four complete turns of the free-vortex wake were used with a relatively coarse 15° azimuth step ($\Delta\psi_b = \Delta\psi_w = 15°$).[6] The explicit scheme initially shows a converging trend but only reaches a minimum error of about 0.05 after 8 iterations. Further iterations cause numerical oscillations in the wake geometry and thereafter slowly drive the solution unstable. In contrast, the pseudo-implicit method essentially shows a monotonically converging solution, which converges below an acceptable error threshold after about 15 iterations.

A comparison of the corresponding wake geometries at the end of the iteration cycles is shown in Fig. 10.33. Note that with the explicit scheme there are several regions of the wake where the numerical errors have propagated, and there are significant signs of developing instabilities that, based on experimental studies of helicopter rotor wakes [e.g., Lehman (1968)], are clearly not of physical origin. These are undesirable characteristics of explicit free-vortex wake schemes that make them unsuitable for rotor flow field predictions.

Comparisons of Vortex Wake Models with Experimental Data

To demonstrate some aspects of the predictive capabilities of free-vortex wake models, comparisons are now shown with experimentally determined wake geometries in hover and forward flight. Figure 10.34 shows a comparison of predicted and experimental wake boundaries at the longitudinal centerline of the rotor (see also Fig. 10.8). It can be

[6] This is probably the largest step size that will give a reasonable physical representation of the rotor wake problem when using straight-line vortex filaments.

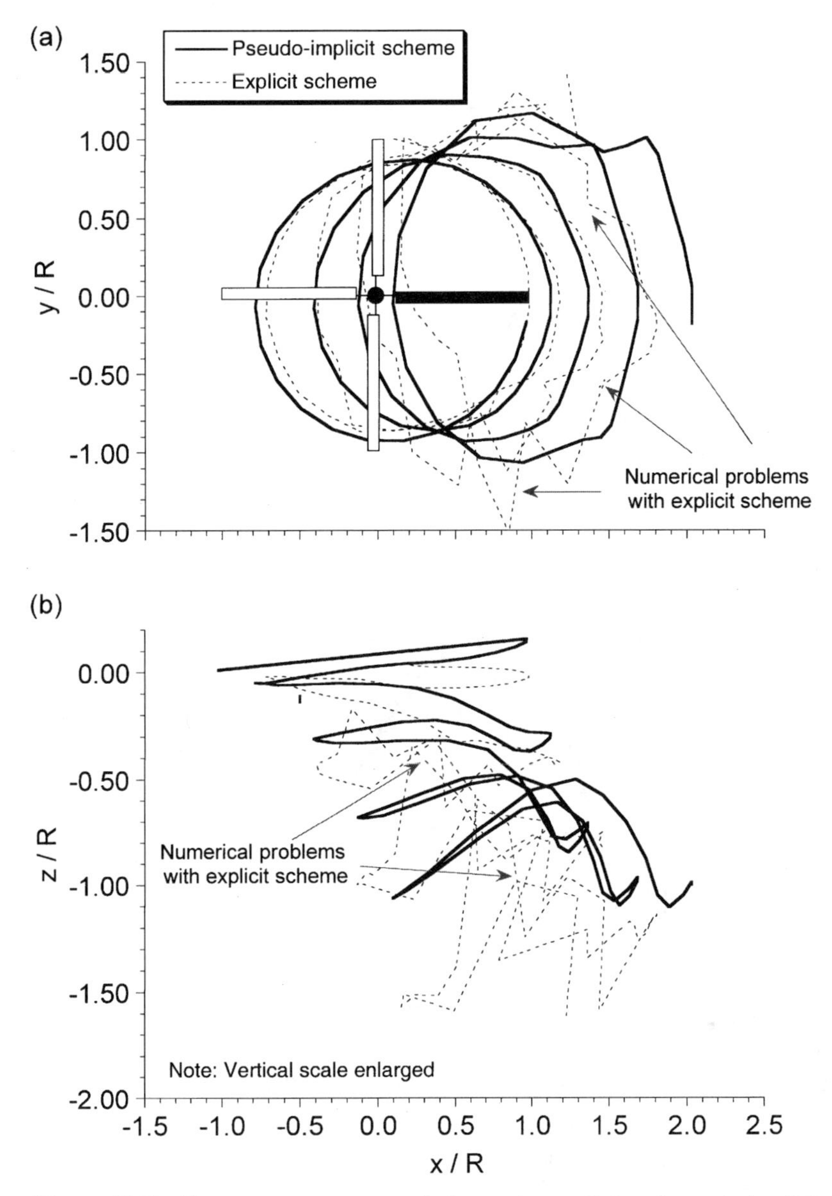

Figure 10.33 Tip vortex geometry predictions using explicit and pseudo-implicit free-vortex wake schemes, with $\mu = 0.05$, $C_T = 0.008$, and $\alpha_s = -6°$. (a) Top view. (b) Side view. Four-bladed rotor, with results for only one blade shown for clarity.

seen that during the transition from hover to forward flight, there are significant changes to the wake geometry, and so these cases provide a good general test of the capabilities of any vortex wake model, free or prescribed.

In hover, the wake is (theoretically) axisymmetric; so the leading- and trailing-edge wake boundaries are identical. For a variety of reasons, this is not necessarily the case for the experimental measurements, but the overall differences are small. As the advance ratio is increased, note that the wake is quickly skewed back by the free-stream flow. Figure 10.34 shows that as the newer tip vortices at the front of the rotor disk move downstream, they

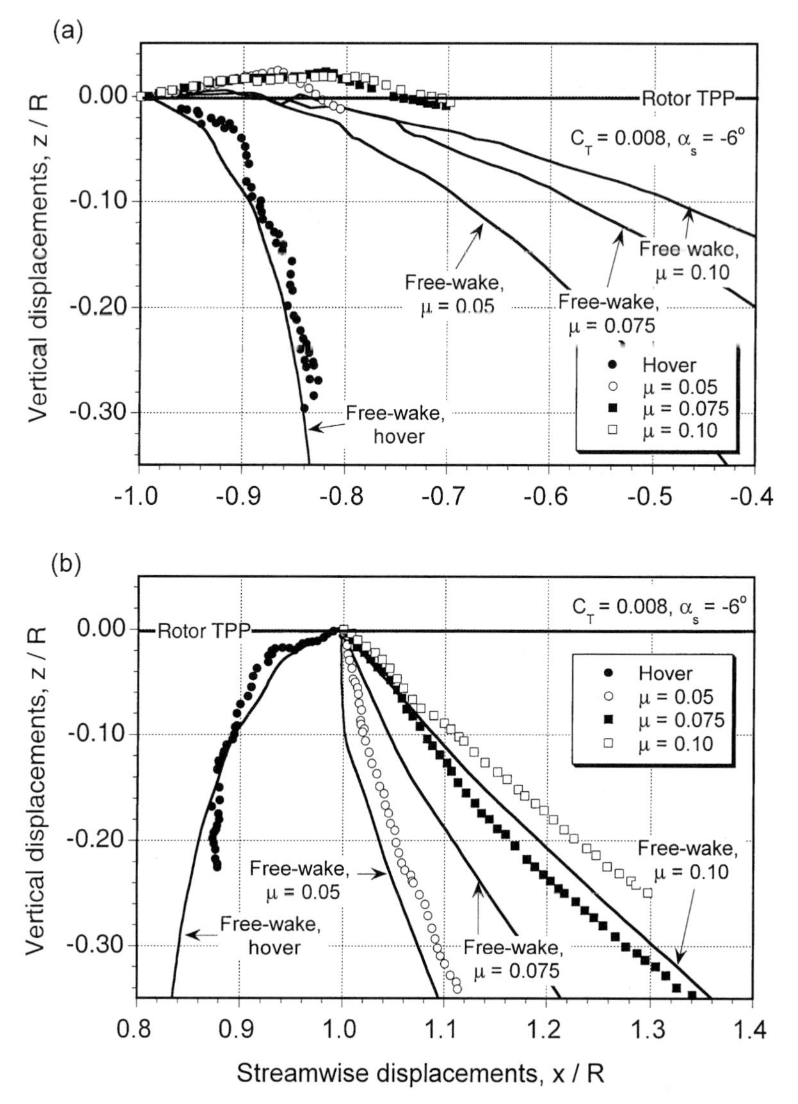

Figure 10.34 Predicted wake boundaries in hover and forward flight using free-vortex wake scheme compared to measured data. (a) Front of rotor. (b) Rear of rotor. Four-bladed rotor, with $C_T = 0.008$ and $\alpha_s = -3^\circ$. Data source: Leishman & Bagai (1991).

are convected over the top of the following blade. As the vortices are convected further downstream, they begin to descend toward the TPP and are ultimately intersected by a blade producing a perpendicular type of BVI, a phenomenon that has been detailed previously in Section 10.4.2.

At the rear of the rotor disk, however, there is a higher downward induced velocity component, and the vorticies remain well below the TPP. As a consequence, the wake skew angle at the rear of the disk is much smaller than at the leading edge of the rotor. It can also be seen from Fig. 10.34 that while the free-vortex wake model shows a high sensitivity to small changes in advance ratio, the streamwise displacements are somewhat underpredicted compared to experimental results. Some of these differences, however, can be attributed to experimental uncertainties in the measured data. Overall, the results from the free-vortex wake model are found to be in much better agreement with the measurements than are those of the rigid wake representation.

Further comparisons of the free-vortex wake model with measured tip vortex displacements in the x–z plane are shown in Figs. 10.35 and 10.36 for two advance ratios and for four longitudinal planes through the wake. The data have been taken from Ghee & Elliott (1995), who have measured the displacements of the tip vortices using a planar laser sheet technique. Data were measured at four longitudinal planes at $y/R = \pm 0.3$ and $y/R = \pm 0.8$. At the inner locations, some gaps occur in the data because the fuselage and rotor hub blocked access for visualization. Free-wake predictions have been made using both a full-span vortex sheet and just with the tip vortex alone. We see that for both advance ratios, the free-vortex wake prediction are in good agreement with the measured tip vortex

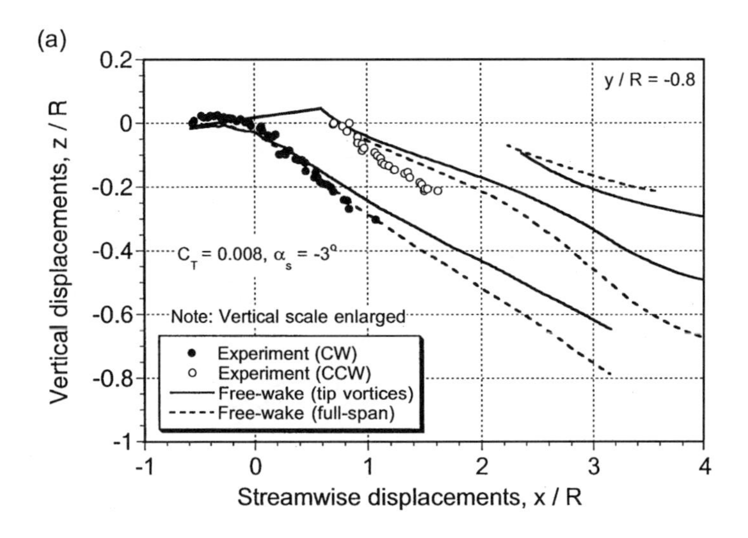

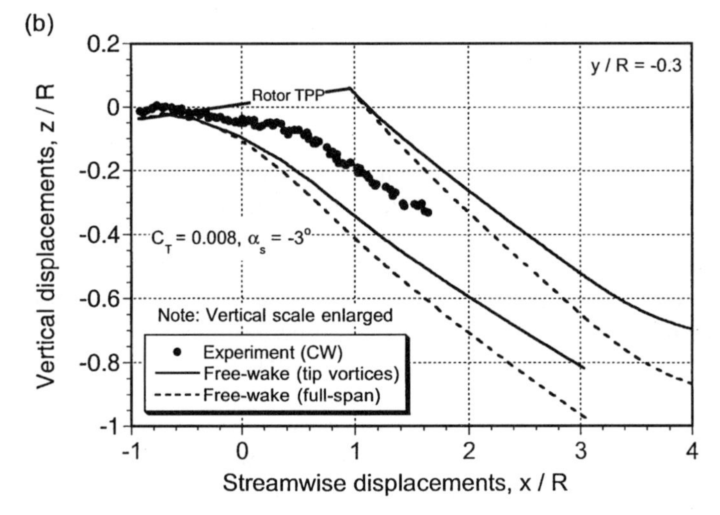

Figure 10.35 Predicted wake boundaries in forward flight compared to measurements. $\mu = 0.15$; $C_T = 0.008$; $\alpha_s - 3^\circ$. (a) $y/R = -0.8$. (b) $y/R = -0.3$. (c) $y/R = 0.3$. (d) $y/R = 0.8$. Measured data taken from Ghee & Elliott (1995).

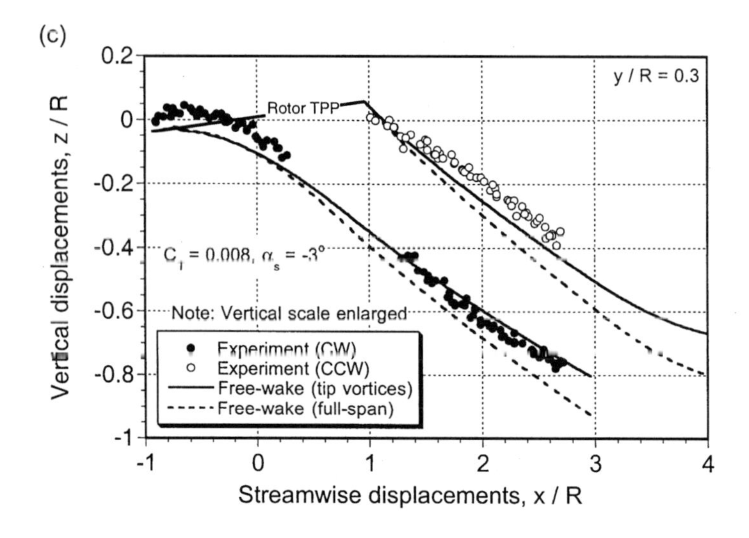

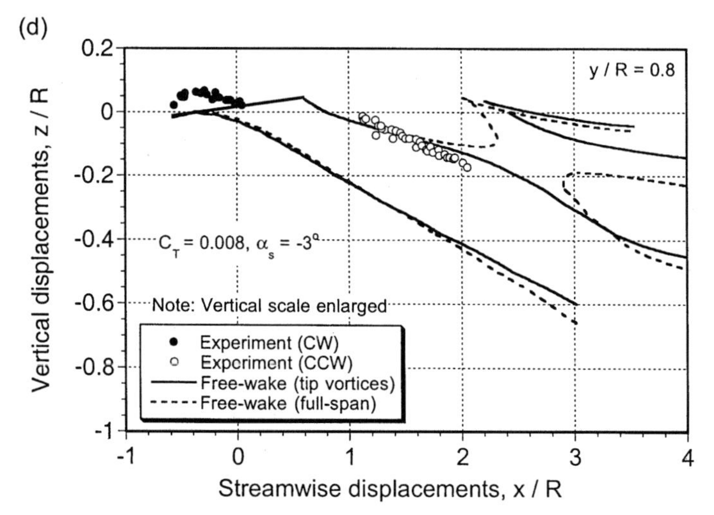

Figure 10.35 (*Continued*)

displacements. Clearly there are some difference between the predictions and experiment, but the overall quality of the results are good, bearing in mind the complexity of the problem. The effect of including a full-span wake compared to a single-tip vortex is relatively small, and so its inclusion may not be justified given that the numerical cost of this approach is greater by about one order of magnitude. However, the inclusion of the inner wake sheet may be required to ensure proper predictions of rotor performance.

Figures 10.37 and 10.38 show the plan and rear views of the rotor wake at an advance ratio of 0.15. The tip vortex trajectories trailed from each blade have been plotted separately to ensure clarity. Again, the results are compared with the measured tip vortex displacements as the vortices intersect the four longitudinal planes. When viewed from above, the trajectories are relatively undistorted from their epicycloidal forms. It is only at the lateral edges of the wake that distortions occur because of the roll-up between individual vortices. The rear views of the wake shown in Fig. 10.38 vividly show this roll-up process and the formation of vortex "bundles" or "super vortices." In these regions, the small radius of curvature of the

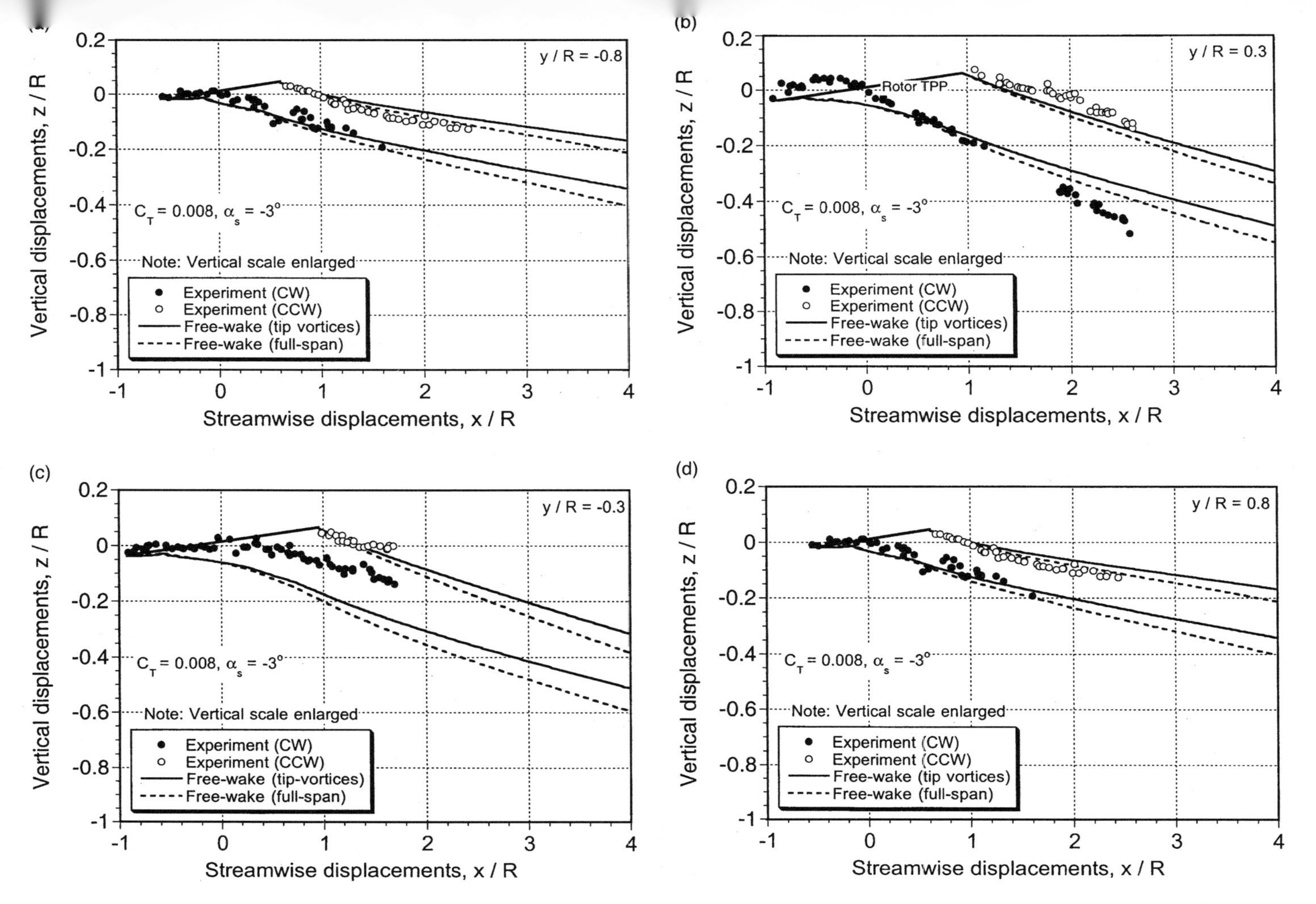

Figure 10.36 Predicted wake boundaries in forward flight compared to measurements. $\mu = 0.23$; $C_T = 0.008$; $\alpha_s - 3^\circ$. (a) $y/R = -0.8$. (b) $y/R = -0.3$, (c) $y/R = 0.3$, (d) $y/R = 0.8$. Measured data taken from Ghee & Elliott (1995).

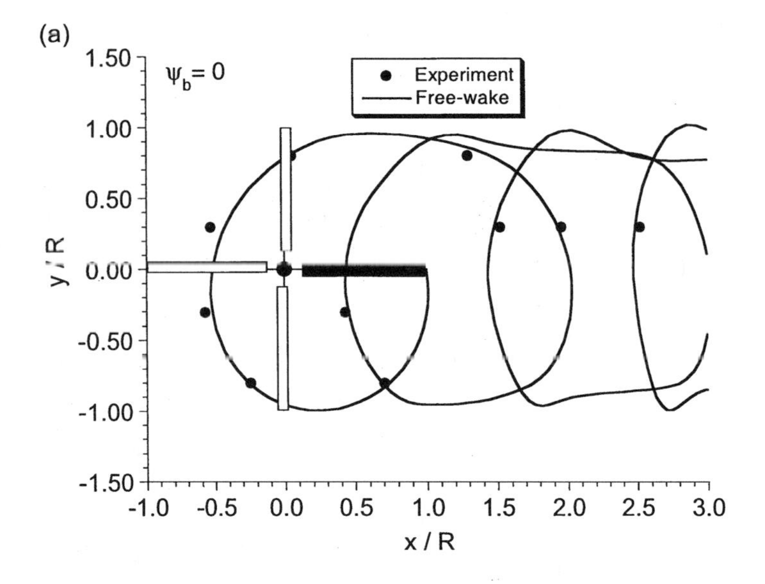

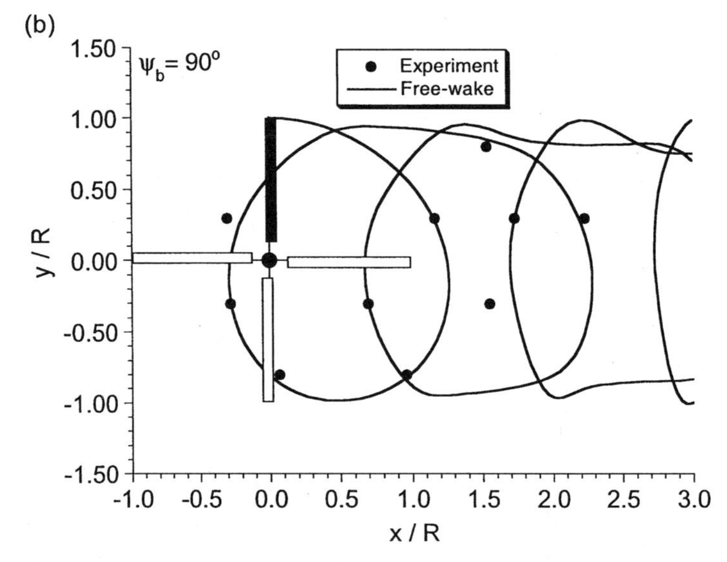

Figure 10.37 Predicted plan view of rotor tip vortex geometry compared to measurements. Results for each blade are shown separately. (a) $\psi_b = 0°$. (b) $\psi_b = 90°$. (c) $\psi_b = 180°$. (d) $\psi_b = 270°$. $\mu = 0.15$; $C_T = 0.008$; $\alpha_s - 3°$. Measured data taken from Ghee & Elliott (1995).

vortex filaments requires a relatively small level of discretization; the present calculation has used $\Delta\psi_b = \Delta\psi_w = 5°$, and the good correlation obtained with the measured data in these regions confirms that, at least in this case, this is an adequate level of discretization.

10.8 Effects of Maneuvers

The problem of modeling helicopter rotor wakes under maneuvering flight conditions is not a new one [see Sadler (1972)]. The maneuver often sets a limit to the normal flight envelope of a helicopter, so the prediction of the rotor airloads under these conditions

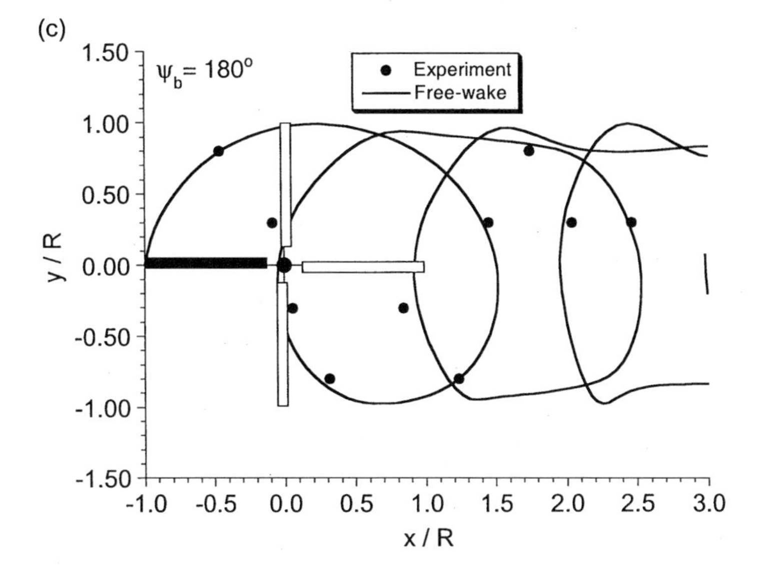

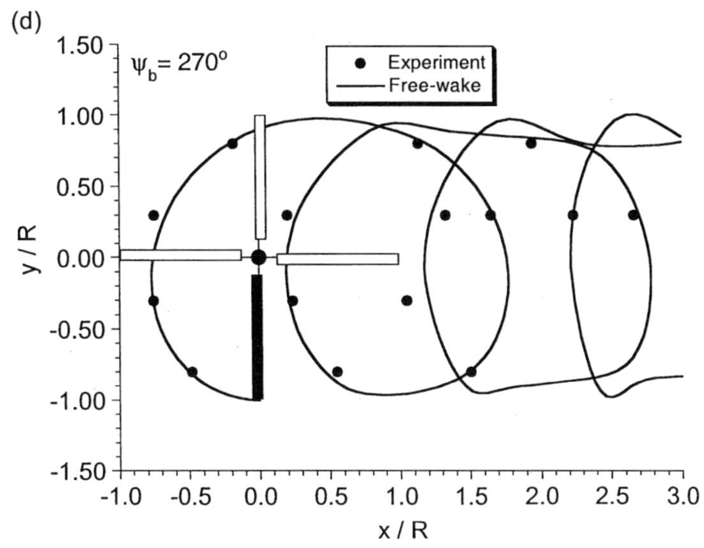

Figure 10.37 (*Continued*)

forms one part of the design process. More recently, the effect on the inflow resulting from rotor wake distortion during maneuvering flight has been suggested as one of the contributing sources of the so-called off-axis response problem of helicopters. This problem manifests as a pitch response resulting from a lateral cyclic input and/or a roll response resulting from a longitudinal cyclic input in pitch – see, for example, Tischler et al. (1994), Mansur & Tischler (1996), and Keller & Curtiss (1996). A maneuver, whether it be pitching or rolling or a combination of such, clearly provides an additional potential source of aerodynamic forcing into the rotor problem and would be expected to produce some further distortion effects into the rotor wake compared to that obtained under straight-and-level flight conditions. The exact understanding of how this effect occurs, however, is the subject of ongoing research. Clearly, if a modified wake topology is produced by the maneuver, this will change the induced inflow, which in turn will affect the blade loads, the rotor

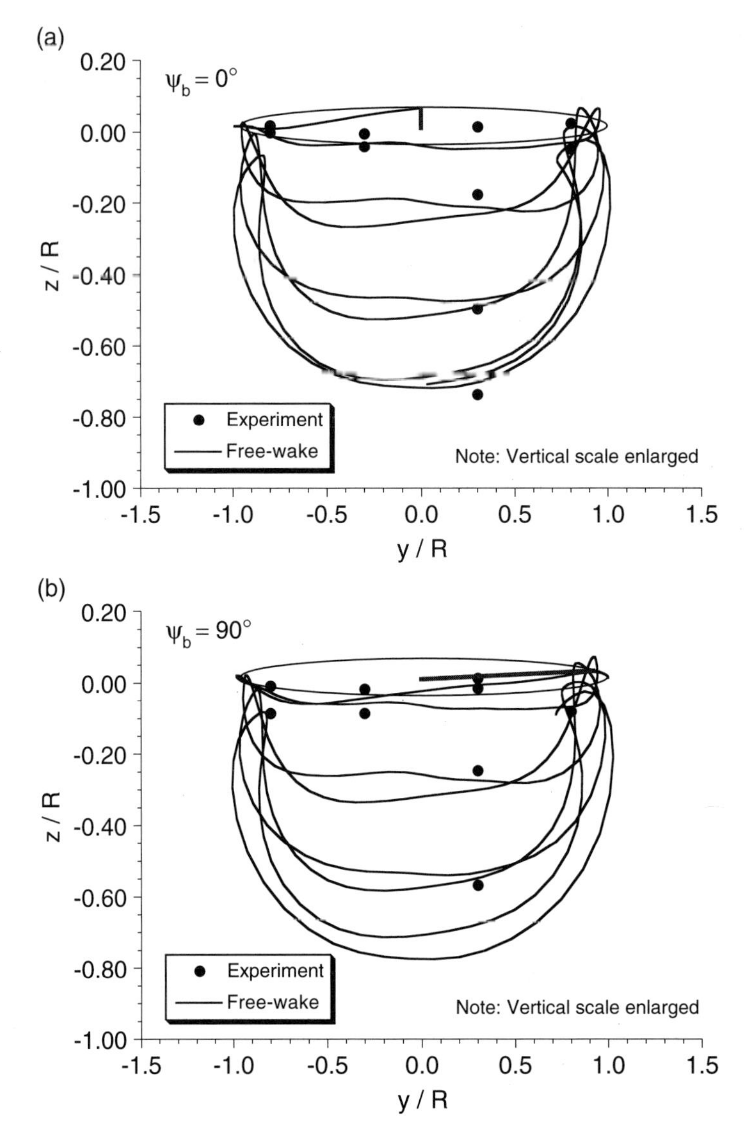

Figure 10.38 Predicted rear view of rotor tip vortex geometry compared to measurements. Results for each blade are shown separately. (a) $\psi_b = 0°$. (b) $\psi_b = 90°$. (c) $\psi_b = 180°$. (d) $\psi_b = 270°$. $\mu = 0.15$; $C_T = 0.008$; $\alpha_s - 3°$. Measured data taken from Ghee & Elliott (1995).

flapping response, and, therefore, the control inputs required to execute the maneuver. The problem is how to calculate such highly coupled effects, which will require a fully integrated aerodynamic, rotor dynamic, and flight mechanics simulation. While comprehensive flight mechanics models are available, the fidelity of the aerodynamic models that can be viably included has been limited so far because of the high computational costs.

There have been several recent developments of rotor wake models directed toward the understanding and prediction of the "off-axis" response problem. Using a relatively simple prescribed wake with assumed (prescribed) maneuver-induced distortion effects, Rosen & Isser (1995), Arnold et al. (1995), Keller & Curtiss (1996), and Barocela et al. (1997) showed improved correlations of the off-axis rotor response when compared to

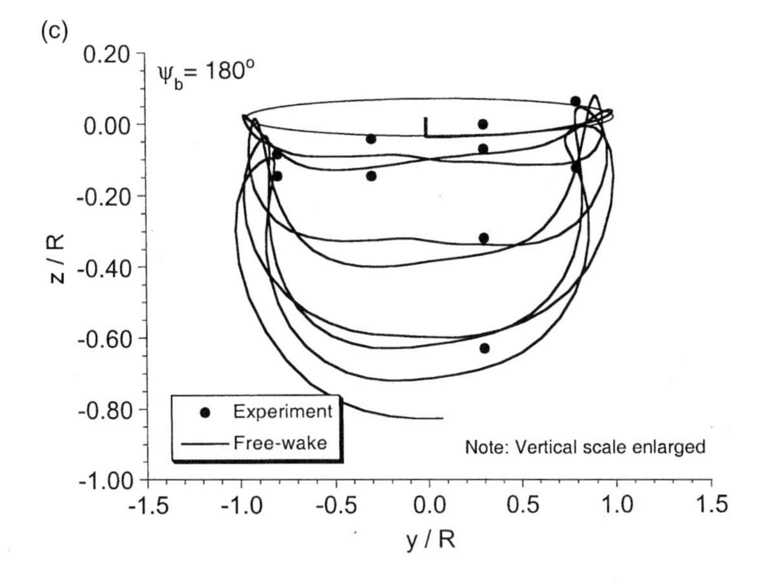

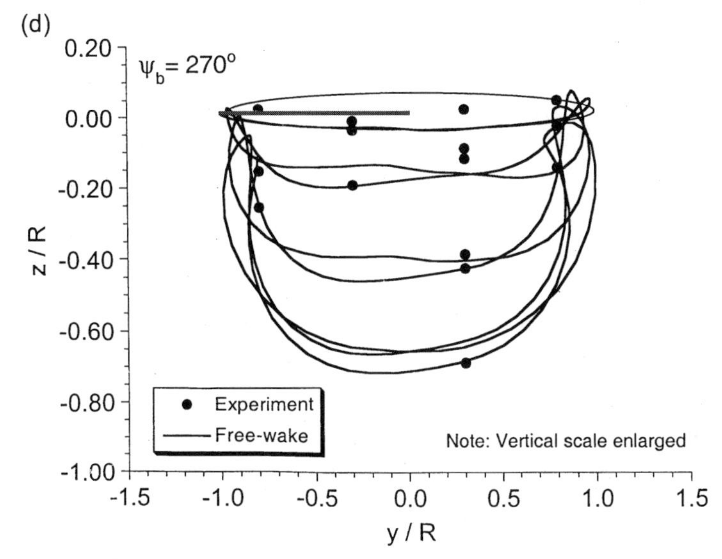

Figure 10.38 *(Continued)*

flight test data. Rosen & Isser's approach, like most vortex wake models that may have the fidelity necessary to predict maneuver-induced wake distortion effects, is computationally expensive to include in flight mechanics simulations. The techniques of Keller & Curtiss are based on "extended" momentum theory to represent the effects of the rotor wake, and is in a form easily included in such simulations. Both these techniques, however, assume that the maneuver effects on wake distortion can be preassumed. This is done by correlating the results with flight test data and including factors into the wake model that produce the desired response. While these empirical approaches can give good results, they give limited insight into the understanding of the aerodynamic mechanisms that may contribute to the off-axis response problem.

The effect of maneuvers on the rotor wake and induced velocity field can be simulated by adding additional "source" terms to the right-hand side of the governing equations

(Eq. 10.8). The source term in Eq. 10.43 can be represented, in general, as $\vec{V}_{\text{loc}}\left(\vec{r}(\psi_b, \psi_w)\right) = \vec{V}_\infty + \vec{V}_{\text{ind}} + \vec{V}_{\text{man}}$, where $\vec{V}_\infty$ is the external velocity field; $\vec{V}_{\text{ind}}$ is the velocity field induced by the rotor blades and their respective vortices from both steady and unsteady aerodynamic effects; and $\vec{V}_{\text{man}}$ is the additional maneuver-induced velocity perturbation, which will be a function of the maneuver rate as well as the positions of the wake filaments relative to the maneuver axis. For simple pitching and rolling motions about the rotor hub axes, the maneuver rate vector can be written as $p\hat{i} + q\hat{j}$, where it is assumed that the rotor rolls about the fixed x axis and pitches about the fixed y axis. Furthermore, because the position of a vortex element in the Cartesian coordinate system is defined by three spatial locations, its position vector can be represented as $\vec{r}(\psi, \zeta) = x\hat{i} + y\hat{j} + z\hat{k}$. The cross-product of the two vectors defines the additional maneuvering velocity field encountered by the free-vortex filaments in the wake, which is

$$\vec{V}_{\text{man}}(\psi, \zeta) = (qz)\hat{i} - (pz)\hat{j} + (py - qx)\hat{k}. \tag{10.86}$$

The first term on the right-hand side of Eq. 10.86, $(qz)\hat{i}$, is equivalent to a velocity component in the x direction in the hub coordinate system resulting from a pitch rate. Similarly, $(pz)\hat{j}$ in Eq. 10.86 represents a y velocity component resulting from rolling. The term $(py - qx)\hat{k}$ corresponds to an axial velocity component resulting from either pitching and rolling motions; that is, the tip-vortex filaments will be either axially closer or further away from the rotor. The $(qz)\hat{i}$ and $(pz)\hat{j}$ velocity components produce a form of asymmetry or curvature to the wake.

Examples of predicted free-vortex wake geometries under idealized (nose-up, positive) pitch rate conditions are shown in Figs. 10.39 and 10.40, for hover and forward flight, respectively. The imposed angular rates are relatively high for a helicopter, but they have been selected to exaggerate the basic effects obtained on the wake geometry. Note the general characteristics manifest as a curvature-type distortion and an axial-type distortion. These are physical effects that have been labeled by Keller & Curtiss (1996) as "bending" and "stretching" of the wake, respectively. Notice that for a positive (nose-up) pitch rate, the vortex elements (tip vortices) at the front of the rotor are further way from the TPP compared to straight-and-level flight conditions. At the back of the disk, the tip vortices are much closer

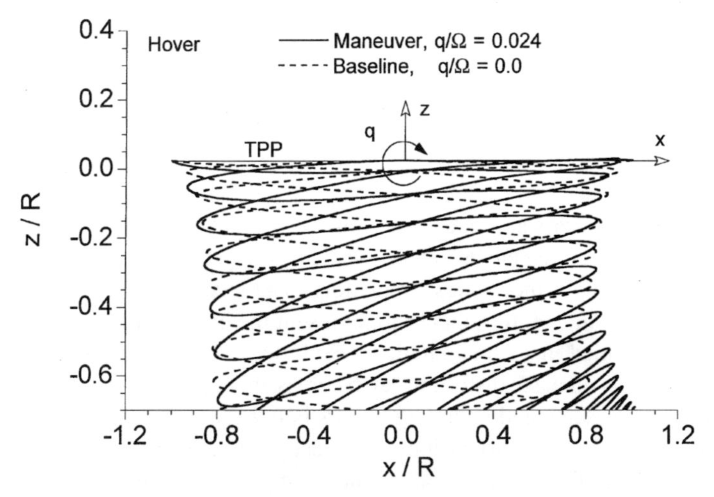

Figure 10.39 Effect of maneuver pitch rate on predicted rotor wake geometry in hover. Four-bladed hovering rotor, with $\bar{q} = 0.024$ and $C_T = 0.008$. Side view.

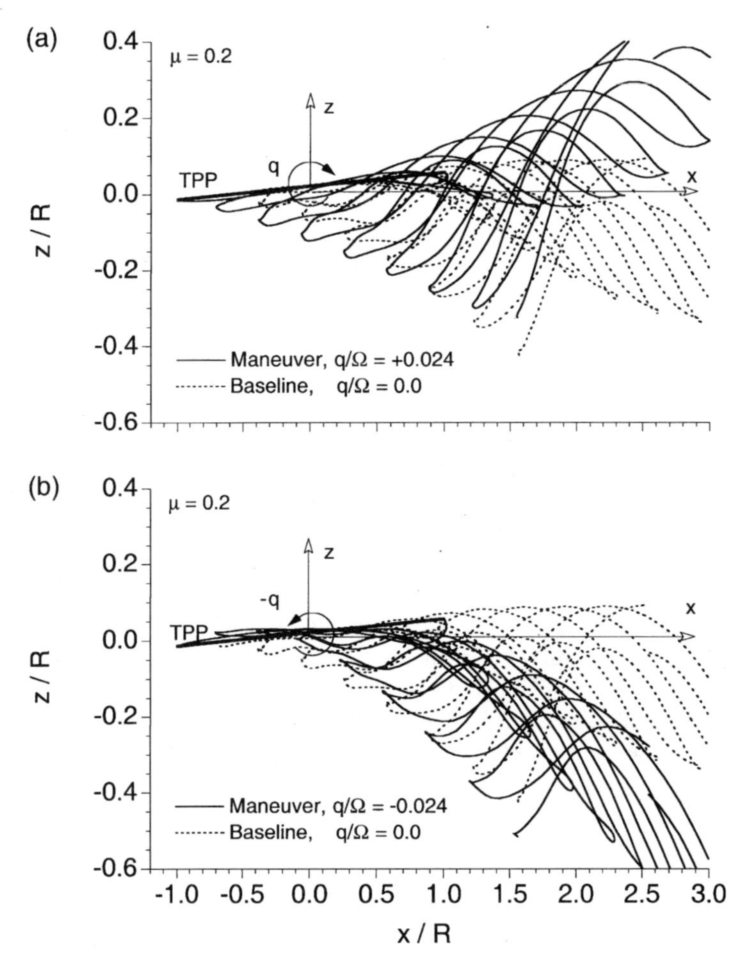

Figure 10.40 Effects of maneuver pitch rate on predicted rotor wake geometries in forward flight. Four-bladed rotor, with $\bar{q} = \pm 0.024$, $C_T = 0.008$, and $\mu = 0.2$. (a) Nose-up pitch rate; $\bar{q} = 0.024$. (b) Nose-down pitch rate; $\bar{q} = -0.024$.

to the TPP. In many ways, these maneuver-induced effects on the wake distortion are similar to an increase in skewness of the wake relative to the rotor TPP. Therefore, the resulting wake distortion must manifest as a change in the longitudinal inflow, which will be expected to be more biased toward the back of the disk.

Theodore & Celi (1998) have incorporated dynamic inflow and free-vortex wake models into a flight mechanics analysis. The results showed some improvements in predicting overall trends in the response of the helicopter to cyclic inputs, but the quantitative predictions were not substantially improved. The results from this study suggest that deficiencies are still present in the aerodynamic models and that other mechanisms may play a role in predicting maneuver airloads and the off-axis response problem. These mechanisms potentially include unsteady aerodynamic effects and transient effects associated with the development of the wake itself. Although the time scales associated with typical helicopter maneuvers are usually long compared to the time scales associated with the development of the wake, the problem of unsteady or transient wake effects is still the subject of research.

10.9 Advanced Computational Models

Supercomputers have enabled new advances to be made in the field of helicopter aerodynamics using so-called computational fluid dynamics (CFD). Here, finite-difference approximations to the governing flow equations are used to solve for the complex flow field about the rotor. The choice of the governing equations affects the level of physics captured by the CFD scheme, as well as the computational time. For the rotor problem, documented approaches have ranged from the transonic small disturbance equations, through full-potential, Euler, and Navier–Stokes equations, each with an increasing level of computational effort. The field is reviewed in McCroskey (1995), which contains many references documenting present CFD capabilities applied to helicopters.

The enormous computer resources required with CFD methods have so far limited their use for many problems in helicopter work. Unlike a fixed-wing aircraft, a helicopter rotor induces significant velocities at large distances from the rotor. Therefore, the computational boundaries, which must be typically small for economy, need to be much larger than for corresponding fixed-wing problems to avoid the possibilities of artificial flow recirculation within the computational domain. Furthermore, the various rotating and nonrotating aerodynamic components of a helicopter make generating an appropriate finite-difference grid on which to solve the equations a challenging problem in itself. Despite these problems, CFD has helped to provide new insight into complex rotor problems.

While the flows about the rotor blades themselves have received the most attention, there has also been some progress in CFD modeling of the rotor wake and flow field around the entire rotorcraft, but it remains a problem of formidable complexity. In some cases, the blade (rotor) and wake can be represented using a hybrid scheme, where a CFD-based solution for the blade aerodynamics is coupled to one of the vortex wake models discussed previously. The inflow from the rotor wake then forms a boundary condition to the computational domain, which can be iterated back and forth until convergence is obtained. Full-potential methods have been used for this purpose and have produced reasonable agreement with experimental measurements of the blade pressure distributions, including the critical area near the tip.

The treatment of the rotor wake itself has proved a more daunting task for CFD methods. Euler methods have been attempted, but because the vortex formation near a blade tip is a result of complex three-dimensional separated flow, Euler-based methods cannot adequately capture the physics of the tip vortex formation. Nevertheless, like full-potential methods, Euler methods have also been used to successfully predict the chordwise pressure distributions and spanwise blade loading. The increasing use of swept tips on helicopter rotors will eventually mean that the Euler-based methods will become the lowest order set of equations that will enable accurate flow field predictions. One particularly promising approach is called the transonic rotor Navier–Stokes (TURNS) method – see Srinivasan & Baeder (1993). In this method, which can be executed in Euler or Navier–Stokes mode, the blades and computational domain are enveloped in body conforming grids, which are constructed using special grid generation programs. This approach is based on the use of an upwind differencing scheme for the convective terms in the governing equations. Upwinding eliminates the addition of explicit numerical dissipation and has been demonstrated to produce less dissipative numerical solutions.

The results from these types of CFD methods have provided considerable physical insight into the detailed flow physics that occur on a rotor blade and have been able to resolve shock induced separation near the blade tip and the roll-up of the tip vortex itself. The proper modeling of the tip vortex formation and the subsequent convection of the vorticity through the flow field is probably the most important step in the calculation of the entire helicopter flow field. The ability to preserve this trailed vorticity from numerical diffusion in the CFD schemes continues to be the subject of much research – see, for example, Tang & Baeder (1999). Most CFD methods also still exhibit grid-dependent solutions and are not yet validated to the level that they can be used confidently for helicopter design purposes.

10.10 Interactions between the Rotor and the Airframe

An additional complication on an actual helicopter is that the rotor cannot be considered to operate in isolation. Under most flight conditions, the rotor wake will envelope significant parts of the airframe. This is especially the case in hovering and low speed forward flight, where the wake skew angle is small. The rotor wake, therefore, may produce average airloads on the airframe that are substantially different to those obtained when it is operating in isolation – see Wilson & Mineck (1975), Sheridan & Smith (1979), Smith & Betzina (1986), Komerath et al. (1985), and Leishman & Bi (1990a,b). What is more of concern, however, is that these airloads on the airframe may change suddenly and unpredictably when the helicopter transitions to or from forward flight and thus may adversely influence aircraft handling qualities (see also the discussion in Chapter 6, Section 6.5.1).

A reciprocal interactional effect may occur on the rotor, whereby the fuselage aerodynamics may affect the rotor loads and performance – see Wilby et al. (1979) and Crouse & Leishman (1992). Such effects are compounded by trends toward the use of smaller and lighter rotors with higher disk loadings and smaller rotor–fuselage spacings. The flow over the fuselage can affect the angles of attack over the front and rear of the rotor, affecting loads, performance, and rotor trim. Although smaller rotor–fuselage spacings cut down hub drag, even without considering the wake effects, the relative proximity of the rotating blades to the airframe can result in large unsteady pressure pulses being produced on both the rotor and airframe. These pulses occur in phase with each blade passage over the airframe surface and may lead to low frequency airframe vibrations with considerable intensity. Several prototype helicopters have been designed with low rotor–fuselage spacings, and these have encountered insurmountable vibration problems that could only be solved by increasing the rotor–fuselage spacing.

Existing predictive models for rotor–airframe interactions are based mostly on extensions of classical panel methods (see Section 6.4.1), with typical methods developed by Freeman (1980), Clark & Maskew (1985, 1988), Egolf & Lorber (1987), Quackenbush et al. (1990, 1994), Lorber & Egolf (1990), and Crouse & Leishman (1992). Predictive capabilities are not entirely satisfactory, and considerable empiricism is required to get predictions of the airloads to correlate with those actually measured. For many flight situations there is a need to retain unsteady terms in the governing equations (at considerably greater computational cost), and also it may be required to represent 3-D separated flows by extending the panels off the airframe surface as free shear layers.

Aside from the effects on the rotor itself, the interactions of the rotor wake vortices with the airframe surface have perhaps received the most attention from a fundamental perspective. These interactions can range from almost benign interactions, where the tip vortex trajectories just glance the airframe or empennage surfaces, to direct impingement and associated large-scale reorganization of the flow topology. However, even though the encounter may vary in terms of its severity, in almost all cases the vortices induce large unsteady airloads on the airframe surface – see, for example, Brand et al. (1989), Leishman & Bi (1990a,b), Crouse et al. (1992), Conlisk & Affes (1993), Conlisk et al. (1993), and

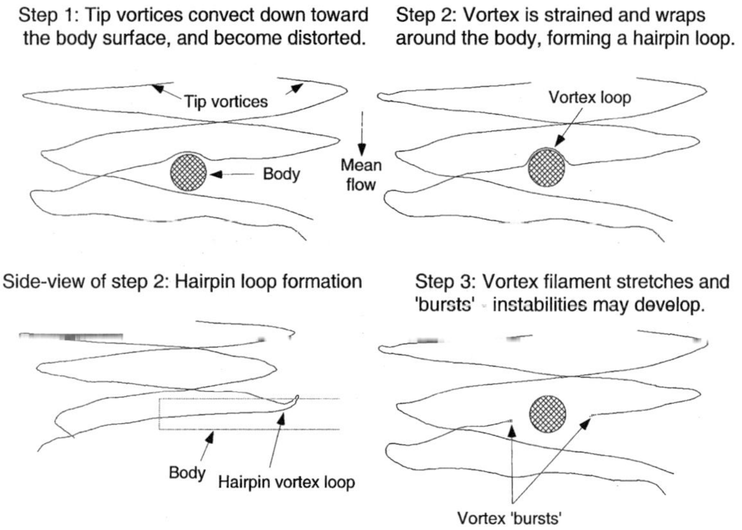

Figure 10.41 Schematic showing the interaction of rotor tip vortices with a cylindrical body.

Moedersheim & Leishman (1998). In many cases, the unsteady pressure fluctuations are much larger than the steady (mean) pressures.

The schematic in Fig. 10.41 shows a typical behavior as the tip vortex filaments come close to and interact with an airframe surface. As the filaments reach the body surface, they distort to form a loop or hairpin vortex structure. These vortex–surface interactions and their effects on the surface pressure pose a complex problem in vortex dynamics, but significant recent progress has been made in modeling the behavior numerically by means of vortex methods – see, for example, Conlisk & Affes (1993a), Conlisk et al. (1993), and Quackenbush et al. (1994).

While the complexity of the rotor–airframe interaction problem provides a good challenge for both the theoretician and experimentalist, it provides a particularly good challenge for more advanced finite-difference numerical methods based on Navier–Stokes or Reynolds-averaged Navier–Stokes equations. Ambitious attempts with entire rotor and airframe configurations have been attempted using millions of grid points in the flow (and many tens of hours of time on a supercomputer), but these have so far provided limited new insight to these types of interactional aerodynamic problems. One general limitation with finite-difference methods is the efficient generation of grids, and especially the proper coupling of any structured and unstructured grids. Better insight has been achieved with less ambitious CFD approaches, such as that of Zori & Rajagopalan (1995), who have developed an incompressible Navier–Stokes solution to the airframe flow, with the effects of the rotor being treated simply as an actuator disk or source term in the governing equations. Such methods take advantage of finding a fundamental solution to the flow about the airframe, but without the expense of modeling the details of the aerodynamics over each blade. However, since 1994, rapid progress in grid generation techniques has been made [see, for example, Chaffin & Berry (1994), Duque & Dimanlig (1994), and Duque (1994)] and it is likely that significant inroads toward an improved understanding of airframe aerodynamics and rotor–airframe interaction problems using these new methods will occur within the next ten years.

10.11 Chapter Review

This chapter has described some of the significant problems associated with an improved understanding of the wakes generated by helicopter rotors. Problems such as tip vortex formation, blade–vortex interaction, and rotor–airframe interference are better understood, and systematic experimental measurements have provided significant results that will help validate numerical predictions of the various phenomena. Recent advances in experimental techniques for flow visualization and three-dimensional velocity field measurements have been significant, and further developments will likely continue. Ambitious experimental tests with rotors have generated large amounts of important results that have provided new insight into old problems and remarkable insight into other problems that were previously considered intractable. However, the complexity of the helicopter rotor wake, together with the sensitivity of its topology to the flight condition, makes all types of rotor wake experiments time consuming and expensive to undertake. The challenge to the rotor analysts of the future is to balance needs with experimental capabilities and to use complementary flow diagnostic techniques wherever possible to provide an improved understanding of the various wake phenomena.

One of the biggest challenges still remaining in helicopter rotor analysis is to reveal more completely the intricate structure of the blade tip vortices. This includes issues such as the vortex strengths, the viscous core size, and velocity field near to and inside the core, as well as their relationships to the loading on the blades. Fundamental issues such as diffusion and core growth of the vortices are not yet fully understood. This behavior is compounded by interactions of vortices and blades. Such problems cannot yet be examined using computational methods based on first principles and require careful experimental studies with rotors in both hover and forward flight. This research may result in new ideas for tip vortex control and possibly the alleviation of several of the complex problems involving vortex dynamics that hinder the development of quieter helicopters with better performance and lower vibration.

The long-term goal of the aerodynamic modeling efforts is to be able to predict the entire flow field around the helicopter rotor and its airframe when operating in any flight condition. To this end, the prediction of the rotor wake is key. However, increasing emphasis is being placed on rotor airframe interaction problems because of their importance in determining performance, vibration levels, and also handling qualities of the helicopter. Rapid advances in computer power mean that prescribed vortex wake models are rapidly giving way to free-vortex models for many routine levels of helicopter analysis and design work. These methods are relatively powerful and can give good results; yet caution should be used. Quantitative errors can only be determined through correlation studies with experiments, and much further validation of vortex wake models is required, particularly in improving the confidence in predicting the combined effects of the rotor flow and the airframe. First principles–based methods, in the form of finite-difference Navier–Stokes computations, are still too computationally intensive other than for basic research purposes. While ambitious problems have been attempted, these methods have not yet been validated enough to be assigned confidence levels suitable for design purposes. With bigger and faster computers, however, CFD methods will ultimately prevail and will become increasingly integrated into the helicopter design process.

10.12 Questions

10.1. For some purposes, it is possible to approximate the rotor wake by means of a series of stacked vortex rings. Find an analytic expression for the velocity induced by a single vortex ring, and show the result for the vertical velocity in the plane of the ring. Determine the induced velocity for a desingularized vortex ring using discretized segmentation of the ring into straight line elements. Show your results for various levels of angular discretization.

10.2. Write a computer program to model a rigid vortex wake and generate the coordinates of the tip vortices. For the case where $\psi_b = 0$, show the top, side, and rear views of the wake relative to the TPP for a two-bladed rotor operating at $C_T = 0.008$ and advance ratios of 0.05, 0.1, and 0.2. Assume the shaft angle is zero.

10.3. Blade vortex interaction or BVI is an important contributor to the unsteady airloads and noise generated by a rotor. Assuming an undistorted wake model in forward flight, show that the potential blade–vortex interaction locations over the rotor disk can be obtained from the equations

$$\psi_b = \sin^{-1}\left\{\left(-\cos\Delta\sin(\psi_b - \Delta) \pm \sin\Delta\sqrt{\mu^2\psi_w^2 - \sin^2(\psi_w - \Delta)}\right)\Big/\mu\psi_w\right\}$$

$$r = \sin(\psi_b - \psi_w)/\sin(\psi_b - \Delta) \quad \text{and} \quad \Delta = \frac{2\pi(i-1)}{N_b},$$

where all the symbols have their usual meaning. Plot the result for a four-bladed rotor operating at $\mu = 0.1, 0.2, 0.3$, and 0.4.

10.4. Use the undistorted wake model to compute the tip vortex trajectories trailed from a four-bladed rotor in forward flight at advance ratios of 0.1, 0.2, 0.3, and 0.4. If it is assumed that $z_w = 0$, that is, all the tip vortices lie in the rotor plane, calculate by means of a suitable numerical method the locus of all the blade vortex interaction locations over the rotor disk.

10.5. Write a computer program for the Beddoes prescribed wake model. Compare the results for a four-bladed rotor operating at $C_T = 0.008$ to those obtained with a rigid wake model. Plot your results in terms of x, y, z coordinates.

10.6. The tangential (swirl) component of velocity in the Oseen-Lamb vortex model is given by

$$V_\theta(r) = \frac{\Gamma}{2\pi r}\left[1 - \exp\left(\frac{-r^2}{4\nu t}\right)\right],$$

where r is the distance from the vortex axis, ν is the kinematic velocity, and t is time. Show that for this model, the viscous core radius of the vortex grows according to the expression $r_c = 1.1209\sqrt{4\nu t}$.

10.7. In a discretized lifting-line model of a rotor blade, it is often assumed that all of the "bound" circulation outboard of the maximum bound circulation rolls up into the tip vortex. Prove that this result is equivalent to the maximum bound circulation on the blade.

10.8. Two rectilinear vortices of unit strength are initially positioned at rest two units of distance apart. By means of time-marching integration using the Biot–Savart law, calculate the subsequent trajectories of the vortex pair. Use both Euler explicit and predictor–corrector integration schemes. Check on the reversibility of the scheme. Comment on your results.

10.9. The ratio Γ_v/ν (where Γ_v is the vortex strength and ν is the kinematic viscosity) is encountered frequently in viscous vortex dynamics, and referred to as a vortex Reynolds number. Explain.

Bibliography

Akimov, A. I., Butov, V. P., Bourtsev, B. N., and Selemenev, S. V. 1994. "Flight Investigation of Coaxial Rotor Tip Vortex Structure," 50th Annual Forum of the American Helicopter Soc., Washington DC, May 11–13.

Althoff, S. L., Elliott, J. W., and Sailey, R. H. 1988. "Inflow Measurements Made with a Laser Velocimeter on a Helicopter Model in Forward Flight," NASA Technical Memorandum 100544.

Arnold, U. T. P., Keller, J. D., Curtiss, H. C., and Reichert, G. 1995. "The Effect of Inflow Models on the Dynamic Response of Helicopters," 21st European Rotorcraft Forum, Saint-Petersburg, Russia, Aug. 30–Sept. 1.

Bagai, A. and Leishman, J. G. 1992a. "Improved Wide-Field Shadowgraph Set-Up for Rotor Wake Visualization," *J. American Helicopter Soc.*, 37 (3), pp. 86–92.

Bagai, A. and Leishman, J. G. 1992b. "A Study of Rotor Wake Developments and Wake/Body Interactions in Hover Using Wide-Field Shadowgraphy," *J. American Helicopter Soc.*, 37 (4), pp. 48–57.

Bagai, A. and Leishman, J. G. 1994. "Flow Visualization of Compressible Vortex Structures using Density Gradient Techniques," *Expts. in Fluids*, 15, pp. 431–442.

Bagai, A. and Leishman, J. G. 1995a. "Rotor Free-Wake Modeling Using a Pseudo-Implicit Technique Including Comparisons with Experiment," *J. American Helicopter Soc.*, 40 (3), pp. 29–41.

Bagai, A. and Leishman, J. G. 1995b. "Rotor Free-Wake Modeling Using a Pseudo-Implicit Relaxation Algorithm," *AIAA J. of Aircraft*, 32 (6), pp. 1276–1285.

Bagai, A. and Leishman, J. G. 1996. "Free-Wake Analysis of Tandem, Tilt-Rotor and Coaxial Rotor Configurations," *J. American Helicopter Soc.*, 41 (3), pp. 196–207.

Bagai, A. and Leishman, J. G. 1998. "Adaptive Grid Sequencing and Interpolation Schemes for Rotor Free-Wake Analyses," *AIAA J.*, 36 (9), pp. 1593–1602.

Barocela, E., Peters, D. A., Krothapalli, K. R., and Prasad, J. V. R. 1997. "The Effect of Wake Distortion on Rotor Inflow Gradients and Off-Axis Coupling," AIAA Flight Mechanics Conf., San Diego, CA, July 29–31.

Batchelor, G. K. 1967. *Introduction to Fluid Dynamics*, Cambridge University Press, Cambridge, UK, p. 87.

Beddoes, T. S. 1985. "A Wake Model for High Resolution Airloads," 2nd Int. Conf. on Basic Rotorcraft Research, Triangle Park, NC.

Bertin, J. H. and Smith, M. L. 1989. *Aerodynamics for Engineers*, 2nd ed., Prentice-Hall, Inc., Englewood Cliffs, NJ, Chapter 7.

Biggers, J. C., Lee, A., Orloff, K. L., and Lemmer, O. J. 1977a. "Measurements of Helicopter Rotor Tip Vortices," 33rd Annual National Forum of the American Helicopter Soc., Washington DC, May 9–11.

Biggers, J. C., Lee, A., Orloff, K. L., and Lemmer, O. J. 1977b. "Laser Velocimeter Measurements of Two-Bladed Helicopter Rotor Flow Fields," NASA TM X-73238.

Biggers, J. C. and Orloff, K. L. 1975. "Laser Velocimeter Measurements of the Helicopter Rotor-Induced Flow Field," *J. American Helicopter Soc.*, 20 (1), pp. 2–10.

Brand, A. G., Komerath, N. M., and McMahon, H. M. 1988. "Results from a Laser Sheet Visualization of a Periodic Rotor Wake," Paper 88-0192, AIAA Aerospace Sci. Meeting, Reno, NV, Jan.

Brand, A. G., McMahon, H. M., and Komerath, N. M. 1989. "Surface Pressure Measurements on a Body Subject to Vortex-Wake Interaction," *AIAA J.*, 27 (5), pp. 569–574.

Campbell, J. F. and Chambers, J. R. 1994. "Patterns in the Sky – Natural Visualization of Aircraft Flow Fields," NASA SP-514.

Caradonna, F. X. and Tung, C. 1981. "Experimental and Analytical Studies of a Model Helicopter Rotor in Hover," 7th European Rotorcraft Forum, Garmisch-Partenkirchen, Germany, Sept. 22–25.

Castles, W., Jr. and De Leeuw, J. H. 1954. "The Normal Component of the Induced Velocity in the Vicinity of a Lifting Rotor and Some Examples of Its Application," NACA Report 1184.

Chaffin, M. S. and Berry, J. D. 1994. "Navier–Stokes and Potential Theory Solutions for a Helicopter Fuselage and Comparison with Experiment," NASA Technical Memorandum 4566, ATCOM Technical Report 94-A-013.

Chigier, N. A. and Corsiglia, V. R. 1971. "Tip Vortices – Velocity Distributions," 27th Annual Forum of the American Helicopter Soc., Washington DC, May.

Clark, D. R. and Leiper, A. C. 1970. "The Free Wake Analysis, A Method for the Prediction of Helicopter Rotor Hovering Performance," *J. American Helicopter Soc.*, 15 (1), pp. 3–11.

Clark, D. R. and Maskew, B. 1985. "Study for Prediction of Rotor/Wake/Fuselage Interference," NASA CR-177340.

Clark, D. R. and Maskew, B. 1988. "Calculation of Unsteady Rotor Blade Loads and Blade/Fuselage Interference," 2nd Int. Conf. on Rotorcraft Basic Research, College Park, MD, Feb.

Coleman, R. P., Feingold, A. M., and Stempin, C. W. 1945. "Evaluation of the Induced Velocity Fields of an Idealized Helicopter Rotor," NACA ARR L5E10.

Conlisk, T. A. and Affes, H. 1993. "A Model for Rotor Tip Vortex–Airframe Interaction, Part 1: Theory," *AIAA J.*, 31 (12), pp. 2263–2273.

Conlisk, T. A., Affes, H., Kim, J. M., and Komerath, N. M. 1993. "A Model for Rotor Tip Vortex–Airframe Interaction, Part 2: Comparison with Experiment," *AIAA J.*, 31 (12), pp. 2263–2273.

Cook, C. V. 1972. "The Structure of the Rotor Blade Tip Vortex," AGARD CP-111.

Crouse, G. L. and Leishman, J. G. 1992. "Aerodynamic Interactional Effects on Rotor Loads and Performance," 48th Annual Forum of the American Helicopter Soc., Washington DC, June 3–5.

Crouse, G. L., Leishman, J. G., and Bi, Nai-pei. 1992. "Theoretical and Experimental Study of Unsteady Rotor/Body Aerodynamic Interactions," *J. American Helicopter Soc.*, 37 (1), pp. 55–65.

Crouse, G. L. and Leishman, J. G. 1993. "A New Method for Improved Rotor Free-Wake Convergence," 31st AIAA Aerospace Sci. Meeting and Exhibit, Reno, NV, Jan. 11–14.

Dadone, L. U. 1970. Unpublished photographs from model tests performed in the 20-by-20 foot wind tunnel at Boeing Helicopters, Philadelphia.

Dadone, L. U. 1995. Technical Evaluation Report, AGARD Fluid Dynamics Panel Symposium on Aerodynamics and Aeroacoustics of Rotorcraft, AGARD-CP-552.

Devenport, W. J., Rife, M. C., Liapis, S. I., and Follin, G. J. 1996. "The Structure and Development of a Wing-Tip Vortex," *J. of Fluid Mech.*, 312, pp. 67–106.

Dingeldein, R. C. 1954. "Wind-Tunnel Studies of the Performance of Multirotor Configurations," NACA TN 3236.

Dosanjh, D. S., Gasparek, E. P., and Eskinazi, S. 1962. "Decay of Viscous Trailing Vortex," *The Aeronautical Quarterly*, 13, pp. 167–188.

Duque, E. P. N. 1994. "A Structured/Unstructured Embedded Grid Solver for Helicopter Rotor Flows," 50th Annual Forum of the American Helicopter Soc., Washington, DC, May 11–13.

Duque, E. P. N. and Dimanlig, A. C. B. 1994. "Navier–Stokes Simulation of the RAH-66 (Comanche) Helicopter," American Helicopter Soc. Aeromechanics Specialists' Conf., San Francisco, CA, Jan. 19–21.

Egolf, T. A. and Landgrebe, A. J. 1983. "Helicopter Rotor Wake Geometry and Its Influence in Forward Flight, Vol. I – Generalized Wake Geometry and Wake Effect on Rotor Airloads and Performance," NASA CR-3726.

Egolf, E. T. and Lorber, P. F. 1987. "An Unsteady Rotor/Fuselage Interaction Method," National Specialists' Meeting on Aerodynamics and Aeroacoustics Arlington, TX, Feb. 25–27.

Elliott, J. W., Althoff, S. L., and Sailey, R. H. 1988. "Inflow Measurements Made with a Laser Velocimeter on a Helicopter Model in Forward Flight," Vols. II and III, Rectangular Blades – Advance Ratios of 0.23 and 0.30, NASA TM-100542, TM-100543.

Felker, F. F., Maisel, M. D., and Betzina, M. D. 1986. "Full-Scale Tilt-Rotor Hover Performance," *J. American Helicopter Soc.*, 31 (2), pp. 10–18.

Freeman, C. E. 1980. "Development and Validation of a Combined Rotor-Fuselage Induced Flow-Field Computational Method," NASA TP-1656.

Gilmore, D. C. and Gartshore, I. S. 1972. "The Development of an Efficient Hovering Propeller/Rotor Performance Prediction Method," AGARD CP-111.

Ghee, T. A. and Elliott, J. W. 1995. "The Wake of a Small-Scale Rotor in Forward Flight Using Flow Visualization," *J. American Helicopter Soc.*, 40 (3), pp. 52–65.

Gray, R. B. 1956. "An Aerodynamic Analysis of a Single Bladed Rotor in Hovering and Low Speed Forward Flight as Determined from Smoke Studies of the Vorticity Distribution in the Wake," Princeton University, Report No. 356.

Gupta, B. P. and Loewy, R. G. 1973. "Analytical Investigation of the Aerodynamic Stability of Helical Vortices Shed from a Hovering Rotor," USAAMRDL 73-84.

Gupta, B. P. and Loewy, R. G. 1974. "Theoretical Analysis of the Hydrodynamic Stability of Multiple Interdigitated Helical Vortices," *AIAA J.*, 12 (10), pp. 1381–1387.

Han, Y., Leishman, J. G., and Coyne, A. 1997. "Measurements of the Velocity and Turbulence Stucture of a Rotor Tip Vortex," *AIAA J.*, 35 (3), pp. 477–485.

Heyson, H. H. and Katzoff, S. 1957. "Induced Velocities near a Lifting Rotor with Non-Uniform Disk Loading," NACA TR 1319.

Hildebrand, F. B. 1976. *Advanced Calculus for Applications*, Prentice-Hall, Inc., Englewood Cliffs, NJ, p. 403.

Hoad, D. R. 1983. "Preliminary Rotor Wake Measurements with a Laser Velocimeter," NASA TM 83246.

Holder, D. W. and North, R. J. 1963. "Schlieren Methods," Notes on Applied Science, No. 31, Dept. of Scientific and Industrial Research, National Physical Laboratory, HMSO, London.

Horner, M. H., Stewart, J. N., Galbraith, R. A. McD., Grant, I., and Coton, F. N. 1994. "An Examination of Vortex Deformation During Blade Vortex Interaction Using Particle Image Velocimetry," 19th ICAS Meeting, Anaheim, CA, June 1994.

Iversen, J. D. 1976. "Correlation of Turbulent Trailing Vortex Decay Data," *J. of Aircraft*, 13 (5), pp. 338–342.

Jenks, M., Dadone, L., and Gad-el-Hak, M. 1987. "Towing Tank Flow Visualization of a Scale Model H-34 Rotor," 43rd Annual Forum of the American Helicopter Soc., St. Louis, MO, May 18–20.

Keller, J. D., Curtiss, H. C. 1996. "Modeling the Induced Velocity of a Maneuvering Helicopter," 52nd Annual Forum of the American Helicopter Soc., Washington, DC, June 4–6.

Kitaplioglu, C. and Caradonna, F. X. 1994. "Aerodynamics and Acoustics of Blade Vortex Interaction Using an Independently Generated Vortex," American Helicopter Soc. Aeromechanics Specialists' Conf., San Francisco, CA, Jan. 19–21.

Kocurek, J. D. and Tangler, J. L. 1976. "A Prescribed Wake Lifting Surface Hover Performance Analysis," *J. American Helicopter Soc.*, 21 (1), pp. 24–35.

Kocurek, J. D. and Berkowitz, L. F. 1982. "Velocity Coupling: A New Concept for Hover and Axial Flow Wake Analysis and Design," AGARD CP-334.

Komerath, N. M., McMahon, H. M., and Hubbard, J. E. 1985. "Aerodynamic Interactions between a Rotor and Airframe in Forward Flight," AIAA Paper 85-1606, Fluid Dynamics, Plasmadynamics & Lasers Conf., 18th, Cincinnati, OH, July 16–18.

Lamb, H. 1932. *Hydrodynamics*, Cambridge University Press, Cambridge, UK, pp. 592–593, 668–669.

Landgrebe, A. J. 1969. "An Analytical Method for Predicting Rotor Wake Geometry," *J. American Helicopter Soc.*, 14 (4), pp. 20–32.

Landgrebe, A. J. 1971."An Analytical and Experimental Investigation of Helicopter Rotor Performance and Wake Geometry Characteristics," USAAMRDL TR 71-24.

Landgrebe, A. J. and Bellinger, E. D. 1971. "An Investigation of the Quantitative Applicability of Model Helicopter Rotor Wake Patterns Obtained from a Water Tunnel," UARL K910917-23.

Landgrebe, A. J. 1972. "The Wake Geometry of a Hovering Rotor and Its Influence on Rotor Performance," *J. American Helicopter Soc.*, 17 (4), pp. 2–15.

Landgrebe, A. J. and Johnson, B. V. 1974. "Measurement of Model Helicopter Rotor Flow Velocities with a Laser Doppler Velocimeter," *J. American Helicopter Soc.*, 19 (3), pp. 39–43.

Landgrebe, A. J. 1988. "Overview of Helicopter Wake and Airloads Technology," 2nd. Int. Conf. on Rotorcraft Basic Research, College Park, MD, Feb.

Larin, A. V. 1973. "Vortex Wake behind a Helicopter," *Aviatsiya i Kosmonavtika*, 3, pp. 32–33. Avail. NTIS as translation ADA005479.

Larin, A. V. 1974. "Vortex Formation in Oblique Flow around a Helicopter Rotor," *Uchenyye Zapiski TSAGI*, 1 (3), pp. 115–122. Avail. NTIS as translation AD781245.

Lee, D. J. and Na, S. U. 1995. "High Resolution Free Vortex Blob Method for Highly Distorted Vortex Wake Generated from a Slowly Starting Rotor Blade in Hover," 21st European Rotorcraft Forum, St. Petersburg, Russia, Aug. 30–Sept. 1.

Lee, D. J. and Na, S. U. 1998. "Numerical Simulations of Wake Structure Generated by Rotating Blades Using a Time Marching, Free Vortex Blob Method," *European J. Fluid Mech. - B/Fluids*, 17 (4), pp. 1–13.

Lehman, A. F. 1968. "Model Studies of Helicopter Rotor Patterns," 24th Annual National Forum of the American Helicopter Soc., Washington DC.

Leishman, J. G. and Bi, Nai-pei. 1990a. "Measurements of a Rotor Flow Field and the Effects of a Fuselage in Forward Flight," *Vertica*, 14 (3), pp 401–415.

Leishman, J. G. and Bi, Nai-pei. 1990b. "Aerodynamic Interactions between a Rotor and a Fuselage in Forward Flight," *J. American Helicopter Soc.*, 35 (3), pp. 22–31.

Leishman, J. G. and Bagai, A. 1991. "Rotor Wake Visualization in Low Speed Forward Flight," Paper 91-3232, AIAA 9th Applied Aerodynamics Conf. Baltimore, MD, Sept. 23–25.

Leishman, J. G., Baker, A. M., and Coyne, A. J. 1995. "Measurements of Rotor Tip Vortices Using Three-Component Laser Doppler Velocimetry," *J. American Helicopter Soc.*, 41 (4), pp. 342–353.

Leishman, J. G. 1996. "Seed Particle Dynamics in Tip Vortex Flows," *J. of Aircraft*, 33 (4), July–Aug., pp. 823–825.

Leishman, J. G. 1998. "Measurements of the Aperiodic Wake of a Hovering Rotor," *Expts. in Fluids*, 25, pp. 352–361.

Leishman, J. G. and Bagai, A. 1998. "Challenges in Understanding the Vortex Dynamics of Helicopter Rotor Wakes," *AIAA J.*, 36 (7), pp. 1130–1140.

Leishman, J. G. 1999. "Acoustic Focusing Effects During Parallel and Oblique Blade Vortex Interaction," *J. of Sound and Vibration*, 221 (3), pp. 415–441.

Levy, H. and Forsdyke, A. G. 1928. "The Steady Motion and Stability of a Helical Vortex," *Proc. of the Royal Soc. of London*, Series A, 120.

Lewis, R. I. 1991. *Vortex Element Methods for Fluid Dynamic Analysis of Engineering Systems*, Cambridge University Press, Cambridge, UK.

Light, J. S., Frerkin, A., and Norman, T. R. 1990. "Application of the Wide-Field Shadowgraph Technique to Helicopters in Forward Flight," 46th Annual Forum of the American Helicopter Soc., Washington DC, May 21–23.

Lorber, P. F. and Egolf, T. A. 1990. "An Unsteady Rotor Fuselage Interaction Analysis," *J. American Helicopter Soc.*, 35 (1), pp. 32–42.

Lorber, P. F., Stauter, R. C., Hass, R. J., Torok, M. S., and Kohlhepp, F. W. 1994. "Techniques for Comprehensive Measurement of Model Helicopter Rotor Aerodynamics," 50th Annual Forum of the American Helicopter Soc., Washington DC, May 11–14.

Lowson, M. V. 1996. "Focusing of Helicopter BVI Noise," *J. of Sound and Vibration*, 190 (3), pp. 477–494.

Mahalingam. R. and Komerath N. M. 1998."Measurements of the Near Wake of a Rotor in Forward Flight," Paper 98-0692, 36th Aerospace Sci. Meeting & Exhibit, Reno, NV, Jan. 12–15.

Mansur, M. H. and Tischler, M. B. 1996. "An Empirical Correction for Improving Off-Axes Response in Flight Mechanics Helicopter Models," AGARD Flight Vehicle Integration Panel Symp. on Advances in Rotorcraft Technology, Ottawa, Canada, May 27–30.

Martin, P. B., Bhagwat, M. B., and Leishman, J. G. 1999. "Strobed Laser Sheet Visualization of a Helicopter Rotor Wake," Proc. of the PSFVIP-2, Honolulu, HI, May 18–20.

McAlister, K. W., Schuler, C. A., Branum, L., and Wu, J. C. 1995. "3-D Wake Measurements near a Hovering Rotor for Determining Profile and Induced Drag," NASA TP 3577.

McAlister, K. W. 1996. "Measurements in the Near Wake of a Hovering Rotor," Paper 96-1958, 27th AIAA Fluid Dynamics Conf., New Orleans, LA, June 18–20.

McCroskey, W. J. 1995. "Vortex Wakes of Rotorcraft," Paper 95-0530, 33rd AIAA Aerospace Sci. Meeting and Exhibit, Reno, NV, Jan. 9–12.

McVeigh, M. A., Liu, J., and O'Toole, S. J. 1997. "V-22 Osprey Aerodynamic Development – A Progress Review," *The Aeronaut J.*, 101 (1006), pp. 231–244.

Mercker, E. and Pengel, K. 1992. "Flow Visualization of Helicopter Blade Tip Vortices," 18th European Rotorcraft Forum, Avignon, France, Sept. 15–18.

Merzkirch, W. 1981. "Density Sensitive Flow Visualization," *Methods of Experimental Physics*, Vol. 18A, pp. 345–375.

Miller, W. O. and Bliss, D. B. 1993. "Direct Periodic Solutions of Rotor Free Wake Calculations ," *J. American Helicopter Soc.*, 38 (2), pp. 53–60.

Miller, W. O. 1993. "A Fast Adaptive Resolution Method for Efficient Free Wake Calculations," 49th Annual Forum of the American Helicopter Soc., St. Louis, MO, May 19–21.

Moedersheim, E., Daghir, M., and Leishman, J. G. 1994. "Flow Visualization of Rotor Wakes Using Shadowgraph and Schlieren Techniques," 20th European Rotorcraft Forum, Amsterdam, The Netherlands, 4–6 Oct.

Moedersheim, E. and Leishman, J. G. 1998. "Investigation of Aerodynamic Interactions between a Rotor and a T-Tail Empennage," *J. American Helicopter Soc.*, 43 (1), pp. 37–46.

Müller, R. H. G. 1990a. "Special Vortices at a Helicopter Rotor Blade," *J. American Helicopter Soc.*, 35 (3), pp. 16–22.

Müller, R. H. G. 1990b. "Winglets on Rotor Blades in Forward Flight – A Theoretical and Experimental Investigation," *Vertica*, 14 (1), pp. 31–46.

Müller, R. H. G. 1994. "Tip Vortex Development and Structure at BVI for a Hovering Rotor with Swept Back Tip Shapes Using the Flow Visualization Gun Technique," 20th European Rotorcraft Forum, Amsterdam, The Netherlands, 4–6 Oct.

Newman, B. G. 1959. "Flow in a Viscous Trailing Vortex," *Aeronautical Quarterly*, May, pp. 167–188.

Norman, T. R. and Light, J. S. 1987. "Rotor Tip Vortex Geometry Measurements Using the Wide-Field Shadowgraph Technique," *J. American Helicopter Soc.*, 32 (2), pp. 40–50.

Ogawa, A. 1993. *Vortex Flow*, CRC Series on Fine Particle Science and Technology, CRC Press Inc., Chapter 12.

Oseen, C. W. 1912. "Uber Wirbelbewegung in Einer Reibenden Flussigkeit," *Ark. J. Mat. Astrom. Fys.*, 7, pp. 14–21.

Parthasarthy, S. P., Cho, Y. I., and Black, L. H. 1985. "Wide-Field Shadowgraph Flow Visualization of Tip Vortices Generated by a Helicopter Rotor," Paper 85-1557, AIAA 18th Fluid Dynamics and Plasmadynamics and Lasers Conf., Cincinnati, OH.

Piziali, R. and Trenka, A. 1970. "An Experimental Study of Blade Tip Vortices," Report AC-2647-S-1, Cornell Aeronautical Laboratory, Buffalo, NY.

Quackenbush, T. R., Bliss, D. B., and Wachspress, D. A. 1988. "Computational Analysis of Hover Performance Using a New Free Wake Method," 2nd. Int. Conf. on Rotorcraft Basic Research, Triangle Park, NC.

Quackenbush, T. R., Bliss, D. B., Lam, C.-M. G., and Katz, A. 1990. "New Vortex/Surface Interaction Methods for the Prediction of Wake-Induced Airframe Loads," 46th Annual Forum of the American Helicopter Soc., Washington DC, May.

Quackenbush, T. R. Lam, C.-M. G., and Bliss, D. B. 1994. "Vortex Methods for the Computational Analysis of Rotor/Body Interaction," *J. American Helicopter Soc.*, 39 (4), pp. 14–24.

Raddatz, J. and Pahlke, K. 1994. "3D Euler Calculations of Multibladed Rotors in Hover: Investigation of the Wake Capturing Properties," AGARD CP-552.

Rorke, J. B., Moffitt, R. C., and Ward, J. F. 1972. "Wind Tunnel Simulation of Full-Scale Vortices," 28th Annual Forum of the American Helicopter Soc., Washington DC, May.

Rorke, J. B. and Moffitt, R. C. 1977. "Measurement of Vortex Velocities over a Wide Range of Vortex Age, Downstream Distance, and Free Stream Velocity," NASA CR 145213.

Rosen, A. and Isser, A. 1995. "A New Model of Rotor Dynamics During Pitch and Roll of a Hovering Helicopter," *J. American Helicopter Soc.*, 30 (3), pp. 17–28.

Rule, J. A. and Bliss, D. B. 1998. "Prediction of Viscous Trailing Vortex Structure from Basic Loading Parameters," *AIAA J.*, 36 (2), pp. 208–218.

Sadler, S. G. 1971a. "A Method for Predicting Helicopter Wake Geometry, Wake Induced Flow, and Wake Effects on Blade Airloads," 27th Annual National V/STOL Forum of the American Helicopter Soc., Washington DC, May.

Sadler, S. G. 1971b. "Development and Application of a Method for Predicting Rotor Free Wake Positions and Resulting Rotor Blade Airloads," NASA CR 1911 and CR 1912.

Sadler, S. G. 1972. "Main Rotor Free Wake Geometry Effects on Blade Air Loads and Response for Helicopters in Steady Maneuvers," NASA CR 2110 and CR 2111.

Sarpkaya, T. 1989. "Computational Methods with Vortices," *J. of Fluids Eng.*, 111, pp. 5–52.

Schmitz, F. H. 1991. "Rotor Noise," *Aeroacoustics of Flight Vehicles: Theory and Practice*, Vol. 1, Chap. 2. NASA Reference Publication 1258.

Scully, M. P. and Sullivan, J. P. 1972. "Helicopter Rotor Wake Geometry and Airloads and Development of Laser Doppler Velocimeter for Use in Helicopter Rotor Wakes," Massachusetts Institute of Technology, AL TR 179.

Scully, M. P. 1975. "Computation of Helicopter Rotor Wake Geometry and Its Influence on Rotor Harmonic Airloads," Massachusetts Institute of Technology, ASRL TR 178-1.

Seelhorst, U., Beesten, B., and Butefisch, K. A. 1994. "Flowfield Investigation of a Rotating Helicopter Rotor Blade by Three-Component Laser Doppler Velocimetry," 75th AGARD Fluid Dynamic Panel Symp., Berlin, Germany, Oct.

Seelhorst, U., Raffel, M., Willert, C., Vollmers, K. A., Butefische, K. A., and Kompenhans, J. 1996. "Comparison of Vortical Structures of a Helicopter Rotor Model Measured by LDV and PIV," 22nd European Rotorcraft Forum, Brighton, UK, Sept. 17–19.

Sheridan, P. and Smith, R. 1979. "Interactional Aerodynamics – A New Challenge to Helicopter Technology," 35th Annual Forum of the American Helicopter Soc., Washington DC, May 21–23.

Srinivasan, G. R. and Baeder, J. D. 1993. "TURNS: A Free-Wake Euler-Navier-Stokes Numerical Method for Helicopter Rotors," *AIAA J.*, 31 (5), pp. 959–962.

Simons, I. A., Pacifico, R. R., and Jones, J. P. 1966. "The Movement, Structure and Breakdown of Trailing Vortices from a Rotor Blade," CAL/USAAVLABS Symp. on Aerodynamic Problems Associated with V/STOL Aircraft, Vol. 1 – Propeller and Rotor Aerodynamics, Buffalo, NY, June 22–24.

Smith, C. A. and Betzina, M. D. 1986. "Aerodynamic Loads Induced by a Rotor on a Body of Revolution," *J. American Helicopter Soc.*, 31 (1), pp. 29–36.

Spencer, R. H. 1969. "Application of Vortex Visualization Test Techniques to Rotor Noise Research," 25th Annual Forum of the American Helicopter Soc., Washington DC, May 14–16.

Squire, H. B. 1965. "The Growth of a Vortex in Turbulent Flow," *Aeronautical Quarterly*, 16, Aug., pp. 302–306.

Steinhoff, J. S. 1985. "A Simple Efficient Method for Flow Measurement and Visualization," in *Flow Visualization III*, ed. W. J. Yang, Hemisphere Publishing, New York, pp. 19–24.

Sternfeld, H. and Schairer, J. O. 1969. "Study of Rotor Blade Tip Vortex Geometry for Noise and Airfoil Applications," Boeing Helicopters D8-2464-1A, Philadelphia, PA.

Sullivan, J. P. 1973. "An Experimental Investigation of Vortex Rings and Helicopter Rotor Wakes Using a Laser Doppler Velocimeter," Massachusetts Institute of Technology Aerophysics Laboratory, Technical Report 183, MIT DSR No. 80038.

Swanson, A. A. 1993. "Application of the Shadowgraph Flow Visualization Technique to a Full-Scale Helicopter Rotor in Hover and Forward Flight," Paper 93-3411-CP, AIAA 11th Applied Aerodynamics Conf., Monterey, CA.

Tang, L. and Baeder, J. D. 1999. "Improved Euler Simulation of Hovering Rotor Tip Vortices with Validation," 55th Annual Forum of the American Helicopter Soc. Int., Montreal, Canada, May 25–27.

Tangler, J. L., Wohlfeld, R. W., and Miley, S. J. 1973. "An Experimental Investigation of Vortex Stability, Tip Shapes, and Noise for Hovering Model Rotors," NASA CR 2305.

Tangler, J. L. 1977. "Schlieren and Noise Studies of Rotors in Forward Flight," 33rd Annual Forum of the American Helicopter Soc., Washington DC, May 9–11.

Taylor, M. K. 1950. "A Balsa-Dust Technique for Air-Flow Visualization and Its Application to Flow through Model Helicopter Rotors in Static Thrust," NACA TN 2220.

Theodore, C. and Celi, R. 1998. "Flight Dynamic Simulation of Hingeless Rotor Helicopters Including a Maneuvering Free Wake Model," 54th Annual Forum of the American Helicopter Soc. Int., Washington DC, May 20–22, 1998.

Thomson, T. L., Komerath, N. M., and Gray, R. B. 1988. "Visualization and Measurement of the Tip Vortex Core of a Rotor Blade in Hover," *J. of Aircraft*, 25 (2), pp. 1113–1121.

Tischler, M. B., Driscoll, J. T., Cauffman, M. G., and Freedman, C. J. 1994. "Study of Bearingless Main Rotor Dynamics from Frequency Response Wind Tunnel Test Data," American Helicopter Soc. Aeromechanics Specialists' Conf., San Francisco, CA, Jan.

Tung, C., Pucci, S. L., Caradonna, F. X., and Morse, H. A. 1983. "The Structure of Trailing Vortices Generated by Model Helicopter Rotor Blades," *Vertica*, 7 (1), pp. 33–43.

Vatistas, G. H., Kozel, V., and Mih, W. C. 1991. "A Simpler Model for Concentrated Vortices," *Expts. in Fluids*, 11, pp. 73–76.

Walters, R. E. and Skujins, O. 1972. "A Schlieren Technique Applied to Rotor Wake Studies," American Helicopter Soc. Symp. of Status of Testing and Modeling Techniques for V/STOL Aircraft, Essington, PA, Oct. 26–28.

Werlé, H. and Armand, C. 1969. "Measures et Visualisation Instationnaires sur les Rotor," ONERA TP 777.

Widnall, S. E. 1972. "The Stability of a Helical Vortex Filament," *J. Fluid Mech.*, 54 (4), pp. 641–663.

Wilby, P. G., Young, C., and Grant, J. 1979. "An Investigation of the Influence of Fuselage Flow Field on Rotor Loads and the Effects of Vehicle Configuration," *Vertica*, 3, pp. 79–94.

Wilson, J. C. and Mineck, R. E. 1975. "Wind-Tunnel Investigation of Helicopter Rotor Wake Effects on Three Helicopter Fuselage Models," NASA TM X-3185.

Winburn, S., Baker, A., and Leishman, J. G. 1996. "Angular Response Properties of Retroreflective Screen Materials Used in Wide-Field Shadowgraphy," *Expts. in Fluids*, 20, pp. 227–229.

Young, C. 1974. "The Prediction of Helicopter Rotor Hover Performance Using a Prescribed Wake Analysis," ARC C & P No. 1341.

Zori, L. A. J. and Rajagopalan, R. G. 1995. "Navier–Stokes Calculations of Rotor–Airframe Interaction in Forward Flight," *J. American Helicopter Soc.*, 40 (2), pp. 57–67.

Appendix

Typical Main Rotor Disk and Power Loadings

Table A.1. *Disk Loadings and Ideal Power Loadings for a Selection of Rotating-Wing Aircraft*

Rotorcraft	Mass, M, kg	Rotor Radius, R, m	Disk Area, A, m^2	DL, N m^{-2}	Ideal PL, N kW^{-1}	DL, lb ft^{-2}	Ideal PL, lb hp^{-1}
AS 332 Puma	8,600	7.79	190.64	442.53	74.41	9.24	12.48
SA 365 Dauphin	4,000	5.97	111.78	351.04	83.54	7.33	14.02
AS 350B Squirrel	1,950	5.34	89.75	213.14	107.21	4.45	17.99
Agusta A109	2,605	5.50	95.03	268.91	95.45	5.62	16.01
Bell 206	1,886	5.64	99.93	185.14	115.04	3.87	19.30
Bell 412	5,409	7.01	154.38	343.72	84.43	7.18	14.16
Bell AH-1 Cobra	4,545	6.71	141.45	315.22	88.16	6.58	14.79
Bell 222B	3,750	6.40	128.68	285.88	92.57	5.97	15.53
Boeing CH-46*	9,706	7.62	182.41	260.99	96.89	5.45	11.49
Boeing CH-47C*	14,969	9.15	262.73	279.15	93.68	5.84	11.11
MBB BO 105	2,600	4.91	75.80	336.49	85.33	7.03	14.32
MBB BK 117	3,200	5.50	95.03	330.33	86.12	6.90	14.45
Hughes 500E	1,364	4.02	50.85	263.17	96.49	5.50	16.19
AH-64 Apache	8,022	7.32	168.10	468.14	72.34	9.78	12.14
Robinson R22	623	3.83	46.18	132.34	136.06	2.76	22.83
Schweizer 300	932	4.09	52.55	173.98	118.67	3.63	19.91
Sikorsky UH-60	10,000	8.18	210.11	446.90	72.44	9.75	12.15
Sikorsky CH-53E	33,409	12.04	455.41	720.74	58.30	15.03	9.78
V-22 Osprey**	27,500	5.79	105.32	1280.75	34.73	26.75	5.19

* Tandem design. ** Tilt-rotor.

Typical Helicopter Tail Rotor Parameters

Table A.2. *Geometric Characteristics of Tail Rotors for Several Helicopters*

Helicopter	TR Radius, m	Chord, m	N_b	MR/TR Diameter
AS 332 Puma	1.525	0.2	5	5.10
SA 365 Dauphin*	0.45	0.0435	11	13.26
AS 350B Squirrel	0.930	0.185	2	5.75
Agusta A109	1.015	0.201	2	5.41
Bell 206	0.826	0.134	2	6.85
Bell 412	1.295	0.297	2	5.41
Bell AH-1 Cobra	1.295	0.293	2	5.18
Bell 222B	1.05	0.244	2	6.10

(continued)

Table A.2 *(continued)*

Helicopter	TR Radius, m	Chord, m	N_b	MR/TR Diameter
MBB BO 105	0.95	0.18	2	5.18
MBB BK 117	0.97	0.2	2	5.68
500E	0.70	0.134	2	5.75
AH-64 Apache	1.4	0.25	4	5.21
Robinson R22	0.533	0.10	2	7.19
Schweizer/Hughes 300	0.646	0.119	2	6.33
Sikorsky UH-60	1.67	0.247	4	4.88
Sikorsky CH-53E	3.05	0.390	4	3.75
Sikorsky S-76A	1.219	0.164	4	5.49

* Fenestron.

Conversion factors

Table A.3. *Conversion Factors*

Multiply	By	To Get
Pounds (lb)	4.448	Newtons (N)
Feet (ft)	0.3048	Meters (m)
Slugs	14.59	Kilograms (kg)
Slugs per cubic foot (slugs/ft^3)	515.4	Kilograms per cubic meter (kg/m^3)
Horsepower (hp)	0.7457	Kilowatts (kW)
Pounds per square inch (psi)	6895.0	Newtons per square meter (N/m^2)
Pounds per square foot (psf)	47.88	Newtons per square meter (N/m^2)
Miles per hour (mph)	0.4471	Meters per second (m/s)
Knots (kt)	0.5151	Meters per second (m/s)
Knots (kt)	1.8518	Kilometers per hour (km/h)
Pounds per square foot (lb/ft^2)	47.893	Newtons per square meter (N/m^2)
Pounds per horsepower (lb/hp)	5.959	Newtons per kilowatt (N/kW)

Main Results from Thin-Airfoil Theory

The general form of chordwise loading is

$$\Delta C_p = 4\alpha \left\{ A_0 \left(\frac{1+\cos\theta}{\sin\theta} \right) + \sum_1^{\infty} A_n \sin n\theta \right\},$$

where $\theta = \cos^{-1}(1 - 2\bar{x})$ and $\bar{x} = x/c$. The coefficients A_0 and A_n are given by

$$A_0 = \alpha - \frac{1}{\pi}\int_0^{\pi} \frac{dy}{dx} d\theta, \qquad A_n = \frac{2}{\pi}\int_0^{\pi} \frac{dy}{dx} \cos n\theta \, d\theta,$$

where $y = y(x)$ is the camberline. For a flat plate

$$\Delta C_p = 4\alpha \left(\frac{1+\cos\theta}{\sin\theta} \right) = 4\alpha\sqrt{\frac{1-\bar{x}}{\bar{x}}}.$$

The lift and pitching moments are given by

$$C_l = 2\pi\left[A_0 + \frac{A_1}{2}\right], \quad C_{m_{LE}} = -\frac{\pi}{2}\left[A_0 + A_1 - \frac{A_2}{2}\right], \quad C_{m_{1/4}} = -\frac{\pi}{4}[A_1 - A_2].$$

Index